Bacterial Population Genetics in Infectious Disease

Bacterial Population Genetics in Infectious Disease

Edited by

D. Ashley Robinson

Daniel Falush

Edward J. Feil

WILEY-BLACKWELL

A John Wiley & Sons, Inc., Publication

Published by John Wiley & Sons, Inc., Hoboken, New Jersey. Published simultaneously in Canada.

Library of Congress Cataloging-in-Publication Data

Bacterial population genetics in infectious disease / editors, D. Ashley Robinson, Daniel Falush, and Edward J. Feil.

p. ; cm.

Includes bibliographical references and index.

ISBN 978-0-470-42474-2 (cloth)

1. Pathogenic bacteria. 2. Bacterial genetics. 3. Population genetics. I. Robinson, D. Ashley. II. Falush, Daniel. III. Feil, Edward J.

[DNLM: 1. Bacteria–genetics. 2. Bacterial Infections–genetics. 3. Genetic Variation. 4. Genetics, Population. QW 51 B1327 2010]

QR201.B34B355 2010

616.07–dc22

2009039211

Printed in the United States of America.

10 9 8 7 6 5 4 3 2 1

In memory of Thomas S. Whittam

Contents

Foreword

When I joined the faculty at the University of Sussex in 1980 the distinguished evolutionary biologist John Maynard Smith despaired of microbiologists for their lack of interest in the population genetics of bacteria. At the time I had little idea what my new colleague was talking about and like many microbiologists was enjoying the opportunities that had just become available through the development of gene manipulation. I suspect that most microbiologists in 1980 saw a bacterial pathogen as a disease-causing entity rather than a population. Where diversity was recognized within bacterial pathogens it was usually based on subdivision into a small number of serological types, and in some cases (for example, the pneumococcus), clear differences in the ability of serotypes to cause disease were recognized. The lack of interest in the population genetics of bacteria among microbiologists was perhaps not surprising since pre-1980 only one paper on the topic had been published, on the genetic diversity of *E. coli*, and that was to address a question of no apparent relevance to microbiology, being a test of the neutral theory. In 2010 the situation has changed radically and high levels of interest are evident among infectious disease and environmental microbiologists, microbial ecologists, taxonomists, and theoreticians.

Progress has been, and continues to be, driven by technology and has been closely allied to our need to characterize isolates of bacterial pathogens and to identify particular strains that cause disease. The first of the approaches to strain characterization that was amenable to population genetic analysis was multi-locus enzyme electrophoresis (MLEE), which was first applied to bacteria in 1973 by Roger Milkman in his test of the neutral theory and, subsequently, by Bob Selander and Bruce Levin. It was the pioneering work of Bob Selander and his laboratory who introduced MLEE into clinical microbiology and demonstrated fundamental aspects of pathogen populations—notably that isolates from disease were genetically diverse but that, worldwide, a small number of strains (electrophoretic types) caused a large proportion of disease. The multi-locus genotypes assigned using MLEE allowed Selander's laboratory to identify strong linkage disequilibrium in pathogen populations which, together with the isolation of indistinguishable genotypes from different countries and different decades, argued that bacterial populations were strongly clonal and that recombination must be rare.

From the mid-1980s the analysis of the sequences of various genes from multiple isolates of bacterial species showed increasing evidence of a history of recombination and a tension developed between the conclusions derived by MLEE and from gene sequencing about the extent of recombination in bacterial populations. This tension was resolved in 1993 when John Maynard Smith and colleagues demonstrated how linkage disequilibrium was compatible with relatively high rates of recombination. A key sub-plot of this 1993 paper was appropriate sampling of pathogen populations, as one cause of linkage disequilibrium was the over-sampling of genotypes that are particularly associated with disease. For many pathogens the vast proportion of the natural population is present as harmless colonizers of the intestines, skin, nasopharynx, and so on, and a representative sample of the natural population would include only a tiny fraction of isolates from cases of disease.

The sampling problem is still very much with us today, as, for most human pathogens, it is straightforward to obtain from clinicians or public health laboratories a collection of isolates from disease, but it can be problematic to obtain large samples from colonization or carriage. Consequently, the population samples that are available tend not to be ideal for the question being addressed. Of course this sampling problem does not arise for some pathogens (for example, those where disease is acquired from an environmental reservoir), and for environmental species, where in many cases the sampling frame can be precisely that required to address the question of interest.

Much of the focus of bacterial population genetics during the 1990s was centered around the extent of recombination in natural populations, and further growth of the field was largely restricted by the nature of the data produced by MLEE. The conversion in 1998 of MLEE into a sequence-based method—multi-locus sequence typing (MLST)—removed this block and opened up the range of topics that can be explored, as amply witnessed by the diverse body of work on bacterial population genetics described in the chapters of this book. The availability of the sequences of seven house-keeping gene fragments from thousands of isolates of several bacterial species has also stimulated theoretical work, with the use of both coalescent approaches and forward simulations where models can be formulated in terms of multi-locus genotypes, allowing model predictions to be compared against real data.

Another area where the improved data richness available from multi-locus sequence analysis is making an impact is the apportioning of bacterial diversity into species, which is a central topic in population biology, although it has not until recently been seen in this light by microbial taxonomists. In parallel, theoretical studies have started to explore the concept of species and the impact of recombination on the processes that lead to irreversible lineage splitting and eventual speciation.

The comfortable view that bacteria evolved by the accumulation of point mutations and short homologous recombinational replacements was shattered in the late 1990s by the appearance of the genome sequences of several bacteria, which unexpectedly showed the rapid acquisition (and loss) of regions of DNA typically from unknown sources. These findings have added a layer of complexity which has yet to be integrated with our current knowledge of bacterial population biology. The next change in technology relevant to bacterial population biology is undoubtedly going to be driven by the opportunities stemming from the continuing rapid development of the capabilities of the new sequencing platforms, which promise the ability to characterize large bacterial populations using complete or nearly complete genome sequences. The integration of gene acquisition and loss with much higher resolution characterization of the core genome will undoubtedly improve our understanding of important areas of bacterial population genetics, particularly ecological differentiation, biogeography, and speciation. These developments will undoubtedly involve contributions from experimentalists, modelers, theoreticians, and ecologists, and closer communication between those who work on pathogens and environmental microbiologists.

Until recently environmental microbiologists have largely used rRNA sequences to monitor and quantify diversity within natural environments, and there has been little meeting of minds between those in this discipline and those working with pathogens. More recently, multilocus sequencing approaches have been used for studying the population dynamics, ecological differentiation, and biogeography of environmental species, and some aspects of the population genetics of bacteria may be more tractable with these organisms than with pathogens. Although, for understandable reasons, this book focuses on the population genetics of pathogens, perhaps by the next edition there will be sufficient integration of approaches to produce a book that covers the population genetics of all bacteria.

The greatly increased interest in bacterial population genetics has not been accompanied by a slew of books on the subject—indeed, there are I think no recent books—and the present excellent and authoritative volume is very timely and greatly welcomed.

Brian G. Spratt
London

Preface

Population genetics is concerned with the causes and effects of genetic variation. It is an observational science, which makes inferences about the processes that shape genetic variation. This book is focused on the population genetics of bacterial pathogens. Infectious diseases caused by bacteria continue to afflict humans even in this era of antibiotics and vaccines, in part because of the genetic flexibility of bacterial populations. However, for all species of bacteria, only a portion of their genetic variation will be relevant to disease processes. If properly deciphered, genetic variation can provide fundamental insights into the pathogens' biology, such as the precise identity of the strains and genes that are dangerous to public health, the elusive nature of bacterial species, and the basis of bacterial adaptations. In turn, this knowledge can assist in the development of new diagnostics, therapeutics, and preventive strategies for combating these terribly common causes of human sickness and death worldwide.

The field of bacterial population genetics developed alongside the methodological improvements that provided a more direct reading of the historical information contained within DNA sequences. In the near future, the field is expected to be inundated with volumes of data that are being generated from rapid, low-cost methods of DNA sequencing. In addition, computationally intensive methods for extracting precise pieces of historical information from sequences are becoming available. Thus, it is an opportune time to summarize progress in the field.

Early population genetics studies of *Escherichia coli*, from the 1970s and 1980s, showed that this model bacterial species exhibited two to five times more genetic variation than that observed in eukaryotic species. Since the early studies, the role of extensive bacterial genetic variation in infectious diseases has attracted the attention of diverse scientists and funding agencies alike. Although the early studies were carried out by population geneticists interested in bacteria, rather than by microbiologists interested in population genetics, scientists from a variety of backgrounds currently work in the field. Perhaps as a consequence of its interdisciplinary character, some of the concepts and methods used in the field are still rather vaguely defined.

Fantastic books are available that give detailed treatments of modern population genetics, but, for those who study bacterial pathogens, one often has to ask "How does this apply to bacteria?" The seminal book edited by Baumberg et al. (1995) from Cambridge University Press was among the first to focus on bacterial population genetics, but it was published prior to the advent of the modern methods that make direct use of DNA sequences. Numerous other books are available that explain how genetic variation originates at the molecular level, but these processes are only part of what is relevant to population genetics. What has been lacking is a book that describes the fate of genetic variation in bacterial populations and that shows how one can generate and analyze bacterial genetic data from a population genetics perspective. We hope that this book will help to synthesize the field and that it will help to train current and future generations of scientists. Here, emphasis has been given to the genetic and population processes that shape genetic variation in bacterial populations and to the methods of analysis that provide a basis for sound inference. An audience for this book could be found among both students and professionals who work in the intersecting fields of genetics, microbiology, infectious diseases, epidemiology, and evolutionary biology.

This book has been subdivided into two sections. The first section covers major concepts and methods of analysis. This section begins with an overview of the coalescent model of population genetics. The next two chapters deal with two different types of data analyses—those that use alleles and those that use DNA sequences. The next three chapters describe different types of genetic and population processes, including recombination, selection, and demographic effects, and the cutting-edge statistical techniques that are used to study their contributions to bacterial population structure. The first section ends with two chapters that describe the modern laboratory techniques that are used to measure bacterial genetic variation from the level of complete genome sequences down to the level of individual nucleotides.

The second section of the book provides a genus-by-genus coverage of some important bacterial pathogens; this is intended to be the applied section of the book. These genera were selected over others because they include species whose study has contributed something unique to the field. For example, some chapters include species where particular laboratory or statistical techniques have been used to great effect, while other chapters include species that illustrate particular clinical or population dynamics. Since many bacterial genera consist of both pathogenic and nonpathogenic species, population genetic data from both types of species have been compared and contrasted in some chapters.

D. Ashley Robinson
Daniel Falush
Edward J. Feil

Contributors

Francois Balloux, MRC Centre for Outbreak Analysis and Modelling, Department of Infectious Disease Epidemiology, Imperial College, London, UK

Robert G. Beiko, Faculty of Computer Science, Dalhousie University, Halifax, Nova Scotia, Canada

Stephen J. Bent, Yale School of Medicine, Yale University, New Haven, CT, USA

Debra E. Bessen, Department of Microbiology and Immunology, New York Medical College, Valhalla, NY, USA

Naiel Bisharat, Department of Medicine, Ha'Emek Medical Center, Afula, Israel

Kristen Butela, Department of Biological Sciences, University of Pittsburgh, Pittsburgh, PA, USA

Erick Denamur, INSERM U722 and Université Denis Diderot, Paris, France

Xavier Didelot, Department of Statistics, University of Warwick, Coventry, UK

Johannes Elias, Institute for Hygiene and Microbiology, University of Würzburg, Würzburg, Germany

Daniel Falush, Department of Microbiology, University College Cork, National University of Ireland, Cork Ireland

Edward J. Feil, Department of Biology and Biochemistry, University of Bath, Bath, UK

Yun-Xin Fu, Human Genetics Center, School of Public Health, University of Texas at Houston, Houston, TX , USA

David S. Guttman, Centre for the Analysis of Genome Evolution and Function, University of Toronto, Toronto, Ontario, Canada

Anne Gatewood Hoen, Children's Hospital Informatics Program, Harvard-MIT Division of Health Sciences and Technology, Boston, MA, USA

Paul J. Jackson, Lawrence Livermore National Laboratory, Livermore, CA, USA

Paul Keim, Department of Biological Sciences, Northern Arizona University, Flagstaff, AZ, USA

Leo J. Kenefic, Department of Biological Sciences, Northern Arizona University, Flagstaff, AZ, USA

Klaus Kurtenbach, Department of Biology and Biochemistry, University of Bath, Bath, UK

Jeffrey Lawrence, Department of Biological Sciences, University of Pittsburgh, Pittsburgh, PA, USA

Xiaoming Liu, Human Genetics Center, School of Public Health, University of Texas at Houston, Houston, TX, USA

Martin C. J. Maiden, Department of Zoology, University of Oxford, Oxford, UK

Gabriele Margos, Department of Biology and Biochemistry, University of Bath, Bath, UK

Darren P. Martin, Institute of Infectious Diseases and Molecular Medicine, University of Cape Town, Rondebosch, South Africa

Nicholas H. Ogden, Public Health Agency of Canada, Centre for Food-borne, Environmental and Zoonotic Infectious Diseases, Université de Montréal, Saint-Hyacinthe, Quebec, Canada

Richard T. Okinaka, Department of Biological Sciences, Northern Arizona University, Flagstaff, AZ, USA

Bertrand Picard, INSERM U722 and Université Denis Diderot, Paris, France

D. Ashley Robinson, Department of Microbiology, University of Mississippi Medical Center, Jackson, MS, USA

Mikkel H. Schierup, Bioinformatics Research Center, Institute of Biology, University of Aarhus, Aarhus, Denmark

Christoph Schoen, Institute for Hygiene and Microbiology, University of Würzburg, Würzburg, Germany

Samuel K. Sheppard, Department of Zoology, University of Oxford, Oxford, UK

Davida S. Smyth, Department of Microbiology, University of Mississippi Medical Center, Jackson, MS, USA

John Stavrinides, Department of Ecology and Evolutionary Biology, University of Arizona, Tucson, Arizona, USA

Olivier Tenaillon, INSERM U722 and Université Denis Diderot, Paris, France

Ulrich Vogel, Institute for Hygiene and Microbiology, University of Würzburg, Würzburg, Germany

Stephanie A. Vollmer, Department of Biology and Biochemistry, University of Bath, Bath, UK

Rob J. Willems, Department of Medical Microbiology, University Medical Center Utrecht, Utrecht, the Netherlands

Carsten Wiuf, Bioinformatics Research Center, Institute of Biology, University of Aarhus, Aarhus, Denmark

Part I

Concepts and Methods in Bacterial Population Genetics

Chapter 1

The Coalescent of Bacterial Populations

Mikkel H. Schierup and Carsten Wiuf

1.1 BACKGROUND AND MOTIVATION

Recent years have seen an explosion in the number of available DNA sequences from many different species. Whereas small genomic regions routinely have been sequenced for more than 20 years and have improved our knowledge of genetic variation at the species and the population levels, new high-throughput techniques have made possible the sequencing of whole genomes and genomic regions for many individuals at an affordable price and in a realistic time frame. This offers unprecedented opportunities for studying genetic variation within and between species and the effects of variation on transcription, regulation, and expression. So, for example, population data sets for bacteria are now expected to consist of full genomes rather than single genes, and the limitations to evolutionary inference are more likely to be found in the analysis rather than in the generation of sequence data (see Chapter 7 of this book).

In the following, we will discuss a mathematical model—*the coalescent*—that describes the process of generating genetic data, with special reference to bacterial populations. For simplicity, we assume the data are in the form of DNA sequences; however, other forms of genetic markers can likewise be modeled. The sequences (or genes) are all homologous copies of the same genetic region in the genome of a species. The relevance of such a model becomes clear when we want to infer/learn details about the evolutionary processes that generated and shaped a sample of present-day sequences. This process may include inferring the mutation rate or demographic parameters, or assessing the age of mutations or common ancestors of sequences. The inferential analysis is retrospective; we seek to understand the evolutionary past of the sample (or population) through analysis of the present-day sequences.

Coalescent theory is the most widespread statistical framework for retrospective statistical analysis of genetic data. The term was coined by Kingman (1982a), who described the genealogy of a sample of n sequences and denoted the genealogical process the coalescent. In subsequent papers, Kingman (1982b,c) developed the theory further and within a few years, it was being studied widely. Kingman's (1980) work built on his own research

Bacterial Population Genetics in Infectious Disease, Edited by D. Ashley Robinson, Daniel Falush, and Edward J. Feil

as well as that of others, for example, Ewens (1972) and Watterson (1974). The coalescent was also independently discovered by Hudson (1983a,b) and Tajima (1983), and in unpublished notes by Bob Griffiths.

In this study, we will first show how simple models of reproduction can be formulated and will discuss their relationship to real bacterial populations. The simple models of reproduction underlie the basic (or standard) coalescent process, which is often used as a null model for statistical analysis. Subsequently, we will introduce some extensions of the basic model that allow for demography and recombination/gene conversion. The extensions predict measurable effects on a sample of sequence data, effects that in turn provide a means for interpreting the data. For further background on the coalescent, see the books by Wakeley (2008) and Hein et al. (2005).

1.2 POPULATION REPRODUCTION MODELS

A simple model of population reproduction was first suggested by Wright (1931) and Fisher (1930). This basic model provides the description of an idealized population and the transmission of genes from one generation to the next. In this study, we consider this model and two other similar models that might be useful for describing bacterial evolution. However, as our exposition is adapted to haploid populations, it may differ slightly from other examples in the literature.

A population of constant size N of haploid individuals forms the basis for our study. At time (generation) $t + 1$, N individuals are drawn from the population at time t—we then consider three different ways that each mimics reproduction in a true physical population (see Fig. 1.1). We use the terms "individuals," "sequences," and "genes" interchangeably in this section since for a haploid, nonrecombining organism, the history of any gene is the same as the history of the bacterial cells. The models we refer to in the study include the following:

Wright–Fisher (WF) model: N individuals are drawn randomly with replacement from the population at time t. The number of descendants of one individual in one time step is approximately Poisson distributed $P(k) = \exp(-1)/k!$.

Moran model: At time t, one individual is chosen randomly to reproduce and one individual is chosen to die. The same individual can be chosen to reproduce and then die. Thus, an individual has either zero, one, or two descendants. Zero and two with equal probability $p_0 = p_2 = (N-1)/N^2$, and one with probability $p_1 = 1 - 2p_2$.

Fission model: At time t, each individual has zero, one, or two descendants with probabilities p_0, p_1, and p_2, respectively. For the population to remain of constant size, we must have $p_0 = p_2 \leq 0.5$.

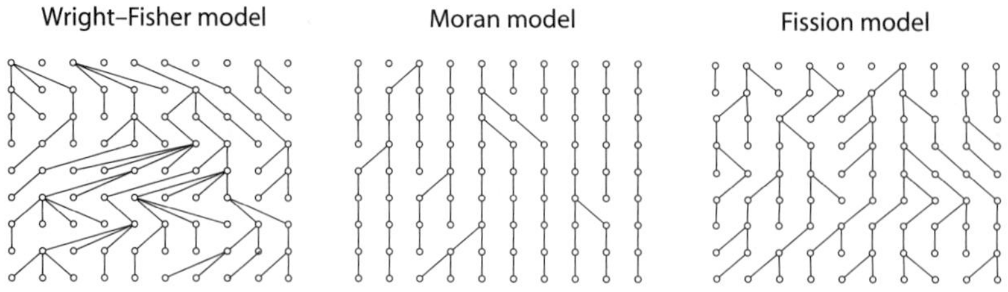

Figure 1.1 Eight generations of reproduction in the Wright–Fisher model, the Moran model, and the fission model, which have properties intermediate between the other two models (see text).

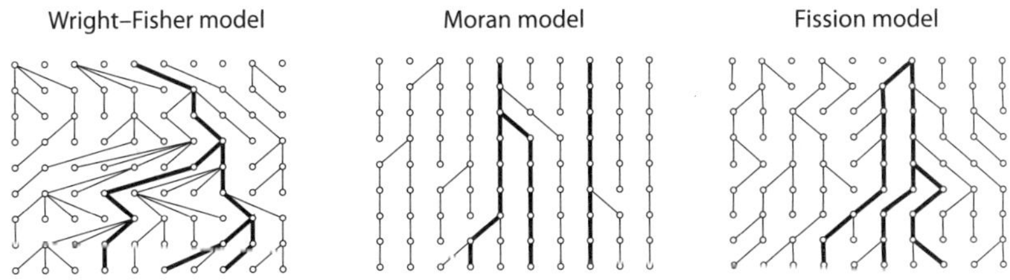

Figure 1.2 The genealogy of a sample of $n = 3$ genes in each of the three reproduction models. Note that coalescent events occur more rapidly in the Wright–Fisher model than in the Moran model. In this example, the three genes coalesce and find an MRCA five generations back, whereas in the Moran model, the three genes have not found an MRCA after eight generations, where there are still two ancestors to the sample. The fission model shows an intermediate pattern.

In the WF model, the entire population is replaced in each time step, whereas in the Moran model, it takes in the order of N time steps before the population is replaced by new individuals. The WF model is often referred to as a nonoverlapping generation model, while the Moran model is referred to as an overlapping generation model, because an individual that does not die continues to the next generation. Figure 1.1 shows eight time steps for each of the three models.

All these models rely on a number of essential, simplifying assumptions: (i) The population is selectively neutral; all alleles are equally fit; (ii) the population has no demographic structure; (iii) the genes are not recombining. We will later discuss how to incorporate recombination and demography, but not selection.

In the study under discussion, we could use these models to trace the genealogical relationship of a sample of n genes backward in time. In Fig. 1.2, this relationship is shown for a sample of size 3 for each of the three reproduction models. In the WF model, the first two genes find a common ancestor two generations back, whereas all three genes share a common ancestor five generations back. The first ancestor of the complete sample is called the *most recent common ancestor* (MRCA) to distinguish it from other ancestors of the sample further back in time. In the Moran model, the three genes have not yet found a common ancestor after eight time steps, but if we progressed far enough back in time, they would eventually find one, since in each time step there is a positive probability for this to happen. The fission model is intermediate between the WF and the Moran model in that coalescent events happen at a slower rate in the fission model than in the WF model, and at a faster rate in the fission model than in the Moran model. In Fig. 1.2, an MRCA is found after eight time steps for the fission model.

1.3 TIME AND THE EFFECTIVE POPULATION SIZE

As the above description suggests, the genealogical history depends on the reproductive model. However, for a large population (large N), all three models show remarkable similarities (Kingman, 1982a–c). To demonstrate this, we first describe the coalescent structure of a sample of size n, taken from the WF model. The probability that *none* of the n genes find a common ancestor in the previous generation is

$$\frac{N-1}{N}\frac{N-2}{N}\cdots\frac{N-n+1}{N}=\left(1-\frac{1}{N}\right)\left(1-\frac{2}{N}\right)\cdots\left(1-\frac{n-1}{N}\right)\approx 1-\frac{n(n-1)}{2N}. \quad (1.1)$$

The latter approximation holds for large N only. The first gene chooses a parent at random; the second can choose among the remaining $N - 1$ genes, the third among $N - 2$ genes, and so on. Consequently, the probability that none of the n genes have found common ancestors in the previous t time steps is

$$P(T_n^N > t) \approx \left(1 - \frac{n(n-1)}{2N}\right)^t, \tag{1.2}$$

where T_n^N denotes the waiting time until the first common ancestor event (superscript N refers to the dependency on population size N). The probability that more than two genes coalesce in the same generation becomes negligible for large N, and henceforth it is ignored in our exposition. The coalescing pair of genes is chosen randomly among all genes in the sample.

Equation 1.2 depends on the population size N. However, if time is scaled in units of N generations, Equation 1.2 takes the approximate form

$$P(T_n > v) \approx \exp\left(-\frac{n(n-1)}{2}v\right), \tag{1.3}$$

where now $T_n = T_n^N / N$. The argument that changes the product in Equation 1.2 into an exponential term in Equation 1.3 relies on N being large and n being relatively small. The right side can be recognized as an exponential variable with rate $n(n - 1)/2$. Consequently, the genealogy of a sample is described by a series of waiting times $T_n, T_{n-1}, \ldots, T_2$ between successive coalescent events; each waiting time is an exponential variable with rate depending on the current number of ancestors. Equation 1.3 has the further important consequence that the genealogy of the sample depends on N only through a scaling of time. For the WF model, the scaling is linear in population size N.

Kingman (1982c) showed that for a variety of reproductive models, including the models discussed here, time can be scaled such that the time between coalescent events is approximately an exponential variable with rate $k(k - 1)/2$, where k is the number of current ancestors (Fig. 1.3). At each coalescent event, two genes are chosen randomly to coalesce.

The scaling factor is known as the *effective population size*, N_e; see Ewens (2005) for discussion and formal definitions. The number N_e depends on N and on the reproductive mechanism in the following way:

$$N_e = \frac{N}{\sigma^2}, \tag{1.4}$$

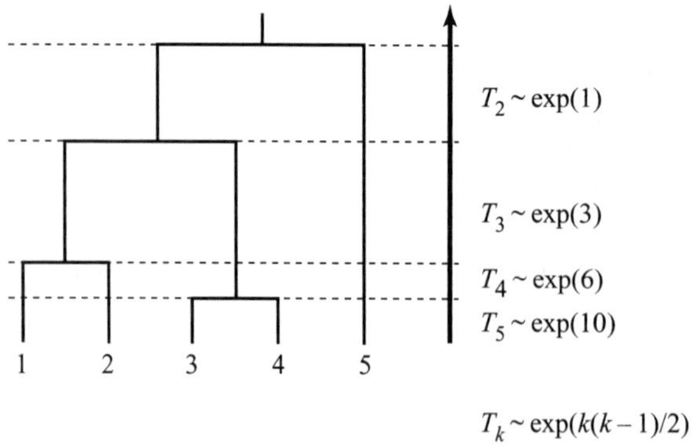

Figure 1.3 The genealogy is described by a series of coalescent events. The waiting times between coalescent events are exponentially distributed with intensities shown in the figure. The intensities depend on the squared number of sequences and therefore grow dramatically with an increasing number of sequences (see also Fig. 1.4). The coalescing pair is choosing randomly among all possible pairs of genes.

where σ^2 is the variance in offspring number (number of lineages subtending an individual in the next generation). For the three models discussed here, we have for large N $N_e^{(WF)} = N$ in the WF model (as already stated), $N_e^{(M)} = N^2/2$ in the Moran model, and $N_e^{(F)} = N/(2p_2)$ in the fission model. Note that if $p_2 = 1/N$, then the fission model is similar to the Moran model, and if $p_2 = 0.5$, then the fission model is similar to the WF model. Hence, in this sense, the fission model embraces both other models, though all of the models differ at the detailed level.

One interpretation of the effective population size is that it is the corresponding size of a similar WF model. For example, a Moran model with population size N corresponds to a WF model with population size $N^2/2$. (Sometimes, the effective population size is defined differently for overlapping generation models; see Ewens, 2005 and below.) Also, if a real physical population has effective population size N_e, then it is similar, with respect to time by generations, to a WF model also with size N_e.

The fission model most closely resembles an idealized bacterial population where individuals divide by fission. In each time step, a certain proportion of cells divide (<50%), a proportion does not divide, and a proportion dies (<50%) in order for the population size to remain constant (growing populations are treated below).

1.3.1 Algorithm 1

Based on the exposition above, an algorithm for simulating the genealogy of a sample of n genes is:

1. Start with $k = n$ genes.
2. Simulate an exponential variable with rate $k(k-1)/2$.
3. Choose two genes randomly among the k genes to coalesce.
4. Put k equal to $k - 1$.
5. If $k > 1$, go to 1; otherwise, stop.

To calculate time in terms of generations, multiply all coalescent times by N_e. This algorithm was used for an initial $n = 50$ genes in order to generate Fig. 1.4.

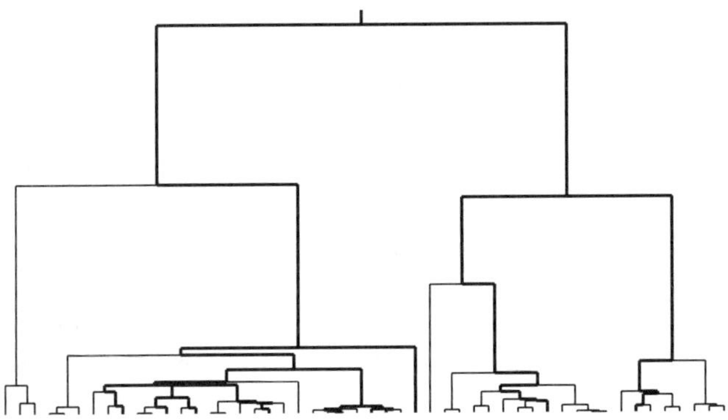

Figure 1.4 An example genealogy of 50 genes under the basic coalescent process. Thick lines track the genealogy of a subsample of size 10. Two features are noteworthy: (i) Coalescent events occur rapidly with many sequences; and (ii) the subsample shares most of the deep branches in the genealogy and the MRCA with the entire sample.

1.4 THE GENEALOGY OF A SAMPLE OF SIZE *n*

In this section, we draw some conclusions from the results of the previous section. The mean and the variance of the (scaled) waiting time while there are k ancestors, $k = 2, \ldots, n$, are, respectively,

$$E(T_k) = \frac{2}{k(k-1)} \tag{1.5}$$

$$Var(T_k) = \frac{4}{k^2(k-1)^2}. \tag{1.6}$$

Thus, more time is spent on average when there are few ancestors than when there are many ancestors (see also Fig. 1.3), and the variance in coalescence times is dominated by the variance when there are few ancestors. The time W_n until the MRCA is found is just the sum of the waiting times T_k; that is, $W_n = \sum_{k=2}^{n} T_k$, which has mean and variance given by

$$E(W_n) = 2\left(1 - \frac{1}{n}\right) \tag{1.7}$$

$$Var(W_n) = \sum_{k=2}^{n} \frac{4}{k^2(k-1)^2} \approx 1.16. \tag{1.8}$$

The latter approximation holds for large sample sizes n. We note some immediate consequences of Equations 1.7 and 1.8: (i) The mean depth of the genealogy of any sample is bounded by 2; hence, an MRCA will *always* be reached, even for very large samples; (ii) even in a large sample, about half of the time is spent while the sample has two ancestors, since $E(T_2) = 1$; (iii) the time while there are two ancestors is much more variable than the remaining time, since $Var(T_2) = 1$, but also $Var(W_n) \approx 1.16$. Thus, unlinked genes might by chance have very different times until their MRCA.

Another quantity of interest is the total size of the genealogy L_n. It is given by $L_n = \sum_{k=2}^{n} kT_k$, because each of the k ancestors contributes T_k to the total size (see Fig. 1.3). It has mean and variance given by

$$E(L_n) = \sum_{k=2}^{n} \frac{2}{k-1} \approx 2\log(n) \tag{1.9}$$

$$Var(L_n) = \sum_{k=2}^{n} \frac{4}{(k-1)^2} \approx 6.58. \tag{1.10}$$

The approximations hold for large sample sizes n. In contrast to the mean depth of the genealogy, the mean of the total size grows without bounds for increasing sample size. However, it grows very slowly, and adding a few more genes only adds a little to the total branch length.

Figure 1.4 shows a sample of size 10 embedded in a larger sample of size 50. In a typical genealogy, the deep branches are shared between the two samples, and adding more genes mainly results in small twigs on the coalescent tree. Consequently, there is high probability that the MRCA of the large sample is also the MRCA of the embedded sample. With the sample sizes of Fig. 1.4, the probability that the embedded sample shares the MRCA with the large sample is 85%. If the larger sample is the entire population (or bacterial species), the probability of MRCA sharing is $(n - 1)/(n + 1)$, where n is the size of the embedded sample (Hein et al., 2005). For $n = 20$, the probability is above 90%, and for $n = 100$, the probability becomes 98%. Thus, the genealogy of a few genes shares important features with the genealogy of the entire population.

1.5 FROM COALESCENT TIME TO REAL TIME

In the above exposition, time is measured in generations or in units of the effective population size N_e. However, it is often of interest to be able to infer the actual physical time in a genealogy. This is possible from sequence data if the mutation rate per time step is known (see below) or if there is an independent estimate of the effective population size. As an example, in *Escherichia coli*, the effective population size may be as large as 50 million (Charlesworth and Eyre-Walker, 2006; Charlesworth, 2009), and if we assume 200 generations per year in the wild, the expected coalescence time for two randomly picked bacteria (if clonal reproduction) would be $N_e = 50$ million generations or 250,000 years. This might be contrasted to humans where the generally agreed numbers are an effective diploid population size of 10,000 and a generation time of 20 years, implying an expected coalescent time of $2N_e = 20{,}000$ generations or 400,000 years, which is surprisingly close to the coalescent time in years in *E. coli*.

The corresponding WF model for the *E. coli* population has $N = 50$ million, whereas the corresponding Moran model has $N = \sqrt{2N_e} \approx 10{,}000$. For the fission model, N depends on the probability of leaving two descendants, p_2.

We note that the above calculations rest entirely on the mathematical formalism set up in Section 1.3 and the desire to equate models with each other. The three models all have different features and capture different aspects of a biological reality. Hence, it is not reasonable per se to say that a certain number of time steps in the Moran model correspond to a number of times steps in the WF model.

1.6 MUTATIONS

Under neutrality, mutations do not affect the number of offspring produced by an individual, and we can impose mutations onto the genealogy after having generated the genealogy, rather than doing it at the same time as generating the genealogy. Figure 1.5 shows the occurrence of three mutations placed at random on the branches under the WF model of reproduction. Only two of these mutations make it to the present generation. Here we assume that mutations happen at a constant rate of u per gene per time steps,

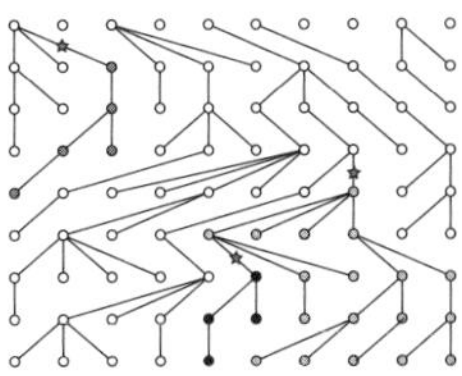

Figure 1.5 The basic coalescent with mutations (shown with stars) imposed. In this example, mutations occurred in generations 1, 4, and 6. The first mutation was lost from the population after three generations, while the third mutation is nested into the second mutation. Thus, there are three types of sequences at the present time (bottom generation)—the original plus two mutated sequences. In the example, they have population frequencies of 10%, 40%, and 50%, respectively.

irrespective of the underlying model. This corresponds to mutations arriving according to a Poisson process on individual lineages.

Since the three models are different, the rate u might be interpreted differently in the three models. In particular, in the Moran and fission models, genes mutate also outside reproduction (see Sniegowski, 2004), where this is suggested as a reasonable scenario for bacterial populations.

With the above definition, the length of the genealogy is directly proportional to the expected number of mutations in a sample; in each time step, there is probability u that the gene mutates; hence, the expected number of mutations is simply the total number of time steps (branch length) times the probability of a mutation. Consequently,

$$E(S_n) = \frac{\theta}{2}E(L_n) = \theta\sum_{k=2}^{n}\frac{1}{k-1} \approx \theta \log(n), \tag{1.11}$$

where S_n denotes the number of mutations in the history of a sample of size n and $\theta = 2N_e u$ is the scaled mutation rate. Thus, if the effective population size is doubled and the mutation rate is halved, then θ remains the same, and we are not able to estimate u and N_e separately from the sample. Equation 1.9 has the further consequence that adding further sequences from other individuals to the sample is not expected to add many more mutations to the data set because the logarithm is a slowly growing function. In contrast, the expected number of mutations increases linearly with sequence length. If mutations happen only during replication, then Equation 1.11 is true for the Moran model and the fission model with $\theta = Nu$, that is, taking the effective size to be $N/2$ in both cases.

These considerations have consequences for parameter inference. For example, for demographic inference, one should aim for longer sequences (potentially from different areas of the genome) rather than for large samples size. Doubling the sample size from 100 to 200 will only increase the expected number of mutations by 13%, whereas doubling the sequence length doubles the expected number of mutations.

A commonly reported estimator of the mutation rate is Watterson's (1975) estimator,

$$\hat{\theta}_W = S_n \Big/ \sum_{k=2}^{n}\frac{1}{k-1}, \tag{1.12}$$

which directly utilizes Equation 1.11 by replacing the expected number of mutations with the observed number. Another estimator, which also has found common support, is Tajima's (1989) estimator,

$$\hat{\theta}_T = \frac{2}{n(n-1)}\sum_{i<j}\pi_{ij}, \tag{1.13}$$

where π_{ij} denotes the number of nucleotide differences between sequences i and j in the sample. This estimator exploits the fact that the number of mutations between a pair of sequences is expected to be θ (Eq. 1.11 with $n = 2$) and considers the average of differences among all possible pairs.

The estimators $\hat{\theta}_T$ and $\hat{\theta}_W$ put a different weight on the mutations in a genealogy. Figure 1.6 shows an example genealogy of five sequences where four mutations have occurred. Watterson's estimator puts equal weight to these mutations, whereas Tajima's estimator puts a larger weight on mutations further up in the genealogy. For instance, in the present example, a mutation carried by two sequences is counted in six comparisons, whereas a mutation carried by only one sequence is counted in four comparisons. Thus,

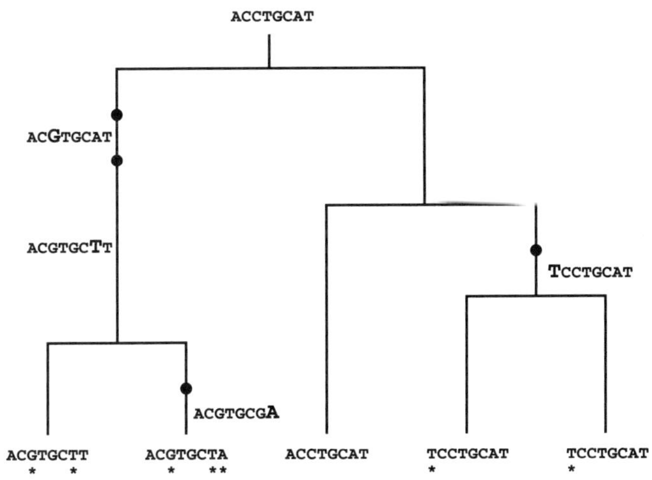


```
ACGTGCTT
ACGTGCTA      Sn = 4  => θw = 4/(1 + 1/2 + 1/3) = 2.3
ACCTGCAT
TCCTGCAT      θT = (1 + 2 + 3 + 3 + 3 + 4 + 4 + 1 + 1 + 0)/10 = 2.2
TCCTGCAT
```

Figure 1.6 An example data set. The effect of mutations is shown in the DNA sequences; at each mutation event (marked by a circle), a nucleotide changes. Sequences 4 and 5 are identical. Below the tree, the calculations leading to Watterson's and Tajima's estimators of the mutation rate are shown. Note that in this example, the two estimators are very similar, and we would not reject the basic coalescent model using Tajima's *D*. The asterisks indicate positions that have changed compared to the root sequence.

if a genealogy has longer inner branches than expected, Tajima's estimator will exceed Watterson's. This fact can be exploited to devise a statistical test for whether sequence data fit the basic coalescent. Tajima (1989) proposed the statistic

$$D = \frac{\hat{\theta}_T - \hat{\theta}_W}{Std\left(\hat{\theta}_T - \hat{\theta}_W\right)}, \tag{1.14}$$

now commonly known as Tajima's *D*, which standardizes the difference of the two estimators (*std* denotes the standard deviation). The distribution of *D* is not known explicitly but can be evaluated by simulation. However, it is sufficiently close to a standard normal distribution, and a rule of thumb is that a Tajima's *D* value >2 or <–2 can be considered significant. This might be used to draw demographic inferences (see the next section).

1.7 DEMOGRAPHY

It is in fact very rare for a population of any species to be of constant size and to mate randomly, as is assumed in the coalescent model. Bacterial populations, for example, have the capacity to very rapidly change population size from a few cells to billions. They can go through dramatic population bottlenecks due to, for example, drugs or during shifts from one host to the next for pathogenic or commensal species. Some bacterial species confined to specific hosts are mainly transmitted from mother to offspring, and they will therefore display a type of population subdivision. Prominent examples of the latter

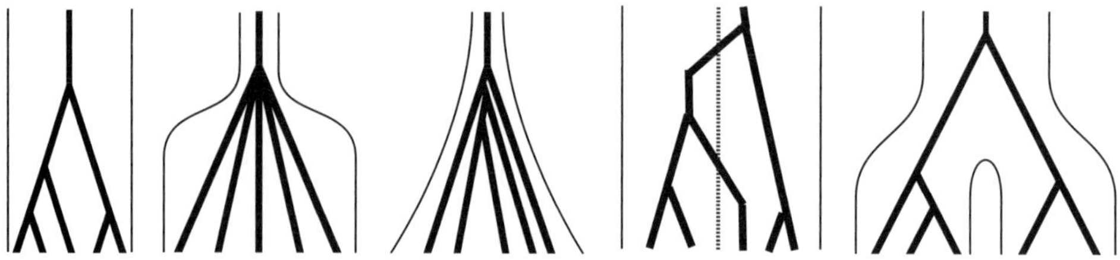

Figure 1.7 Four common demographic scenarios that create deviations from the basic reproduction model with a stable population size. Bottleneck: The rate of coalescence increases dramatically at the time of the bottleneck. Growing: The coalescent rate increases gradually back in time. Subdivision: Initial coalescent events occur preferentially within subpopulations, whereas the last coalescent events need to wait for migration between demes to occur. Splitting: Coalescence can only occur within populations until the time where the two populations merge (viewed backward in time). Subdivision and splitting lead to decreasing coalescent rates back in time, resulting in positive values of Tajima's D. In contrast, growth and bottleneck result in negative values of Tajima's D.

include studies of human migration patterns inferred from the population structure of bacterial species (Falush et al., 2003; Moodley et al., 2009). Even free-living bacterial species are not necessarily very mobile, and one would expect that bacterial cells close to each other are related by fewer cell divisions than bacterial cells far apart. This leads to many different types of population subdivision that are not reflected in the basic coalescent model (see Chapter 6 of this book). Figure 1.7 shows a cartoon of four different demographic population stratifications that deviate from the basic coalescent model.

Growing population: In a growing population, the rate of coalescence increases back in time because the chance of finding a common ancestor is larger in a small population than in a large one. Indeed, the coalescence rate is proportional to the population size. Thus, if the population size has been growing exponentially, then the coalescent rate measured in the present population size will be exponentially increasing back in time. This implies that the last coalescent events (those farthest away from the present) occur relatively faster than the first coalescent events compared with the basic coalescent. Consequently, the internal branches of the coalescent tree are comparatively shorter, which in turn implies that Tajima's D (Eq. (1.14) should be negative. Large negative values of Tajima's D have indeed been interpreted as evidence for population growth in many studies (see, e.g., Venkatesan et al., 2007).

Population bottleneck: A population bottleneck viewed back in time is a fast and dramatic decrease in population size. During the bottleneck, the coalescent rate is therefore much higher than outside the bottleneck. Therefore, the effect on Tajima's D will often resemble that of population growth. If the bottleneck lasts for a very short while, it is possible that not all ancestral lineages coalesce during the bottleneck. In that case, the coalescent genealogy would have a time interval where many coalescent events occurred at almost the same time.

Population subdivision: When bacteria occupy separated habitats, for example, distinct hosts, they can have a stable pattern of subdivision with cell division occurring within each subpopulation and occasional migration between subpopulations. This situation is modeled by equilibrium models, among these the popular n-island model. Population subdivision implies that lineages can only coalesce within demes, so lineages from different demes will need to migrate to the same subpopulation before coalescence can occur. The number of migration events (viewed back in time) is proportional to the number of lineages, whereas the coalescent rate is proportional to the square of the number of lineages (Eq. 1.3). The consequence is that for a sample of individuals, the first coalescent events will be relatively fast because they occur between pairs of lineages in the same subpopulation. The last coalescent events often need to wait for migration events to bring together lineages in the same subpopulation, so if the migration rate is low, we expect that the last coalescent events take a comparatively long time. This implies that the resulting coalescent tree has longer internal branches than the basic coalescent tree and that Tajima's D is expected to be positive.

Population splitting (and merging): It is not possible in general to predict the effect of nonequilibrium population subdivision on the coalescent tree; hence, more subtle ways than Tajima's D are required to detect this scenario. Much progress has been made in predicting population splitting for human populations in the past using single nucleotide polymorphism (SNP) data (Li et al., 2008).

1.8 RECOMBINATION AND GENE CONVERSION

Many bacterial species are very amenable to coalescent-based analysis because they reproduce clonally. However, exchange of genetic material between cells of the same species is also prevalent in many species. This can occur in different ways (see Chapter 4 of this book), but the main effect is the same, namely, that there will no longer be a single coalescent tree describing the fate of the complete genome (or genomic region). This complicates analysis, but it is also the basis for association mapping of a phenotype of interest to a particular loci.

The effect of recombination is described in Fig. 1.8. Forwards in time, a genetic element from one individual is exchanged with the homologous element in a recipient individual. Backward in time, this has the consequence of splitting the genetic material of one individual onto two ancestral individuals, and the genealogical histories of two positions sitting close to each other but on different sides of one of the black bars in Fig. 1.8 will differ. The backward process depends on the rate of recombination and on the length of the exchanged segments (Hudson, 1983a, 1994; Wiuf, 2001; Hein et al., 2005).

Assuming an individual undergoes recombination with probability r, we find that the number of time steps until a lineage has experienced recombination has a probability distribution,

$$P\left(T_{\text{Rec}}^{N_e} > t\right) = (1-t)^t. \tag{1.15}$$

Assuming as in the previous sections that time is scaled in the effective population size, then

$$P(T_{\text{Rec}} > v) = \left(1 - \frac{N_e r}{N_e}\right)^{N_e v} \approx \exp(-v\rho/2), \tag{1.16}$$

where $\rho = 2N_e r$ is the scaled recombination rate (similar to the scaled mutation rate) and $T_{\text{Rec}} = T_{\text{Rec}}^{N_e}/N_e$. Thus, lineages wait for recombination and coalescence to occur, and the ancestral sample is modified according to whatever happens first. The total rate of recombination is $n\rho/2$, while that of coalescence is $n(n-1)/2$. This gives the following algorithm for simulating a sample history (see also Fig. 1.9).

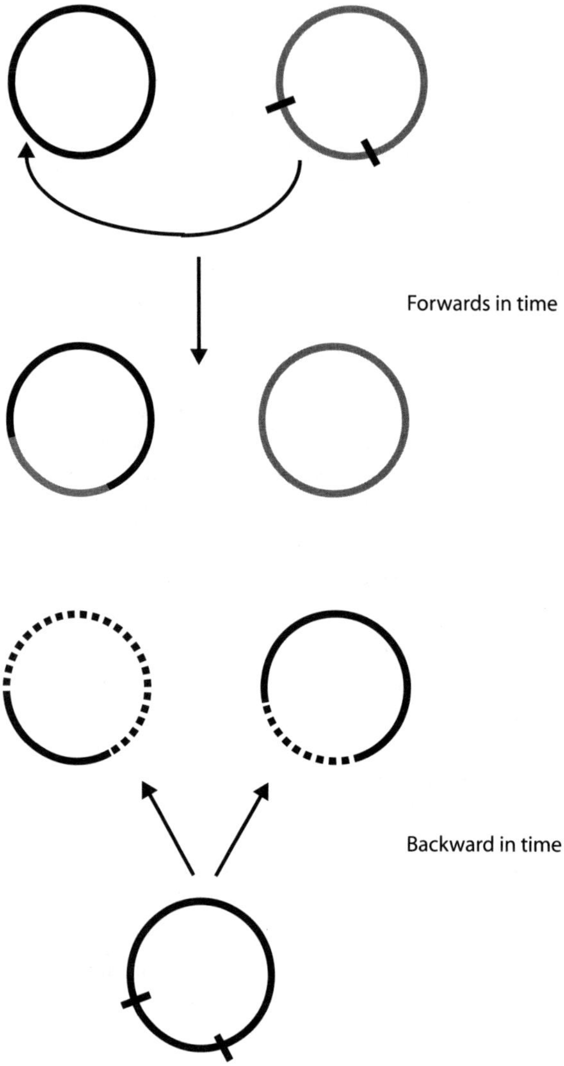

Figure 1.8 Schematic representations of bacterial recombination forwards and backward in time.

1.8.1 Algorithm 2

The algorithm is a modification of Algorithm 1. Do the following:

1. Start with $k = n$ genes.
2. Simulate an exponential variable with rate $k\rho/2 + k(k-1)/2$ (the sum of the rates for coalescence and recombination).
3. With probability $\dfrac{k\rho/2}{k(k-1)/2+k\rho/2}=\dfrac{\rho}{k-1+\rho}$, perform a recombination event; otherwise, with probability $\dfrac{k-1}{k-1+\rho}$, perform a coalescent event

   a. If the result is a recombination event, choose a sequence at random and split it into two. This can be accomplished in different ways; for example, one could

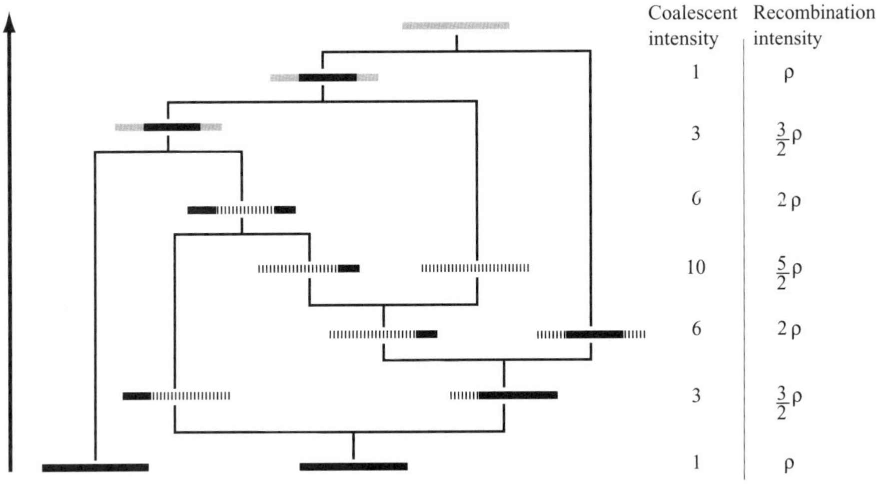

Figure 1.9 The coalescent process with recombination for a sample of two sequences. The intensities of coalescence and recombination at each time point are shown to the right, assuming that each sequence has a recombination rate of ρ/2. The third recombination event (counted from the present time) creates a sequence that is not ancestral to the present-day sample (the dotted black sequences). The solid black lines represent material ancestral to the present-day sample. Finally, the solid gray lines represent common ancestral material. Figure adapted from Hein et al. (2005).

choose two points at random, or one could choose one point at random and the other in a fixed distance from it.

b. If the result is a coalescent event, choose two sequences randomly among the k genes to coalesce.

4. If the result is a recombination event, put k equal to $k + 1$; if a coalescent event, put k equal to $k - 1$.

5. If $k > 1$, go to 1; otherwise, stop.

To get time in generations, multiply all times by N_e. During bacterial conjugation, the F factor is transferred, which can only happen once per replication cycle. In that case, it is reasonable to scale the recombination rate by $N/2$ rather than by N_e (i.e., similar to the discussion of Eq. 1.11).

This algorithm is illustrated in Fig. 1.9. A sample size of two waits for recombination and coalescence to occur. Here we only look at a small (linear) segment of the entire (circular) genome. The first two events are both recombination events and spread the ancestral material of the right sequence onto three ancestors. The third event is also a recombination event, but it creates an "empty" sequence in the sense that the recombination break point is in the part of the sequence that does not carry material ancestral to the present-day sequence. Hence, this sequence might be ignored. After the three recombination events, the first coalescent event happens, which brings together two pieces of ancestral material (see Fig. 1.9). The next event is also a coalescent event and at this event, some positions in the sample find an MRCA (shown in gray in the figure).

It is worth noticing that different positions might have different genealogies and MRCAs. Also, positions far apart might share some history; in Fig. 1.9, the leftmost and the rightmost positions share MRCA, but they do not share their entire genealogical history. Also, and in contrast to recombination in linear genomes, each recombination

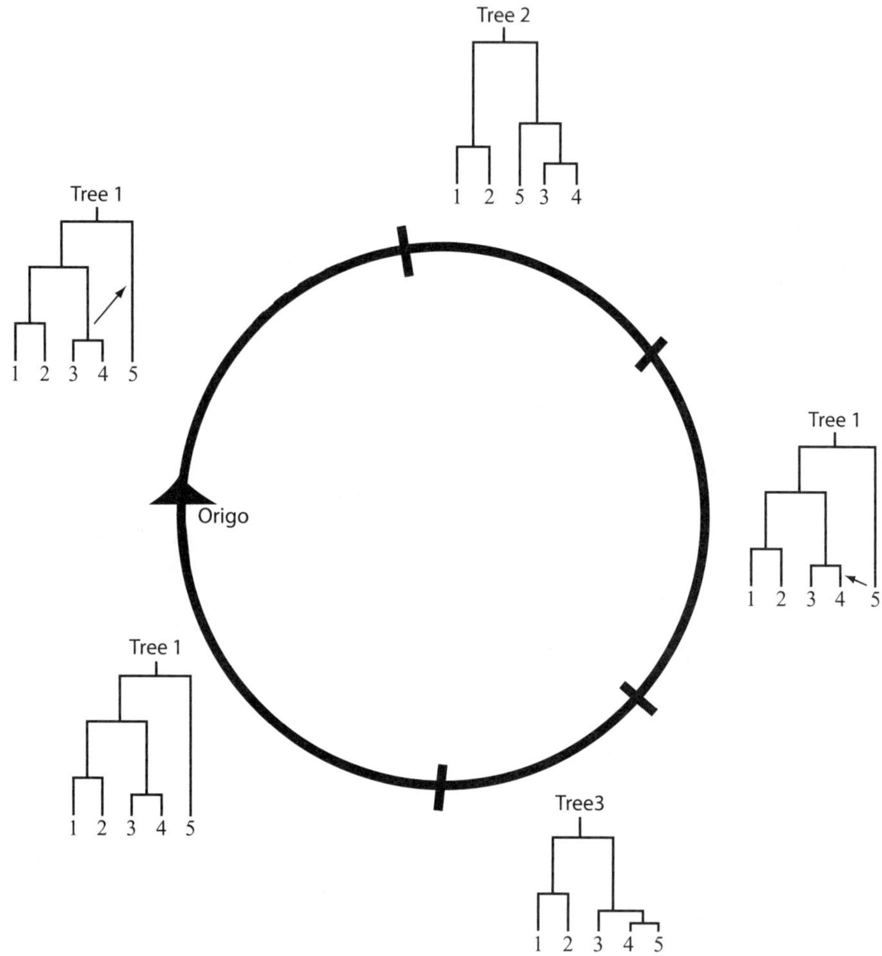

Figure 1.10 The consequences of the recombination process for a bacterial genome. Since the genome is circular, all recombination events resemble gene conversion events. Starting from the origin of replication (origo) moving in the direction of the arrow, a sample of five genes is related through coalescent tree 1. A recombination/gene conversion break point at the top results in a subtree transfer (indicated by an arrow) of the tree carrying sequences 3 and 4 to a different branch leading to coalescent tree 2. At the next break point (right break point of the gene conversion event), we return to tree 1. The next break point results in a subtree transfer of sequence 5 to the branch leading to sequence 4. This also leads to a different time of the MRCA in coalescent tree 3. At the final break point, we again return to coalescent tree 1.

event requires two break points (a beginning and an end of the segment being exchanged), and hence the recombination in circular genomes resembles gene conversion in linear genomes (Wiuf and Hein, 2000; Wiuf, 2001). Computationally, it is important to note that the coalescent with recombination is much more difficult to handle because the number of ancestral sequences might go up or go down, whereas the number of ancestral sequences in the pure coalescent process always goes down by one at each event.

Figure 1.10 illustrates further some of the consequences of a circular genome. At the origin of replication, the sample is related through a single coalescent tree. Moving away from the origin, a recombination break point is encountered in a branch. The effect of the recombination event is to move the subtree subtending the branch to a different location

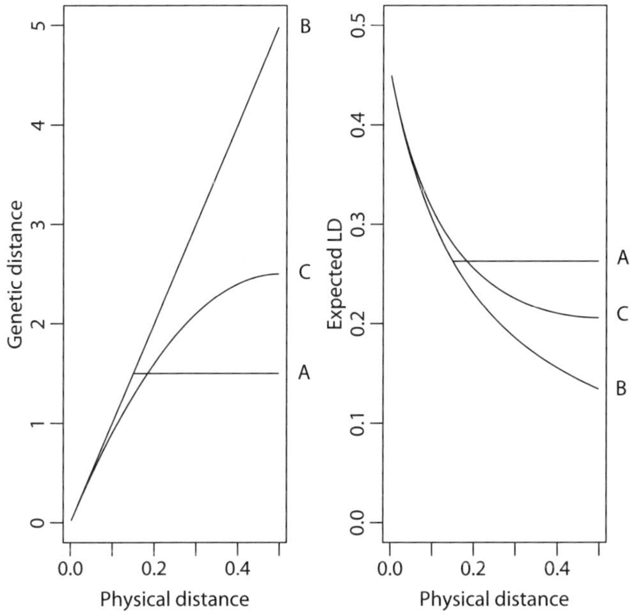

Figure 1.11 The left part of the figure shows the relationship between the physical and the genetic distances. Here the entire circular genome has length 1; hence, the maximal distance between two positions is 0.5. The genetic distance here is scaled to 10 (corresponding to $\rho/2 = 10$). Break points are chosen in the following way: (A) The first position is chosen randomly, the second at distance $L = 0.15$ away; (B) the first position is chosen randomly, the second at distance $L = 0.5$ away; (C) both positions are chosen randomly. The right figure shows the expected LD for each of the three models.

in the original tree. In Fig. 1.9, the sample size is only two and each recombination event potentially moves the common ancestor up or down (the subtree is just a single lineage). In Fig. 1.10, the subtree consisting of sequences 3 and 4 is moved to a different location, thereby creating a new tree. Moving further away from the origin, another recombination break point is encountered and we end up with the first tree again. This can likewise happen in linear genomes but is less frequent since the recombination process does not have the same similarity to gene conversion as in circular genomes.

For linear genomes, the linkage disequilibrium (LD) is expected to decay to zero over long distances because the genetic distance is roughly proportional to the physical distance. This is not the case for circular genomes. Figure 1.11 shows, for two different models, the relationship between the genetic and the physical distance, and the expected LD (measured by the quantity r^2) in a large sample,

$$E(r^2) \approx \frac{10+g}{22+13g+g^2}, \tag{17}$$

where g is the genetic distance between two positions in the genome. It is noteworthy that LD decays faster in a linear genome than in a circular genome.

1.9 SUMMARY

We have presented a basic powerful framework for modeling population variation data. In this framework, it appears that many properties are shared by apparently different reproductive models and that the approximating coalescent process is a very robust approximation. Genetic processes, such as mutation and recombination, as well as demographic effects can easily be incorporated into the coalescent; the consequences of these additions/changes can be studied by simulation and can be compared to real data. In this chapter, we have focused on describing a variety of different models and processes and have ignored the statistical analysis of real data. For further background on statistical inference in population genetics, we refer to Balding et al. (2007).

REFERENCES

BALDING, D. J., BISHOP, M., and CANNINGS, C., eds. (2007) *Handbook of Statistical Genetics*, 3rd ed. Wiley, New York.

CHARLESWORTH, B. (2009) Fundamental concepts in genetics: Effective population size and patterns of molecular evolution and variation. *Nat Rev Genet*. Advanced online publication.

CHARLESWORTH, J. and EYRE-WALKER, A. (2006) The rate of adaptive evolution in enteric bacteria. *Mol Biol Evol* **23,** 1348–1356.

EWENS, W. (2005) *Mathematical Population Genetics*, 2nd ed. Springer, New York.

EWENS, W. J. (1972) Sampling Theory Of Selectively Neutral Alleles. *Theor Popul Biol* **3,** 87–112.

FALUSH, D., WIRTH, T., LINZ, B. et al. (2003) Traces of human migrations in *Helicobacter pylori* populations. *Science* **299,** 1582–1585.

FISHER, R. (1930) *The Genetical Theory of Natural Selection*. Clarendon Press, Oxford.

HEIN, J., SCHIERUP, M. H., and WIUF, C. (2005) *Gene Genealogies, Variation and Evolution: A Primer in Coalescent Theory*. Oxford University Press, Oxford.

HUDSON, R. R. (1983a) Properties of a neutral allele model with intragenic recombination. *Theor Popul Biol* **23,** 183–201.

HUDSON, R. R. (1983b) Testing the constant-rate neutral allele model with protein-sequence data. *Evolution* **37,** 203–217.

HUDSON, R. R. (1994) Analytical results concerning linkage disequilibrium in models with genetic-transformation and conjugation. *J Evol Biol* **7,** 535–548.

KINGMAN, J. F. C. (1980) *Mathematics of Genetic Diversity*. SIAM, Philadelphia, PA.

KINGMAN, J. F. C. (1982a) The coalescent. *Stoch Process Appl* **13,** 235–248.

KINGMAN, J. F. C. (1982b) *Exchangeability and the Evolution of Large Populations. Exchangeability in Probability and Statistics*, pp. 97–112. North-Holland, Amsterdam.

KINGMAN, J. F. C. (1982c) On the genealogy of large populations. *J Appl Probab* **19A,** 27–43.

LI, J. Z., ABSHER, D. M., TANG, H. et al. (2008) Worldwide human relationships inferred from genome-wide patterns of variation. *Science* **319,** 1100–1104.

MOODLEY, Y., LINZ, B., YAMAOKA, Y. et al. (2009) The peopling of the Pacific from a bacterial perspective. *Science* **323,** 527–530.

SNIEGOWSKI, P. (2004) Evolution: Bacterial mutation in stationary phase. *Curr Biol* **14,** R245–R246.

TAJIMA, F. (1983) Evolutionary relationship of DNA sequences in finite populations. *Genetics* **105,** 437–460.

TAJIMA, F. (1989) Statistical method for testing the neutral mutation hypothesis by DNA polymorphism. *Genetics* **123,** 585–595.

VENKATESAN, M., WESTBROOK, C. J., HAUER, M. C., and RASGON, J. L. (2007) Evidence for a population expansion in the West Nile virus vector *Culex tarsalis*. *Mol Biol Evol* **24,** 1208–1218.

WAKELEY, J. (2008) *Coalescent Theory: An Introduction*. Roberts & Co., Greenwood Village.

WATTERSON, G. (1974) The sampling theory of selectively neutral alleles. *Adv Appl Probab* **6,** 463–488.

WATTERSON, G. A. (1975) Number of segregating sites in genetic models without recombination. *Theor Popul Biol* **7,** 256–276.

WIUF, C. (2001) Recombination in human mitochondrial DNA? *Genetics* **159,** 749–756.

WIUF, C. and HEIN, J. (2000) The coalescent with gene conversion. *Genetics* **155,** 451–462.

WRIGHT, S. (1931) Evolution in Mendelian populations. *Genetics* **16,** 97–159.

Chapter 2

Linkage, Selection, and the Clonal Complex

EDWARD J. FEIL

2.1 INTRODUCTION—HISTORICAL OVERVIEW

There are two reasons one may wish to distinguish one bacterial strain from another. The first is that assaying genetic variation within natural populations is a prerequisite to addressing key evolutionary questions such as the molecular processes underlying the emergence of variation (mutation and recombination) or the processes (selection and drift) that determine the subsequent fate of this variation. The second reason is that the assignment of pathogenic bacteria to one of a number of "types" provides information of relevance to the clinician on virulence or resistance properties, and facilitates epidemiological surveillance (and, ideally, management) at both local and global scales (Turner and Feil, 2007). Multilocus sequence typing (MLST) was born out of an appreciation that these two perspectives are not independent and that the principals of population genetics should inform on the generation and interpretation of epidemiological typing data (Maiden et al., 1998).

Much is made of the advantages of MLST as a nucleotide sequenced-based approach, and rightly so for these advantages are obvious. The establishment of MLST databases on the Internet provided the first opportunity for the meaningful and instantaneous comparisons of typing data from independent studies (Spratt and Maiden, 1999; Maiden, 2006). There is also no doubt that the high information content of nucleotide sequence data provides the most illuminating evidence for detailed population and evolutionary analysis. However, there is a second quality that distinguishes MLST from phenotypic methods (serotyping, antibiotyping, or phage-resistance typing) or gel-based methods (pulsed field gel electrophoresis [PFGE], amplified fragment length polymorphism [AFLP], and random amplification of polymorphic DNA [RAPD]), that is, the simultaneous use of multiple, known housekeeping loci. This basic concept was directly borrowed from the predecessor to MLST, multilocus enzyme electrophoresis (MLEE), in which enzyme variation is detected via differences in electrophoretic mobility (Selander et al., 1986). The organization of MLST data into strings of allele identifiers, and the assignment of each allelic combination as a distinct "sequence type" (ST), is a faithful facsimile of the MLEE approach; "ST" evolved from "electrophoretic type" ("ET").

Bacterial Population Genetics in Infectious Disease, Edited by D. Ashley Robinson, Daniel Falush, and Edward J. Feil

There was a sound logic behind the adoption of the principles of MLEE for sequence-based typing. The use of multiple housekeeping loci acts as a buffer against the effects of recombination or atypical selection pressures on single loci. Moreover, MLEE data had provided the engine for many of the key advances in population genetics throughout the 1960s, 1970s, and 1980s. Although the potential of studying enzyme polymorphism in bacteria was noted as early 1963 (Norris, 1963), the technique was much more rapidly adopted as the standard method for eukaryotic populations following its initial use on *Drosophila pseudoobscura* (Lewontin and Hubby, 1966) and on humans in 1966 (Harris, 1966). While these and subsequent MLEE studies on eukaryotic populations laid the foundations for the formulation of the non-Darwinian concepts of drift and neutrality (Kimura and Crow, 1964; Kimura, 1968a), the rich intraspecies variation revealed in prokaryotic populations simply muddied the water for the microbial taxonomists of the time. Besides an exceptional study by Milkman in 1973, which aimed to test predictions of Kimura's (1979) radical new neutral theory, it was not until the 1980s that MLEE was used in earnest to shed light on bacterial population structure. A succession of large-scale MLEE studies throughout this decade, spearheaded by Selander and colleagues, led to the first major wave of population-level data for microbes, thus stoking the nascent field of bacterial population genetics into life (Whittam et al., 1983; Ochman and Selander, 1984; Caugant et al., 1987; Selander et al., 1987; Musser et al., 1988).

These pioneering large-scale MLEE studies defined a number of key debates, many of which—in one form or another—rumble on in the present day. Many of the analytical concepts and tools were also laid down during this period, or at least were modified for prokaryotic populations. I will discuss how allele-based approaches have informed on some of the key debates in bacterial population genetics, and the continued relevance of these methods in the face of the ever-strengthening nucleotide "data storm" (Strous, 2007). The chapter is divided into three overlapping themes: (i) recombination, linkage, and substructure; (ii) neutrality versus selection; and (iii) clustering techniques.

2.2 RECOMBINATION, LINKAGE, AND SUBSTRUCTURE

The core debates concerning the impact of recombination, or horizontal gene transfer, predate the genomic era and, in fact, can be traced back to the earliest studies on bacterial evolution and population genetics. Indeed, the rapid dissemination of drug-resistant microbes was the key motivation for early studies on bacterial evolution, and these quickly led to an appreciation of the potential importance of horizontal gene transfer, in particular via plasmids (Orskov and Orskov, 1973). I aim to provide the briefest of sketches of recombination in this chapter (for an excellent recent review addressing the mechanisms of recombination and its evolutionary significance, see Vos, 2009).

Many sophisticated tests are now available to detect recombination from nucleotide sequence data (see the comprehensive discussion in Chapter 4 and the catalog at http://www.bioinf.manchester.ac.uk/recombination/programs.shtml). Allele-based data also provide a simple means to gauge the impact of recombination within a given population sample, and this approach remains influential. The term "linkage" simply refers to the probability that alleles at different loci are found together in the same genome. If there is no linkage, and all alleles are randomly assorted, the presence of a given allele at a given locus will not provide any predictive information concerning the presence of alleles at any other loci. This case is described as linkage equilibrium, a term that is doubly confusing as it refers to the absence of linkage, and there is no reason to suppose such a population

Loci:	A	B	C	D	E	F	A	B	C	D	E	F
Strain 1	1	3	1	1	2	4	1	2	1	2	2	4
Strain 2	1	3	1	1	2	4	1	3	1	1	6	2
Strain 3	1	3	1	1	2	4	2	4	2	2	6	3
Strain 4	1	3	1	1	3	4	1	3	1	2	3	4
Strain 5	2	1	2	3	1	1	4	1	2	3	3	3
Strain 6	2	1	2	3	6	1	2	2	2	3	6	1
Strain 7	2	1	2	3	1	1	1	1	1	2	1	1
Strain 8	2	1	2	3	1	1	3	1	2	4	1	4
Strain 9	3	2	1	2	2	2	3	4	1	4	2	4
Strain 10	3	4	1	2	2	2	4	4	1	2	3	1
	Linkage disequilibrium						Linkage equilibrium					

Figure 2.1 The extremes of population structure as revealed by MLEE/MLST data. In this cartoon example, 10 strains are characterized at six loci, and all unique alleles are assigned an identifier. On the left-hand side, the 10 strains fall into three clusters, where variation is limited within clusters, but the clusters are unrelated to each other. In this case, only a small fraction of all possible allelic combinations are observed, and the population is said to be in a state of linkage disequilibrium. On the right-hand side, the alleles are randomly assorted, and strains do not fall into clusters. This is an example of linkage equilibrium.

is resting at equilibrium. At the opposite extreme, the recovery of only a small fraction of all possible allelic combinations within a population (but each of these at a high frequency) is expected if alleles have been inherited together over time and the population is in a state of linkage disequilibrium (Fig. 2.1).

A popular means to quantify the degree of linkage in a given MLEE/MLST data set is to use Brown's index of association (I_A) (Brown et al., 1980). This index compares the variance in pairwise differences in the data, in terms of allele mismatches, compared to that expected under a null hypothesis of linkage equilibrium, which would give a value of around zero (Smith et al., 1993). However, there are some difficulties with this index as originally used. While a higher I_A broadly reflects stronger linkage (thus lower rates of recombination), its value also depends on the number of loci in the input data. To address this, and to facilitate comparisons between data sets, Haubold and Hudson (2000) proposed a standardized index (I^S_A). Additionally, complications in determining whether the computed I_A represents a significant departure from linkage equilibrium led to a revised empirical method for calculating significance. Resampling from randomized data sets remains a commonly used, and perfectly valid, alternative. Tools to calculate I_A are available on both of the U.K.-based MLST websites (http://www.mlst.net/ and http://pubmlst.org/).

MLEE data for most bacterial species point to high levels of linkage disequilibrium, and such populations are commonly referred to as "clonal." While the use of the word clonal in this context broadly implies very modest rates of recombination, the term is also used for highly uniform species, such as *Bacillus anthracis*, where all isolates appear identical by MLEE/MLST. As recombination between identical sequences will not leave any genetic footprint, so it is not possible to reliably gauge how much recombination occurs within such populations, only that DNA does not appear to be imported from elsewhere. "Clonal complex" is a related term, and this will be discussed at more length elsewhere in the chapter. The clonality of bacterial populations has broadly been confirmed by MLST data. As recombination might be expected to lower the degree of linkage between genes, these data led Selander and others to conclude that recombination is likely to occur at modest rates in most bacterial populations, a hypothesis promoted to the "clonal paradigm." The use of I_A as described above supports this position by testing against a null hypothesis of linkage equilibrium. Given that bacteria are asexual, and that rates of recombination approximately 20-fold higher than those of mutation are required to achieve random assortment (Smith et al., 1993, Hudson, 1994), it is perhaps not surprising that this index is typically "significant" and that reports of linkage equilibrium are rather rare. Further, while the absence of linkage ($I_A \sim 0$) is difficult to explain without invoking very high levels of recombination, linkage disequilibrium may be introduced into populations, or more precisely into multilocus data sets, even when recombination is profoundly impacting on diversification and adaptation.

A perennial difficulty in bacterial population genetics lies first in defining a single "population" then in drawing from it a strictly representative sample of isolates. The over-representation of certain genotypes within a sample will introduce misleadingly high levels of linkage. A well-cited example is that of occasional human pathogens, such as *Neisseria meningitidis*, where isolates from cases of disease may be overrepresented and the bulk of the population that is carried asymptomatically may be underrepresented. Alternatively, a given sample may actually encompass two or more freely recombining populations that are isolated from each other either geographically, ecologically, or temporally. The pooling of multiple populations will necessarily introduce linkage disequilibrium if there is little or no gene flow between them.

Interpretations of linkage should therefore be made within the context of as much clinical, geographic, ecological, and temporal metadata as are available. While a simple means to counter these confounders is to analyze a single example of each genotype, or even one example of each cluster of genotypes, it is unclear how much the subsequent reduction in I_A is due to a loss of power in the data (Lenski, 1993). Furthermore, just as genotypic clusters are best interpreted in the light of ecological and other information, so the presence of such clusters may well inform on the ecology of the organism. One must be careful not to overlook such possibilities in the pursuit of a single clean figure describing "recombination rate," which will in any case almost certainly be an oversimplification. The same caveat should also be applied to other summary statistics commonly computed for allele-based data, most notably diversity (heterozygosity, H), measures of selection, or F-statistics, which provide a measure of gene flow between populations and are reviewed elsewhere (Weir and Hill, 2002).

2.2.1 Recent Applications

A simple means by which to gauge the continuing importance of allele-based linkage analysis is by examining the citation record for key work by Maynard Smith et al. in 1993. This paper has been cited about once a week during 2008/2009, and recent works continue to illustrate both the utility and the potential pitfalls of drawing inferences on population dynamics and diversification from estimates of allele linkage. Kaiser et al. (2009) recently presented MLST data for 70 strains of the opportunistic pathogen *Stenotrophomonas maltophilia*. They noted significant linkage disequilibrium within the data set as a whole (as gauged by a resampling procedure), and argued that this observation was consistent with other lines of evidence pointing to a basically clonal population structure. They went on to compare isolates from clinical and environmental origins, and noted significant disequilibrium for the clinical isolates but not for the environmental isolates. Although it would be tempting to speculate from this that the environmental reservoir constitutes a freely recombining pool, from which clinically relevant genotypes occasionally emerge and expand, the authors took a cautious line by pointing out that the environmental sample may be too small to detect significant linkage disequilibrium.

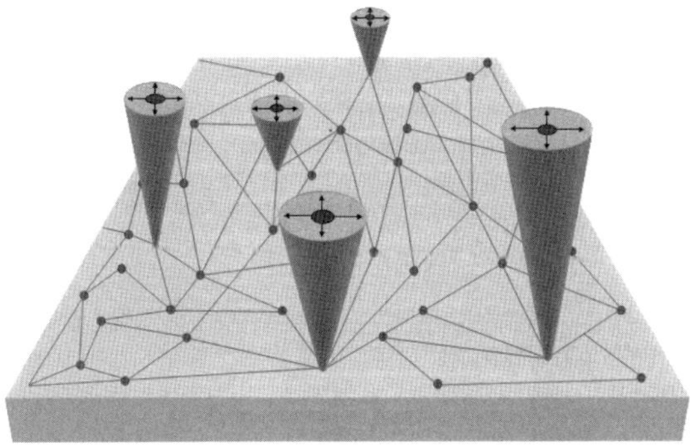

Figure 2.2 The "epidemic" bacterial population structure. The background population is composed of a large number of relatively rare and unrelated genotypes (small circles) that are recombining at a high frequency. Superimposed upon this background are clusters of closely related genotypes (clonal complexes), illustrated as cones. These emerge from a single, highly adaptive ancestral genotype (the large circles).The diversification of these clones by recombination and mutation is indicated by the arrows (from Smith et al., 2000). See color insert.

The authors also noted that the linkage disequilibrium was lost for the clinical strains when only one example of each ST was used, an observation consistent with the rather inappropriately named "epidemic" model of clonal expansion proposed by Maynard Smith et al. Under this model, nature is implicated in playing a role in generating sampling bias through the clonal expansion of specific lineages (Smith et al., 2000) (Fig. 2.2). In terms of consequence, this process is indistinguishable from the human-induced sampling bias leading to the overrepresentation of "interesting" genotypes as described above. Indeed, in cases such as the clinical lineages of *N. meningitidis*, it is likely these genotypes possess selective advantage both to flourish in nature and within our culture collections. Whether selection or stochastic processes are important for natural expansion events is discussed in the next section, although of course, there is no need for an either/or approach.

In addition to sampling artifacts and clonal expansion, the question of geographic structuring in bacterial populations has received a great deal of attention in recent years (Whitaker et al., 2003; Martiny et al., 2006; Ramette and Tiedje, 2007). As discussed above, the pooling of multiple isolated populations into a single sample can be an important source of linkage disequilibrium. MLST has been used to address this question for a number of eubacterial and archaeal species, and sequence-based analyses are generally presented in concert with those based on comparisons of allele frequencies between populations, gene flow (*F*-statistics), and linkage. These works encompass both pathogens and environmental species occupying terrestrial and aquatic habitats, and range from intercontinental comparisons (Vesaratchavest et al., 2006) down to comparisons between samples taken from a single square meter (Vos and Velicer, 2008). In a nutshell, these studies address the issue of whether what we call a single population in fact represents a metapopulation of isolated foci, either geographically, genetically, or both. If strong structure is revealed by a combination of allele-based and phylogenetic analyses, then one is free to speculate on the role of selection and drift on the emergence and maintenance of this structure, as discussed below. If so inclined, one may go on to speculate as to what this tells us about "species" (Doolittle, 2009; Fraser et al., 2009).

While it is perhaps not surprising that the field has yet to find a consensus, it is striking that many of the best-designed studies point to a seemingly intangible and probably highly dynamic combination of processes for single taxa. Papke et al. (2007) aimed to examine the relative roles of local adaptation and geographic distance on halophilic archaea. They chose three sites, two of which were adjacent but differed in salinity; the third was distant but similar in salinity to one of the first two. The question was simple: Which two sites should be more similar—the pair in close proximity, implicating geography, or the distant sites with similar salinity, implicating ecology? Unfairly, phylogenetic analysis weakly supported the third possibility—of the five markers used, three clustered the two sites, which were distinct both in terms of geography and salinity (although see Pagaling et al., 2009 for a more recent study). However, the geographically close sites shared more identical alleles than either of the other two pairs of sites. Thus, while gene flow may be more common between the geographically close sites, at least over the short term, the long-term consequences of this in terms of adaptation and divergence are far from clear.

Studies on spatial structuring have also been carried out for pathogenic species and may prove particularly pertinent for zoonotic diseases (see Chapter 12). Goethert et al. (2009) have recently presented data on the tick-borne species *Francisella tularensis*, which is the causative agent of type A tularaemia. The last nine years have witnessed a heightened frequency of this disease on Martha's Vineyard (MA, USA). These authors used variable number of tandem repeat (VNTR) to characterize bacteria from two samples of questing dog ticks from this small island from two distinct locations approximately 15 km apart. This technique is based on hypervariable repeat loci and is useful in cases where there is limited sequence diversity. The use of multiple VNTR loci is known by the nested acronym multiple loci VNTR analysis (MLVA, pronounced "mulva"). All commonly used allele-based methods are applicable to MLVA data sets.

In the first site, the percentage of infected ticks had been consistently high in the preceding 9 years (~5%), whereas the infection prevalence in ticks in the other site had been steadily increasing from 0.4% in 2003 to 3.9% in 2006. Thus, while one site is thought to be "stable," the other is "emerging." VNTR data for bacteria from the two sites revealed that the stable site was essentially clonal (low-diversity, high-linkage disequilibrium), while the emerging site consisted of many unrelated genotypes. This study therefore highlights how local populations, even when in close proximity, may vary not only in terms of genotype/allele frequencies but also in terms of diversity and overall structure. Such differences may reflect the clonal spread of a single genotype within a given locale, which removes much of the local diversity, or alternatively the frequent import of diverse genotypes into a locale from elsewhere.

Have such studies shed any light on the question of whether "everything is everywhere" for prokaryotes? As discussed above, the evidence will not only be mixed between taxa but also between sites within single studies. Moreover, the results depend greatly on the resolution of the typing methods used; it is unclear what exactly it is that we might expect to find everywhere (Nesbo et al., 2006). Clonal expansion must affect all bacterial populations at some level, as the recombination rate per cell must be many orders of magnitude lower than one event per generation, and each event is likely to affect a tiny proportion of the genome. It follows that, given sufficient discriminatory power and an appropriate spatial scale, geographic structuring must occur in all populations. There should therefore be no surprises when future population studies based on full genome sequences reveal striking geographic structure within species previously considered to be globally homogenized. The interesting part will not be the detection of geographic structure per se but the level of discrimination required for its detection, and the range over which endemic domains extend (Ruimy et al., 2009). While the former could fall anywhere between a single highly conserved 16S rRNA sequence, and a full genome sequence, it is clear that as the discriminatory power of routine typing methods increase, so will the importance of spatial statistics.

2.3 NEUTRALITY VERSUS SELECTION

Perhaps the single most important realization to come from early MLEE studies on eukaryotes was the finding of a much higher level of diversity within natural populations than had previously been supposed (Kimura, 1968b). Many investigators considered it implausible that selection could account for the maintenance of such high levels of diversity, so the evolutionary significance of purely stochastic processes was reconsidered. Through a comparative analysis of the amino acid sequences of mammalian hemoglobin genes, Kimura concluded that the rate of mutation is so high that most changes must have no adaptive relevance (Kimura, 1968b, Gillespie, 1987). Kimura laid down the elegant mathematical predictions for neutral evolution, whereby heterozygosity (diversity) is simply a function of the neutral mutation rate μ and the effective population size N_e. A problem soon arose with testing the theory, as the model assumed an equilibrium condition rarely met for eukaryotic populations. In order to fit neutral expectations, a population was required to have maintained a very large population size for a very long time. It was this condition, along with short generation times and a global distribution, that first led Milkman in 1973 to consider *Escherichia coli* as a "no excuse" organism for testing Kimura's theory. The utilization of a haploid organism also ruled out the possibility of heterosis, which was popular with selectionists at the time as it offered a selective basis for the maintenance of polymorphism. *E. coli* subsequently became the model organism for bacterial population genetics.

Milkman's study revealed that at least 90% of loci are variable in natural populations of *E. coli*; the equivalent figure for eukaryotes being around 30%. Despite this diversity, Milkman rejected the neutral theory on the grounds that the *effective* number of alleles per locus fell way short of neutral predictions. The effective number of alleles equals the reciprocal of the sum of the squared frequencies of all alleles, and was used to correct for the excess of rare alleles in the population. Unfortunately, this pioneering study did nothing to resolve the debate. Observed levels of polymorphism fell awkwardly for both camps, too much to be explained by balancing selection and too little to be explained by drift. Nevertheless, one thing was clear: unprovable selective scenarios could be proposed to explain nearly anything, whereas neutrality was far easier to parameterize and thus test. It is for this reason, rather than being more parsimonious, that neutrality has been widely adopted as the "gigantic null hypothesis" in molecular evolution (Avise, 1994; Hahn, 2008).

The effective population size is notoriously difficult to estimate from bacterial populations, and MLST data have underpinned a number of studies that have sought to reevaluate neutral predictions by reestimating (lowering) N_e (Feil, 2004; Fraser et al., 2009). For pathogenic bacteria, which are able to colonize or infect a new host from a very few pioneering cells, a simple approximation is to equate N_e with the number of infected hosts. Appreciating that this may still not account for the paucity of variation with respect to neutral expectations, Fraser et al. (2005) went one stage further in proposing a "microepidemic" model. By considering the effects of localized transmission chains, this model reduces N_e still further and appears to bring the observed level of diversity in line with neutral expectations. However, although it was raised by the authors, one crucial test was not convincingly performed. Regardless of how one estimates N_e, any neutral model will predict an increase in diversity with increasing population size. In practice, this means that diversity (heterozygosity, *H*) should increase when epidemiologically unlinked samples are combined. A preliminary analysis on MLST data for asymptomatically carried *Staphylococcus aureus* suggests that combining carefully sampled data sets from Europe, Asia, and Africa has little effect on *H* (data not shown). If confirmed for other species, such analyses should help to reinvigorate efforts to reexamine the effects of selection in shaping bacterial population structures.

One widely cited mode of selection in bacteria is that of "periodic selection"(Levin, 1981). This is the process whereby an adaptive mutant emerges and rises sufficiently in frequency to become an observable "clone" or clonal complex (if minor variants are detected). Such lineages are thought to remain coherent in the face of mutation, and possibly recombination, by regular "selective sweeps," whereby newly emerged variants of high fitness outcompete their less fit clone mates, thus purging the lineage of diversity. Such a model has intuitive appeal for explaining the existence of discrete genotypic clusters in many bacterial populations, some of which can be attributed specific metabolic, resistance, or virulence properties.

A large body of theoretical work spearheaded by Cohan examines how ecologically adaptive clusters, ecotypes, might be maintained in nature in the face of frequent migration, recombination, and drift (Cohan, 2002; Gevers et al., 2005; Cohan and Perry, 2007; Koeppel et al., 2008). For pathogenic bacteria, such as *N. meningitidis*, biotic factors such as immune evasion and interstrain competition are likely to play a key role in the maintenance of discrete lineages (Buckee et al., 2008). In the case of *S. aureus*, there is also evidence that gene flow between lineages may be limited by lineage-specific restriction/modification systems, and resistance to different types of phage may play a major role in shaping the selective landscape (Waldron and Lindsay, 2006). Although there are good reasons to believe that many, if not most, clonal complexes in bacterial populations represent fitness peaks of one sort or another, this need not necessarily be case. There are certain conditions—most notably allopatry—by which clusters can form and diverge by entirely neutral processes (Fraser et al., 2007).

Periodic selection will certainly result in a dramatic reduction in the effective population size and in fact has been used to reconcile low levels of variation within neutral expectations (Levin, 1981; Fraser et al., 2009). While it may seem slightly odd to evoke selection to support a neutral position, the point is that the variation assayed, within the various MLEE electromorphs or MLST alleles, is likely to be neutral; the adaptive mutation is somewhere else on the genome. Periodic selection thus greatly raises the power of very occasional strong adaptive changes in shaping population structure, and in purging neutral diversity, via the hitchhiking effect. Although this means there is no need to surrender the position that almost all variation is effectively neutral, the hugely disproportionate impact of rare adaptive changes should also be of comfort to the selectionist. As it was originally conceived for sexual populations, this represents a departure from Kimura's neutral theory, and perhaps could be distinguished from it as the "essentially neutral but a little selection goes a long way" theory.

There remain two theoretical caveats. First, the efficiency of periodic selection in purging neutral diversity is reliant upon tight linkage between adaptive and neutral loci, which implies that rates of recombination should be modest (Levin, 1981). However, there is a good deal of evidence that recombination rates are high for many natural populations, particularly between closely related strains (Feil, 2004). Second, neutrally diversifying ecotypes that are reliant on periodic selection for genetic cohesiveness should have a finite shelf life. If all diversity generated within an adaptive lineage was strictly neutral, there should be no loss of fitness between sweeps. If so, one might imagine that the time between

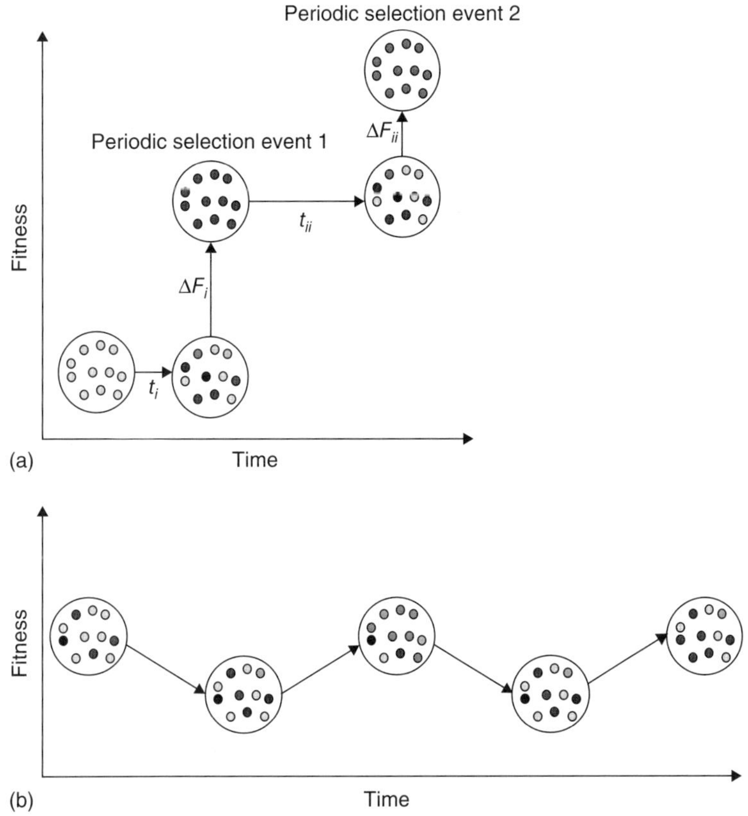

Figure 2.3 (a) The maintenance of ecotypes by repeated periodic selection. If all accumulated diversity within an ecotype is assumed to be strictly neutral, and the niche is assumed to be strictly stable, then there should be no loss of fitness between periodic selection events. This means that each periodic selection event would reflect a fitness gain over the previous event. As each subsequent periodic selection event becomes "harder" (as the ecotype ascends the fitness peak), so the time between selection events should increase (i.e., $t_{ii} > t_i$), and the fitness gain at each event should decrease (i.e., $\Delta F_i > \Delta F_{ii}$). (b) The consolidation of fitness within an ecotype through purging of slightly deleterious changes and partial selective sweeps. In this case, fitness decreases over time as mutations confer an overall fitness cost. The selective purging of these mutations will partially constrain divergence. The emergence of adaptive mutations may help to counter the fitness cost, but this assumes high rates of recombination, as complete selective sweeps (as in Fig. 2.3a) would lead to the fixation of deleterious changes and an overall loss of fitness over time. Adaptive mutations will be more likely if the niche is liable to change. See color insert.

selective sweeps should increase incrementally, and the increase in fitness at each successive sweep should decrease (Fig. 2.3a). Assuming they are only permitted to occupy a single, stable niche, it follows that there will eventually come a point where they have reached the summit of the fitness peak.

It is worth recalling that the neutral predictions of diversity rest on two parameters, the effective population size and the neutral mutation rate. Difficulties estimating the former are well documented as discussed above, but the neutral mutation rate may also require reevaluation. It has recently been shown that a high proportion of recent mutations (such as those apparent within clonal lineages) are not neutral at all but are slightly deleterious. This is evident by the observation that a higher proportion of nonsynonymous changes (which are likely to be slightly deleterious) are observed between sequences at the very initial stages of divergence (Rocha et al., 2006; Balbi and Feil, 2007; Balbi et al., 2009). Although the neutrality of synonymous changes is a widely held approximation, is this really valid? Codon bias may impose fitness costs on maladapative synonymous changes (i.e., those to unpreferred codons) (Hershberg and Petrov, 2008), and synonymous GC->AT changes, which are enriched over the short term by mutation bias, have also been shown to be preferentially purged by selection over time in *E. coli* (Balbi et al., 2009). Wilson recently estimated that approximately 60% of nascent molecular variation is purged by purifying selection in *Campylobacter jejuni* (Wilson et al., 2009).

Purifying selection of deleterious mutations offers a simple means by which the accumulation of standing variation within clonal complexes could be partially constrained. Furthermore, a gradual loss of fitness due to the stochastic fixation of deleterious mutations within a clonal complex means it is easier to imagine repeated adaptive mutations arising, as these would be necessary just to retain fitness rather than to increase it continually (Fig. 2.3b). The high rate of recombination observed in many populations is also consistent with such a model. While recombination may mean that selective sweeps only partially purge diversity (as the adaptive change is not tightly linked to all other genes on the genome), this would be beneficial as it would prevent the fixation of some of the deleterious changes hitchhiking with the adaptive change (Rocha et al., 2006). It should also be noted that uncertainties concerning the effective population size have a knock-on effect for estimates of the *effective* neutral mutation rate. As the strength of purifying selection is far greater in large populations than in small ones, these two parameters should be inversely proportional (Ohta, 1973). In many respects, the processes described above fit those classically reserved for "species," and it may be illuminating to compare intra- versus inter "population" patterns of diversity on the basis of clonal complex.

The effect of purifying selection on deleterious changes is distinct from positive selection on adaptive change in that it is not incompatible with the neutral theory. A problem may have arisen due to a rather casual acceptance that all synonymous changes are neutral and that deleterious changes are likely to have been removed prior to sampling. There is surprisingly little basis for this assumption, and much of the synonymous variation observed between closely related strains may well be destined for selective removal over time. Advocates of MLST, myself included, have repeatedly claimed that the variation within MLST alleles is likely to be neutral, a claim occasionally tempered by the scarcely adequate disclaimer "or nearly so." While this may do to justify the phylogenetic and typing utility of MLST data, it is less obvious that slightly deleterious changes can be safely ignored for detailed simulations of population dynamics, in which I include the coalescent (Chapter 1). The emergence and subsequent purification of slightly deleterious changes will result in an apparently higher rate of mutation toward the tips of the tree, thus mimicking the effect of recent population growth.

Finally, it is worth considering the likely stability of the niche occupied by a single ecotype. For environmental taxa, abiotic factors such as pH, temperature, and nutrient availability are most often assumed relevant, and one might imagine these to be relatively stable. In contrast, for pathogenic bacteria, biotic factors, such as immune pressure, are most often cited (Fraser et al., 2009). However, there is little biological basis for this distinction. Biotic factors, in the form of interstrain/species competition, predation, phage infection, and transient associations with eukaryotes, are likely to form key dynamic components of the selective landscape in the environment (Vos et al., 2009). Thus, it may be that the primary role of selective sweeps in nature, even when partial, is in preventing catastrophic loss of fitness due to the accumulation of deleterious mutations and a dynamic selective landscape.

2.4 CLUSTERING TECHNIQUES

In the 10 years since the publication of the original MLST paper by Maiden et al. (1998), the use of allele-based data for clustering isolates has remained one of the most contentious and, for many, anachronistic issues. Dissenting voices argue that the proportion of allelic mismatches between isolates, the distance measure upon which clustering techniques are based, provides only a fraction of the information contained within the sequences themselves. Treating alleles simply as either the same or different ignores the fact that some may be different by a single base, whereas others may be different by 20% of sites. Phylogenetic analyses will of course capture this information easily, so why throw it all away?

A single recombination event may introduce one, or many, changes into a sequence. This means that the amount of sequence divergence between strains is not necessarily tightly coupled to the number of events each strain has encountered since sharing a common ancestor, and that the number of events, rather than the overall sequence divergence, is a more reliable estimator of relatedness. A recombination event that changes 20 sites within an MLST allele may radically affect its position on a tree, particularly if the imported allele is present in unrelated strains in the data set. By considering only alleles as the same or different, clustering procedures will score this as a single event, regardless of how many bases are affected.

An important caveat of this logic is that it only really makes sense when considering very closely related strains differing at one, or maybe two, of seven MLST loci. In such cases, the high level of allelic identity is unlikely to be due to convergence, even in freely recombining populations, and thus reflects identity by descent. In contrast, stochastic effects mean there is little (or no) basis for assuming that two strains differing at, say, five loci are any more closely related than two strains differing at six loci. Such strains may well have experienced multiple events at single loci. Further, the effect of convergence by recombination between more distantly related isolates may be profound, particularly in cases where a small number of alleles at a given locus are very common, and a large number of alleles are very rare (the typical situation for most multilocus data sets, and one consistent with the emergence of a high proportion of slightly deleterious mutations). It is also important to consider that the "lumpiness" (presence of clonal complexes) in bacterial populations means that the relationship between divergence time and the number of allelic mismatches is completely nonlinear. Whereas isolates differing at zero, one, or two MLST alleles may belong to the same clonal cluster, hence are very closely related, more distant comparisons are likely to be between clonal complexes, corresponding to a significant jump in divergence time (Feil, 2004). Finally, the majority of pairwise comparisons within multilocus data sets are likely to be different at all seven alleles, meaning that there is no information by which such isolates may be clustered.

Traditional clustering methods such as unweighted pair group method with arithmetic mean (UPGMA) (Sneath, 1973), as well as the currently popular minimum spanning tree as implemented in Bionumerics™, should therefore be interpreted with caution. Whereas these methods may be used to identify clonal complexes—indeed UPGMA was the standard clustering tool used for MLEE data sets—they also attempt to depict relationships between them. For modestly recombining populations, or perhaps even for nonrecombining populations, the links between clonal complexes as assigned by clustering procedures are likely to be completely arbitrary. For example, Didelot et al. (2009) recently demonstrated that robust relationships between clonal complexes of *N. meningitidis* could not be reconstructed, even using state-of-the-art Bayesian approaches designed to account for recombination events, and sequence data incorporating 20 loci (Chapter 3). Given this, it seems foolhardy indeed to imagine that simple clustering tools will be able to do any better.

2.4.1 eBURST as a Simple Exploratory Tool

As the early MLST data sets grew, so two things became apparent: (i) The data confirmed the basic population structure evident from MLEE data that clonal complexes were present in many populations, and (ii) UPGMA was not ideal for exploring these data because it attempted to link the clusters together. The based upon related sequence types (BURST) algorithm was an attempt to address these issues simply by only focusing on relationships within clonal complexes and by ignoring those between them. This approach immediately presented a second advantage, that by presenting the data as a forest of discrete trees, it became possible to represent very large data sets in a single figure. eBURST (http://eburst.mlst.net/), the JAVA implementation of BURST, remains the only clustering procedure capable of representing thousands of genotypes within a single figure (Feil et al., 2004) (Fig. 2.4).

A full description of the BURST algorithm is given in the eBURST instructions (http://eburst.mlst.net/), but briefly, the approach is as follows. The first step is to divide the data into mutually exclusive groups approximating to clonal complexes. These groups are defined on the basis that each member of a group must share at least a theshold number of alleles in common with at least one other member of the group. The default setting for eBURST is the most conservative definition; every member of the group must be exhibit no more than a single allelic mismatch from at least one other member. Once the groups are assigned, the algorithm examines the degree to which the variation within the groups corresponds to a simple model of radial diversification from a clonal founder. Founders

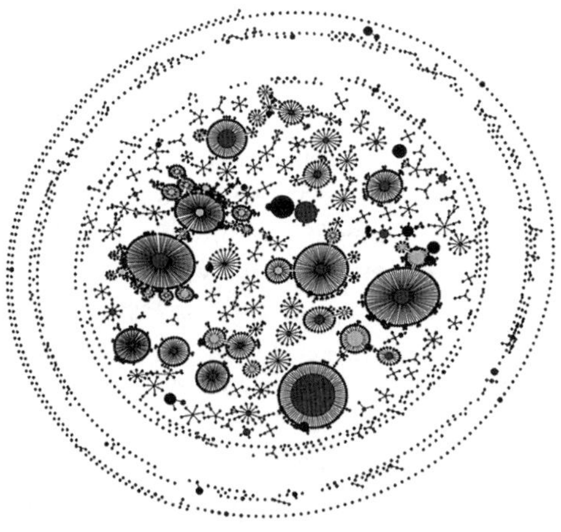

Figure 2.4 eBURST representation of the MLST data set for *N. meningitidis*. Each circle represents an ST. The frequency of the ST is indicated by the size of the circle. STs that only differ by one locus are linked (see text). Clonal founders are shown in blue; subfounders are shown in yellow. The pattern shows large clonal complexes against a background of diversity, as predicted by the "epidemic model" (e.g., Figure 2.2, as seen from the top). See color insert.

are assigned on the basis that they define the largest number of single-locus variants (SLVs) when examined against all other genotypes in the group. These SLVs are then linked to the founder. The largest subgroup is then defined in the same way, and SLVs are again preferentially linked to this subgroup founder. The process continues until all strains in the group are linked.

The BURST algorithm provides a single acyclic solution by preferentially assigning founders as "centers of gravity" according to the number of SLVs they define. It should, however, be borne in mind that there will be a large number of alternative ways by which STs within a single complex could be linked, and for this reason, eBURST was designed to be exploratory. As Prim's algorithm does not distinguish between different optimal solutions, which may be extremely numerous, the Bionumerics implementation of the minimum spanning tree borrows the BURST rules for closely related strains. Hence, the two approaches give identical results for intracomplex patterns. The interested reader is invited to explore their data using the minimum spanning tree tool at http://pubmlst.org/ without invoking BURST rules.

eBURST provides the utility to display all possible SLV (and double-locus variant [DLV]) links within a group, and this illustrates the myriad of ways in which the STs could potentially be linked. As an example, consider two strains that have diverged from a founder at a single locus. These will both be SLVs of the founder, but if by chance the same locus was affected in both cases, they will also be SLVs of each other, and so could potentially be linked. Similarly, it is also common that an ST is an SLV of more than one founder within a group.

The assignment of one overall clonal founder is often a close call, and may be due to one founder defining one or two SLVs more than another founder. eBURST thus provides the option to manually change the group founder. It also provides bootstrap scores, estimated by resampling, in order to gauge the confidence of a founder assignment, along with other lines of evidence such as the frequency of the assigned founders (these are more likely to be, and often are, the most frequent genotypes in the group) and the average distance from all other members of the group. Other lines of evidence, including clinical, phenotypic, epidemiological, and genetic, should also be used to interpret eBURST outputs.

2.4.2 BURST Aid

A basic eBURST diagram provides only three types of information: whether two strains are linked or not, the frequency of a given ST (illustrated by the diameter of the circle), and the assignment of group (blue) or subgroup (yellow) founders. Comparative eBURST can be used to compare two data sets. The lengths and angles of the links mean nothing; neither does the relative positioning of the groups or singletons. "Population snapshots," which show all the groups within the data set (e.g., Fig. 2.4), are generated by setting the group definition to zero alleles in common. eBURST is designed to deal with very large data sets, and it is much more informative to analyze one's own data set within the context of the entire MLST database, if available. Small samples on their own may not give a very representative picture of the overall clonal structure of the species, often producing a series of unconnected dots with the occasional doublet. An alternative implementation of the BURST approach, goeBURST (http://goeburst.phyloviz.net/), has recently been developed, which incorporates double-locus links that may be meaningful under some circumstances (Francisco et al., 2009). This paper is also recommended for an excellent discussion of the underlying network theory.

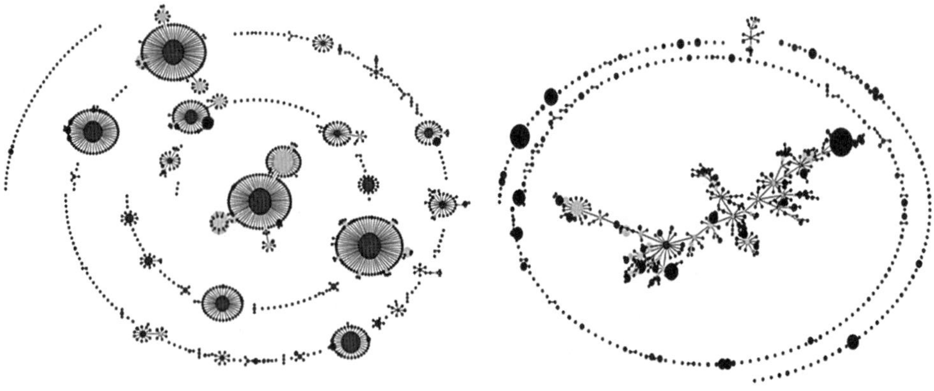

Figure 2.5 Interpreting eBURST diagrams with respect to recombination rate. The *S. aureus* figure illustrates discrete clonal lineages, diversifying in a radial fashion from well-supported clonal founders. The output from *Burkholderia pseudomallei* illustrates that many STs have merged into a single "straggly" group, with less clear evidence of radial diversification and poorly supported founders. This difference is likely due to much higher rates of recombination in the latter species. See color insert.

As discussed above, eBURST can be used to examine how closely real data corresponds to a simple model of radial diversification from a small number of clonal founders. For populations with low rates of recombination, such as *S. aureus*, the population snapshot corresponds closely to such a model. Founders and SLVs can be defined with a high level of confidence in this species, and these are often supported by other lines of genetic and phenotypic evidence (e.g., antibtiotic resistance). Other species, such as *Burkholderia pseudomallei*, do not produce clean radial groups but instead produce large "straggly" groups containing many STs, but large numbers of these are not clearly descended from well-supported founders or subfounders. This pattern is likely to be the result of high rates of recombination, which can result in the merging of discrete lineages and the loss of radial structure (Fig. 2.5).

Turner et al. (2007) tested the effect of recombination on eBURST groups by simulation. They noted that the accuracy of links inferred by eBURST was good for low (>90%) or moderate (>85%) rates of recombination, a range that encompasses the majority of MLST data sets. However, the accuracy of eBURST was found to drop off rapidly when rates of recombination were very high (~60%). They also suggested that the percentage of STs in the largest group can be used as a reasonable proxy for recombination rates and that, as a rule of thumb, if >25% of the STs in a given representative data set belong to a single large straggly group, then the reliability of the eBURST links is low. However, this does not mean that eBURST has not told you something useful; on the contrary, it has pointed to the possibility of high rates of recombination, which can then be explored using other methods. The species with perhaps the highest rate of recombination, *Helicobacter pylori*, does not produce a single straggly group but a series of unconnected dots. This reflects the fact that nearly every strain is a unique, and unrelated, ST, and that levels of allelic diversity are extremely high.

eBURST can also be used to estimate the relative contributions of recombination (r) and mutation (m) to clonal divergence by comparing variant alleles within SLVs with the equivalent alleles in the assigned founders. This approach, which was inspired by a paper by Guttman and Dykhuizen (1994) on closely related *E. coli* sequences and is described in full elsewhere (Feil et al., 1999, 2000), has been used to confirm high rates of recombination in *N. meningitidis* and in *Streptococcus pneumoniae*, and much lower rates in *S. aureus*. The r/m ratio estimated by this approach for *N. meningitidis* (~4.5 : 1.0) is consistent with a recent analysis based on 20 gene loci using ClonalFrame (Didelot et al., 2009). Finally, it should be noted that clustering procedures are only ever as good as the input data. MLST data are based on a small proportion of the genome; thus, inferences drawn from them using eBURST, or any other method, may not be reflective of the whole genome.

REFERENCES

Avise, J. C. (1994) *Molecular Markers, Natural History and Evolution*. Chapman and Hall, New York.

Balbi, K. J. and Feil, E. J. (2007) The rise and fall of deleterious mutation. *Res Microbiol* **158,** 779–786.

Balbi, K. J., Rocha, E. P., and Feil, E. J. (2009) The temporal dynamics of slightly deleterious mutations in *Escherichia coli* and *Shigella* spp. *Mol Biol Evol* **26,** 345–355.

Brown, A. H., Feldman, M. W., and Nevo, E. (1980) Multilocus structure of natural populations of *Hordeum spontaneum*. *Genetics* **96,** 523–536.

Buckee, C. O., Jolley, K. A., Recker, M., Penman, B., Kriz, P., Gupta, S., and Maiden, M. C. (2008) Role of selection in the emergence of lineages and the evolution of virulence in *Neisseria meningitidis*. *Proc Natl Acad Sci U S A* **105,** 15082–15087.

Caugant, D. A., Zollinger, W. D., Mocca, L. F., Frasch, C. E., Whittam, T. S., Froholm, L. O., and Selander, R. K. (1987) Genetic relationships and clonal population structure of serotype 2 strains of *Neisseria meningitidis*. *Infect Immun* **55,** 1503–1512.

Cohan, F. M. (2002) What are bacterial species? *Annu Rev Microbiol* **56,** 457–487.

Cohan, F. M. and Perry, E. B. (2007) A systematics for discovering the fundamental units of bacterial diversity. *Curr Biol* **17,** R373–R386.

Didelot, X., Urwin, R., Maiden, M. C., and Falush, D. (2009) Genealogical typing of *Neisseria meningitidis*. *Microbiology* **155**(10), 3176–3186.

Doolittle, W. F. (2009) Eradicating typological thinking in prokaryotic systematics and evolution. *Cold Spring Harb Symp Quant Biol*, Aug 10 [Epub ahead of print].

Feil, E. J. (2004) Small change: Keeping pace with microevolution. *Nat Rev Microbiol* **2,** 483–495.

Feil, E. J., Li, B. C., Aanensen, D. M., Hanage, W. P., and Spratt, B. G. (2004) eBURST: Inferring patterns of evolutionary descent among clusters of related bacterial genotypes from multilocus sequence typing data. *J Bacteriol* **186,** 1518–1530.

Feil, E. J., Maiden, M. C., Achtman, M., and Spratt, B. G. (1999) The relative contributions of recombination and mutation to the divergence of clones of *Neisseria meningitidis*. *Mol Biol Evol* **16,** 1496–1502.

Feil, E. J., Smith, J. M., Enright, M. C., and Spratt, B. G. (2000) Estimating recombinational parameters in *Streptococcus pneumoniae* from multilocus sequence typing data. *Genetics* **154,** 1439–1450.

Francisco, A. P., Bugalho, M., Ramirez, M., and Carrico, J. A. (2009) Global optimal eBURST analysis of multilocus typing data using a graphic matroid approach. *BMC Bioinformatics* **10,** 152.

Fraser, C., Alm, E. J., Polz, M. F., Spratt, B. G., and Hanage, W. P. (2009) The bacterial species challenge: Making sense of genetic and ecological diversity. *Science* **323,** 741–746.

Fraser, C., Hanage, W. P., and Spratt, B. G. (2005) Neutral microepidemic evolution of bacterial pathogens. *Proc Natl Acad Sci U S A* **102,** 1968–1973.

Fraser, C., Hanage, W. P., and Spratt, B. G. (2007) Recombination and the nature of bacterial speciation. *Science* **315,** 476–480.

Gevers, D., Cohan, F. M., Lawrence, J. G., Spratt, B. G., Coenye, T., Feil, E. J., Stackebrandt, E., Van de Peer, Y., Vandamme, P., Thompson, F. L., and Swings, J. (2005) Opinion: Re-evaluating prokaryotic species. *Nat Rev Microbiol* **3,** 733–739.

Gillespie, J. H. (1987) Molecular evolution and the neutral allele theory. *Oxf Surv Evol Biol* **4,** 10–37.

Goethert, H. K., Saviet, B., and Telford, S. R. III (2009) Metapopulation structure for perpetuation of *Francisella tularensis tularensis*. *BMC Microbiol* **9,** 147.

Guttman, D. S. and Dykhuizen, D. E. (1994) Clonal divergence in *Escherichia coli* as a result of recombination, not mutation. *Science* **266,** 1380–1383.

Hahn, M. W. (2008) Toward a selection theory of molecular evolution. *Evolution* **62,** 255–265.

Harris, H. (1966) Enzyme polymorphisms in man. *Proc R Soc Lond B Biol Sci* **164,** 298–310.

Haubold, B. and Hudson, R. R. (2000) LIAN 3.0: Detecting linkage disequilibrium in multilocus data. Linkage analysis. *Bioinformatics* **16,** 847–848.

Hershberg, R. and Petrov, D. A. (2008) Selection on codon bias. *Annu Rev Genet* **42,** 287–299.

Hudson, R. R. (1994) Analytical results concerning linkage disequlibrium in models with genetic transformation and conjugation. *J Evol Biol* **7,** 535–548.

Kaiser, S., Biehler, K., and Jonas, D. (2009) A *Stenotrophomonas maltophilia* multilocus sequence typing scheme for inferring population structure. *J Bacteriol* **191,** 2934–2943.

KIMURA, M. (1968a) Evolutionary rate at the molecular level. *Nature* **217,** 624–626.

KIMURA, M. (1968b) Genetic variability maintained in a finite population due to mutational production of neutral and nearly neutral isoalleles. *Genet Res* **11,** 247–269.

KIMURA, M. (1979) The neutral theory of molecular evolution. *Sci Am* **241,** 98–100, 102, 108 passim.

KIMURA, M. and CROW, J. F. (1964) The number of alleles that can be maintained in a finite population. *Genetics* **49,** 725–738.

KOEPPEL, A., PERRY, E. B., SIKORSKI, J., KRIZANC, D., WARNER, A., WARD, D. M., ROONEY, A. P., BRAMBILLA, E., CONNOR, N., RATCLIFF, R. M., NEVO, E., and COHAN, F. M. (2008) Identifying the fundamental units of bacterial diversity: A paradigm shift to incorporate ecology into bacterial systematics. *Proc Natl Acad Sci U S A* **105,** 2504–2509.

LENSKI, R. E. (1993) Assessing the genetic structure of microbial populations. *Proc Natl Acad Sci U S A* **90,** 4334–4336.

LEVIN, B. R. (1981) Periodic selection, infectious gene exchange and the genetic structure of *E. coli* populations. *Genetics* **99,** 1–23.

LEWONTIN, R. C. and HUBBY, J. L. (1966) A molecular approach to the study of genic heterozygosity in natural populations. II. Amount of variation and degree of heterozygosity in natural populations of *Drosophila pseudoobscura*. *Genetics* **54,** 595–609.

MAIDEN, M. C. (2006) Multilocus sequence typing of bacteria. *Annu Rev Microbiol* **60,** 561–588.

MAIDEN, M. C., BYGRAVES, J. A., FEIL, E., MORELLI, G., RUSSELL, J. E., URWIN, R., ZHANG, Q., ZHOU, J., ZURTH, K., CAUGANT, D. A., FEAVERS, I. M., ACHTMAN, M., and SPRATT, B. G. (1998) Multilocus sequence typing: A portable approach to the identification of clones within populations of pathogenic microorganisms. *Proc Natl Acad Sci U S A* **95,** 3140–3145.

MARTINY, J. B., BOHANNAN, B. J., BROWN, J. H., COLWELL, R. K., FUHRMAN, J. A., GREEN, J. L., HORNER-DEVINE, M. C., KANE, M., KRUMINS, J. A., KUSKE, C. R., MORIN, P. J., NAEEM, S., OVREAS, L., REYSENBACH, A. L., SMITH, V. H., and STALEY, J. T. (2006) Microbial biogeography: Putting microorganisms on the map. *Nat Rev Microbiol* **4,** 102–112.

MILKMAN, R. (1973) Electrophoretic variation in *Escherichia coli* from natural sources. *Science* **182,** 1024–1026.

MUSSER, J. M., KROLL, J. S., MOXON, E. R., and SELANDER, R. K. (1988) Evolutionary genetics of the encapsulated strains of *Haemophilus influenzae*. *Proc Natl Acad Sci U S A* **85,** 7758–7762.

NESBO, C. L., DLUTEK, M., and DOOLITTLE, W. F. (2006) Recombination in *Thermotoga*: Implications for species concepts and biogeography. *Genetics* **172,** 759–769.

NORRIS, J. R. (1963) Esterases of cyrstalliferous bacteria pathogenic for insects: Epizootological application. *J Insect Pathol* **5,** 460–472.

OCHMAN, H. and SELANDER, R. K. (1984) Evidence for clonal population structure in *Escherichia coli*. *Proc Natl Acad Sci U S A* **81,** 198–201.

OHTA, T. (1973) Slightly deleterious mutant substitutions in evolution. *Nature* **246,** 96–98.

ORSKOV, I. and ORSKOV, F. (1973) Plasmid-determined H2S character in *Escherichia coli* and its relation to plasmid-carried raffinose fermentation and tetracycline resistance characters. Examination of 32 H2S-positive strains isolated during the years 1950 to 1971. *J Gen Microbiol* **77,** 487–499.

PAGALING, E., WANG, H., VENABLES, M., WALLACE, A., GRANT, W. D., COWAN, D. A., JONES, B. E., MA, Y., VENTOSA, A., and HEAPHY, S. (2009) Microbial biogeography of six salt lakes in Inner Mongolia China and a salt lake in Argentina. *Appl Environ Microbiol* **75**(18), 5750–5760.

PAPKE, R. T., ZHAXYBAYEVA, O., FEIL, E. J., SOMMERFELD, K., MUISE, D., and DOOLITTLE, W. F. (2007) Searching for species in haloarchaea. *Proc Natl Acad Sci U S A* **104,** 14092–14097.

RAMETTE, A. and TIEDJE, J. M. (2007) Biogeography: An emerging cornerstone for understanding prokaryotic diversity, ecology, and evolution. *Microb Ecol* **53,** 197–207.

ROCHA, E. P., SMITH, J. M., HURST, L. D., HOLDEN, M. T., COOPER, J. E., SMITH, N. H., and FEIL, E. J. (2006) Comparisons of dN/dS are time dependent for closely related bacterial genomes. *J Theor Biol* **239,** 226–235.

RUIMY, R., ARMAND-LEFEVRE, L., BARBIER, F., RUPPE, E., COCOJARU, R., MESLI, Y., MAIGA, A., BENKALFAT, M., BENCHOUK, S., HASSAINE, H., DUFOURCQ, J. B., NARETH, C., SARTHOU, J. L., ANDREMONT, A., and FEIL, E. J. (2009) Comparisons between geographically diverse samples of carried *Staphylococcus aureus*. *J Bacteriol* **191,** 5577–5583.

SELANDER, R. K., CAUGANT, D. A., OCHMAN, H., MUSSER, J. M., GILMOUR, M. N., and WHITTAM, T. S. (1986) Methods of multilocus enzyme electrophoresis for bacterial population genetics and systematics. *Appl Environ Microbiol* **51,** 873–884.

SELANDER, R. K., MUSSER, J. M., CAUGANT, D. A., GILMOUR, M. N., and WHITTAM, T. S. (1987) Population genetics of pathogenic bacteria. *Microb Pathog* **3,** 1–7.

SMITH, J. M., FEIL, E. J., and SMITH, N. H. (2000) Population structure and evolutionary dynamics of pathogenic bacteria. *Bioessays* **22,** 1115–1122.

SMITH, J. M., SMITH, N. H., O'ROURKE, M., and SPRATT, B. G. (1993) How clonal are bacteria? *Proc Natl Acad Sci U S A* **90,** 4384–4388.

Sneath (1973) *Numerical Taxonomy*. W.H. Freeman and Company, San Francisco.

SPRATT, B. G. and MAIDEN, M. C. (1999) Bacterial population genetics, evolution and epidemiology. *Philos Trans R Soc Lond B Biol Sci* **354,** 701–710.

STROUS, M. (2007) Data storm. *Environ Microbiol* **9,** 10–11.

TURNER, K. M. and FEIL, E. J. (2007) The secret life of the multilocus sequence type. *Int J Antimicrob Agents* **29,** 129–135.

TURNER, K. M., HANAGE, W. P., FRASER, C., CONNOR, T. R., and SPRATT, B. G. (2007) Assessing the reliability of eBURST using simulated populations with known ancestry. *BMC Microbiol* **7,** 30.

Vesaratchavest, M., Tumapa, S., Day, N. P., Wuthiekanun, V., Chierakul, W., Holden, M. T., White, N. J., Currie, B. J., Spratt, B. G., Feil, E. J., and Peacock, S. J. (2006) Nonrandom distribution of *Burkholderia pseudomallei* clones in relation to geographical location and virulence. *J Clin Microbiol*, **44,** 2553–7.

Vos, M. (2009) Why do bacteria engage in homologous recombination? *Trends Microbiol* **17,** 226–232.

Vos, M., Birkett, P. J., Birch, E., Griffiths, R. I., and Buckling, A. (2009) Local adaptation of bacteriophages to their bacterial hosts in soil. *Science* **325,** 833.

Vos, M. and Velicer, G. J. (2008) Isolation by distance in the spore-forming soil bacterium *Myxococcus xanthus*. *Curr Biol* **18,** 386–391.

Waldron, D. E. and Lindsay, J. A. (2006) Sau1: A novel lineage-specific type I restriction-modification system that blocks horizontal gene transfer into *Staphylococcus aureus* and between *S. aureus* isolates of different lineages. *J Bacteriol* **188,** 5578–5585.

Weir, B. S. and Hill, W. G. (2002) Estimating F-statistics. *Annu Rev Genet* **36,** 721–750.

Whitaker, R. J., Grogan, D. W., and Taylor, J. W. (2003) Geographic barriers isolate endemic populations of hyperthermophilic archaea. *Science* **301,** 976–978.

Whittam, T. S., Ochman, H., and Selander, R. K. (1983) Multilocus genetic structure in natural populations of *Escherichia coli*. *Proc Natl Acad Sci U S A* **80,** 1751–1755.

Wilson, D. J., Gabriel, E., Leatherbarrow, A. J., Cheesbrough, J., Gee, S., Bolton, E., Fox, A., Hart, C. A., Diggle, P. J., and Fearnhead, P. (2009) Rapid evolution and the importance of recombination to the gastroenteric pathogen *Campylobacter jejuni*. *Mol Biol Evol* **26,** 385–397.

Chapter 3

Sequence-Based Analysis of Bacterial Population Structures

XAVIER DIDELOT

3.1 INTRODUCTION

The aim of this chapter is to introduce the sequence-based methods of analysis of bacterial population structures used in subsequent chapters. By "sequence" is usually meant a fragment of the chain of bases adenine (A), guanine (G), cytosine (C), and thymine (T) that make up DNA. The methods we describe can, however, be equally applied to other types of biological sequences such as the chains of amino acids that make up proteins.

The modern methodology for sequencing DNA was first introduced by Sanger et al. (1977), and its efficiency was later greatly improved by the discovery of the polymerase chain reaction (PCR) by Saiki et al. (1985, 1988). Sequencing has since become increasingly cheap and easy to perform. The first bacterial genomes to be sequenced were those of *Haemophilus influenzae* (Fleischmann et al., 1995) followed by *Escherichia coli* (Blattner et al., 1997). At the time of writing, more than 700 complete bacterial genomes have been published in the Genomes OnLine Database (GOLD; Liolios et al., 2006). Several species have multiple genomes (e.g., there are at least 15 genomes for both *Salmonella enterica* and *E. coli*), which opens up the possibility to investigate bacterial population structures based on whole genomes. The recent development of high-throughput sequencing methods such as 454 sequencing (Margulies et al., 2005) or Solexa sequencing (Bennett et al., 2005) is making DNA sequencing even cheaper, so that there may soon be hundreds of genomes available for bacterial species of interest.

Genomic sequences clearly contain a lot of information about bacterial population structures. Extracting this information typically involves the identification of homologous nucleotides (i.e., nucleotides from the different sequences that are derived from the same ancestor) and then the study of the patterns of diversity they exhibit. The first step in this process is called "alignment" and is the subject of our next section, whereas the analysis of alignments is treated in subsequent sections.

Bacterial Population Genetics in Infectious Disease, Edited by D. Ashley Robinson, Daniel Falush, and Edward J. Feil

3.2 ALIGNMENTS

3.2.1 The Need for Alignments

Here we consider three basic processes altering a DNA sequence over time: substitution, insertion, and deletion. When a substitution occurs, a given nucleotide is replaced by another one. When an insertion happens, a number of nucleotides are inserted in a given location. When a deletion takes place, a number of adjacent nucleotides are removed from the sequence. For example, the following DNA sequence is a fragment from the *abcZ* gene in *Neisseria meningitidis*:

Sequence 1:

```
         1         2
12345678901234567890
TTTGATACTGTTGCCGAAGG
```

If a substitution A → C occurred at site 5, followed by an insertion of ACT at site 10 and a deletion of two nucleotides at site 15 (which is all very unlikely since *abcZ* is a housekeeping gene!), we would obtain the following sequence:

Sequence 2:

```
         1         2
123456789012345678901
TTTGCTACTACTGTCCGAAGG
```

Note that since insertions and deletions are of arbitrary lengths, sequence 2 is not necessarily of the same length as sequence 1. Yet, since sequence 2 was derived from sequence 1 through the action of substitution, insertion, and deletion, they are called homologous sequences.

We now consider that a bacteria carrying the first sequence gave birth (by binary fission) to two daughter cells 1 and 2, each carrying the first sequence, but that cell 2 later endures the substitution, insertion, and deletion described above so that it now carries sequence 2. Cells 1 and 2 are then sequenced and are found to carry sequences 1 and 2, respectively. It is impossible to reconstruct from these sequences the exact list of events (substitutions, insertions, and deletions) that happened to the cells. For example, the fact that cell 1 carries an A at site 5 and cell 2 carries a C is due to the substitution A → C that happened to cell 2, but could just as well be explained by a substitution C → A happening to cell 1. Similarly, the insertion in cell 2 has the same effect as a deletion in cell 1 and vice versa. This problem is made more difficult by the possibility of overlapping events, such as a substitution that disappears in a deletion, or an insertion in one sequence and a deletion in the other at the same site, which may look like a single event. It becomes even harder when considering more than two sequences. For these reasons, reconstituting the list of events that gave rise to a set of observed homologous sequences is never attempted. Instead, we limit ourselves to building an alignment of the sequences, which means finding the groups of nucleotides from the different sequences that are homologous (i.e., derived from the same nucleotide in their most recent common ancestor).

In our example, sites 1–9 in sequence 1 are homologous to sites 1–9 in sequence 2. Sites 10–12 in sequence 2 are not homologous to any site in sequence 1. Sites 10 and 11 in sequence 1 are homologous to sites 13 and 14 in sequence 2. Sites 12 and 13 in sequence 1 are not homologous to any site in sequence 2. Sites 14–20 in sequence 1 are homologous to sites 15–21 in sequence 2. An alignment of sequences is traditionally represented by inserting gaps in the sequences for the sites where it has no homologue to the other sequences. In our example, the alignment of sequences 1 and 2 would therefore be represented as

First alignment of sequences 1 and 2:

```
             1         2
123456789---01234567890 Position in Sequence 1
TTTGATACT---GTTGCCGAAGG
TTTGCTACTACTGT--CCGAAGG
12345678901234--5678901 Position in Sequence 2
         1         2
```

In the example above, we have used our knowledge of the events that happened to cells 1 and 2 to find out which nucleotides were homologous and therefore build the alignment of sequences 1 and 2 that we know to be correct. The alignment above uses one gap of size 3, one gap of size 2, and one substitution (at site 5). If we did not know the events that happened to cells 1 and 2, and that we only knew sequences 1 and 2, there would be other possibilities of alignments, for example,

Second alignment of sequences 1 and 2:

```
           1         2
123456789-01234567890 Position in Sequence 1
TTTGATACT-GTTGCCGAAGG
TTTGCTACTACTGTCCGAAGG
123456789012345678901 Position in Sequence 2
         1         2
```

The alignment above uses one gap of size 1 and four substitutions (at sites 5, 10, 12, and 13). Given that several alignments are compatible with a set of sequences, how can we decide which one is correct? To do so, we give each possible alignment a score. Elaborate scoring functions are required for amino acid chains, the two most popular ones being the blocks substitution matrix (BLOSOM) score (Henikoff and Henikoff, 1992) and the point accepted mutation (PAM) score (Dayhoff et al., 1978). For DNA sequences, simple scoring functions are often used, which involve the number of matches, mismatches, gaps, and their lengths. For example, the default score of ClustalW (Thompson et al., 1994) is equal to the number of matches, minus a penalty for each gap equal to 10 plus 0.1 times its length after the first site. Using this scoring function, we find a score of $17 - (10 + 2 \times 0.1) - (10 + 1 \times 0.1) = -3.3$ for the first alignment and a score of $16 - 10 = 6$ for the second one. Since the second alignment has a higher score, we conclude that it is more likely to be correct than the first one.

With this formulation, the problem of aligning a set of homologous sequences requires us to explore the set of all possible alignments in search for the one with the highest score. We will first describe how this is done for pairwise alignments (i.e., alignments of two sequences) and then how it can be extended to multiple alignments (i.e., alignments of more than two sequences).

3.2.2 Pairwise Alignment

Constructing alignments for pairs of sequences may not seem a very important application since we often want to align more than two sequences, but the methodology we describe here to build a pairwise alignment underpins all multiple aligners. So let us consider a set of two homologous sequences. The number of possible alignments is very large for sequences of a realistic size, and it is therefore not feasible to enumerate them all. Instead, there exists a powerful dynamic programming procedure to quickly find the optimal alignment: the Needleman–Wunsch algorithm (Needleman and Wunsch, 1970). The details of this algorithm are fairly complex, but since it represents the foundation for any alignment work, we will describe broadly how it works.

The idea is to recursively build the matrix $F(i, j)$ of the score of the best alignment between the first sequence up to the ith nucleotide against the second sequence up to the jth nucleotide. If i is the length of the first sequence and j is the length of the second sequence, then clearly, $F(i, j)$ is equal to the score of the best alignment of the two whole sequences. Here we assume for simplicity that the score is equal to the number of matching nucleotides minus the sum of the lengths of the gaps, although any scoring function can be used.

We start by noticing that an empty alignment has a score of 0, so that $F(0, 0) = 0$. We put this value at the top left of the matrix F. We then fill in the first row $F(0, j)$ with the scores corresponding to a gap of size j (this is equal to $-j$ with our scoring function) because the only way that sequence 1 up to site 0 aligns against sequence 2 up to site j is if we use a gap of size j. We also fill in the first column $F(i, 0)$ with the scores corresponding to a gap of size i (here $-i$) for the same reasons.

For the rest of the entries of the matrix F, we notice that there are three configurations for sequence 1 up to i to align against sequence 2 up to j:

1. Sequence 1 up to i is aligned against sequence 2 up to $j - 1$, and the jth nucleotide in sequence 2 aligns against a gap in sequence 1. The score of this configuration is therefore $F(i, j - 1) - 1$.
2. Sequence 1 up to $i - 1$ is aligned against sequence 2 up to j, and the ith nucleotide in sequence 1 aligns against a gap in sequence 2. The score of this configuration is therefore $F(i - 1, j) - 1$.
3. Sequence 1 up to $i - 1$ is aligned against sequence 2 up to $j - 1$, and the ith nucleotide in sequence 1 aligns against the jth nucleotide in sequence 2. The score of this configuration is therefore $F(i - 1, j - 1)$ if sites i and j mismatch, or $F(i - 1, j - 1) + 1$ if they match.

Since $F(i, j)$ is the score of the best alignment up to (i, j), we take $F(i, j)$ equal to the maximum of the score of these three configurations. This allows filling in the matrix row after row, from left to right. As we fill in each cell of the matrix, we use an arrow to indicate which of the three possibilities above gave the best score. We eventually reach the bottom-right cell, which corresponds to the best score of the alignment of sequences 1 and 2 in their entirety.

An example of the algorithm being run on the sequences "CTGTTGCCG" and "CTACTGTCCG" (these are the same as sequences 1 and 2 from the previous example except that the first seven and last four nucleotides of each sequence have been removed for simplicity) is shown in Fig. 3.1.

The best score for an alignment of these two sequences is equal to 6. By tracing back the arrow from the bottom-right corner to the top-left corner, we reconstitute the alignment that gives this score:

```
CTGTTG-CCG
CTACTGTCCG
```

		C	T	A	C	T	G	T	C	C	G
	0	← -1	← -2	← -3	← -4	← -5	← -6	← -7	← -8	← -9	← -10
C	-1	↖ 1	← 0	← -1	← -2	← -3	← -4	← -5	← -6	← -7	← -8
T	-2	↑ 0	↖ 2	← 1	← 0	← -1	← -2	← -3	← -4	← -5	← -6
G	-3	↑ -1	↑ 1	↖ 2	← 1	← 0	↖ 0	← -1	← -2	← -3	← -4
T	-4	↑ -2	↑ 0	↑ 1	↖ 2	↖ 2	← 1	↖ 1	← 0	← -1	← -2
T	-5	↑ -3	↑ 1	↑ 0	↑ 1	↖ 3	← 2	↖ 2	← 1	← 0	← -1
G	-6	↑ -4	↑ -2	↑ -1	↑ 0	↑ 2	↖ 4	← 3	← 2	← 1	↖ 1
C	-7	↑ -5	↑ -3	↑ -2	↖ 0	↑ 1	↑ 3	↖ 4	↖ 4	← 3	← 2
C	-8	↑ -6	↑ -4	↑ -3	↑ -1	↑ 0	↑ 2	↑ 3	↖ 5	↖ 5	← 4
G	-9	↑ -7	↑ -5	↑ -4	↑ -2	↑ -1	↑ 1	↑ 2	↑ 4	↖ 5	↖ 6

Figure 3.1 Alignment of "CTGTTGCCG" and "CTACTGTCCG" using the Needleman–Wunsch algorithm.

It is easy to check that the score of this alignment is indeed 6 (since we have seven matches and one gap). The Needleman–Wunsch algorithm above finds the best scoring global alignment, meaning an alignment of the two sequences in their entirety. A similar dynamic procedure exists, called the Smith–Waterman algorithm (Smith and Waterman, 1981), to find the best local alignment, meaning the best scoring alignment for subsequences of the two sequences.

If we assume that the two sequences are approximately of the same length n, then the algorithm requires filling in a matrix with n^2 entries. The computational cost and the memory requirement of this algorithm therefore scale as n^2. For this reason, the algorithm is unsuitable to align very long sequences such as whole bacterial genomes. We will return to this problem in Section 3.2.4.

3.2.3 Multiple Alignment

We now consider the problem of aligning a set of $N > 2$ sequences. The first approach is to extend the dynamic procedure of the previous section, with the matrix F becoming N-dimensional. Unfortunately, the time taken by the resulting algorithm scales as n^N (Carrillo and Lipman, 1988), which is too slow for application to a reasonable number of sequences. Instead, we need to use a heuristic, which is a method that does not explore the complete space of all possible alignments but that is meant to focus on the high-scoring ones. For this reason, these methods are not guaranteed to give the optimal alignment unlike in the previous section. The most popular heuristic is to build the alignment "progressively" using a guide tree. This method was first conceived by Feng and Doolittle (1987) and follows a three-step procedure:

1. Every pair of sequences is aligned in a pairwise fashion and is given a score of homology. These values are stored into a matrix of pairwise homology.
2. A guide tree is built from the matrix of pairwise homology.
3. The sequences are aggregated into a multiple alignment following the branching order of the guide tree.

Many multiple aligners exist based on this principle, depending on how steps 2 and 3 are performed, the most popular one being ClustalW (Higgins and Sharp, 1988; Higgins et al., 1992; Thompson et al., 1994). In ClustalW, the guide tree is built using the neighbor-joining method of Saitou and Nei (1987) (cf. Section 3.3.3 for a description of the neighbor-joining algorithm). We then explore the guide tree from the leaves up to the root. When two sequences find a common ancestor in the guide tree, we create a new pairwise alignment using the dynamic procedure from the previous section. When a sequence finds a common ancestor in the guide tree with a group of sequences (for which there is already an alignment), a consensus sequence (also called "profile") is built for the preexisting alignment and is aligned with the new sequence using the dynamic procedure. A new alignment that contains the added sequence can thus be deduced. Finally, when two groups of sequences find a common ancestor in the guide tree, a consensus sequence is built for each group; they are aligned using the dynamic procedure, and a new alignment is deduced, which contains the sequences from both groups.

A limitation of the progressive method is that mistakes in the alignment introduced as we explore the guide tree cannot be corrected later. For example, in an alignment of three sequences, two sequences are first aligned, then a profile is created for this alignment and the third sequence is aligned against the profile. However, it is possible that the third sequence gives important clues as to how the first two sequences should be aligned, and yet the progressive method does not allow for the alignment of the first two sequences to be changed when aligning them against the third. For this reason, a new iterative methodology has been recently introduced, which starts with a progressive alignment, and modifications are repeatedly applied to it, accepted only if the overall score is improved. Examples of iterative aligners include PRRP (Gotoh, 1996) and MUSCLE (Edgar, 2004).

3.2.4 Genomic Alignment

The progressive and iterative approaches described above are well suited for the alignment of even large numbers (e.g., hundreds) of short sequences. However, application to long sequences of more than 10 Kbp would be extremely expensive in terms of time and memory and therefore not feasible (Ureta-Vidal et al., 2003). The alignment of long sequences can, however, be achieved by first identifying short subsequences (e.g., 10 bp) identical in all sequences and then by using these as anchors when building the alignment, so that only the regions between two anchors need to be aligned. Examples of programs implementing this idea to produce multiple alignments include MAVID (Bray and Pachter, 2004) and Multi-LAGAN (Brudno et al., 2003).

However, the alignment of long sequences in bacteria is made harder by the effect of genomic rearrangements, which shuffle the location of homologous fragments around the genome. All the algorithms and programs we have described so far are unable to deal with rearrangements since they assume that the sequences to be aligned are colinear. One program that does not make that assumption is MAUVE (Darling et al., 2004, 2007; Darling, 2006). For this reason, MAUVE is suitable for the alignment of large genomic regions in bacteria, up to whole genomes.

In order to allow the genomes to be noncolinear, the anchors need to be allowed to occur in a different order in each sequence. This greatly increases the possibilities for anchor choice, especially since bacterial genomes often contain many repeated elements. For this reason, MAUVE uses for anchors the exactly matching subsequences shared by at least two sequences and that occur only once in those sequences. A neighbor-joining guide tree is then built using the number and size of anchors shared by any two sequences as a measure of their relatedness. The anchors are then grouped into colinear groups according to their level of agreement. The anchors that are not found to belong to a significant group are discarded. Each resulting group of anchors therefore corresponds to

Figure 3.2 MAUVE alignment of the Typhi (top) and Paratyphi A (bottom) genomes, shown in the Artemis Comparison Tool (Rutherford et al., 2000; Carver et al., 2005).

a region present in at least two of the sequences and that occurs colinearly in each sequence where it is present. Finally, a multiple alignment is built for each such region using either ClustalW (Thompson et al., 1994) or MUSCLE (Edgar, 2004).

Figure 3.2 illustrates the 44 colinear regions found by MAUVE when aligning the genomes of *Salmonella enterica* Typhi and Paratyphi A (Parkhill et al., 2001; McClelland et al., 2004; Didelot et al., 2007). MAUVE can be used to align tens of whole genomes of bacteria. The fact that it identifies regions occurring colinearly in two or more of the sequences is useful to study the action of genomic rearrangements (Darling et al., 2008). The regions found to be present in several genomes from the same species constitute the core genome of that species, and the regions present in one but not all genomes represent the dispensable genome (Medini et al., 2005). MAUVE also enables the study of the genomic flux that occurred during the evolution of a group of genomes by analysis of the regions found in subsets of the genomes (Didelot et al., 2009).

3.3 PHYLOGENETIC METHODS

3.3.1 Introduction

Having shown in the previous section how alignments can be built, we now turn to the problem of analyzing bacterial population structures given an alignment of sequences. By far the most commonly used object to describe population structure is phylogeny. If one can reconstruct the correct phylogeny for a given sample, the evolutionary relationships between its members become apparent.

Id	Isolate	Year	Country	Serogroup	ST
61	393	1968	Greece	A	1
120	F4698	1987	Saudi	A	5
299	80049	1963	China	A	5
314	D1	1989	Mali	C	11
420	NG F26	1988	Norway	B	14
421	NG H15	1988	Norway	B	43
422	NG H41	1988	Norway	B	27
423	NG H38	1988	Norway	B	36
424	NG E31	1988	Norway	B	15
425	NG G40	1988	Norway	B	25
426	NG E28	1988	Norway	B	26
427	NG E30	1988	Norway	B	44
428	NG H36	1988	Norway	B	47
442	297-0	1987	Chile	B	49
655	E32	1988	Norway	Z	31
656	E26	1988	Norway	X	39
659	A22	1986	Norway	W-135	22

Figure 3.3 Isolates in the example data set.

We illustrate all the methods described below using the multi-locus sequence typing (MLST) data from the 17 carrier isolates of *N. meningitidis* sequenced by Maiden et al. (1998). Figure 3.3 gives some details about these 17 isolates. We consider the seven loci of the standard MLST scheme for *N. meningitidis*, which are *abcZ*, *adk*, *areE*, *fumC*, *gdh*, *pdhC*, and *pgm*. Except for two isolates of sequence type (ST)-5, each isolate has a unique ST. This tiny data set is unlikely to contain much information about the population structure of carrier *N. meningitidis*, but its smallness is useful to clearly illustrate and contrast the different methods of analysis. This data set is available online from http://pubmlst.org/neisseria/. All trees were drawn using FigTree (Rambaut, 2008) unless otherwise stated.

3.3.2 Unweighted Pair Group Method Using Arithmetic Averages (UPGMA)

The UPGMA (Fitch and Margoliash, 1967) is the simplest method of construction of a phylogenetic tree. The UPGMA algorithm first requires building a distance matrix that contains the genetic distance between every pair of individuals. The simplest measure of distance between two isolates is the proportion of sites at which they differ. This, however, has the inconvenience not to converge to infinity for completely unrelated sequences, and so is often corrected using a logarithmic scale following the work of Jukes and Cantor (1969). This assumes that all substitutions are equally likely, and more complex distance measures can be used to relax this assumption (e.g., Kimura, 1980; Felsenstein, 1981; Hasegawa et al., 1985).

The UPGMA algorithm starts with each isolate in a cluster of its own. The two clusters with the smallest distance are then found and grouped into a single cluster. The distance of this new cluster to the others is the average of the distance of their members. The process is then repeated until all members are clustered together. The order of the clustering defines the topology of a tree, and every time two clusters are grouped into one, their distance defines the age of their common ancestor.

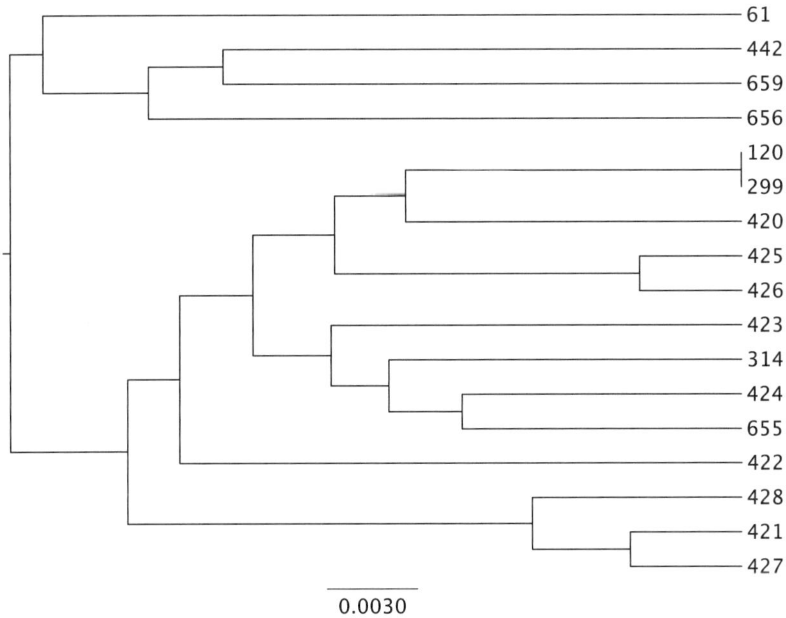

Figure 3.4 UPGMA tree for the example data set, built using PHYLIP (Felsenstein, 1989) with Jukes–Cantor distances (Jukes and Cantor, 1969).

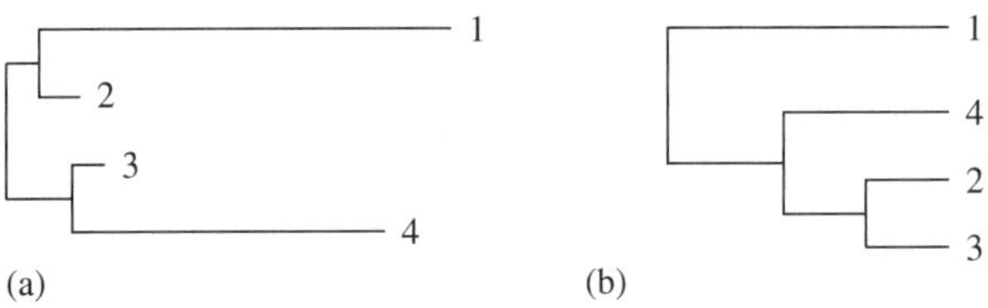

Figure 3.5 Illustration of the limitations of UPGMA when the molecular clock assumption does not hold: (a) real phylogeny with branch lengths proportional to the number of substitutions introduced and (b) UPGMA reconstruction based on data from (a).

A number of variants of the UPGMA algorithm exist, such as single linkage, where the distance between two clusters is the minimum (rather than the average) of the distances of their members, or complete linkage, where it is the maximum. Figure 3.4 shows the UPGMA tree corresponding to our example data set. Almost every phylogenetic software includes the UPGMA method, including PHYLIP (Felsenstein, 1989), PAUP (Swofford, 2002), START (Jolley et al., 2001), and MEGA (Tamura et al., 2007).

The UPGMA algorithm has the advantage to be simple and fast, even for large numbers of isolates, which is the reason why it is still very popular in spite of its limitations compared to the more elaborate algorithms described below. An important assumption behind the UPGMA algorithm is that differences between sequences are accumulated according to a molecular clock. If the molecular clock exists and large amounts of sequence data are available (so that the pairwise distances are very precisely measured), then UPGMA is guaranteed to reconstruct the correct tree.

However, the UPGMA procedure offers little robustness against deviations from the molecular clock assumption. To illustrate this problem, consider the tree shown in Fig. 3.5a where the lengths of the branches are proportional to the number of substitutions introduced in the sequence. The smallest distance between two sequences is found for 2 and 3, and the UPGMA will therefore cluster these two sequences together first, and then sequences 4 and 1 as shown in Fig. 3.5b. Thus, not only is the UPGMA algorithm unable to reconstruct the lengths of branches correctly, but the long branches also confuse the topology reconstruction so that the reconstructed clustering order is wrong.

3.3.3 Neighbor Joining

The neighbor-joining method (Saitou and Nei, 1987) is another hugely popular method of phylogeny reconstruction. Like the UPGMA algorithm, it uses a matrix of pairwise distances between the sequences. Sequences are iteratively clustered together as in the UPGMA algorithm until all sequences are grouped. The main difference is that in order to find which clusters to group, a modification of the distance matrix is used, which corrects for the possibility of having different rates of evolution on the different branches of the tree. The details of this modification are not important here. Furthermore, when a new cluster is formed, it is not assigned an age as in the UPGMA algorithm, but a distance to the two clusters it is made of and the clusters that are yet to group. For this reason, the trees produced by the neighbor-joining method are unrooted, unlike those produced by UPGMA.

Figure 3.6 shows the neighbor-joining algorithm applied to our example data set. Comparison with the UPGMA reconstruction of Fig. 3.4 reveals some similarities in the clustering order (e.g., concerning the sequences 61, 442, 659, and 656) but also some differences (e.g., the sequences 120/399 are most closely related to 420 in the UPGMA reconstruction and to 425/426 in the neighbor-joining tree).

Because it does not assume a molecular clock, the neighbor-joining algorithm is usually regarded as superior to UPGMA. It is almost as quick to apply and is widely available in many phylogenetic software (cf. list for UPGMA). However, UPGMA can be preferable if one wants to enforce a molecular clock in order to reconstruct a rooted ultrametric tree (i.e., where all distances from root to leaves are the same).

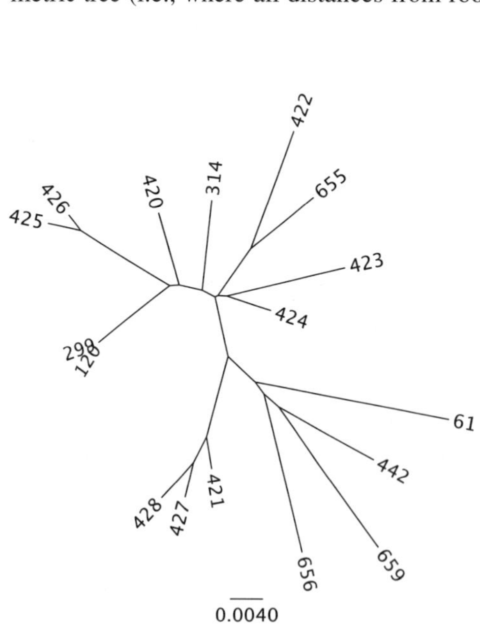

Figure 3.6 Neighbor-joining tree for the example data set, built using PHYLIP (Felsenstein, 1989) with Jukes–Cantor distances (Jukes and Cantor, 1969).

The neighbor-joining method is based on an assumption of additivity, whereby if two successive branches have lengths *A* and *B*, then the distance from the beginning of the first branch to the end of the second one is *A* + *B*. This assumption is correct, for example, under the infinite site model of Kimura (1969). If additivity holds and sufficient data are used (so that the pairwise distances are precisely known), then the neighbor-joining technique is guaranteed to reconstruct the right phylogeny. However, if two mutations occur on different branches and on the same site, then the additivity assumption is violated. In bacteria, the effect of recombination is likely to seriously compromise the additivity rule. More generally, distance-based methods such as UPGMA or neighbor joining are limited by the fact that they use a distance matrix to summarize the sequence data, thus potentially discarding a significant part of the information contained about the evolutionary history of the sample.

3.3.4 Parsimony

Parsimony (Camin and Sokal, 1965) is another method of phylogeny reconstruction that, unlike UPGMA and neighbor joining, is not based on pairwise distance summaries but uses the sequence data directly. The idea is to find the unrooted tree topology, which minimizes the cost, defined as the number of substitutions required to explain how the data arose. Calculating the cost of a tree would be easy if the ancestral genotypes of the internal nodes were known: it would simply be the sum over all branches of the number of sites that differ on each side of the branch. To calculate the cost of a tree when the genotypes of the internal nodes is unknown requires considering all possibilities for these ancestral genotypes. Luckily, there exists a dynamic procedure that allows doing this quickly (Fitch, 1971), the details of which are not important here.

All that remains to be done is therefore to explore the space of possible trees. Unfortunately, the number of trees becomes very large with the number of sequences (Felsenstein, 1978a), so that it is impossible to consider all trees for much more than 10 sequences. The first solution is to use a method known as branch and bound (Hendy and Penny, 1982), which cuts down the number of trees to consider while guaranteeing that it finds the best tree. This method is, however, too slow to be used for much more than 25 sequences. Larger data sets require the use of heuristics that are not certain to find the best tree. One such method consists in adding the sequences one by one to the tree at the place where it minimizes the cost (Felsenstein, 1981). Another possibility is to propose small modifications to an existing tree and accepting them only if they reduce the cost.

Maximum parsimony techniques are implemented in many phylogenetic software, including PHYLIP (Felsenstein, 1989), PAUP (Swofford, 2002), and MEGA (Tamura et al., 2007). Figure 3.7 shows the result of applying maximum parsimony to the example data set. The tree agrees with neighbor joining rather than with UPGMA concerning the phylogenic relationships of 120, 299, 425, 426, and 420. However, it disagrees with both distance-based methods in that it finds a more recent common ancestor of 61 with 656 than with 442/659.

The idea behind maximum parsimony is simple and appealing. Its increased computational cost compared to distance-based methods has long been an issue, but heuristic approaches to finding the best tree have greatly improved its usability, so that it can now be applied to thousands of sequences. Yet, the method has met criticism, most famously from Felsenstein (1978b), because it is unable to reconstruct the correct tree when different branches evolved at different rates. The problem known as “long branch attraction” arises,

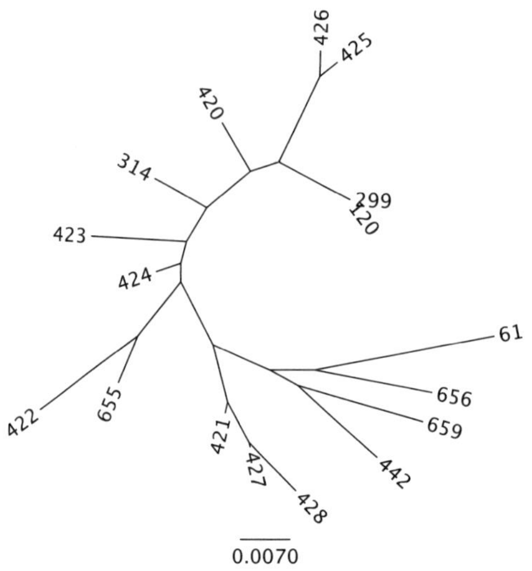

Figure 3.7 Parsimony tree for the example data set, built using PHYLIP (Felsenstein, 1989).

for example, if four sequences are related as shown in Fig. 3.5a with very long branches leading to 1 and 4 and short branches everywhere else. In that case, 2 and 3 will be identical for many sites, whereas 1 and 4 will often by substituted, and when they both are at the same site, there is a chance that they will become equal to one another but different from 2 and 3, thus pointing toward a wrong tree where they would have a more recent common ancestor than with 2 and 3.

Maximum parsimony, like distance-based methods, but in contrast with other methods described below, is not based on an explicit model of sequence evolution (Sanderson and Kim, 2000). This is perceived as an advantage by supporters of the parsimony method since it means that no simplistic assumption is made. However, it also means that its usage is limited purely to tree reconstruction, and that it is of no use for statistical hypothesis testing.

3.3.5 Maximum Likelihood

The maximum likelihood method of phylogeny reconstruction was first introduced by Edwards and Cavalli-Sforza (1964) but only became applicable in practice thanks to the theoretical work of Felsenstein (1981). The idea is to find the unrooted tree, which maximizes the probability of the data under some evolutionary model. The model commonly assumed is that substitutions are introduced at a constant rate along the branches of the tree. The simplest model is that all substitutions are equally likely to occur (Jukes and Cantor, 1969), although more sophisticated models can easily be used too (as discussed in Section 3.3.2).

As for the parsimony cost, calculating the probability of the data if the genotypes of the ancestral nodes were known would be easy, but doing it without this knowledge requires considering all possibilities of the ancestral genotypes. This can be done efficiently using the pruning algorithm described by Felsenstein (1981). The main difficulty is therefore to find the tree that maximizes the likelihood. This is done using similar heuristics as for maximum parsimony, with the added difficulty that the likelihood (unlike the parsimony cost) is a function of branch lengths. Although the use of heuristics is only

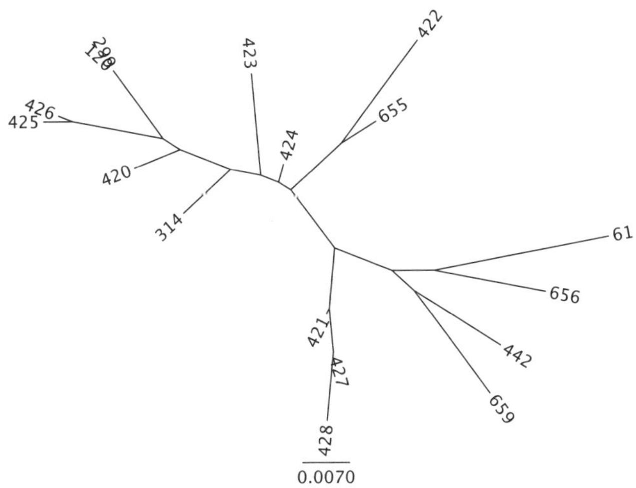

Figure 3.8 Maximum-likelihood tree for the example data set, built using PHYLIP (Felsenstein, 1989).

approximate, Kuhner and Felsenstein (1994) showed using simulated data that the true phylogeny is very often reconstructed.

The maximum likelihood method is implemented in PHYLIP (Felsenstein, 1989) and in PAML (Yang, 2007). Figure 3.8 shows the maximum likelihood phylogeny reconstruction for our example data set. The tree structure is exactly the same as that of the parsimony reconstruction shown in Fig. 3.7.

The maximum likelihood method has many similarities with the maximum parsimony approach, at least when a simple substitution model is assumed (Tuffley and Steel, 1997). It is the only method we described that really uses all the information contained in the data set (maximum parsimony does not use the nonpolymorphic sites). The maximum likelihood approach is guaranteed to reconstruct the best tree if the model assumed is correct and if large amounts of data are being used. In practice, the model is never absolutely correct, which can result in mistakes in the estimated branching structure and branch lengths. Maximum likelihood is the most computationally expensive of the four methods we described so far, although efficient implementations that can handle large data sets exist.

3.4 MEASURES OF UNCERTAINTY

The previous section described four methods to reconstruct a single phylogeny given an aligned sequence data set. However, they do not offer the possibility to directly assess the reliability of the reconstruction. The first approach to this issue is to use a bootstrapping procedure, whereas recent years have seen the development of Bayesian approaches to phylogeny, which offer more reliable measures of confidence.

3.4.1 Bootstrapping

Bootstrapping (Felsenstein, 1985) is a method that can be applied to any of the methods of reconstruction described in the previous section in order to assess how well supported the reconstruction is. The idea is to generate a fake data set, of the same size as the real

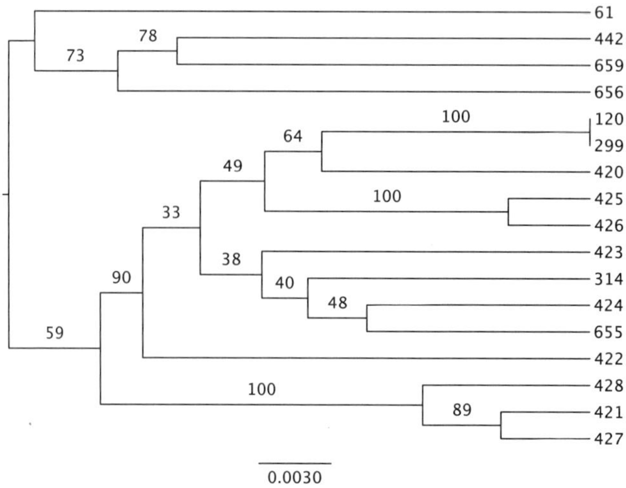

Figure 3.9 Bootstrapped UPGMA tree for the example data set, built using MEGA (Tamura et al., 2007) with Jukes–Cantor distances (Jukes and Cantor, 1969).

one, and where each site corresponds to a randomly chosen site of the real one. Thus, some sites of the real data set may be found more than once in the new data set, and others not at all. The fake data set is then analyzed using the same method as the real one. The procedure is repeated a large number of times (at least 1000 times). For each cluster in the tree made from the real data set, the percentage of times that this cluster appears in the trees made from fake data sets is counted. This value is called the bootstrap support of a cluster.

Figure 3.9 shows the UPGMA reconstruction of Fig. 3.4 for our example data set, but with the bootstrap support of each cluster indicated. Whereas some clusters are very strongly supported (e.g., the cluster made of 421, 427, and 428, which was found in all fake data sets), many others are only weakly supported, with values of the bootstrap support going as low as 33%. Typically, clusters with support below 50% are considered untrustworthy and are often removed from the tree representation.

The bootstrapping method has the advantage to be reasonably quick to apply, at least for distance-based methods such as UPGMA or neighbor joining. One common mistake, however, is to consider that the bootstrap support of a cluster is equal to the probability that this cluster is correct given the data (Soltis and Soltis, 2003). This interpretation is wrong as only a Bayesian method can reveal this probability.

3.4.2 Bayesian Inference

The Bayesian approach to phylogeny reconstruction was first outlined by Yang and Rannala (1997) and by Mau and Newton (1997) and has proved very popular since. The Bayesian approach is similar to the maximum likelihood approach (cf. Section 3.3.5) in that it is based on a probabilistic model of how the tree and the data are interrelated, but it presents two very important differences with the maximum likelihood approach.

First, the suitability of a tree is not evaluated on the basis of its likelihood (i.e., the probability of the data given the tree, as in the maximum likelihood approach) but based on its posterior probability (i.e., the probability of the tree given the data). Evaluating this

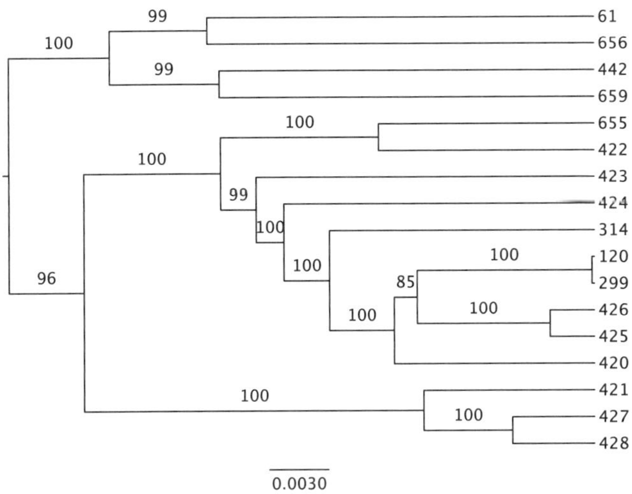

Figure 3.10 Consensus tree based on the output of BEAST (Drummond and Rambaut, 2007) for the example data set.

probability involves the likelihood, but also the prior probability of a tree. This prior probability is given by a model that describes our expectation of what trees look like, before observing the data. One very popular prior model is the coalescent model (Kingman, 1982).

Second, the Bayesian approach returns a sample of trees from the posterior rather than a single optimal tree as in the maximum likelihood approach. Consequently, the probability of any feature of interest (e.g., whether a subsample of the isolates has a most recent common ancestor than with the rest of the sample) can be evaluated on the basis of the proportion of the trees that support it. Such sampling from the posterior distribution typically involves the use of a Markov chain Monte Carlo (MCMC; Hastings, 1970).

The fact that the Bayesian approach returns a sample of trees is very useful to evaluate the probability of any specific clustering, but it does pose the problem of how to present such results. One commonly used method is to draw a consensus tree (Adams, 1972), for example, a tree that contains only the clusters found in at least 50% of the sampled trees (otherwise known as a 50% majority-rule consensus tree).

Bayesian phylogeny reconstruction is available in MrBayes (Huelsenbeck and Ronquist, 2001) and BEAST (Drummond and Rambaut, 2007). In order to apply BEAST to our example data set, a coalescent prior with constant population size was assumed, as well as a Jukes–Cantor model of nucleotide substitution, and the program was run for 10 million iterations, leaving all other settings as default. The first half of the iterations was discarded to allow for the MCMC to converge to the posterior distribution, and a consensus tree was built based on the trees sampled in the second half. Two runs were compared and showed good convergence using Tracer (Rambaut and Drummond, 2007). Each run took approximately 10 min. Figure 3.10 shows a consensus tree based on our results. The tree structure is the same as that of parsimony and maximum likelihood, except that isolates 423 and 424 have been inverted in relationship with the rest of the sample.

One difficulty with Bayesian approaches to phylogeny reconstruction is that they are highly demanding computationally, which makes them impractical for very large data sets. Yet, they are the only way to obtain a complete measure of uncertainty, which makes them highly desirable. Inference is always only as good as the model assumed, and if this model

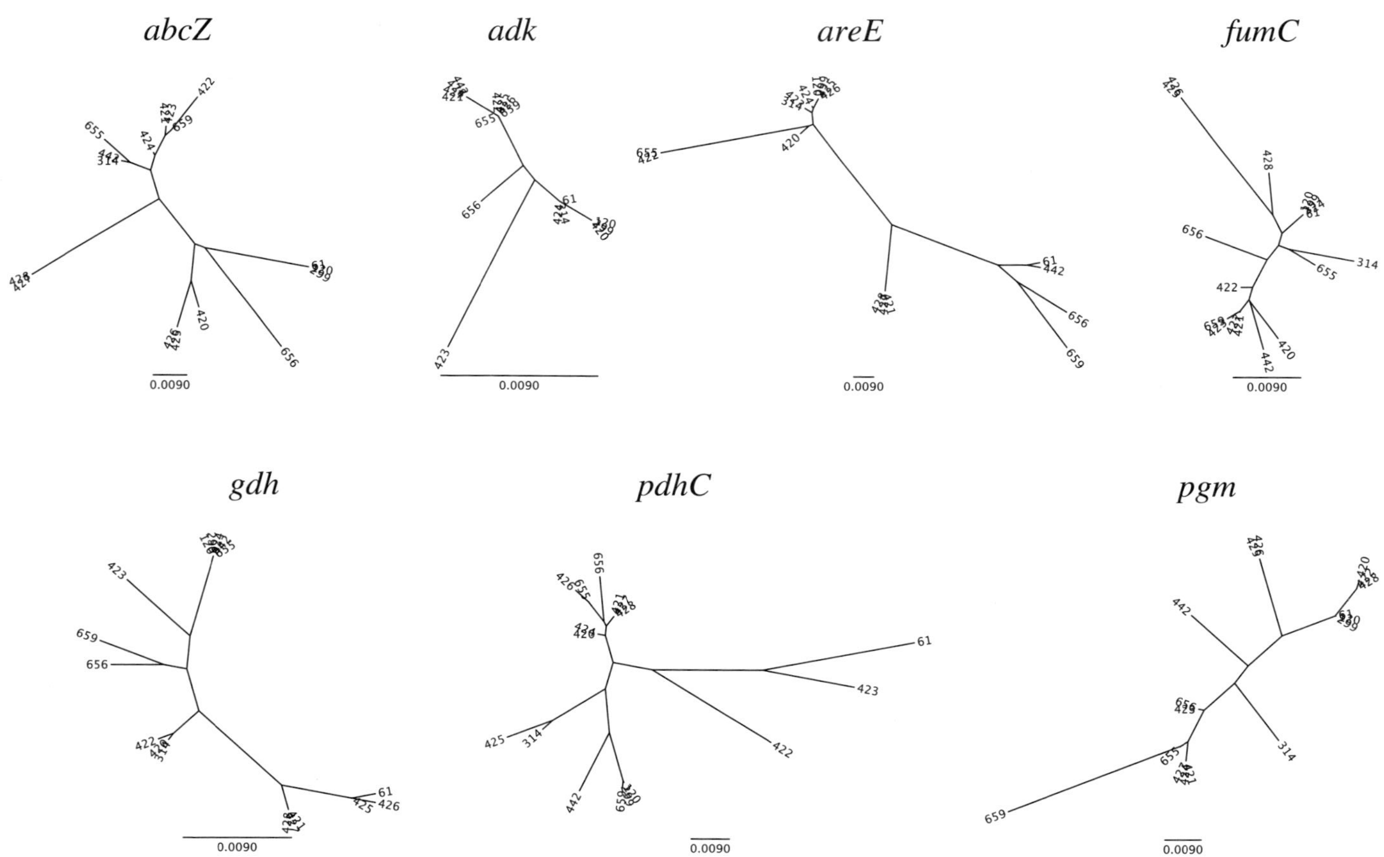

Figure 3.11 Neighbor-joining trees for each of the seven genes in the example data set. Each tree was built using PHYLIP (Felsenstein, 1989) with Jukes–Cantor distances (Jukes and Cantor, 1969).

is violated, the reconstruction can be wrong, including the confidence ascertainment. Extensions of the Bayesian methods have therefore been developed in order to deal with complex evolutionary scenarios such as changes in demography (Drummond et al., 2002), relaxed molecular clock (Drummond et al., 2006), or even recombination (cf. Section 3.5.3).

3.5 BEYOND THE TREE MODEL

3.5.1 Introduction

The methods described in the previous section can be used to assess the support in a phylogeny reconstruction. They still assume that there exists a unique tree that gave rise to the data, but they attempt to quantify the uncertainty that there is when inferring this tree from given genetic sequences. Such a tree might, however, not exist at all. For example, let us consider three isolates, A, B, and C, such that A and B are more closely related to each other than to C. If we now consider that A has recently imported a gene from C through recombination, then it means that for this gene, A and C are more closely related to each other than to B. In other words, recombination can cause different parts of the genomes to have different phylogenies.

We can illustrate this on our data set by estimating a different tree for each of the seven housekeeping genes. Figure 3.11 shows the result using the neighbor-joining method (cf. Section 3.3.3). The seven trees have very little in common, apart from the fact that 120 and 299 are always clustered together and that 421, 427, and 428 often cluster together (except for genes *abcZ* and *fumC*). These are the only two clusters found by building a consensus tree from the seven trees. We can therefore conclude that recombination has played a very important role on our data set, which is consistent with the fact that *N. meningitidis* is highly recombinogenic (Feil et al., 1996, 1999; Jolley et al., 2000, 2005). More clonal species (e.g., *Staphylococcus aureus*; Feil et al., 2003) show a much better agreement between gene trees.

Because all the methods described so far assumed that evolution followed a tree, they can be mistaken when applied on data that have been affected by recombination. Therefore, even if the Bayesian reconstruction of Fig. 3.10 was given a very strong support, it may not have much evolutionary relevance. Building a tree for each of the seven genes as we did in Fig. 3.11 is useful to illustrate the effect of recombination, but it does not take into account the possibility that each gene may not have evolved according to a single tree (i.e., recombination may have affected only parts of a gene) and cannot be applied to data sets where there is no unit of sequencing (e.g., whole genome data).

3.5.2 Split Networks

The concept of a split network is a natural extension of phylogenetic trees: if the data support not just one but several trees, then it makes sense to attempt building a network that contains all those trees. There exist many different methods to build split networks, reviewed by Huson and Bryant (2006). The most famous one is called the split decomposition (Bandelt and Dress, 1992) and is implemented in the program SplitsTree (Huson, 1998). Here we use an improvement called neighbor net (Bryant and Moulton, 2004) and which is implemented in SplitsTree4 (Huson and Bryant, 2006).

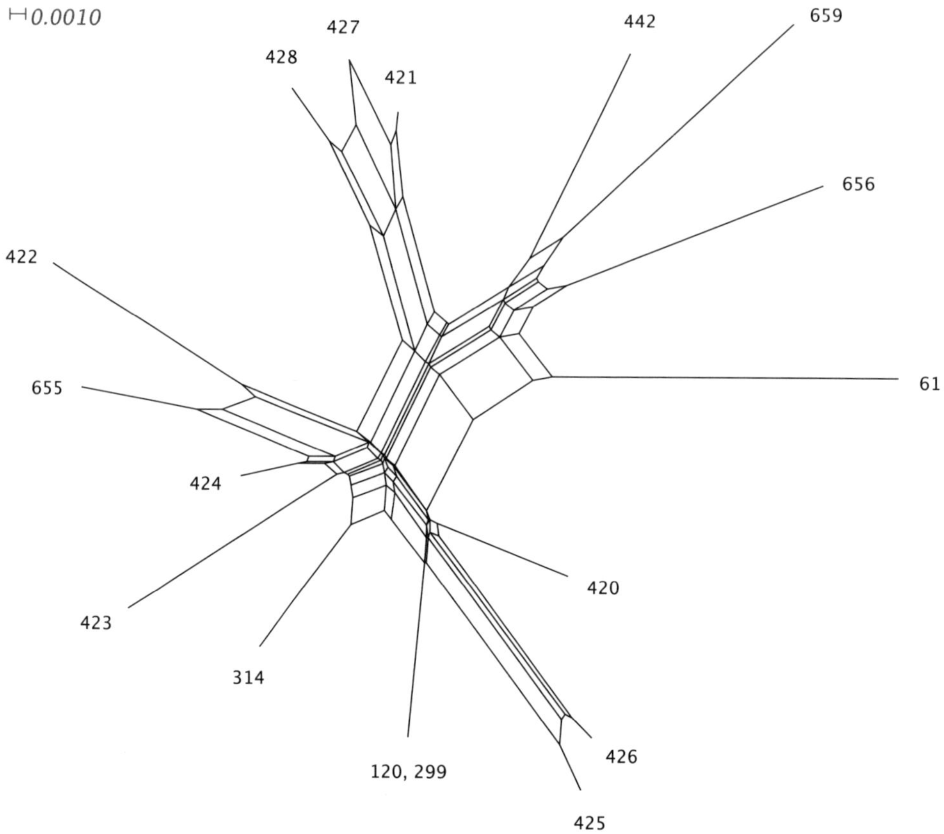

Figure 3.12 Neighbor-net graph (Bryant and Moulton, 2004) for the example data set, drawn using SplitTree4 (Huson and Bryant, 2006).

Figure 3.12 shows the result of applying the neighbor-net algorithm to our example data set. The network contains many cycles that are indicators of conflicting signals of phylogeny, and therefore of a potential effect of recombination. Isolates 120 and 299 are on the same branch because they share the same ST. A few clear clusters exist, such as 425 and 426, or 421, 427, and 428. These correspond to the clusters that we often find in the gene trees shown in Fig. 3.11. The deep phylogeny is, however, completely unresolved.

Split networks are useful to represent large data sets suspected to have been affected by recombination. Like distance-based methods, their main advantage is their speed. The neighbor-net method in particular is very fast and can be applied to data sets containing hundreds of isolates in a few minutes. Split networks are, however, unable to distinguish whether recombination really took place or whether the data are simply unclear about which tree is correct. When tree-based methods give strong support to a single tree (as we saw in the Bayesian analysis of our example data set; cf. Section 3.4.2), then we can strongly suspect recombination to be the cause of the cycles observed in a split network. In that case, the exact position of the cycles in the network may even give indication as to which recombination events happened. However, split networks only ever give indications as to what may have happened and are mostly useful as a guide for further analysis (Huson and Bryant, 2006).

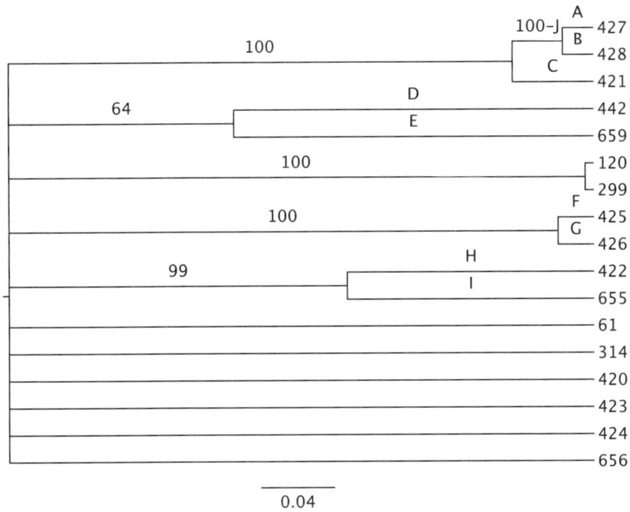

Figure 3.13 Consensus tree based on the output of ClonalFrame (Didelot and Falush, 2007) for the example data set.

3.5.3 ClonalFrame

ClonalFrame (Didelot and Falush, 2007) follows the Bayesian approach to reconstructing phylogenies outlined in Section 3.4.2: It is based on an explicit evolutionary model and uses an MCMC to sample from the distribution of evolutionary scenarios likely to have given rise to the observed data. The main difference with the previous Bayesian methods we described is that the evolutionary model at the heart of ClonalFrame accounts for the effect of recombination. Thus, it can reconstruct the clonal genealogy, as well as the mutation and recombination events that took place on the branches of the genealogy. Explicitly modeling recombination presents two advantages over methods that do not: The reconstruction of the phylogeny is more accurate since the recombination events are likely to confuse other methods, and the analysis reveals some information about the recombination process itself.

ClonalFrame is able to infer the rates at which mutation and recombination events occur over time, as well as the average size of recombination events. However, since our example data set is very small and is unlikely to contain enough information to estimate all these parameters, we decided not to estimate the mutation rate in ClonalFrame but instead to use the Watterson estimator (Watterson, 1975), which was equal to 118.61. For the same reason, we did not attempt to estimate the mean recombination tract length, but we set it equal to 1000 bp, in agreement with previous findings based on a much larger data set (Jolley et al., 2005). The remainders of the ClonalFrame parameters were left as default, and four instances of 100,000 iterations each were run to evaluate convergence. Each run took approximately an hour.

Figure 3.13 shows a consensus tree based on the output of ClonalFrame for our example data set. The recent phylogeny is similar to that found by BEAST (cf. Fig. 3.10, with clusters inferred with high probability for 421/427/428, 120/299, 425/426, and 422/655). One further potential relationship was found between 442 and 659, although ClonalFrame is a lot less assertive than BEAST (posterior probability of 0.64 vs. 0.99). ClonalFrame was, however, unable to reconstruct any deep phylogeny at all, unlike all the

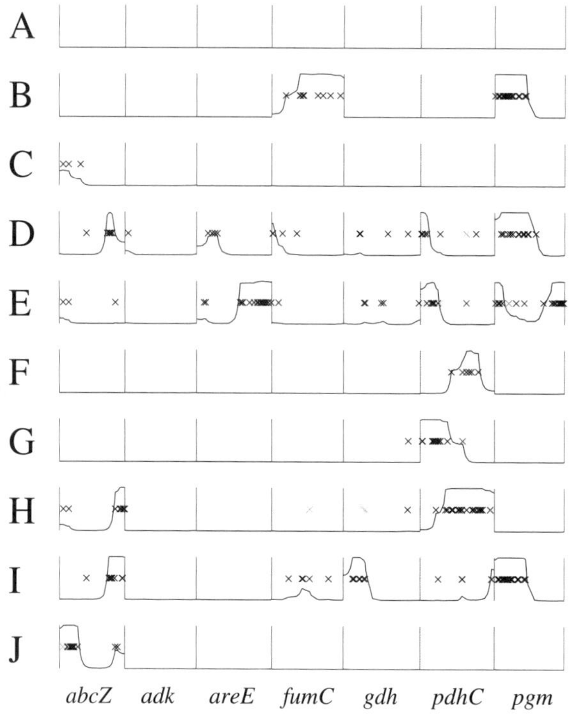

Figure 3.14 Graphical representation of the mutation and recombination events inferred by ClonalFrame (Didelot and Falush, 2007) on the branches labeled from A to J in Fig. 3.13.

previous methods we have seen so far. This is because ClonalFrame takes recombination into account and infers that most of the phylogenic signal found for the deep clusters comes from recombination rather than clonal relationship, and that it is therefore impossible to reconstruct the deep clonal relationships between the members of the sample.

One important advantage of ClonalFrame over other methods is that it can be used to analyze the mutation and recombination events that happened on the branches of the clonal genealogy. Figure 3.14 shows the events found on 10 of the branches from the consensus tree, as labeled from A to J in Fig. 3.13. Each row corresponds to one of the branches, and each column corresponds to one of the seven MLST gene fragments. Crosses indicate genetic differences between the genotype at the top and at the bottom of the branch, while the height of the lines in each box indicates the probability that recombination took place on that branch for each site. For example, line A contains no crosses and the line remains at the bottom of the boxes, meaning that no mutation or recombination event happened on branch A. On branch B, however, two recombination events have been detected affecting the second half of *fumC* and the first half of *pgm*. Three substitutions have been found on gene *abcZ* on branch C, which may be caused either by recombination (with probability of approximately a third as indicated by the height of the line) or by mutation. Branches D and E exhibit many more mutation and recombination events, which is consistent with the length of these branches. Branch F shows a single recombination event, whereas branch G shows a mutation on gene *gdh* and a recombination event on gene *pdhC*. Branches H and I show several events, and finally, branch J shows that recombination affected the first half of gene *abcZ* and that the second half of this gene was affected either by recombination or by a couple of mutations.

As ClonalFrame identifies the mutation and recombination events that may have taken place, it can be used to estimate the relative contributions of those two forces on the observed patterns of genetic diversity. One measure for this is the ratio r/m of the rates at which recombination and mutation introduce substitutions over time. This measure was first introduced by Guttman and Dykhuizen (1994) and has since been extensively used by Feil et al. (1999, 2000, 2001). A survey of previously estimated values can be found in Didelot and Falush (2008), whereas Vos and Didelot (2008) used ClonalFrame to estimate and to compare the values of r/m in over 40 bacterial species.

One difficulty with ClonalFrame is that it is based on a rather complex model, so that it can be slow to converge. This needs to be tested using multiple runs as described above. For very large MLST data sets, containing more than 500 isolates, ClonalFrame is unlikely to give an answer in a reasonable amount of time. One way around this problem, if possible, is to split the data into clades of, at most, a few hundred isolates each, and to run ClonalFrame on each clade separately. For example, in *N. meningitidis*, there exist seven well-defined hypervirulent lineages (Wang et al., 1992, 1993; Seiler et al., 1996; Maiden et al., 1998), which could be analyzed separately with no loss of power (in fact, the power to detect interlineage recombination events is increased).

3.5.4 A Nonphylogenic Model of Population Structure

Since recombination occasionally erases the clonal signal in its entirety, it may be better not to attempt to reconstruct the whole phylogeny in species that are highly recombinogenic. For example, in the ClonalFrame results shown in Fig. 3.13, a lot of time is spent trying to infer the deep phylogeny only to conclude that this is not possible. It can therefore be interesting to use models of population structure that are not based on a phylogeny. One example is the model within Structure (Pritchard et al., 2000), which assumes that the individuals from a sample come from a number of populations, but makes no attempt at modeling the relationships of these populations with one another or the relationships of the individuals within each population.

Three versions of the STRUCTURE population model have been implemented. The no-admixture version assumes that each individual in the data set comes entirely from one of the populations. In the admixture version, each individual is made of a mixture of each population. These two versions of the model assume complete linkage equilibrium between loci, which is a very strong assumption, not very well suited to the analysis of sequence data where we know that the linkage between two sites decreases approximately linearly with the distance that separates them. In order to relax this assumption, the linkage version of the model was implemented (Falush et al., 2003a), where each site of each sequence originates from one of the populations. The linkage version assumes a correlation between the origins of close sites, which decreases linearly as their distance on the genome increases. The linkage version of the STRUCTURE model can therefore be used to identify individuals that are entirely or mostly from one population, and others that are hybrids, that is, have imported fragments of DNA from different populations.

Our example data set is too small for STRUCTURE to be run effectively. Interesting applications of the STRUCTURE method can, however, be found for species as diverse as *Helicobacter pylori* (Falush et al., 2003b), *E. coli* (Wirth et al., 2006), *S. enterica* (Falush et al., 2006), or *Campylobacter jejuni* (Sheppard et al., 2008).

REFERENCES

Adams, E. N. (1972) Consensus techniques and the comparison of taxonomic trees. *Systematic Zoology* **21,** 390–397.

Bandelt, H. and Dress, A. (1992) A canonical decomposition theory for metrics on a finite set. *Advances in Mathematics* **92,** 47–105.

Bennett, S., Barnes, C., Cox, A., Davies, L., and Brown, C. (2005) Toward the $1000 human genome. *Pharmacogenomics* **6,** 373–382.

Blattner, F., Plunkett, G. III, and Bloch, C. et al. (1997) The complete genome sequence of *Escherichia coli* K-12. *Science* **277,** 1453–1474.

Bray, N. and Pachter, L. (2004) MAVID: Constrained ancestral alignment of multiple sequences. *Genome Research* **14,** 693–699.

Brudno, M., Do, C., Cooper, G. et al. (2003) LAGAN and Multi-LAGAN: Efficient tools for large-scale multiple alignment of genomic DNA. *Genome Research* **13,** 721–731.

Bryant, D. and Moulton, V. (2004) Neighbor-net: An agglomerative method for the construction of phylogenetic networks. *Molecular Biology and Evolution* **21,** 255–265.

Camin, J. H. and Sokal, R. R. (1965) A method for deducing branching sequences in phylogeny. *Evolution* **19,** 311–326.

Carrillo, H. and Lipman, D. (1988) The multiple sequence alignment problem in biology. *SIAM Journal on Applied Mathematics* **48,** 1073–1082.

Carver, T. J., Rutherford, K. M., Berriman, M., Rajandream, M. A., Barrell, B. G., and Parkhill, J. (2005) ACT: The Artemis Comparison Tool. *Bioinformatics* **21,** 3422–3423.

Darling, A., Miklós, I., and Ragan, M. (2008) Dynamics of genome rearrangement in bacterial populations. *PLoS Genetics* **4,** e1000128.

Darling, A., Treangen, T., Messeguer, X., and Perna, N. (2007) Analyzing patterns of microbial evolution using the mauve genome alignment system. *Methods in Molecular Biology* **396,** 135–152.

Darling, A. C., Mau, B., Blattner, F. R., and Perna, N. T. (2004) Mauve: Multiple alignment of conserved genomic sequence with rearrangements. *Genome Research* **14,** 1394–1403.

Darling, A. E. (2006) Computational analysis of genome evolution. PhD thesis. University of Wisconsin.

Dayhoff, M., Schwartz, R., and Orcutt, B. (1978) A model of evolutionary change in proteins. *Atlas of Protein Sequence and Structure* **5,** 345–352.

Didelot, X., Achtman, M., Parkhill, J., Thomson, N. R., and Falush, D. (2007) A bimodal pattern of relatedness between the *Salmonella* Paratyphi A and Typhi genomes: Convergence or divergence by homologous recombination? *Genome Research* **17,** 61–68.

Didelot, X., Darling, A., and Falush, D. (2009) Inferring genomic flux in bacteria. *Genome Research* **19,** 306–317.

Didelot, X. and Falush, D. (2007) Inference of bacterial microevolution using multilocus sequence data. *Genetics* **175,** 1251–1266.

Didelot, X. and Falush, D. (2008) Bacterial recombination in vivo. In *Horizontal Gene Transfer in the Evolution of Pathogenesis* (ed. M. Hensel and H. Schmidt), pp. 23–48. Cambridge University Press.

Drummond, A. and Rambaut, A. (2007) BEAST: Bayesian evolutionary analysis by sampling trees. *BMC Evolutionary Biology* **7,** 214.

Drummond, A. J., Ho, S. Y., Phillips, M. J., and Rambaut, A. (2006) Relaxed phylogenetics and dating with confidence. *PLoS Biology* **4,** 699–710.

Drummond, A. J., Nicholls, G. K., Rodrigo, A. G., and Solomon, W. (2002) Estimating mutation parameters, population history and genealogy simultaneously from temporally spaced sequence data. *Genetics* **161,** 1307–1320.

Edgar, R. C. (2004) MUSCLE: Multiple sequence alignment with high accuracy and high throughput. *Nucleic Acids Research* **32,** 1792–1797.

Edwards, A. and Cavalli-Sforza, L. (1964) Reconstruction of evolutionary trees. *Phenetic and Phylogenetic Classification* **6,** 67–76.

Falush, D., Stephens, M., and Pritchard, J. K. (2003a) Inference of population structure using multilocus genotype data: Linked loci and correlated allele frequencies. *Genetics* **164,** 1567–1587.

Falush, D., Wirth, T., Linz, B. et al. (2003b) Traces of human migrations in *Helicobacter pylori* populations. *Science* **299,** 1582–1585.

Falush, D., Torpdahl, M., Didelot, X., Conrad, D. F., Wilson, D. J., and Achtman, M. (2006) Mismatch induced speciation in *Salmonella*: Model and data. *Philosophical Transactions of the Royal Society B* **361,** 2045–2053.

Feil, E., Maiden, M., Achtman, M., and Spratt, B. (1999) The relative contributions of recombination and mutation to the divergence of clones of *Neisseria meningitidis*. *Molecular Biology and Evolution* **16,** 1496–1502.

Feil, E., Zhou, J., Maynard Smith, J., and Spratt, B. G. (1996) A comparison of the nucleotide sequences of the *adk* and *recA* genes of pathogenic and commensal *Neisseria* species: Evidence for extensive interspecies recombination within *adk*. *Journal of Molecular Evolution* **43,** 631–640.

Feil, E. J., Cooper, J. E., Grundmann, H. et al. (2003) How clonal is *Staphylococcus aureus*? *Journal of Bacteriology* **185,** 3307–3316.

Feil, E. J., Holmes, E. C., Enright, M. et al. (2001) Recombination within natural populations of pathogenic bacteria: Short-term empirical estimates and long-term phylogenetic consequences. *Proceedings of the National Academy of Sciences of the United States of America* **98,** 182–187.

Feil, E. J., Smith, J. M., Enright, M. C., and Spratt, B. G. (2000) Estimating recombinational parameters in

Streptococcus pneumoniae from multilocus sequence typing data. *Genetics* **154,** 1439–1450.

FELSENSTEIN, J. (1978a) The number of evolutionary trees. *Systematic Zoology* **27,** 27–33.

FELSENSTEIN, J. (1978b) Cases in which parsimony or compatibility methods will be positively misleading. *Systematic Zoology* **27,** 401–410.

FELSENSTEIN, J. (1981) Evolutionary trees from DNA sequences: A maximum likelihood approach. *Journal of Molecular Evolution* **17,** 368–376.

FELSENSTEIN, J. (1985) Confidence limits on phylogenies: An approach using the bootstrap. *Evolution* **39,** 783–791.

FELSENSTEIN, J. (1989) PHYLIP—Phylogeny Inference Package (Version 3.2). *Cladistics* **5,** 164–166.

FENG, D. F. and DOOLITTLE, R. F. (1987) Progressive sequence alignment as a prerequisite to correct phylogenetic trees. *Journal of Molecular Evolution* **25,** 351–360.

FITCH, W. (1971) Toward defining the course of evolution: Minimum change for a specific tree topology. *Systematic Zoology* **20,** 406–416.

FITCH, W. and MARGOLIASH, E. (1967) Construction of phylogenetic trees. *Science* **155,** 279–284.

FLEISCHMANN, R., ADAMS, M., WHITE, O. et al. (1995) Whole-genome random sequencing and assembly of *Haemophilus influenzae* Rd. *Science* **269,** 496–512.

GOTOH, O. (1996) Significant improvement in accuracy of multiple protein sequence alignments by iterative refinement as assessed by reference to structural alignments. *Journal of Molecular Biology* **264,** 823–838.

GUTTMAN, D. S. and DYKHUIZEN, D. E. (1994) Clonal divergence in *Escherichia coli* as a result of recombination, not mutation. *Science* **266,** 1380–1383.

HASEGAWA, M., KISHINO, H., and YANO, T. (1985) Dating of the human-ape splitting by a molecular clock of mitochondrial DNA. *Journal of Molecular Evolution* **22,** 160–174.

HASTINGS, W. K. (1970) Monte Carlo sampling methods using Markov chains and their applications. *Biometrika* **57,** 97–109.

HENDY, M. D. and PENNY, D. (1982) Branch and bound algorithms to determine minimal evolutionary trees. *Mathematical Biosciences* **59,** 277–290.

HENIKOFF, S. and HENIKOFF, J. (1992) Amino acid substitution matrices from protein blocks. *Proceedings of the National Academy of Sciences of the United States of America* **89,** 10915–10919.

HIGGINS, D. and SHARP, P. (1988) CLUSTAL: A package for performing multiple sequence alignment on a microcomputer. *Gene* **73,** 237–244.

HIGGINS, D. G., BLEASBY, A. J., and FUCHS, R. (1992) CLUSTAL V: Improved software for multiple sequence alignment. *Computer Applications in the Biosciences* **8,** 189–191.

HUELSENBECK, J. P. and RONQUIST, F. (2001) MrBayes: Bayesian inference of phylogenetic trees. *Bioinformatics* **17,** 754–755.

HUSON, D. and BRYANT, D. (2006) Application of phylogenetic networks in evolutionary studies. *Molecular Biology and Evolution* **23,** 254–267.

HUSON, D. H. (1998) SplitsTree: Analyzing and visualizing evolutionary data. *Bioinformatics* **14,** 68–73.

JOLLEY, K. A., FEIL, E. J., CHAN, M.-S., and MAIDEN, M. C. J. (2001) Sequence Type Analysis and Recombinational Tests (START). *Bioinformatics* **17,** 1230–1231.

JOLLEY, K. A., KALMUSOVA, J., FEIL, E. J., GUPTA, S., MUSILEK, M., KRIZ, P., and MAIDEN, M. C. (2000) Carried meningococci in the Czech Republic: A diverse recombining population. *Journal of Clinical Microbiology* **38,** 4492–4498.

JOLLEY, K. A., WILSON, D. J., KRIZ, P., MCVEAN, G., and MAIDEN, M. C. J. (2005) The influence of mutation, recombination, population history, and selection on patterns of genetic diversity in *Neisseria meningitidis*. *Molecular Biology and Evolution* **22,** 562–569.

JUKES, T. H. and CANTOR, C. R. (1969) Evolution of protein molecules. In *Mammalian Protein Metabolism* (ed. H. N. Munro), pp. 21–132. Academic Press, New York.

KIMURA, M. (1969) The number of heterozygous nucleotide sites maintained in a finite population due to steady flux of mutations. *Genetics* **61,** 893–903.

KIMURA, M. (1980) A simple method for estimating evolutionary rates of base substitutions through comparative studies of nucleotide sequences. *Journal of Molecular Evolution* **16,** 111–120.

KINGMAN, J. F. C. (1982) The coalescent. *Stochastic Processes and Their Applications* **13,** 235–248.

KUHNER, M. and FELSENSTEIN, J. (1994) A simulation comparison of phylogeny algorithms under equal and unequal evolutionary rates. *Molecular Biology and Evolution* **11,** 459–468.

LIOLIOS, K., TAVERNARAKIS, N., HUGENHOLTZ, P., and KYRPIDES, N.C. (2006) The Genomes OnLine Database (GOLD) v.2: A monitor of genome projects worldwide. *Nucleic Acids Research* **34,** 332–334.

MAIDEN, M. C. J., BYGRAVES, J. A., FEIL, E. et al. (1998) Multilocus sequence typing: A portable approach to the identification of clones within populations of pathogenic microorganisms. *Proceedings of the National Academy of Sciences of the United States of America* **95,** 3140–3145.

MARGULIES, M., EGHOLM, M., ALTMAN, W. et al. (2005) Genome sequencing in microfabricated high-density picolitre reactors. *Nature* **437,** 376–380.

MAU, B. and NEWTON, M. (1997) Phylogenetic inference for binary data on dendograms using Markov chain Monte Carlo. *Journal of Computational and Graphical Statistics* **6,** 122–131.

MCCLELLAND, M., SANDERSON, K. E., CLIFTON, S. W. et al. (2004) Comparison of genome degradation in Paratyphi A and Typhi, human-restricted serovars of *Salmonella enterica* that cause typhoid. *Nature Genetics* **36,** 1268–1274.

MEDINI, D., DONATI, C., TETTELIN, H., MASIGNANI, V., and RAPPUOLI, R. (2005) The microbial pan-genome. *Current Opinion in Genetics and Development* **15,** 589–594.

NEEDLEMAN, S. B. and WUNSCH, C. D. (1970) A general method applicable to the search for similarities in the amino acid sequence of two proteins. *Journal of Molecular Biology* **48,** 443–453.

PARKHILL, J., DOUGAN, G., JAMES, K. D. et al. (2001) Complete genome sequence of a multiple drug resistant *Salmonella enterica* serovar Typhi CT18. *Nature* **413,** 848–852.

PRITCHARD, J., STEPHENS, M., and DONNELLY, P. J. (2000) Inference of population structure using multilocus genotype data. *Genetics* **155,** 945–959.

RAMBAUT, A. (2008) FigTree, a graphical viewer of phy logenetic trees. http://tree.bio.ed.ac.uk/software/figtree/ (accessed September 1, 2008).

RAMBAUT, A. and DRUMMOND, A. (2007) Tracer v1.4. http://beast.bio.ed.ac.uk/Tracer (accessed September 1, 2008).

RUTHERFORD, K., PARKHILL, J., CROOK, J., HORSNELL, T., RICE, P., RAJANDREAM, M. A., and BARRELL, B. (2000) Artemis: Sequence visualization and annotation. *Bioinformatics* **16,** 944–945.

SAIKI, R., GELFAND, D., STOFFEL, S. et al. (1988) Primer-directed enzymatic amplification of DNA with a thermostable DNA polymerase. *Science* **239,** 487–491.

SAIKI, R., SCHARF, S., FALOONA, F. et al. (1985) Enzymatic amplification of beta-globin genomic sequences and restriction site analysis for diagnosis of sickle cell anemia. *Science* **230,** 1350–1354.

SAITOU, N. and NEI, M. (1987) The neighbor-joining method: a new method for reconstructing phylogenetic trees. *Molecular Biology and Evolution* **4,** 406–425.

SANDERSON, M. J. and KIM, J. (2000) Parametric phylogenetics? *Systematic Biology* **49,** 817–829.

SANGER, F., NICKLEN, S., and COULSON, A. (1977) DNA sequencing with chain-terminating inhibitors. *Proceedings of the National Academy of Sciences of the United States of America* **74,** 5463–5467.

SEILER, A., REINHARDT, R., SARKARI, J., CAUGANT, D. A., and ACHTMAN, M. (1996) Allelic polymorphism and site-specific recombination in the *opc* locus of *Neisseria meningitidis*. *Molecular Microbiolog* **19,** 841–856.

SHEPPARD, S., MCCARTHY, N., FALUSH, D., and MAIDEN, M. (2008) Convergence of *Campylobacter* species: Implications for bacterial evolution. *Science* **320,** 237–239.

SMITH, T. F. and WATERMAN, M. S. (1981) Identification of common molecular subsequences. *Journal of Molecular Biology* **147,** 195–197.

SOLTIS, P. S. and SOLTIS, D. E. (2003) Applying the bootstrap in phylogeny reconstruction. *Statistical Science* **18,** 256–267.

SWOFFORD, D. (2002) *PAUP*. Phylogenetic Analysis Using Parsimony (*and Other Methods). Version 4.* Sinauer Associates, Sunderland, MA.

TAMURA, K., DUDLEY, J., NEI, M., and KUMAR, S. (2007) MEGA4: Molecular Evolutionary Genetics Analysis (MEGA) Software Version 4.0. *Molecular Biology and Evolution* **24,** 1596–1599.

THOMPSON, J. D., HIGGINS, D. G., and GIBSON, T. J. (1994) CLUSTAL W: Improving the sensitivity of progressive multiple sequence alignment through sequence weighting, position-specific gap penalties and weight matrix choice. *Nucleic Acids Research* **22,** 4673–4680.

TUFFLEY, C. and STEEL, M. (1997) Links between maximum likelihood and maximum parsimony under a simple model of site substitution. *Bulletin of Mathematical Biology* **59,** 581–607.

URETA-VIDAL, A., ETTWILLER, L., and BIRNEY, E. (2003) Comparative genomics: Genome-wide analysis in metazoan eukaryotes. *Nature Reviews. Genetics* **4,** 251–262.

VOS, M. and DIDELOT, X. (2008) A comparison of homologous recombination rates in bacteria and archaea. *ISME Journal* **3,** 199–208.

WANG, J. F., CAUGANT, D. A., LI, X. et al. (1992) Clonal and antigenic analysis of serogroup a *Neisseria meningitidis* with particular reference to epidemiological features of epidemic meningitis in the People's Republic of China. *Infection and Immunity* **60,** 5267–5282.

WANG, J. F., CAUGANT, D. A., MORELLI, G., KOUMARÉ, B., and ACHTMAN, M. (1993) Antigenic and epidemiologic properties of the et-37 complex of *Neisseria meningitidis*. *Journal of Infectious Diseases* **167,** 1320–1329.

WATTERSON, G. A. (1975) On the number of segregating sites in genetical models without recombination. *Theoretical Population Biology* **7,** 256–276.

WIRTH, T., FALUSH, D., LAN, R. et al. (2006) Sex and virulence in *Escherichia coli*: An evolutionary perspective. *Molecular Microbiology* **60,** 1136–1151.

YANG, Z. (2007) PAML 4: Phylogenetic Analysis by Maximum Likelihood. *Molecular Biology and Evolution* **24,** 1586–1591.

YANG, Z. and RANNALA, B. (1997) Bayesian phylogenetic inference using DNA sequences: A Markov chain Monte Carlo method. *Molecular Biology and Evolution* **14,** 717–724.

Chapter 4

Genetic Recombination and Bacterial Population Structure

Darren P. Martin and Robert G. Beiko

4.1 INTRODUCTION

The transmission of genetic information can be achieved via means other than parent-to-offspring inheritance. The lateral or horizontal transfer of genes can occur between closely or distantly related organisms, facilitated by any of a number of different mechanisms. Transfers that occur between closely related organisms (e.g., individuals within the same species) are often termed *recombination* events if they are driven by homologous processes. Conversely, de novo gene acquisition can occur through illegitimate, or nonhomologous, recombination processes; such events are typically called horizontal or lateral gene transfers (LGTs). For the purposes of this chapter, we use LGT to refer to any transfer of genetic information between a donor and a recipient cell, regardless of the mechanism by which the genetic material is integrated into the recipient genome or of the degree of relatedness between donor and recipient.

LGT can facilitate the spread of adaptive mutations through a population in a manner similar to sexual recombination, and in so doing, it has the potential to disrupt the strictly vertical patterns of inheritance that are usually thought to dominate the population genetics of clonally reproducing organisms. By shuffling alleles within a population, homologous recombination-driven LGT can yield novel genetic combinations that can facilitate, among other things, the evasion of host immunity, the emergence of multiple drug resistance, host range modification, adaptation to new environments, and the evolution of increased pathogenicity (Arnold et al., 2008). While LGT can enhance the genetic and phenotypic diversity within populations, it can also yield an effect similar to that of concerted evolution, in which these populations are continuously purged of deleterious mutations (Michod et al., 2008). Finally, illegitimate recombination allows the acquisition of novel genes and operons, and consequently, completely novel phenotypes. Such gains of function are evident in the gene distribution patterns seen in closely related genomes such as members of the genus *Pseudomonas* (Schmidt et al., 1996).

While a recipient organism can potentially gain a selective advantage from LGT, many introduced sequences including viral genes such as the cII protein from

Bacterial Population Genetics in Infectious Disease, Edited by D. Ashley Robinson, Daniel Falush, and Edward J. Feil

bacteriophage λ (Kedzierska et al., 2003), β-lactamase enzymes in isolation (Morosini et al., 2000; Hossain et al., 2004), and restriction enzymes (unless accompanied by a protective methyltransferase gene; Jeltsch and Pingoud, 1996) can be deleterious. Indeed, most evidence suggests that none of the processes that facilitate the acquisition of genes initially arose for that specific purpose: instead, transformation processes typically degrade DNA upon its introduction into the cell, while other transfer mechanisms may arise as a side effect of the propagation of selfish elements (Redfield, 2001). DNA repair has been suggested as a primary purpose for the evolution of transformation (Michod et al., 1988), but this hypothesis is controversial and is contradicted by the regulatory processes that govern competence (Redfield, 1993). However, once a molecule of DNA has crossed into a recipient cell, genes can potentially be recruited into the genome and expressed in integrons (Mazel, 2006). These structures, which include an enzyme that facilitates the recombination of novel elements into the genome, have been found associated with both genes of unknown function and genes conferring antibiotic resistance and environmental adaptations.

While LGT as a phenomenon has been known for many decades (Jones and Sneath, 1970; Heinemann and Sprague, 1989), the importance of LGT as an evolutionary process has risen to greater prominence with the advent of large-scale whole-genome sequencing. When the first few genomes of cellular organisms became available for comparison, it was quickly recognized that many genes had either unusual nucleotide compositional properties or displayed discordant phylogenetic relationships that set them apart from the principal "genomic signature" (Karlin et al., 1997) of the genomes in which they resided. This was particularly true for the archaea and thermophilic bacteria for which early analyses indicated that >20% of genes in some genomes may have had xenologous origins (Koonin et al., 1997; Nelson et al., 1999). While controversies surrounded the interpretation of discordant phylogenies and the predictions of composition-based approaches (Ragan, 2001; Kurland et al., 2003), even a comparison of the homologous gene contents of different *Escherichia coli* strains (Welch et al., 2002) revealed the existence of many genes that were unique to specific strains. The distributions of homologous and orthologous genes across all prokaryotic life also indicated an important role for LGT.

It is important to point out that the apparent pervasiveness of LGT events discovered in comparative genome scans could, at least in part, be explained by gene loss rather than by gene acquisition through LGT. If, for example, a gene present in a distant common ancestor was lost in multiple descendant lineages, it could appear as though the gene was laterally transferred to the lineages that retained it. However, since invoking a gene loss hypothesis for every potential instance of LGT would require an impossibly large cenancestral "genome of Eden" (Doolittle et al., 2003; Dagan and Martin, 2007), it is almost certain that a large proportion of apparent "gain-of-function" LGT events are genuine instances of gene acquisition.

Besides examining the evolutionary value and biological consequences of LGT, this chapter also deals with the technical issues of both analyzing signals of genetic recombination using nucleotide sequence data and accounting for the potentially misleading effects that LGT can have on various evolutionary analyses.

4.2 CONSTRAINTS ON LGT

Gogarten et al. (2002) enumerated five factors that influence LGT between organisms: two of these (propinquity and gene transfer mechanisms) reflect the opportunity for an organism to acquire a gene from a potential donor, while the other three (metabolic compatibility, compatibility of gene expression mechanisms, and environmental adaptation) will influence the probability that an acquired and integrated gene will spread to fixation in the recipient population. Constraints on opportunity have been considered extensively elsewhere (Kurland et al., 2003; Thomas and Nielsen, 2005); we note here that individuals within the same population are likely to share compatible LGT vectors such as transducing phages, conjugation mechanisms, or uptake sequences for transformation (Fitzmaurice et al., 1984). Propinquity may not be necessary if DNA can be transferred between habitats, either as naked DNA (Levy-Booth et al., 2007; Vlassov et al., 2007) or packaged in a vector (Snyder et al., 2007).

Although precise definitions vary, genetic exchange between organisms can be driven by homology-dependent or homology-independent recombination processes (e.g., nonhomologous or illegitimate recombination), or through the transmission of self-replicating genetic elements such as plasmids. Closely related genomes can potentially use homologous recombination as the integrative process due to higher average sequence similarity. However, homologous recombination frequency drops off exponentially with decreasing sequence identity (Vulić et al., 1997). Unless recombination can rapidly purge all genetic variants from a population, mutational drift and nonhomologous recombination events may raise barriers to homologous recombination within the genome and may serve as speciation "seeds" that lead to strain diversification (Lawrence, 2002). Vetsigian and Goldenfeld (2005) used simulation techniques to show that homologous recombination mechanisms that require sequence identity at both ends of the recombined region can give rise to "propagating fronts" following the introduction into genomes of initial seed sequences by nonhomologous recombination. Homologous recombination in the genus *Bacillus* is known to require high degrees of sequence identity at both ends of transferred sequences to operate (Majewski and Cohan, 1999), and examination of several *Bacillus* genomes has revealed a steplike pattern of sequence similarity that suggests the existence of diversifying regions within these genomes.

Once acquired DNA has been integrated into a recipient genome, either as a plasmid or via recombination, its fate in population genetic terms will depend on its contribution to the fitness of the organism. Genes may be lethal to the cell due to their disruption of important processes. For instance, Sorek et al. (2007) claimed that some transfers between closely related genomes were less likely due to the potentially disruptive effects of modified gene copy number. If two homologous copies of a given gene (one "native," one xenologous) are both expressed within a cell, these products may indeed disrupt multisubunit complexes (such as the ribosome) that require specific ratios of constituent proteins for correct assembly. However, this situation may not arise for many introgressed genes (or gene fragments) if they go unexpressed because they possess promoters or other regulatory elements that are unrecognized by the host transcription and/or translation machinery. Expression of such genes would require their being brought under the control of endogenous promoters. This process could in turn also involve homologous recombination with the homologous locus in the host genome, in which case the original copy might be lost. Conversely, fortuitous nonhomologous recombination immediately downstream of unsuitably strong promoters could lead to the expression of the xenologous gene products at toxic concentrations.

There is no mechanistic barrier to homologous recombination occurring within gene boundaries. Gene conversion, in which recombination occurs within the coding sequences of homologous genes, is a process in many ways analogous to homologous recombination-driven LGT, and has been shown to occur extensively in both prokaryotic and eukaryotic systems (Zangenberg et al., 1995; Morris and Drouin, 2007; Palmer and Brayton, 2007).

Santoyo and Romero (2005) identified several important roles played by gene conversion, including the reversion of mutants, concerted evolution (the spread of potentially beneficial mutations among paralogous sequences), and the generation of increased diversity among positively selected genes such as those encoding antigenic proteins. Although potentially confounded by other factors such as positive selection, an analysis of a type IV secretion system in *Bartonella* has shown evidence for extensive homologous recombination within certain genes (such as *trwJ* and *trwL*) both within and among strains. The roles of these genes in pilus formation and host cell attachment suggest that homologous recombination within and between species may be an important generator of sequence diversity (Nystedt et al., 2008).

Even though they are not constrained by mechanistic barriers, gene conversion events may be selectively unfavorable if they disrupt conserved domains or the interdomain interactions of an encoded protein. Lefeuvre et al. (2007) found such an effect in virus genomes, with recombination break points underrepresented in gene regions that separate encoded amino acids whose interactions are important for the 3-D structure of proteins. Conversely, Chan et al. (2009) found that inferred break points in a bacterial data set did not appear to be preferentially associated with domain boundaries. The extent to which protein fold disruption influences patterns of LGT in natural populations likely varies from gene to gene according to the structural and selective constraints on the proteins they encode.

Since nonhomologous recombination generally involves the integration of foreign genetic material at any site within a genome, this process can potentially disrupt any gene. This is especially significant in prokaryotic genomes where protein-coding genes typically constitute 70–90% of the genome sequence. Examples of open reading frames (ORFs) disrupted by nonhomologous recombination have been identified in various sequenced genomes: for instance, in the *Mycobacterium tuberculosis* H37Rv genome, gene Rv2353c has been disrupted by a retrotransposition event (Betts et al., 2000).

The complexity hypothesis (Jain et al., 1999) predicts that the transfer of "informational" genes involved in conserved processes will, during evolution, be selectively disfavored over the transfer of metabolic and regulatory "operational" genes (Rivera et al., 1998) due to the tendency of the former to participate in large, multisubunit complexes. Informational genes have been used to define a nontransferrable "core" of genomic content that is either immune or extremely recalcitrant to successful propagation following LGT. This assumption and the presence of homologous informational genes in most genomes have underpinned the use of concatenated sets of these genes (or their encoded amino acid sequences) to infer organismal histories (Baldauf et al., 2000; Brochier et al., 2004).

The complexity hypothesis has been tested many times in large-scale analyses using both homology-independent (Ragan, 2001; Nakamura et al., 2004) and homology-dependent (Charlebois et al., 2004; Beiko et al., 2005; Zhaxybayeva et al., 2006) approaches to identify LGT events. Categories of genes have typically been defined using homology-based assignments of function and classification schemes such as the NCBI Clusters of Orthologous Groups based on the groupings of Riley (1993). Statistical contrasts of the transfer of these two classes of genes have often, but not always (see, e.g., Zhaxybayeva et al., 2006), shown significantly fewer transfers for informational genes than for operational ones. Wellner et al. (2007) explored the relationship between connectivity (which includes a protein's interactions within and outside of its complex), and found that genes encoding proteins with high connectivity were not more recalcitrant to successful transfer, except in cases where they formed part of an essential complex. In spite of these differences, it is clear that informational genes can be successfully transferred, a hypothesis borne out indirectly by the presence of informational genes such as 16S rDNA on phage vectors (Beumer and

Robinson, 2005), and directly by observations of phylogenetic discordance in informational gene and protein phylogenies (Garcia-Vallvé et al., 2002; Gogarten et al., 2002).

While discordance can potentially be remedied by filtering out these genes or proteins from concatenated sets (e.g., Ciccarelli et al., 2006), what remains after filtering in such cases is typically a small "rump" of genes that, due to weakened phylogenetic signals, potential evolutionary model violations, and the possibility of further undetected LGT among the remaining genes, might not defensibly constitute the history of the represented organisms (Dagan and Martin, 2006). Furthermore, the operational versus informational distinction is an imperfect surrogate for interaction complexity. For example, informational proteins such as aminoacyl-tRNA synthetases, which interact only with a single tRNA and one amino acid moiety, show considerable evidence of historical transfers, even between distantly related taxonomic groups (Woese et al., 2000).

Within species and genera, the relatively slow evolutionary rate of informational proteins (Wuchty et al., 2003; Aris-Brosou, 2005) may slow the emergence of mutational barriers to homologous recombination between closely related organisms. The exchange of proteins with relatively low amino acid sequence divergence is also less likely to disrupt the protein–protein interactions that underpin complex formation, minimizing the effects reported by Lefeuvre et al. (2007). Detecting homologous recombination in informational genes transferred between closely related organisms may also be difficult, since the lower degrees of sequence divergence commonly found in these genes will diminish the power of recombination or phylogenetic analyses (see, e.g., Lefébure and Stanhope, 2007). This effect might significantly bias the relative amounts of recombination observed within "core" versus noncore genes. On the other hand, the low degrees of sequence divergence among closely related individuals may mean that gene transfers between them are likely to be selectively neutral and dependent only on rates of acquisition and recombination.

Beyond the possible negative effects on metabolic networks and protein complexes, introgressed genes may disrupt important structural aspects of the genome and may consequently reduce the fitness of the recipient organism. Such structural properties can include supercoiling domains, leading/lagging strand gene organization (reviewed in Rocha, 2008), codon preference (Medrano-Soto et al., 2004), and features involved in chromosome partitioning during replication (Hendrickson and Lawrence, 2006). However, such effects may be small when examined at the level of introgression of single genes or operons, and a comprehensive model of how such disruptions might impact on the survival probabilities of a recipient genome has not yet been articulated.

4.3 INFLUENCES OF LGT ON SEQUENCE ANALYSES

When not accounted for, LGT can potentially influence a number of inferences one can make when considering nucleotide sequence data. Primary among these is the expression of evolutionary relationships within phylogenetic trees. The combined evolutionary histories of laterally transferred sequences cannot be accurately portrayed using conventional bifurcating phylogenetic trees (Posada and Crandall, 2002), and, as a result, any nucleotide sequence analysis that derives power from the correct inference of phylogenetic tree topologies and branch lengths will be at least mildly confounded by recombination. Such analyses include the detection of positive selection (Scheffler et al., 2006), ancestral sequence prediction (Posada and Crandall, 2002), molecular clock estimates of evolution rates (Schierup and Hein, 2000), identification of coevolving sites (Shapiro et al., 2006), and nested clade-based phylogeographic tracing of migration routes or transmission pathways.

By influencing the linkage between nucleotide polymorphisms, frequent homologous recombination between the sequences of closely related individuals (such as occurs during parasexual reproduction) will also influence a variety of population genetic analyses that do not necessarily rely on the correct inference of phylogenetic trees. By a process called "genetic hitchhiking," neutral nucleotide polymorphisms (or alleles) can be driven to high frequencies within populations due to their occurring within close proximity to high fitness polymorphisms. Relative to other genome regions, this process will result in decreased genetic variability in those regions surrounding high fitness polymorphisms. Conversely, in genome regions that are distantly separated (or "unlinked") from high fitness polymorphisms, frequent homologous recombination might extensively mix neutral polymorphisms and, in so doing, might create increased genetic diversity in these regions. In the presence of recombination, effective population size estimates will therefore be alternatively biased downward when they are derived using genome loci carrying strongly selected polymorphisms and upward when they are derived using loci that are distantly separated from positively selected polymorphisms. Such biases will strongly influence summary statistics that rely on accurate estimates of population sizes. These include Tajima's (1989) D-statistic and Fu and Li's (1993) F-statistic that describe the relative neutrality of evolution. Unless carefully estimated recombination rates are explicitly accounted for when interpreting the values of these statistics, it could result in the false inference of either selective sweeps or population expansion (both of which can be inferred from a departure from neutral expectations; for a review, see Awadalla, 2003).

4.4 THE DETECTION OF INDIVIDUAL LGT EVENTS

In general, the detection of individual LGT events involves the analysis of nucleotide sequence data sampled from a number of taxa within either the same or different species. Although the nucleotide sequence analysis tools that have been devised to perform this task are mostly geared toward detecting homologous recombination, they can also be used to identify potential nonhomologous recombination events. This chapter will, however, focus on their use for detecting signals of homologous recombination, which for the sake of simplicity will simply be referred to as "recombination." Also, as the tools that will be described are largely blind to the conceptual differences between whole gene transfers (traditionally known as LGT events) and partial gene transfers (traditionally referred to as gene conversion events), all detectable sequence exchanges will simply be referred to here as "recombination events."

Many, if not the vast majority, of the recombination events that have taken place during the evolutionary history of a sampled data set of DNA molecules will have left no detectable trace of their occurrence (Posada and Crandall, 2001). These undetectable recombination events will include those that have occurred either between identical sequences or prior to the last common ancestor(s) of the sampled DNA sequences. Even recombination events that have occurred between quite distantly related sequences might never be detectable if either only a small fragment of sequence was exchanged or sequences resembling the recombinant's parents remain unsampled.

Identification of recombination within a group of DNA sequences usually requires one or more statistical or phylogenetic tests to assess whether individual members of the group have evolutionary histories that vary depending on the genome or gene region considered. The power of such recombination tests (see Table 4.1 for examples) is heavily

Table 4.1 Recombination Analysis Methods

Program	Method(s) implemented	Analysis approach: Partition	Analysis approach: Scan	Gives p value	Gives breakpoint positions	Identifies recombinants	Reference
3Seq	3Seq	DM	E	+	+	–	Boni et al. (2007)
Barce	Barce	DM	E/Q	+/–	+	–	Husmeier and McGuire (2003)
Chimaera	Chimaera	MW	E	+	+	–	Posada and Crandall (2001)
ClonalFrame	ClonalFrame	DM	E/Q	–	+	+	Didelot and Falush (2007)
DualBrotherS	DualBrothers	DM	Q	–	+	+	Minin et al. (2005)
EEEP	EEEP	RS	Q	–	–	+	Beiko and Hamilton (2006)
GARD	GARD	DM	E	+	+	–	Kosakovsky Pond et al. (2006)
GENECONV	GENECONV	DM	E	+	+	–	Sawyer (1989)
Homoplasy	Homoplasy test	DM	E	+	–	–	Maynard Smith and Smith (1998)
HorizStory	HorizStory	RS	Q	–	–	+	McLeod et al. (2005)
jpHMM	jpHMM	DM	Q	–	+	+	Schultz et al. (2006)
LARD	LARD	MP/DM	E/Q	+/–	+	–	Holmes et al. (1999)
MaxChi	MaxChi	MW	E/Q	+	+	–	Maynard Smith (1992)
SplitsTree	Phi test	DM + MW	E	+	–	–	Bruen et al. (2006)
PhylPro	PhylPro	MP	E	–	+	+	Weiller (1998)
PIST	PIST	MW	Q	+	–	–	Grassly and Holmes (1997)
PLATO	PLATO	MW	Q	+	+	–	Worobey (2001)
RAT	RAT	MW	E/Q	–	+	–	Etherington et al. (2005)
RDP3	RDP, GENECONV, 3Seq, BootScan, MaxChi, Chimaera, Dss, SiScan, PhylPro, LARD, VisRD	MW/DM	E	+	+	+	Martin et al. (2005b)
RecPars	RecPars	MW	E	–	+	–	Hein (1990)
Rega	Rega Bootscan	MW	Q	+	+	+	de Oliveira et al. (2005)
RIP	RIP	MW	Q	+	+	+	Siepel et al. (1995)
SimPlot	SimPlot Bootscan	MW	Q	–	+	+	Lole et al. (1999) and Salminen et al. (1995)
SiScan	SiScan	MW	E	+/–	+	–	Gibbs et al. (2000)
TOPAL	Dss	MW	E	+	+	–	McGuire and Wright (2000)
TOPALi	Dss, Barce, Jambe	MW/DM	E/Q	+	+	–	Milne et al. (2004)
VisRD	VisRD	MW	E	+	+	+	Forslund et al. (2004)

DM = multiple dynamically placed partitions; MW = moving window-based partitioning; MP = single moving partition; RS = rigid single user-specified partition; E = exploratory scanning approach with no prior specification of a nonrecombinant set of reference sequences; Q = query versus reference scanning approach in which a potential recombinant (the query) is scanned against a set of known nonrecombinant sequences (the references).

dependent on both the nature of the recombination events being detected and the thoroughness of sampling. Most tests will fail unless all of the following criteria are met:

1. Along with the recombinant sequence, at least one sequence resembling one of its parents must be sampled.
2. Recombination must have occurred between parental sequences carrying enough polymorphisms that the origin of recombinant sequences can be unambiguously traced back to at least one parental lineage.
3. The distribution of polymorphisms observed within recombinant sequences must not be credibly attributable to other evolutionary processes such as convergent point mutations or mutation rate variation.

The process of detecting recombination can be illustrated by depicting the evolutionary histories of recombinant sequences using phylogenetic trees (Fig. 4.1). Given a multiple sequence alignment containing one recombinant and its two parental sequences (respectively labeled recombinant, "major parent," and "minor parent" in Fig. 4.1), two phylogenetic trees can be constructed from the two nonoverlapping segments of the nucleotide sequence alignment that correspond to the two tracts of the recombinant sequence that were inherited from its different parents. When these trees are compared, the recombinant sequence apparently "jumps" between clades. While many recombination detection methods directly apply such phylogenetic approaches, others rely on different measures of sequence relatedness. Without exception, however, all recombination detection methods work by

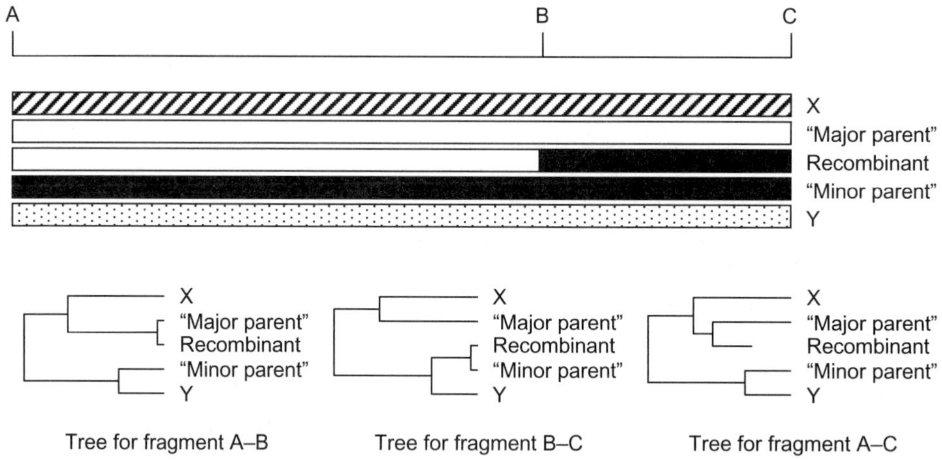

Figure 4.1 Phylogenetic signals betraying the presence of recombination in a sample of five nucleotide sequences. The patterned lines represent aligned nucleotide sequences of five individuals sampled from nature. The phylogenetic trees depict the probable evolutionary histories of the five sequences as inferred from different portions of the nucleotide sequence alignment. Notice how the recombinant sequence "jumps" between the "major parent"—sequence X clade and the "minor parent"—sequence Y clade in the A–B and B–C fragment trees. The recombinant sequence has inherited genome fragment A–B from a sequence resembling that labeled "major parent" and fragment B–C from a sequence resembling that labeled "minor parent." Note that neither of these "parental" sequences are the actual parents of the recombinant but are simply sequences in the sample that most closely resemble the recombinant's actual parent. The probability of sampling a recombinant and its two actual parents is very small.

identifying shifting relationships between sequences in different genomic regions. Such shifts or jumps in sequence relatedness are often referred to as "recombination signals."

Recombination signals might be detected in a number of ways. Most recombination detection methods have two basic components: (i) a mechanism for partitioning sequences into two or more tracts and (ii) a test statistic that can be used to compare the relatedness of sequences in different partitions.

4.4.1 Partitioning Schemes

The most simple form of alignment partitioning is that involving the comparison of different genes where partitions are placed at gene boundaries. For example, methods such as EEEP and HorizStory (Table 4.1) detect recombination signals by either comparing phylogenetic trees constructed from different genes (often called gene trees) or comparing phylogenetic trees for individual genes with that obtained using either full genome sequences, concatenated informational gene trees, or some other preferably LGT-free (or at least mostly LGT-free) genomic portion (with these latter trees often claimed to be species trees).

By dynamically optimizing the location of sequence partitions, many recombination detection methods can also infer the locations of recombination break points. The simplest of the dynamic partitioning approaches use a so-called sliding partition to split a sequence alignment into two pieces (e.g., those implemented in programs such as LARD and PhylPro; Table 4.1). In this approach, a partition is moved one or more nucleotides at a time along the length of the alignment with the tracts of sequence on the left-hand side of the partition being compared to those on the right. The partition position at which the two segments of the alignment are most different is taken to be the most probable recombination breakpoint position.

More sophisticated dynamic partitioning schemes involve two or more moving partitions. Having more than one partition makes good sense when one considers that most recombination events in long DNA molecules, and all recombination events in circular molecules, involve two recombination break points. However, even when considering just two moving partitions, comparing every possible tract of sequence with every other one can be quite computationally intensive. Therefore, many methods test only a subset of all the possible partitions. Most commonly, a pair of partitions called a sliding window is moved one or more nucleotides at a time along the alignment. The piece of the alignment within the window is then either compared to the remainder of the alignment or to immediately adjacent windows.

Although some sliding window recombination detection methods incorporate a window-size optimization step (e.g., the MaxChi and Chimaera implementations in the program RDP3), most of these methods are strongly influenced by the arbitrary choice of an initial window size. The key issue is that the optimal window size for identifying any particular recombination signal will be the distance between the pair of recombination break points comprising the signal. When a data set contains a variety of recombination signals produced by the exchange of both large and small pieces of DNA, it can be quite difficult to select a suitable window size. Some of the most sophisticated recombination breakpoint detection methods such as GARD and DualBrothers (Table 4.1) therefore employ quite complex "windowless" partitioning schemes. In these, large numbers of dynamically placed partition sets (often involving 10 or more partitions) are evaluated and compared to one another to yield the most probable locations of recombination break points.

4.4.2 Test Statistics

There are many possible statistics that could be used to compare the relatedness of sequences in different genome locations. The simplest and most intuitively obvious of these are based on genetic distances, codon usage biases, or GC content. These simple tests are based on the reasonable (but not always true) assumption that any given sequence will be most similar to whichever sequence it shares a most recent common ancestor. Such tests will often involve moving a sliding window along an alignment and calculation of the relative differences/similarities between each pair of sequences in each window along the alignment. Whereas, for example, the genetic distances between nonrecombinant sequences should remain consistent across all alignment partitions, this should not be the case with recombinant sequences. The relative distances between a recombinant sequence and sequences closely related to its parents should shift at one or more points along the alignment. The points at which such shifts occur should correspond with recombination breakpoint positions.

Although evolutionary relatedness usually correlates well with pairwise genetic distances, similarities in GC content, and similarities in codon usage biases, this is not always the case. Therefore, using phylogenetic methods to more accurately determine the relative relatedness of sequences has been extensively explored in the context of recombination signal detection (examples include the Dss, DualBrothers, PLATO, RecPars, GARD, Barce, ClonalFrame, and BootScan methods; Table 4.1). The popular BootScan method uses a sliding window partitioning scheme in which bootstrapped neighbor-joining trees are constructed for segments of the alignment within each window. The relative relatedness of the sequences in each window is then expressed in terms of bootstrap support for the phylogenetic clusters within which they occur in the tree. Recombination signals are detectable as taxa "jumping" between different branches of the tree. Recombination break points are, in turn, detectable as the points along the alignment where there is a sudden change in bootstrap support grouping the potential recombinant with sequences resembling its potential parental sequences.

A major drawback of the pure phylogenetic methods is that they tend to be a lot slower than, for example, genetic distance-based methods. As a result of this, a diverse group of methods has been devised that, while primarily genetic distance based, also take some phylogenetic information into account. Methods such as VisRD, RDP, and SiScan (Table 4.1), for example, will note the relationships between sequences in a phylogenetic tree and, when calculating genetic distances, only consider nucleotide sequence differences that map to specific branches of the tree.

Regardless of whether phylogenetic, genetic distance, GC content, or codon bias-based measures of relatedness are used to detect potential recombination signals, it is usually desirable to determine the probability that the observed signals could have arisen in the absence of recombination. To achieve this, many methods will use a specific statistical test such as one based on the binomial, normal, or chi-squared distributions. As the validity of these tests is not entirely obvious (the actual underlying probability distribution is invariably unknown and some tests may be undermined by evolutionary processes such as selection), many methods such as Dss, GENECONV, VisRD, and LARD (Table 4.1) additionally employ either "permutation" or "parametric bootstrap" tests. Although these tests can better account for underlying probability distributions, they are generally very slow in that they involve reanalysis of hundreds or thousands of data sets. Whereas in permutation tests alignment columns in the real data set are simply reshuffled to yield "permuted data sets" (reshuffling usually destroys the patterns of sites that are detectable as recombination signals), for parametric bootstrapping, data sets that superficially resemble the actual data set are simulated in the absence of recombination.

4.4.3 Advantages and Disadvantages of Different Recombination Analysis Approaches

Very little is currently known about the relative merits of different recombination analysis methods. The main problem with comparing different methods is that they often require different types of data and provide information on different aspects of the recombination process. For example, a method might quite accurately identify evidence of intraspecies recombination but will suffer an intolerably high false-positive rate if used to infer interspecies recombination (e.g., homoplasy test; Posada and Crandall, 2001). Alternatively, a method might simply give an estimate of the most likely recombination breakpoint positions but provide no indication of the probability of recombination having occurred or not (e.g., the SimPlot Bootscan, DualBrothers, and LARD methods; Table 4.1).

Although the relative abilities of 21 recombination analysis methods to detect recombination have been tested using the same simulated and real data sets (Posada and Crandall, 2001; Posada, 2002; Martin et al., 2005a; Bruen et al., 2006; Carvajal-Rodríguez et al., 2006; Kosakovsky Pond et al., 2006; Boni et al., 2007), nearly nothing is known about how accurately these methods identify recombination break points and recombinant sequences. There is, however, some indication that methods vary quite widely in their abilities to detect both the presence of recombination (Posada and Crandall, 2001) and the positions of recombination break points (Chan et al., 2006).

Whereas the more sophisticated, computationally intense methods tend to perform slightly better than the simple methods with respect to recombination detection power, relatively simple methods such as Phi test and MaxChi seem to be only slightly less powerful than these but are capable of analyzing much bigger and more complex data sets. Also, while the breakpoint detection accuracies of sophisticated methods such as GARD, jpHMM, and DualBrothers are apparently very high relative to some simple methods (Kosakovsky Pond et al., 2006; Schultz et al., 2006; Chan et al., 2006), they have never been thoroughly compared either to one another or to the most promising simple breakpoint identification methods such as MaxChi and Chimaera.

The accuracy with which available recombination analysis methods can be used to identify recombinant sequences is currently completely unknown. In fact, it is unlikely that without either prior knowledge of a set of nonrecombinant parental sequences or a reference "nonrecombinant" phylogeny, any of the most widely used recombination analysis methods other than VisRD, PhylPro, EEEP, and HorizStory could be productively used to identify recombinant sequences. Given a set of known nonrecombinant sequences or a nonrecombinant phylogeny, most methods that identify recombination break points could also be used to identify recombinant sequences using a so-called query versus reference analysis approach (see below).

Methods such as VisRD and PhylPro do not require any help identifying recombinants because they effectively scan every sequence in a data set against all others and in so doing identify the sequences in an alignment that generate the strongest recombination signals. EEEP and HorizStory compare phylogenetic trees constructed from different genome regions and attempt to infer the minimum number of unique recombination events that could account for changes in topology between the trees. By directly inferring the tree permutations needed to create the observed recombination signals, these methods also identify a credible subset of recombination events in the tree and in so doing identify the probable recombinants.

It is important to point out, however, that despite their promise, VisRD, PhylPro, EEEP, and HorizStory probably still have a high failure rate when identifying recombinants. With the PhylPro and VisRD methods, for example, nonrecombinant sequences will often generate the strongest recombination signals when parental sequences have remained unsampled or when reciprocal or nearly reciprocal recombinants (recombinants that are mirror images of one another) have been sampled. Also, the most parsimonious set of recombination events proposed by the HorizStory and EEEP methods is likely to be neither unique (there may be many equally parsimonious sets of events) nor entirely accurate for even modest-sized data sets (i.e., those containing 20 or more sequences) and as a result, at least a small percentage of all recombinant designations are likely to be incorrect. Different approaches can be used to choose from a set of equally parsimonious recombination scenarios (Beiko and Ragan, 2008), but these rely on assumptions about which lineages are more likely to participate in transfer, thereby potentially conflating the hypothesis with the method used to aggregate results.

One of the reasons that there are so many methods with which to detect individual recombination events is that no single general approach, let alone any one method, has yet emerged as being best under all possible analysis conditions. The powerful, very sophisticated methods such as GARD, DualBrothers, and Barce are also extremely slow and are currently only applicable to relatively small or simplified analysis problems. Some simpler methods, while slightly less powerful, can easily handle enormous extremely complex data sets. Also, unlike phylogenetic methods, simple genetic distance-based methods can detect recombination events that do not alter tree topologies. Probably as a consequence of this ability, however, these methods also probably suffer from a higher false-positive rate than phylogenetic methods when the assumption that smaller genetic distances equate with more recent common ancestry is violated—such as often occurs when different lineages in a data set are evolving at vastly different rates.

The difficulty of rationally choosing an appropriate analysis method has spawned recombination analysis tools such as RDP3 (Martin et al., 2005b) and TOPALi (Milne et al., 2004), which provide access to multiple recombination signal detection methods that can be used in conjunction with one another. Not only can these methods be used for cross-checking recombination signals identifiable by individual methods but, in the case of RDP3, they can be collectively combined to analyze large data sets for evidence of recombination. While this may seem like a good idea, it is still unclear how much extra credibility should be given to recombination signals that are detectable by multiple recombination detection methods.

4.4.4 Recombination Analysis Using Exploratory and "Query versus Reference" Methods

There are two basic approaches that can be used to detect and characterize individual recombination events. Choosing which to use depends primarily on the availability of reliable "nonrecombinant" reference sequences or recombination-free phylogenies. Some recombination analysis tools such as SimPlot, DualBrothers, RIP, jpHMM, and many others implemented in the program RDP3 allow the comparison of potentially recombinant query sequences with a set of known nonrecombinant, or reference, sequences and can potentially identify (i) recombination events during the evolutionary history of the query sequence, (ii) the approximate locations of recombination break points, and (iii) the parental sequences (or at least sequences closely related to the parental sequences if these are among the chosen references). Other tools such as HorizStory and EEEP allow one to compare phylogenetic trees constructed using sequences derived from specific genomic loci (called gene trees) with a reference phylogeny (often representing a credible "species" history of vertical descent) that can, for example, be derived using full genome sequences. Although these methods will not provide information on recombination breakpoint positions, they will identify (i) genes that have been obtained through recombination, (ii) the origins of these genes, and (iii) relatively unbiased estimates of recombination event numbers in the history of the sequences examined. Methods such as HorizStory, EEEP, jpHMM, and DualBrothers might be broadly referred to as query versus reference methods.

Another class of methods, called exploratory methods, adopt a different analysis approach in that, unlike the query versus reference methods, they can be used to identify individual recombination events without any nonrecombinant reference sequences or recombination-free reference phylogenies. Examples of these methods include RDP, MaxChi, Chimaera, GENECONV, 3Seq, SiScan, RecPars, VisRD, GARD, and PhylPro (Table 4.1). It should also be pointed out that methods that compare phylogenetic trees (such as HorizStory and EEEP) might be used in an exploratory way. An example of how this could be achieved is described in Hickey et al. (2008), where two unrooted trees are compared using a search procedure that originates from both trees simultaneously: an approach that trades accurate identification of recombinant sequences for unbiased recombination analysis in the absence of a reference phylogeny. Although these exploratory methods are fundamentally more objective than the query versus reference methods (which rely on the subjective and often flawed assembly of a recombination-free reference data set), it is important to point out that they have two serious drawbacks that diminish their appeal.

To blindly enumerate the recombination signals evident within a data set, the exploratory methods will often perform millions of sequence comparisons. This creates serious multiple testing problems that must be accounted for when assessing the statistical significance of potential recombination signals. For large data sets, such as those containing hundreds of sequences, multiple testing corrections can erode statistical power to the point that even relatively obvious recombination signals are missed.

The second problem with exploratory recombination detection methods is that they usually provide no real indication of which sequences in a data set are recombinant. What they will usually provide are either pairs or triplets of sequences within which individual recombination signals are evident. Although the PhylPro method identifies recombinant sequences and the program RDP3 uses this and other accessory methods (based largely on the PhylPro, HorizStory, and EEEP methods) to identify recombinants, the failure rate of these approaches is currently unknown. As a result, the high degrees of objective analysis automation afforded by exploratory recombination detection methods are counterbalanced by the largely subjective and time-consuming manual process of figuring out which sequences are recombinant. This can be a particularly difficult to do by, for example, identifying recombinants as those sequences that "jump" around on phylogenetic trees (the approach generally used to identify recombinant sequences), because in many cases, exploratory methods will detect recombination events between parental sequences that are themselves recombinant—that is, two or three of the sequences used to detect a recombination signal might all jump around in a tree.

Another consideration when choosing a recombination analysis method is that some query versus reference approaches such as those implemented in EEEP and HorizStory rely on extrinsically inferred phylogenetic trees rather than on the sequences themselves.

As noted above, they cannot be used to detect recombination break points or to identify recombined regions: Typically, such methods are applied to gene or protein trees with the implicit assumption that the source data were not subjected to recombination within the blocks of sequence used to infer trees. Also, these and other phylogenetic methods (such as Bootscan, RecPars, DualBrothers, and Dss; Table 4.1) generally consider only the branching order of phylogenetic trees and not their branch lengths. Consequently, recombination events that do not change tree topologies might be detectable by nonphylogenetic methods such as MaxChi, Chimaera, GENECONV, and 3Seq (Table 4.1), but would not be identified by phylogenetic methods. An advantage of these approaches, however, is that both homologous and illegitimate recombinations can produce discordant phylogenies, and both types of event can potentially be identified.

Given a reasonably reliable reference phylogeny or a set of known nonrecombinant sequences, one should therefore always consider using a hybrid exploratory and query versus reference approach that utilizes both phylogenetic and nonphylogenetic recombination detection methods. For, example, exploratory analyses can be used either to test the "nonrecombinant" status of reference sequences (recombinants may have inadvertently been identified as being nonrecombinant) or to derive a credible set of potentially nonrecombinant reference sequences. Exploratory methods could also be used to identify tracts of sequence or genes that should be excluded during the construction of reference phylogenies. Following the exploratory identification of recombination signals, query versus reference analyses with carefully curated reference data sets could then be used to identify the recombinant sequences that are responsible for detectable recombination signals.

4.5 THE ESTIMATION OF HOMOLOGOUS RECOMBINATION RATES

Whereas identifying and characterizing individual LGT events can give qualitative information on gross patterns of sequence exchange between and within species, it cannot provide quantitative estimates of how frequently recombination events occur. The main reason for this is that most recombination events that occur between very similar sequences (such as LGT between members of the same species) are completely undetectable by even the most powerful methods used to characterize individual recombination events. Although simply counting detectable recombination events and mapping their locations can yield valuable information on relative recombination frequencies in different parts of a genome, there is no way that this approach can yield estimates of absolute recombination rates.

Absolute recombination rates can perhaps be best calculated within the framework of traditional population genetics. The concept of linkage disequilibrium in population genetics refers at least in part to the tendency of adjacent sites along a eukaryote chromosome to be inherited together from the same parent following meiotic recombination during sexual reproduction. Generally, the closer two genes are on a chromosome, the smaller will be the probability that a recombination break point will occur between them during meiosis. When closely linked genes are not inherited independently of one another, they are said to be in linkage disequilibrium. Although not all genes in linkage disequilibrium are physically linked on a chromosome (i.e., there are causes of linkage disequilibrium other than low probabilities of recombination occurring between linked sites), tools devised for the analysis of linkage disequilibrium are ideally suited to the estimation of recombination rates.

Given any population with some degree of neutral genetic diversity (i.e., a genetically diverse population of equally fit individuals) and a constant recombination rate, one would expect a specific degree of linkage disequilibrium within the population. Randomly sampling individuals from the population and estimating average degrees of linkage disequilibrium at different pairs of linked polymorphic sites can therefore yield an estimate of linkage disequilibrium that is easily translatable into a recombination rate.

There are various programs that will estimate such "population-scaled" recombination rates. These include LDHAT (McVean et al., 2002), DnaSP (Rozas et al., 2003), LAMARC (Kuhner, 2006), and SITES (Hey and Wakeley, 1997). The recombination rates that these programs calculate are probably better described as recombination frequencies as they have no easily definable time component and are not simply estimates of recombination break points per site per generation. Such recombination frequency estimates can only be translated into genuine recombination rate estimates if the neutral mutation rate is known. However, even without this conversion, recombination frequency estimates can be very useful in that they can be used to compare relative recombination rates within populations known to have similar neutral mutation rates even when these neutral rates are unknown.

It is important to realize, however, that besides physical genetic linkage, degrees of linkage disequilibrium (and hence recombination rate estimates) that are detectable within a sample of nucleotide sequences drawn from a population can be influenced by various other factors including (i) natural selection disfavoring the disruption of intragenome interaction networks that is expected to occur following recombination, (ii) mutation rates, and (iii) biological (e.g., varying migration rates or positive assortative mating) and environmental (e.g., geographic barriers or niche variation across a geographic range) factors that influence the randomness of genetic exchange within the population. Meaningful use of population-scaled recombination rates or recombination frequency estimates is therefore heavily contingent on the quality of sequence sampling, and great care must therefore be taken to assemble appropriate data sets. As a rule, sequences should be sampled as randomly as possible, and these methods should only be applied in comparative studies where the data sets being compared contain evidence of very similar neutral mutation rates (estimated, e.g., using Tajima's [1989] *D*-statistic or Fu and Li's [1993] *F*-statistic in the program DnaSP) and degrees of nonrandom mating and population structure (e.g., estimated with programs such as STRUCTURE [Falush et al., 2007] or LAMARC [Kuhner, 2006]).

4.6 PROPERLY ACCOUNTING FOR LGT DURING SEQUENCE ANALYSES

There are a number of ways in which LGT can be factored into the evolutionary analysis of nucleotide sequences. From the perspective of graphically depicting the evolution of recombining sequences, network-based representations of nucleotide sequence relationships can prove more meaningful than standard bifurcating phylogenetic trees. While programs such as SplitsTree (Huson, 1998), CombineTrees (Cassens et al., 2005), and NETWORK (Forster et al., 2007) can be used to produce such network graphs, it should be pointed out that, strictly speaking, these graphs are not simply phylogenetic trees that represent LGT. There are currently no methods that will construct true phylogenetic network graphs in which branch lengths accurately represent evolutionary distances and internal "cycles" (the branches between branches that make the networks) represent actual LGT events (Woolley et al., 2008). Therefore, rather than being used as an alternative to conventional phylogenetic tree construction and recombination analysis, these network construction tools should always be used in conjunction with these other methods.

LGT can also be accounted for in most other sequence analysis methods that require the inference of phylogenetic trees. While the most obvious way in which this might be achieved is for the methods to use phylogenetic networks rather than bifurcating trees, the lack of proper phylogenetic network construction algorithms has meant that other simpler approaches have been used to deal with LGT.

The most widely used of these do not deal with LGT directly but rather mitigate the influences of LGT on the accuracy of phylogenetic tree construction by adopting an "averaging over tree space" methodology. Put simply, LGT is expected to have a largely unpredictable impact on the topology and branch lengths of phylogenetic trees, and these average-over-tree-space methods minimize this impact by considering large numbers of possible alternative tree topologies and branch lengths. Also called Markov chain Monte Carlo (MCMC) methods, they are expected to be more robust to the influences of LGT than other phylogenetic-based sequence analysis approaches. The programs that apply these methods can be used to infer ancestral sequences (MrBayes; Ronquist and Huelsenbeck, 2003), to identify the occurrence of purifying or diversifying selection (MrBayes), to estimate evolution rates (BEAST; Drummond and Rambaut, 2007), to infer demographic processes (BEAST), and to study migration patters (BEAST).

Other approaches deal with recombination directly. One of these relies on the use of standard recombination analysis methods to identify recombination break points and then uses separately inferred phylogenetic trees for the different "nonrecombinant" sections of the data set. While accounting for LGT in this way has only proven successful in the detection of positive selection in coding sequences (Scheffler et al., 2006), it should be applicable to any other phylogenetic-based methods. In fact, given a set of recombination breakpoint positions, methods implemented in programs such as MrBayes, BEAST, and HyPhy (Pond et al., 2005) can be set up to directly account for recombination breakpoint positions.

LGT can also be directly accounted for in many population genetic-based analyses by first inferring the population-scaled recombination rate from a sequence data set using a program such as LDHAT (McVean et al., 2002) and then by using this rate in subsequent analyses. For example, given a recombination rate and the DNA sequences used to infer this rate, a program such as DnaSP (Rozas et al., 2003) can be used to test whether estimates of Tajima's D-statistic or Fu and Li's F-statistic represent significant departures from neutrality.

4.7 QUESTIONS RELATING DIRECTLY TO LGT

While LGT can be treated as potentially confounding in analyses seeking to quantify evolutionary processes such as selection and speciation, LGT is an important evolutionary phenomenon in its own right. The acquisition of novel functions via laterally transferred genes can provide new metabolic and ecological opportunities, but also poses a risk to the acquiring organism. The analysis of gene flow patterns within a population of microorganisms can also reveal a great deal about both the pressures faced by that population and the barriers to free sharing of DNA among close relatives. These questions have been reviewed extensively (Ochman et al., 2000; Lawrence, 2002; Gogarten and Townsend, 2005; Ragan and Beiko, 2009); here we review some of the major themes of LGT research and address recent work in microbial population dynamics.

4.7.1 What Are the Global Patterns of Gene Sharing among Lineages?

Questions about the frequency of LGT typically focus on the relative rarity of such events. Even if LGT is sufficiently frequent to obscure the relationships among major microbial lineages, in generational terms, the acquisition of a xenologous gene and its subsequent fixation may be extremely rare. Furthermore, the patterns that emerge from large phylogenomic analyses suggest that gene trees depicting most evolutionary time scales have definite nonrandom structures that apparently represent the "mostly" vertical descent of genes. It is not the *consistency* of phylogenomic results across many analyses (e.g., the splitting of bacteria from archaea or the apparent monophyly of proteobacteria) that argues against a dominant role for LGT but rather the *internal support* from any given data set. For instance, Creevey et al. (2004) used permutation tests to show that the observed phylogenetic support for the monophyly of relatively recent groups is considerably greater than random, although the deepest relationships did not reject a null hypothesis of no phylogenetic signal (i.e., rampant LGT). Heat map analyses of phylogenetic signals in several prokaryotic groups (Bapteste et al., 2005) have shown that for many genes, there are insufficient data to achieve the statistical power needed to reject a substantial number of alternative tree topologies, thus calling into question the use of gene trees to address relationships among these groups.

If a majority or plurality phylogenetic signal can be recovered from a set of genomes, then this signal might be taken to represent the "vertical" history of genes that have been inherited without LGT (but see Doolittle and Bapteste, 2007). Using the taxon relationships represented by such trees as a null hypothesis, other topologies that reject the "reference" tree can be examined to see whether they support a dominant evolutionary role for LGT. Of particular interest is testing the possibility that preferential sharing of genes within discrete biomes is a fundamental cohesive force shaping the emergence and evolution of novel bacterial communities, such that otherwise distantly related taxa (e.g., members of different phyla) might share a disproportionately large number of laterally transferred genes.

Several types of extremophilic organisms show considerable evidence for gene sharing within their preferred habitats. The phylogenetic position of hyperthermophilic bacterial lineages such as Aquificae and Thermotogae remains controversial (Cavalier-Smith, 2002; Gupta and Griffiths, 2002; Beiko et al., 2005; Boussau et al., 2008), but there is considerable evidence that these groups have shared genes extensively with other thermophilic lineages such as Pyrococcus (Noll et al., 2008). Similarly, halophiles such as *Salinibacter ruber* possess numerous genes that have apparently been derived from species within halophilic genera such as *Halobacterium*. Confounding the inference of such "environmental LGT highways," however, is the tendency for extremophilic organisms to display similar biases in protein composition, with halophiles having higher frequencies of acidic residues (Mongodin et al., 2006). Nonetheless, phylogenetic inferences of LGT in such bacteria are supported by distributional analyses of homologous and orthologous proteins, which have shown that distributions of particular protein sets are all similarly restricted to the same sets of cohabiting extremophilic taxa.

Among mesophiles, the genomes of many soil bacteria appear to have been shaped by extensive LGT. The Rhizobiales (a group of α-proteobacteria) and the β-proteobacterial genus *Ralstonia* appear to have exchanged a large number of genes (Kaneko et al., 2002; Kunin et al., 2005). Plant pathogens such as the gamma-proteobacterial Xanthomonadales also appear to have acquired many genes via LGT (van Sluys et al., 2002; Comas et al.,

2006), thus contributing to the unstable positioning of this lineage in genome phylogenies (Beiko et al., 2005). The γ-proteobacterium *Pseudomonas aeruginosa* is a remarkable generalist that can survive in many environments and infect many eukaryotic organisms: its diverse lifestyle appears to be supported by extensive within-genus gene sharing and acquisition of genes from other bacterial lineages (Shen et al., 2006).

The extents to which potential pathways of LGT are constrained by DNA acquisition mechanisms are not well understood. Prokaryotes that can develop competence can theoretically acquire any DNA from the environment, while mechanisms that require vectors such as transduction by phage or conjugation will only shuttle genes between their potential hosts (Thomas and Nielsen, 2005). Although the detection and characterization of LGT can identify potential gene sharing pathways, it cannot directly propose mechanisms for individual transfer events. Conversely, so-called surrogate (sensu Ragan, 2001) and genomic context analyses (Hsiao et al., 2005; Nakamura et al., 2004) can potentially suggest mechanisms of transfer but may not be able to precisely identify donor lineages.

Given the apparently nonrandom nature of LGT, it is difficult to meaningfully express a global rate of gene sharing, particularly if one is primarily interested only in those LGT events that are "successful," that is, those events that are not immediately purged by natural selection but instead are spread through a population either through neutral drift or due to their providing some selective advantage. In addition to biased patterns of sharing between organisms, there is very likely variation in LGT rates over time, as the physical properties and microbiota of habitats change. Possibly as a result of this, it is apparent that many LGT events are transient in that many laterally transferred genes are subsequently lost within a few generations of acquisition (Berg and Kurland, 2002; Hao and Golding, 2006).

4.7.2 What Is the Metabolic and Ecological Significance of Apparent Biases in the Types of Gene That Are Transferred?

Genes acquired via LGT can replace existing homologous sequences, typically through homologous recombination, or can confer a new function if no existing homologue was present in the recipient genome. The latter genes often cluster into genomic "islands" that are patchily distributed across closely related genomes (Hacker and Kaper, 2000). Given that the greatest bacterial sequencing effort has thus far been focused on the genomes of closely related pathogens, it is not surprising that many of the genomic islands examined thus far confer adaptations that relate directly to host interactions. While viral genes and transposable elements are frequent in genomic islands (suggesting mechanisms by which they arise; see, e.g., Zaneveld et al., 2008), also common are genes encoding antibiotic resistance, secretion systems, and toxin production (Dobrindt et al., 2004). Such genes can apparently be transferred between very distantly related taxa: for instance, the *ermB* gene encoding a methylase modification system appears to have originated in gram-positive bacteria such as *Streptococcus* or *Clostridium* and then has been transferred to members of the gram-negative genus *Bacteroides* (Shoemaker et al., 2001).

Environmental genomic islands are similar in nature to pathogenicity islands: indeed, the two are the same if one considers a host to be a highly specialized environment. While different strains of the marine picocyanobacterium *Prochlorococcus marinus* differ by >3% in the sequence of their 16S ribosomal DNA, they show considerably greater diversity in niche adaptation and genome content. A great deal of these differences are due to the presence of genomic islands enriched in nutrient uptake and stress response functions

(Coleman et al., 2006). Other common functions associated with such islands include xenobiotic degradation and the expression of toxins (Dobrindt et al., 2004).

Comprehensive genome analyses using phylogenetic or nonphylogenetic methods support an overrepresentation within genomic islands of genes encoding specific types of functions (e.g., Nakamura et al., 2004; Beiko et al., 2005; Hsiao et al., 2005). Phylogenetic analysis has revealed that novel metabolic pathways can be assembled from genes laterally transferred even between bacteria and archaea. For example, Fournier and Gogarten (2008) have shown that acetoclastic methanogenesis with the *Methanosarcina* was initiated by the ancestral acquisition of two clostridial genes, *ackA* and *pta*. The successful transfer of informational genes (sensu Rivera et al., 1998) is generally less common than that of other gene categories (Wellner et al., 2007), but evidence is still seen for transfer of even core informational genes including elongation factor 1α (Inagaki et al., 2006) and 16S rDNA (Yap et al., 1999).

Several studies have assessed the role played by transferred genes in metabolic networks. Pál et al. (2005) found that whereas LGT plays a much greater role than gene duplication in producing metabolic innovations in *E. coli*, acquired genes were much more likely to map to the periphery of metabolic networks. Lercher and Pál (2008) extended this observation by noting that transferred genes appear to remain at the fringes of networks and that they probably acquire only a few additional interaction partners over time. Such observations support the notion of a metabolic "core" that is highly connected by protein–protein interactions and whose predominant mode of inheritance is vertical rather than lateral. However, such core metabolic genes might still be susceptible to replacement by orthologous copies from other genomes. Transferred genes need to be expressed in the recipient cell to provide any sort of selective advantage: This requires the presence of *cis*-acting sequences such as promoters and operators as well as *trans*-acting transcription factors. Accordingly, the majority of genes acquired by *E. coli* appear to be peripheral in regulatory as well as in metabolic terms (Cosentino Lagomarsino et al., 2007), although Price et al. (2008) distinguished between "global" regulators in *E. coli* such as *crp*, which are inherited vertically, and "neighbor" *trans*-acting regulators that are adjacent to their regulatory targets and show evidence of acquisition via LGT.

Although these studies are compelling, it is worth highlighting the drawbacks of generalizing to a broad spectrum of prokaryotic lifestyles when the studies have dealt largely with *E. coli*. Kreimer et al. (2008) examined the metabolic network structure and modularity of over 300 different prokaryotic genomes and showed that the connectivity patterns as well as the proteins comprising these networks varied dramatically both between pathogenic and nonpathogenic taxa, and between different types of pathogen—it is very likely that these differences will strongly influence global patterns of LGT.

4.7.3 Does LGT Support, or Detract from, the Notion of Microbial Species?

Bacterial species have historically been defined in operational terms, that is, based on their phenotypic traits of greatest interest. DNA–DNA hybridization dynamics have been used since the 1970s as a quantitative method to define microbial species, with 70% hybridization as the de facto minimum standard for the assignment of two organisms to the same species. More recently, molecular techniques such as marker gene (e.g., 16S ribosomal DNA) analysis and multilocus sequence typing (MLST) have been used to quickly assign species memberships. These genetic differences do not, however, reflect an underlying philosophical species concept (Gevers et al., 2005) and furthermore can be undermined by LGT. Another challenge to microbial species concepts is the need to deal with temporal aspects of speciation: For instance, Retchless and Lawrence (2007) have proposed that different genes that are shared by *Escherichia* and *Salmonella* likely diverged at different times over a span of 70 million years. Consequently, rather than the "clean break" that might be expected during speciation (such as, e.g., that occurring with eukaryotic allopatry), *Escherichia* and *Salmonella* required tens of millions of years to speciate. What terminology would have been appropriate to describe the relationship between these two groups over this long time span?

A species concept can be applied to different prokaryotic lineages to varying degrees, depending on the particular evolutionary properties of each lineage. A critical question is the extent to which genomes within a lineage are homogeneous: Tettelin et al. (2005) proposed that both group A and group B *Streptococcus* have "open" pan-genomes, since every lineage sequenced contributed a substantial number of new genes (>25) to the pool for that genus. A similar analysis of 12 *P. marinus* genomes also identified that the members of this species possess open pan-genomes (Kettler et al., 2007). Conversely, four examined strains of *Bacillus anthracis* showed very similar gene content, suggesting a "closed" pan-genome. Furthermore, within lineages such as group A *Streptococcus*, genome composition does not correlate with MLST data (which characterize relationships among "core" housekeeping elements of the genome) or pathogenicity (Medini et al., 2005; McMillan et al., 2006). While a pan-genome can theoretically be defined for any set of genomes at any taxonomic level, there might exist "natural" groupings of organisms that share a common pan-genome, which is qualitatively different from the genomic composition of their next closest relatives. If such pan-genomes can be discovered, then a natural microbial species concept may exist that does not depend on the drawing of arbitrary lines in a series of gradations of evolutionary relatedness. However, if the capacity for gene sharing among distantly related taxa is sufficiently high, then it may be impossible to define species based on their propensity toward gene sharing (Gogarten et al., 2002).

REFERENCES

Aris-Brosou, S. (2005) Determinants of adaptive evolution at the molecular level: The extended complexity hypothesis. *Molecular Biology and Evolution* **22,** 200–209.

Arnold, M. L., Sapir, Y., and Martin, N. H. (2008) Genetic exchange and the origin of adaptations: Prokaryotes to primates. *Philosophical Transactions of the Royal Society of London. Series B, Biological Sciences* **363,** 2813–2820.

Awadalla, P. (2003) The evolutionary genomics of pathogen recombination. *Nature Reviews. Genetics* **4,** 50–60.

Baldauf, S. L., Roger, A. J., Wenk-Siefert, I., and Doolittle, W. F. (2000) A kingdom-level phylogeny of eukaryotes based on combined protein data. *Science* **290,** 972–977.

Bapteste, E., Susko, E., Leigh, J., MacLeod, D., Charlebois, R. L., and Doolittle, W. F. (2005) Do orthologous gene phylogenies really support tree-thinking? *BMC Evolutionary Biology* **5,** 33.

Beiko, R. G. and Hamilton, N. (2006) Phylogenetic identification of lateral genetic transfer events. *BMC Evolutionary Biology* **6,** 15.

Beiko, R. G., Harlow, T. J., and Ragan, M. A. (2005) Highways of gene sharing in prokaryotes. *Proceedings of the National Academy of Sciences of the United States of America* **102,** 14332–14337.

Beiko, R. G. and Ragan, M. A. (2008) Detecting lateral genetic transfer: A phylogenetic approach. *Methods in Molecular Biology* **452,** 457–469.

Berg, O. G. and Kurland, C. G. (2002) Evolution of microbial genomes: Sequence acquisition and loss. *Molecular Biology and Evolution* **19,** 2265–2276.

Betts, J. C., Dodson, P., Quan, S., Lewis, A. P., Thomas, P. J., Duncan, K., and McAdam, R. A. (2000) Comparison of the proteome of *Mycobacterium tuberculosis* strain H37Rv with clinical isolate CDC 1551. *Microbiology* **146,** 3205–3216.

Beumer, A. and Robinson, J. B. A. (2005) Broad-host-range, generalized transducing phage (SN-T) acquires 16S rRNA genes from different genera of bacteria. *Applied and Environmental Microbiology* **71,** 8301–8304.

BONI, M. F., POSADA, D., and FELDMAN, M. W. (2007) An exact nonparametric method for inferring mosaic structure in sequence triplets. *Genetics* **176,** 1035–1047.

BOUSSAU, B., GUÉGUEN, L., and GOUY, M. (2008) Accounting for horizontal gene transfers explains conflicting hypotheses regarding the position of aquificales in the phylogeny of bacteria. *BMC Evolutionary Biology* **8,** 272.

BROCHIER, C., FORTERRE, P., and GRIBALDO, S. (2004) Archaeal phylogeny based on proteins of the transcription and translation machineries: Tackling the *Methanopyrus kandleri* paradox. *Genome Biology* **5,** R17.

BRUEN, T. C., PHILIPPE, H., and BRYANT, D. (2006) A simple and robust statistical test for detecting the presence of recombination. *Genetics* **172,** 2665–2681.

CARVAJAL-RODRÍGUEZ, A., CRANDALL, K. A., and POSADA, D. (2006) Recombination estimation under complex evolutionary models with the coalescent composite-likelihood method. *Molecular Biology and Evolution* **23,** 817–827.

CASSENS, I., MARDULYN, P., and MILINKOVITCH, M. C. (2005) Evaluating intraspecific "network" construction methods using simulated sequence data: Do existing algorithms outperform the global maximum parsimony approach? *Systematic Biology* **54,** 363–372.

CAVALIER-SMITH, T. (2002) The neomuran origin of archaebacteria, the negibacterial root of the universal tree and bacterial megaclassification. *International Journal of Systematic and Evolutionary Microbiology* **52,** 7–76.

CHAN, C. X., DARLING, A. E., BEIKO, R. G., and RAGAN, M. A. (2009) Are protein domains modules of lateral genetic transfer? *PLoS ONE* **4,** e4524.

CHAN, C. X., BEIKO, R. G., and RAGAN, M. A. (2006) Detecting recombination in evolving nucleotide sequences. *BMC Bioinformatics* **7,** 412.

CHARLEBOIS, R. L., BEIKO, R. G., and RAGAN, M. A. (2004) Genome phylogenies. In *Organelles, Genomes and Eukaryote Phylogeny: An Evolutionary Synthesis in the Age of Genomics* (eds. R. P. Hirt and D. S. Horne), pp. 189–206. CRC Press, Boca Raton, FL.

CICCARELLI, F. D., DOERKS, T., VON MERING, C., CREEVEY, C. J., SNEL, B., and BORK, P. (2006) Toward automatic reconstruction of a highly resolved tree of life. *Science* **311,** 1283–1287.

COLEMAN, M. L., SULLIVAN, M. B., MARTINY, A. C., STEGLICH, C., BARRY, K., DELONG, E. F., and CHISHOLM, S. W. (2006) Genomic islands and the ecology and evolution of *Prochlorococcus*. *Science* **311,** 1768–1770.

COMAS, I., MOYA, A., AZAD, R. K., LAWRENCE, J. G., and GONZALEZ-CANDELAS, F. (2006) The evolutionary origin of Xanthomonadales genomes and the nature of the horizontal gene transfer process. *Molecular Biology and Evolution* **23,** 2049–2057.

COSENTINO LAGOMARSINO, M., JONA, P., BASSETTI, B., and ISAMBERT, H. (2007) Hierarchy and feedback in the evolution of the *Escherichia coli* transcription network. *Proceedings of the National Academy of Sciences of the United States of America* **104,** 5516–5520.

CREEVEY, C. J., FITZPATRICK, D. A., PHILIP, G. K., KINSELLA, R. J., O'CONNELL, M. J., PENTONY, M. M., TRAVERS, S. A., WILKINSON, M., and MCINERNEY, J. O. (2004) Does a tree-like phylogeny only exist at the tips in the prokaryotes? *Proceedings. Biological sciences/The Royal Society* **271,** 2551–2558.

DAGAN, T. and MARTIN, W. (2006) The tree of one percent. *Genome Biology* **7,** 118.

DAGAN, T. and MARTIN, W. (2007) Ancestral genome sizes specify the minimum rate of lateral gene transfer during prokaryote evolution. *Proceedings of the National Academy of Sciences of the United States of America* **104,** 870–875.

DE OLIVEIRA, T., DEFORCHE, K., CASSOL, S., SALMINEN, M., PARASKEVIS, D., SEEBREGTS, C., SNOECK, J., VAN RENSBURG, E. J., WENSING, A. M., VAN DE VIJVER, D. A., BOUCHER, C. A., CAMACHO, R., and VANDAMME, A. M. (2005) An automated genotyping system for analysis of HIV-1 and other microbial sequences. *Bioinformatics* **21,** 3797–3800.

DIDELOT, X. and FALUSH, D. (2007) Inference of bacterial microevolution using multilocus sequence data. *Genetics* **175,** 1251–1266.

DOBRINDT, U., HOCHHUT, B., HENTSCHEL, U., and HACKER, J. (2004) Genomic islands in pathogenic and environmental microorganisms. *Nature Reviews. Microbiology* **2,** 414–424.

DOOLITTLE, W. F. and BAPTESTE, E. (2007) Pattern pluralism and the tree of life hypothesis. *Proceedings of the National Academy of Sciences of the United States of America* **104,** 2043–2049.

DOOLITTLE, W. F., BOUCHER, Y., NESBØ, C. L., DOUADY, C. J., ANDERSSON, J. O., and ROGER, A. J. (2003) How big is the iceberg of which organellar genes in nuclear genomes are but the tip? *Philosophical Transactions of the Royal Society of London. Series B, Biological Sciences* **358,** 39–57.

DRUMMOND, A. J. and RAMBAUT, A. (2007) BEAST: Bayesian evolutionary analysis by sampling trees. *BMC Evolutionary Biology* **7,** 214.

ETHERINGTON, G. J., DICKS, J., and ROBERTS, I. N. (2005) Recombination Analysis Tool (RAT): A program for the high-throughput detection of recombination. *Bioinformatics* **21,** 278–281.

FALUSH, D., STEPHENS, M., and PRITCHARD, J. K. (2007) Inference of population structure using multilocus genotype data: Dominant markers and null alleles. *Molecular Ecology Notes* **7,** 574–578.

FITZMAURICE, W. P., BENJAMIN, R. C., HUANG P. C., and SCOCCA, J. J. (1984) Characterization of sites on DNA segments from bacteriophage HP1c1 which interact with specific DNA recognition system of transformable *Haemophilus influenzae* Rd. **Gene 31,** 187–196.

FORSLUND, K., HUSON, D. H., and MOULTON, V. (2004) VisRD-Visual recombination detection. *Bioinformatics* **20,** 3654–3655.

FORSTER, M., FORSTER, P., and WATSON, J. (2007). *Network Version 4.2.0.1: A Software for Population Genetics Data*

Analysis, 4.2.0.1 ed., pp. 1999–2007. Fluxus Technology Ltd.

FOURNIER, G. P. and GOGARTEN, J. P. (2008) Evolution of acetoclastic methanogenesis in *Methanosarcina* via horizontal gene transfer from cellulolytic Clostridia. *Journal of Bacteriology* **190,** 1124–1127.

FU, Y. X. and LI, W. H. (1993) Statistical tests of neutrality of mutations. *Genetics* **133,** 693–709.

GARCIA-VALLVÉ, S., SIMÓ, F. X., MONTERO, M. A., AROLA, L., and ROMEU, A. (2002) Simultaneous horizontal gene transfer of a gene coding for ribosomal protein l27 and operational genes in *Arthrobacter* sp. *Journal of Molecular Evolution* **55,** 632–637.

GEVERS, D., COHAN, F. M., LAWRENCE, J. G., SPRATT, B. G., COENYE, T., FEIL, E. J., STACKEBRANDT, E., VAN DE PEER, Y., VANDAMME, P., THOMPSON, F. L., and SWINGS, J. (2005). Re-evaluating prokaryotic species. *Nature Reviews. Microbiology* **3,** 733–739.

GIBBS, M. J., ARMSTRONG, J. S., and GIBBS, A. J. (2000) Sister-scanning: A Monte Carlo procedure for assessing signals in recombinant sequences. *Bioinformatics* **16,** 573–582.

GOGARTEN, J. P., DOOLITTLE, W. F., and LAWRENCE, J. G. (2002) Prokaryotic evolution in light of gene transfer. *Molecular Biology and Evolution* **19,** 2226–2238.

GOGARTEN, J. P., and TOWNSEND, J. P. (2005) Horizontal gene transfer, genome innovation and evolution. *Nature Reviews. Microbiology* **3,** 679–687.

GRASSLY, N. C. and HOLMES, E. C. (1997) A likelihood method for the detection of selection and recombination using nucleotide sequences. *Molecular Biology and Evolution* **14,** 239–247.

GUPTA, R. S. and GRIFFITHS, E. (2002) Critical issues in bacterial phylogeny. *Theoretical Population Biology* **61,** 423–434.

HACKER, J. and KAPER, J. B. (2000) Pathogenicity islands and the evolution of microbes. *Annual Review of Microbiology* **54,** 641–679.

HAO, W. and GOLDING, G. B. (2006) The fate of laterally transferred genes: Life in the fast lane to adaptation or death. *Genome Research* **16,** 636–643.

HEIN, J. (1990) Reconstructing evolution of sequences subject to recombination using parsimony. *Mathematics in the Biosciences* **98,** 185–200.

HEINEMANN, J. A. and SPRAGUE, G. F. Jr. (1989) Bacterial conjugative plasmids mobilize DNA transfer between bacteria and yeast. *Nature* **340,** 205–209.

HENDRICKSON, H. and LAWRENCE, J. G. (2006) Selection for chromosome architecture in bacteria. *Journal of Molecular Evolution* **62,** 615–629.

HEY, J. and WAKELEY, J. (1997) A coalescent estimator of the population recombination rate. *Genetics* **145,** 833–846.

HICKEY, G., DEHNE, F., RAU-CHAPLIN, A., and BLOUIN, C. (2008) SPR distance computation for unrooted trees. *Evolutionary Bioinformatics* **4,** 17–27.

HOLMES, E. C., WOROBEY, M., and RAMBAUT, A. (1999) Phylogenetic evidence for recombination in dengue virus. *Molecular Biology and Evolution* **16,** 405–409.

HOSSAIN, A., REISBIG, M. D., and HANSON, N. D. (2004) Plasmid-encoded functions compensate for the biological cost of AmpC overexpression in a clinical isolate of *Salmonella typhimurium*. *Journal of Antimicrobials and Chemotherapy* **53,** 964–970.

HSIAO, W. W., UNG, K., AESCHLIMAN, D., BRYAN, J., FINLAY, B. B., and BRINKMAN, F. S. (2005) Evidence of a large novel gene pool associated with prokaryotic genomic islands. *PLoS Genetics* **1,** e62.

HUSMEIER, D. and MCGUIRE, G. (2003) Detecting recombination in 4-taxa DNA sequence alignments with Bayesian hidden Markov models and Markov chain Monte Carlo. *Molecular Biology and Evolution* **20,** 315–337.

HUSON, D. H. (1998) SplitsTree: Analyzing and visualizing evolutionary data. *Bioinformatics* **14,** 68–73.

INAGAKI, Y., SUSKO, E., and ROGER, A. J. (2006) Recombination between elongation factor 1alpha genes from distantly related archaeal lineages. *Proceedings of the National Academy of Sciences of the United States of America* **103,** 4528–4533.

JAIN, R., RIVERA, M. C. and LAKE, J. A. (1999) Horizontal gene transfer among genomes: The complexity hypothesis. *Proceedings of the National Academy of Sciences of the United States of America* **96,** 3801–3806.

JELTSCH, A. and PINGOUD, A. (1996) Horizontal gene transfer contributes to the wide distribution and evolution of type II restriction-modification systems. *Journal of Molecular Evolution* **42,** 91–96.

JONES, D. and SNEATH, P. H. (1970) Genetic transfer and bacterial taxonomy. *Bacteriology Reviews* **34,** 40–81.

KANEKO, T., NAKAMURA, Y., SATO, S., MINAMISAWA, K., UCHIUMI, T., SASAMOTO, S., WATANABE, A., IDESAWA, K., IRIGUCHI, M., KAWASHIMA, K., KOHARA, M., MATSUMOTO, M., SHIMPO, S., TSURUOKA, H., WADA, T., YAMADA, M., and TABATA, S. (2002) Complete genomic sequence of nitrogen-fixing symbiotic bacterium *Bradyrhizobium japonicum* USDA110. *DNA Research* **9,** 189–197.

KARLIN, S., MRÁZEK, J., and CAMPBELL, A. M. (1997) Compositional biases of bacterial genomes and evolutionary implications. *Journal of Bacteriology* **179,** 3899–3913.

KEDZIERSKA, B., GLINKOWSKA, M., IWANICKI, A., OBUCHOWSKI, M., SOJKA, P., THOMAS, M. S., and WEGRZYN, G. (2003) Toxicity of the bacteriophage lambda cII gene product to *Escherichia coli* arises from inhibition of host cell DNA replication. *Virology* **313,** 622–628.

KETTLER, G. C., MARTINY, A. C., HUANG, K., ZUCKER, J., COLEMAN, M. L., RODRIGUE, S., CHEN, F., LAPIDUS, A., FERRIERA, S., JOHNSON, J., STEGLICH, C., CHURCH, G. M., RICHARDSON, P., and CHISHOLM, S. W. (2007) Patterns and implications of gene gain and loss in the evolution of *Prochlorococcus*. *PLoS Genetics* **3,** e231.

KOONIN, E. V., MUSHEGIAN, A. R., GALPERIN, M. Y., and WALKER, D. R. (1997) Comparison of archaeal and bacterial genomes: Computer analysis of protein sequences predicts novel functions and suggests a chimeric origin for the archaea. *Molecular Microbiology* **25,** 619–637.

KOSAKOVSKY POND, S. L., POSADA, D., GRAVENOR, M. B., WOELK, C. H., and FROST, S. D. (2006) Automated phylogenetic detection of recombination using a genetic algorithm. *Molecular Biology and Evolution* **23,** 1891–1901.

KREIMER, A., BORENSTEIN, E., GOPHNA, U., and RUPPIN, E. (2008) The evolution of modularity in bacterial metabolic networks. *Proceedings of the National Academy of Sciences of the United States of America* **105,** 6976–6981.

KUHNER, M. K. (2006) LAMARC 2.0: Maximum likelihood and Bayesian estimation of population parameters. *Bioinformatics* **22,** 768–770.

KUNIN, V., GOLDOVSKY, L., DARZENTAS, N., and OUZOUNIS, C. A. (2005) The net of life: Reconstructing the microbial phylogenetic network. *Genome Research* **15,** 954–959.

KURLAND, C. G., CANBACK, B., and BERG, O. G. (2003) Horizontal gene transfer: A critical view. *Proceedings of the National Academy of Sciences of the United States of America* **100,** 9658–9662.

LAWRENCE, J. G. (2002) Gene transfer in bacteria: Speciation without species? *Theoretical Population Biology* **61,** 449–460.

LEFEUVRE, P., LETT, J. M., REYNAUD, B., and MARTIN, D. P. (2007) Avoidance of protein fold disruption in natural virus recombinants. *PLoS Pathogens* **3,** e181.

LEFÉBURE, T. and STANHOPE, M. J. (2007) Evolution of the core and pan-genome of Streptococcus: positive selection, recombination, and genome composition. *Genome Biology* **8,** R71.

LERCHER, M. J. and PÁL, C. (2008) Integration of horizontally transferred genes into regulatory interaction networks takes many million years. *Molecular Biology and Evolution* **25,** 559–567.

LEVY-BOOTH, D. J., CAMPBELL, R. G., GULDEN, R. H., HART, M. M., POWELL, J. R., KLIRONOMOS, J. N., PAULS, K. P., SWANTON, C. J., TREVORS, J. T., and DUNFIELD, K. E. (2007) Cycling of extracellular DNA in the soil environment. *Soil Biology and Biochemistry* **39,** 2977–2991.

LOLE, K. S., BOLLINGER, R. C., PARANJAPE, R. S., GADKARI, D., KULKARNI, S. S., NOVAK, N. G., INGERSOLL, R., SHEPPARD, H. W., and RAY, S. C. (1999) Full-length human immunodeficiency virus type 1 genomes from subtype C-infected seroconverters in India, with evidence of intersubtype recombination. *Journal of Virology* **73,** 152–160.

MAJEWSKI, J. and COHAN, F. (1999) DNA sequence similarity requirements for interspecific recombination in *Bacillus*. *Genetics* **153,** 1525–1533.

MARTIN, D. P., POSADA, D., CRANDALL, K. A., and WILLIAMSON, C. (2005a) A modified bootscan algorithm for automated identification of recombinant sequences and recombination breakpoints. *AIDS Research and Human Retroviruses* **21,** 98–102.

MARTIN, D. P., WILLIAMSON, C., and POSADA, D. (2005b) RDP2: Recombination detection and analysis from sequence alignments. *Bioinformatics* **21,** 260–262.

MAYNARD SMITH, J. (1992) Analyzing the mosaic structure of genes. *Journal of Molecular Evolution* **34,** 126–129.

MAYNARD SMITH, J. and SMITH, N. H. (1998) Detecting recombination from gene trees. *Molecular Biology and Evolution* **15,** 590–599.

MAZEL, D. (2006) Integrons: Agents of bacterial evolution. *Nature Reviews. Microbiology* **4,** 608–620.

MCGUIRE, G. and WRIGHT, F. (2000) TOPAL 2.0: Improved detection of mosaic sequences within multiple alignments. *Bioinformatics* **16,** 30–134.

MCLEOD, D., CHARLEBOIS, R. L., DOOLITTLE, F., and BAPTESTE, E. (2005) Deduction of probable events of lateral gene transfer through comparison of phylogenetic trees by recursive consolidation and rearrangement. *BMC Evolutionary Biology* **5,** 27–37.

MCMILLAN, D. J., BEIKO, R. G., GEFFERS, R., BUER, J., SCHOULS, L. M., VLAMINCKX, B. J., WANNET, W. J., SRIPRAKASH, K. S., and CHHATWAL, G. S. (2006) Genes for the majority of group a streptococcal virulence factors and extracellular surface proteins do not confer an increased propensity to cause invasive disease. *Clinical and Infectious Disease* **43,** 884–891.

MCVEAN, G., AWADALLA, P., and FEARNHEAD, P. (2002) A coalescent-based method for detecting and estimating recombination from gene sequences. *Genetics* **160,** 1231–1241.

MEDINI, D., DONATI, C., TETTELIN, H., MASIGNANI, V., and RAPPUOLI, R. (2005) The microbial pan-genome. *Current Opinions in Genetics and Development* **15,** 589–594.

MEDRANO-SOTO, A., MORENO-HAGELSIEB, G., VINUESA, P., CHRISTEN, J. A., and COLLADO-VIDES, J. (2004) Successful lateral transfer requires codon usage compatibility between foreign genes and recipient genomes. *Molecular Biology and Evolution* **21,** 1884–1894.

MICHOD, R. E., BERNSTEIN, H., and NEDELCU, A. M. (2008) Adaptive value of sex in microbial pathogens. *Infection Genetics and Evolution* **8,** 267–285.

MICHOD, R. E., WOJCIECHOWSKI, M. F., and HOELZER, M. A. (1988) DNA repair and the evolution of transformation in the bacterium *Bacillus subtilis*. *Genetics* **118,** 31–39.

MILNE, I., WRIGHT, F., ROWE, G., MARSHALL, D. F., HUSMEIER, D., and MCGUIRE, G. (2004) TOPALi: Software for automatic identification of recombinant sequences within DNA multiple alignments. *Bioinformatics* **20,** 1806–1807.

MININ, V. N., DORMAN, K. S., FANG, F., and SUCHARD, M. A. (2005) Dual multiple change-point model leads to more accurate recombination detection. *Bioinformatics* **21,** 3034–3042.

MONGODIN, E. F., NELSON, K. E., DAUGHERTY, S., DEBOY, R. T., WISTER, J., KHOURI, H., WEIDMAN, J., WALSH, D. A., PAPKE, R. T., SANCHEZ PEREZ, G., SHARMA, A. K., NESBØ, C. L., MACLEOD, D., BAPTESTE, E., DOOLITTLE, W. F., CHARLEBOIS, R. L., LEGAULT, B., and RODRIGUEZ-VALERA, F. (2006) The genome of *Salinibacter ruber*: Convergence and gene exchange among hyperhalophilic bacteria and archaea. *Proceedings of the National Academy of Sciences of the United States of America* **102,** 18147–18152.

MOROSINI, M. I., AYALA, J. A., BAQUERO, F., MARTINEZ, J. L., and BLAZQUEZ, J. (2000) Biological cost of AmpC production for *Salmonella enterica* serotype Typhimurium. *Antimicrobial Agents and Chemotherapy* **44,** 3137–3143.

MORRIS, R. T. and DROUIN, G. (2007) Ectopic gene conversions in bacterial genomes. *Genome* **50,** 975–984.

NAKAMURA, Y., ITOH, T., MATSUDA, H., and GOJOBORI, T. (2004) Biased biological functions of horizontally transferred genes in prokaryotic genomes. *Nature Genetics* **36,** 760–766.

NELSON, K. E., CLAYTON, R. A., GILL, S. R., GWINN, M. L., DODSON, R. J., HAFT, D. H., HICKEY, E. K., PETERSON, J. D., NELSON, W. C., KETCHUM, K. A., MCDONALD, L., UTTERBACK, T. R., MALEK, J. A., LINHER, K. D., GARRETT, M. M., STEWART, A. M., COTTON, M. D., PRATT, M. S., PHILLIPS, C. A., RICHARDSON, D., HEIDELBERG, J., SUTTON, G. G., FLEISCHMANN, R. D., EISEN, J. A., WHITE, O., SALZBERG, S. L., SMITH, H. O., VENTER, J. C., and FRASER, C. M. (1999) Evidence for lateral gene transfer between archaea and bacteria from genome sequence of *Thermotoga maritima. Nature* **399,** 323–329.

NOLL, K. M., LAPIERRE, P., GOGARTEN, J. P. and NANAVATI, D. M. (2008) Evolution of mal ABC transporter operons in the Thermococcales and Thermotogales. *BMC Evolutionary Biology* **8,** 7.

NYSTEDT, B., FRANK, A. C., THOLLESSON, M., and ANDERSSON, S. G. (2008) Diversifying selection and concerted evolution of a type IV secretion system in *Bartonella. Molecular Biology and Evolution* **25,** 287–300.

OCHMAN, H., LAWRENCE, J. G., and GROISMAN, E. A. (2000) Lateral gene transfer and the nature of bacterial innovation. *Nature* **405,** 299–304.

PÁL, C., PAPP, B., and LERCHER, M. J. (2005) Adaptive evolution of bacterial metabolic networks by horizontal gene transfer. *Nature Genetics* **37,** 1372–1375.

PALMER, G. H. and BRAYTON, K. A. (2007) Gene conversion is a convergent strategy for pathogen antigenic variation. *Trends in Parasitology* **23,** 408–413.

POND, S. L., FROST, S. D., and MUSE, S. V. (2005) HyPhy: Hypothesis testing using phylogenies. *Bioinformatics* **21,** 676–679.

POSADA, D. (2002) Evaluation of methods for detecting recombination from DNA sequences: Empirical data. *Molecular Biology and Evolution* **19,** 708–717.

POSADA, D. and CRANDALL, K. A. (2001) Evaluation of methods for detecting recombination from DNA sequences: Computer simulations. *Proceedings of the National Academy of Sciences of the United States of America* **98,** 13757–13762.

POSADA, D. and CRANDALL, K. A. (2002) The effect of recombination on the accuracy of phylogeny estimation. *Journal of Molecular Evolution* **54,** 396–402.

PRICE, M. N., DEHAL, P. S., and ARKIN, A. P. (2008) Horizontal gene transfer and the evolution of transcriptional regulation in *Escherichia coli. Genome Biology* **9,** R4.

RAGAN, M. A. (2001) On surrogate methods for detecting lateral gene transfer. *FEMS Microbiology Letters* **201,** 187–191.

RAGAN, M. A. and BEIKO, R. G. (2009) Lateral genetic transfer: open issues. *Philosophical Transactions of the Royal Society of London. Series B. Biological Sciences* **364**, 2241–2251.

REDFIELD, R. J. (1993) Evolution of natural transformation: Testing the DNA repair hypothesis in *Bacillus subtilis* and *Haemophilus influenzae. Genetics* **133,** 755–761.

REDFIELD, R. J. (2001) Do bacteria have sex? *Nature Reviews. Genetics* **2,** 634–639.

RETCHLESS, A. C. and LAWRENCE J. G. (2007) Temporal fragmentation of speciation in bacteria. *Science* **317,** 1093–1096.

RILEY, M. (1993) Functions of the gene products of *Escherichia coli. Microbiology Reviews* **57,** 862–952.

RIVERA, M. C., JAIN, R., MOORE, J. E., and LAKE, J. A. (1998) Genomic evidence for two functionally distinct gene classes. *Proceedings of the National Academy of Sciences of the United States of America* **195,** 6239–6244.

ROCHA, E. P. (2008) The organization of the bacterial genome. *Annual Review of Genetics* **42,** 211–233.

RONQUIST, F. and HUELSENBECK, J. P. (2003) MrBayes 3: Bayesian phylogenetic inference under mixed models. *Bioinformatics* **19,** 1572–1574.

ROZAS, J., SÁNCHEZ-DELBARRIO, J. C., MESSEGUER, X., and ROZAS, R. (2003) DnaSP, DNA polymorphism analyses by the coalescent and other methods. *Bioinformatics* **19,** 2496–2497.

SALMINEN, M. O., CARR, J. K., BURKE, D. S., and MCCUTCHAN, F. E. (1995) Identification of breakpoints in intergenotypic recombinants of HIV type 1 by bootscanning. *AIDS Research and Human Retroviruses* **11,** 1423–1425.

SANTOYO, G. and ROMERO, D. (2005) Gene conversion and concerted evolution in bacterial genomes. *FEMS Microbiology Reviews* **29,** 169–183.

SAWYER, S. (1989) Statistical tests for detecting gene conversion. *Molecular Biology and Evolution* **6,** 526–538.

SCHEFFLER, K., MARTIN D. P., and SEOIGHE C. (2006) Robust inference of positive selection from recombining coding sequences. *Bioinformatics* **22,** 2493–2499.

SCHIERUP, M. H. and HEIN, J. (2000) Recombination and the molecular clock. *Molecular Biology and Evolution* **17,** 1578–1579.

SCHMIDT, K. D., TUMMLER, B., and ROMLING, U. (1996) Comparative genome mapping of *Pseudomonas aeruginosa* PAO with *P. aeruginosa* C, which belongs to a major clone in cystic fibrosis patients and aquatic habitats. *Journal of Bacteriology* **178,** 85–93.

SCHULTZ, A. K., ZHANG, M., LEITNER, T., KUIKEN, C., KORBER, B., MORGENSTERN, B., and STANKE, M. (2006) A jumping profile hidden Markov model and applications to recombination sites in HIV and HCV genomes. *BMC Bioinformatics* **7,** 265.

SHAPIRO, B., RAMBAUT, A., PYBUS, O. G. and HOLMES, E. C. (2006) A phylogenetic method for detecting positive epistasis in gene sequences and its application to RNA virus evolution. *Molecular Biology and Evolution* **23,** 1724–1730.

SHEN, K., SAYEED, S., ANTALIS, P., GLADITZ, J., AHMED, A., DICE, B., JANTO, B., DOPICO, R., KEEFE, R., HAYES, J., JOHNSON, S., YU, S., EHRLICH, N., JOCZ, J., KROPP, L., WONG, R., WADOWSKY, R. M., SLIFKIN, M., PRESTON, R. A., ERDOS, G., POST, J. C., EHRLICH, G. D., and HU, F. Z. (2006) Extensive genomic plasticity in *Pseudomonas aeruginosa* revealed by identification and distribution studies of novel genes among clinical isolates. *Infection and Immunity* **74,** 5272–5283.

Shoemaker, N. B., Vlamakis, H., Hayes, K., and Salyers, A. A. (2001) Evidence for extensive resistance gene transfer among *Bacteroides* spp. and among *Bacteroides* and other genera in the human colon. *Applied and Environmental Microbiology* **67,** 561–568.

Siepel, A. C., Halpern, A. L., Macken, C., and Korber, B. T. (1995) A computer program designed to screen rapidly for HIV type 1 intersubtype recombinant sequences. *AIDS Research and Human Retroviruses* **11,** 1413–1416.

Snyder, J. C, Wiedenheft, B., Lavin, M., Roberto, F. F., Spuhler, J., Ortmann, A. C., Douglas, T., and Young, M. (2007) Virus movement maintains local virus population diversity. *Proceedings of the National Academy of Sciences of the United States of America* **104,** 19102–19107.

Sorek, R., Zhu, Y., Creevey, C. J., Francino, M. P., Bork, P., and Rubin, E. M. (2007) Genome-wide experimental determination of barriers to horizontal gene transfer. *Science* **318,** 1449–1452.

Tajima, F. (1989) Statistical method for testing the neutral mutation hypothesis by DNA polymorphism. *Genetics* **123,** 585–595.

Tettelin, H., Masignani, V., Cieslewicz, M. J., Donati, C., Medini, D., Ward, N. L., Angiuoli, S. V., Crabtree, J., Jones, A. L., Durkin, A. S., Deboy, R. T., Davidsen, T. M., Mora, M., Scarselli, M., Margarit, Y. Ros, I., Peterson, J. D., Hauser, C. R., Sundaram, J. P., Nelson, W. C., Madupu, R., Brinkac, L. M., Dodson, R. J., Rosovitz, M. J., Sullivan, S. A., Daugherty, S. C., Haft, D. H., Selengut, J., Gwinn, M. L., Zhou, L., Zafar, N., Khouri, H., Radune, D., Dimitrov, G., Watkins, K., O'Connor, K. J., Smith, S., Utterback, T. R., White, O., Rubens, C. E., Grandi, G., Madoff, L. C., Kasper, D. L., Telford, J. L., Wessels, M. R., Rappuoli, R., and Fraser, C.M. (2005) Genome analysis of multiple pathogenic isolates of *Streptococcus agalactiae*: Implications for the microbial "pan-genome." *Proceedings of the National Academy of Sciences of the United States of America* **102,** 13950–13955.

Thomas, C. M. and Nielsen, K. M. (2005) Mechanisms of, and barriers to, horizontal gene transfer between bacteria. *Nature Reviews. Microbiology* **3,** 711–721.

van Sluys, M. A, Monteiro-Vitorello, C. B., Camargo, L. E. A., Menck, C. F. M., da Silva, A. C. R, Ferro, J. A., Oliveira, M. C., Setubal, J. C., Kitajima, J. P., and Simpson, A. J. (2002) Comparative genomic analysis of plant-associated bacteria. *Annual Review of Phytopathology* **40,** 169–189.

Vetsigian, K. and Goldenfeld, N. (2005) Global divergence of microbial genome sequences mediated by propagating fronts. *Proceedings of the National Academy of Sciences of the United States of America* **102,** 7332–7337.

Vlassov, V. V., Laktionov, P. P., and Rykova, E. Y. (2007) Extracellular nucleic acids. *BioEssays* **29,** 654–667.

Vulić, M., Dionisio, F., Taddei, F., and Radman, M. (1997) Molecular keys to speciation: DNA polymorphism and the control of genetic exchange in enterobacteria. *Proceedings of the National Academy of Sciences of the United States of America* **94,** 9763–9767.

Weiller, G. F. (1998) Phylogenetic profiles: A graphical method for detecting genetic recombinations in homologous sequences. *Molecular Biology and Evolution* **15,** 326–335.

Welch, R. A., Burland, V., Plunkett, G., Redford, P., Roesch, P., Rasko, D., Buckles, E. L., Liou S. R., Boutin, A., Hackett, J., Stroud, D., Mayhew, G. F., Rose, D. J., Zhou, S., Schwartz, D. C., Perna, N. T., Mobley, H. L., Donnenberg, M. S., and Blattner, F. R. (2002) Extensive mosaic structure revealed by the complete genome sequence of uropathogenic *Escherichia coli*. *Proceedings of the National Academy of Sciences of the United States of America* **99,** 17020–17024.

Wellner, A., Lurie, M. N., and Gophna, U. (2007) Complexity, connectivity, and duplicability as barriers to lateral gene transfer. *Genome Biology* **8,** R156.

Woese, C. R., Olsen, G. J., Ibba, M., and Söll, D. (2000) Aminoacyl-tRNA synthetases, the genetic code, and the evolutionary process. *Microbiology and Molecular Biology Reviews* **64,** 202–236.

Woolley, S. M., Posada, D., and Crandall, K. A. (2008) A comparison of phylogenetic network methods using computer simulation. *PloS One* **3,** e1913.

Worobey, M. (2001) A novel approach to detecting and measuring recombination: New insights into evolution in viruses, bacteria, and mitochondria. *Molecular Biology and Evolution* **18,** 1425–1434.

Wuchty, S., Oltvai, Z. N., and Barabási, A. L. (2003) Evolutionary conservation of motif constituents in the yeast protein interaction network. *Nature Genetics* **35,** 176–179.

Yap, W. H., Zhang, Z., and Wueng, Y. (1999) Distinct types of rRNA operons exist in the genome of the actinomycete *Thermomonospora chromogena* and evidence for horizontal transfer of an entire rRNA operon. *Journal of Bacteriology* **181,** 5201–5209.

Zaneveld, J. R., Nemergut, D. R., and Knight, R. (2008) Are all horizontal gene transfers created equal? Prospects for mechanism-based studies of HGT patterns. *Microbiology* **154,** 1–15.

Zangenberg, G., Huang, M. M., Arnheim, N., and Erlich, H. (1995) New HLA-DPB1 alleles generated by interallelic gene conversion detected by analysis of sperm. *Nature Genetics* **10,** 407–414.

Zhaxybayeva, O., Gogarten, J. P., Charlebois, R. L., Doolittle, W. F., and Papke, R. T. (2006) Phylogenetic analyses of cyanobacterial genomes: quantification of horizontal gene transfer events. *Genome Research* **16,** 1099–1108.

Chapter 5

Statistical Methods for Detecting the Presence of Natural Selection in Bacterial Populations

YUN-XIN FU AND XIAOMING LIU

5.1 INTRODUCTION

Natural selection is one of the most powerful mechanisms determining the fate of a population and thus, elucidating the impact of natural selection has always been an important aspect of population study. Natural selection in the evolution of a population often leaves traces at the molecular level. Therefore, samples from a population or from multiple related populations or species can be used to reveal the signature of natural selection. Many times, various sampling strategies are required to detect the presence of natural selection at different evolutionary time scales. Molecular sequences sampled from sufficiently divergent populations or species are needed for dissecting natural selection that persists for long periods of time. In such situations, many nucleotide sites have accumulated multiple mutations so that the relative tendency of nucleotide changes can be evaluated, which provide the basis for judging if natural selection is an important evolutionary force and if so, the type of natural selection. A large body of literature exists in this area (e.g., Nielsen and Yang, 1998; Suzuki and Gojobori, 1999; Yang and Nielsen, 2002).

When one is interested in the evolutionary forces that govern the recent significant events of a population, including natural selection that is operating on the extant population, samples from individuals within the sample populations or within closely related populations are necessary. One main characteristic of such samples is that most of the observed polymorphic sites have experienced few mutations or just one mutation. When studying a pathogen population, it is often necessary to take multiple samples from the evolving population over a period of time so that molecular changes can be tracked. To reveal the presence of natural selection from a sample within a population and from longitudinal samples, a different statistical approach than those used when studying long-term evolution is needed.

This chapter focuses on statistical methods for detecting the presence of recent natural selection that can be revealed from within population samples. The review is not meant to

Bacterial Population Genetics in Infectious Disease, Edited by D. Ashley Robinson, Daniel Falush, and Edward J. Feil

be comprehensive as we are more interested in statistical tests that are applicable to bacterial population studies. We will start with the general predictions of the outcome of natural selection, then describe a few widely used statistical methods and end with the discussion of some statistical approaches that are specific to the study of bacterial populations.

5.2 NATURAL SELECTION

Due to the haploidy of a bacterial genome, the type of natural selection in a bacterial population is limited and the outcome is relatively easy to predict: The better allele will ultimately win. A straightforward classical demonstration of the prediction is as follows.

Consider two alleles, A and a, at a locus of a bacterial population, with fitness W_A and W_a, respectively. Suppose the frequency of the two alleles are p_{t-1} and q_{t-1} at generation $t - 1$, then in generation t at reproduction time,

$$p_t = \frac{p_{t-1}W_A}{p_{t-1}W_A + q_{t-1}W_a}. \tag{5.1}$$

Therefore, the ratio of p_t and q_t is

$$\frac{p_{t-1}W_A}{q_{t-1}W_a} = \ldots = \left(\frac{p_0}{q_0}\right)\left(\frac{W_A}{W_a}\right)^t. \tag{5.2}$$

Suppose A is the fitter allele ($W_A > W_a$), then the ratio will approach infinity as t goes to infinity, which indicates that A will be fixed in the population eventually.

The above demonstration assumes that population size is sufficiently large so that the formula for p_t is accurate. Since typically bacterial population size is large (at least census size), it is traditionally thought that random drift is not a significant factor. This may be appropriate when dealing with a situation in which the new allele is sufficiently advantageous over the existing one; however, in reality, it is often difficult to identify an advantageous allele from a snapshot of the population in the form of a sample of DNA sequences, in which many polymorphic sites are present. Although collectively a bacterial species is usually large indeed, its population is often geographically structured, and selection may proceed differently in different local populations in which random genetic drift can become a significant factor. The observation of many mutations of various frequencies in a sample of DNA sequences of reasonable length from a bacterial population is a strong indication that random genetic drift is important in dealing with the recent evolution of bacteria.

When the locus under study is subject to purifying selection, that is, some mutations are deleterious, the prediction by Equation 5.2 is that their frequencies should decrease each generation when the population size is infinitely large. However, random genetic drift has played a significant role in keeping some deleterious mutations in the population longer, and the signature of such mutations can be found by the pattern of polymorphism in a sample.

5.3 STATISTICAL METHODS FOR DETECTING THE PRESENCE OF NATURAL SELECTION

5.3.1 Summary Statistics of Polymorphism

There has been a long history in both biological sciences as well as in the statistical field to gain insight into a scientific query through the use of quantities that summarize impor-

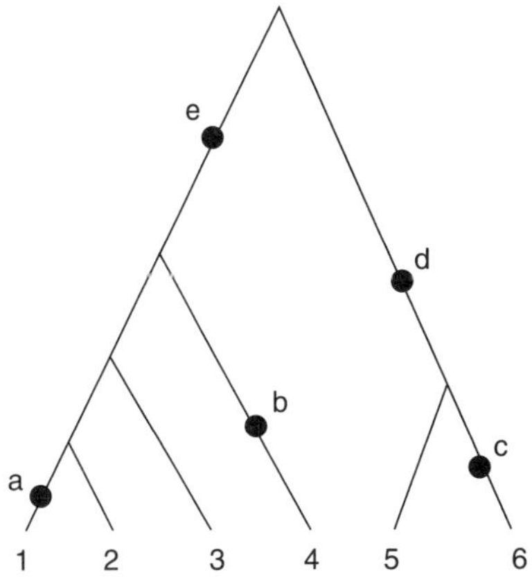

Figure 5.1 A genealogy of six sequences with five mutations (a–e) since the most recent common ancestor, three of which (a, b, and c) are of size 1, one of which (d) is of size 2, and one of which (e) is of size 4.

tant features of the data. These quantities are commonly known as summary statistics. The polymorphism in a sample of DNA sequences from a population can be summarized by a number of summary statistics from which a number of statistical tests have been developed. Three most widely known summary statistics are the number of distinct alleles (k), the number of segregating sites (K), and the mean number of nucleotide differences between two sequences in a sample (Π). A mutation observed in a sample must have occurred in the genealogy of the sample and can be further classified into size classes, which are the number of sequences that carry the mutant nucleotide. For a sample of n sequences, a mutation is thus of size from 1 to $n - 1$. Most summary statistics can be expressed as linear functions of the number of mutations of various classes (ξ_i, $i = 1, \dots,$ $n - 1$). To illustrate the definitions of various summary statistics, consider the following hypothetical sample of six sequences of 15 bps:

1: GAGGCTCTGATCCCA
2: AAGGCTCTGATCCCA
3: AAGGCTCTGATCCCA
4: AAGGTTCTGATCCCA
5: AAGGCTCTGATCTCG
6: AAGGCTCAGATCTCG

which resulted from the genealogy in Fig. 5.1. By direct counting, it is found that $k = 5$, and from the genealogy, it follows that $\xi_1 = 3$, $\xi_2 = 1$, and $\xi_4 = 1$, while $\xi_2 = \xi_2 = 0$. The number of segregating sites K is equal to five, which is also the number of mutations in the genealogy. Let d_{ij} represent the number of nucleotide differences between sequences i and j, then it is easy to see, for example, $d_{12} = d_{13} = 1$, $d_{14} = 2$, and the average of all the d_{ij} leads to $\Pi = 2.07$.

Under the infinite allele model, that is, every mutation in the population creates a new allele, the number of distinct alleles, k, in a sample of n sequences has the following distribution (Ewens, 1972; Karlin and McGregor, 1972):

$$Pr(k|\theta) = \frac{|S_k|\theta^k}{S_n(\theta)}, \tag{5.3}$$

where $S_n(\theta) = \theta(\theta - 1). \dots (\theta - n + 1)$ and S_k is the coefficient of θ^k when $S_n(\theta)$ is expanded to the polynomial of θ, also known as the Stirling number of the first kind (Abramowitz and Stegun, 1965). Furthermore, if n_i is the occurrence of allele type $i(i = 1, \ldots, k)$, we have

$$Pr(n_1, n_2, \ldots, n_k | k) = \frac{n!}{|S_n^k| k! n_1 n_2 \ldots n_k}. \tag{5.4}$$

Under the infinite site model (i.e., every mutation occurs in a new site),

$$K = \xi_1 + \ldots + \xi_{n-1}, \tag{5.5}$$

which in this example leads to $K = 3 + 1 + 0 + 1 + 0 = 5$. By definition, $\Pi = \frac{2}{n(n-1)} \sum_{i<j} d_{ij}$. When there is no recombination, Π can be expressed as

$$\Pi = \frac{2}{n(n-1)} \sum_{i=1}^{n-1} i(n-i)\xi_i. \tag{5.6}$$

In this example, we can compute Π from the above formula, which results in Π equal to 2.07. Many potentially useful linear functions of $\xi_i(i = 1, \ldots, n-1)$ can be defined, for example,

$$m = \frac{1}{n-1}(\xi_1 + 2\xi_2 + \cdots + (n-1)\xi_{n-1}), \tag{5.7}$$

which is the mean number (or more appropriately corrected) of mutations in a sequence since the most recent common ancestor (MRCA). Another interesting quantity is

$$h = \frac{2}{n(n-1)}\left(\xi_1 + 2^2\xi_2 + \ldots + (n-1)^2\xi_{n-1}\right), \tag{5.8}$$

which places a heavier weight on mutations of large sizes. For the example, $m = 1.8$ and $h = 1.53$.

When the sample is taken from a population evolving according to the Wright–Fisher model with constant population size, and all mutations are selectively neutral (the so-called neutral Wright–Fisher model), it follows (Fu, 1995) that $E(\xi_i) = \frac{\theta}{i}$, where $E(\xi_i)$ represents the mathematical expectation of ξ_i, from which it is easy to show that

$$E(K) = a_n\theta, \tag{5.9}$$

$$E(\Pi) = \theta, \tag{5.10}$$

$$E(m) = \theta, \text{ and} \tag{5.11}$$

$$E(h) = \theta, \tag{5.12}$$

where $a_n = 1 + \frac{1}{2} + \cdots + \frac{1}{n-1}$. Historically, $E(K)$ and $E(\Pi)$ (as well as their variances) were first derived without referring to ξ_i by Watterson (1975) and Tajima (1983), respectively. The variances of these summary statistics are as follows:

$$\mathrm{Var}(K) = a_n\theta + b_n\theta^2 \tag{5.13}$$

$$\mathrm{Var}(\Pi) = \frac{n+1}{3(n-1)}\theta + \frac{2(n^2+n+3)}{9(n(n-1))}\theta^2, \tag{5.14}$$

where $b_n = 1 + \frac{1}{2^2} + \ldots + \frac{1}{(n-1)^2}$ and Var stands for variance. The variances of all the summary statistics described above as well as their covariance can be derived from those of $\xi_i(i = 1, \ldots, n-1)$. The case of ξ_1 is of special interest since ξ_1 is a widely used quantity. Fu and Li (1993) shows that

$$\mathrm{Var}(\xi_1) = \theta + 2\left[na_n - \frac{2(n-1)}{(n-1)(n-2)}\right]\theta^2, \tag{5.15}$$

while from Fu (2009) (also see Zeng et al., 2006), we have

$$Var(m) = \frac{n}{2(n-1)}\theta + \left(2\left(\frac{n}{n-1}\right)^2(b_{n+1}-1)-1\right)\theta^2. \tag{5.16}$$

Although it is more powerful to use summary statistics that are linear functions of $\xi_i (i = 1, \ldots, n-1)$, sometimes the values of ξ_i may be difficult to identify. In such cases, it is preferable to use $\eta_i = \xi_i + \xi_{n-1}$ $(i = 1, \ldots, n/2)$ as the building blocks for summary statistics. Two such summary statistics are K and Π. The variance and covariance between each pair of η_i are given by Fu (1995). For bacterial and viral population studies, often longitudinal samples are taken. In addition to the summary statistic described above, there are some new informative quantities. One such quantity is the number of private (unique) mutations to a sample taken at a specific time.

5.3.2 Statistical Test of Neutrality Based on Summary Statistics

In the presence of natural selection, for example, some of the mutations in the sample are deleterious, or some are advantageous, or the sequences are from a locus that is linked to another one, which is the target of natural selection (genetic hitchhiking). The expectation is that almost all the summary statistics described in the previous section will be changed, but the extent of change varies from situation to situation, and more importantly from statistic to statistic.

When there are some deleterious mutations in the locus being sequenced, it is expected that most of these deleterious mutations are either quickly removed from the population or are kept in low frequencies. Therefore, summary statistics that are influenced strongly by low-frequency mutations are expected to be inflated in the presence of deleterious mutations. To be more specific, in the presence of many deleterious mutations, the number of singleton mutations (ξ_1) will be high. Therefore, summary statistics that weigh heavily on ξ_1, such as ξ_1, will be inflated severely. The number of segregating sites (K) is also expected to be inflated, even more pronounced when compared with Π, which gives much less weight on mutations of low frequencies.

A similar effect to most summary statistics will be observed if the sample is taken from a locus that has experienced (or tightly linked to one) a recent fixation of an advantageous allele. This is because when an advantageous allele reaches fixation from an initial low frequency, it mimics a population whose size is expanding relatively fast; since ξ_1 represents mutations that are on average young in age, a large population size corresponds to a large value for ξ_1. Similarly, K is expected to be inflated more severely than the value of Π. However, if the sample is taken at a time before fixation is completed, the sequences in the sample may fall into two types: one carries the advantageous alleles (or linked to), and another does not carry the advantageous allele (or linked to). If the advantageous allele is nearly fixed, the class that carries the advantageous allele will be in high frequency, and the mutations that separate the two classes of sequences will be in relatively high frequency as well. Summary statistics that weigh heavily on high-frequency mutant classes will be inflated; among the summary statistics described above, h, as defined by Equation 5.8, is one such statistic.

It should be pointed out that altered expectations of summary statistics are not always due to the presence of natural selection. As we have mentioned above, a rapidly growing population can lead to inflated numbers of low-frequency mutants. A structured population, on the other hand, will lead to the presence of an excess of mutations of intermediate frequencies. Therefore, summary statistics such as Π are likely inflated in the presence of population structure.

The responses of summary statistics to departure from neutrality (i.e., evolving according to the Wright–Fisher model with constant population size and all mutations are selectively neutral) lead to a class of statistical tests that is of the form

$$\frac{L_1 - L_2}{\sqrt{Var(L_1 - L_2)}} \tag{5.17}$$

where $E(L_1) = E(L_2) = \theta$ under neutrality, but are likely to be different in the presence of natural selection. Therefore, a significant departure from zero is taken as evidence against neutrality. The reason for the denominator is to standardize the test statistic so that it is not affected by or at least not sensitive to unknown values of θ. Note that

$$\text{Var}(L_1 - L_2) = \text{Var}(L_1) + \text{Var}(L_2) - 2\text{Cov}(L_1, L_2), \tag{5.18}$$

where $\text{Cov}(L_1, L_2)$ stands for the covariance between L_1 and L_2.

Therefore, as long as the variance of two summary statistics and their covariance are known, the variance of their difference can be computed. Even with the standardization, such a statistic does not usually follow some standard distribution. Therefore, its critical values are normally determined from simulated samples, which can be performed easily using efficient coalescent algorithms.

The first such statistical test was proposed by Tajima (1989) and was known as Tajima's D test, which has $L_1 = \Pi$ and $L_2 = \frac{K}{a_n}$. The covariance between L_1 and L_2 is found by Tajima (1989) as

$$\text{Cov}(L_1, L_2) = \frac{\theta}{a_n} + \frac{n+2}{2na_n}\theta^2. \tag{5.19}$$

To compute the value of the variance, an estimate of θ is required. In the case of Tajima's test, θ is estimated by $\frac{K}{a_n}$, which is known as Watterson's estimator (Watterson, 1975), and θ^2 is estimated by $\frac{K(K-1)}{a_n^2 + b_n}$.

Fu and Li (1993) proposed several tests utilizing rare mutants. Their D test correspondsto $L_1 = \frac{K}{a_n}$ and $L_2 = \xi_1$. For this test, the covariance between L_1 and L_2 is

$$\text{Cov}\left(\frac{K}{a_n}, \xi_1\right) = \theta + \frac{a_n}{n-1}\theta^2. \tag{5.20}$$

Again, to evaluate the variance, θ is estimated by Watterson's estimator. Fay and Wu (2000) proposed a test using $L_1 = \Pi$ and $L_2 = h$, but it turns out that this is equivalent to a test with $L_1 = \Pi$ and $L_2 = m$. The covariance between Π and m is given by Fu (2009) (also, see Zeng et al., 2006) as

$$\text{Cov}(\Pi, m) = \frac{n+1}{3(n-1)}\theta + \frac{7n^2 + 3n - 2 - 4n(n+1)b_{n+1}}{2(n-1)^2}\theta^2. \tag{5.21}$$

Another line of statistical tests is to utilize Ewens' sampling formula (Ewens, 1972; Karlin and McGregor, 1972). The first well-known test of this type is Watterson's (1978) homozygosity test, which was motivated by that conditional on the number of alleles in a sample, the frequencies of each allele are independent of population parameter θ. The test statistic is as follows:

$$H = \sum_i \left(\frac{n_i}{n}\right)^2. \tag{5.22}$$

Although Watterson's test is appropriate when detailed DNA sequence variation is not available, it is in general less powerful when compared with statistical tests that utilize such detailed patterns of DNA variation. One way to utilize Ewens' sampling formulas is to compare the number of distinct alleles in a sample with a predicted number under certain assumptions. Given the value of θ, too many k and too few k can be taken as evidence against neutrality. Fu (1997) proposed to substitute θ by Π, and this led to the following test:

$$F_S = \log\left(\frac{s}{1-s}\right), \tag{5.23}$$

where $s = \Sigma_{l=1}^{k} Pr(k|\theta = \Pi)$ and $Pr(k|\theta = \Pi)$ is computed by Equation 5.3. This test is found to be particularly powerful for detecting the access of rare alleles; it is also sensitive to the presence of recombination.

Most of the statistical tests described above as well as the test to be described in the next section can be found in a number of popular softwares used for analyzing population data. These include DnaSP (Librado and Rozas, 2009), Arlequin (Excoffier et al., 2005), and NeutralityTest (Li and Fu, 2009).

5.3.3 Statistical Test Utilizing Both within and between Population Variations

The statistical tests described in the previous section utilize only within-sample polymorphism; often, samples from multiple closely related species are available, and it is desirable to utilize interspecific variation as well. There are two well-known tests of the kind known as the MK test (McDonald and Kreitman, 1991) and the HKA test (Hudson et al., 1987). We shall describe the MK test in this section.

Consider two closely related populations from each of which a sample of DNA sequences of a protein-coding region is taken. In the total sample, a polymorphic site may be such that all the sequences in one sample possess one particular nucleotide, while all the sequences in the other sample possess another different nucleotide. This type of polymorphism is called between-sample variation; otherwise, it is called within-sample variation. Since the sequences are from a protein-coding region, each polymorphism will be either a synonymous change or a nonsynonymous change. The pattern of polymorphism in the total sample can thus be summarized in a 2 × 2 table:

	Within sample	Between sample
Synonymous	a	b
Nonsynonymous	c	d

where a, for example, is the number of polymorphic sites that are both within-sample variation and synonymous change. When mutations are selectively neutral, it is expected that the ratio of nonsynonymous and synonymous changes (dN/dS) remains constant over time. That is, under neutrality, $\frac{a}{c}=\frac{b}{d}$. This equality can be tested statistically by a chi-square test, which results in the following test statistic:

$$X^2=\frac{n(ad-bc)^2}{[(a+b)(a+c)(b+d)(c+d)]} \tag{5.24}$$

where $n = a + b + c + d$ is the total number of polymorphic sites. When n is sufficiently large, X^2 can be approximated by a X^2 variable with one degree of freedom (df). Significantly large values of X^2 are taken as evidence against neutrality. Alternatively, the G test or Fisher's exact test can be used when n is small. Note that such a test can easily be extended to more than two species.

Typically, a significant departure in the MK test is caused by an excess of nonsynonymous between-sample variation, which is taken as evidence of positive selection in favor of some amino acid changes. Since MK is a widely applicable and powerful test, its validity has been a subject of debate. Early debate partially stemmed from confusion of terminology. A discussion can be found in Fu (2000). Eyre-Walker (2002) found that existence of some deleterious mutations and increasing population sizes can lead to a significant MK test. Rocha et al. (2006) suggested that for closely related populations, d*N*/d*S* depends on their separation time and that a lag in the removal of slightly deleterious mutations may explain the change of d*N*/d*S* over time. Therefore, caution is also needed for inferences of selection based on the MK test.

5.4 STATISTICAL METHODS FOR BACTERIAL POPULATIONS

5.4.1 Longitudinal Samples

Longitudinal samples are samples taken at different time points from the same population (Fig. 5.2). For genetic studies of most organisms with relatively low mutation rates, longitudinal samples can be pooled together as a single sample taken at the same time, which simplifies the analysis. The justification of such convention is that the sampling interval is so small that the possible mutations accumulated on the sequences studied within the sampling intervals are negligible. However, for fast-evolving organisms, including some bacteria, some sampling intervals (in years) may be sufficiently long to allow for the observation of significant genetic change within samples. Although new statistical methods have been developed for analyzing longitudinal DNA samples (see review by Drummond et al., 2003), few methods are available for detecting the presence of natural selection. On the other hand, if longitudinal samples can be safely pooled as a single sample, more methods for selection detection can be applied (see Sections 5.3.2 and 5.3.3). For longitudinal samples taken from fast-evolving bacterial populations, it is suggested to test whether there are significant genetic changes between longitudinal samples as the first step. If not, the samples can be pooled as a single sample and methods for selection detection for single samples can be used. Otherwise, the methods designed for longitudinal samples should be applied.

Liu and Fu (2007) proposed several methods for testing genetical isochronism or for detecting significant genetical heterochronism in longitudinal samples. Here we introduce a test based on the number of private mutations within samples. Suppose there are two samples taken from an evolving haploid population at time t_0 and $t_0 + t$, respectively, where t is the sampling interval in generations. Let n_1 and n_2 be the sizes of samples taken at t_0

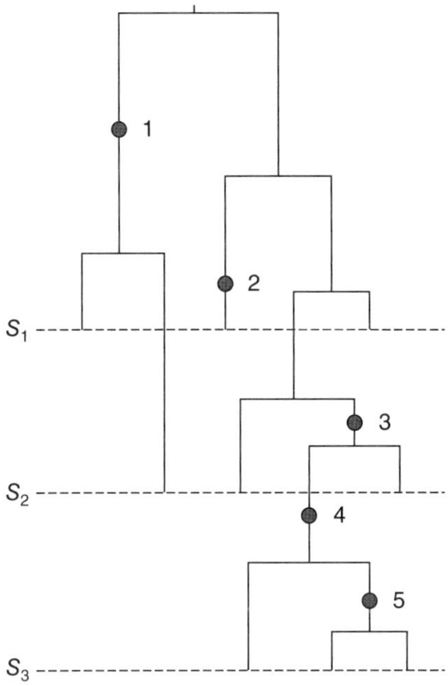

Figure 5.2 A genealogy of three longitudinal samples, s_1, s_2, and s_3, sampled at three different time points. Each sample has three sequences.

and $t_0 + t$, respectively. The number of private mutations within a sample is the number of sites that are not only polymorphic in that sample but are monomorphic in the other samples. Let $K_p(i)(i = 1, 2)$ be the number or private mutations of sample i. Then the test statistic is

$$T_c = \frac{c(K_p(1) - E(K_p(1))) + (1-c)(K_p(2) - E(K_p(2)))}{\sqrt{\mathrm{Var}(cK_p(1) + (1-c)K_p(2))}}. \tag{5.25}$$

The detailed computation of mean and variance can be found in Liu and Fu (2007), in which several values of c were compared using simulation. It was found that the test statistic with $c_2 = \frac{n_2}{n_1 + n_2}$ or $c_3 = \frac{n_2^2}{n_1^2 + n_2^2}$ has the highest power. The significance level of the test can be determined by either permutation or coalescent simulation.

If significant genetical heterochronism between longitudinal samples is suggested by the test, then the samples should not be pooled and analyzed as a single sample. Unfortunately, there are only a very few studies on detecting potential selection with longitudinal samples.

Goode et al. (2008) extended Nielsen and Yang's (1998) codon model for protein-coding sequences to apply to longitudinal samples. Using their model, nucleic sites can be assigned to different selection categories (negatively selected, positively selected, and neutral). However, their method is based on an inferred phylogenetic tree and does not take into account the uncertainty of phylogeny reconstruction.

Edwards et al. (2006) and later Drummond and Suchard (2008) tried to overcome this shortcoming and to take into account the uncertainty of the gene genealogy and the parameters of the mutation model and the demographic model at the same time. To do so, their methods sample gene genealogies of the sequences, along with model parameters using a Markov chain Monte Carlo (MCMC) framework (Drummond et al., 2002). More specifically, the genealogy and parameters are sampled according to the posterior probability distribution $Pr(G, \Omega, \theta|Y)$, where G is the gene genealogy, θ is the mutation parameters, and Ω is the parameter for an exponential growth model. Given each sampled genealogy, six different statistics are calculated. They include one classic neutrality test statistic (Fu and Li's [1993] D-statistic), two measures of branch length distribution (age of the MRCA and total tree length), and three measures of tree imbalance (Kirkpatrick and Slatkin's [1993] B_1, McKenzie and Steel's [2000] C_n, and Colless's [1982] I_c). With a large number of genealogies sampled, empirical null distributions of these statistics can be obtained. If the same statistics observed in the original sample are significantly unlikely to be observed from the null distributions, the null hypothesis of neutrality is rejected. This method is theoretically promising, although several caveats need to be considered. First, it assumes no recombination in the sequences and so far, it is unknown how robust the method is when recombination cannot be ignored. Second, a large parameter space needs to be explored, so the power and reliability of the conclusion may be a concern if the sample size is not large. Third, there are always some technical issues for the application of MCMC, such as prior choices and convergence detection.

5.4.2 Selection Based on DNA Fingerprints

In the area of studying bacterial populations, DNA fingerprints are often used to determine the polymorphic level or diversity of the population. Typically, bacteria (or clones, or strains) were sampled from a population. Then, a specified locus was amplified using polymerase chain reaction (PCR) for each bacterium. Finally, certain DNA fingerprinting methods (such as PCR-SSCP and PCR-DGGE) were used to identify the alleles of the locus (Nocker et al., 2007). If each bacterium can be independently genotyped using DNA fingerprinting, the frequencies of different alleles $n_i(i = 1, \ldots, k)$ can be directly counted. Then, Watterson's homozygosity test (Equation 5.22) can be applied.

Sometimes, the smallest sampling unit may still consist of multiple bacteria. For example, a host was infected by multiple strains of pathogens, and a sample taken from that host may contain more than one strain. Targeting this problem, Rannala et al. (2000) used a Poisson distribution to model the number of strains in each sample and a multinomial distribution to model the number of allele copies carried by these strains. Based on this model, they proposed a maximum likelihood estimator of the allele frequency given the observed presence/absence frequency of each allele. Anderson and Scheet (2001) derived another estimator of allele frequency from the same model, and their estimator is supposed to be less biased when compared to Rannala et al.'s (2000) original estimator. After the allele frequency is estimated, a similar test based on Ewens' sampling formulas can be conducted. However, it is not clear to what extent uncertainty in the allelic frequencies will affect the neutrality test.

5.4.3 Selection Based on the Presence/Absence of Certain Genomic Islands (GIs)

One important mechanism of bacterial genome evolution is horizontal gene transfer. Many of the accessory genes transferred by this mechanism form a distinct DNA segment called GI (Juhas et al., 2009). Studies have shown that GIs may be associated with many important adaptive functions, such as pathogenicity, symbiosis, sucrose, aromatic compound metabolism, mercury resistance, and siderophore synthesis (Juhas et al., 2009). However, since GIs typically carry various novel genes (no detectable homologues in other species)

(Hsiao et al., 2005), a statistical test of association between a GI and a phenotype is needed to detect adaptive GIs. Instead of testing the correlation between the concentration of the presence/absence of one binary character with the presence/absence of another binary character, as Maddison (1990)'s concentrated changes test, here we propose a test of nonrandom copresence/coabsence of two binary characters (one GI, one phenotype, or two GIs) on branches of a given phylogeny. It is possible to further extend the method by taking gene genealogy uncertainty into account, as Edwards et al. (2006) and Drummond and Suchard (2008) did. We assume that in addition to the presence/absence data of GIs of interest, each strain in the sample is also assayed by other markers, such as multilocus sequence typing (MLST).

First, a phylogenetic tree is reconstructed using MLST. Then, given the presence/absence states of the GIs on the external nodes of the tree and the phylogeny, the presence/absence states of the internal nodes of the tree (ancestors of the sample) can be inferred using available phylogeny reconstruction programs, such as PAUP* (Swofford, 2003; http://paup.csit.fsu.edu/), PHYLIP (Felsenstein, 1989; http://evolution.genetics.washington.edu/phylip/), or PAML (Yang, 2007; http://abacus.gene.ucl.ac.uk/software/paml.html). Similarly, the presence/absence of certain adaptive phenotypes of the internal nodes can also be inferred. After the states of the GI or phenotype on internal nodes are inferred, the number of state change (i.e., presence to absence or absence to presence) events on the branches of the phylogeny can be counted.

Then some statistical measures of the correlation between the (inferred) phenotype and the (inferred) GI states are calculated, such as Gini impurity (Breiman et al., 1984). For example, if we use 0 and 1 to represent the absence or the presence of a particular trait (GI or phenotype), then n_{00}, n_{01}, n_{10}, and n_{11} are the counts of nodes that have both traits absent, the first absent and the second present, the first present and the second absent, and both present, respectively. Let $n_0 = n_{00} + n_{01}$, $n_1 = n_{10} + n_{11}$, and $n = n_0 + n_1$. Then, the Gini impurity is calculated as

$$G = \left[1 - \left(\frac{n_{01}+n_{11}}{n}\right)^2 - \left(\frac{n_{00}+n_{10}}{n}\right)^2\right] - \frac{n_1}{n}\left[1 - \left(\frac{n_{11}}{n_1}\right)^2 - \left(\frac{n_{10}}{n_1}\right)^2\right] - \frac{n_0}{n}\left[1 - \left(\frac{n_{01}}{n_0}\right)^2 - \left(\frac{n_{00}}{n_{10}}\right)^2\right]. \tag{5.26}$$

A larger Gini impurity measure means a better correlation. The Gini impurity measure can be calculated for internal nodes only, external nodes only, and all nodes combined on the phylogenetic tree.

To test the significance of correlation between a GI and a phenotype, a Monte Carlo simulation can be used by superimposing the state change events onto the phylogeny while fixing their number according to their inferred counts using original data (see details below). For each replication, Gini impurity measures are compared to those calculated with original data, which are designated as G_0. After a large number of replications, the percent of the replications with a larger Gini impurity measure than G_0 was counted. This is the empirical p value for the significance of correlation between the GI and the phenotype, with the null hypothesis that the presence/absence events of each trait independently occur on the phylogeny.

To reasonably simulate the horizontal transfer of the GI, the superimposing process used in the simulation needs to be carefully designed. The process begins with all nodes having the same states (0/1) as the root. If the root state is "1," an absence (deletion) event is randomly superimposed onto a branch with a probability that equals to its branch length divided by the total length of the branches with state "1." If the root state is "0," then a presence (insertion) event is superimposed onto the phylogeny. After an event is superimposed onto a branch, all the descendant nodes of the branch change their states accordingly. After an event is superimposed, the probability of whether an insertion or deletion is the next event to superimpose is determined by the relative ratio of the total length of the remaining branches with state "0" and state "1." There are two restrictions of the above process. One is that the number of insertions/deletions to be superimposed needs to be fixed to the number of events as inferred using the original data. If an internal node has undetermined states (due to their equality according to the criteria used in the phylogenetic algorithm, such as maximum parsimony) within each replication, its state is randomly assigned while fixing the total number of event changes. The other restriction is that to superimpose an event, there must be some eligible branches to be superimposed. If we encounter a situation, such as an insertion event that needs to be superimposed because the deletion quotas have already dried up, but there are no eligible remaining branches with state "0," then we have to stop the process and restart from the beginning. So this superimposing process is a trial and error simulation. Another limitation of the process is that it does not allow two events to be superimposed onto the same branch. Further developments addressing these problems are needed.

5.5 AN EXAMPLE

A recent example of applying various statistical tests described in this chapter is given by Zhao and Qin (2007). The data set was originally from a study of the phycoerythrin (*ppe*) gene in two ecotypes of *Prochlorococcus*, which are specifically adapted to high light (HL) or low light (LL) conditions (Steglich et al., 2003). In its original analysis, the authors only did phylogenetic analysis and found the monophyletic origin of the HL and LL sequences. They concluded *ppe* is suitable as a sensitive molecular marker to study *Prochlorococcus* populations. Zhao and Qin (2007) reanalyzed the data and applied multiple methods for selection detection.

Zhao and Qin (2007) applied different intraspecific neutrality tests, including Tajima's (1989) D-statistic and Fu and Li's (1993) D^*- and F^*-statistics on the HL- and LL-*ppeB* locus. They found significant negative values for the tests on HL-*ppeB* ($D = -1.9542$, $p < 0.01$, $D^* = -4.2708$, $p < 0.01$, $F^* = -4.1726$, $p < 0.01$), which suggests an excess of rare variants probably due to directional selection or population bottleneck. As to the LL sequences, Tajima's D showed a marginally significant positive value ($D = 3.4205$, $p < 0.05$), suggesting an excess of intermediate variants possibly due to balancing selection or population subdivision. However, Fu and Li's D^*- and F^*-statistics did not show significant departure from the expectation under neutrality, although their values are also positive. Considering the possibility of mutation rate heterogeneity along sites, they also applied Misawa and Tajima's D^+ test (Misawa and Tajima, 1997) on the same data, which is a modified version of Tajima's D under the finite site model. D^+ also showed significant negative values in the HL-*ppeB* sequences but no significant departure from neutrality in the LL-*ppeB* sequences. A likelihood ratio test for neutrality based on phylogeny (Yang and Nielsen, 2002) was then conducted. The result confirmed the hypothesis of positive selection on the HL-*ppeB* sequences. Besides intraspecific tests, Zhao and Qin (2007) also conducted interspecific neutrality tests, including the MK test (McDonald and Kreitman, 1991) and the likelihood ratio test mentioned above. The MK test showed an excess of nonsynonymous fixed substitutions in the *ppeB* and *ppeA* loci (Fisher's exact test,

$p < 0.001$), which suggests a positive selection on those loci since the divergence of *Prochlorococcus* and *Synechococcus*. This hypothesis was confirmed using the likelihood ratio test. By inferring the selection pressures acting on the *ppeB* loci along with the functional structural information, the authors conclude that HL- and LL-*ppeB* should be under different selective pressures, and positive selection may drive HL-*ppeB* to obtain a new function.

5.6 DISCUSSION AND PERSPECTIVE

The theory and statistical methods for detecting the presence of natural selection using samples from within a population or within closely related populations are reviewed in this chapter, and several new statistical approaches specifically designed for bacterial populations, such as for fast-evolving bacterial pathogens and for GIs, are also presented. While many statistical approaches developed earlier can be applied to bacterial populations, there is also the need for methods that are more specific to microorganisms, including bacterial populations. Longitudinal samples from pathogen populations present some challenges that are not found in the traditional one-sample analysis. New summary statistics as well as new statistical methods for detecting natural selection for longitudinal samples likely will be developed in the future. We note that even for a single sample analysis, there is still considerable room for developing new and useful summary statistics, some perhaps in the form described in Fu (2009).

Bacterial genomes often evolve by acquiring novel and foreign genomic elements or sometimes by eliminating some existing segments; such a mechanism many times creates a pattern of presence/absence of a certain element (GI). How to evaluate the importance of the presence or absence event is not trivial. Although we have presented a method to do so, further analysis of this method as well as developing more powerful methods is desirable.

As far as a statistical approach is concerned, most of the methods described are based on comparisons between two summary statistics. One useful extension is to consider such multiple tests simultaneously (e.g., Innan, 2006; Zeng et al., 2006, 2007). Also note that all such tests of natural selection are based on the comparison of data to the prediction of the null model, which usually assumes a variation of neutrality. Such an approach in many ways is desirable since a neutral model is well accepted as the starting point of the data analysis and can be clearly defined. Because of the nature of such analysis, a significant departure from the null model should be interpreted by noting that natural selection is only, albeit important, one of the possible causes. Other causes include population structure, population growth or shrinkage, and even sampling bias. Biased sampling may be even more pronounced in studying bacterial pathogens since samples are often based on opportunity rather than on design. An alternative statistical approach, such as the maximum likelihood approach or even the Bayesian test (e.g., Drummond and Suchard, 2008), may be desirable when the situation warrants. This is typically true when a particular alternative model of evolution can be identified and justified. Statistical tests that take the specific alternative model into consideration may be more powerful, but one must also be cautious in the interpretation because a number of different evolutionary models may all fit the data adequately.

As far as studying infectious diseases is concerned, it is in an exciting stage since new sequencing technologies (such as pyrosequencing) are capable of generating large amounts of data. However, analyzing such data also presents a considerable challenge (e.g., Eriksson et al., 2008; Rodrigo et al., 2008), which is not unique to the study of bacterial populations. This is due to the large intrinsic error as well as other uncertainties in the sequencing. Developing statistical tests based on such data will be desirable as part of the effort to meet the challenge.

REFERENCES

ABRAMOWITZ, M. and STEGUN, I. (1965) *Handbook of Mathematical Functions*. New York, Dover Publications, Inc.

ANDERSON, E. C. and SCHEET, P. A. (2001) Improving the estimation of bacterial allele frequencies. *Genetics* **158(3),** 1383–1386.

BREIMAN, L., FRIEDMAN, J. H., OLSHEN, R. A., and STONE, C. J. (1984) *Classification and Regression Trees*. Kluwer Academic Publishers, Dordrecht.

COLLESS, D. H. (1982) Review of "Phylogenetics: The theory and practice of phylogenetic systematics." *Systematic Zoology* **31(1),** 100–104.

DRUMMOND, A. J., NICHOLLS, G. K., RODRIGO, A. G., and SOLOMON, W. (2002) Estimating mutation parameters, population history and genealogy simultaneously from temporally spaced sequence data. *Genetics* **161(3),** 1307–1320.

DRUMMOND, A. J., PYBUS, O. G., RAMBAUT, A. et al. (2003) Measurably evolving populations. *Trends in Ecology and Evolution* **18,** 481–488.

DRUMMOND, A. and SUCHARD, M. A. (2008) Fully Bayesian tests of neutrality using genealogical summary statistics. *BMC Genetics* **9(1),** 68.

EDWARDS, C. T. T., HOLMES, E. C., PYBUS, O. G. et al. (2006) Evolution of the human immunodeficiency virus envelope gene is dominated by purifying selection. *Genetics* **174(3),** 1441–1453.

ERIKSSON, N., PACHTER, L., MITSUYA, Y. et al. (2008) Viral population estimation using pyrosequencing. *PLoS Computational Biology* **4(5),** e1000074.

EWENS, W. J. (1972) The sampling theory of selectively neutral alleles. *Theoretical Population Biology* **3(1),** 87–112.

EXCOFFIER, L., LAVAL, G., and SCHNEIDER S. (2005) Arlequin (version 3.0): An integrated software package for population genetics data analysis. *Evolutionary Bioinformatics Online* **1,** 47–50.

EYRE-WALKER, A. (2002) Changing effective population size and the McDonald-Kreitman test. *Genetics* **162(4),** 2017–2024.

FAY, J. C. and WU, C. I. (2000) Hitchhiking under positive Darwinian selection. *Genetics* **155,** 1405–1413.

FELSENSTEIN, J. (1989) PHYLIP—Phylogeny inference package (version 3.2). *Cladistics* **5,** 164–166.

FU, Y. X. (1995) Statistical properties of segregating sites. *Theoretical Population Biology* **48,** 172–197.

FU, Y. X. (1997) Statistical tests of neutrality of mutations against population growth, hitchhiking and background selection. *Genetics* **147**, 915–925.

FU, Y. X. (2000) Neutrality and selection in molecular evolution: Statistical tests. In *Encyclopedia of Life Sciences* http://www.els.net/ (accessed January 6, 2010)

FU, Y. X. (2009) Variances and covariances of linear summary statistics of segregating sites (manuscript in preparation).

FU, Y. X. and LI, W. H. (1993) Statistical tests of neutrality of mutations. *Genetics* **133(3),** 693–709.

GOODE, M., GUINDON, S., and RODRIGO, A. (2008) Modelling the evolution of protein coding sequences sampled from measurably evolving populations. *Genome Informatics* **21,** 150–164.

HSIAO, W. W. L., UNG, K., AESCHLIMAN, D. et al. (2005) Evidence of a large novel gene pool associated with prokaryotic genomic islands. *PLoS Genetics* **1(5),** e62.

HUDSON, R. R., KREITMAN, M., and AGUADE, M. (1987) A test of neutral molecular evolution based on nucleotide data. *Genetics* **116,** 153–159.

INNAN, H. (2006) Modified Hudson-Kreitman-Aguade test and two-dimensional evaluation of neutrality tests. *Genetics* **173,** 1725–1733.

JUHAS, M., VAN DER MEER, J. R., GAILLARD, M. et al. (2009) Genomic islands: Tools of bacterial horizontal gene transfer and evolution. *FEMS Microbiology Reviews* **33(2),** 376–393.

KARLIN, S. and MCGREGOR, J. L. (1972) Addendum to a paper of W. Ewens. *Theoretical Population Biology* **5,** 95–105.

KIRKPATRICK, M. and SLATKIN, M. (1993) Searching for evolutionary patterns in the shape of a phylogenetic tree. *Evolution* **47(4),** 1171–1181.

LI, H. and FU, Y. X. (2009) NeutralityTest: A Novel Software for Testing Neutrality. http://xfiles.uth.tmc.edu/xythoswfs/webview/_xy-1789858_1 (manuscript in preparation).

LIBRADO, P. and ROZAS, J. (2009) DnaSP v5: A software for comprehensive analysis of DNA polymorphism data. *Bioinformatics*, 2009 Apr 3. [Epub ahead of print] doi:10.1093/bioinformatics/btp187

LIU, X. and FU, Y. X. (2007) Test of genetical isochronism for longitudinal samples of DNA sequences. *Genetics* **176,** 327–342.

MADDISON, W. P. (1990) A method for testing the correlated evolution of two binary characters: Are gains or losses concentrated on certain branches of a phylogenetic tree? *Evolution* **44(3),** 539–557.

MCDONALD, J. H. and KREITMAN, M. (1991) Adaptive protein evolution at the Adh locus in *Drosophila*. *Nature* **351,** 652–654.

McKenzie, A. and Steel, M. (2000) Distributions of cherries for two models of trees. *Mathematical Biosciences* **164(1),** 81–92.

Misawa, K. and Tajima, F. (1997) Estimation of the amount of DNA polymorphism when the neutral mutation rate varies among sites. *Genetics* **147,** 1959–1964.

Nielsen, R. and Yang, Z. (1998) Likelihood models for detecting positively selected amino acid sites and applications to the HIV-1 envelope gene. *Genetics* **148(3),** 929–936.

Nocker, A., Burr, M., and Camper, A. K. (2007) Genotypic microbial community profiling: A critical technical review. *Microbial Ecology* **54,** 276–289.

Rannala, B., Qiu, W. G., and Dykhuizen, D. E. (2000) Methods for estimating gene frequencies and detecting selection in bacterial populations. *Genetics* **155(2),** 499–508.

Rocha, E. P. C., Maynard Smith, J., Hurst, L. D. et al. (2006) Comparisons of dN/dS are time dependent for closely related bacterial genomes. *Journal of Theoretical Biology* **239,** 226–235.

Rodrigo, A., Bertels, F., Heled, J. et al. (2008) The perils of plenty: What are we going to do with all these genes? *Philosophical Transactions of the Royal Society of London. Series B, Biological Sciences* **363**(15**12),** 3893–3902.

Steglich, C., Post, A. F., and Hess, W. R. (2003) Analysis of natural populations of *Prochlorococcus* spp. in the northern Red Sea using phycoerythrin gene sequences. *Environmental Microbiology* **5,** 681–690.

Suzuki, Y. and Gojobori T. (1999) A method for detecting positive selection at single amino acid sites. *Molecular Biology and Evolution* **16,** 1315–1328.

Swofford, D. L. (2003) PAUP*. Phylogenetic Analysis Using Parsimony (*and Other Methods) Version 4. http://paup.csit.fsu.edu/ (accessed January 5, 2010).

Tajima, F. (1983) Evolutionary relationship of DNA sequences in finite populations. *Genetics* **105,** 437–460.

Tajima, F. (1989) Statistical method for testing the neutral mutation hypothesis by DNA polymorphism. *Genetics* **123,** 585–595.

Watterson, G. A. (1975) On the number of segregating sites. *Theoretical Population Biology* **7,** 256–276.

Watterson, G. A. (1978) The homozygosity test of neutrality. *Genetics* **88(2),** 405–417.

Yang, Z. (2007) PAML 4: Phylogenetic analysis by maximum likelihood. *Molecular Biology and Evolution* **24(8),** 1586–1591

Yang, Z. and Nielsen, R. (2002) Codon-substitution models for detecting molecular adaptation at individual sites along specific lineages. *Molecular Biology and Evolution* **19,** 908–917.

Zeng, K., Fu, Y. X., Shi, S., and Wu, C-I. (2006) Statistical tests for detecting positive selection by utilizing high frequency variants. *Genetics* **174,** 1431–1439.

Zeng, K., Shi, S., and Wu, C-I. (2007) Compound tests for the detection of hitchhiking under positive selection. *Molecular Biology and Evolution* **24(8),** 1898–1908.

Zhao, F. and Qin, W. (2007) Comparative molecular population genetics of phycoerythrin locus in *Prochlorococcus*. *Genetica* **129,** 291–299.

Chapter 6

Demographic Influences on Bacterial Population Structure

Francois Balloux

6.1 BACTERIAL POPULATION SIZE

The population structures of pathogenic bacteria are extraordinarily diverse. This variation is due to the wide range in life histories and niches exploited by different species, but is also due to the extensive variation in the dynamics of transfer of genes from one generation to the next. In particular, bacteria occupy the full continuum between strict parthenogenesis and essentially free genetic recombination, a factor crucial in conditioning the distribution of genes in time and space. As a consequence, there are few commonalities between populations of different bacteria. Possibly, the only overarching characteristic shared between essentially all bacterial populations is their large census sizes. Population size is a key factor in gene dynamics and population structure as it conditions genetic diversity at neutral markers, the efficacy and pace of natural selection, as well as the evolvability of populations.

Bacterial population sizes are indeed mind-boggling. The number of bacteria on Earth has been estimated around 5×10^{33} (Whitman et al., 1998), over a trillion (10^{12}) times the number of stars in the universe or 10 trillion times the number of grains of sand on Earth. Equally extraordinary is the figure of 10^{14} bacteria carried by each healthy human mainly in the gut, a figure outnumbering human body cells by ten to one (Berg, 1996). Such population sizes remain gigantic even after accounting for the fact that these bacteria belong to multiple species. Several hundred species have been found on the skin (Gao et al., 2007), and there have been estimates of up to 1000 species in the human gut (Hooper and Gordon, 2001), even if 30–40 dominant species represent 99% of the intestinal flora (Savage, 1977). Most of the species routinely found in the human body represent commensals or even symbionts. Opportunistic pathogens are also expected to form immense population sizes at least in species where an important fraction of strains are potentially pathogenic. This seems to be the case in *Staphylococcus aureus*, *Neisseria meningitides*, and *Streptococcus mutans*, which are all present in a large proportion of the human population and where many strains seem to be capable to evolve toward pathogenicity merely accidentally (Herczegh et al., 2008; van Belkum et al., 2009).

Bacterial Population Genetics in Infectious Disease, Edited by D. Ashley Robinson, Daniel Falush, and Edward J. Feil

In some facultative pathogens (e.g., *Escherichia coli*), only a small proportion of strains are harmful (Wirth et al., 2006). Thus, the size of the infective population is expected to be much reduced compared to the global bacterial population. The proportion of strains capable of pathogenicity—together with their capacity to maintain a sustained epidemic in humans—is also expected to be a major determinant of the population size in animal pathogens or free-living bacteria for which humans represent only a secondary niche. For example, *Vibrio cholera* (the agent of cholera) occurs naturally in large numbers in the plankton of fresh, brackish, and salt water, and the global population has been estimated around 10^{20} cells (Thompson et al., 2004; Fraser et al., 2009a); however, only a small subset of strains is pathogenic (Waldor and Mekalanos, 1996; Karaolis et al., 1998). Finally, the population sizes of obligate pathogens are directly constrained by the number of infected carriers and are expected to be smaller except for the most widespread diseases. But even such populations can be extraordinarily large. For example, there were an estimated 14.4 million active cases of tuberculosis in 2006 (WHO, 2008), with each carrier harboring millions, if not billions, of bacteria.

6.2 MEASURES OF GENETIC DIVERSITY

A central tenet of population genetics is that larger populations are expected to maintain higher neutral genetic diversity (e.g., synonymous polymorphisms). Genetic diversity is generally measured as the variation in the sequence polymorphism of genes shared between all individuals in the population under study. The simplest measure of genetic diversity is that two genes drawn at random from a population are of different allelic types. It is generally referred to as *gene diversity* or *expected heterozygosity*, even if the concept of heterozygosity is obviously not biologically meaningful in haploid organisms:

$$H_E = \frac{n}{n-1}\left(1-\sum_{i=1}^{k} p_i^2\right). \tag{6.1}$$

The expression simply reads as one minus the sum of the squared frequencies p_i for the k alleles. The sample size is denoted as n, and the $(n/n - 1)$ term is a correction for small sample size accounting for sampling without replacement. This measure of genetic diversity can be applied to single polymorphisms (e.g., point mutations or indels) or haplotypes (stretches of DNA sequences including several polymorphic sites). When analyzing multiple markers, gene diversity is simply averaged over loci. Several extensions of the formula above have been devised specifically for the estimation of genetic diversity from sequence data. One can measure the average number of differences between pairs of DNA sequences as

$$\pi = \frac{n}{n-1}\sum_{i=1}^{k}\sum_{j\neq i}^{k} p_i p_j \pi_{ij}, \tag{6.2}$$

where p_i and p_j are the frequencies of sequences i and j, and π_{ij} is the number of nucleotide differences between two sequences. A similar expression can be written for the proportion of shared nucleotides ν_{ij}:

$$\nu = \frac{n}{n-1}\sum_{i=1}^{k}\sum_{j\neq i}^{k} p_i p_j \nu_{ij}. \tag{6.3}$$

In practice, the average genetic distance over all pairs of individuals within a population is generally used, as this approach allows specifying an underlying model of molecular evolution. Variable probabilities can be assigned to the different transitions between nucleotidic states, and variation in mutation rate among sites can be accounted for (e.g., Tamura and Nei, 1993).

A complication arises in bacteria because of the highly dynamic nature of bacterial genomes. For instance, only a small fraction of core genes are shared between different strains of *E. coli* (Welch et al., 2002). The recent annotation of 20 *E. coli* genomes showed that out of ~18,000 genes observed, only ~2000 were common to all strains (Touchon et al., 2009). This separation into an essential core and variable "pan-genome" challenges traditional measures of genetic diversity. While such variation can be modeled in the context of genome evolution (Didelot et al., 2009), population structuring analyses have so far been restricted to the variation of genes in the core genome. Population genetics data will probably be available for pan-genomic variation for some species in the near future. At that stage, it would be relatively straightforward to consider variation in the gene complement for instance by measuring the proportion of genes shared between pairs of individual strains. While this may provide interesting insight into the ecology of various strains, it will not replace the study of orthologous genes, which allows making inference on the past genealogy of a population.

6.2.1 Expected and Observed Genetic Diversity

The basic expression of genetic diversity given in Equation 6.1 allows making some rough quantitative predictions. Indeed, the expected genetic diversity H_E at equilibrium of a neutral genetic marker with mutation rate per generation μ in an idealized random mating population of size N reads

$$H_E = \frac{2N_\mu}{1+2N_\mu}. \tag{6.4}$$

Using this relation, we can make some predictions on the expected genetic diversity of neutral mutations for hypothetical bacterial populations of different sizes at equilibrium. We can, for instance, work out the expected genetic diversity of a stretch of 1000 base pairs (bp) of coding DNA assuming that synonymous mutations are neutral and that all nonsynonymous mutations are deleterious and will be lost immediately. Estimates for the mutation rate per nucleotide per division for bacteria lie around 10^{-10} (Drake et al., 1998; Ochman et al., 1999; Tago et al., 2005). The probability of a random mutation leading to a nonsynonymous change has been estimated at 0.761 using the complete genome of the K12 strain of *E. coli* (Zhang, 2005). Assuming all nonsynonymous mutations to be deleterious and the synonymous ones to be neutral, this would lead to a realized mutation rate around 2.5×10^{-8} per nucleotide per division for a coding sequence of 1000 bp. We can conservatively assume that the dominant species have average population sizes of at least 10^{12} per human host. Considering this within-host population in isolation already leads to an estimate of $H_E \sim 1$, implying that no pairs of strains are expected to share exactly the same 1000 bp DNA sequence.

The amount of genetic diversity is highly variable between different bacterial species. Among the well-studied species, the most genetically diverse is the gut bacterium *Helicobacter pylori*. Out of 3850 nucleotides located in housekeeping genes, 1418 turned out to be polymorphic (Falush et al., 2003b), and essentially any isolate genotyped from unrelated hosts turned out to be unique (Schwarz et al., 2008). This diversity is clearly exceptional and most pathogenic bacteria have moderate genetic diversities. For 34 species, MultiLocus Sequence Typing (MLST) schemes based on the sequencing of seven housekeeping genes are sufficient to identify well-defined sets of strains represented by multiple isolates (Maiden, 2006). Then there are species characterized by very low genetic diversity. Such cases include several human pathogens and have been termed "genetically monomorphic pathogens" by Achtman (2008), as they display very little or essentially no genetic diversity. This heterogeneous group includes several important pathogens such as *Escherichia coli* O157:H7 (Zhang et al., 2006), the agent of plague *Yersinia pestis* (Achtman et al., 1999), tuberculosis (the *Mycobacterium tuberculosis* complex) (Sreevatsan et al., 1997), the agent of leprosy (*Mycobacterium leprae*) (Monot et al., 2005), *Salmonella enterica* serovar Typhi (Kidgell et al., 2002), and anthrax (*Bacillus anthracis*).

6.3 THE CONCEPT OF EFFECTIVE POPULATION SIZE

Populations with large census sizes but with low genetic diversity are not unique to bacteria. It has been recognized a long time ago that essentially any deviation from an ideal population model leads to a decrease in genetic diversity. To account for this, Wright introduced the concept of effective population size nearly 80 years ago (Wright, 1931). It is defined as the parameter summarizing the amount of genetic drift to which a population is subjected and quantified as the number of idealized randomly mating individuals, which experience the same amount of random fluctuations at neutral loci as the population under scrutiny. The dynamics of idealized randomly mating individuals is described by the Wright–Fisher model, whose well-studied properties lead to different definitions of the effective population size depending on whether the quantities of interest are the variance of change in allelic frequencies or genetic diversity (Whitlock and Barton, 1997).

An alternative way to understand effective population size is to consider the mean coalescence time instead (the average time taken for a random pair of alleles to coalesce in a common ancestor), as this quantity is intimately linked to the effective population size (Slatkin, 1991). In an ideal population size of N individuals, the mean coalescence time will be N generations back in time (Fig. 6.1). Deviations from the random mating pattern (e.g., higher variance in reproductive success), population size fluctuations, or population structuring will have a parallel effect on effective size and mean coalescence times. Thus, the fluctuations in mean coalescence times along a phylogeny can be used as a proxy for past demographic fluctuations. Moreover, the concept of mean coalescence

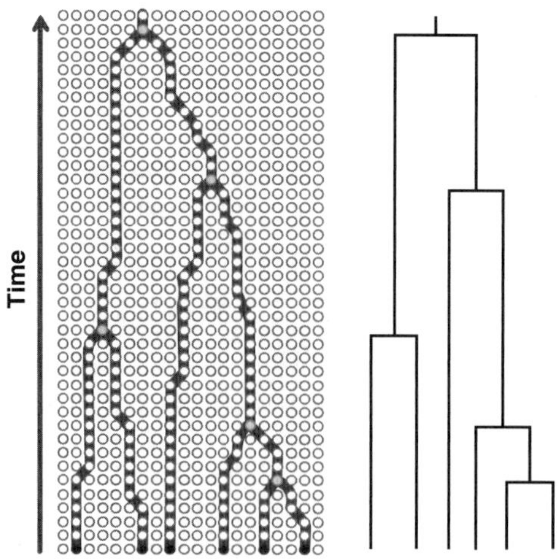

Figure 6.1 Coalescence dynamic in a hypothetical population of constant size (left) and the resulting phylogenetic tree (right).

time is probably more intuitive than effective population size and leads to interesting cross simplification in analytical work (Balloux et al., 2003).

The effective size of a population is generally much smaller than its census size. For most plant and animal populations, census size is expected to be one or two orders of magnitude larger than effective population size (Frankham, 1996). For pathogens such as viruses, bacteria, or protozoa, this ratio is expected to be much larger. Several of the features of pathogen populations are expected to reduce effective population drastically. These include a recent origin due to host shift, variation in population size over time due to epidemic bursts, the bottlenecks inherent to host-to-host transmission, and the strong selective pressures induced by host immunity in particular when combined with limited genomic recombination. In the following, I will explore the main factors that can reduce effective population size.

6.3.1 Recent Origin

A recent origin would be the most straightforward explanation for reduced genetic diversity in pathogens (Achtman, 2008). Following a host shift, the founding population adapting to the new host is expected to undergo a bottleneck or founder event that could involve only a few or even possibly a single bacterium, thus drastically reducing the genetic diversity of the population. The population will eventually go back to the equilibrium genetic diversity corresponding to its new effective size. However, as this buildup of genetic diversity relies on the accumulation of new mutations, this is a very slow process. Thus, if a founding event happened fairly recently, there would not have been enough time to replenish the genetic diversity of the population.

A recent origin could have played a role in all the aforementioned examples of bacteria with reduced genetic diversity. There is evidence for a possible founder effect accompanying a change in ecological niche associated with the acquisition of two plasmids by the progenitor of *Y. pestis*, the agent of plague (Achtman et al., 1999). Another possible example of recent ancestry is *E. coli* O157:H7, which is represented by closely related serotypes and is believed to have arisen as a result of horizontal gene transfer of virulence factors, some 40,000 years ago (Zhang et al., 2006). Other pathogens with low diversity are believed to have arisen relatively recently with an estimated age of ~17,000 years (Van Ert et al., 2007) for anthrax, 10,000–71,000 years (Roumagnac et al., 2006) for Typhi, and with the origin of the *M. tuberculosis* complex (*M. tuberculosis*, *Mycobacterium bovis*, *Mycobacterium africanum*, and *Mycobacterium microti*) dated at 15,000–20,000 years (Kapur et al., 1994; Sreevatsan et al., 1996, 1997). While all these dates are indicative of a relatively recent origin, these figures have to be taken with some caution as there is considerable uncertainty around the mutation rates used behind these calculations (Achtman, 2008). Moreover, while the relatively recent ancestry of these pathogens is likely to have played a role in their reduced genetic diversity, a recent origin does not provide a sufficient explanation for their low genetic diversity. For example, *H. pylori*, possibly the most genetically diverse bacterium, has been estimated to share a similar age (54,000–62,000 years; Linz et al., 2007).

6.3.2 Variable Population Size

Few species maintain stable population sizes over extended time periods; this is likely to be particularly true for pathogens. Population size fluctuations will be affected by changes in the host population as well as by environmental conditions. Changes in population sizes are expected to be particularly dramatic in pathogens characterized by epidemic dynamics. The effective population (and hence the mean coalescence time) is very sensitive to the smallest census sizes of a population over time and can be approximated by the harmonic mean of the census population sizes over generations, which reads

$$N_S \cong \frac{t}{\sum_{i=0}^{t-1} \frac{1}{N_i}}, \tag{6.5}$$

where t stands for the time in number of generations and N_i is an index for the population sizes from the first generation considered ($N_{i=0}$) to the last parental generation(N_{t-1}). The formula given above is an approximation, and it additionally makes the assumption that population sizes are uncorrelated over time. However, it helps visualizing the dramatic effect of small census sizes a lineage may have experienced in the past irrespective of its average (arithmetic mean) population size. For instance, if we assumed that a population had the following census size over successive generations: 10,000, 10,000, 10,000, 10, 20, and 10,000, the effective population (harmonic mean) would be just below 40, despite the average census size lying around 6672 individuals.

The disproportionate effect of small population sizes on the observed genetic diversity can be understood more easily when thinking in terms of mean coalescence times. The probability that two lineages coalesce in the immediately preceding generation is the probability that they share the same parent. In a haploid population, there are N_{t-1} "potential parents" in the previous generation, so the probability that two alleles sampled at times t share a parent is $1/(2N_{t-1})$.

Variation in population size will also affect the shape of phylogenetic trees. The shape of a phylogeny is defined by the topology (the branching order of the taxa) and the length of the branches, which will be conditioned by the distribution of mutational events along the tree. As such, the shape of a tree will be affected by past demography. For instance, a population that underwent a recent demographic expansion will be characterized by a phylogeny with short branches (Fig. 6.2) and numerous mutations at low frequency on the tips of the tree. Conversely, populations that have been stable over long evolutionary time scales will lead to phylogenies with longer branches and mutations that will be shared by a larger number of strains.

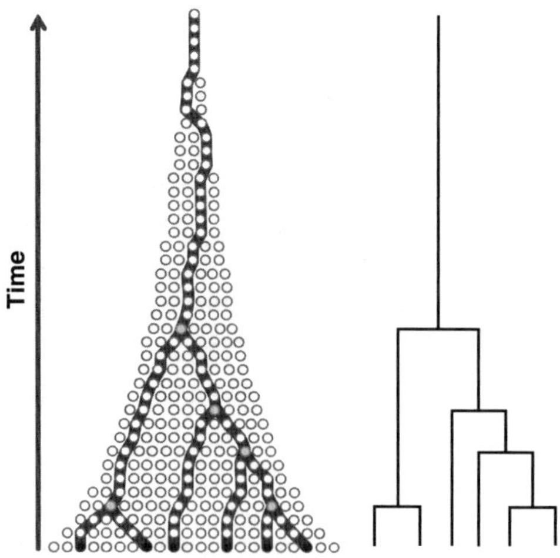

Figure 6.2 Coalescence dynamic in a hypothetical population increasing in size (left) and the resulting phylogenetic tree (right).

6.3.3 Outbreaks and Selective Sweeps

The simplest scenario for a rapid population expansion in pathogens is generally referred to as an outbreak. However, there are also more complex situations where, despite the rapid expansion of a strain, the population size as a whole remains essentially unchanged. In such a case, it is the relative frequency of strains that is changing, with a more successful strain replacing others. These could be considered as a form of cryptic outbreak. Such events of periodic selection, in which a strain recurrently displaces the resident bacterial population, have been well documented in the lab (Atwood et al., 1951; Notley-McRobb and Ferenci, 2000). The same phenomenon is believed to be happening frequently in natural populations and has been invoked to explain the low genetic diversity of bacteria (Cohan, 2005). Note that similar arguments have been put forward to explain the lower than expected genetic diversities in organisms with extensive recombination (Gillespie, 1987, 2001). An important feature is that these selective sweeps well known to microbiologists concern entire bacteria rather than genes (i.e., total replacement by one chromosomal linkage group due to complete lack of recombination).

In contrast, a selective sweep has a different meaning for eukaryote geneticists, where the concept captures the increase in frequency of a specific allele. In the presence of sufficient recombination, the loss of genetic diversity will be limited to the locus under positive selection and its vicinity. Under extensive recombination, a further signature will be a progressive loss of genetic diversity with increasing physical distance along the chromosome from the locus under selection. This pattern arises because the nucleotides physically closest to the site under selection had the highest probability to sweep through the population together with the positively selected allele, a phenomenon generally referred to as genetic hitchhiking (Maynard Smith and Haig, 1974). These patterns represent the signal used by many statistical tests aiming at detecting natural selection in organisms with some level of genetic recombination (Sabeti et al., 2002; Voight et al., 2006). While this decay of genetic diversity from the locus under selection can extend over hundreds of kilobases in eukaryotes, similar patterns will be far more localized under homologous recombination. Interestingly, selective sweeps specific to particular regions of the genome, similar to the ones described in eukaryotes, also exist in bacteria. The emergence of BRO-1 and BRO-2 beta-lactamases in *Moraxella catarrhalis* is believed to have followed an import by horizontal gene transfer and to have swept through a large part of the species within a couple of decades due to antibiotic selection (Bootsma et al., 2000). There are also several documented examples of antibiotic-driven locus-specific selective sweeps in *E. coli* (Milkman et al., 2003; Lescat et al., 2009).

6.3.4 Genetic Recombination and Selection

This brief overview of the concept of outbreaks and selective sweeps exemplifies the complexity of bacterial population genetics and the crucial role of recombination in the dynamics of genes. Recombination is generally assumed to be rare in most bacterial species (e.g., Cohan, 2005). This is likely to be generally true, but the situation is actually more complex. Given the immense population sizes of bacteria, even minute rates of recombination can have a dramatic impact on gene dynamics. It is also important to separate the potential for genetic recombination with its actual effect on the dynamics of genes; an event of genetic recombination does only leave a genetic trace when it happened between individuals that are genetically sufficiently differentiated. As an instructive parallel, we can consider the situation in human mitochondrial DNA. Human mitochondria have all the required molecular machinery to recombine (Kraytsberg et al., 2004), and they probably do so at high rates. However, there is only one case of actual recombination having been detected (Schwartz and Vissing, 2002), and mitochondrial phylogenies do not bear the classical hallmarks of recombining DNA sequences (frequent homoplasies and increasing linkage disequilibrium with physical distance along DNA sequences). The reason is that mitochondria are inherited strictly maternally and leakage through the paternal line is excessively rare. Moreover, there is a severe bottleneck in the mitochondrial population passed on from one generation to the next. As such, mitochondrial genetic diversity within individuals is negligible, and even sustained recombination between the mitochondrial copies within an individual is unlikely to leave any trace.

The situation is similar in pathogenic bacteria. Homologous genetic recombination will only leave a trace if it happened between strains that are genetically sufficiently differentiated. For strict pathogens, this will require multiple independent infections within a host. The frequency of such events will vary between species but will probably be uncommon for most virulent epidemic pathogens. The genetic diversity of coexisting strains is expected to be particularly low in epidemic species. Strains collected from within the same outbreak (or sweep) will be genetically very homogeneous due to their recent common ancestry. If strains from several outbreaks are analyzed jointly, the apportionment of genetic diversity will be a so-called clonal structure, where sequences form a finite number of well-defined clusters. This is also a signature of the absence of genetic recombination (Maynard Smith et al., 1993).

Besides, host-to-host transmission will also generally create a strong population bottleneck. The extent of this population reduction will vary dramatically from species to species depending on the number of bacteria required for an infection. A single tuberculosis bacterium is believed to be sufficient to initiate an infection in a healthy host (Ratcliffe, 1952; Nyka, 1962). Conversely, the infectious dose for cholera ranges from one million bacteria in certain foods to over one billion bacteria in contaminated water. Irrespective of this initial transmission bottleneck, the host will harbor a large bacterial population once the infection has started. Thus, the bacteria introduced through a secondary infection will be outnumbered and are unlikely to contribute much to the bacterial gene pool within the host. Furthermore, the rare recombinants produced may not pass through the next transmission bottleneck unless they benefit from a fitness advantage. Thus, the situation where there is a potential for recombination between differentiated lineages is relatively rare in pathogenic bacteria. Moreover, recombination events will not systematically affect the evolutionary dynamics of genes because recombination took place between closely related strains and/or the recombinants did not leave any progeny themselves.

This impact of recombination is greatly increased when recombinants benefit from some fitness advantage over the resident population. It is probably no coincidence that the most striking examples in bacteria of extensive homologous recombination over short time periods are associated with the spread of antibiotic resistance (Bootsma et al., 2000; Milkman et al., 2003; Hanage et al., 2009; Lescat et al., 2009). There is also some evidence that recombination is more widespread in the most pathogenic strains; this may be due to the higher selective pressure exerted by host immunity on virulent lineages (Wirth et al., 2006). Outside genomic regions experiencing extraordinary strong selective pressure, homologous recombination will tend to affect bacterial phylogenies only over longer evolutionary time periods (Maynard Smith et al., 1993; Feil et al., 2001, 2003).

6.4 INFERRING PAST DEMOGRAPHY FROM GENETIC SEQUENCE DATA

As mentioned before, the past demography of a population will affect genetic diversity and the shape of phylogenetic trees. These signatures can be exploited to infer the past demography of a genetic sample by fitting an underlying coalescent-based demographic model to a within-species phylogeny (e.g., Griffiths and Tavare, 1994; Kuhner et al., 1998; Wilson and Balding, 1998; Beaumont, 1999; Drummond et al., 2002). Over the past years, the most widely used software for inferences on past demography is BEAST (Drummond and Rambaut, 2007). It has been extensively applied to the reconstruction of the past demography of RNA viruses but also of animal mitochondria. Its use in bacteria has been very limited to date (Eppinger et al., 2006; Roumagnac et al., 2006; Jaenike and Dyer, 2008). The main limitation to the application of such methods is the low genetic diversity within many bacterial populations. This problem is likely to be alleviated in the near future, thanks to considerable progress in high-throughput sequencing methods, and there is no reason to believe methods like the one implemented in BEAST will not be more widely used in bacterial population genetics.

Despite this potential, there are also reasons why the success in bacterial population will probably prove more limited than in RNA viruses. These coalescent-based methodologies are particularly adapted to the reconstruction of simple demographies over short evolutionary time spans. The most straightforward applications are to disease outbreaks, where the main questions of interest are the age of the pathogen's most recent common ancestor (MRCA) and the basic reproductive rate (R_0; the average number of descendents left by each individual). There are parallel situations in bacteria that would be highly adequate to such coalescent-based methodologies.

Disease outbreaks are fairly common across nosocomial and foodborne bacteria. For instance, it might be possible to obtain an even finer picture of the frequency at which methicillin-resistant *Staphylococcus aureus* (MRSA) strains arise de novo by mutation (Enright et al., 2002; Robinson and Enright, 2003; Nubel et al., 2008; Witte et al., 2008) and to get further insight in the subsequent spread of MRSA lineages using coalescence-based modeling. However, such methods also have limitations. Foremost, sufficient polymorphisms must be available. Another limitation is recombination. In the presence of extensive homologous recombination, different loci will be characterized by different phylogenies and there will be no single consensus genealogy to recover. Coalescence-based simulation also do not allow making any inference on the past spatial dynamics (i.e., they do not allow reconstructing the spatial spread of the lineages). Finally, as for any quantitative inference in population genetics, great care should be exercised when selecting genetic markers.

For bacteria with very limited genetic polymorphism, it is tempting to take advantage of mutation discovery approaches. A limited number of strains are sequenced in depth and genetic markers are defined on this panel. These polymorphisms are then typed on a larger sample of strains. While this may sound like a perfectly sensible approach, it is bound to lead to serious biases affecting later population genetics inferences. The first problem is that the number of strains used for initial screening is generally small compared to the final sample and do not contain representatives of all studied populations. As such, this approach will be limited to the discovery of polymorphisms represented in the discovery panel. The second problem is that genetic markers are rarely selected at random but represent a subset that satisfies specific criteria, in particular prior knowledge of them having high genetic diversity.

These insidious biases inherent to mutation discovery have been studied extensively in the field of human genetics where the phenomenon is generally referred to as ascertainment bias (Rogers and Jorde, 1996; Kuhner et al., 2000; Wakeley et al., 2001; Akey et al., 2003; Bustamante et al., 2005; Romero et al., 2009). An ascertained set of genetic markers will overestimate the diversity of the strains represented in the discovery panel and will bias all subsequent population genetics inferences. As there is no satisfying way to correct for such biases, mutation discovery approaches should be avoided whenever the objective of the study includes quantitative population genetics inference. If there is really no alternative to such a procedure, the biases should be minimized by using the largest and most representative possible panel in the initial stages. Polymorphism selected for subsequent typing should also be picked up randomly rather than as the most variable set.

6.5 POPULATION SUBDIVISION

There are probably very few species where individuals are randomly distributed over the entire range or niche. Pathogenic bacteria are no exception to this general rule that related individuals tend to be clustered in time and space. This pattern will arise through a variety of factors. First, the geographic distribution of susceptible individuals will generally be heterogeneous itself. For instance, nosocomial pathogens can spread rapidly within hospital wards due to the high frequency of immunocompromised individuals but are unable to spread through the general population. If the potential range or niche is geographically structured, strains will preferentially infect hosts within the same location, and the number of dispersal events to susceptible individuals in different locations will be low. The relatedness of strains found within a locality will be particularly high for species undergoing periodic clonal outbreaks, as all local strains are likely to share a recent common ancestor under such epidemic dynamics. An extreme example of epidemic clonal structure as found in *S. aureus* is represented in Fig. 6.3a.

When susceptible hosts are distributed more homogeneously, the bacterial population structure will not necessarily translate into well-defined clusters of related strains. However, as long as the dispersal capacity of a pathogen is considerably smaller than the entire distribution range, there may still be a correlation between pairwise relatedness of strains and geographic proximity, a pattern referred to as isolation by distance (IBD). The strength of this correlation will depend on a variety of factors linked to the transmission dynamics but also to the genetics of the species. There can be a complete absence of IBD as in *Salmonella* Typhi (Roumagnac et al., 2006), where the genetic relatedness of different strains seems completely unrelated to their continental origin (Fig. 6.3b). Conversely, extreme patterns of IBD can be observed for *H. pylori* (Fig. 6.3c). In the latter case, they have been generated by the striking similarities between the spatial distribution of genetic diversity between *H. pylori* and its human host, where native populations display very strong IBD worldwide (Linz et al., 2007). Similar yet less striking correlations between the genetic structure of human and bacterial populations have been described for tuberculosis (*M. tuberculosis*) (Hershberg et al., 2008; Wirth et al., 2008) and leprosy (Monot et al., 2005). The general interpretation behind this covariation is that the structure of the bacterial populations has been shaped by past migration events in human settlement history.

The differences between the three data sets summarized in Fig. 6.3 run much deeper than the differences in the large-scale geographic structure. The *S. aureus* data set in panel a was analyzed with the eBURST algorithm, which divides an MLST data set into groups

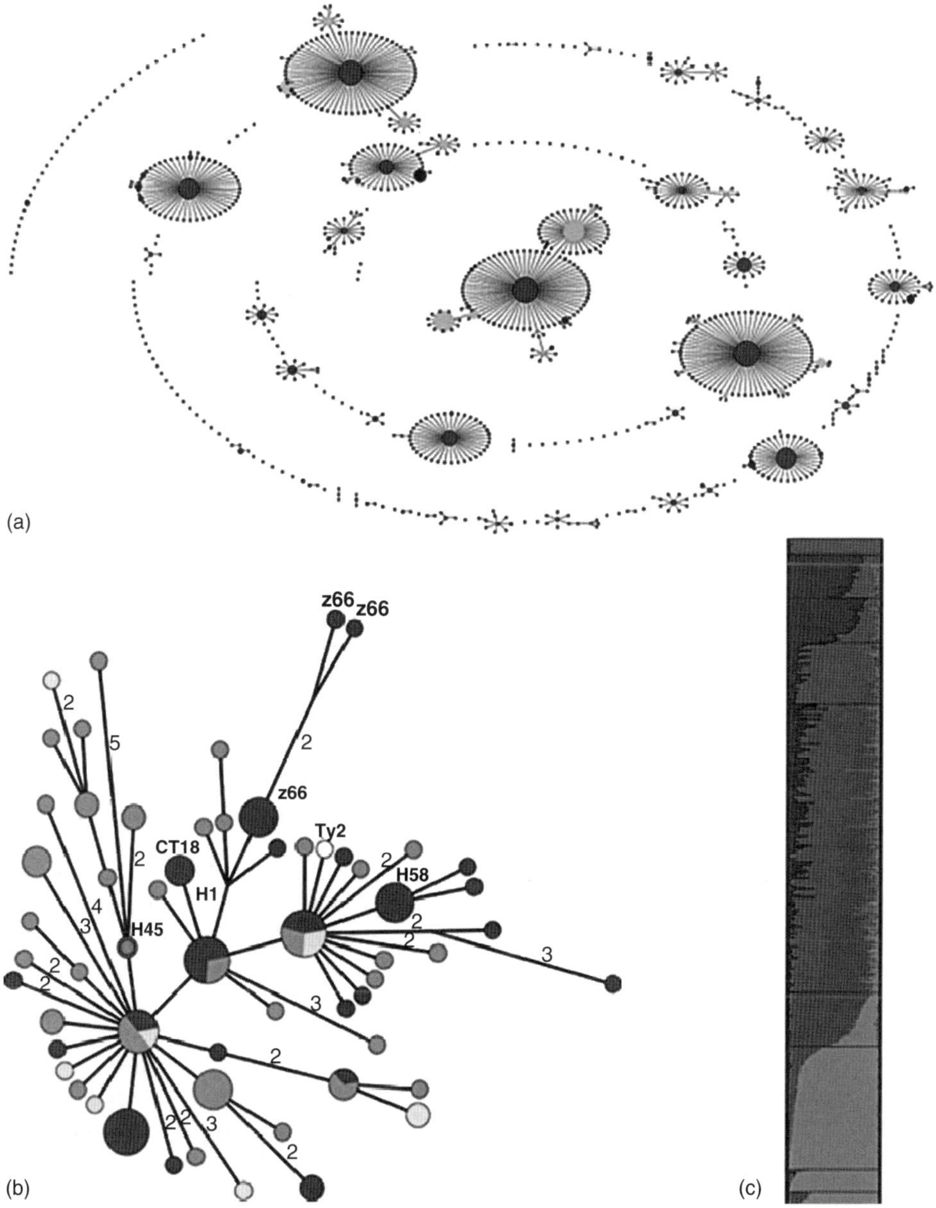

Figure 6.3 Representation of three bacterial population structures. (a) eBURST analysis of *S. aureus* (courtesy of David Aanensen); (b) minimum spanning tree of worldwide Typhi isolates (reprinted with permission from Roumagnac et al., 2006, *Science* 314, 1301–1304); (c) structure output for worldwide strains of *H. pylori* (reproduced with permission from Linz et al., 2007, *Nature* 445, 915–918).

of related isolates and clonal complexes and infers the most likely founding strain for each clonal complex based on the proportion of identical gene fragments shared between pairs of strains (Feil et al., 2004; Turner et al., 2007). The Typhi data set was analyzed with a minimum spanning tree with parsimony-based distances (Fig. 6.3b). Finally, the *H. pylori* data set in Fig. 6.3c was analyzed with the software Structure (Pritchard et al., 2000; Falush et al., 2003a), which infers clusters as random mating populations but allows for individual isolates with mixed ancestry to be assigned to multiple clusters. The strong IBD pattern in *H. pylori* leads to a high frequency of intermediate genotypes represented by striking bleeding patterns between clusters.

While in each of these cases the analyses produced striking and biologically meaningful representations of the population subdivision, none of the three data sets would be amenable to the methodology applied to the two other ones. The eBURST algorithm applied to *S. aureus* was specifically devised for epidemic clonal complexes analyzed with MLST schemes. The minimum spanning tree is an elegant methodological solution for the analysis of the Typhi data set thanks to the most unusual absence of any homoplasy in the 88 single-nucleotide polymorphisms (SNPs) analyzed. Finally, the Structure algorithm, which is highly popular in eukaryotic population genetics, could be applied to *H. pylori* due to the extensive genetic recombination in this organism. However, it is unlikely to be adequate for the analysis of population structure in many other bacterial species. There have been some applications of the Structure algorithm to clonal bacteria such as tuberculosis or *Listeria* (e.g., Filliol et al., 2006; Ragon et al., 2008; Wirth et al., 2008). However, strong correlations in allele frequencies generated by an absence of recombination clearly violate the underlying assumptions of the methodology and thus lead to questionable inferences.

6.6 WHAT IS A BACTERIAL POPULATION?

A commonality between the three methods employed to analyze the data sets represented in Fig. 6.3 is that they do not require a priori grouping of individuals into populations. Most classical population genetics requires individuals to be assigned into populations. Ideally, the basic level of population subdivision should represent a level of structure at which mating is assumed to happen at random. Inference on gene flow can then be reached using (often implicitly) a model of population structure. The by-default model is the Island model where random mating populations of equal size are exchanging migrants at a constant rate with all other populations (Wright, 1931). A further assumption is that the population is at demographic equilibrium. There are other alternative models such as the continuous IBD model (Wright, 1943; Malecot, 1948). There are also more complex models including metapopulation models with local populations arising and going extinct over time (Levins, 1969), which seem conceptually more adequate for pathogenic bacteria.

The main problem for the application of any population genetics model is the difficulty to define a bacterial population. The issue is parallel to the problem of the definition of bacterial species, which has recently received considerable attention in the literature (Gevers et al., 2005; Achtman and Wagner, 2008; Cohan and Koeppel, 2008; Fraser et al., 2009a). In the absence of extensive recombination, it is impossible to define populations as random mating groups. This obviously does not preclude grouping strains into arbitrary populations depending on the research question investigated. There are a large number of possible meaningful hierarchical levels at which the grouping may be performed. Bacterial populations can be structured within hosts (Grant et al., 2008), at microgeographic levels (Chantratita et al., 2008), and at larger geographic scales (Achtman et al., 2004; Linz et al., 2007; Hershberg et al., 2008).

The level of grouping that will be chosen will depend on the structure of the population and the scientific questions. Meaningful subdivision levels could include strains collected from human ethnic groups, hospitals, and cities. If the subdivision is arbitrary, this largely precludes quantitative population genetics inference. However, the comparison of the genetic diversity and the pairwise genetic distances between these groups will still be informative on the genetic similarity and genetic exchanges between the groups. Additional information can be obtained by correlating the pairwise genetic differentiation with various environmental factors using Mantel tests. The most classical spatially explicit analysis is the correlation between genetic distance and geographic isolation indicative of IBD. However, more sophisticated friction routes can be computed including information on human travel data to represent connectivity for epidemic species (Fraser et al., 2009b).

One important determinant for population genetic analyses is the quality of the sampling. The availability of strains can be problematic for pathogenic species, and sampling is often realized in an opportunistic way by analyzing all strains that are available at a given time. While this is often unavoidable, such data sets are often suboptimal in particular when sample sizes between the different populations are highly imbalanced. There is no absolute rule for the ideal sampling strategies of population genetics data sets, but there are a few rules of thumb.

The populations should be defined with a strict set of invariant criteria. Care should also be taken to collect a similar number of strains in each population. A hierarchical sampling strategy (e.g., patients within hospitals within regions within the continent) offers the most flexibility and the highest chances to detect the important levels of structuring. The first hierarchical level of sampling should ideally be performed at the lowest possible biologically meaningful scale, as it is always possible to pool isolates belonging to different subsamples at lower hierarchical levels if no population structuring was detected at that scale.

The ideal sampling scheme should apportion the genetic variance at all relevant levels, from the smallest spatial units where most individuals are expected to be clone mates to larger areas encompassing a greater genetic diversity. Sampling at the relevant scale is crucial for obtaining accurate inferences, and ill-defined a priori sampling units constitute a major source of misleading results. This will be true regardless of which genetic markers are assayed and which statistical tests are applied. The sampling strategy should ensure that there is no hidden genetic structuring within the units defined as subpopulations to avoid a Wahlund effect, which is known to strongly influence parameter estimates, such as F-statistics and linkage disequilibrium, and tends to mimic the signal of clonal reproduction.

Some methods enabling a posteriori partitioning of samples into appropriate biological sampling units are available, such as maximization of total genetic variance (Dupanloup et al., 2002). Alternatively, hierarchical levels that explain a negligible amount of variance (Excoffier et al., 1992) can be removed. In this context, it can be mentioned that the Bayesian clustering method implemented in the widely used software Structure (Falush et al., 2003a) is not ideally suited for organisms reproducing mainly asexually (Drummond et al. 2003). In addition to spatial subdivision, it is also important to consider possible temporal structuring. Samples from the geographic locations collected over long periods of time cannot necessarily be pooled as allele frequencies evolve both over space and time.

6.7 CONCLUSION

Depending on the species, the genomic region, the past demography of the population analyzed, and the time span over which inferences are made, different methodological tools are needed. Thus, there is probably no such thing as a unified field of bacterial population genetics. When recombination is rare, phylogenetic-based approaches and their coalescent-based extensions will be the tools of choice. Conversely, in the presence of recombination, the toolbox of eukaryote geneticists, largely based on summary statistics, will generally be more adequate. The need to adapt the analytical methodology to the specific question and the bacterial species under study makes bacterial population genetics one of the most challenging but also exciting fields.

REFERENCES

Achtman, M. (2008) Evolution, population structure, and phylogeography of genetically monomorphic bacterial pathogens. *Annual Review of Microbiology* **62,** 53–70.

Achtman, M., Morelli, G., Zhu, P. X., Wirth, T., Diehl, I., Kusecek, B., Vogler, A. J., Wagner, D. M., Allender, C. J., Easterday, W. R., Chenal-Francisque, V., Worsham, P., Thomson, N. R., Parkhill, J., Lindler, L. E., Carniel, E., and Keim, P. (2004) Microevolution and history of the plague bacillus, *Yersinia pestis. Proceedings of the National Academy of Sciences of the United States of America* **101,** 17837–17842.

Achtman, M. and Wagner, M. (2008) Microbial diversity and the genetic nature of microbial species. *Nature Reviews. Microbiology* **6,** 431–440.

Achtman, M., Zurth, K., Morelli, C., Torrea, G., Guiyoule, A., and Carniel, E. (1999) *Yersinia pestis,* the cause of plague, is a recently emerged clone of *Yersinia pseudotuberculosis. Proceedings of the National Academy of Sciences of the United States of America* **96,** 14043–14048.

Akey, J. M., Zhang, K., Xiong, M., and Jin, L. (2003) The effect of single nucleotide polymorphism identification strategies on estimates of linkage disequilibrium. *Molecular Biology and Evolution* **20,** 232–242.

Atwood, K. C., Schneider, L. K., and Ryan, F. J. (1951) Periodic selection in *Escherichia coli. Proceedings of the National Academy of Sciences of the United States of America* **37,** 146–155.

Balloux, F., Lehmann, L., and De Meeus, T. (2003) The population genetics of clonal and partially clonal diploids. *Genetics* **164,** 1635–1644.

Beaumont, M. A. (1999) Detecting population expansion and decline using microsatellites. *Genetics* **153,** 2013–2029.

Berg, R. D. (1996) The indigenous gastrointestinal microflora. *Trends in Microbiology* **4,** 430–435.

Bootsma, H. J., Van Dijk, H., Vauterin, P., Verhoef, J., and Mooi, F. R. (2000) Genesis of bro beta-lactamase-producing *Moraxella catarrhalis*: Evidence for transformation-mediated horizontal transfer. *Molecular Microbiology* **36,** 93–104.

Bustamante, C. D., Fledel-Alon, A., Williamson, S., Nielsen, R., Hubisz, M. T., Glanowski, S., Tanenbaum, D. M., White, T. J., Sninsky, J. J., Hernandez, R. D., Civello, D., Adams, M. D., Cargill, M., and Clark, A. G. (2005) Natural selection on protein-coding genes in the human genome. *Nature* **437,** 1153–1157.

Chantratita, N., Wuthiekanun, V., Limmathurotsakul, D., Vesaratchavest, M., Thanwisai, A., Amornchai, P., Tumapa, S., Feil, E. J., Day, N. P., and Peacock, S. J. (2008) Genetic diversity and microevolution of *Burkholderia pseudomallei* in the environment. *PLoS Neglected Tropical Diseases* **2,** 6.

Cohan, F. M. (2005) Periodic selection and ecological diversity in bacteria. In *Selective Sweep* (ed. Nurminsky, D.), pp. 78–93. Kluwer Academic, New York.

Cohan, F. M. and Koeppel, A. F. (2008) The origins of ecological diversity in prokaryotes. *Current Biology* **18,** R1024–U17.

Didelot, X., Darling, A., and Falush, D. (2009) Inferring genomic flux in bacteria. *Genome Research* **19,** 306–317.

Drake, J. W., Charlesworth, B., Charlesworth, D., and Crow, J. F. (1998) Rates of spontaneous mutation. *Genetics* **148,** 1667–1686.

Drummond, A. J., Nicholls, G. K., Rodrigo, A. G., and Solomon, W. (2002) Estimating mutation parameters, population history and genealogy simultaneously from temporally spaced sequence data. *Genetics* **161,** 1307–1320.

Drummond, A. J., Pybus, O. G., Rambaut, A., Forsberg, R., and Rodrigo, A. G. (2003) Measurably evolving populations. *Trends in Ecology & Evolution* **18,** 481–488.

Drummond, A. J. and Rambaut, A. (2007) BEAST: Bayesian evolutionary analysis by sampling trees. *BMC Evolutionary Biology* **7,** 8.

Dupanloup, I., Schneider, S., and Excoffier, L. (2002) A simulated annealing approach to define the genetic structure of populations. *Molecular Ecology* **11,** 2571–2581.

Enright, M. C., Robinson, D. A., Randle, G., Feil, E. J., Grundmann, H., and Spratt, B. G. (2002) The evolutionary history of methicillin-resistant *Staphylococcus aureus* (MRSA). *Proceedings of the National Academy of Sciences of the United States of America* **99,** 7687–7692.

Eppinger, M., Baar, C., Linz, B., Raddatz, G., Lanz, C., Keller, H., Morelli, G., Gressman, H., Achtman, M., and Schuster, S. C. (2006) Who ate whom? Adaptive *Helicobacter* genomic changes that accompanied a host jump from early humans to large felines. *PLoS Genetics* **2,** 1097–1110.

Excoffier, L., Smouse, P., and Quattro, J. (1992) Analysis of molecular variance inferred from metric distances among DNA haplotypes—Application to human mitochondrial DNA restriction data. *Genetics* **131,** 479–491.

Falush, D., Stephens, M., and Pritchard, J. (2003a) Inference of population structure using multilocus genotype data: Linked loci and correlated allele frequencies. *Genetics* **164,** 1567–1587.

Falush, D., Wirth, T., Linz, B., Pritchard, J. K., Stephens, M., Kidd, M., Blaser, M. J., Graham, D. Y., Vacher, S., Perez-Perez, G. I., Yamaoka, Y., Megraud, F., Otto, K., Reichard, U., Katzowitsch, E., Wang, X. Y., Achtman, M., and Suerbaum, S. (2003b) Traces of human migrations in *Helicobacter pylori* populations. *Science* **299,** 1582–1585.

Feil, E. J., Cooper, J. E., Grundmann, H., Robinson, D. A., Enright, M. C., Berendt, T., Peacock, S. J., Smith, J. M., Murphy, M., Spratt, B. G., Moore, C. E., and Day, N. P. J. (2003) How clonal is *Staphylococcus aureus*? *Journal of Bacteriology* **185,** 3307–3316.

Feil, E. J., Holmes, E. C., Bessen, D. E., Chan, M. S., Day, N. P. J., Enright, M. C., Goldstein, R., Hood, D. W., Kalla, A., Moore, C. E., Zhou, J. J., and Spratt, B. G. (2001) Recombination within natural populations of pathogenic bacteria: Short-term empirical estimates and long-term phylogenetic consequences. *Proceedings of the National Academy of Sciences of the United States of America* **98,** 182–187.

Feil, E. J., Li, B. C., Aanensen, D. M., Hanage, W. P., and Spratt, B. G. (2004) eBURST: Inferring patterns of evolutionary descent among clusters of related bacterial genotypes from multilocus sequence typing data. *Journal of Bacteriology* **186,** 1518–1530.

Filliol, I., Motiwala, A. S., Cavatore, M., Qi, W., Hazbon, M. H., Bobadilla Del Valle, M., Fyfe, J., Garcia-Garcia, L., Rastogi, N., Sola, C., Zozio, T., Guerrero, M. I., Leon, C. I., Crabtree, J., Angiuoli, S., Eisenach, K. D., Durmaz, R., Joloba, M. L., Rendon, A., Sifuentes-Osornio, J., Ponce De Leon, A., Cave, M. D., Fleischmann, R., Whittam, T. S., and Alland, D. (2006) Global phylogeny of *Mycobacterium tuberculosis* based on single nucleotide polymorphism (SNP) analysis: Insights into tuberculosis evolution, phylogenetic accuracy of other DNA fingerprinting systems, and recommendations for a minimal standard SNP set. *Journal of Bacteriology* **188,** 759–772.

Frankham, R. (1996) Relationship of genetic variation to population size in wildlife. *Conservation Biology* **10,** 1500–1508.

Fraser, C., Alm, E. J., Polz, M. F., Spratt, B. G., and Hanage, W. P. (2009a) The bacterial species challenge: Making sense of genetic and ecological diversity. *Science* **323,** 741–746.

Fraser, C., Donnelly, C. A., Cauchemez, S., Hanage, W. P., Van Kerkhove, M. D., Hollingsworth, T. D., Griffin, J., Baggaley, R. F., Jenkins, H. E., Lyons, E. J., Jombart, T., Hinsley, W. R., Grassly, N. C., Balloux, F., Ghani, A. C., Ferguson, N. M., Rambaut, A., Pybus, O. G., Lopez-Gatell, H., Apluche-Aranda, C. M., Chapela, I. B., Zavala, E. P., Guevara, D. M. E., Checchi, F., Garcia, E., Hugonnet, S., Roth, C., and The, W. H. O. R. P. A. C. (2009b) Pandemic potential of a strain of influenza A (H1N1): Early findings. *Science* 1557–1561.

Gao, Z., Tseng, C.-H., Pei, Z., and Blaser, M. J. (2007) Molecular analysis of human forearm superficial skin bacterial biota. *Proceedings of the National Academy of Sciences of the United States of America* **104,** 2927–2932.

Gevers, D., Cohan, F. M., Lawrence, J. G., Spratt, B. G., Coenye, T., Feil, E. J., Stackebrandt, E., Van De Peer, Y., Vandamme, P., Thompson, F. L., and Swings, J. (2005) Re-evaluating prokaryotic species. *Nature Reviews. Microbiology* **3,** 733–739.

Gillespie, J. H., ed. (1987) Molecular evolution and the neutral allele theory. *Oxford Surveys in Evolutionary Biology* **4,** 10–37.

Gillespie, J. H. (2001) Is the population size of a species relevant to its evolution? *Evolution* **55,** 2161–2169.

Grant, A. J., Restif, O., Mckinley, T. J., Sheppard, M., Maskell, D. J., and Mastroeni, P. (2008) Modelling within-host spatiotemporal dynamics of invasive bacterial disease. *PLoS Biology* **6,** E74.

Griffiths, R. C. and Tavare, S. (1994) Sampling theory for neutral alleles in a varying environment. *Philosophical Transactions of the Royal Society of London. Series B, Biological Sciences* **344,** 403–410.

Hanage, W. P., Fraser, C., Tang, J., Connor, T. R., and Corander, J. (2009) Hyper-recombination, diversity, and antibiotic resistance in *Pneumococcus*. *Science* **324,** 1454–1457.

Herczegh, A., Ghidan, A., Deseo, K., Kamotsay, K., and Tarjan, I. (2008) Comparison of *Streptococcus mutans* strains from children with caries-active, caries-free and gingivitis clinical diagnosis by pulsed-field gel electrophoresis. *Acta Microbiologica et Immunologica Hungarica* **55,** 419–427.

Hershberg, R., Lipatov, M., Small, P. M., Sheffer, H., Niemann, S., Homolka, S., Roach, J. C., Kremer, K., Petrov, D. A., Feldman, M. W., and Gagneux, S. (2008) High functional diversity in *Mycobacterium tuberculosis* driven by genetic drift and human demography. *PLoS Biology* **6,** E311.

Hooper, L. V. and Gordon, J. I. (2001) Commensal host-bacterial relationships in the gut. *Science* **292,** 1115–1118.

Jaenike, J. and Dyer, K. A. (2008) No resistance to male-killing *Wolbachia* after thousands of years of infection. *Journal of Evolutionary Biology* **21,** 1570–1577.

Kapur, V., Whittam, T. S., and Musser, J. M. (1994) Is *Mycobacterium tuberculosis* 15,000 years old? *Journal of Infectious Diseases* **170,** 1348–1349.

Karaolis, D. K. R., Johnson, J. A., Bailey, C. C., Boedeker, E. C., Kaper, J. B., and Reeves, P. R. (1998) A *Vibrio cholerae* pathogenicity island associated with epidemic and pandemic strains. *Proceedings of the National Academy of Sciences of the United States of America* **95,** 3134–3139.

Kidgell, C., Reichard, U., Wain, J., Linz, B., Torpdahl, M., Dougan, G., and Achtman, M. (2002) *Salmonella*

Typhi, the causative agent of typhoid fever, is approximately 50,000 years old. *Infection, Genetics and Evolution* **2,** 39–45.

KRAYTSBERG, Y., SCHWARTZ, M., BROWN, T. A., EBRALIDSE, K., KUNZ, W. S., CLAYTON, D. A., VISSING, J., and KHRAPKO, K. (2004) Recombination of human mitochondrial DNA. *Science* **304,** 981.

KUHNER, M. K., BEERLI, P., YAMATO, J., and FELSENSTEIN, J. (2000) Usefulness of single nucleotide polymorphism data for estimating population parameters. *Genetics* **156,** 439–447.

KUHNER, M. K., YAMATO, J., and FELSENSTEIN, J. (1998) Maximum likelihood estimation of population growth rates based on the coalescent. *Genetics* **149,** 429–434.

LESCAT, M., CALTEAU, A., HOEDE, C., BARBE, V., TOUCHON, M., ROCHA, E., TENAILLON, O., MEDIGUE, C., JOHNSON, J. R., and DENAMUR, E. (2009) A module located at a chromosomal integration hot spot is responsible for the multidrug resistance of a reference strain from *Escherichia coli* clonal group A. *Antimicrobial Agents and Chemotherapy* **53,** 2283–2288.

LEVINS, R. 1969. Some demographic and genetic consequences of environmental heterogeneity for biological control. *Bulletin of the Entomological Society of America* **15,** 237–240.

LINZ, B., BALLOUX, F., MOODLEY, Y., MANICA, A., LIU, H., ROUMAGNAC, P., FALUSH, D., STAMER, C., PRUGNOLLE, F., VAN DER MERWE, S. W., YAMAOKA, Y., GRAHAM, D. Y., PEREZ-TRALLERO, E., WADSTROM, T., SUERBAUM, S., and ACHTMAN, M. (2007) An African origin for the intimate association between humans and *Helicobacter pylori*. *Nature* **445,** 915–918.

MAIDEN, M. C. J. (2006) Multilocus sequence typing of bacteria. *Annual Review of Microbiology* **60,** 561–588.

MALECOT, G. (1948) *Les Mathematiques de L'heredite*. Masson, Paris.

MAYNARD SMITH, J. and HAIG, J. (1974) The hitch-hiking effect of a favourable gene. *Genetical Research* **23,** 23–35.

MAYNARD SMITH, J., SMITH, N. H., O'ROURKE, M., and SPRATT, B. G. (1993) How clonal are bacteria? *Proceedings of the National Academy of Sciences of the United States of America* **90,** 4384–4388.

MILKMAN, R., JAEGER, E., and MCBRIDE, R. D. (2003) Molecular evolution of the *Escherichia coli* chromosome. VI. Two regions of high effective recombination. *Genetics* **163,** 475–483.

MONOT, M., HONORE, N., GARNIER, T., ARAOZ, R., COPPEE, J. Y., LACROIX, C., SOW, S., SPENCER, J. S., TRUMAN, R. W., WILLIAMS, D. L., GELBER, R., VIRMOND, M., FLAGEUL, B., CHO, S. N., JI, B. H., PANIZ-MONDOLFI, A., CONVIT, J., YOUNG, S., FINE, P. E., RASOLOFO, V., BRENNAN, P. J., and COLE, S. T. (2005) On the origin of leprosy. *Science* **308,** 1040–1042.

NOTLEY-MCROBB, L. and FERENCI, T. (2000) Experimental analysis of molecular events during mutational periodic selections in bacterial evolution. *Genetics* **156,** 1493–1501.

NUBEL, U., ROUMAGNAC, P., FELDKAMP, M., SONG, J. H., KO, K. S., HUANG, Y. C., COOMBS, G., IP, M., WESTH, H., SKOV, R., STRUELENS, M. J., GOERING, R. V., STROMMENGER, B., WELLER, A., WITTE, W., and ACHTMAN, M. (2008) Frequent emergence and limited geographic dispersal of methicillin-resistant *Staphylococcus aureus*. *Proceedings of the National Academy of Sciences of the United States of America* **105,** 14130–14135.

NYKA, W. (1962) Studies on the infective particle in airborne tuberculosis. I. Observations in mice infected with a bovine strain of *M. tuberculosis*. *American Review of Respiratory Disease* **85,** 33–39.

OCHMAN, H., ELWYN, S., and MORAN, N. A. (1999) Calibrating bacterial evolution. *Proceedings of the National Academy of Sciences of the United States of America* **96,** 12638–12643.

PRITCHARD, J. K., STEPHENS, M., and DONNELLY, P. (2000) Inference of population structure using multilocus genotype data. *Genetics* **155,** 945–959.

RAGON, M., WIRTH, T., HOLLANDT, F., LAVENIR, R., LECUIT, M., Le MONNIER, A., and BRISSE, S. (2008) A new perspective on *Listeria monocytogenes* evolution. *PLoS Pathogens* **4,** E1000146.

RATCLIFFE, H. L. (1952) Tuberculosis induced by droplet nuclei infection. *American Journal of Hygiene* **55,** 36–48.

ROBINSON, D. A. and ENRIGHT, M. C. (2003) Evolutionary models of the emergence of methicillin-resistant *Staphylococcus aureus*. *Antimicrobial Agents and Chemotherapy* **47,** 3926–3934.

ROGERS, A. R. and JORDE, L. B. (1996) Ascertainment bias in estimates of average heterozygosity. *American Journal of Human Genetics* **58,** 1033–1041.

ROMERO, I. G., MANICA, A., GOUDET, J., HANDLEY, L. L., and BALLOUX, F. (2009) How accurate is the current picture of human genetic variation? *Heredity* **102,** 120–126.

ROUMAGNAC, P., WEILL, F. X., DOLECEK, C., BAKER, S., BRISSE, S., CHINH, N. T., LE, T. A. H., ACOSTA, C. J., FARRAR, J., DOUGAN, G., and ACHTMAN, M. (2006) Evolutionary history of *Salmonella Typhi*. *Science* **314,** 1301–1304.

SABETI, P. C., REICH, D. E., HIGGINS, J. M., LEVINE, H. Z., RICHTER, D. J., SCHAFFNER, S. F., GABRIEL, S. B., PLATKO, J. V., PATTERSON, N. J., MCDONALD, G. J., ACKERMAN, H. C., CAMPBELL, S. J., ALTSHULER, D., COOPER, R., KWIATKOWSKI, D., WARD, R., and LANDER, E. S. (2002) Detecting recent positive selection in the human genome from haplotype structure. *Nature* **419,** 832–837.

SAVAGE, D. C. (1977) Microbial ecology of gastrointestinal tract. *Annual Review of Microbiology* **31,** 107–133.

SCHWARTZ, M. and VISSING, J. (2002) Paternal inheritance of mitochondrial DNA. *New England Journal of Medicine* **347,** 576–580.

SCHWARZ, S., MORELLI, G., KUSECEK, B., MANICA, A., BALLOUX, F., OWEN, R. J., GRAHAM, D. Y., VAN DER

Merwe, S., Achtman, M., and Suerbaum, S. (2008) Horizontal versus familial transmission of *Helicobacter pylori*. *PLoS Pathogens* **4**.

Slatkin, M. (1991) Inbreeding coefficients and coalescence times. *Genetical Research* **58,** 167–175.

Sreevatsan, S., Escalante, P., Pan, X., Gillies, D. A., Siddiqui, S., Khalaf, C. N., Kreiswirth, B. N., Bifani, P., Adams, L. G., Ficht, T., Perumaalla, V. S., Cave, M. D., Vanembden, J. D. A., and Musser, J. M. (1996) Identification of a polymorphic nucleotide in oxyr specific for *Mycobacterium bovis*. *Journal of Clinical Microbiology* **34,** 2007–2010.

Sreevatsan, S., Pan, X., Stockbauer, K. E., Connell, N. D., Kreiswirth, B. N., Whittam, T. S., and Musser, J. M. (1997) Restricted structural gene polymorphism in the *Mycobacterium tuberculosis* complex indicates evolutionarily recent global dissemination. *Proceedings of the National Academy of Sciences of the United States of America* **94,** 9869–9874.

Tago, Y., Imai, M., Ihara, M., Atofuji, H., Nagata, Y., and Yamamoto, K. (2005) *Escherichia coli* mutator delta pola is defective in base mismatch correction: The nature of in vivo DNA replication errors. *Journal of Molecular Biology* **351,** 299–308.

Tamura, K. and Nei, M. (1993) Estimation of the number of nucleotide substitutions in the control region of mitochondrial DNA in humans and chimpanzees. *Molecular Biology and Evolution* **10,** 512–526.

Thompson, J. R., Randa, M. A., Marcelino, L. A., Tomita-Mitchell, A., Lim, E., and Polz, M. F. (2004) Diversity and dynamics of a North Atlantic coastal *Vibrio* community. *Applied and Environmental Microbiology* **70,** 4103–4110.

Touchon, M., Hoede, C., Tenaillon, O., Barbe, V., Baeriswyl, S., Bidet, P., Bingen, E., Bonacorsi, S., Bouchier, C., Bouvet, O., Calteau, A., Chiapello, H., Clermont, O., Cruveiller, S., Danchin, A., Diard, M., Dossat, C., Karoui, M. E., Frapy, E., Garry, L., Ghigo, J. M., Gilles, A. M., Johnson, J., Le Bouguenec, C., Lescat, M., Mangenot, S., Martinez-Jehanne, V., Matic, I., Nassif, X., Oztas, S., Petit, M. A., Pichon, C., Rouy, Z., Ruf, C. S., Schneider, D., Tourret, J., Vacherie, B., Vallenet, D., Medigue, C., Rocha, E. P. C., and Denamur, E. (2009) Organised genome dynamics in the *Escherichia coli* species results in highly diverse adaptive paths. *PLoS Genetics* **5,** E1000344.

Turner, K., Hanage, W., Fraser, C., Connor, T., and Spratt, B. (2007) Assessing the reliability of eBURST using simulated populations with known ancestry. *BMC Microbiology* **7,** 30.

van Belkum, A., Melles, D. C., Nouwen, J., Van Leeuwen, W. B., Van Wamel, W., Vos, M. C., Wertheim, H. F. L., and Verbrugh, H. A. (2009) Co-evolutionary aspects of human colonisation and infection by *Staphylococcus aureus*. *Infection Genetics and Evolution* **9,** 32–47.

Van Ert, M. N., Easterday, W. R., Huynh, L. Y., Okinaka, R. T., Hugh-Jones, M. E., Ravel, J., Zanecki, S. R., Pearson, T., Simonson, T. S., U'ren, J. M., Kachur, S. M., Leadem-Dougherty, R. R., Rhoton, S. D., Zinser, G., Farlow, J., Coker, P. R., Smith, K. L., Wang, B., Kenefic, L. J., Fraser-Liggett, C. M., Wagner, D. M., and Keim, P. (2007) Global genetic population structure of *Bacillus anthracis*. *PloS One* **2,** E461.

Voight, B. F., Kudaravalli, S., Wen, X., and Pritchard, J. K. (2006) A map of recent positive selection in the human genome. *PLoS Biology* **4,** E72.

Wakeley, J., Nielsen, R., Liu-Cordero, S. N., and Ardlie, K. (2001) The discovery of single-nucleotide polymorphisms—And inferences about human demographic history. *American Journal of Human Genetics* **69,** 1332–1347.

Waldor, M. K. and Mekalanos, J. J. (1996) Lysogenic conversion by a filamentous phage encoding cholera toxin. *Science* **272,** 1910–1914.

Welch, R. A., Burland, V., Plunkett, G., Redford, P., Roesch, P., Rasko, D., Buckles, E. L., Liou, S. R., Boutin, A., Hackett, J., Stroud, D., Mayhew, G. F., Rose, D. J., Zhou, S., Schwartz, D. C., Perna, N. T., Mobley, H. L. T., Donnenberg, M. S., and Blattner, F. R. (2002) Extensive mosaic structure revealed by the complete genome sequence of uropathogenic *Escherichia coli*. *Proceedings of the National Academy of Sciences of the United States of America* **99,** 17020–17024.

Whitlock, M. and Barton, N. (1997) The effective size of a subdivided population. *Genetics* **146,** 427–441.

Whitman, W. B., Coleman, D. C., and Wiebe, W. J. (1998) Prokaryotes: The unseen majority. *Proceedings of the National Academy of Sciences of the United States of America* **95,** 6578–6583.

WHO (2008) Global tuberculosis control—Surveillance, planning, financing. World Health Organization, Geneva. http://www.who.int/tb/publications/global_report/2008/en/index.html.

Wilson, I. J. and Balding, D. J. (1998) Genealogical inference from microsatellite data. *Genetics* **150,** 499–510.

Wirth, T., Falush, D., Lan, R. T., Colles, F., Mensa, P., Wieler, L. H., Karch, H., Reeves, P. R., Maiden, M. C. J., Ochman, H., and Achtman, M. (2006) Sex and virulence in *Escherichia coli*: An evolutionary perspective. *Molecular Microbiology* **60,** 1136–1151.

Wirth, T., Hildebrand, F., Allix-Beguec, C., Wolbeling, F., Kubica, T., Kremer, K., Van Soolingen, D., Rusch-Gerdes, S., Locht, C., Brisse, S., Meyer, A., Supply, P., and Niemann, S. (2008) Origin, spread and demography of the *Mycobacterium tuberculosis* complex. *PLoS Pathogens* **4,** E1000160.

Witte, W., Cuny, C., Klare, I., Nuebel, U., Strommenger, B., and Werner, G. (2008) Emergence and spread of antibiotic-resistant gram-positive bacterial pathogens. *International Journal of Medical Microbiology* **298,** 365–377.

Wright, S. (1931) Evolution on Mendelian populations. *Genetics* **16,** 97–159.

Wright, S. (1943) Isolation by distance. *Genetics* **28,** 114–138.

ZHANG, J. (2005) On the evolution of codon volatility. *Genetics* **169,** 495–501.

ZHANG, W., QI, W. H., ALBERT, T. J., MOTIWALA, A. S., ALLAND, D., HYYTIA-TREES, E. K., RIBOT, E. M., FIELDS, P. I., WHITTAM, T. S., and SWAMINATHAN, B. (2006) Probing genomic diversity and evolution of *Escherichia coli* O157 by single nucleotide polymorphisms. *Genome Research* **16,** 757–767.

Chapter 7

Population Genomics of Bacteria

David S. Guttman and John Stavrinides

7.1 INTRODUCTION

From a single gene taken from a dozen strains in 1990 to a single genome in 2000 to whole genomes from dozens of strains in 2010, the progress made in genome sequencing over the past 20 years far exceeds even the most optimistic of expectations. Not surprisingly, microbial genomics has been on the forefront of this advance, and consequently, our understanding of microbial ecology and evolution has matured immeasurably. These extraordinary advances permit us to seriously ask what was once only a rhetorical question, "What would you do if you could sequence everything?" (Kahvejian et al., 2008).

The advances of the genomics era have brought about dramatic changes in the way we study and view microbes and microbial populations. The development of cost-effective and high-throughput sequencing technologies has paved the way for addressing long-standing and fundamental questions in new and innovative ways. Traditional population genetic analyses have been expanding from the detailed analysis of only a few specific loci to whole genome exploration, giving rise to the newly emerging field of population genomics (Gulcher and Stefansson, 1998; Black et al., 2001; DeLong, 2002, 2004; Whitaker and Banfield, 2006). Gulcher and colleagues (Gulcher and Stefansson, 1998) defined population genomics as the study of the evolutionary processes that influence variation across populations using whole genome data. Population genomics applies established population genetic theories and methodologies to whole genome sequences obtained from multiple individuals that share the potential to exchange genetic material (populations). It permits the separation of evolutionary forces that affect individual loci (e.g., mutation, recombination, selection) from those forces that influence the genome as a whole (e.g., population bottlenecks, genetic drift) (Black et al., 2001). Genome-wide changes are more likely to reflect population demographic patterns, while single-gene effects are more informative for deciphering the selective pressures underlying bacterial adaptation.

We are now well into the post-genomic era dominated by next-generation (next-gen) sequencing technology. Next-gen platforms can interrogate tens to hundreds of billions of bases on a single run. A single bacterial genome can be sequenced to high coverage in less than a day, and soon it will be possible to generate a full bacterial genome sequence during the time it takes to get an overpriced coffee. If dozens or hundreds of bacterial genomes can be sequenced in a week, we are now in the remarkable situation where data

Bacterial Population Genetics in Infectious Disease, Edited by D. Ashley Robinson, Daniel Falush, and Edward J. Feil

acquisition is effectively a trivial step in population genomic studies, raising a number of very interesting questions. Is next-gen bacterial population genomics just classical population genetics writ large, or can we now address fundamentally different questions than we could 10 years ago? In other words, has the next-gen revolution made qualitative change in the way we study and understand microbes, or has it merely provided a quantitative advance in our science? Are we seeing an evolutionary or revolutionary change in population genomics? Jacques Monod once famously stated, "What's true for *E. coli* is true for elephants, only more so." Perhaps now we must ask, "Is what's true for Sanger also true for Solexa, only more so?"

This chapter will examine the history, power, application, and potential future of bacterial population genomics. We will begin by describing the principles of classical population genetics, focusing on its major applications, including its use in characterizing the fundamental evolutionary forces, the study of bacterial population structure, strain typing, and experimental evolution. We will then examine some of the advances made through "classical" population genomics, including a vastly more sophisticated understanding of bacterial genome structure and dynamics, and will briefly look at the potential for microarray-based population genomic approaches. We will then move on to next-gen population genomics and discuss its tremendous potential. Unfortunately, next-gen population genomics is such a fledgling field that there are very few examples to discuss, so part of the section will be devoted to potential future directions and prospects for the field. Finally, we provide a brief survey of the next-gen genomics technology and analytical methods and tools as a guide for those interested in pursuing bacterial population genomics. We hope to illustrate that a $100 bacterial genome not only complements and carries forward 75 years of population genetics theory and study but also moves the field into areas barely even envisioned at the time when the first complete bacterial genome sequence was published.

7.2 CLASSICAL BACTERIAL POPULATION GENETICS

The methodologies and conceptual advances of microbial population genomics are seeded in the basic principles, applications, and limitations of classical population genetics. Population genetics as a discipline is concerned with how evolutionary forces drive and distribute genetic variation within and among individuals, populations, and species. Conversely, the patterns of genetic variation found among extant populations can also be used to infer the evolutionary history of populations. Although the majority of population genetics theory was founded on diploid, eukaryotic models, this framework was readily extended to haploid, prokaryotic systems by Roger Milkman and Robert Selander (Milkman, 1973; Selander and Levin, 1980), giving rise to modern microbial population genetics. Their work and others' (Maynard Smith, 1991; Maynard Smith et al., 1991; Baumberg et al., 1995; Spratt and Maiden, 1999; Rozenfeld et al., 2007) have strengthened our ability to study bacterial populations in light of the biological features that distinguish most eukaryotes and prokaryotes, namely, haploidy, clonality, obscure and porous species boundaries, nonreciprocal genetic exchange among distantly related individuals, and a propensity to undergo very rapid fluctuations in population size.

7.2.1 Evolutionary Forces

Genetic diversity is introduced and molded within populations by the four primary evolutionary forces: mutation, gene flow, natural selection, and genetic drift. Mutation is the ultimate source of all genetic variation, and while it is often assumed to be a constant, mutation rates can vary considerably between genes and species, and can even fluctuate under different environmental conditions (Drake et al., 1998). While mutation is a stochastic process that affects DNA irrespective of its function, we, as geneticists, can only observe and measure those mutations that have made it through the sieve of genetic drift or natural selection.

Genetic drift describes changes in allele frequency over time as a result of random sampling (survival) from generation to generation. This stochastic process is dependent on the effective population size, with smaller populations being more strongly influenced by genetic drift than larger populations. Unlike the stochastic drift process, natural selection is a deterministic process that describes the differential survival and propagation of genetic variants within a population. Genetic drift and selection can oppose each other, with the dominant force being largely determined by the effective size of the population.

Mutation, selection, and drift are intimately intertwined and interdependent. Most mutations observed in essential and phylogenetically conserved "core" genes, which encode conserved proteins and RNAs with critical cellular functions in processes such as metabolism, transcription, and translation, tend to be synonymous and largely neutral. This is due to the simple fact that nonsynonymous mutations are more likely to result in deleterious amino acid substitutions, which are likely to be purged from the population by negative (purifying) selection. In contrast, it is not uncommon for mutations in genes associated with niche adaptation (e.g., disease-associated genes) to cause nonsynonymous, amino acid substitutions that are driven to high frequencies by positive selection. Finally, it is also possible for natural selection to favor the selective maintenance of multiple alleles within a population through a variety of mechanisms collectively called balancing or diversifying selection.

While mutation generates all variation, gene flow, in the form of recombination or horizontal gene transfer (HGT), can reassort that variation, introduce it into new genetic backgrounds, and shuffle it among populations. Following acquisition of genetic information, it may be incorporated into the genomic repertoire through either homologous or nonhomologous recombination. Homologous recombination requires an existing region of sufficient similarity and is more likely to result in the integration of incoming DNA when the donor and recipient are closely related. Consequently, homologous recombination can affect variation in housekeeping or core genes. Nonhomologous recombination can introduce entirely novel DNA into the genome and is almost always associated with some form of selfish/mobile genetic element such as plasmids, phages, or transposons. It plays a critical role in structuring the bacterial "flexible" genome, which is composed of genes that vary between strains, genomic islands that may encode determinants for pathogenicity and resistance, mobile genetic elements, and genes that aid in niche-specific adaptation.

Recombination also has a more subtle, but perhaps even more important role in breaking down the linkage between mutations. Linkage disequilibrium (LD), or the extent to which alleles at distinct loci are associated, is inversely related to the rate of recombination between those loci. Therefore, selection (whether it be positive or negative) acting on one polymorphism will likewise influence surrounding polymorphisms unless their evolutionary trajectories are separated by recombination. This genetic hitchhiking can result in the fixation of deleterious mutations simply due to their proximity to a positively selected polymorphism, or vice versa (the Hill–Robertson effect [Hill and Robertson, 1966]). A selective sweep is the most dramatic example of genetic hitchhiking whereby an entire genome can rapidly sweep to fixation in a population due to its linkage to a positively selected site. Given the relatively low rate of recombination in many bacterial core genomes, the Hill–Robertson dynamic between positively and negatively selected polymorphisms must be intense. This is a topic perfectly suited to population genomic study but which has not yet received adequate attention.

A recent study by Ma and colleagues (Ma and Guttman, 2008) reveals how evolutionary forces can interact in one family of virulence-associated proteins within the *Pseudomonas syringae* species complex and how these changes impact host–pathogen interactions. The *Pseudomonas syringae* HopZ family of type III secreted effector proteins, a member of the YopJ family of effectors found in both plant and animal pathogens, is present in three major homology groups widely distributed among *P. syringae* strains. The HopZ1 homologue appears to be ancestral in *P. syringae* and has evolved at least three functional allelic classes and a number of degenerate forms through the mutational process. The HopZ2 and HopZ3 homologues, on the other hand, were brought into *P. syringae* via horizontal transfer from other plant pathogenic bacteria. The introduction of the HopZ allele that is most similar to the ancestral allele (*hopZ1a*) into strains harboring alternate or degenerate *hopZ* alleles results in a resistance protein-mediated defense response in their respective hosts, which is not observed with the endogenous allele. All of the homologues, whether derived through the mutational process or HGT, seem to retain similar biochemical functions, although they show very different patterns of recognition in different hosts. Some of these host-specificity changes may be due to the very extensive positively selected genetic variation found in the family. The apparent maintenance of similar biochemical function in the functional HopZ homologues illustrates a possible constraint on effector evolution. If an effector performs an essential virulence function that is recognized by a subset of hosts, there are a number of avenues of adaptive change that could permit the evasion of host detection. An effector could be lost or may degenerate if the function can be assumed by a different effector. Alternatively, an effector could modify its specificity through the mutational process and through positive selection, or could acquire a homologous virulence factor through HGT. All of these routes have been taken in the *P. syringae* HopZ family. The particular path taken likely depends on the availability of novel virulence factors and the plasticity of host targets.

Population genetic analyses can be used to correlate the presence of new genetic determinants or changes within existing determinants that enable a given lineage to persist and thrive in a population. *Escherichia coli*, which normally lives as a harmless commensal in the animal gut, has diversified into at least five genotypically distinct pathogenic types: enterohemorrhagic, enteropathogenic, enteroinvasive, enterotoxigenic, and enteroaggregative (Welch et al., 2002). A separate extraintestinal group has also surfaced, which includes uropathogenic strains. Almost all these *E. coli* strains express type 1 fimbriae, which are attachment structures composed of two subunits, FimA and FimH. FimH is an adhesive subunit situated on the tip of the fimbrial tip, and can normally bind trimannosyl receptors. A survey of uropathogenic *E. coli* isolates revealed that these pathogens have selectively accumulated mutations in the adhesive subunit FimH that enhance the binding efficiency of FimH to monomannose receptors found in the uroepithelium (Sokurenko et al., 1998, 1999; Weissman et al., 2006). These positively selected mutations enhancing tissue tropism resulted in a 15-fold greater ability to colonize the urinary tract. In contrast to *E. coli*, antigenic variability in the pili used by pathogenic *Neisseria gonorrhoeae* for attachment to host epithelial cells is generated through recombination (Hagblom et al., 1985). The pilin subunits comprise a conserved N terminus and a hypervariable C terminus. The *Neisseria* genome carries *pilE*, an intact pilin gene with its own promoter, along with multiple truncated silent pilin loci (Seifert et al., 1988). Following transformation with DNA from lysed cells, recombination occurs between one of the silent truncated pilin loci and the expressed *pilE*, resulting in the formation of a mosaic pilin gene with the potential for an altered antigenicity.

7.2.2 Bacterial Population Structure

While studies of natural selection and the other evolutionary forces are an important component of population genetics, the study of bacterial population structure and dynamics dominates the field due to its immediate importance for understanding medically and agriculturally important microbes. Bacteria display a broad continuum of population structures, ranging from strictly clonal to freely recombining (Maynard Smith et al., 2000) (Fig. 7.1).

The varied bacterial population structures are the manifestation of the underlying ecological and evolutionary dynamics of the organism. An idealized clonal bacterium reproduces strictly by binary fission of the mother cell into two daughter cells, which will be genetically identical to the mother except for the rare mutation. These mutations descend vertically through time, resulting in the progressive accumulation of mutations and clonal divergence of descendant lineages. An analysis of the distribution of diversity within this idealized population would reveal nested descendant clades being defined by shared derived traits (mutations). Since each mutation would occur within the existing genomic context defined by previously occurring mutations, their association would be entirely nonrandom, and thus would be in LD. From a phylogenetic context, the entire clonal genome would share a common evolutionary history, with the exception of the sequentially acquired mutations; therefore, all genomic regions will display the same phylogenetic relationships (Fig. 7.2).

While it is easiest to envision and analyze an idealized clonal bacterium, most species have multiple mechanisms to permit and facilitate genetic exchange. A population undergoing frequent recombination would display low LD because of the reshuffling of alleles into different genetic backgrounds. In these cases, strains do not have a single evolutionary history, but instead, each recombining unit has a potentially unique line of descent. Consequently, the evolutionary and phylogenetic relationships among recombining strains cannot be appropriately captured with a traditional bifurcating phylogeny. Instead, a network approach, such as split decomposition, must be used to visualize the phylogenetic reticulations that result from recombination events (Bandelt and Dress, 1992; Huson, 1998).

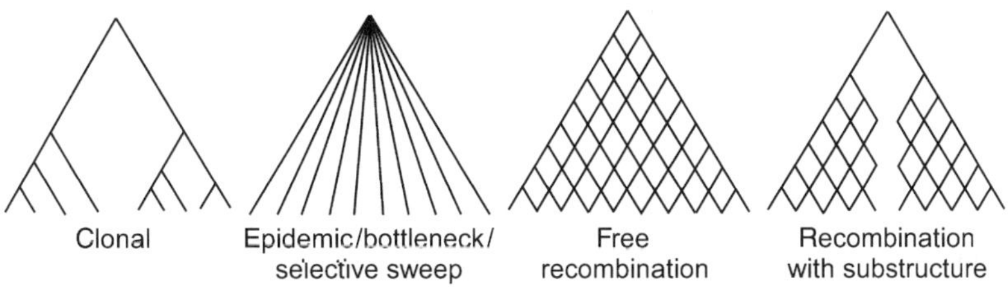

Figure 7.1 Bacterial population structures. A clonal population results from the accumulation and vertical inheritance of mutations via binary fission with no exchange of genetic material. Epidemic populations, or those having undergone population bottlenecks, or selective sweeps coalesce to a relatively recent common ancestor, such that their genealogical structure appears as a star radiation of all isolates from a single ancestral node. In freely recombining populations, strains can potentially exchange genetic material with any other member of that population, resulting in a net-like or reticulated genealogical structure. Populations that are ecologically or geographically isolated recombine mainly within their subpopulation.

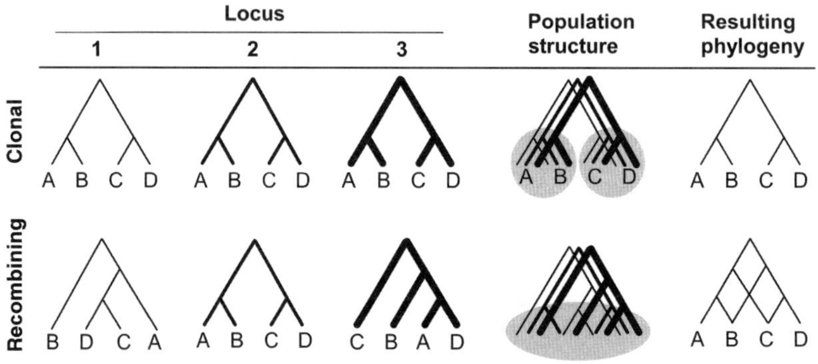

Figure 7.2 Principle of multilocus sequence typing (MLST). Independent gene genealogies for four taxa at three loci will be congruent if the population is clonal, resulting in a single phylogeny representing the common evolutionary relationships among strains at all loci. In populations with high rates of recombination, each locus can have a unique evolutionary history, resulting in incongruent gene genealogies, and a reticulated net species phylogeny.

An epidemic population structure has a background population composed of recombining clones that have a reticulated relationship as described above. Epidemic clonal lineages emerge out of this recombining population after the formation of an adaptive genotype. These epidemic clones may then increase in frequency and persist for long periods of time.

Given the broad spectrum of population structures exhibited by different bacterial species, efforts have been made to quantify the level of LD and the relative evolutionary significance of recombination. LD is often benchmarked by the "index of association," with a value significantly deviating from zero indicating low LD (frequent recombination) (Maynard Smith et al., 1993). Another approach indexed recombination by determining how much variation was introduced into populations by recombination versus mutation (Guttman and Dykhuizen, 1994; Maynard Smith, 1999). An analysis of four loci from 12 natural isolates of *E. coli* revealed a recombination rate ~50 times higher than the mutation rate (Guttman and Dykhuizen, 1994), suggesting that recombination, which is traditionally considered a process that homogenizes genetic variation within populations, in fact plays a more important role in generating variation than mutation. These findings were supported in a more extensive multilocus sequence typing (MLST, discussed below) study from Wirth and colleagues (2006), who analyzed seven housekeeping genes from a collection of 462 *E. coli* isolated from a variety of sources (humans and domesticated, captive, and wild mammals, birds, and reptiles). The authors concluded that homologous recombination was a more important source of variation than mutation, that population bottlenecks reduced the diversity of *E. coli* approximately 10–30 million years ago, and that the extant genetic diversity has accumulated in this species only in the last 5 million years. Furthermore, an analysis of the evolutionary histories of the housekeeping genes revealed that traditional phylogenetic approaches could not be applied for establishing population structure due to insufficient signal, rapid diversification, and frequent homologous recombination.

This last study provides an important reminder that while LD is an extremely powerful indicator of allele association, it is not an absolute indicator of clonality. LD can arise from a poor or biased sample that does not reflect the true population diversity, or when samples are taken from multiple populations that are geographically or ecologically isolated, such that the opportunity for recombination is restricted. Similarly, populations having undergone epidemic expansion or selective sweeps may also exhibit LD. LD may also indicate the presence of epistatic interactions among linked loci that confer a selective advantage to the organism.

7.2.3 Strain and Population Typing

Given the importance of strain typing for a wide range of applications, it is not surprising that a diversity of approaches have been developed for this task. One critical assumption underlying all of these approaches, however, is that isolates are collected without bias. The sample collection must represent the breadth of natural diversity and must not over-represent those isolates with specific traits or phenotypes, particularly when it is that very phenotype that is the focus of the study. This is a particularly difficult problem for studies of pathogenic species, where strains only come to the attention of pathologists when they cause disease. This, so-called iceberg sampling (Tibayrenc, 1999) frequently overlooks virulence-attenuated strains, those strains that have evolved different modes of interaction with their hosts, or closely related strains living in different niches. While this may not be a significant issue for mechanistic studies of virulence and avirulence, it can profoundly distort our understanding of the ecology and evolution of potential pathogens. Since ecology and evolution are intimately tied to epidemiology, host specificity, antibiotic resistance, and the emergence of new infectious agents, we ignore this issue at our peril.

The bipartite nature of the bacterial genome (core verses flexible) poses a special challenge for typing strains. Strain typing is ultimately concerned with reconstructing the evolutionary relationships among strains of interest, and assumes that a single evolutionary history actually exists. In reality, genes that have been introduced by HGT have, by definition, evolutionary histories and relationships different from those genes that have been faithfully transmitted in a vertical manner from mother cell to daughter cell. In population genetics terms, core gene genealogies will very likely not be congruent with genealogies from the flexible genome. Consequently, if we are interested in characterizing the clonal (vertical) evolutionary relationships of strains, we need to focus on those genes that most likely have undergone clonal evolutionary descent.

Housekeeping genes are functionally essential and evolutionary constrained genes involved in replication, transcription, translation, and the central metabolism (Medini et al., 2008). Given their essential cellular role, most variation in these genes will be neutral due to the strong negative selection against those amino acid substitutions that disrupt function. For example, synonymous substitutions are primarily found in the essential replication initiation factor gene *dnaA* since these do not alter the function of this essential protein. In contrast, surface-exposed proteins involved in virulence can be subject to strong positive or diversifying selection (Endo et al., 1996) due to pressures imposed by host immunity or surveillance systems. These positively selected genes may not be the best choices for inferring population structure, since selection can distort evolutionary relationships.

Multilocus Enzyme Electrophoresis (MLEE) Typing

A wide range of methods have been used for quantifying evolutionary relationships and for measuring genetic diversity within bacterial populations. One method, popular in the early 1980s, is MLEE, a typing technique that uses the differential starch gel electrophoretic mobility of housekeeping enzymes as a measure of allelic variation among individuals within a population (Selander et al., 1986). The identification of allozymes allows distinct alleles of a given locus to be assigned electrophoretic types, with the characterization of multiple loci permitting the generation of a profile for each individual strain. The genetic distance between profiles can then be used as a phenetic measure of strain similarity. This information can also be used as an index of population structure, where LD among the MLEE alleles is used to infer the degree of clonal population structure. MLEE was used to characterize the population structure of pathogenic bacteria such as *Neisseria* (Caugant et al., 1987), *Salmonella* (Selander et al., 1990), *Legionella* (Selander et al., 1985), *Haemophilus* (Musser et al., 1988, 1990), and *Bordetella* (Musser et al., 1987). An MLEE analysis of 688 clinical strains of *Neisseria meningitidis* collected over a 14-year period from 20 countries revealed a clonal or possibly an epidemic population structure that may have been the result of multiple selective sweeps (Caugant et al., 1987). Further, several of the electrophoretic profiles were also found to be more prominent among isolates from specific countries. In contrast to *N. meningitidis*, a MLEE analysis of 2209 worldwide isolates of *Haemophilus influenzae* identified 280 distinct electrophoretic types, and because specific electrophoretic types were correlated with specific serotypes with limited geographical distributions, *H. influenzae* was determined to be clonal (Musser et al., 1988).

DNA-Based Typing Methods

The gradual progression toward molecular typing led to DNA-based diversity measures such as restriction fragment length polymorphism (RFLP) analyses. RFLP analyses were applied in a number of ways, including the use of specific regions of DNA such as 16S and 23S ribosomal RNAs (rRNAs) (ribotyping), targeted loci amplified prior to restriction digestion (amplified fragment length polymorphism [AFLP]), and even restriction digests of whole genome extracts that are separated by pulsed-field gel electrophoresis (PFGE) (Mueller and Wolfenbarger, 1999). PFGE analysis is the most common approach used in clinical microbiology laboratories. For example, PFGE typing of 239 *Staphylococcus aureus* isolates from 142 patients produced 26 fingerprints, with substantially higher resolution than that obtained by ribotyping (Prevost et al., 1992). A larger analysis of 957 oxacillin-resistant *S. aureus* isolates collected from the United Kingdom, the United States, Canada, and Australia revealed a geographically restricted population subdivision (McDougal et al., 2003). Eight lineages (US100-800) were defined, with the US100 lineage being found to comprise the majority of multidrug-resistant strains US200, US500, and US600 comprising epidemic strains from Europe and Australia, and the US300 and US400 lineages containing community isolates resistant only to beta-lactam drugs and erythromycin.

Another DNA-based typing method assesses the overall genomic similarity by measuring DNA–DNA hybridization kinetics between two genomes. This approach provides a standardized method for the taxonomical classification of bacterial species; however, it not only requires the pairwise comparison of all strains but also completely ignores the significance of horizontally acquired genes and the flexible genome, which can account for over 50% of the genomic complement (Welch et al., 2002).

The development of cost-effective DNA sequencing techniques permitted the rapid and efficient classification of bacterial species based on universally distributed rRNA sequence. Pioneered by Woese and colleagues (1975; Fox et al., 1977), 16S rRNA has become the standard for the taxonomic classification of new bacterial isolates, with a threshold of 97–98% sequence identity defining species boundaries. The suitability of 16S rRNA sequence for identifying species lies in its cellular essentiality and, thus, its reduced likelihood of horizontal transfer (Rossello-Mora and Amann, 2001). Unfortunately, 16S rRNA evolves too slowly to adequately resolve intraspecific relationships and is not immune from recombination (Ueda et al., 1999).

MLST

MLST is the most recent and sophisticated technique developed to characterize intraspecific strain relationships. This approach uses the combined information from ~500 bp of DNA sequence data from a set of each of (typically) seven housekeeping genes (Enright and Spratt, 1998; Maiden et al., 1998). MLST offers increased resolution over its predecessor, MLEE, and data collected by MLST are readily comparable between laboratories (Achtman, 2004). Since its first application for the dissection of the population dynamics of *N. meningitidis* (Maiden et al., 1998), MLST has been used to study over two dozen bacterial species (see http://www.mlst.net/).

Salmonella enterica serovar Typhi provides a nice example of the power of MLST analysis. This globally distributed, human-adapted pathogen is the causal agent of typhoid fever and poses a major health problem in Asia, in Africa, and in South America (Parry et al., 2002; Roumagnac et al., 2006; Chau et al., 2007; Achtman, 2008). An analysis of genetic diversity among hundreds of isolates collected between 1958 and 2004 revealed highly congruent MLST gene genealogies, strongly supporting a highly clonal population structure. Of the 85 haplotypes identified by Roumagnac et al., only 8 were found on multiple continents. Consequently, it appears that each haplotype arose only once, and there have been at least eight instances of global spread.

Some bacterial species exhibit such high levels of recombination that most or all MLST loci exhibit unique phylogenetic histories (Fig. 7.2). Examples of highly recombinogenic bacterial species include *N. gonorrhoeae* and *Helicobacter pylori* (Spratt et al., 1995; Salaun et al., 1998; Suerbaum et al., 1998; Solca et al., 2001; Suerbaum and Josenhans, 2007). One approach used in *H. pylori* to circumvent the issue of high recombination rates was to evaluate the diversity of three nonhousekeeping gene fragments, *flaA*, *flaB*, and *vacA*, from 50 Canadian, South African, and German strains (Suerbaum et al., 1998). FlaA and FlaB are flagellar filament proteins covered by a flagellar sheath. The authors claim that these proteins are unlikely to be under positive selection since they are not directly exposed to the external milieu. The third protein, VacA, is a secreted vacuolating toxin thought to be involved in the formation of ulcers (Cover, 1996) and is therefore more likely to be under positive selection. All three of these loci showed patterns consistent with high levels of recombination via a homoplasy test, which examines the probability of a single site change occurring two or more times in the evolutionary history of these sequences.

Another *H. pylori* study examined the population structure of 78 strains collected from a limited geographic range by surveying four housekeeping genes (Solca et al., 2001). Three other virulence-associated loci (including *vacA*) and antibiotic resistance phenotypes were also included in this study to provide additional resolving power and to identify possible associations among virulence-associated loci and disease phenotypes. As in previous studies, the evolutionary histories of the different housekeeping genes proved to be incongruent. In addition, no significant correlation was observed between allelic variants, virulence, and antibiotic resistance for any group of strains.

Most natural bacterial populations are believed to fall somewhere between clonality and freely recombining, with limited recombination interspersed among long periods of clonal reproduction and expansion. *N. meningitidis*, a bacterial pathogen responsible for most cases of meningococcal disease, retains a diverse population characterized by clonal subgroups that emerge following recombination (Achtman, 2004). Hyperinvasive lineages show an epidemic population structure with clades of closely related strains radiating out from an ancestral clone following a selective sweep or population bottleneck (Zhu et al., 2001).

7.2.4 Experimental Evolution

Natural population studies provide tremendous retrospective insight into evolutionary processes but are still only a single snapshot of one realization of the descent process. Experimental evolution studies, on the other, hand make it possible to directly follow evolution in real time, and to effectively rewind history and restart or replicate the evolutionary process. In theory, these studies permit population geneticists to observe the dynamics of evolutionary change under carefully controlled environmental conditions, with sufficient replication to separate selective versus stochastic events.

Richard Lenski et al. (1991) initiated the most notable experimental evolution program using 12 replicate populations of *E. coli* B/6 maintained by serial transfer in minimal medium with glucose as the growth-limiting carbon source. Mutation was predicted to be the main source of adaptation in these populations since the ancestor was plasmid free, and recombination was presumed to be infrequent. This amazing resource has been used to study a wide variety of evolutionary and ecological questions. For example, Souza and colleagues (1997) addressed the importance of recombination by mixing clonal strains that had undergone 7000 generations of experimental evolution with recombining strains. The populations were propagated for 1000 generations, with periodic mating cycles when they were mixed with a genetically distinct high-frequency recombination (HFR) donor *E. coli* strains. This HFR strain could transfer genetic material to donors via F plasmid-mediated conjugation but could not replicate under the particular culturing conditions. After 1000 generations, the 12 control populations that were not exposed to the HFR strains retained their ancestral alleles at 14 tested loci, whereas populations exposed to recombination treatment exhibited substantially more genetic variation. Surprisingly though, fitness assays against a common competitor found that populations subjected to recombination treatment did not gain any significant improvement in their rate of adaptive evolution relative to the control population. In other words, a higher level of genetic diversity did not provide a net improvement in mean fitness. Subsequent experiments revealed complex selection dynamics, where recombinant genotypes in fact did show a selective advantage, but only in the presence of other genotypes from the same experimental population. This adaptive evolution was based on frequency-dependent, genotype-by-genotype interactions that were not captured in simple competition experiments.

Experimental evolutionary approaches have also been used to study the evolution of antibiotic resistance. One study examined the ability of bacteria to develop resistance to the cationic antimicrobial peptide pexiganan, a class of drug to which resistance was claimed to be less likely to evolve (Perron et al., 2006). Both mutator and nonmutator lines of *E. coli* and *Pseudomonas fluorescens* were passaged for >600 generations in the presence of pexiganan, and in the end, both mutator and nonmutator lines acquired resistance. This indicates that the naturally occurring mutation rate is sufficient for generating the variation needed to respond to this selective pressure. Ultimately, these results reinforced the intense selective pressures imposed on microbial populations through the use of any antibiotic.

7.3 THE GENOMICS ERA

The 10 years following the 1995 publication of the first complete genome sequence from a free-living organism (*H. influenzae*) (Fleischmann et al., 1995) could be considered the classical age of genomics. Extraordinarily scientific, methodological, and technological advances were made during this period, culminating with the publication of the human genome sequence in 2001 (Lander et al., 2001; Venter et al., 2001) and the release of the "finished" human genome by the Human Genome Project in 2003 and then again in 2006.

One of the most exciting and important discoveries arising from the first bacterial genome studies was that many virulence-associated or niche-specific genes are clustered on the genome and exhibit signatures of having been horizontally inherited as a single unit. These regions are referred to as "genomic islands" or "pathogenicity islands" (PAIs) when the encoded genes are implicated in the disease process (Ziebuhr et al., 1999; Hacker and Kaper, 2000; Hacker and Carniel, 2001). PAIs are typically virulence-associated gene clusters situated adjacent to tRNA genes, flanked by direct repeats or mobile elements, and having a G + C content that differs from the rest of the bacterial host genome (Ziebuhr et al., 1999; Hacker and Kaper, 2000; Hacker and Carniel, 2001). PAIs with these explicit characteristics have been identified in numerous pathogens of both plants and animals (Arnold et al., 2003).

The significance of PAIs is made clear in the study of vancomycin-resistant *Enterococcus faecium*, where a specific lineage has adapted to the nosocomial environment, and is now associated with the majority of hospital outbreaks (Leavis et al., 2006). An AFLP analysis of vancomycin-resistant commensal and outbreak isolates collected from both humans and animals showed that human commensal isolates clustered with those of pigs, while outbreak isolates formed their own clonal "CC17" complex. A subsequent MLST analysis of 411 isolates from animal and human sources collected from different continents revealed that the CC17 clone is responsible for the emergence of vancomycin-resistant enterococcal outbreaks worldwide. Genetic and phenotypic characterization the CC17 clonal complex found broad resistance to ampicillin and quinolones. CC17 isolates also typically carry the genes *esp* and *hyl*, which encode a surface protein and hyaluronidase enzyme, respectively. Both of these genes are involved in enterococcal colonization (Leavis et al., 2004) and have been found to cluster on a 150-kb PAI that also carries multiple mobile elements and numerous virulence determinants in the closely related species *Enterococcus faecalis*. Much like other PAIs, it exhibits a much lower G + C content than the rest of the genome (Shankar et al., 2002). The *esp* gene itself encodes a domain containing 13 amino acid repeats that constitute approximately 50% of the entire 1873 amino acid protein (Shankar et al., 1999). The number of repeats present in this gene varies among isolates, implicating homologous recombination in the formation of new gene variants, which may contribute to the evasion of host immunity.

Another beautiful example of the influence of PAI on bacterial evolution is provided by *Streptococcus pneumoniae*. This species has been categorized into 91 capsular polysaccharide serotypes, with specific serotypes being correlated with geographical location and tissue tropism (Henriques-Normark et al., 2008). Serotypes 1, 4, 5, and 7F are associated with invasive disease but are only rarely found in the nasopharynx of healthy carriers. In contrast, isolates of serotypes 3, 6A, 19F, and 23F are less likely to cause invasive disease and are commonly found in healthy carriers. Pathogenic nasopharyngeal *S. pneumoniae* strains carry the *rlrA* islet, a 12-kb island carrying *rlrA* and *srtD* encoding surface proteins and flanked on both sides by mobile element IS1167. RlrA and SrtD have been implicated in the survival of *S. pneumoniae* during lung infection (Hava and Camilli, 2002; Hava et al., 2003). The former is a pneumococcal pilin protein that enhances adhesion to human lung epithelial cells, and which elicits an inflammatory response (Hava and Camilli, 2002). Surprisingly, only 20–30% of isolates surveyed carried the *rlrA* islet, possibly implicating population-level host immunity factors in determining its frequency (Henriques-Normark et al., 2008). A second islet was identified that is associated with serotypes 1, 2, 7F, 19A, and 19F, implicating another locus not only in adhesion but also in lung infection and general invasiveness (Bagnoli et al., 2008).

Pitman and colleagues (2005) demonstrated how exposure to the host immune system can select for the loss of a virulence- and avirulence-associated PAI. In this study, the bean pathogen *Pseudomonas syringae* pv. phaseolicola 1302, which carries the type III secreted effector *hopAR1*, was infiltrated into the bean cultivar Tendergreen, which expresses the cognate R3 resistance protein that induces host defenses in response to HopAR1. After five serial passages *in planta*, the interaction changed from a typical incompatible interaction with a very rapid and strong defense response to water-soaking lesions typically seen during compatible disease interactions with virulent strains. This phenotypic change was brought about via the excision of the ~106-kb genomic island PPHGI-1, which contains 100 predicted open reading frames, including the *hopAR1* effector, other putative pathogenicity factors, genes involved in transcriptional regulation, chemotaxis, signaling, and plasmid-associated sequences responsible for replication, partitioning, conjugal transfer, and type IV pilus biosynthesis. The entire genomic island is surrounded by 52-bp direct repeats associated with a tRNAlys gene, and its excision is controlled by a XerC integrase encoded just inside the right border. XerC expression was found to be upregulated 50-fold *in planta*.

7.4 BACTERIAL POPULATION GENOMICS

For most of the genomics era, population genomics *sensu stricto* was not feasible for many research groups due to the expense, expertise, and resources required to generate even one representative genome sequence from a species. The development and application of second-generation, or "next-gen" genomic technology rapidly changed this by dramatically reducing the cost, time, and expertise needed to produce a genome sequence. It effectively moved genomics out of the genome center and into the research laboratory, allowing for some of the first true population genomic studies.

Microbiologists have known as far back as the work of Frederick Griffith (1928) that bacteria are capable of exchanging genetic material, yet the evolutionary and ecological significance of genetic exchange was still a matter of significant debate well into the 1990s (Maynard Smith et al., 1993; Guttman, 1997; Maynard Smith, 1999). Welch and colleagues (2002) used comparative genomics to shed new light on the significance of recombination-associated genomic variation in the first and, arguably, the most important bacterial population genomic study. They compared the genomic complement of the newly sequenced *Escherichia coli* CFT073, a pathogenic strain isolated from the blood of a patient with acute pyelonephritis, to the enterohemorrhagic strain EDL933 and the nonpathogenic laboratory strain MG1655. Using a fairly stringent criterion for orthology, they concluded that while the three strains shared a highly conserved core genome, the clonal backbone was interspersed with a remarkably large amount of DNA acquired via HGT. In fact, only 39.2% of the genome was common to all three strains, while an average of 15.5% of the genome complement of each of the three strains was unique to that strain (Welch et al., 2002) (Fig. 7.3). This remarkable finding has been observed in nearly every

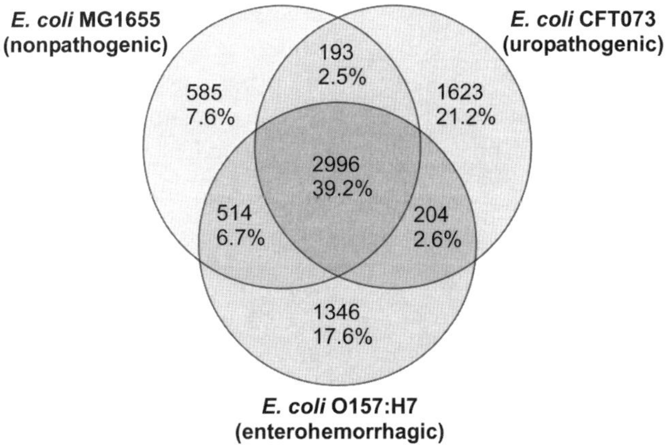

Figure 7.3 Venn diagram showing the *Escherichia coli* core and flexible genome. Orthologous coding sequences shared among nonpathogenic (MG1655), uropathogenic (CFT073), and enterohemorrhagic (O157 : H7) *E. coli* isolates. A total of 7638 proteins were analyzed, with orthologous coding sequences being counted only once. Adapted from Welch et al. (2002).

bacterial species examined, and has led to a more sophisticated view of the dynamic and complex nature of bacterial genomes. As discussed above, we now understand that many bacterial species have a conserved core genome, and a highly dynamic and horizontally transferable flexible genome (Dobrindt and Hacker, 2001; Ochman and Davalos, 2006). Being that the flexible genome may account for as much as half of the entire genome of any one isolate, the entire pan-genome (defined as the total suite of core, flexible, and unique genes in a species) (Medini et al., 2005) must clearly dwarf the core genome in size and scope. Further, the evolution and ecology of the mobilome, defined as the collection of transposons, plasmids, and phages that are largely responsible for the horizontal movement of bacterial genes (Koonin and Wolf, 2008), must be taken into consideration if we are to understand the drivers and constraints imposed on bacterial populations.

Another extensive population genomic study was performed by Tettelin and colleagues (2005, 2008) on *Streptococcus agalactiae* (group B *Streptococcus* [GBS]), which is the leading cause of illness and death among newborns. The authors sequenced six pathogenic strains via traditional Sanger sequencing and determined the number of shared and unique genes for each strain. They then used an exponential decaying function to estimate the number of shared core genomes in GBS and the size of the GBS pan-genome. While the average number of predicted genes per strain was 2245, the estimated core genome size was 1806 genes. Consequently, each strain had a flexible genome of ~439 genes, but remarkably, the regression analysis also showed that the GBS pan-genome increased by over 33 genes with each additional strain examined. The authors also showed that classical serotyping is a poor tool for taxonomic classification and does not reflect the true diversity of the strains. Genome comparisons found that two strains with very closely related serotypes were in fact the least-conserved strains in the comparison set, while the two most closely related strains as determined by DNA-level genomic comparisons had distinct serotypes.

Similar studies have been performed in *Neisseria* (Maiden, 2008; Sheppard et al., 2008). One particularly fascinating finding from this system came through the analysis of DNA uptake sequences (DUSs), which are required for efficient DNA uptake in Pasteurellaceae. Treangen and colleagues (2008) found DUSs to be overrepresented in core genes, underrepresented in highly polymorphic genes, and absent in genes recently acquired or lost. They suggest that this correlation implicates DUSs (and by extension, natural transformation) in facilitating genome stability via some regenerative process, rather than in promoting hypervariation and adaptive diversification.

Genomic studies have led to important advances in our understanding of the evolution and diversification of the plague pathogen *Yersina pestis* from the enteropathogenic *Yersinia pseudotuberculosis* (Hinchliffe et al., 2003; Chain et al., 2004; Pouillot et al., 2008). MLST analysis supported the emergence of the *Yersinia pestis* clone within the last 1500–20,000 years, while full genome comparisons found that 13% of *Y. pseudotuberculosis* genes are degenerate in *Y. pestis*, mostly due to the action of insertion sequences. Surprisingly, five of nine chromosomal regions deleted in *Y. pestis* are important for survival, growth, or virulence of *Y. pseudotuberculosis*.

A similarly important role for mobile elements was found in the evolution of the acidophilic archaeon *Ferroplasma acidarmanus* ferl (Allen et al., 2007). The 1.94-Mb genome of this strain isolated out of the Richmond Mine in Northern California was compared to ~103 Mb of sequence data obtained from a natural biofilm community at the same site. The environmental sample recapitulated 92% of the ferl genome but also showed extensive genomic heterogeneity in the form of gain and loss of novel gene blocks, largely mediated by transposase and phage movement. Comparison of homologous sequences revealed evidence for strong negative (purifying) selection and extensive recombination. The authors conclude that recombination and transposition are much stronger drivers of population diversification than mutation. The high rate of recombination also results in a highly mosaic gene pool that may promote population cohesion (Fig. 7.4).

7.4.1 Microarray-Based Population Genomics

While whole genome sequencing is the most obvious approach to population genomics, microarray-based methods have also been used to assess genome-wide population diversity. In contrast to whole genome sequencing, microarrays are comparatively cheap and accessible. Depending on their design, DNA microarrays can be used not only to determine the presence or the absence of entire coding sequences but also to identify single nucleotide polymorphisms (SNPs). Sung and colleagues (2008) pursued the former approach to identify host-specific genes from *S. aureus*. They spotted all 3623 coding sequences from seven completely sequenced *S. aureus* strains and probed the array with DNA from 56 strains isolated from cows, horses, goats, sheep, and a camel, as well as 161 strains isolated from healthy and diseased humans. They found human and (nonhuman) animal isolates tended to cluster in multiple but distinct lineages, although a small number of sequence types were found in humans and in multiple animal hosts. Horse isolates proved to have a particularly high likelihood of also being carried by humans. Further, a number of instances of animal diseases were caused by strains from the human lineages. The authors conclude that despite the presence of host-specific colonization and virulence phenotypes among *S. aureus* isolates, there are relatively few conserved genetic differences between strains that colonize human and animal hosts.

A higher-resolution microarray approach was taken by Zwick and colleagues (2005) who used Affymetrix high-density oligonucleotide resequencing arrays to interrogate 56 *Bacillus anthracis* strains. This array design permitted the authors to get SNP-level variation for 92.6% of all bases in the genome with a discrepancy rate of 7.4×10^{-7}, which is over threefold better than what is obtainable through conventional Sanger sequencing.

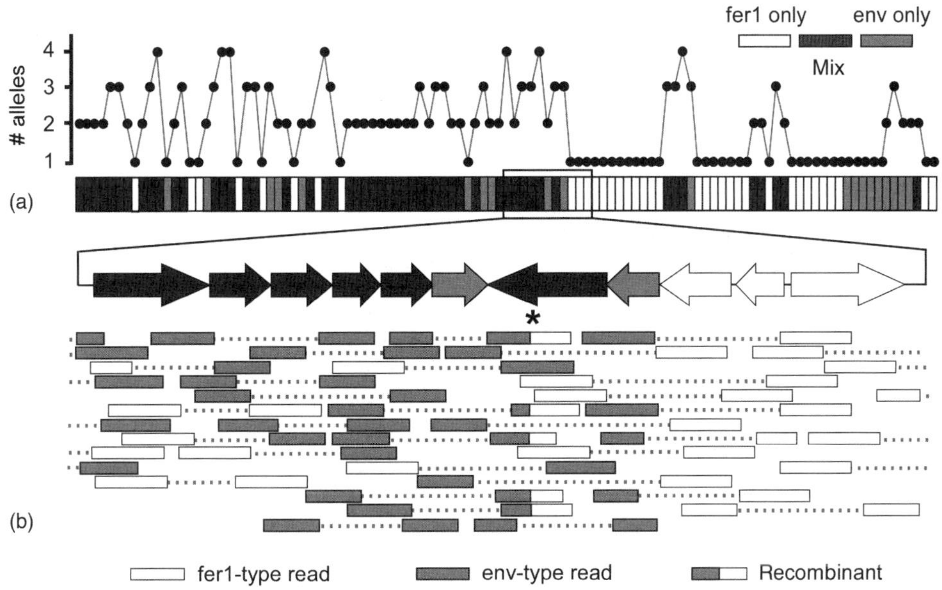

Figure 7.4 Population structure of the archaeon *Ferroplasma acidarmanus* fer1. (a) An analysis of a genomic region spanning 109 genes shows the highly mosaic structure of an environmental population (env) of the same species and collected from the same site as *F. acidarmanus* fer1. The population carries more than one allele for nearly all loci, with the fer1 allele being most common. White indicates the fer1 allele; light gray indicates an environmental (env) allele; and black indicates a mix of the two. (b) Individual sequencing of 11 genes within the 109 gene cluster reveals evidence of recombination within the population (indicated with an asterisk). Adapted from Allen et al. (2007).

Remarkably, they found only 37 SNPs out of >1.5 Mb of chromosomal and plasmid DNA. Only nine of these SNPs resulted in a replacement substitution. This excess of rare SNPs is indicative of either recent population expansion or a selective sweep. Since there was no evidence for recent recombination or plasmid exchange, the authors concluded that the extant *B. anthracis* population is recently derived from a single ancestral clone.

While microarrays hold great promise for characterizing gene composition and identifying strain-specific difference within bacterial populations (Ochman and Santos, 2005; Sarkar et al., 2006), they also bring some very significant challenges. Specifically, all microarray studies suffer from a built-in ascertainment bias. Microarrays require a priori knowledge of genome content and variation during their design and cannot provide any information for those regions not encoded on the array. In addition, the physical interaction inherent in microarray hybridization means that there is a positive association between the similarity of the probe and target sequences, and the binding efficiency of the two. Consequently, probes coming from more divergent strains will bind more poorly to the array and will produce a weaker signal. This limitation can be overcome with the use of high-density oligonucleotide resequencing arrays, but these require the greatest amount of a priori knowledge for their construction.

7.5 NEXT-GEN BACTERIAL POPULATION GENOMICS

The impact of next-gen sequencing is just beginning to be felt in population genomics. Perhaps the first such study was published by Velicer and colleagues (2006), who used experimental evolution to identify mutations influencing social cooperation in the bacterium *Myxococcus xanthus*. They evolved an ancestral social cooperator for 1000 generations and isolated an obligate social cheater. They then changed the selection regime and identified a competitively dominant social cooperator. The authors sequenced the 9.14-Mb genome to a depth of 19× using a combination of Sanger sequencing and 454 pyrosequencing, and uncovered only 15 polymorphisms (six transversions, eight transitions, and one deletion) between the ancestral and evolved lines. Fourteen of the mutations occurred during the first transition, while only one occurred during the second transition. One of the 14 mutations leading to social cheating was in the *pilQ* gene, which encodes a protein required for the biosynthesis of type IV pili required for social motility in *M. xanthus*. The single mutation that resulted in the re-evolution of social cooperation occurred in a 7-base homo-cytosine stretch located 128 bp upstream of the start codon of a predicted GNAT-family acetyltransferase. The function of this protein is unknown.

More recently, next-gen sequencing on the Illumina Solexa Genome Analyzer was used to sequence two quasi-independent isolates of the *Bacillus subtilis* laboratory strain 168 in addition to 14 other related *B. subtilis* laboratory strains (Srivatsan et al., 2008). The authors were able to use the >35× sequence coverage to identify errors in the published sequence of strain 168, differences between the two isolates, and mutations responsible for important laboratory-associated phenotypes. For example, they were able to identify unmappable, second-site mutations in *yjbM* and *ywaC* that suppress stringent response-associated growth deficits. YjbM and YwaC were shown to work with RelA in the modulation of (p)ppGpp, a critical signaling molecule that controls bacterial response to environmental change.

Holt and colleagues (2008) used both 454 and Illumina Solexa sequencing (discussed below) to generate 19 genome sequences that span the phylogenetic diversity of the *S. enterica* serovar Typhi group, which, as discussed above, is a human-restricted pathogen that is the causal agent of typhoid fever. They found only very weak evidence of either purifying or positive (adaptive) selection, and 72% of all genes were monomorphic across all strains. Twenty-six genes did show signals consistent with positive selection, and half of these encoded surface-exposed, exported, or secreted proteins, or proteins involved in these processes. Very little evidence of recombination or antigenic variation was observed. The authors also identified 42 genes with deletion events and another 55 with nonsense mutations that resulted in a premature stop codon. Overall, 4.5% of Typhi genes were pseudogenes, which is much higher than the 0.9% observed in *Salmonella enterica* serovar Typhimurium or the 0.7% seen in *E. coli*. Consequently, gene loss seems to be an important force in the evolution of Typhi, which is consistent with genetic drift due to the small effective population size. The lack of evidence for selection-driven antigenic variation is in contrast to what is seen among other human pathogens. The observed genomic patterns of diversity are consistent with a small and human-restricted population, where asymptomatic carriers make up the primary reservoir for this important pathogen.

7.5.1 The Future of Bacterial Population Genomics

As next-gen technology permeates microbial population genetics, not only will new life be infused into long-standing questions, but a more sophisticated understanding of microbial form, function, and diversity will inevitably arise, and from this, entirely new questions will emerge. A perfect example of this is the recognition of the bipartite nature of the bacterial genome with interdigitated flexible and core element—a finding coming out of the first microbial comparative genomics studies. This revelation provided fresh perspective on such contentious issues as the relative importance and role of recombination, the reason for very significant differences in genome size among closely related strains, taxonomic discrepancies between different strain typing approaches, and even questions related to the nature and very existence of bacterial species. While it is notoriously difficult to forecast the future, it is clear that next-gen genomics is already moving the field in new and sometimes unexpected directions. The use of whole genome scans for genes showing strong selective signatures is an extraordinarily powerful means to identify functionally significant genes. The coupling of metabolic modeling and whole genome sequencing of unculturable microbes and microbial consortia provides remarkable biological perspectives on organisms that were hereto complete unknowns. The integration of multiple "omic" and interaction data sets (e.g., genomic, transcriptomic, proteomic, metabolomic, Yeast-2-Hybrid, and synthetic lethal) using the methodologies of information theory and systems biology has the potential to formalize the function of cells, populations, and communities to an unprecedented degree. The probing of environmental gene nurseries and the unimaginably diverse "phage world" provides an exciting glimpse into the quasi-living world underlying and even potentiating the evolution of microbes. Finally, the interrogation of microbial communities via metagenomics, particularly with whole genome sequencing of environmentally extracted DNA, and even community-level whole transcriptome analysis, has tremendous potential beyond basic research, as these communities directly influence the function, health, and stability of the entire global ecosystem.

7.6 NEXT-GEN GENOMICS TECHNOLOGY

The advent of next-gen genomics is as significant a step forward for population genetics as Lewontin and Hubbys's (1966) classic paper describing the first application of starch gels to study population genetic variation. While starch gel technology enabled any population genetics laboratory to study genotype (or at least at a level quite close to genotype) rather than inferring it from select phenotypes in a very few model systems, next-gen technologies permit any population genetics laboratory to study full genomic variation rather than inferring relationships from a few relative well-characterized genes. Next-gen technologies remove the incredibly laborious and expensive processes of library construction and physical mapping that are the hallmarks of hierarchical genomic sequencing. Whereas 5 or 10 years ago it would have taken hundreds of individuals months of work and millions of dollars to obtain a genome sequence, today, the same sequence can be obtained by a single individual in a couple of weeks for a few thousand dollars. Next-gen technology truly brings genomics to the masses by bringing the cost and effort down to a level that can be supported by a moderate grant. It also effectively knocks model systems off their pedestal by making it possible to perform sophisticated comparative and functional genomic studies on nearly any organism or natural community.

The first next-gen sequencer was the Roche 454 Genome Sequencer (originally, the 454 Life Sciences GS-20; Fig. 7.5) (Rothberg and Leamon, 2008). This revolutionary technology hit the commercial market in 2005 and produced ~20 Mb of data with read lengths of ~110 bp during an 8-h run. As of January 2009, the platform (GS FLX Titanium series) produced reads of >450 bp with a total throughput of 400–600 Mb/10-h run. This pyrosequencing platform first binds millions of sheared single-stranded DNA fragments to agarose beads. Each fragment is amplified in parallel in its own microreaction vessel, produced by mixing oil with the aqueous reagents in what is called emulsion PCR.

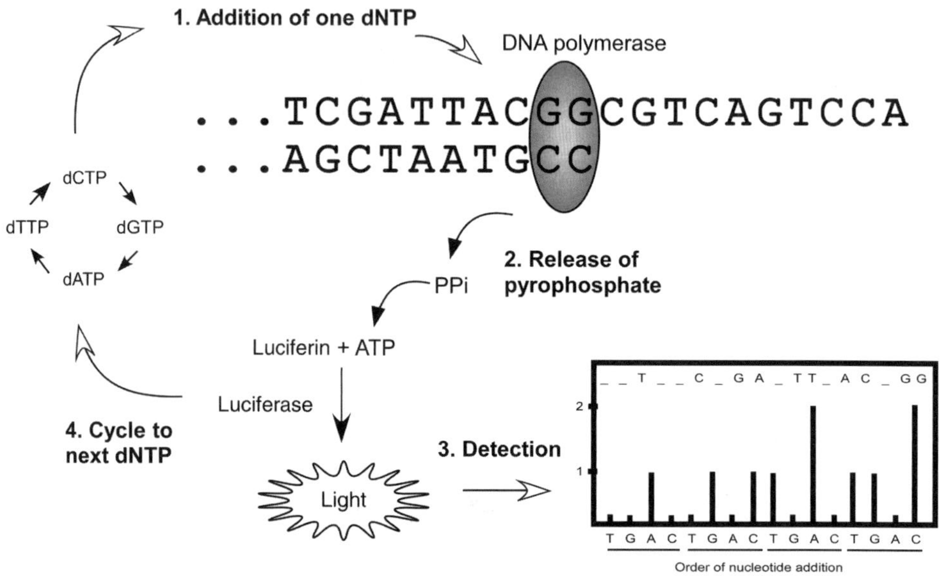

Figure 7.5 Pyrosequencing/454. (1) Sequencing begins with the addition of one of the four dinucleotide triphosphates (dNTPs) to the sequencing reaction. (2) The successful incorporation of the specific dNTP results in the release of a pyrophosphate (PPi), which is subsequently converted to ATP. The resulting ATP drives the luciferase-mediated conversion of luciferin to oxyluciferin, producing light that is detected (3). The light produced is proportional to the amount of ATP (and thus the amount of PPi) produced, such that the incorporation of two or three of the same base will produce double and triple the amount of light, respectively. Extra dNTPs not used in the reaction are then eliminated with apyrase; the next dNTP is added (4); and the cycle is repeated.

The beads, which are now each holding millions of copies of one DNA fragment, are then loaded onto a picotiter plate with wells sized to hold single beads and the enzymes needed for DNA sequencing. The actual sequencing process proceeds by monitoring light flashes produced by a luciferase-based reaction with a phosphate molecule released after the incorporation of one or more nucleotides into the complementary DNA strand. The sequential addition of each nucleotide to the picotiter plate makes it possible to track the identity of the incorporated nucleotide.

The Illumina (Solexa) Genome Analyzer became widely available at the beginning of 2007, producing ~1 Gb of data in 36-bp reads (Fig. 7.6). As of January 2009, it is in its second major iteration and can produce >12 Gb of data with read lengths of 72 bp or longer during a 1.5-week run. Some genome centers have even reportedly pushed the machine to >120 bp reads, but these protocols are not widely available. Solexa sequencing starts by ligating adapters to sheared DNA fragments for PCR amplification. The library is then fixed to a glass flow cell, and each single molecule amplified over 1000-fold into a clonal cluster through a process called bridge amplification. The clusters are then sequenced by successive single-base extensions from a sequencing primer. Each nucleotide carries one of four fluorescent labels and is blocked with a 3′-OH terminator to ensure only one base is incorporated per cycle. The slide is imaged via laser excitation to determine which base has been incorporated and then deblocked for the next cycle. The sequence of each fragment is tracked based on the absolute position of the clonal cluster on the flow cell.

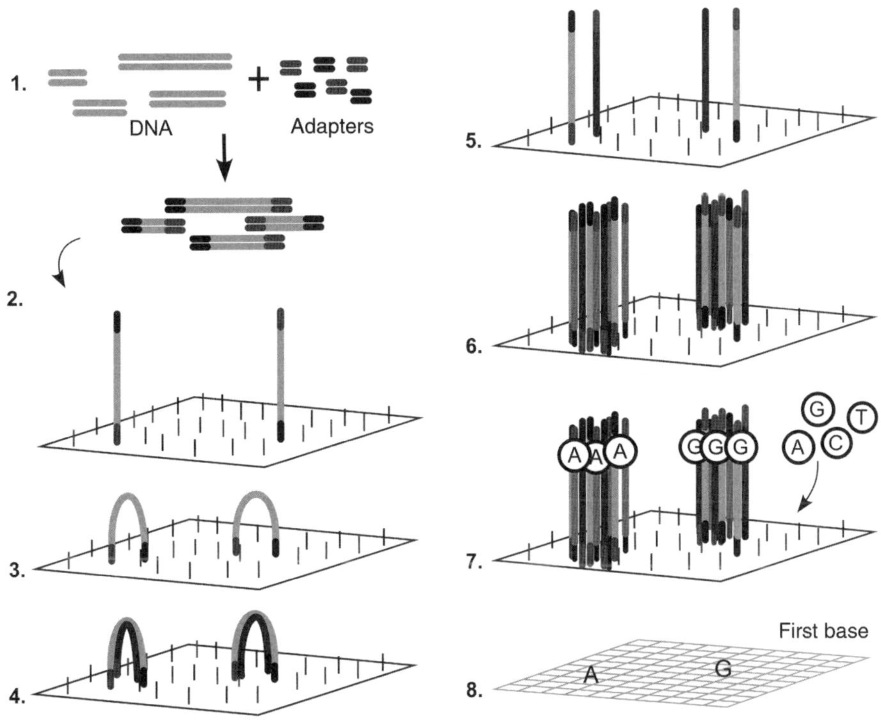

Figure 7.6 Solexa/Illumina sequencing. (1) Adapters are ligated to the template DNA. (2) Single-stranded fragments are affixed onto the inside surface of flow cells. The surface also contains a lawn of primers that are complementary to the adapters. (3) Unlabeled dinucleotide triphosphates (dNTPs) and enzyme are added, initiating bridge amplification. (4) Double-stranded fragments are created. (5) Denaturation creates single-stranded fragments, which serve as the template for the next round of amplification. (6) Clusters of double-stranded DNA are generated in each channel, and (7) are then subjected to a reaction containing all four labeled reversible terminators. Laser excitation captures the fluorescence emitted by the incorporated base, and (8) the identity of the base is recorded. This process is repeated until the sequence is complete.

The Applied Biosciences Sequencing by Oligo Ligation and Detection (SOLiD) system was released near the end of 2007 and uses a unique ligation-based approach for next-gen sequencing (Fig. 7.7). As of January 2009, the SOLiD system can produce >20 Gb of data with 35-bp read lengths. The higher throughput compared to the Solexa system is largely due to the ability of this platform to interrogate two flow cells simultaneously. As with the 454 systems, a SOLiD library is prepared by emulsion PCR amplification of DNA fragments fixed to microbeads. These beads are randomly deposited on a glass flow cell where they are sequenced through the addition of partly degenerate 8-mer oligos carrying 5′ linked fluorescent tags that uniquely correspond to the two bases of the 8-mer. When an 8-mer hybridizes to the sequence adjacent to the primer, the two are ligated together, and the 5′ bases of the 8-mer are chemically cleaved along with the fluorescent tag that is laser detected and recorded. The process proceeds in steps so that every fifth base is interrogated. At the end of the run, the synthesized fragment is removed and a second round begins with a primer complementary to the $n - 1$ position. This continues for a total of five rounds. A major advantage of the SOLiD system is that each base is effectively interrogated twice due to the 2-base encoding, so there is inherent error checking, which reduces the error rate.

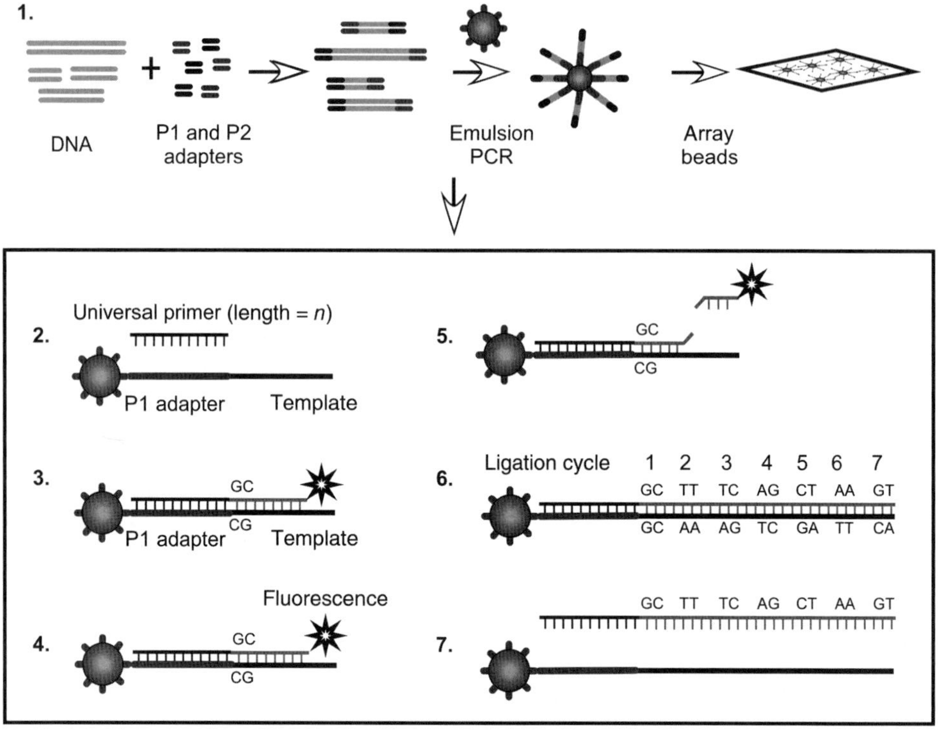

Figure 7.7 ABI SOLiD sequencing. (1) Preparation of the library begins with the ligation of P1 and P2 adapters to the template DNA. Emulsion PCR follows, in which a single-stranded DNA molecule is bound to a bead, and amplified using adapter primers. Each bead, which holds only one specific nucleic acid fragment, is then arrayed in flow cells, and is processed as a separate sequencing reaction. (2) The labeling reaction of each nucleic acid species involves annealing a primer, and (3) hybridizing and ligating with a mixture of fluorescent oligonucleotides. The oligonucleotides have one of 16 specific dinucleotides in the last two 3′ end bases and hold one of four fluorescent dyes. (4) Successful ligation of one of these oligos is followed by detection of the specific fluor bound. (5) Removal of the fluor proceeds via cleavage of the three 5′ bases, leaving a 5-base probe. (6) The entire process is repeated five to seven times, after which (7) the initial primer and all ligated fragments are removed. New primers having lengths of $n-1$, $n-2$, $n-3$, and $n-4$ are used in succession, allowing each base to be queried by two different oligonucleotides.

The newcomer in the field could rightfully be called the first third-generation (or next-next gen) sequencer—the Helicos Genetic Analysis Systems using the HeliScope Single Molecule Sequencer. While information on this system is still limited as of January 2009, company literature claims that a routine run will produce 21–28 Gb of data with an average read length of 30–35 bp, during the course of an 8-day run. Expectations for this platform are very high, with the assumption that throughput will increase dramatically. What Helicos terms "true single-molecule sequencing" involves fixing fragmented DNA to a surface and directly sequencing without amplification using a sequencing-by-synthesis process. Fluorescently labeled nucleotides are added sequentially and are incorporated into the growing complementary strand where they are interrogated by laser excitation. The fluorescent label is removed before the start of each cycle, and nucleotide incorporation is track based for each individual DNA molecule.

A number of other emerging technologies are also moving forward. One of the most hotly anticipated is the Pacific Biosystems Single-Molecule Real-Time (SMRT) DNA sequencing technology, in which the DNA synthesis of a single DNA molecule by a polymerase tethered to a chip is monitored. This technology has been made possible by developments in semiconductor manufacturing, where a hole that is only tens of nanometers in diameter is made in a metal film deposited on a silicon dioxide surface. The hole acts as a reaction and detection vessel holding a single DNA polymerase in a volume of only 20 zL (10^{-21} L). In theory, fluorescently labeled nucleotides can be incorporated at speeds of tens per second, resulting in extremely long reads produced in minutes. Pacific Biosystems claims that its system can be run to maximize throughput or read length, with individual reads exceeding those obtainable through Sanger sequencing. The SMRT system is scheduled for a commercial launch in the second half of 2010.

A novel approach is being pursued by ZS Genetics, an organization working on an electron microscopy-based sequencing system whereby single DNA strands labeled with single-atom labels such as iodated or brominated nucleotides are stretched on a slide and are visualized directly. By using this approach, they have obtained read lengths of >10 kb (W. Glover, pers. comm.).

Genomics innovation is advancing so rapidly that this discussion will be far out of date by the time it is published. Nevertheless, one very exciting point is very clear—the age of next-gen genomics moves genome analysis out of the genome center and makes it available to nearly any researcher. Relatively low-cost genome analyzers are becoming more common, and as we move into third-generation and even look forward to fourth-generation technology, it is apparent that data acquisition will become less and less of a stumbling block. The real challenge moves from data collection to data management and analysis.

7.7 NEXT-GEN GENOMIC DATA ANALYSIS

7.7.1 Data, Data Everywhere

One of the most challenging and persistent issues facing next-gen sequencing is what to do with all the data. Each run of an Illumina Solexa GA-II produces over a terabyte of raw data and gigabases of raw sequences typically in the format of a multi-fasta file with tens of millions of short sequence entries. In this section, we will briefly discuss some of the tools used for next-gen sequence analysis. We will not venture into the world of population genetics data analysis software since to our knowledge, no tools have been specifically designed to work with population genomic (as opposed to population genetic) data. Instead, we will focus on the tools needed to work directly with the next-gen data up to the point of extracting information on genetic variation. We will highlight some of the most popular applications but unfortunately do not have the space to do justice to all of the powerful and sophisticated tools available. Further, new applications and methods are being developed at a tremendous pace, making it an exciting, yet difficult field to cover comprehensively. We also refer the reader to an excellent summary of next-gen sequence analysis software curated on the SEQanswer.com next-gen sequencing forum (http://seqanswers.com/forums/showthread.php?t=43).

It is illustrative to compare the growth in DNA sequence data in GenBank versus the increase in computer power and storage over the same time period. Moore's law describes the projected and realized doubling in computing power (and hard drive storage space) every 2 years. Figure 7.8 shows the total number of base pairs stored in GenBank from 1982 through 2008 on a semilog scale (combining both the GenBank data with the whole genome sequence (WGS) data) plotted along with the expectation given Moore's law

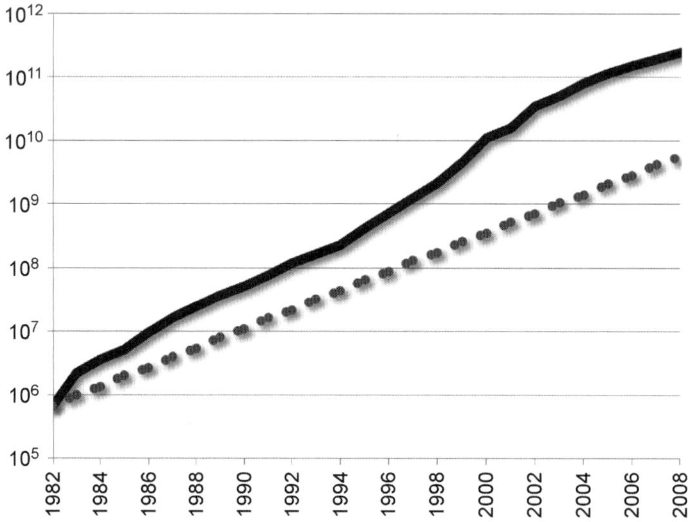

Figure 7.8 Comparison between the growth of Genbank and the growth of computing power. The solid line is the total number of base pairs (including whole genomes) stored in Genbank over the 26-year period spanning 1982–2008, excluding data in the short read archive. The dashed line indicates the projected and realized growth in computing power based on the number of transistor per chip as described by Moore's law. While computing power doubles every 2 years, GenBank doubles approximately every 1.5 years.

starting at the same initial value. While computing power and storage double every 2 years, GenBank doubles approximately every 1.5 years, and certain periods have seen GenBank increase over fivefold in a 2-year period. It is important to realize that these GenBank statistics are largely pre-next-gen data collection and include none of the data currently deposited in the NCBI Short Read Archive. As more and more laboratories collect gigabases of data per run, the discrepancy between computing power and data generation will grow at a tremendous rate.

One never wants to complain about having data, but what do you do with files so large that they take a full day to download and, once downloaded, cannot be opened on a typical desktop machine? Clearly, the next-gen revolution has created nearly as many information technology (IT) problems as biological advances. The first challenge is simple data management, since next-gen data requires access to computing clusters for data analysis and disk arrays for storage. Even if the primary data files, which can exceed 1 Tb in size, are not kept in long-term storage, significant computational infrastructure is needed for primary data extraction and analysis, and robust databases and laboratory information management systems are required for data management and retrieval. All of these demand a significant investment in and commitment to IT infrastructure and personnel.

While dealing with data management issues can be expensive, its implementation is relatively straightforward. But data management is just the beginning, as it makes little sense to collect biological data if it will not be used to address a biological question. Lincoln Stein (2008) describes some of the issues and potential solutions to biological cyberinfrastructure problems with reference to the data, computational, communication, and human infrastructure. He proposes that the best long-term solution involves the integration of community annotation systems with grid systems that provide access to common web-based databases, services, and resource that can be invoked programmatically. These need to be linked via web service links that speak a common language and use common semantics so that disparate and diverse databases and services can be integrated. An example of just such a web service is BioMOBY, which is an open source, object-driven query system with defined object and query ontologies (Wilkinson and Links, 2002; Wilkinson et al., 2005). BioMOBY aims to overcome the non-interoperability of different analytical tools by permitting users to link diverse data sets and analyses through an intuitive interface. Assuming that all the systems speak the same language, these analyses can be performed irrespective of the nature or location of the data sets or analysis tools.

7.7.2 Genome Assembly

Bacterial population genomic studies, by definition, assess genomic variation among multiple strains. Often these new sequenced genomes are first aligned against a well-validated and annotated reference genome prior to SNP calling. This task can be performed in one of two ways. First, the reads (often as short as 25–36 bp) can be aligned against the reference genome directly, an approach known as read mapping. Alternatively, the sequence reads can be assembled de novo first, with the resulting contigs then being aligned against the reference genome.

Alignment to a Reference Genome (Read Mapping)

Read mapping is relatively straightforward and well supported for next-gen sequence data. In fact, it was the only approach that was supported upon the release of the first sequencing platforms. Short sequence alignment is a relatively simple algorithmic problem but becomes computationally intensive when attempting to align millions or tens of millions of short reads to the reference genome. While many applications can perform this task, the specific algorithm used can determine whether the assembly will be completed in hours or years. Another issue is whether the aligner can exploit paired-end data, where two sequencing reads are generated from the same template, since paired-end data can enhance the accuracy of genome assembly substantially (Miller et al., 2008). As of January 2009, almost two dozen applications have been developed for the task, although only a few will be described here.

One of the first applications developed for read mapping was MUMmer (Schatz et al., 2007), which uses a suffix tree data structure for very efficient mapping of short exact matches that are extended with a seed-and-extend strategy similar to BLAST. MUMmer is a very feature-rich package with modules optimized for matching highly similar sequences, highly divergent sequences, draft sequences to a finished reference sequence, substitution detection, and repeat identification. MUMmer can work with both nucleotide and protein sequences and has some visualization capabilities.

Efficient Large-Scale Alignment of Nucleotide Databases (ELAND), which uses Dirichlet's principle or the pigeonhole principle, was developed by Solexa and is implemented in the Illumina Solexa GA analysis pipeline. In addition to having been benchmarked as one of the fastest assemblers available, this application reports the number of alternative matches for repetitive reads, aligns reads with one or two mismatches, and can handle paired-end data. The downside of ELAND is that it cannot handle indels and cannot align reads >32 bp in its native form. Similar approaches are taken by SeqMap (Jiang and Wong, 2008), RMAP (Smith et al., 2008), and SOAP (Li et al., 2008), which can map longer reads with greater numbers of substitutions or indels. A new player is Bowtie (http://bowtie-bio.sourceforge.net), which is reportedly substantially faster than other aligners and is capable of supporting reads of up to 1024 bp. Unfortunately, Bowtie does not currently support indel or paired-end mapping, or ABI SOLiD "color space" alignments.

De Novo Assembly

While aligning raw reads against a reference is unquestionably useful, most population genomic studies will not have access to a reference genome requiring entirely de novo assembly of the raw reads. This is a difficult computational task even for "simple" bacterial genomes due to the presence of repetitive sequences and regions of low information content. Nevertheless, a number of de novo assemblers have been developed that can rapidly generate robust draft genomes. Draft bacterial genomes may be encoded on hundreds or even thousands of contigs, with the ends of the contigs commonly falling in repetitive insertion element sequences. While these assemblies cannot make a closed and polished genome, they are acceptable for the vast majority of evolutionary and functional analyses assuming the median site coverage is high enough to ensure reliable base calling. Of course while polymorphism discovery and analysis is possible with draft genomes, questions related to gene synteny and overall genome structure can be much more difficult to address.

One of the most successful de novo assemblers is Velvet (Jeck et al., 2007; Warren et al., 2007; Zerbino and Birney, 2007; Salzberg et al., 2008; Zerbino and Birney, 2008), which builds contigs from k-mer length word overlaps among reads using de Bruijn graph methods. This assembler works with both short and long reads as well as with single and paired-end data. It permits the user to balance specificity with sensitivity by adjusting the k-mer hash length. Exact DE Novo Assembler (Edena) is another very promising de novo assembler developed strictly for very short read data (Hernandez et al., 2008). This program uses an overlap-layout-assembly approach similar to that used in traditional assembly algorithms, but speeds up processing by accepting only exact matches. Velvet and Edena seem to perform similarly, although Edena does not yet assemble paired-end data. A newcomer in the next-gen assembly arena is Mimicking Intelligent Read Assembly (MIRA), which can perform hybrid de novo assemblies using reads generated from Roche 454, Illumina Solexa, and even traditional Sanger sequencing (http://chevreux.org/projects_mira.html).

A unique "gene-boosted" approach that essentially falls between read-mapping alignment and de novo assembly has been proposed by Salzberg and colleagues (2008). The approach uses related genomes for an initial comparative read-mapped assembly and then fills in gaps by identifying genes that span two or more contigs. The translated protein sequences of these spanning genes are then queried by translated reads that have not been already assigned to a contig to find reads to fill the assembly gaps. The authors also were able to map Velvet-generated contigs (de novo assembly) to their reference genome prior to gene boosting with excellent results. In principle, there is no reason this approach cannot be applied to strict de novo assembly by BLASTing all contig ends to identify those that have syntenic relationships on a database sequence. In our experience with bacterial genomes, the vast majority of contig ends correspond to insertion element-associated sequences.

7.7.3 Genome Annotation and Visualization

Genomes are generally of very little use until they have been annotated. Fortunately, a number of new and largely automated systems have been developed to facilitate this often extremely challenging and time-consuming task. The Rapid Annotation using Subsystems Technology (RAST) server automatically generates rapid (usually within a couple of days) and high-quality bacterial genome annotations (Aziz et al., 2008). This open source, cooperative, and curated project is structured around a comparative genomics environment called the SEED (http://www.theseed.org/) and is largely run by groups at Argonne National Laboratory and the University of Chicago. RAST enhances its annotations with curated subsystems that organize genes involved in common functional roles or biological processes. The annotation, which can be performed on both finished as well as draft genomes, provides a mapping of genes to subsystems and a metabolic reconstruction. The site also provides basic visualization and analysis tools for further refining the annotation.

GenColors (Romualdi et al., 2005, 2007) is a web-based analysis and database system designed for improving the annotation of prokaryotic genomes using strong comparative tools. It permits the integration of user data with annotated genomic sequences from GenBank, and provides excellent import and export tools to assist the preparation of GenBank files. Some of the comparison and visualization tools available include reciprocal BLAST analysis, synteny comparison, COG analysis, HGT prediction, and genome plots.

System for Automatic Bacterial (genome) Integrated Annotation (SABIA) is a freely distributed web server that provides both contig assembly and genome annotation, although the contig assembly does not appear to support next-gen data at this point. The annotation pipeline uses most of the major publicly available tools and databases, including BLAST, GO, and PSORT analyses, and the KEGG and COG databases to name just a few.

Many of the tools mentioned above have integrated genome browsers, but there are also many other outstanding options for viewing and working with genomes. Perhaps the best known is the UCSC Genome Browser (Kent et al., 2002), which provides a highly flexible framework for genome visualization via its incorporation and management of standard and custom annotation tracks. Tracks can display almost any type of additional data and can be linked to supplementary off-site databases. Strong searching and filtering tools are also incorporated. Another option is the Java-based Argo Genome Browser, which can be run as an applet or webstart application (http://www.broad.mit.edu/annotation/argo/). This open source browser permits user annotations and supports a wide variety of annotation tracks. It also comes with a comparative viewer that permits pairwise alignments between one or more query genomes and a reference genome (Engels et al., 2006).

The Generic Genome Browser (GBrowser) is a web server application designed for the Generic Model Organism Database (GMOD) project (Stein et al., 2002). This collection of open source tools supports the management and analysis of large-scale genome projects and is used and supported by a large number of organism-centric research consortia. As such, this highly configurable system has a large and active developer and user base, and a wide variety of associated tools. A stand-alone option for genome viewing is EagleView, a next-gen assembly viewer combining data integration and visualization (Huang and Marth, 2008). EagleView can view mixed read data from multiple next-gen platforms, can handle large data sets and annotation features, has strong navigation and viewing options, and can run on Windows, Mac, or Linux systems.

7.7.4 Comparative Genome Alignment

Many analyses require accurately aligned whole genome sequences, and a number of groups have developed powerful tools to perform this computationally difficult analysis.

Mauve is a multiple genome alignment tool that rapidly identifies and displays conserved genomic regions, rearrangements, inversions, and sequence breakpoints across multiple genomes (Darling et al., 2004). It does this by identifying locally colinear blocks, which are homologous sequence regions that contain no rearrangements in two or more of the query genomes under study. Mauve is enhanced by a number of software tools that have been designed to further analyze Mauve-based alignment, such as tools for identifying genomic islands and mobile DNA.

The SynView comparative genome analyzer is a visualization tool developed for the GBrowse framework (Pan et al., 2005; Brendel et al., 2007). This tool has two components: a web-based front end and a relational database that stores precomputed alignments of the query genome against a reference genome as well as genome annotation information. CoreAligner is a new dynamic programming-based algorithm designed to identify the core bacterial genome structure by finding the maximal genomic regions with conserved synteny, which are indicative of vertical inheritance (Uchiyama, 2008). This tool will likely be particularly useful for identifying regions useful for population genomic analysis.

7.7.5 Polymorphism Calling (SNP Discovery)

Polymorphism detection and calling is a critical step in population genomic and association-based studies (Buckingham, 2008), and the ease with which comparative data can be produced with next-gen approaches has spurred the development of many new methods and applications. ssahaSNP, from the Wellcome Trust Sanger Institute, detects SNPs and indels by read mapping to a reference genome. It filters out highly repetitive reads by ignoring high-frequency k-mer words. Pairwise alignment scores are used to find the best match for unique and low repetitive reads, and SNPs are identified based on the quality value of the variable bases as well as the quality values in the neighboring bases. Indels are checked to determine if they are mapped to multiple reads to increase call confidence.

Maq is primarily a read-mapping tool but can also measure the alignment error probability for individual reads, call SNP and indel polymorphisms with associated Phred quality scores, find large deletions and translocations, and even evaluate copy number variation based on read coverage (Li, 2008). Gobor Marth has also updated his PolyBayes SNP discovery tools for working with next-gen data. PbShort (http://bioinformatics.bc.edu/marthlab/PbShort) is only currently been released in beta form but will presumably integrate multiple sequence alignment, paralog detection, and SNP detection into one tool similar to the original PolyBayes application (Marth et al., 1999).

7.8 CONCLUSIONS/FUTURE PROSPECTS

So, we now return to our initial question: Is what's true for Sanger also true for Solexa, only more so? In other words, is next-gen population genomics simply classical population genetics writ large, or have we moved into a fundamentally new age where we can address questions not even envisioned earlier? We are certainly not foolish enough to pretend to have the answer to this question, but it is clear that having the ability to interrogate full genomes in real time will change the face of population analyses. We have dispensed with beanbag genetics, but that does not necessarily mean that Fisher, Wright, and Haldane no longer have relevance and are destined for the bin as well. The fundamental questions of population genetics still stand, but for the first time in the history of the field, the ability to collect data has far outstripped the development of theory to frame and interpret that data. A field driven for so long by theory now is being driven by sequencers on steroids and databases that are shooting past the terabyte range into petabytes, exabytes, and zettabytes. The transition from theory-rich/data-poor science to theory-poor/data-rich science certainly does pose new challenges, but are there other dangers? Some will say that an elegant and sophisticated science is being replaced by brute force sample grinding and data crunching. Nevertheless, the question really should be: Are we making progress? Do we understand microbes better today than we did yesterday? And the answer to this is certainly yes.

REFERENCES

Achtman, M. (2004) Population structure of pathogenic bacteria revisited. *International Journal of Medical Microbiology* **294**(2-3), 67–73.

Achtman, M. (2008) Evolution, population structure, and phylogeography of genetically monomorphic bacterial pathogens. *Annual Review of Microbiology* **62**, 53–70.

Allen, E. E., Tyson, G. W. et al. (2007) Genome dynamics in a natural archaeal population. *Proceedings of the National Academy of Sciences of the United States of America* **104**(6), 1883–1888.

Arnold, D. L., Pitman, A. et al. (2003) Pathogenicity and other genomic islands in plant pathogenic bacteria. *Molecular Plant Pathology* **4**(5), 407–420.

Aziz, R. K., Bartels, D. et al. (2008) The RAST server: Rapid Annotations using Subsystems Technology. *BMC Genomics* **9**, 75.

Bagnoli, F., Moschioni, M. et al. (2008) A second pilus type in *Streptococcus pneumoniae* is prevalent in emerging serotypes and mediates adhesion to host cells. *Journal of Bacteriology* **190**(15), 5480–5492.

Bandelt, H. J. and Dress, A. W. (1992) Split decomposition: A new and useful approach to phylogenetic analysis of distance data. *Molecular Phylogenetics and Evolution* **1**(3), 242–252.

Baumberg, S., Young, J. P. W. et al., eds. (1995) Population genetics of bacteria. Symposium of the Society for General Microbiology, Cambridge University Press, Cambridge, UK.

Black, W. C., Baer, C. F. et al. (2001) Population genomics: Genome-wide sampling of insect populations. *Annual Review of Entomology* **46**, 441–469.

Brendel, V., Kurtz, S. et al. (2007) Visualization of syntenic relationships with SynBrowse. *Methods in Molecular Biology* **396**, 153–163.

Buckingham, S. D. (2008) Scientific software: Seeing the SNPs between us. *Nature Methods* **5**(10), 903–908.

Caugant, D. A., Mocca, L. F. et al. (1987) Structure of *Neisseria meningitidis* populations in relation to serogroup, serotype, and outer membrane protein pattern. *Journal of Bacteriology* **169**(6), 2781–2792.

Chain, P. S., Carniel, E. et al. (2004) Insights into the evolution of *Yersinia pestis* through whole-genome comparison with *Yersinia pseudotuberculosis*. *Proceedings of the National Academy of Sciences of the United States of America* **101**(38), 13826–13831.

Chau, T. T., Campbell, J. I. et al. (2007) Antimicrobial drug resistance of *Salmonella enterica* serovar Typhi in Asia and molecular mechanism of reduced susceptibility to the fluoroquinolones. *Antimicrobial Agents and Chemotherapy* **51**(12), 4315–4323.

Cover, T. L. (1996) The vacuolating cytotoxin of *Helicobacter pylori*. *Molecular Microbiology* **20**(2), 241–246.

Darling, A. C. E., Mau, B. et al. (2004) Mauve: Multiple alignment of conserved genomic sequence with rearrangements. *Genome Research* **14**(7), 1394–1403.

DeLong, E. F. (2002) Microbial population genomics and ecology. *Current Opinion in Microbiology* **5**(5), 520–524.

DeLong, E. F. (2004) Microbial population genomics and ecology: The road ahead. *Environmental Microbiology* **6**(9), 875–878.

Dobrindt, U. and Hacker, J. (2001) Whole genome plasticity in pathogenic bacteria. *Current Opinion in Microbiology* **4**(5), 550–557.

Drake, J. W., Charlesworth, B., et al. (1998) Rates of spontaneous mutation. *Genetics* **148**(4), 1667–1686.

Endo, T., Ikeo, K. et al. (1996) Large-scale search for genes on which positive selection may operate. *Molecular Biology and Evolution* **13**(5), 685–690.

Engels, R., Yu, T. et al. (2006) Combo: A whole genome comparative browser. *Bioinformatics* **22**(14), 1782–1783.

Enright, M. C. and Spratt, B. G. (1998) A multilocus sequence typing scheme for *Streptococcus pneumoniae*: Identification of clones associated with serious invasive disease. *Microbiology* **144**(Pt 11), 3049–3060.

Fleischmann, R. D., Adams, M. D. et al. (1995) Whole-genome random sequencing and assembly of *Haemophilus influenzae* Rd. *Science* **269**(5223), 496–512.

Fox, G. E., Pechman, K. R. et al. (1977) Comparative cataloging of 16S ribosomal ribonucleic acid-molecular approach to procaryotic systematics. *International Journal of Systematic Bacteriology* **27**(1), 44–57.

GRIFFITH, F. (1928) The significance of pneumococcal types. *Journal of Hygiene* **27,** 113–159.

GULCHER, J. and STEFANSSON, K. (1998) Population genomics: Laying the groundwork for genetic disease modeling and targeting. *Clinical Chemistry and Laboratory Medicine* **36**(8), 523–527.

GUTTMAN, D. S. (1997) Recombination and clonality in natural population of *Escherichia coli. Trends in Ecology and Evolution* **12**(1), 16–22.

GUTTMAN, D. S. and DYKHUIZEN, D. E. (1994) Clonal divergence in *Escherichia coli* as a result of recombination, not mutation. *Science* **266,** 1380–1383.

HACKER, J. and CARNIEL, E. (2001) Ecological fitness, genomic islands and bacterial pathogenicity—A Darwinian view of the evolution of microbes. *EMBO Reports* **2**(5), 376–381.

HACKER, J. and KAPER, J. B. (2000) Pathogenicity islands and the evolution of microbes. *Annual Review of Microbiology* **54,** 641–679.

HAGBLOM, P., SEGAL, E. et al. (1985) Intragenic recombination leads to pilus antigenic variation in *Neisseria gonorrhoeae. Nature* **315**(6015), 156–158.

HAVA, D. and CAMILLI A. (2002) Large-scale identification of serotype 4 *Streptococcus pneumoniae* virulence factors. *Molecular Microbiology* **45**(5), 1389–1405.

HAVA, D. L., HEMSLEY, C. J. et al. (2003) Transcriptional regulation in the *Streptococcus pneumoniae* RlrA pathogenicity islet by RlrA. *Journal of Bacteriology* **185**(2), 413–421.

HENRIQUES-NORMARK, B., BLOMBERG, C. et al. (2008) The rise and fall of bacterial clones: *Streptococcus pneumoniae. Nature Reviews. Microbiology* **6**(11), 827–837.

HERNANDEZ, D., FRANCOIS, P. et al. (2008) *De novo* bacterial genome sequencing: Millions of very short reads assembled on a desktop computer. *Genome Research* **18**(5), 802–809.

HILL, W. G. and ROBERTSON, A. I. (1966) Effect of linkage on limits to artificial selection. *Genetical Research* **8**(3), 269–294.

HINCHLIFFE, S. J., ISHERWOOD, K. E. et al. (2003) Application of DNA microarrays to study the evolutionary genomics of *Yersinia pestis* and *Yersinia pseudotuberculosis. Genome Research* **13**(9), 2018–2029.

HOLT, K. E., PARKHILL, J. et al. (2008) High-throughput sequencing provides insights into genome variation and evolution in *Salmonella typhi. Nature Genetics* **40**(8), 987–993.

HUANG, W. and MARTH, G. (2008) EagleView: A genome assembly viewer for next-generation sequencing technologies. *Genome Research* **18**(9), 1538–1543.

HUSON, D. H. (1998) SplitsTree: Analyzing and visualizing evolutionary data. *Bioinformatics* **14**(1), 68–73.

JECK, W. R., REINHARDT, J. A. et al. (2007) Extending assembly of short DNA sequences to handle error. *Bioinformatics* **23**(21), 2942–2944.

JIANG, H. and WONG, W. H. (2008) SeqMap: Mapping massive amount of oligonucleotides to the genome. *Bioinformatics* **24**(20), 2395–2396.

KAHVEJIAN, A., QUACKENBUSH, J. et al. (2008) What would you do if you could sequence everything? *Nature Biotechnology* **26**(10), 1125–1133.

KENT, W. J., SUGNET, C. W. et al. (2002) The human genome browser at UCSC. *Genome Research* **12**(6), 996–1006.

KOONIN, E. V. and WOLF, Y. I. (2008) Genomics of bacteria and archaea: The emerging dynamic view of the prokaryotic world. *Nucleic Acids Research* **36**(21), 6688–6719.

LANDER, E. S., LINTON, L. M. et al. (2001) Initial sequencing and analysis of the human genome. *Nature* **409**(6822), 860–921.

LEAVIS, H., TOP, J. et al. (2004) A novel putative enterococcal pathogenicity island linked to the *esp* virulence gene of *Enterococcus faecium* and associated with epidemicity. *Journal of Bacteriology* **186**(3), 672–682.

LEAVIS, H. L., BONTEN, M. J. M. et al. (2006) Identification of high-risk enterococcal clonal complexes: Global dispersion and antibiotic resistance. *Current Opinion in Microbiology* **9**(5), 454–460.

LENSKI, R. E., ROSE, M. R. et al. (1991) Long-term experimental evolution in *Escherichia coli.* I. Adaptation and divergence during 2,000 generations. *American Naturalist* **138,** 1315–1341.

LEWONTIN, R. C. and HUBBY, J. L. (1966) A molecular approach to the study of genic heterozygosity in natural populations. II. Amount of variation and degree of heterozygosity in natural populations of *Drosophila pseudoobscura. Genetics* **54**(2), 595–609.

LI, H. (2008) Maq: Mapping and Assembly with Qualities. Software for Short-read Mapping Assemblies.

LI, R., LI, Y. et al. (2008) SOAP: Short oligonucleotide alignment program. *Bioinformatics* **24**(5), 713–414.

MA, W. and GUTTMAN, D. S. (2008) Evolution of prokaryotic and eukaryotic virulence effectors. *Current Opinion in Plant Biology* **11**(4), 412–419.

MAIDEN, M. C. J. (2008) Population genomics: Diversity and virulence in the *Neisseria. Current Opinion in Microbiology* **11**(5), 467–471.

MAIDEN, M. C. J., BYGRAVES, J. A. et al. (1998) Multilocus sequence typing: A portable approach to the identification of clones within populations of pathogenic microorganisms. *Proceedings of the National Academy of Sciences of the United States of America* **95**(6), 3140–3145.

MARTH, G. T., KORF, I. et al. (1999) A general approach to single-nucleotide polymorphism discovery. *Nature Genetics* **23**(4), 452–456.

MAYNARD SMITH, J. (1991) The population genetics of bacteria. *Proceedings of the Royal Society of London. Series B, Biological Sciences* **245,** 37–41.

MAYNARD SMITH, J. (1999) The detection and measurement of recombination from sequence data. *Genetics* **153**(2), 1021–1027.

MAYNARD SMITH, J., DOWSON, C. G. et al. (1991) Localized sex in bacteria. *Nature* **349,** 29–31.

MAYNARD SMITH, J., FEIL, E. J. et al. (2000) Population structure and evolutionary dynamics of pathogenic bacteria. *Bioessays* **22**(12), 1115–1122.

Maynard Smith, J., Smith, N. H. et al. (1993) How clonal are bacteria? *Proceedings of the National Academy of Sciences of the United States of America* **90,** 4384–4388.

McDougal, L. K., Steward, C. D. et al. (2003) Pulsed-field gel electrophoresis typing of oxacillin-resistant *Staphylococcus aureus* isolates from the United States: Establishing a national database. *Journal of Clinical Microbiology* **41**(11)**,** 5113–5120.

Medini, D., Donati, C. et al. (2005) The microbial pan-genome. *Current Opinion in Genetics & Development* **15**(6)**,** 589–594.

Medini, D., Serruto, D. et al. (2008) Microbiology in the post-genomic era. *Nature Reviews. Microbiology* **6**(6)**,** 419–430.

Milkman, R. (1973) Electrophoretic variation in *Escherichia coli* from natural sources. *Science* **182,** 1024–1026.

Miller, J. R., Delcher, A. L. et al. (2008) Aggressive assembly of pyrosequencing reads with mates. *Bioinformatics* **24**(24)**,** 2818–2824.

Mueller, U. G. and Wolfenbarger, L. L. (1999) AFLP genotyping and fingerprinting. *Trends in Ecology & Evolution* **14**(10)**,** 389–394.

Musser, J. M., Bemis, D. A. et al. (1987) Clonal diversity and host distribution in *Bordetella bronchiseptica. Journal of Bacteriology* **169**(6)**,** 2793–2803.

Musser, J. M., Kroll, J. S. et al. (1988) Clonal population structure of encapsulated *Haemophilus influenza. Infection and Immunity* **56,** 1837–1845.

Musser, J. M., Kroll, J. S. et al. (1990) Global genetic-structure and molecular epidemiology of encapsulated *Haemophilus influenzae. Reviews of Infectious Diseases* **12**(1)**,** 75–111.

Ochman, H. and Davalos, L. M. (2006) The nature and dynamics of bacterial genomes. *Science* **311**(5768)**,** 1730–1733.

Ochman, H. and Santos, S. R. (2005) Exploring microbial microevolution with microarrays. *Infection, Genetics and Evolution* **5**(2)**,** 103–108.

Pan, X. K., Stein, L. et al. (2005) SynBrowse: A synteny browser for comparative sequence analysis. *Bioinformatics* **21**(17)**,** 3461–3468.

Parry, C. M., Hien, T. T. et al. (2002) Typhoid fever. *New England Journal of Medicine* **347**(22)**,** 1770–1782.

Perron, G. G., Zasloff, M. et al. (2006) Experimental evolution of resistance to an antimicrobial peptide. *Proceedings of the Royal Society. Series B, Biological Sciences* **273**(1583)**,** 251–256.

Pitman, A. R., Jackson, R. W. et al. (2005) Exposure to host resistance mechanisms drives evolution of bacterial virulence in plants. *Current Biology* **15**(24)**,** 2230–2235.

Pouillot, F., Fayolle, C. et al. (2008) Characterization of chromosomal regions conserved in *Yersinia pseudotuberculosis* and lost by *Yersinia pestis. Infection and Immunity* **76**(10)**,** 4592–4599.

Prevost, G., Jaulhac, B. et al. (1992) DNA fingerprinting by pulsed-field gel-electrophoresis is more effective than ribotyping in distinguishing among methicillin-resistant *Staphylococcus aureus* isolates. *Journal of Clinical Microbiology* **30**(4)**,** 967–973.

Romualdi, A., Felder, M. et al. (2007) GenColors: Annotation and comparative genomics of prokaryotes made easy. *Methods in Molecular Biology* **395**, 75–96.

Romualdi, A., Siddiqui, R. et al. (2005) GenColors: Accelerated comparative analysis and annotation of prokaryotic genomes at various stages of completeness. *Bioinformatics* **21**(18)**,** 3669–3671.

Rossello-Mora, R. and Amann, R. (2001) The species concept for prokaryotes. *FEMS Microbiology Reviews* **25**(1)**,** 39–67.

Rothberg, J. M. and Leamon, J. H. (2008) The development and impact of 454 sequencing. *Nature Biotechnology* **26**(10)**,** 1117–1124.

Roumagnac, P., Weill, F. X. et al. (2006) Evolutionary history of *Salmonella typhi. Science* **314**(5803)**,** 1301–1304.

Rozenfeld, A. F., Arnaud-Haond, S. et al. (2007) Spectrum of genetic diversity and networks of clonal organisms. *Journal of the Royal Society Interface* **4**(17)**,** 1093–1102.

Salaun, L., Audibert, C. et al. (1998) Panmictic structure of *Helicobacter pylori* demonstrated by the comparative study of six genetic markers. *FEMS Microbiology Letters* **161**(2)**,** 231–239.

Salzberg, S. L., Sommer, D. D. et al. (2008) Gene-boosted assembly of a novel bacterial genome from very short reads. *PLoS Computational Biology* **4**(9)**,** e1000186.

Sarkar, S. F., Gordon, J. S. et al. (2006) Comparative genomics of host-specific virulence in *Pseudomonas syringae. Genetics* **174**(4)**,** 1041–1056.

Schatz, M. C., Trapnell, C. et al. (2007) High-throughput sequence alignment using Graphics Processing Units. *BMC Bioinformatics* **8,** 1–10.

Seifert, H. S., Ajioka, R. S. et al. (1988) DNA transformation leads to pilin antigenic variation in *Neisseria gonorrhoeae. Nature* **336**(6197)**,** 392–395.

Selander, R. K., Beltran, P. et al. (1990) Evolutionary genetic relationships of clones of *Salmonella* serovars that cause human typhoid and other enteric fevers. *Infection and Immunity* **58**(7)**,** 2262–2275.

Selander, R. K., Caugant, D. A. et al. (1986) Methods of multilocus enzyme electrophoresis for bacterial population genetics and systematics. *Applied and Environmental Microbiology* **51,** 873–884.

Selander, R. K. and Levin, B. R. (1980) Genetic diversity and structure in *Escherichia coli* populations. *Science* **210,** 545–547.

Selander, R. K., McKinney, R. M. et al. (1985) Genetic Structure of populations of *Legionella pneumophila. Journal of Bacteriology* **163**(3)**,** 1021–1037.

Shankar, N., Baghdayan, A. S. et al. (2002) Modulation of virulence within a pathogenicity island in vancomycin-resistant *Enterococcus faecalis. Nature* **417**(6890)**,** 746–750.

SHANKAR, V., BAGHDAYAN, A. S. et al. (1999) Infection-derived *Enterococcus faecalis* strains are enriched in *esp*, a gene encoding a novel surface protein. *Infection and Immunity* **67**(1), 193–200.

SHEPPARD, S. K., MCCARTHY, N. D. et al. (2008) Convergence of *Campylobacter* species: Implications for bacterial evolution. *Science* **320**(5873), 237–239.

SMITH, A., XUAN, Z. et al. (2008) Using quality scores and longer reads improves accuracy of Solexa read mapping. *BMC Bioinformatics* **9**(1), 128.

SOKURENKO, E. V., CHESNOKOVA, V. et al. (1998) Pathogenic adaptation of *Escherichia coli* by natural variation of the FimH adhesin. *Proceedings of the National Academy of Sciences of the United States of America* **95**(15), 8922–8926.

SOKURENKO, E. V., HASTY, D. L. et al. (1999) Pathoadaptive mutations: Gene loss and variation in bacterial pathogens. *Trends in Microbiology* **7**(5), 191–195.

SOUZA, V., TURNER, P. E. et al. (1997) Long-term experimental evolution in *Escherichia coli*. 5. Effects of recombination with immigrant genotypes on the rate of bacterial evolution. *Journal of Evolutionary Biology* **10**(5), 743–769.

SPRATT, B. G. and MAIDEN, M. C. J. (1999) Bacterial population genetics, evolution and epidemiology. *Philosophical Transactions of the Royal Society of London. Series B, Biological Sciences* **354**(1384), 701–710.

SPRATT, B. G., SMITH, N. H. et al. (1995) The population genetics of the pathogenic *Neisseria*. In *Population Genetics of Bacteria* (eds. S. Baumberg, J. P. W. Young, E. M. H. Wellington, and S. R. Saunders), p. 52. Cambridge University Press, Cambridge, U.K.

SRIVATSAN, A., HAN, Y. et al. (2008) High-precision, whole-genome sequencing of laboratory strains facilitates genetic studies. *PLoS Genetics* **4**(8), e1000139.

STEIN, L. D. (2008) Towards a cyberinfrastructure for the biological sciences: Progress, visions and challenges. *Nature Reviews. Genetics* **9**(9), 678–688.

STEIN, L. D., MUNGALL, C. et al. (2002) The Generic Genome Browser: A building block for a model organism system database. *Genome Research* **12**(10), 1599–1610.

SUERBAUM, S. and JOSENHANS, C. (2007) *Helicobacter pylori* evolution and phenotypic diversification in a changing host. *Nature Reviews. Microbiology* **5**(6), 441–452.

SUERBAUM, S., SMITH, J. M. et al. (1998) Free recombination within *Helicobacter pylori*. *Proceedings of the National Academy of Sciences of the United States of America* **95**(21), 12619–12624.

SUNG, J. M. L., LLOYD, D. H. et al. (2008) *Staphylococcus aureus* host specificity: Comparative genomics of human versus animal isolates by multi-strain microarray. *Microbiology* **154**, 1949–1959.

TETTELIN, H., MASIGNANI, V. et al. (2005) Genome analysis of multiple pathogenic isolates of *Streptococcus agalactiae*: Implications for the microbial "pan-genome." *Proceedings of the National Academy of Sciences of the United States of America* **102**(39), 13950–13955.

TETTELIN, H., RILEY, D. et al. (2008) Comparative genomics: The bacterial pan-genome. *Current Opinion in Microbiology* **11**(5), 472–477.

TIBAYRENC, M. (1999) Toward an integrated genetic epidemiology of parasitic protozoa and other pathogens. *Annual Review of Genetics* **33**, 449–477.

TREANGEN, T. J., AMBUR, O. H. et al. (2008) The impact of the neisserial DNA uptake sequences on genome evolution and stability. *Genome Biology* **9**(3), R60.

UCHIYAMA, I. (2008) Multiple genome alignment for identifying the core structure among moderately related microbial genomes. *BMC Genomics* **9**, 515.

UEDA, K., SEKI, T. et al. (1999) Two distinct mechanisms cause heterogeneity of 16S rRNA. *Journal of Bacteriology* **181**(1), 78–82.

VELICER, G. J., RADDATZ, G. et al. (2006) Comprehensive mutation identification in an evolved bacterial cooperator and its cheating ancestor. *Proceedings of the National Academy of Sciences of the United States of America* **103**(21), 8107–8112.

VENTER, J. C., ADAMS, M. D. et al. (2001) The sequence of the human genome. *Science* **291**(5507), 1304–1351.

WARREN, R. L., SUTTON, G. G. et al. (2007) Assembling millions of short DNA sequences using SSAKE. *Bioinformatics* **23**(4), 500–501.

WEISSMAN, S. J., CHATTOPADHYAY, S. et al. (2006) Clonal analysis reveals high rate of structural mutations in fimbrial adhesins of extraintestinal pathogenic *Escherichia coli*. *Molecular Microbiology* **59**(3), 975–988.

WELCH, R. A., BURLAND, V. et al. (2002) Extensive mosaic structure revealed by the complete genome sequence of uropathogenic *Escherichia coli*. *Proceedings of the National Academy of Sciences of the United States of America* **99**(26), 17020–17024.

WHITAKER, R. J. and BANFIELD, J. F. (2006) Population genomics in natural microbial communities. *Trends in Ecology & Evolution* **21**(9), 508–516.

WILKINSON, M., SCHOOF, H. et al. (2005) BioMOBY successfully integrates distributed heterogeneous bioinformatics web services. The PlaNet exemplar case. *Plant Physiology* **138**(1), 5–17.

WILKINSON, M. D. and LINKS, M. (2002) BioMOBY: An open source biological web services proposal. *Briefings in Bioinformatics* **3**(4), 331–341.

WIRTH, T., FALUSH, D. et al. (2006) Sex and virulence in *Escherichia coli*: An evolutionary perspective. *Molecular Microbiology* **60**(5), 1136–1151.

WOESE, C. R., FOX, G. E. et al. (1975) Conservation of primary structure in 16S ribosomal-RNA. *Nature* **254**(5495), 83–86.

ZERBINO, D. and BIRNEY, E. (2007) Velvet: Sequence assembler for very short reads.

ZERBINO, D. and BIRNEY, E. (2008) Velvet: Algorithms for de novo short read assembly using de Bruijn graphs. *Genome Research* **18**(5), 821–829.

ZHU, P., VAN DER ENDE, A. et al. (2001) Fit genotypes and escape variants of subgroup III *Neisseria meningitidis* during three pandemics of epidemic meningitis. *Proceedings of the National Academy of Sciences of the United States of America* **98**(9), 5234–5239.

ZIEBUHR, W., OHLSEN, K. et al. (1999) Evolution of bacterial pathogenesis. *Cellular and Molecular Life Sciences* **56,** 719–728.

ZWICK, M. E., MCAFEE, F. et al. (2005) Microarray-based resequencing of multiple *Bacillus anthracis* isolates. *Genome Biology* **6**, R10.

Chapter 8

The Use of MLVA and SNP Analysis to Study the Population Genetics of Pathogenic Bacteria

PAUL J. JACKSON

8.1 INTRODUCTION

The study of the population genetics of pathogenic bacteria has benefited significantly from the increasingly rapid and less costly ability to generate direct DNA sequence information from individual isolates. It is now possible to sequence and assemble a high-quality draft sequence of a bacterial genome in just a few days. However, circumstances that require analyses of many different bacteria or that must be fielded to diagnostic laboratories still rely primarily on methods that interrogate a relatively small sampling of a bacterial genome. Consequently, databases of multiple locus variable number tandem repeat analysis (MLVA) and single-nucleotide polymorphisms (SNPs) have been generated and are still being populated with MLVA and SNP profiles for different infectious bacterial pathogens. Such databases allow a rapid characterization of a new or unknown isolate relative to all other isolates for which such profiles are available. In the following chapter, the methods used to initially identify MLVA profiles and SNPs will be described; examples of the application of these methods to generate phylogenetic information for highly virulent bacterial species will be presented; databases in use to allow rapid comparison among isolates will be introduced; and limitations of the different methods will be discussed.

Prior to applying any method to differentiate among members of a population of bacterial isolates, one must have a sufficiently large and diverse collection of isolates to ascertain whether the method to be used to differentiate among the isolates will provide sufficient resolution to be useful for further studies. In the absence of a validated method to demonstrate diversity, this can be problematic. In some species, diversity has been demonstrated by a number of phenotypic or immunologic methods. In *Yersinia pestis*, the causative agent of bubonic plague, isolates were differentiated by their ability to ferment glycerol and to reduce nitrate (Devignat, 1951). In *Bacillus thuringiensis*, a highly diverse insect pathogen of commercial value, diversity is measured by the presence of different

Bacterial Population Genetics in Infectious Disease, Edited by D. Ashley Robinson, Daniel Falush, and Edward J. Feil

insecticidal toxins and diversity in flagellar H-antigen agglutination reactions (Lecadet and Frachon, 1994; Crickmore et al., 1998). Subspecies differentiation in *Francisella tularensis*, the causative agent of tularemia, was done by biochemical analysis (Johansson et al., 2000). Biovar A ferments glycerol and glucose and produces citrulline ureidase, while biovar B ferments only glucose and does not produce this enzyme (Gurycova, 1998). More recently, immunoassays using monoclonal antibodies have been applied to detect and to differentiate among different subtypes (Grunow et al., 2000). Similar methods did not separate different *Bacillus anthracis* isolates into distinctively different groups.

8.2 MLVA AND OTHER DNA FRAGMENT-BASED METHODS

There are a significant number of other pathogenic bacteria besides those outlined in this chapter that have been analyzed using the MLVA approach. The utility of a technique that could separate isolates into more than two categories by analysis of a single locus is obvious. Studies demonstrate that the "mutation frequency" of variable number tandem repeats (VNTRs) is significantly higher than that of other types of mutations suggesting that MLVAs provide greater resolution than analyses of most other changes while interrogating fewer loci. On a per locus basis, VNTRs generally contain greater discriminatory capacity than any other type of molecular typing system (Richards and Sutherland, 1997; van Belkum et al., 1998). Most pathogenic bacteria contain VNTRs, but, in some highly diverse species, not all isolates contain the same complement of these loci.

There are online databases for those who wish to conduct MLVAs of different bacteria (Grissa et al., 2008). One such site can be found at http://minisatellites.u-psud.fr/. The Institut Pasteur maintains an MLVA database that can be accessed at http://www.pasteur.fr/mlva. However, MLVA profiles are available for only a small number of pathogens. Often, such databases are limited to species of interest to those who constructed the site. The author could not find a single online source containing all MLVA profiles for all the pathogens mentioned in this chapter.

Perhaps the most difficult aspect of using MLVA is the difficulty of comparing MLVAs from different laboratories. Early MLVAs were often conducted using polyacrylamide-based DNA sequencing gels. Such gels and the current capillary-based electrophoresis instruments were designed to resolve single-nucleotide differences between DNA fragments, not to determine specific fragment sizes. Calling the size (in base pairs) of a fragment on a gel-based sequencer is within ±3 nucleotides depending on the number of molecular weight markers included in the analysis. This is based on a comparison of a known number of nucleotides in a DNA fragment to results deduced by analysis on such gels (our own personal observations). MLVA fragment lengths may differ by less than three nucleotides. Moreover, commercially available DNA fragment sizing standards provide different putative lengths because DNA standard fragments migrate in the gel based on both the length of the sequences and their nucleotide content. That is, two fragments of exactly the same nucleotide length may migrate slightly differently on the gel, resulting in a different fragment length call for an MLVA allele relative to the same analysis with a different set of size standards. Consequently, fragment length calls can differ significantly among different laboratories. This problem is magnified when using capillary gel electrophoresis, where DNA fragment length calls vary ±9 nucleotides or more (our own observations). Direct measurement of fragment masses by mass spectrometry can provide very accurate analysis of MLVA results, but such instruments are very expensive and are not readily available to most laboratories conducting such analyses. One approach to solving this

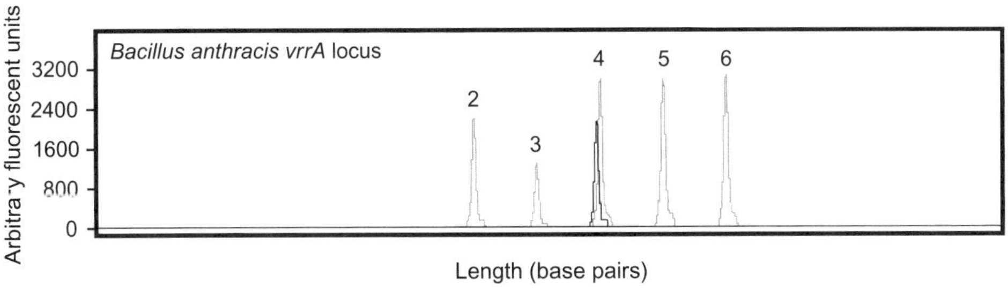

Figure 8.1 Identification of the *vrrA* allele in an unknown sample by direct comparison to a set of differentially labeled *vrrA* amplification fragments. VIC™-labeled *vrrA* PCR primers (Applied Biosystems, Foster City, CA) were used to amplify the *vrrA* allele from five different *B. anthracis* isolates, each containing a different *vrrA* allele. The resulting five allele fragments were purified then mixed to make a molecular weight standard. DNA from an uncharacterized *B. anthracis* isolate was then used as template in a reaction containing FAM™-labeled *vrrA* PCR primers (Applied Biosystems, Foster City, CA). A small amount of the resulting amplicon was mixed with the VIC-labeled *vrrA* standard and was analyzed on an Applied Biosystems 3100 capillary DNA sequencer. Numbers above each peak represent the number of (CAATATCAACAA) repeats present in the VNTR locus. The *vrrA* amplicon from the unknown samples migrates with the $(CAATATCAACAA)_4$ fragment. The slight size difference between the VIC-labeled $(CAATATCAACAA)_4$ molecular weight control fragment and that from the unknown is due to differences in the molecular weights of the two dyes used to label the DNA fragments.

problem is to make DNA fragment size standards directly from a set of different MLVA alleles at a particular VNTR locus. This allows direct comparisons that can be shared among laboratories using the same fragment array for comparison. For example, the *vrrA* VNTR locus of *B. anthracis* has five alleles, containing from two to six repeats, 12 nucleotides in length. Generation of a set of all five alleles differentially labeled relative to the assay allows direct comparison of an assay result to an array of all five possible alleles (Fig. 8.1). This provides a relatively simple determination of the allele present in an unknown sample. However, direct comparison of alleles among laboratories requires that all laboratories use the same sets of size standards, that is, a set of size standards for each MLVA locus that contains fragments representing all of the alleles at that locus. Such specialized molecular weight markers are not generally available. One can also identify MLVA loci and the number of repeats present by analysis of complete or partial genome sequences from different bacterial isolates (Denœud and Vergnaud, 2004). This, of course, assumes that the sequences are already available because the cost and time of sequencing and assembling a genome are still significantly more than MLVAs with developed assays and reagents.

Amplified fragment length polymorphism (AFLP) analysis was first used to demonstrate differences among the many isolates with limited diversity within this species (Keim et al., 1997). Analysis of only the few polymorphic fragments in the AFLP profiles for the different strains analyzed provided the first successful differentiation among multiple isolates of this species. Further analysis of the AFLP profiles revealed several DNA fragments that were mutually exclusive among different isolates. That is, a strain that manifested a particular fragment always lacked a different fragment, often of similar size, while a different strain would manifest the second fragment but not the first. These observations led to the discovery of five VNTRs in *B. anthracis* (Keim et al., 2000). However, the first example of a VNTR in *B. anthracis* was discovered by a much less extensive survey of the *B. anthracis* genome using arbitrarily primed polymerase chain reaction (AP-PCR) (Andersen et al., 1996). A single genetic locus manifested two different alleles in a limited number of *B. anthracis* isolates. Subsequent analyses using a much larger strain collection

(198 isolates) revealed five alleles at this locus (Jackson et al., 1997). Additional MLVA loci were subsequently identified by, first, direct analysis of the available *B. anthracis* pXO1 and pXO2 sequences (Keim et al., 2000), then by analysis of the entire *B. anthracis* genome (Van Ert et al., 2007a). MLVA8 analysis, using eight VNTR loci, separates all tested *B. anthracis* isolates into 89 different genotypes. MLVA15 analysis, using 15 different VNTR loci, in combination with a larger number of isolates analyzed, increased the genotype number from 89 to 221 (Van Ert et al., 2007a).

The first VNTR in *Y. pestis* was a tetranucleotide repeat, $(CAAA)_n$, where $n = 3$–10 (Adair et al., 2000). All possible alleles between 3 and 10 were found in a survey of 35 diverse *Y. pestis* isolates. The $(CAAA)_2$ was found in *Yersinia pseudotuberculosis*, a close relative of *Y. pestis*, but the same VNTR was not found in *Yersinia enterocolitica*, another relatively close relative. Analysis of *Y. pestis* chromosomal DNA sequences and the sequences of two plasmids, pMT1 and pCD1, identified an additional 42 VNTR loci in *Y. pestis* (Klevytska et al., 2001). VNTR-based phylogenetic trees were generally consistent with common biovar evolutionary scenarios and with IS100-based analyses (Motin et al., 2002). Pourcel et al. (2004) used 25 *Y. pestis* MLVA markers to characterize 180 different isolates into 61 different genotypes. The three traditional *Y. pestis* biovars were consistently distributed into three branches with some exceptions, primarily in the Medievalis biovar. Studies of VNTR mutation rates in *Y. pestis* and in other species allow application of this technology to better epidemiological understanding of disease outbreaks and their progression (Vogler et al., 2007).

MLVAs have also been applied to distinguish among different *Escherichia coli* O157:H7 isolates. Twenty-nine putative VNTR loci were identified by interrogation of the *Escherichia coli* genome, and these were validated by analyses of 56 different *Escherichia coli* O157:H7/HN and O55:H7 isolates (Keys et al., 2005). The number of alleles at each locus ranged from 2 to 29, while the diversity index varied from 0.23 to 0.95. Values of this index can range from 0 (no diversity) to 1 (complete diversity) (Nei and Kumar, 2000). A comparison of MLVA typing to pulsed-field gel electrophoresis (PFGE) results showed that both methods provided consistent results, but MLVAs were able to further resolve among sample isolates that were identical by PFGE analysis. Thus, MLVA of an outbreak cluster should generate superior resolution to the more traditional PFGE methods in addition to being somewhat easier and significantly more rapid to execute. MLVA was used to better understand *E. coli* O157:H7 contamination of lettuce and spinach in the Salinas and San Juan valleys of California between 1995 and 2006 (Cooley et al., 2007). A comparison to PFGE results again demonstrated resolution among apparently identical isolates by MLVA. The MLVA of 54 feedlot isolates separated into 12 different MLVA types and suggested that animals entering the feedlot at initial stocking are an important source of this contamination. Once *Escherichia coli* O157 inoculated the feedlot, water troughs, pen bars, pen floor feces, and feed were all found to be means of transmitting this pathogen (Murphy et al., 2008). Seventy-two human and animal strains of Shiga toxin-producing *E. coli* O157 were typed using MLVA assays (Lindstedt et al., 2003), and Ohata et al. (2008) typed Japanese *E. coli* O157:H7 clinical isolates. Comparisons in both studies again demonstrated the superior resolution of MLVA typing relative to PFGE analysis.

F. tularensis, the etiologic agent of tularemia, is found naturally throughout the northern hemisphere in North America, in Asia, and in Europe, although it has also been isolated in Australia. This highly infective, gram-negative intracellular pathogen can infect a large number of different species. There are four subspecies of *F. tularensis*. The subspecies *tularensis* is the most virulent of these. Biochemical studies and 16S rDNA sequence analysis have traditionally been used to distinguish among different subspecies of this pathogen. Six polymorphic VNTRs were initially identified based on canvassing the *F. tularensis* genome followed by analysis of these putative VNTR loci in 55 different *F. tularensis* isolates (Farlow et al., 2001). The allele number in these six loci ranged from 2 to 20. The analysis of an additional 56 samples resulted in the identification of 39 different allele combinations. Unweighted Pair Group Method with Arithmetic Mean (UPGMA) cluster analysis revealed two major clusters of isolates. However, there were no absolute fixed allelic differences between the two clusters. California isolates were found in both major groups. Oklahoma isolates mapped to one of two subgroups within the second cluster, while all of the Arizona isolates analyzed appeared to be identical at the resolution of the MLVA. An additional 19 variable VNTR loci containing between 2 and 31 alleles were used to analyze 192 geographically diverse *F. tularensis* isolates (Johansson et al., 2004a). Nei's diversity values ranged between 0.05 and 0.95 and were correlated with the number of alleles at each locus. *Francisella tularensis* ssp. *tularensis* (type A) isolates showed great diversity, but *Francisella tularensis* ssp. *holarctica* (type B) isolates were much less diverse in spite of a much broader geographic range. Some but not all genetically similar isolates were isolated from geographically proximal locations.

Burkholderia pseudomallei is a genetically diverse pathogen that causes a disease called melioidosis. This disease is endemic throughout Southeast Asia and Northern Australia (White, 2003; Cheng and Currie, 2005), and the number of cases increases significantly during the wet monsoon season. PFGE, AFLP, and multilocus sequence typing (MLST; see below) have been used to distinguish among different isolates of this pathogen. *B. pseudomallei* has numerous VNTRs, some duplicated at more than one site within the genome. Duplicated repeat regions may facilitate genomic rearrangement and, possibly, altered gene expression. However, they can significantly complicate an MLVA and are therefore not usually used in such analyses. U'Ren et al. (2007) used 32 VNTR loci displaying between 7 and 28 alleles, with Nei's diversity values ranging between 0.47 and 0.94 to analyze 66 geographically diverse *B. pseudomallei* and 21 *Burkholderia mallei* isolates. They also applied these assays to 95 lineages of an 18,000-generation passage experiment to better understand the mutation frequencies at the different VNTR loci. MLVA-based phylogenetic analyses of the *Burkholderia* isolates demonstrated that the *B. mallei* isolates were significantly less diverse, clustered tightly relative to all of the *B. pseudomallei* isolates. Similar results were demonstrated using AFLP and MLST analyses. Analysis of isolates from the passage experiment revealed that variation in 12 of the VNTR loci occurred during the passage study. Most of these changes resulted in single repeat changes with a bias toward increases in tandem repeat copy number. More recently, Currie et al. (2009) developed a four-locus MLVA for rapid typing of *B. pseudomallei* based on selection of a subset of informative VNTR markers. This analysis provides resolution similar to PFGE and MLST results and can provide genotyping results within 8 h following receipt of samples.

8.3 SNP AND DNA SEQUENCE-BASED METHODS

Development of alternative DNA-based methods of sample analysis was initially driven by the cost and time required to generate high-quality DNA sequences, and the increased use of SNPs to differentiate among different bacterial isolates has closely paralleled the increased access to low-cost DNA sequencing. Recently, introduction of new, much more rapid methods of generating DNA sequences, especially for comparison to previously sequenced organisms, has made the initial identification of SNPs across multiple isolates of the same species and among multiple, closely related species much more rapid and less expensive than was previously possible. Indeed, potential VNTR loci are now almost exclusively identified by analysis of bacterial genomes (see Dégrange et al., 2009 for a recent example).

Historically, the first extensively used SNP-based method of characterizing bacteria was the analysis of a specific region of the 16S ribosomal RNA (rRNA) gene common to all bacteria (Lane et al., 1985). The sequence of the small subunit (16S) rRNA varies in an orderly manner across phylogenetic lines and contains segments that are conserved at the species, genus, or kingdom level. By designing oligonucleotide primers to prime off sequences conserved throughout the eubacterial kingdom, it is possible to use PCR to amplify DNA fragments encoding phylogenetically informative sections of the 16S RNA gene. Subsequent sequencing of these amplicons provides information that is useful to identify an isolate at least to the genus level. Sometimes, this approach provides differentiation among different isolates of highly diverse species (Collins and East, 1998), although based on differences in 16S rRNA sequences, one could argue that the diverse species in question is actually multiple species, each represented by a different 16S sequence. The 16S rRNA typing approach is still widely used to generate information about previously uncharacterized isolates. The Ribosomal Database Project (http://rdp.cme.msu.edu/) archives over 920,000 16S rRNA sequences with the software and informatics tools for comparison to sequences generated from unknown isolates.

The 16S rRNA sequences vary little among some closely related species. For example, this approach will not differentiate between *B. anthracis* and some closely related *Bacillus cereus* isolates when amplifying the template normally generated when analyzing unknown *Bacillus* isolates. Another method, analyzing selected sequences of highly conserved so-called housekeeping genes, provides higher resolution among different closely related bacterial isolates. MLST was first applied to analyze populations of pathogenic bacteria by Maiden et al. in 1998 in a study of *Neisseria meningitidis*. The study analyzed fragments approximately 470 nucleotides in length from 11 different genes. The amplicon size was selected to provide rapid full sequencing of both amplicon DNA strands using the best available sequencing technology of the time. Most MLST analyses now target portions of seven different conserved genes. Analysis of the *B. cereus* group relies on sequencing portions of the *glpF*, *gmk*, *ilvD*, *pta*, *pur*, *pycA*, and *tpi* genes, encoding the glycerol uptake facilitator protein, guanylate kinase, dihydroxy-acid dehydratase, phosphate acetyltransferase, phosphoribosyl aminoimidazole carboxamide, pyruvate carboxylase, and triosephosphate isomerase, respectively. Hoffmaster et al. (2006) applied MLST and AFLP analysis to *B. cereus* isolates associated with fatal pneumonias. The results were complementary. Isolates that appeared to be closely related by AFLP analysis were also closely linked by MLST analysis. Isolates that appeared to be identical by MLST analysis were also identical within the resolution of the AFLP analysis. Both methods allowed comparison across a large, diverse collection of *B. cereus* and *B. thuringiensis* isolates. Neither method provided significant resolution among different *B. anthracis* isolates, but both methods clearly differentiated all *B. anthracis* isolates from even very closely related *B. cereus* isolates.

Y. pestis and its closest relatives *Y. pseudotuberculosis* and *Y. enterocolitica* have been subjected to MLST analyses by sequencing portions of the *thrA*, *trpE*, *glnA*, *tmk*, *dmsA*, and *manB* genes (Achtman et al., 1999). The MLST results in combination with other information about the genome organization in different isolates of these species support the contention that *Y. pestis* is a recently emerged clone on *Y. pseudotuberculosis*. Another

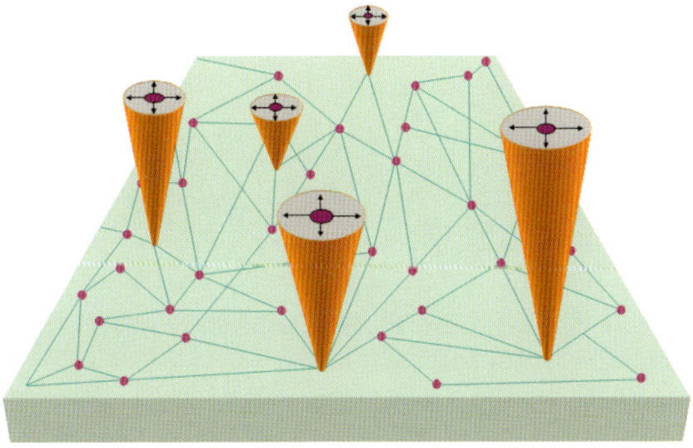

Figure 2.2 The "epidemic" bacterial population structure. The background population is composed of a large number of relatively rare and unrelated genotypes (small circles) that are recombining at a high frequency. Superimposed upon this background are clusters of closely related genotypes (clonal complexes), illustrated as cones. These emerge from a single, highly adaptive ancestral genotype (the large circles).The diversification of these clones by recombination and mutation is indicated by the arrows (from Smith et al., 2000).

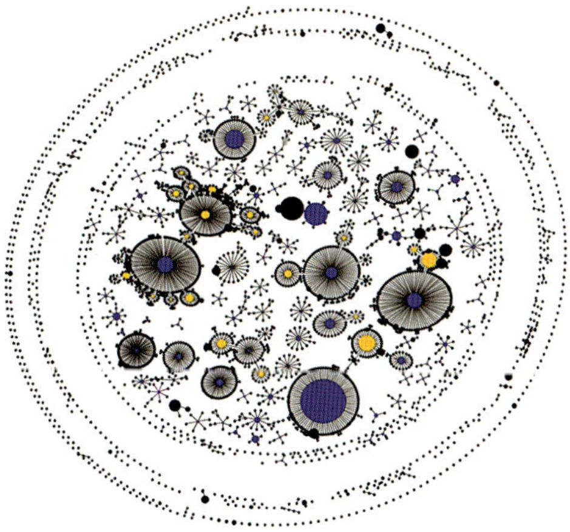

Figure 2.4 eBURST representation of the MLST data set for *N. meningitidis*. Each circle represents an ST. The frequency of the ST is indicated by the size of the circle. STs that only differ by one locus are linked (see text). Clonal founders are shown in blue; subfounders are shown in yellow. The pattern shows large clonal complexes against a background of diversity, as predicted by the "epidemic model" (e.g., Figure 2.2, as seen from the top).

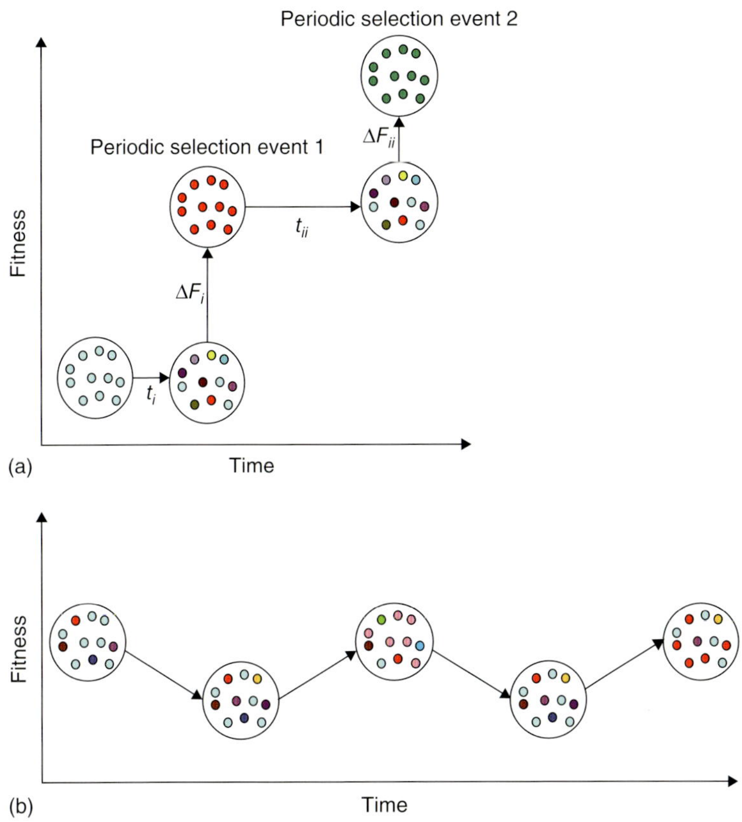

Figure 2.3 (a) The maintenance of ecotypes by repeated periodic selection. (b) The consolidation of fitness within an ecotype through purging of slightly deleterious changes and partial selective sweeps. (See text for full caption.)

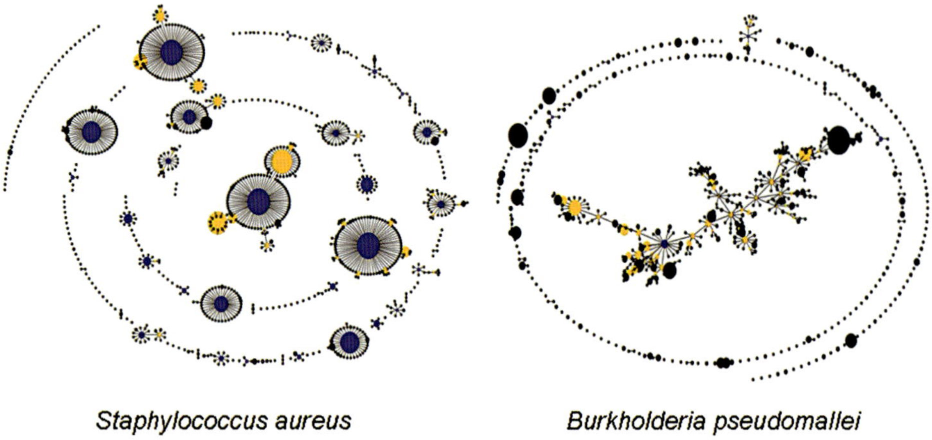

Figure 2.5 Interpreting eBURST diagrams with respect to recombination rate. (See text for full caption.)

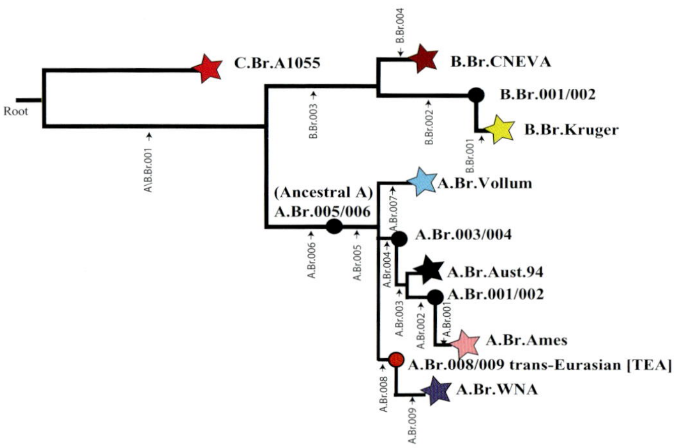

Figure 9.2 CanSNP phylogenetic tree and nomenclature. (See text for full caption.)

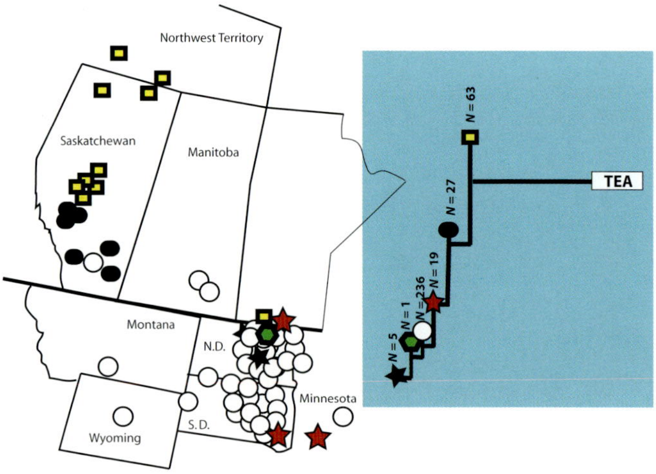

Figure 9.3 Phylogeography of the Western North American clade. (See text for full caption.)

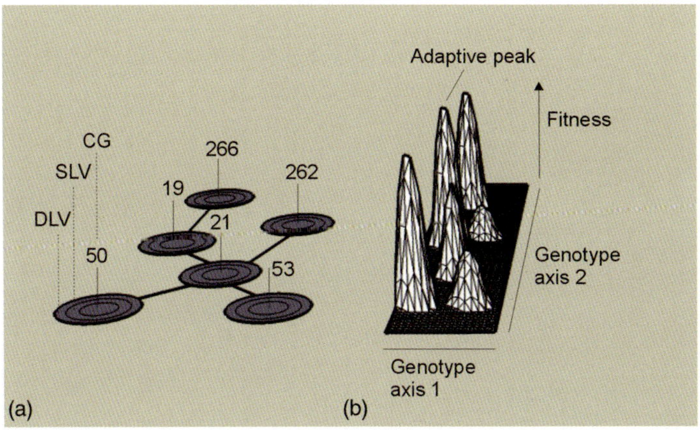

Figure 10.5 Speculation on the adaptive signal in the MLST loci. (a) Examples of *C. jejuni* allelic profiles matching at five loci belonging to the ST-21 clonal complex. The predicted primary founder, ST-21, is defined, and other central genotypes (CGs) are linked to this. Within groups, concentric circles represent single-locus variants (SLVs) and double-locus variants (DLVs). (b) Hypothetical adaptive landscape in relation to clonal complex substructure. Peaks are local fitness optima for genome-wide variation; troughs represent suboptimal genotype combinations that will be selected against impeding evolutionary transitions between peaks.

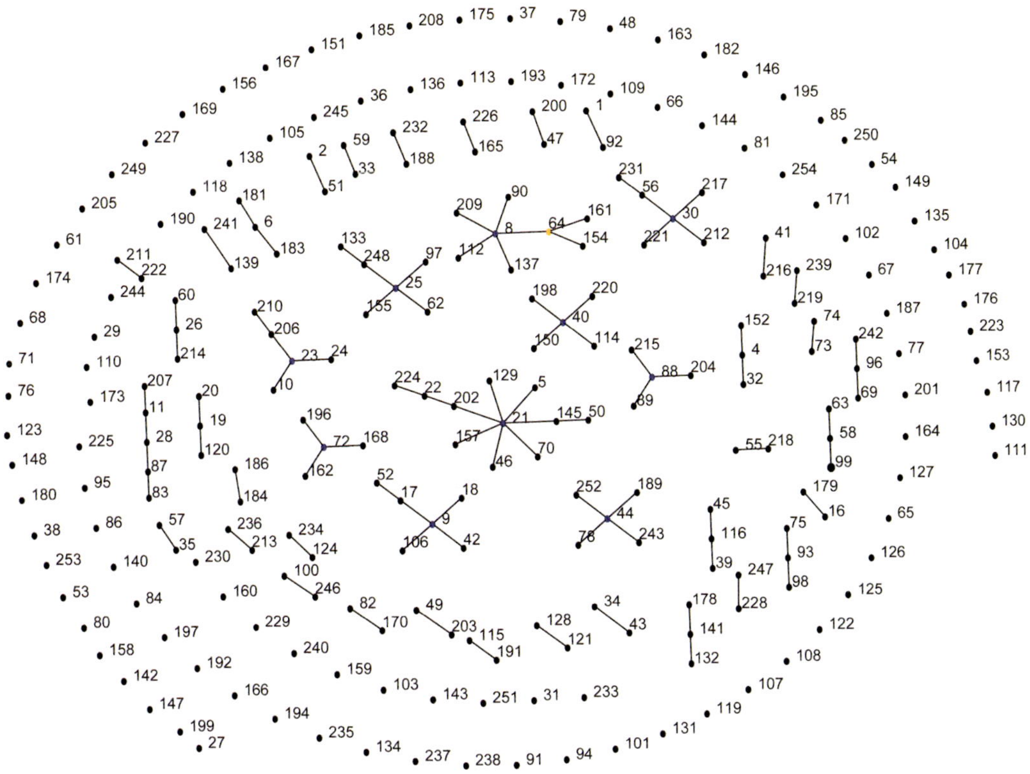

Figure 11.3 Population snapshot of 643 *E. faecalis* isolates (http://efaecalis.mlst.net/) representing 249 STs based on MLST allelic profiles using the eBURST algorithm (Feil et al., 2004). See also Fig. 11.1 for the explanation of the eBURST-based population snapshot.

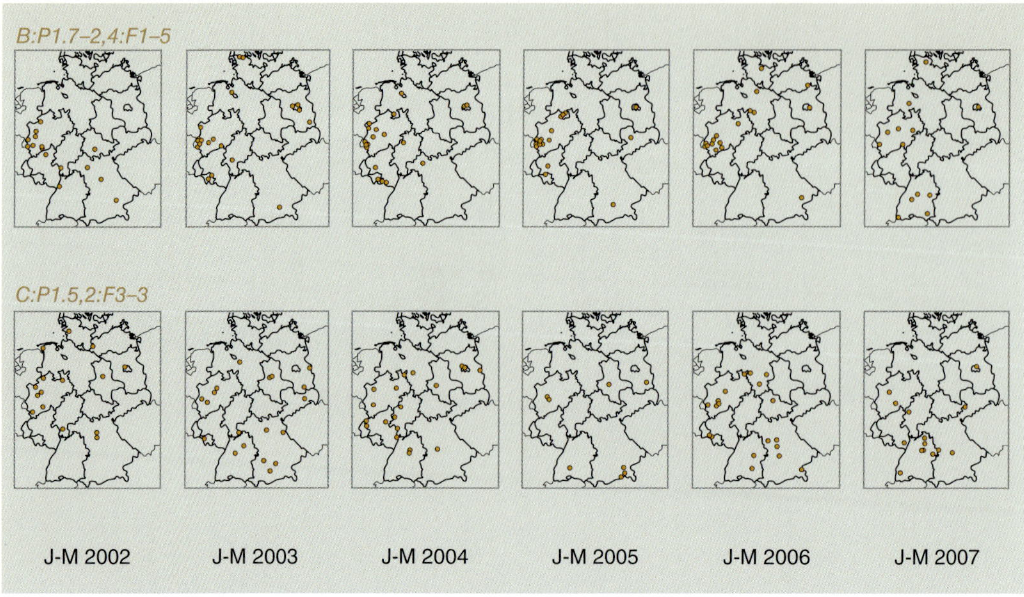

Figure 13.4 Geographic spread of two distinct meningococcal finetypes between 2002 and 2008 from Germany. The months January to March are depicted for each year. The finetype B:P1.7-2,4:F1-5 mostly belongs to the ST-41/44 complex; most strains of the finetype C:P1.5,2:F3-3 are ST-11 complex. B:P1.7-2,4:F1-5 persists in Western Germany, whereas C:P1.5,2:F3-3 appears to occur with a more or less random distribution. The maps were generated with EpiScanGIS (http://www.episcangis.org/) (Reinhardt et al., 2008).

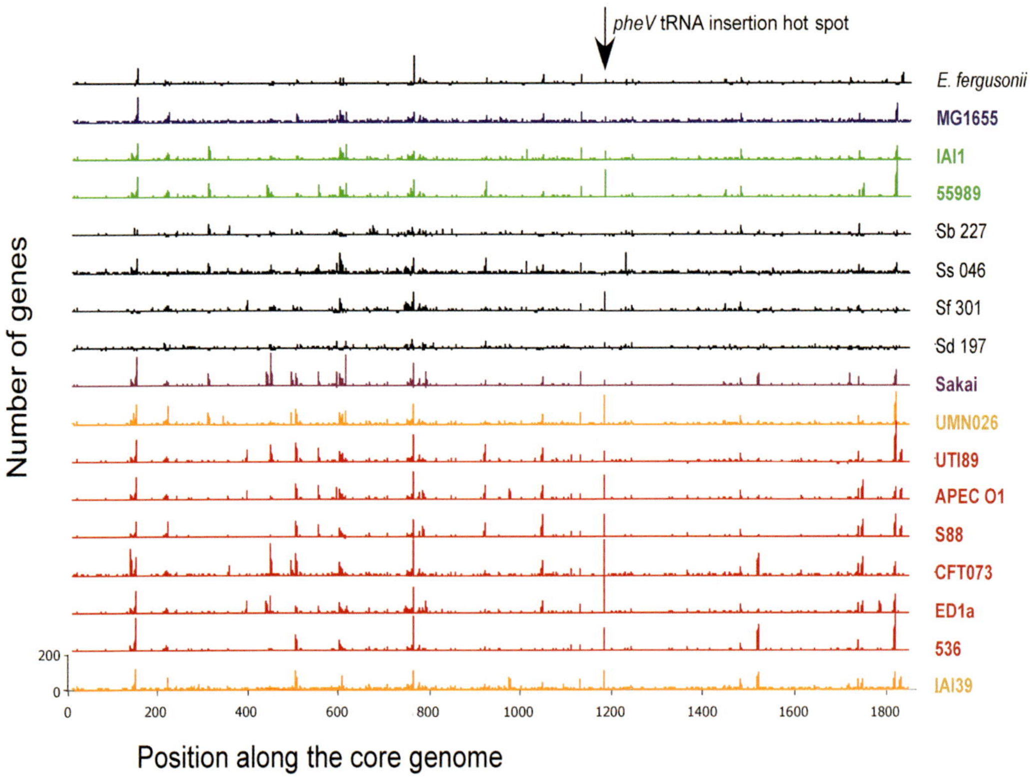

Figure 14.1 Global view of insertion/deletion hot spots on the chromosome of 16 *Escherichia coli*/*Shigella* and *Escherichia fergusonii*. Number of genes (ranging from 0 to 200) in indels along the genomes according to the ancestral gene order of the core genome (Touchon et al., 2009). The numbers on the *x*-axis represent the order of genes in the core genome, as in *E. coli* K-12 MG1655. Sb 227: *Shigella boydii* 4, Ss 046: *Shigella sonnei*, Sf 301: *Shigella flexneri* 2a, Sd 197: *Shigella dysenteriae* 1.

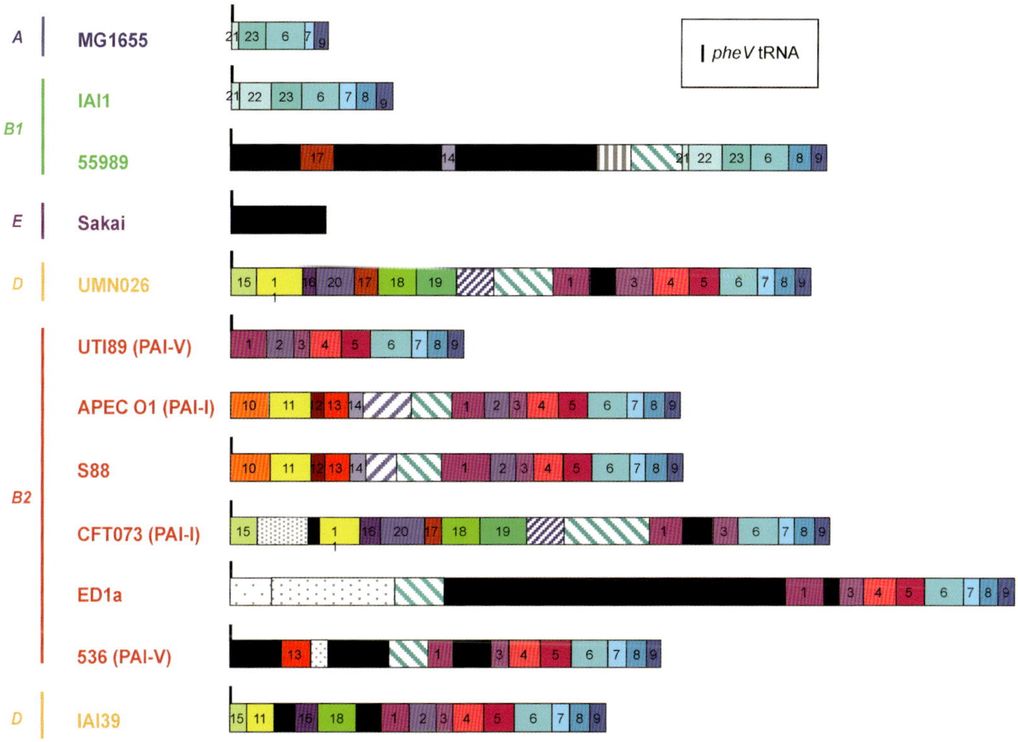

Figure 14.2 The genomic island at the *pheV* tRNA insertion hot spot in 12 different *Escherichia coli* strains. (See text for full caption.)

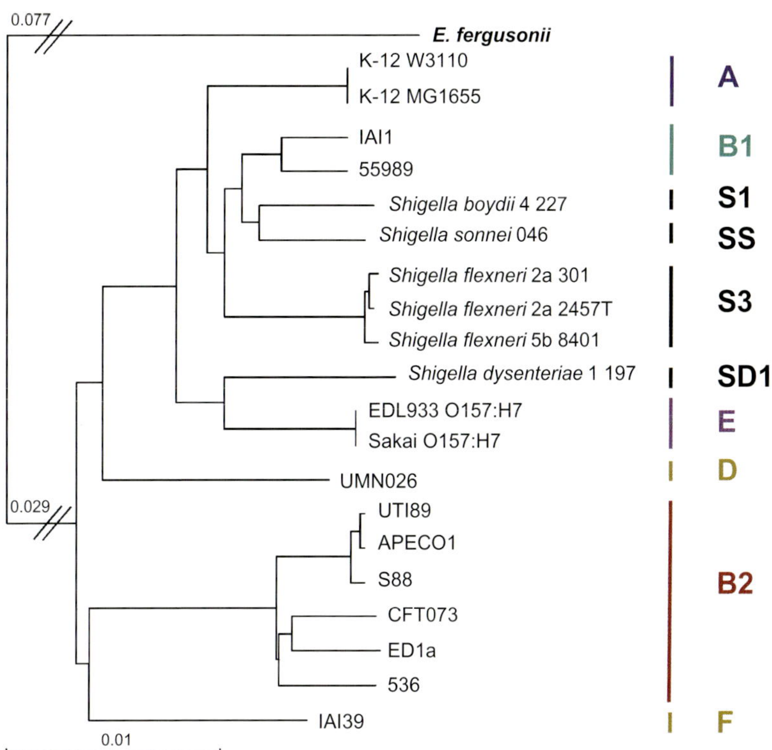

Figure 14.3 Maximum likelihood phylogenetic tree of the 20 *Escherichia coli* and *Shigella* strains as reconstructed from the sequences of the 1878 genes of the *Escherichia* core genome. (See text for full caption.)

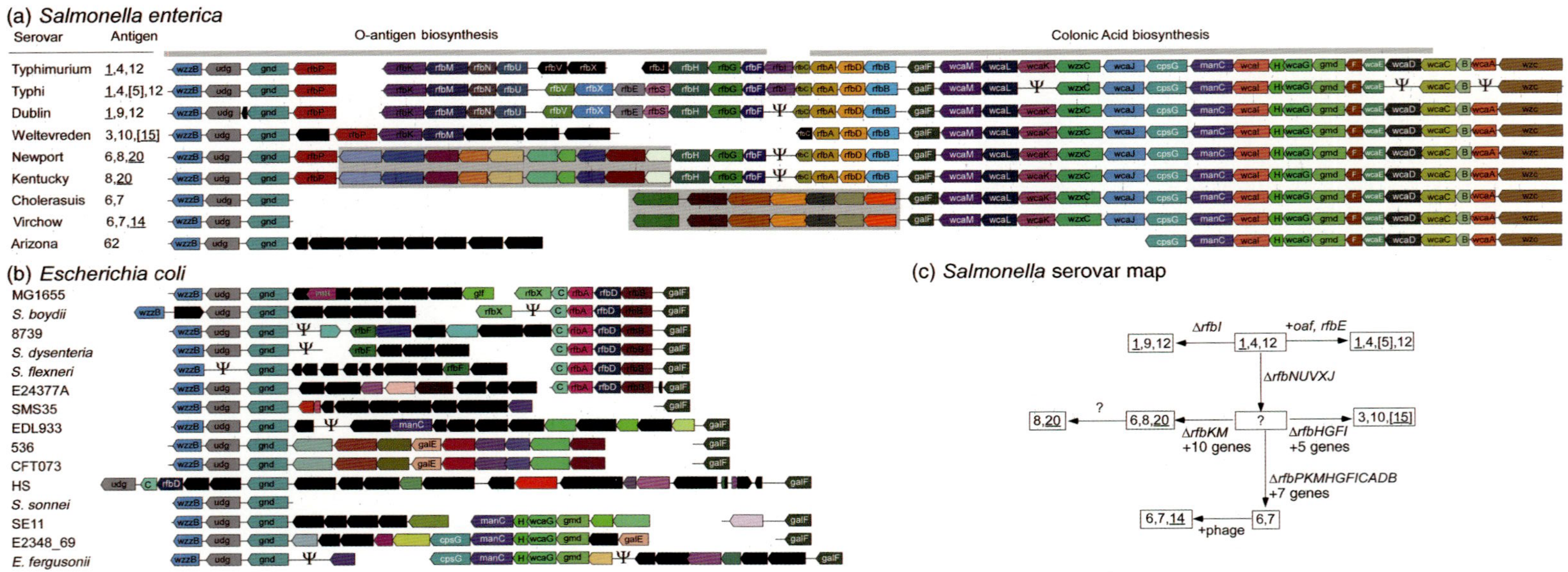

Figure 15.5 (a) Alignment of *rfb* operon regions of *Salmonella* strains. Orthologous genes are shaded with the same color. (b) Alignment of *rfb* operon regions of *E. coli* strains. Genes with *Salmonella* orthologues are shown in cognate colors. (c) Differences between *Salmonella* O antigens. A color version of this figure appears in the center of this volume.

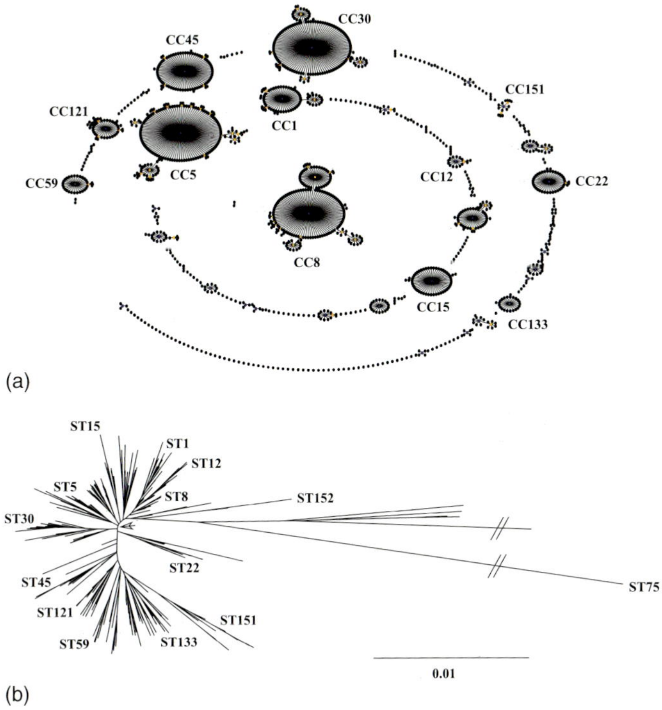

Figure 16.1 Overview of population structure in *S. aureus*. (See text for full caption.)

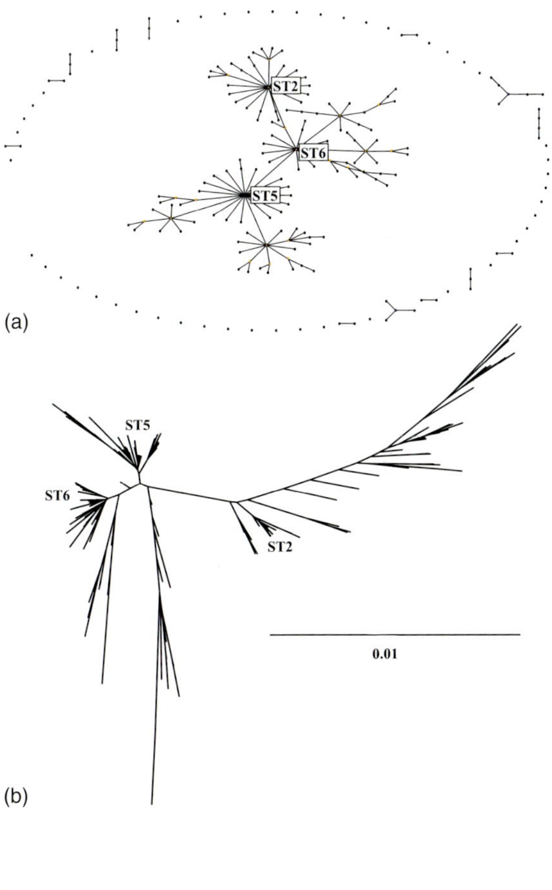

Figure 16.2 Overview of population structure in *S. epidermidis*. (a) eBURST analysis of all 211 STs in the MLST database. Each circle represents a unique ST. Lines connect STs that differ at a single locus, though not all such connections are depicted. Names of various STs within the large, "straggly" CC are indicated. (b) Neighbor-joining phylogenetic tree based on concatenated MLST sequences. The tree shows relationships between 210 STs; one ST with insertion–deletion polymorphisms was dropped from the analysis. Scale is in substitutions per site.

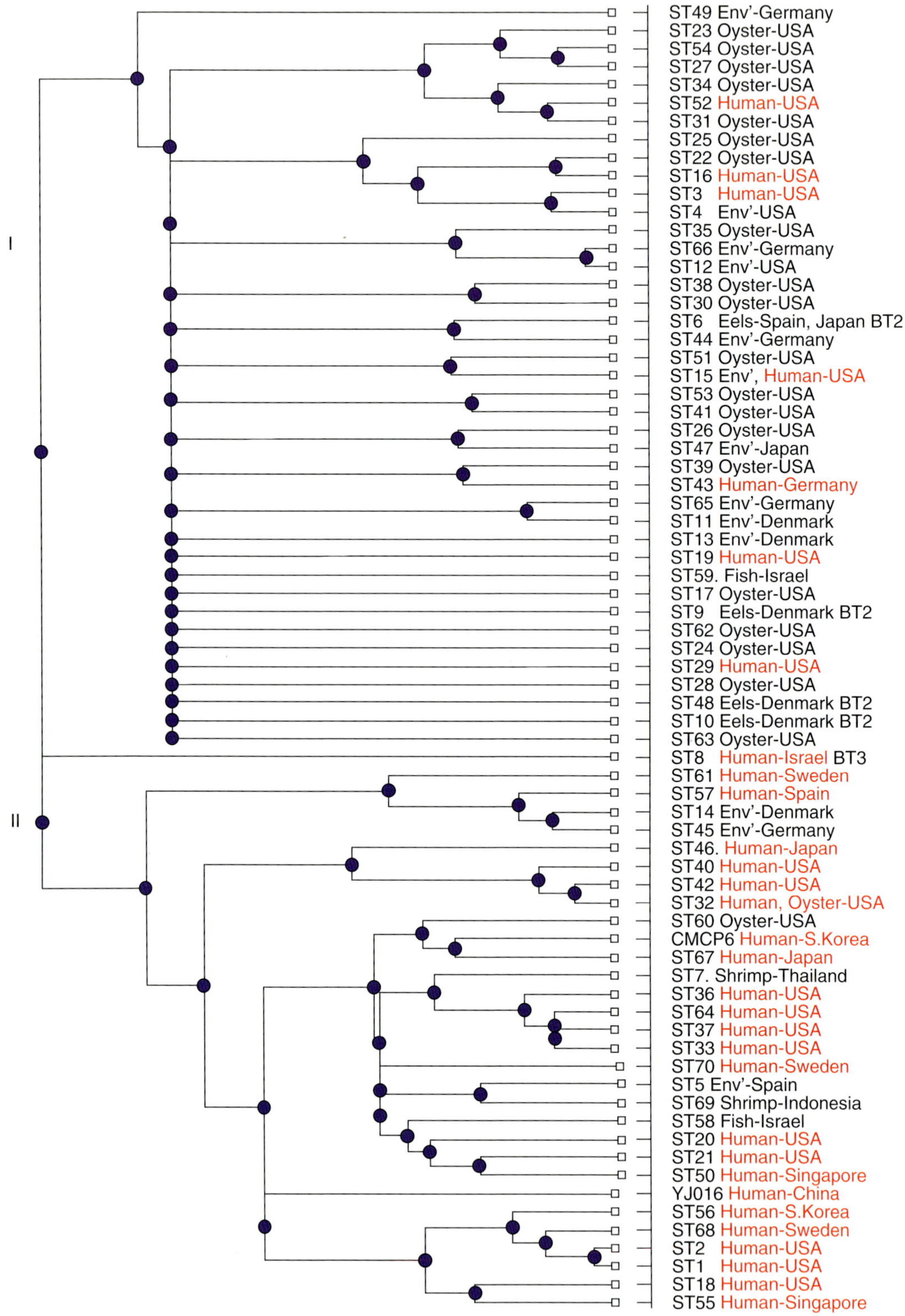

Figure 18.3 Majority-rule consensus tree based on the posterior distribution of genealogies inferred by ClonalFrame. The biotype identity is shown for some isolates; the rest of the isolates belong to biotype 1. Env' = environment.

study of *Y. pestis* isolates from the Republic of Georgia and neighboring former Soviet Union countries applied MLST to differentiate among different isolates (Revazishvile et al., 2008). However, analysis at seven loci (portions of the *hsp60*, *glnA*, *gyrB*, *recA*, *manB*, *thrA*, and *tmk* genes) and the 16S rRNA gene provided little resolution, and the authors found that PFGE discriminated among the *Y. pestis* isolates more effectively than MLST. It appears that, like *B. anthracis*, *Y. pestis* isolates show little diversity based on such analyses. The lack of diversity within the MLST loci analyzed is in direct contrast to methods that interrogate changes in genome organization, suggesting that many differences among different *Y. pestis* isolates may be based on differences in the relative spatial distribution of sequences within the genome (Motin et al., 2002).

MLST-based methods have also been applied to the study of *E. coli* isolates. Noller et al. (2003) found no sequence diversity in the sequenced portions of seven housekeeping genes among 77 *E. coli* O157:H7 isolates shown to be diverse using PFGE. In an attempt to better understand the evolution of new bacterial pathogens, Reid et al. (2000) looked at seven housekeeping genes in enteropathogenic *E. coli* (EPEC), enterohemorrhagic *E. coli* (EHEC), strains of other Shiga toxin-producing *E. coli* serotypes, and the laboratory strain K-12. While MLST analyses provided useful data to draw conclusions about evolutionary mechanisms, the diversity index within the fragments of the housekeeping genes amplified ranged from approximately 2% in the *arcA* gene to only 7.5% in the *mtlD* gene, again suggesting the limited value of the MLST approach to differentiate among even very different *E. coli* isolates.

MLST has also been applied to *F. tularensis* and closely related species. However, Johansson et al. (2004b) showed that the use of seven housekeeping genes of *F. tularensis* distinguished the subspecies but did not provide high-resolution discrimination of individual isolates. In contrast, MLVA was only exceeded by whole genome sequencing in providing resolution among different *F. tularensis* isolates.

There are several published reports describing application of MLST to the study of *B. pseudomallei* and *B. mallei* isolates (Godoy et al., 2003; Currie et al., 2007; Wattiau et al., 2007). MLST analysis of 128 isolates of a geographically diverse collection of *B. pseudomallei* isolates using sequences from the *ace*, *gltB*, *gmhD*, *lepA*, *lipA*, *nark*, and *ndh* genes resolved the collection into 71 sequence types (Godoy et al., 2003). Resolution was improved by the presence of multiple SNPs within the different amplicons. For example, there were 15 different SNPs in the *gmhD* gene fragment and 14 SNPs in the *narK* gene fragment. Specific nucleotides present at five different SNP loci can be used together to unambiguously differentiate between all *B. mallei* and all *B. pseudomallei* isolates tested (R. Okinaka, unpublished data).

Commercially available SNP analysis kits and custom-synthesized primers and probes are now routinely available. As outlined above, SNP analyses can be applied, with varying success that depends on the species in question, to differentiate among different isolates of the same species or across a group of closely related species depending on the targets chosen for the analysis.

Demonstration of the resolution of any typing method requires signatures for a large collection of diverse isolates, and this is sometimes the limiting factor in demonstrating the utility of such methods, especially when analyzing species where there is a lack of genetic diversity among the isolates. There are a significant number of publications describing application of MLST methods to differentiate among different, closely related species or among different isolates of the same species. However, another inherent weakness in this approach is the lack of common databases containing all of the MLST profiles for a particular target species. MLST profiles for the *B. cereus* group and *B. pseudomallei* (also applicable to *B. mallei*) are available at PubMLST (http://pubmlst.org/), but the database contains a relatively small number of isolates relative to the large collections available. While *Y. pestis* and its closest relatives *Y. pseudotuberculosis* and *Y. enterocolitica* have been subjected to MLST analyses by sequencing portions of the *thrA*, *trpE*, *glnA*, *tmk*, *dmsA*, and *manB* genes (Achtman et al., 1999), the available database for MLST analysis of *Y. pseudotuberculosis* (http://mlst.ucc.ie/mlst/dbs/Ypseudotuberculosis) provides profiles for a somewhat different set of genes: *thrA*, *trpE*, *glnA*, *tmk*, *adk*, *argA*, and *aroA*. Two different *E. coli* MLST databases are available (http://www.pasteur.fr/recherche/genopole/PF8/mlst/EColi.html and http://mlst.ucc.ie/mlst/dbs/Ecoli), but they each provide profiles for a different set of target genes with only one common target. Besides containing profiles for a limited number of profiles, the lack of consensus for some species on which genes to target is a significant weakness. Analysis of mutually exclusive sets of target genes does not allow comparison of results. Perhaps a forum could be established to determine which MLST loci provide the highest resolution among all available isolates for particular species. It would clearly be beneficial if consensus on the genes targeted for analysis was reached.

There is also a Web site for comparison of *B. pseudomallei* MLST profiles and those from closely related species that targets seven MLST loci (http://www.bpseudomallei.mlst.net). This Web site provides information about the geographic location (by country) of a large collection of isolates as well as the MLST genotype for the collection of isolates from that country.

MLST analysis is advantageous because it exploits the unambiguous nature of DNA sequences to allow data comparison across multiple laboratories. However, in some species, it provides only very limited resolution among demonstrated diverse isolates. It is also dangerous to make general assumptions about evolutionary and phylogenetic differences based only on analysis of changes in a limited number of gene sequences. Sequence differences identify one kind of diversity but do not capture whole genome changes such as the relative presence or absence of particular sequences or genome reorganization among closely related species or among different isolates of the same species. It has been clearly demonstrated in *Y. pestis* that genome reorganization, probably facilitated by the large number of IS elements present, contributes to phenotype among different isolates of this pathogen and, perhaps, in differentiating between this pathogen and other closely related bacterial species.

Another approach to differentiating among isolates of the same pathogenic species focuses on changes within genes that are unique to that species. In particular, studies have targeted genes encoding known virulence or toxin factors. Price et al. (1999) showed that *B. anthracis* isolates could be differentiated by analysis of the protective antigen gene. Analysis of seven SNPs within this gene in combination with MLVA of different isolates provided a unique signature for an isolate associated with the 1979 Sverdlovsk release. Twelve SNPs have now been identified in the protective antigen gene (P. J. Jackson, unpublished data). Five additional SNPs have been found in the *cya* gene, encoding edema factor also residing, with the protective antigen gene, on pXO1 (R. T. Okinaka, unpublished data). Combinations of SNPs in a single gene have been used to differentiate between a target species and its close relatives. Qi et al. (2001) identified four SNPs within the *rpoB* gene that were reported to be specific for *B. anthracis* relative to other *Bacillus* isolates. However, comparisons were not made between *B. anthracis* and *B. cereus* and *B. thuringiensis* isolates that have been shown by other methods to be very closely related to this pathogen. In the absence of either a discrete genetic change that correlates with the SNP—for example, the expression of a particular gene characteristic of the pathogen—

or a very extensive survey of closely related species, it is dangerous to assume that one or a very limited number of SNPs may, in themselves, represent a species-specific assay.

The availability of fully sequenced genomes from different genetically diverse isolates of the same pathogenic species has led to extensive surveys that identify virtually all of the phylogenetically informative SNPs in a genome. Comparative full-genome sequencing among eight *B. anthracis* strains led to the discovery of approximately 3500 SNPs (Read et al., 2002; Pearson et al., 2004). To some, it might seem that the binary nature of SNPs provides only limited subtyping power, and a large number of SNPs might be required to provide the resolution needed to differentiate among closely related isolates. However, it has been shown that a surprisingly small number of SNPs can be used to provide high-definition resolution among different genetic groups (Van Ert et al., 2007b). Keim et al. (2004) developed this concept further and proposed the "canonical SNP," a SNP that can be used to define a point in the evolutionary history of a species. Such canonical SNPs can be used diagnostically to define major genetic lineages within a species or, more narrowly, to define specific isolates. Moreover, combining canonical SNP and MLVAs provides insights into the evolutionary history of the species. A set of only 12 canonical SNPs representing different points in the evolutionary history of *B. anthracis* was used, in combination with MLVA15 analyses to type a large, diverse, global collection of *B. anthracis* isolates. SNP analyses placed all isolates into 12 conserved groups or lineages (Van Ert et al., 2007b). The analysis of the slowly evolving canonical SNP in combination with the MLVA15 results greatly enhanced the resolution beyond the 221 genotypes resolved by MLVA15 analysis alone. Analysis of slowly evolving canonical SNPs allowed definition of major clonal lineages, while younger, population-level structure was revealed using the more rapidly evolving MLVA markers.

SNP analyses can also detect changes of a more sinister nature. Resistance to ciprofloxacin in *B. anthracis* results from a number of single-nucleotide changes in the *gyrA* and *parC* genes, encoding the proteins targeted by this antibiotic (Price et al. 2003). It is highly unlikely that naturally occurring *B. anthracis* isolates will be resistant to this antibiotic because anthrax is primarily a zoonotic disease and infected animals are seldom provided antibiotic therapy to treat their condition. Therefore, the presence of one or more SNPs in critical positions in these genes suggests that an isolate containing such SNPs may have been intentionally subjected to increasing antibiotic concentrations to select a ciprofloxacin-resistant *B. anthracis* isolate. In particular, initial selection for ciprofloxacin resistance results in a high frequency of mutations at only two specific nucleotides (Price et al., 2003; P. J. Jackson, unpublished data), allowing rapid screening for such isolates with just two assays. It will likely be possible to identify and develop assays for other such phenotypic changes with the increased ease of producing whole genome sequences and comparing these to similar isolates with slightly different phenotypes.

The availability of multiple whole genome sequences also allows design of high-density microarrays that can be used to compare different isolates of the same species or closely related bacterial species (Zwick et al., 2008). In principle, microarrays should allow rapid identification of even minor differences between the array sequence and the challenge DNA. However, the relatively high frequency of "false-positive" SNPs exhibited by microarray data does not allow efficient identification of very minor differences relative to the arrayed sequence. Array technology can be used to demonstrate total, additive differences between a reference genome of a particular isolate and those of other isolates. It can also rapidly screen a large number of isolates to provide information about the relationship of an unknown isolate relative to a reference. When the array hybridization results are compared to the most recent available genome sequences for the same strains,

Strain	Chip A0488	Chip A0465	Chip A0442	Chip Ames	Chip Sterne	Chip A0174	Chip A0193	Chip A0149	Chip Ba-like1	Chip Ba-like2	Chip A0039	Chip A0220	Chip A0382
CNEVA-9066/JCVI 2008	1020	180	589	1066	1068	977	984	929	852	979	1079	937	508
BAM	1096	117	667	1145	1140	1052	1059	1008	932	1058	1159	1018	586
A0465/JCVI 2008	1062	**82**	632	1110	1105	1017	1024	973	897	1023	1124	983	552
KrugerB genotype 87 group B1	1434	756	168	1482	1486	1390	1399	1345	1276	1395	1499	1353	195
A0442/JCVI 2008	1336	656	**68**	1382	1386	1290	1301	1247	1176	1303	1399	1253	**95**
BAH	1396	716	128	1442	1444	1350	1361	1303	1236	1363	1459	1313	155
Kruger B/JCVI 2008	1349	671	83	1397	1401	1303	1314	1260	1189	1316	1412	1266	110
A01055/JCVI 2008	1712	1662	1953	1760	1740	1670	1671	1633	1559	1673	1794	1653	1908
A0149/LANL_BAG	330	977	1237	377	381	242	265	**57**	**145**	318	380	233	1179
Western North America USA6153/JCVI 2008	377	1025	1284	427	428	109	108	257	169	365	425	282	1226
BAQ	536	1188	1441	585	587	265	223	417	329	520	579	442	1387
A0193/JCVI 2008	377	1027	1283	425	428	107	**64**	257	167	363	421	280	1227
BAO	416	1066	1325	464	466	100	170	299	208	402	464	323	1267
A0174/JCVI 2008	370	1020	1279	418	420	**54**	124	251	164	356	416	275	1223
BAS	357	1007	1257	398	405	268	285	236	171	341	400	257	1201
Tsiankovskii-I/JCVI 2008	342	994	1248	385	388	257	274	225	158	330	387	242	1190
Ames	486	1121	1369	73	195	414	437	378	324	472	385	244	1325
BAA	493	1128	1378	82	206	423	444	387	331	479	394	251	1332
BAB	483	1118	1368	72	196	413	434	377	321	469	384	241	1322
Ames Ancestor	477	1112	1362	66	190	407	428	371	315	463	378	235	1316
A0389/JCVI 2008	493	1127	1383	193	205	421	445	394	333	480	393	248	1337
Sterne	470	1108	1357	173	**87**	400	416	367	304	452	363	224	1314
A0039/JCVI 2008	501	1144	1392	389	400	429	447	401	335	486	**37**	**170**	1346
BMA	237	1193	1453	618	616	497	512	464	404	247	611	469	1396
BAV	61	1082	1338	502	502	383	402	352	292	132	504	354	1283
Vollum/JCVI 2008	**42**	1071	1329	493	491	375	393	343	285	125	497	347	1276
BAC	44	1075	1331	495	495	377	395	345	287	129	499	349	1278
BAE	105	1134	1392	554	556	438	456	406	346	188	558	408	1337
A0488/JCVI 2008	43	1074	1330	494	494	376	394	344	286	128	498	348	1277
A0248/RefSeq	477	1112	1362	66	190	407	428	371	315	463	378	235	1316
A2012/NCBI 2008	456	1072	1319	**62**	186	388	413	356	302	441	369	229	1275
CDC 684/RefSeq	168	1124	1384	549	547	429	443	396	336	179	543	401	1327
Afghan/NBACC	92	1022	1270	446	443	328	341	296	233	**81**	451	302	1220

Figure 8.2 Hamming distances between available *B. anthracis* genomes and microarray genotype calls. Hamming distance matrix comparing SNP and insertion/deletion allele calls from microarrays hybridized to 13 *B. anthracis* samples with alleles predicted from 33 finished or draft whole genome sequences. "Ba-Like1" and "Ba-Like2" are uncharacterized isolates; A0220 and A0382 are characterized isolates but have not been sequenced. Numbers in bold represent hybridization of a known isolate's DNA to a microarray containing sequences from that isolate. The lower the Hamming distance number, the more closely matched are the sequences between the two isolates.

the arrays correctly identify strains 100% of the time (Fig. 8.2; S. N. Gardner et al., unpublished data).

8.4 CONCLUSION

The development of different DNA-based methods to interrogate and distinguish among different species and strains of pathogenic bacteria has roughly paralleled development of more rapid, less expensive DNA sequencing technologies. In the absence of easily and rapidly obtained whole genome sequences from multiple isolates of the same species, methods that indirectly detected differences among different species and different isolates of the same species were developed. Early assays were based on methods that used restriction endonucleases and basic PCR methods to demonstrate differences among species or isolates. These included AFLP, single VNTR, then, later, MLVAs. As the cost of generating DNA sequences continued to drop and the speed with which sequences could be generated increased, assay methods that could exploit the newly available direct sequence information were developed. These involved, first, single SNP assays then multiple SNP assays. Finally, as the significance of specific SNPs became more apparent with increased information from multiple isolates of the same species, multiple SNP assays developed to interrogate genetically or phylogenetically significant changes. SNPs represent evolutionarily slow genome changes relative to changes in VNTR loci. Approaches that use a combination of SNP analysis and MLVA can therefore provide significant insights into the definition of major clonal lineages and population-level structure within a species.

REFERENCES

ACHTMAN, M., ZURTH, K., MORELLI, G. et al. (1999) *Yersinia pestis*, the cause of plague, is a recently emerged clone of *Yersinia pseduotuberculosis*. *Proceedings of the National Academy of Sciences of the United States of America* **96,** 14033–14048.

ADAIR, D. M., WORSHAM, P. L., HILL, K. K. et al. (2000) Diversity in a variable-number tandem repeat from *Yersinia pestis*. *Journal of Clinical Microbiology* **38,** 1516–1519.

ANDERSEN, G. L., SIMCHOCK, J. M., and WILSON, K. H. (1996) Identification of a region of genetic variability among *Bacillus anthracis* strains and related species. *Journal of Bacteriology* **178,** 377–384.

CHENG, A. C., and CURRIE, B. J. (2005) Melioidosis: Epidemiology, pathophysiology and management. *Clinical Microbiology Reviews* **18,** 383–416.

COLLINS, M. D., and EAST, A. K. (1998) Phylogeny and taxonomy of the food-borne pathogen *Clostridium botulinum* and its neurotoxins. *Journal of Applied Microbiology* **84,** 5–17.

COOLEY, M., CARYCHAO, D., CRAWFORD-MIKSZA, L. et al. (2007) Incidence and tracking of *Escherichia coli* O157:H7 in a major produce production region in California. *PloS One* **2,** e1159.

CRICKMORE, N., ZEIGLER, D. R., FEITELSON, J. et al. (1998) Revision of the nomenclature for the *Bacillus thuringiensis* pesticidal crystal proteins. *Microbiology and Molecular Biology Reviews* **62,** 807–813.

CURRIE, B. J., HASLEM, A., PEARSON, T. et al. (2009) Identification of melioidosis outbreak by multilocus variable number tandem repeat analysis. *Emerging Infectious Diseases* **15,** 169–174.

CURRIE, B. J., THOMAS, A. D., GODOY, D. et al. (2007) Australian and Thai isolates of *Burkholderia pseudomallei* are distinct by multilocus sequence typing: Revision of a case of mistaken identity. *Journal of Clinical Microbiology* **45,** 3828–2829.

DÉGRANGE, S., CAZANAVE, C., CHARRON, A. et al. (2009) Development of multiple-locus variable-number tandem-repeat analysis for molecular typing of *Mycoplasma pneumoniae*. *Journal of Clinical Microbiology* **47,** 914–923.

DENŒUD, F. and VERGNAUD, G. (2004) Identification of polymorphic tandem repeats by direct comparison of genome sequence from different bacterial strains: A web-based resource. *BMC Bioinformatics* **5,** 4.

DEVIGNAT, R. (1951) Varietes de l'espece *Pasteurella pestis*. Nouvelle hypothese. *Bulletin World Health Organization* **4,** 247–263.

FARLOW, J., SMITH, K. L., WONG, J. et al. (2001) *Francisella tularensis* strain typing using multiple-locus, variable number tandem repeat analysis. *Journal of Clinical Microbiology* **39,** 3186–3192.

GODOY, D., RANDLE, G., SIMPSON, A. J. et al. (2003) Multilocus sequence typing and evolutionary relationships among the causative agents of melioidosis and glanders, *Burkholderia pseudomallei* and *Burkholderia mallei*. *Journal of Clinical Microbiology* **41,** 2068–2079.

GRISSA, I., BOUCHON, P., POURCELA, C., and VERGNAUD, G. (2008) On-line resources for bacterial micro-evolution studies using MLVA or CRISPR typing. *Biochimie* **90,** 660–668.

GRUNOW, R., SPLETTSTOESSER, W., MCDONALD, S. et al. (2000) Detection of *Francisella tularensis* in biological specimens using a capture enzyme-linked immunosorbent assay, an immunochromatographic handheld assay, and a PCR. *Clinical and Diagnostic Laboratory Immunology* **7,** 86–90.

GURYCOVA, D. (1998) First isolation of *Francisella tularensis* subsp. *tularensis* in Europe. *European Journal of Epidemiology* **14,** 797–802.

HOFFMASTER, A. R., HILL, K. K., GEE, J. E. et al. (2006) Characterization of *Bacillus cereus* isolates associated with fatal pneumonias: Strains are closely related to *Bacillus anthracis* and harbor *B. anthracis* virulence genes. *Journal of Clinical Microbiology* **44,** 3352–3360.

JACKSON, P. J., WALTHERS, E. A., KALIF, A. S. et al. (1997) Characterization of the variable-number tandem repeats in *vrrA* from different *Bacillus anthracis* isolates. *Applied and Environmental Microbiology* **63,** 1400–1405.

JOHANSSON, A., BERGLUND, L., ERIKSSON, U. et al. (2000) Comparative analysis of PCR versus culture for diagnosis of ulceroglandular tularemia. *Journal of Clinical Microbiology* **38,** 22–26.

JOHANSSON, A., FARLOW, J., LARSSON, P. et al. (2004a) Worldwide genetic relationships among *Francisella tularensis* isolates determined by multiple-locus variable-number tandem repeat analysis. *Journal of Bacteriology* **186,** 5808–5818.

JOHANSSON, A., FORSMAN, M., and SJÖSTEDT, A. (2004b) The development of tools for diagnosis of tularemia and typing of *Francisella tularensis*. *Acta Pathologica Microbiologica et Immunologica Scandinavica* **112,** 898–907.

KEIM, P., KALIF, A., SCHUPP, J. et al. (1997) Molecular evolution and diversity in *Bacillus anthracis* as detected by amplified fragment length polymorphism markers. *Journal of Bacteriology* **179,** 818–824.

KEIM, P., PRICE, L. B., KLEVYTSKA, A. M. et al. (2000) Multiple-locus variable-number tandem repeat analysis reveals genetic relationships within *Bacillus anthracis*. *Journal of Bacteriology* **182,** 2928–2936.

KEIM, P., VAN ERT, M. N., PEARSON, T. et al. (2004) Anthrax molecular epidemiology and forensics: Using the appropriate marker for different evolutionary scales. *Infection, Genetics and Evolution* **4,** 205–213.

KEYS, C., KEMPER, S., and KEIM, P. (2005) Highly diverse variable number tandem repeat loci in *E. coli* 0157:H7 and O55:H7 genomes for high-resolution molecular typing. *Journal of Applied Microbiology* **98,** 928–940.

KLEVYTSKA, A. M., PRICE, L. B., SCHUPP, J. M. et al. (2001) Identification and characterization of variable number tandem repeats in the *Yersinia pestis* genome. *Journal of Clinical Microbiology* **39,** 3179–3185.

LANE, D. J., PACE, B., OLSEN, G. J. et al. (1985) Rapid determination of 16S ribosomal RNA sequences for phylogenetic analyses. *Proceedings of the National Academy of Sciences of the United States of America* **82,** 6955–6959.

LECADET, M.-M., FRACHON, E., COSMAO-DUMANOIR, V. et al. (1994) An updated version of the *Bacillus thuringiensis* strains classification according to H-serotypes. Presented at the 6th International Colloquium on Invertebrate Pathology and Microbial Control, Montpellier, France, August 28–September 2, 1994.

LINDSTEDT, B.-A., HEIR, E., GJERNES, E., VARDUND, T., and KAPPERUD, G. (2003) DNA fingerprinting of Shiga-toxin producing *Escherichia coli* O157 based on multiple-locus variable-number tandem-repeats analysis (MLVA). *Annals of Clinical Microbiology and Antimicrobials* **2,** 12.

MAIDEN, M. C., BYGRAVES, J. A., FEIL, E. et al. (1998) Multilocus sequence typing: A portable approach to the identification of clones within populations of pathogenic microorganisms. *Proceedings of the National Academy of Sciences of the United States of America* **95,** 3140–3145.

MOTIN, V. L., GEORGESCU, A. M., ELLIOTT, J. M. et al. (2002) Genetic variability of *Yersinia pestis* isolates as predicted by PCR-based IS*100* genotyping and analysis of structural genes encoding glycerol-3-phosphae and dehydrogenase (*glpD*). *Journal of Bacteriology* **184,** 1019–1027.

MURPHY, M., MINIHAN, D., BUCKLEY, J. F. et al. (2008) Multiple-locus variable number of tandem repeat analysis (MLVA) of Irish verocytotoxigenic *Escherichia coli* O157 from feedlot cattle: Uncovering strain dissemination routes. *BMC Veterinary Research* **4,** 2.

NEI, M. and KUMAR, S. (2000) *Molecular Evolution and Phylogenetics*. Oxford University Press, New York.

NOLLER, A. C., MCELLISTREM, M. C., STINE, O. C. et al. (2003) Multilocus sequence typing reveals a lack of diversity among *Escherichia coli* O157:H7 isolates that are distinct by pulsed-field gel electrophoresis. *Journal of Clinical Microbiology* **41,** 675–679.

OHATA, K., SUGIYAMA, K., MASUDA, T. et al. (2008) Molecular typing of Japanese *Escherichia coli* O157:H7 isolates from clinical specimens by multilocus variable-number tandem repeat analysis and PFGE. *Journal of Medical Microbiology* **57,** 58–63.

PEARSON, T., BUSCH, J. D., RAVEL, J. et al. (2004) Phylogenetic discovery bias in *Bacillus anthracis* using single-nucleotide polymorphisms from whole genome sequencing. *Proceedings of the National Academy of Sciences of the United States of America* **101,** 13536–13541.

POURCEL, C., ANDRÉ-MAZEAUD, F., NEUBAUER, H., RAMISSE, F., and VERGNAUD, G. (2004) Tandem repeats analysis for the high resolution phylogenetic analysis of *Yersinia pestis*. *BMC Microbiology* **4,** 22.

PRICE, L. B., HUGH-JONES, M., JACKSON, P. J., and KEIM, P. (1999) Genetic diversity in the protective antigen gene of *Bacillus anthracis*. *Journal of Bacteriology* **181,** 2358–2362.

PRICE, L. B., VOGLER, A., PEARSON, T. et al. (2003) In vitro selection and characterization of *Bacillus anthracis* mutants with high-level resistance to ciprofloxacin. *Antimicrobial Agents and Chemotherapy* **47,** 2362–2365.

QI, Y., PATRA, G., LIANG, X. et al. (2001) Utilization of the *rpoB* gene as a specific chromosomal marker for real-time PCR detection of *Bacillus anthracis*. *Applied and Environmental Microbiology* **67,** 3720–3727.

READ, T. D., SALZBERG, S. L., POP, M. et al. (2002) Comparative genome sequencing for discovery of novel polymorphisms in *Bacillus anthracis*. *Science* **296,** 2028–2033.

REID, S. D., HERBELIN, C. J., BUMBAUGH, A. C., SELANDER, R. K., and WHITTAM, T. S. (2000) Parallel evolution of virulence in pathogenic *Escherichia coli*. *Nature* **406,** 64–67.

REVAZISHVILE, T., RAJANNA, C., BAKANIDZE, L. et al. (2008) Characterisation of *Yersinia pestis* isolates from natural foci of plague in the Republic of Georgia, and their relationship to *Y. pestis* isolates from other countries. *Clinical Microbiology and Infection* **14,** 429–436.

RICHARDS, R. I. and SUTHERLAND, G. R. (1997) Dynamic mutation: Possible mechanisms and significance in human disease. *Trends in Biochemical Sciences* **22,** 432–436.

U'REN, J. M., SCHUPP, J. M., PEARSON, T. et al. (2007) Tandem repeat regions within the *Burkholderia pseudo-*

mallei genome and their application for high resolution genotyping. *BMC Microbiology* **7,** 23.

VAN BELKUM, A., SCHERER, S., VAN ALPHEN, L., and VERBRUGH, H (1998) Short sequence DNA repeats in prokaryotic genomes. *Microbiology and Molecular Biology Reviews* **62,** 275–293.

VAN ERT, M. N., EASTERDAY, W. R., HUYNH, L. Y. et al. (2007a) Global genetic population structure of *Bacillus anthracis*. *PloS One* **5,** e461.

VAN ERT, M. N., EASTERDAY, W. R., SIMONSON, T. S. et al. (2007b) Strain-specific single-nucleotide polymorphism assays for the *Bacillus anthracis* Ames strain. *Journal of Clinical Microbiology* **45,** 47–53.

VOGLER, A. M., KEYS, C. E., ALLENDER, C. et al. (2007) Mutations, mutation rates, and evolution at the hypervariable VNTR loci of *Yersinia pestis*. *Mutation Research* **616,** 145–158.

WATTIAU, P., VAN HESSCHE, M., NEUBAUER, H. et al. (2007) Identification of *Burkholderia pseudomallei* and related bacteria by multiple-locus sequence typing-derived PCR and real time PCR. *Journal of Clinical Microbiology* **45,** 1045–1048.

WHITE, N.J. (2003) Melioidosis. *Lancet* **361,** 1715–1722.

ZWICK, M. E., KILEY, M. P., STEWART, A. C., MATECZUN, A., and READ, T. D. (2008) Genotyping of *Bacillus cereus* strains by microarray-based resequencing. *PloS One*, **3,** e2513.

Part II

Population Genetics of Select Bacterial Pathogens

Chapter 9

Population Genetics of *Bacillus*: Phylogeography of Anthrax in North America

LEO J. KENEFIC, RICHARD T. OKINAKA, AND PAUL KEIM

9.1 INTRODUCTION

Bacillus anthracis, the etiological agent of anthrax, is a facultative, spore-forming, gram-positive bacillus that primarily causes disease in ruminants (e.g., cattle, goats, sheep, and bison) and secondarily in humans and in other animals. There is no recognized reservoir for the disease, but sensitive animals such as cattle, which ingest infectious spores via the soil, can rapidly develop bacteremia and succumb to the disease within a matter of days (Friedlander, 2000). The bacterium has a worldwide geographic distribution (Turnbull, 2002; Van Ert et al., 2007a; WHO, 2008), and there have been numerous documented outbreaks among both domesticated and wildlife bovine species in North America including two prominent regions within the Dakotas/Nebraska and the coastal regions of Texas and Louisiana (Hugh-Jones and De Vos, 2002; Blackburn et al., 2007).

There are other documented incidences of anthrax in North America that resulted from the human handling and processing of spore-laden wool and hides in textile mills and from other activities (Brachman and Fekety, 1958; Brachman et al., 1966; Brachman, 2003). Combined with the more ecological established outbreaks in cattle, bison, goats, and so on, there is a distinct geographic distribution of anthrax in North America (Stein, 1945, 1953; Stein and Van Ness, 1955). This current review establishes correlations between the historical, geographic, and ecological accounts of anthrax outbreaks and incidents in North America and the recent developments in molecular DNA typing and reconstructive inferences that have built an accurate phylogenetic tree for *B. anthracis* (Keim et al., 2004; Pearson et al., 2004; Van Ert et al., 2007a).

9.2 HISTORY OF ANTHRAX IN NORTH AMERICA

Historical accounts of the movement of anthrax into North America have been described in many reports (Stein, 1945; Hanson, 1959; Klemm and Klemm, 1959). Some suggest

Bacterial Population Genetics in Infectious Disease, Edited by D. Ashley Robinson, Daniel Falush, and Edward J. Feil

that early French settlers may have introduced the disease into the Mississippi River Delta as early as the mid 1700s. Others suggest that anthrax may have been introduced into the Gulf Coast states (Louisiana and Texas) from Spanish herds via Veracruz, Mexico sometime in the early 1800s (Hugh-Jones and De Vos, 2002). While these accounts are somewhat obscure and inconsistent, it is clear that by the 1830s, there were widespread outbreaks of the disease reported in the coastal regions of Louisiana and subsequently in Texas (Stein, 1945).

Similarly, early accounts suggest that French-occupied Saint Dominique (modern Haiti) already had an anthrax presence when an earthquake on June 3, 1770 caused a massive outbreak described as intestinal anthrax, which killed ~15,000 people in Port-au-Prince (Morens, 2002, 2003). While these historical diagnoses lack the clinical, microbial, and molecular standards that constitute the modern definition of anthrax (Okinaka et al., 2006), these accounts from the Mississippi Delta and Haiti are supported by the ecological and continual recurrence of anthrax outbreaks in these same areas well into the twentieth century. Stein designated these to be "anthrax districts" where soil infections are confined to specific regions and "in such districts it (anthrax) constitutes a perennial problem" (Stein, 1945).

Soon after the turn of the twentieth century, the United States (United States Department of Agriculture [USDA]) had begun national surveys to account for the distribution and density of anthrax outbreaks in livestock within each of the states (Stein, 1945). In 1916, when the first national survey was completed, there were several "hot spots" for anthrax, which included New York/Pennsylvania, North Dakota/Nebraska, Texas/Louisiana, and California (see Fig. 9.1). By the fourth such survey in 1944, Stein had concluded that there were three of these anthrax districts where there had been continuous recurrences of anthrax in livestock, that is, Texas/Louisiana, North and South Dakota/Nebraska, and Northern California. By this latter survey, the level of outbreaks in certain other regions

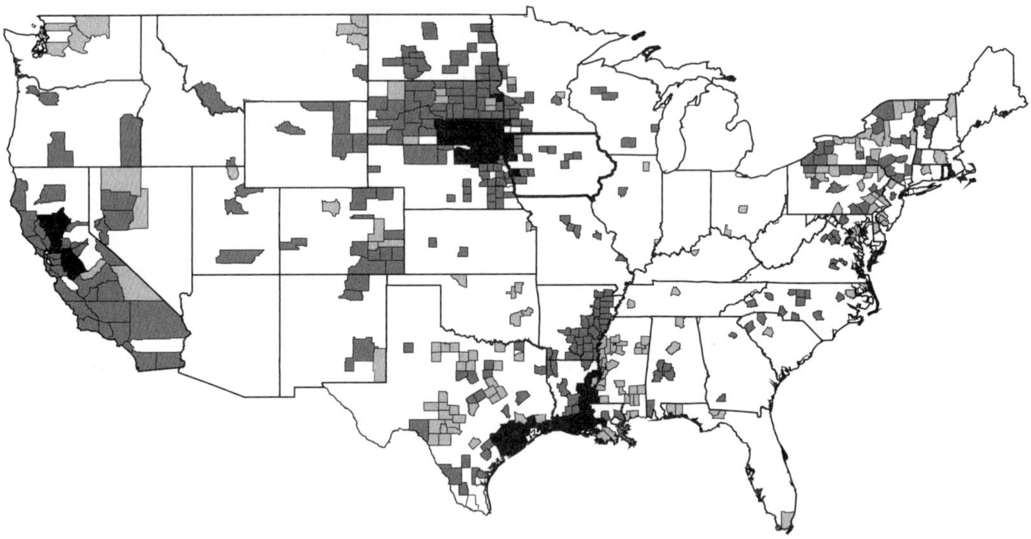

Figure 9.1 U.S. anthrax outbreaks in livestock between 1916 and 1944 (modeled after Stein, 1945). Dark patches indicate counties that Stein labeled as "anthrax districts" where anthrax outbreaks had occurred repeatedly over the periods where these surveys were conducted. Medium-density patches indicate counties where outbreaks occurred between 1934 and 1944. Light patches indicate regions with sporadic outbreaks since 1916.

had begun to decrease, probably due to increased awareness of the disease, the introduction of an effective vaccine, and urbanization along the northeastern states (New York and Pennsylvania) and California.

These geographically based outbreak analyses represent a fairly accurate distribution of ecologically established regions for anthrax on a subcontinental scale in the United States. Surprisingly, the persistence of *B. anthracis* spores and the continued recurrence of anthrax in some of these same regions allow Stein's map not only to hold its historical value but also to provide insights into our perception of the phylogeography of anthrax today. "Modern" isolates of *B. anthracis* have been recovered from sites within Stein's three anthrax districts where previously established lineages continue to cause sporadic outbreaks when conditions become favorable (Dragon et al., 1999; WHO, 2008).

9.3 THE ANTHRAX DISTRICTS AFTER 1944

9.3.1 Texas/Louisiana

Many natural outbreaks in livestock and wildlife have been described over the course of the last 60 years in North America. The majority of these outbreaks have occurred along a corridor that extends from the Mississippi Delta northward through the Dakotas and extending further north into the Northwest Territories in Central Canada. Several of the larger and more recent epizootic events in these regions have been extensively reviewed in a USDA report, "Epizootiology and Ecology of Anthrax" (http://www.aphis.usda.gov/vs/ceah/cei/taf/emerginganimalhealthissues_files/anthrax.pdf). This includes outbreaks in livestock in Louisiana in 1971 (Fox et al., 1973) and in Texas in 1974 (Fox et al., 1977) and 2001 (Hugh-Jones and De Vos, 2002). The history of sporadic outbreaks, especially in Texas, has been a year-by-year surveillance effort and dates back to the middle of the nineteenth century (Stein, 1945, 1953; Stein and Van Ness, 1955; Young, 1975; Fox et al., 1977).

Notably absent from the literature are references to an outbreak in Jim Hogg County, Texas, in 1981 where the infamous "Ames strain" from the 2001 anthrax letter attacks was originally isolated (http://www.albionmonitor.com/0208a/anthrax.html). This strain was initially characterized in vaccine challenge experiments (Little and Knudson, 1986) and has only been recovered once from nature (Kenefic et al., 2008). However, "Ames-like" isolates (Van Ert et al., 2007b) were associated with a large epizootic affecting both livestock and wildlife in Real, Uvalde and in Kinney counties in 1997 and 2001 (Hugh-Jones and De Vos, 2002). Interestingly, this particular region of southwest Texas has recorded sporadic outbreaks in the past but not to the extent of previous outbreaks located in regions much further east in Falls County, Texas (Stein, 1945). More recently, two small outbreaks in this region of Texas lend support to the idea that more intensive surveillance in this region would likely yield more Ames and Ames-like isolates (Kenefic et al., 2008).

9.3.2 North and South Dakota/Nebraska

The upper Midwest, and especially South Dakota and Nebraska, was designated by Stein as an anthrax district, and similar to the conditions in Texas, sporadic outbreaks are a natural and almost yearly burden on the farmers and ranchers in these states. As an example, North Dakota had averaged two farms per year quarantined for anthrax in the 40 years prior to 2000 when conditions became favorable and produced a larger epizootic

(MMWR, 2001; Blackburn et al., 2007). Like Texas and Louisiana, there has been a long and historical presence for anthrax in the Dakotas, in Nebraska, and in the surrounding states including Manitoba and Saskatchewan in Canada (MMWR, 2001).

9.3.3 Canada

Stein's anthrax districts could have included Canada because the recurring epizootics in North and South Dakota are often reflected in outbreaks in neighboring Manitoba and Saskatchewan (MMWR, 2001). Dragon et al. (1999) discuss accounts from the early twentieth century that indicate outbreaks in cattle in the Midwestern provinces, which include Saskatchewan and Manitoba. The first well-documented anthrax outbreak in Northern Canada was in the Slave River Lowlands of the Northwest Territories in 1962 (Dragon and Elkin, 2001) and was discovered by chance during aerial surveillance of bison herds. The origin of this outbreak is unknown, but it has been linked to the unexplained deaths of horses imported into the area from Alberta during the winter of 1960–1961. However, *B. anthracis* was not detected in the horse carcasses and there was no history of anthrax in the region of Alberta from which the animals were imported (Gates et al., 1995). Subsequent outbreaks have expanded the affected area south to include the Mackenzie Bison Sanctuary (MBS) and the Woods Buffalo National Park (WBNP). From 1962 to 1993, nine sporadic outbreaks were reported among bison herds in this region resulting in the deaths of over 1300 animals (Gates et al., 1995; Dragon et al., 1999). Three more recent outbreaks were recorded in 2000, 2001, and 2006 (Dragon et al., 1999, 2005; Nishi et al., 2007), resulting in over 200 additional fatalities. Regions where carcasses have been located are typically remote and, although not documented, anthrax outbreaks in this region prior to 1962 likely went undetected due to sparse human activity and the lack of aerial surveillance.

The seemingly distant relationship between anthrax outbreaks in bison in the northern Northwest Territory and the more recent outbreaks in Manitoba and in the Dakotas in cattle have become the most accurate phylogeographic account for a lineage in *B. anthracis*.

9.4 MOLECULAR GENOTYPING OF *B. ANTHRACIS*

B. anthracis was renowned for its lack of genetic diversity (Harrell et al., 1995), but several developments including the initial use of amplified fragment length polymorphism (AFLP) analysis (Keim et al., 1997) began to change this dilemma. AFLP transitioned into multiple locus variable number tandem repeat analysis (MLVA), and by 2000, the monomorphic *B. anthracis* population structure containing 486 isolates had been resolved into 89 distinct genotypes (Keim et al., 2000). During this time frame, whole genome sequencing had begun to greatly alter the perception for microbial analysis and genotyping and, in the case of *B. anthracis*, the sequencing, assembly, and comparative analysis of five diverse genomes (Pearson et al., 2004).

The initial comparison of these genomes led to the discovery of 3500 single-nucleotide polymorphisms (SNPs). The status of nearly 1000 of these SNPs in 26 diverse *B. anthracis* isolates revealed an extremely conserved, clonal population structure for this species. The polymorphic sites were only on branches created by the five reference strains (Fig. 9.2). The 26 diverse isolates were then aligned along the branches created by these reference strains (lines or lineages, stars in Fig. 9.2) as "collapsed" branch points or nodes (circles).

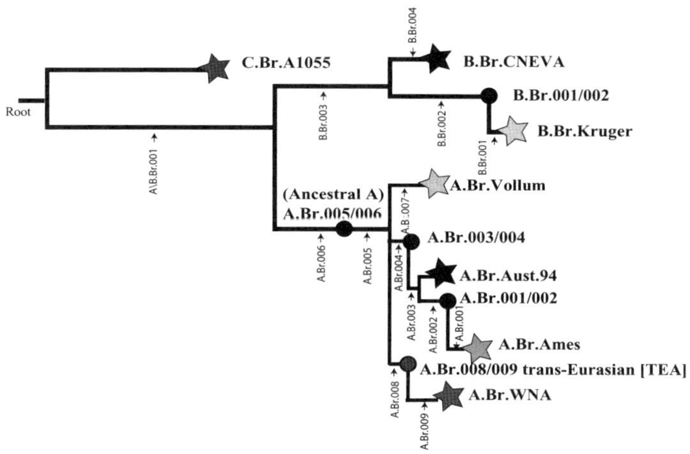

Figure 9.2 CanSNP phylogenetic tree and nomenclature. The tree was originally defined by ~1000 discovery SNPs that resulted from the whole genome sequence of seven *B. anthracis* genomes (stars). Each of these genomes defines a specific branch or lineage. This analysis also resulted in the identification of five new nodes (small circles) that were defined by two canSNPs on either side of the node (e.g., the node labeled A. Br.001/002 falls along the Ames branch). Fourteen canonical SNPs (vertical arrows; e.g., A.Br.001 = canSNP that described the A.Br.Ames lineage) were used to analyze each of 1033 isolates on this tree. All 1033 isolates fell into one and only one of these five nodes (circles) or seven lineages (stars). See references (Pearson et al., 2004; Van Ert et al., 2007a) for more details. See color insert.

The extremely conserved nature of the phylogenetic tree for *B. anthracis* led to a hypothesis that only a few canonical single-nucleotide polymorphisms (canSNPs) at key positions in the SNP-based tree could be used to represent the branches created by the 1000 original SNPs (Keim et al., 2004). A few SNPs could then replace the 1000 original SNPs and still describe the conserved phylogenetic pattern for *B. anthracis.* This hypothesis was confirmed when 14 canSNPs were used to accurately place >1000 worldwide isolates of *B. anthracis* into one and only one of 12 canSNP genotypes (Van Ert et al., 2007a). The addition of two *B. anthracis* whole genome sequences created seven total sublineages (or branches) and five collapsed branch points (nodes depicted as circles), and all 1000 isolates fell into one of the seven sublineages or into one of the five collapsed branch points (called subgroups in Van Ert et al., 2007a). The combination of these canSNP analyses and a more recent MLVA of 15 of these isolates increased the number of the once genetically indistinguishable *B. anthracis* species into 221 separate genotypes (Van Ert et al., 2007a).

It is important to reemphasize that the 12 genotypes include seven sublineages represented by the whole genome sequencing of seven reference genomes (the stars in Fig. 9.2). The SNPs that are unique to each of these reference genomes form a linear branch or lineage that terminates with the reference genome, for example, A.Br.Ames (Pearson et al., 2004). The seven sublineages include four A branch isolates (A.Br.Ames, A.Br.WNA, A.Br.Aust94, and A.Br.Vollum), two B branch isolates (B.Br.CNEVA and B.Br.KrugerB), and one C branch isolate (C.Br.A1055). The global distribution of >1000 *B. anthracis* isolates has been determined using canSNP typing, and all of these isolates fall into one of 12 original canSNP sublineages or subgroups (Van Ert et al., 2007a).

9.5 GENOTYPES WITHIN THE ANTHRAX DISTRICTS IN NORTH AMERICA

The global distribution of *B. anthracis* isolates has several distinctive features, but the most striking are the geographic locations of two groups, the node A.Br.008/009, more commonly referred to as the trans-Eurasian (TEA) subgroup, because these isolates are mostly spread across the whole of Europe, the Middle East, and portions of China and the sublineage A.Br.WNA (for Western North America [WNA]). In North America, the WNA sublineage represents the vast majority of a large number of isolates (currently numbering 350 isolates) that were mostly found in North and South Dakota and in the entire central portions of Canada stretching from Manitoba/Saskatchewan to the Northwest Territories. These results suggest that Stein's anthrax district for this region is considerably larger than the areas in his survey that were limited to the contiguous United States.

Only a small number of isolates (20) have actually been recovered and genotyped from the other prominent anthrax districts (Louisiana and Texas) with the canSNP typing methodology. These data indicate three distinct genotypes with 11 isolates carrying the Ames-like genotype (Simonson et al., 2009), five carrying the Vollum-like genotype (A.Br.Vollum), and four carrying the WNA genotype (A.Br.WNA) (unpublished results). An important unknown is whether all three of these lineages actually represent ecologically established lineages in either Texas or Louisiana. Evidence for this possibility is supported by the fact that all of the isolates involved in these analyses were recovered from bovines, deer, or goats.

The third anthrax district in Stein's map is located in California. Although California experienced severe outbreaks of anthrax in the San Joaquin and Sacramento Valleys in 1937 and 1942 (Stein, 1945), there have been no subsequent outbreaks in this region. Stein's data on incidence of infections during this time suggest that anthrax would be endemic and subsequent infections would be likely. Several reasons may account for this. Rangelands and pasturelands once used for grazing have, in many instances, been developed and urbanized (USDA, 2001). Some of this land has also been converted to prime agricultural lands used for farming. Additionally, some of the historic ranches in this region have been designated as land trust areas and are no longer used for ranching.

Nevertheless, one of the ecologically established lineages in California belongs to the B branch on the *B. anthracis* tree (Fig. 9.2). This is an old but rare lineage, and isolates belonging to this major grouping have only been found in Africa, in limited locations in Europe and in California (Van Ert et al., 2007a). Two questions remain in California: (i) What is the origin of the B Branch lineage? and (ii) Did other genotypes become ecologically established in vast areas that had previously reported sporadic outbreaks?

9.6 PHYLOGENETIC RESOLUTION WITHIN THE WNA LINEAGE

The canSNP genotyping experiments were useful in organizing and in accurately placing each isolate from a large collection into one of 12 phylogenetic clusters. But this data set lacked significant resolution because only 26 diverse isolates were used to construct the first whole genome comparison SNP tree (Pearson et al., 2004) and because only a small subset of 14 SNPs was used in the canSNP typing experiment (Van Ert et al., 2007a). As an initial step to provide further resolution within the WNA lineage, a custom Affymetrix whole genome tiling array using 2850 SNPs was used to genotype 128 diverse *B. anthracis* isolates (Kenefic et al., 2009) including 10 isolates from the TEA node and 21 isolates

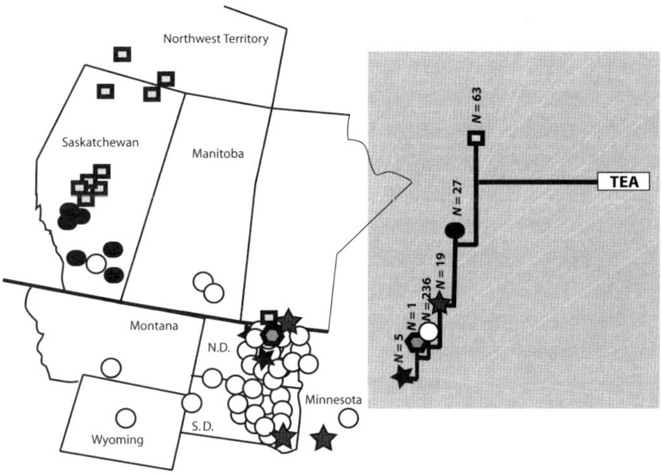

Figure 9.3 Phylogeography of the Western North American clade. The geolocation of isolates along a phylogenetic tree (right panel) illustrates the movement of the ancestral population southward to the most derived populations in North and South Dakota in the United States. Note, for example, that the most ancestral population (yellow squares, $N = 63$) was mostly found in the Northwest Territory and in Saskatchewan with a spurious outbreak to the south, and that the next oldest population (black circles, $N = 27$) is entirely in the southern half of Saskatchewan. In the younger populations, the disease is mostly in domesticated livestock, and thus the geolocation becomes more mixed, probably the result of increased human interactions. See color insert.

from the WNA lineage. This analysis identified 78 unique SNPs that separate the TEA node from the WNA lineage and 28 additional SNPs that subdivide the WNA lineage into six genotypes (Kenefic et al., 2009).

Ten of the WNA-specific SNPs were converted into Taqman^{R} MGB dual-probe real-time PCR SNP assays, and these assays were used to genotype 352 WNA isolates. The genotyping of these 352 individual isolates followed both a phylogenetic and geographic pattern that defines an ancestral to a derived population structure from the Canadian Northwest Territories through Manitoba/Saskatchewan and into North and South Dakota in the United States (Fig. 9.3; also Kenefic et al., 2009). These results indicate a relatively rapid phylogeographic southward expansion from Northern Canada into the Upper Midwestern United States.

These observations coupled to molecular clock calibrations prompted a hypothesis (Kenefic et al., 2009) for a northern and pre-Columbian introduction of anthrax into North America. They suggested that early human migrations included transport of the WNA lineage across the Beringian steppe ecosystem land bridge (Shapiro et al., 2004) between Asia and the North American continent ~13,000 ybp. The large evolutionary gap (78 SNPs) with missing representative taxa between the TEA clade and the WNA sublineage suggests that the evolutionary steps from TEA to WNA are not present in the highly populated regions of Europe. The Beringia land bridge model would suggest that this transformation may have occurred in northern Asia and that the WNA lineage experienced a bottleneck that created the unique modern lineage.

A sidelight to the geography and accounts of anthrax in North America are historical/medical accounts of a massive outbreak in humans for a disease that the French called "charbon" in Saint Dominique (modern-day Haiti) that was associated with a large earthquake on June 3, 1770 (Morens, 2002, 2003). This country has had recurring outbreaks of anthrax for much of the time since this devastating incident. Isolates from Haiti collected from the mid-1990s and earlier have been typed using canSNPs, and these isolates belong to the WNA clade (Van Ert et al., 2007a). The introduction of anthrax into Haiti appears to involve the movement of an endemic North American clade and not the genotypes common to Europe.

9.7 PHYLOGEOGRAPHIC RESOLUTION WITHIN THE AMES LINEAGE

The anthrax district defined in Stein's surveys for Louisiana/Texas has long been thought to be the originating point for anthrax in the United States. And while the recent hypothesis for a pre-Columbian entry into North America would supersede this notion, it is still certain that anthrax outbreaks have been prevalent in these coastal regions since the early 1800s. Despite a number of epizootic events in both Texas and Louisiana since 1957, there appear to be very few isolates from these areas that have been successfully archived in accredited Select Agent facilities. An example would be 179 culture-confirmed cases of anthrax from 39 counties in Texas in the USDA report described earlier (Dr. L. Sneed, pers. comm.).

Despite these deficiencies, there are 20 isolates recovered from animals in Texas and in Louisiana that have been canSNP genotyped. Of these, the 11 Ames-like isolates from Texas have been extensively analyzed using MLVA and SNP markers that are specific to resolving isolates in the Ames lineage (Van Ert et al., 2007b; Kenefic et al., 2008; Simonson et al., 2009). Five of the isolates were isolated in 1997, 2001, 2001, 2006, and 2007 in the Big Bend area of Texas from Terrell, Real, Kinney, and Uvalde Counties where sporadic outbreaks have been reported on a nearly yearly basis over the last several decades. Fine-structure SNP typing using 31 SNPs specific to the Ames branch places these five isolates (plus five isolates from Texas with no county assignment) on a node that is four SNPs removed from the derived position of the original Ames strain. This is significant because the closest relatives to these 10 Ames-like isolates and the original Ames strain are eight isolates recovered from a collection obtained from China (Simonson et al., 2009).

These results indicate that the Ames lineage is very rare (only in the United States and in China, thus far) and that the closest known common ancestor to the Ames-like node and the Ames strain resides in China and is only divergent by 8 and 12 SNPs, respectively (Simonson et al., 2009). And although 8 and 12 SNPs do not represent an extended evolutionary presence of the WNA lineage, they do support Stein's idea that once ecologically established, the Ames lineage could have caused repeated infections over the course of a few hundred years. These observations could also pertain to both the WNA and Vollum-like isolates recovered from cattle, deer, or goats in Texas (unpublished results). Ecologically established *B. anthracis* isolates can cause infections decades after an incident, and this pattern appears to repeat itself in many ecological niches in China (Simonson et al., 2009), in Texas/Louisiana, and in the Dakotas.

An interesting historical aspect to anthrax outbreaks in Texas involves the great Western cattle drives in the 20–25 years following the end of the Civil War. There have been reports that the 1000-mile or more cattle drives from Texas had a significant affect in spreading anthrax along a corridor from Texas through Oklahoma/Kansas and points north (e.g., *The New York Times*, October 29, 2001). And an ecological niche model and historical records provide evidence that validate the notion that anthrax could have spread in this manner (Blackburn et al., 2007). But Stein's anthrax district map implies a disconnect between the outbreaks in the Dakotas/Nebraska and those in Texas/Louisiana.

The recent discovery of the WNA lineage into the Dakotas and Nebraska to the north and the Ames and Vollum lineages only in Texas or in Louisiana is a significant genetic discontinuity along this corridor. These results suggest that while cattle probably did move anthrax along these Western cattle trails, the disease in Texas did not spread much beyond its borders.

9.8 ADDITIONAL *B. ANTHRACIS* GENOTYPES IN NORTH AMERICA

9.8.1 Other Outbreaks

There have been other sporadic outbreaks where additional canSNP genotypes have been identified in North America. These include the recovery of the A.Br.001/002 genotype from isolates in cattle from Ohio in 1952, a cow from Kansas (date uncertain), and a llama in Texas (date uncertain). The A.Br.001/002 subgroup is a rare subgroup that appears to have either originated and/or expanded significantly in China (Simonson et al., 2009).

The A.Br.Australia94 lineage is a branch created by the sequencing of an isolate from Australia, but the origins of the lineage appear to be in Asia and/or in the Middle East, for example, India, China, and Turkey (Van Ert et al., 2007a; Simonson et al., 2009). In North America, there have been two incidents in cattle in Ohio (date uncertain) and in Oklahoma (cow spleen, August 20, 1957) that canSNP type to this lineage. While these appear to be sporadic events without a significant history or precedence, the specific date of the single isolate from Oklahoma obtained from the Centers for Disease Control and Prevention (CDC), Atlanta, corresponds to the height of a major epizootic (August 20, 1957) that occurred on the border between Oklahoma and Kansas in 1957 (Van Ness et al., 1959). The reported losses totaled 1627 farm animals on 741 premises. These correlations are important because the map of anthrax in livestock (1916–1944) generated by Stein (Fig. 9.1) does show sporadic activity in two counties in northeastern Oklahoma that were involved (Rogers) or were immediately adjacent (Tulsa) to the outbreak in 1957. Because this outbreak appears to be a recurrence of previous anthrax outbreaks, a single culture-confirmed isolate from the incident describes at least one new genotype, A.Br. Aust94, which became an ecological established lineage. The origin of this lineage in Oklahoma/Kansas remains a mystery.

Similarly, the A.Br.003/004 lineage is also rare in North America. But two archival isolates in our collection type to this lineage and are representative of an outbreak in Mississippi (1957) and one from a ranch in Florida (1952). These latter genotypes have not become widely established and may suggest a significantly later introduction and/or a lack of opportunities to spread. Outside of North America, the A.Br.003/004 lineage is well established in South America (Argentina, Bolivia, and Chile) with rare isolates from Europe and South Africa as well (Van Ert et al., 2007a).

9.8.2 Industrial Incidents

"Woolsorter's disease" or inhalation anthrax has not been the problem in the United States as it was in Europe, and until the anthrax letters of 2001, less than 20 cases of inhalation anthrax had been reported since 1900 (Brachman et al., 1966). Nevertheless, industrial outbreaks of cutaneous anthrax have been historically associated with textile mills located along the East Coast, that is, New Hampshire, Massachusetts, Rhode Island, Pennsylvania, North Carolina, and South Carolina (Brachman and Fekety, 1958; Brachman et al., 1966;

Suffin et al., 1978; Bell et al., 2002; Plotkin et al., 2002). The sources for spores associated with various human infections (primarily goat hair and wool) include imports from many countries, for example, India, Pakistan, Syria, Turkey, Iraq, Afghanistan, China, and Iran (Brachman et al., 1960; Dahlgren et al., 1960; MMWR, 1988). It is not evident that incidents in textile mills can lead to ecological establishment in wildlife and domesticated animals, but countries in the Middle East and in the Far East involved in importation of hides into the United States are most likely to harbor the A.Br.Vollum, A.Br.Aust94, and the TEA clades. These are not very common lineages in North America.

9.9 CONCLUSIONS

The phylogeographic pattern for the dispersal of anthrax in North America has been determined using historical, geography-based analysis of outbreaks dating back to the turn of the twentieth century coupled to molecular genetic analysis that can accurately establish the phylogenetic structure of *B. anthracis*. These studies include a phylogeographic reconstruction that suggests a pre-Columbian evolution and migration of the WNA lineage from the European/Asian continent to the North American continent. This hypothesis suggests that the migration may have occurred ~13,000 ybp when a land bridge existed in what is now the Bering Straits. These conclusions are based on the distribution of >350 isolates belonging to the WNA lineage along an anthrax corridor that stretches from the Northwest Territories in Canada to South Dakota and Nebraska.

These studies also describe the analysis of isolates belonging to an anthrax district that covers the coastal regions of Texas/Louisiana and the description of the ecologically established, Ames-like lineage that has had a significant presence in Texas. Other potential established lineages in this area include Vollum and WNA. But these studies suffer from sampling and archival issues.

There are 8 of 10 canSNP groups identified in North America: A.Br.WNA, A.Br.Ames, A.Br.Vollum, A.Br.001/002, A.Br.003/004, A.Br.Aust94, B.Br.001/002, and A/B.001.002. The three anthrax districts described geographically by Stein contains the A.Br.WNA lineage in the Midwestern United States, the B.Br.001/002 lineage in California, and the A.Br.Ames lineage in Texas/Louisiana with Vollum and WNA lineages possibly also ecologically established in the latter district.

REFERENCES

Bell, J. H., Fee, E., and Brown, T. M. (2002) Anthrax and the wool trade. 1902. *Am J Public Health* **92,** 754–757.

Blackburn, J. K., Mcnyset, K. M., Curtis, A., and Hugh-Jones, M. E. (2007) Modeling the geographic distribution of *Bacillus anthracis*, the causative agent of anthrax disease, for the contiguous United States using predictive ecologic niche modeling. *Am J Trop Med Hyg* **77,** 1103–1110.

Brachman, P. S. (2003) Infectious diseases–Past, present, and future. *Int J Epidemiol* **32,** 684–686.

Brachman, P. S. and Fekety, F. R. (1958) Industrial anthrax. *Ann N Y Acad Sci* **70,** 574–584.

Brachman, P. S., Kaufman, A. F., and Dalldorf, F. G. (1966) Industrial inhalation anthrax. *Bacteriol Rev* **30,** 646–659.

Brachman, P. S., Plotkin, S. A., Bumford, F. H., and Atchison, M. M. (1960) An epidemic of inhalation anthrax: The first in the twentieth century. II. Epidemiology. *Am J Hyg* **72,** 6–23.

Dahlgren, C. M., Buchanan, L. M., Decker, H. M., Freed, S. W., Phillips, C. R., and Brachman, P. S. (1960) *Bacillus anthracis* aerosols in goat hair processing mills. *Am J Hyg* **72,** 24–31.

Dragon, D. C., Bader, D. E., Mitchell, J., and Woollen, N. (2005) Natural dissemination of *Bacillus anthracis* spores in Northern Canada. *Appl Environ Microbiol* **71,** 1610–1615.

Dragon, D. C. and Elkin, B. T. (2001) An overview of early anthrax outbreaks in Northern Canada: Field reports of the Health of Animals Branch, Agriculture Canada, 1962–1971. *Artic* **54,** 32–40.

DRAGON, D. C., ELKIN, B. T., NISHI, J. S., and ELLSWORTH, T. R. (1999) A review of anthrax in Canada and implications for research on the disease in northern bison. *J Appl Microbiol* **87,** 208–213.

FOX, M. D., BOYCE, J. M., KAUFMANN, A. F., YOUNG, J. B., and WHITFORD, H. W. (1977) An epizootiologic study of anthrax in Falls County, Texas. *J Am Vet Med Assoc* **170,** 327–333.

FOX, M. D., KAUFMANN, A. F., ZENDEL, S. A., KOLB, R. C., SONGY, C. G. Jr., CANGELOSI, D. A., and FULLER, C. E. (1973) Anthrax in Louisiana, 1971: Epizootiologic study. *J Am Vet Med Assoc* **163,** 446–451.

FRIEDLANDER, A. M. (2000) Anthrax: Clinical features, pathogenesis, and potential biological warfare threat. *Curr Clin Top Infect Dis* **20,** 335–349.

GATES, C. C., ELKIN, B. T., and DRAGON, D. C. (1995) Investigation, control and epizootiology of anthrax in a geographically isolated, free-roaming bison population in Northern Canada. *Can J Vet Res* **59,** 256–264.

HANSON, R. P. (1959) The earliest account of anthrax in man and animals in North America. *J Am Vet Med Assoc* **135,** 463–465.

HARRELL, L. J., ANDERSEN, G. L., and WILSON, K. H. (1995) Genetic variability of *Bacillus anthracis* and related species. *J Clin Microbiol* **33,** 1847–1850.

HUGH-JONES, M. E. and DE VOS, V. (2002) Anthrax and wildlife. *Rev Sci Tech* **21,** 359–383.

KEIM, P., KALIF, A., SCHUPP, J., HILL, K., TRAVIS, S. E., RICHMOND, K., ADAIR, D. M., HUGH-JONES, M., KUSKE, C. R., and JACKSON, P. (1997) Molecular evolution and diversity in *Bacillus anthracis* as detected by amplified fragment length polymorphism markers. *J Bacteriol* **179,** 818–824.

KEIM, P., PRICE, L. B., KLEVYTSKA, A. M., SMITH, K. L., SCHUPP, J. M., OKINAKA, R., JACKSON, P. J., and HUGH-JONES, M. E. (2000) Multiple-locus variable-number tandem repeat analysis reveals genetic relationships within *Bacillus anthracis*. *J Bacteriol* **182,** 2928–2936.

KEIM, P., VAN ERT, M. N., PEARSON, T., VOGLER, A. J., HUYNH, L. Y., and WAGNER, D. M. (2004) Anthrax molecular epidemiology and forensics: Using the appropriate marker for different evolutionary scales. *Infect Genet Evol* **4,** 205–213.

KENEFIC, L. J., PEARSON, T., OKINAKA, R., CHUNG, W.-K., MAX, T., VAN ERT, M., MARSTON, C. K., GUTIERREZ, K., SWINFORD, A. K., HOFFMASTER, A., and KEIM, P. (2008) Texas isolates closely related to *Bacillus anthracis* Ames. *Emerg Infect Dis* **14,** 1825.

KENEFIC, L., PEARSON, T., OKINAKA, R. T., SCHUPP, J. M., WAGNER, D. M., TRIM, C. P., CHUNG, W.-K., BEAUDRY, J. A., FOSTER, J. T., MEAD, J. I., and KEIM, P. (2009) Pre-Columbian origins for North American anthrax. *PloS One* **4**(3), E4813.

KLEMM, D. M. and KLEMM, W. R. (1959) A history of anthrax. *J Am Vet Med Assoc* **135,** 458–462.

LITTLE, S. F. and KNUDSON, G. B. (1986) Comparative efficacy of *Bacillus anthracis* live spore vaccine and protective antigen vaccine against anthrax in the guinea pig. *Infect Immun* **52,** 509–512.

MILROY, R. E. (2001) Anthrax hides along cattle trails of the old west. *New York Times*, October 29, 2001, p. A9.

MMWR (1988) Human cutaneous anthrax–North Carolina, 1987. *MMWR Morb Mortal Wkly Rep* **37,** 413–414.

MMWR (2001) Human anthrax associated with an epizootic among livestock–North Dakota, 2000. *MMWR Morb Mortal Wkly Rep* **50,** 677–680.

MORENS, D. M. (2002) Epidemic anthrax in the eighteenth century, the Americas. *Emerg Infect Dis* **8,** 1160–1162.

MORENS, D. M. (2003) Characterizing a "new" disease: Epizootic and epidemic anthrax, 1769–1780. *Am J Public Health* **93,** 886–893.

NISHI, J. S., ELLSWORTH, T. R., LEE, N., DEWAR, D., ELKIN, B. T., and DRAGON, D. C. (2007) Northwest Territories. An outbreak of anthrax (*Bacillus anthracis*) in free-roaming bison in the Northwest Territories, June–July 2006. *Can Vet J* **48,** 37–38.

OKINAKA, R., PEARSON, T., and KEIM, P. (2006) Anthrax, but not *Bacillus anthracis*? *PLoS Pathog* **2,** E122.

PEARSON, T., BUSCH, J. D., RAVEL, J., READ, T. D., RHOTON, S. D., U'REN, J. M., SIMONSON, T. S., KACHUR, S. M., LEADEM, R. R., CARDON, M. L., VAN ERT, M. N., HUYNH, L. Y., FRASER, C. M., and KEIM, P. (2004) Phylogenetic discovery bias in *Bacillus anthracis* using single-nucleotide polymorphisms from whole-genome sequencing. *Proc Natl Acad Sci U S A* **101,** 13536–13541.

PLOTKIN, S. A., BRACHMAN, P. S., UTELL, M., BUMFORD, F. H., and ATCHISON, M. M. (2002) An epidemic of inhalation anthrax, the first in the twentieth century: I. Clinical features. 1960. *Am J Med* **112,** 4–12; discussion 2–3.

SHAPIRO, B., DRUMMOND, A. J., RAMBAUT, A., WILSON, M. C., MATHEUS, P. E., SHER, A. V., PYBUS, O. G., GILBERT, M. T., BARNES, I., BINLADEN, J., WILLERSLEV, E., HANSEN, A. J., BARYSHNIKOV, G. F., BURNS, J. A., DAVYDOV, S., DRIVER, J. C., FROESE, D. G., HARINGTON, C. R., KEDDIE, G., KOSINTSEV, P., KUNZ, M. L., MARTIN, L. D., STEPHENSON, R. O., STORER, J., TEDFORD, R., ZIMOV, S., and COOPER, A. (2004) Rise and fall of the Beringian steppe bison. *Science* **306,** 1561–1565.

SIMONSON, T. S., OKINAKA, R. T., WANG, B., EASTERDAY, R., HUYNH, L., U'REN, J. M., DUKERICH, M., ZANECKI, S. R., KENEFIC, L. J., BEAUDRY, J., JAMESMSCHUPP, J. M., PEARSON, T., WAGNER, D. M., HOFFMASTER, A., RAVEL, J., and KEIM, P. (2009) *Bacillus anthracis* in China and its relationship to worldwide lineages. *BMC Microbiol* **9,** 71.

STEIN, C. D. (1945) The history and distribution of anthrax in livestock in the United States. *Vet Med* **40,** 340–349.

STEIN, C. D. (1953) Anthrax in animals and its relationship to the disease in man. *Tex Rep Biol Med* **11,** 534–546.

STEIN, C. D. and VAN NESS, G. B. (1955) A ten-year survey of anthrax in livestock with special reference to outbreaks in 1954. *Vet Med* **11,** 579–589.

SUFFIN, S. C., CARNES, W. H., and KAUFMANN, A. F. (1978) Inhalation anthrax in a home craftsman. *Hum Pathol* **9,** 594–597.

TURNBULL, P. C. (2002) Introduction: Anthrax history, disease and ecology. *Curr Top Microbiol Immunol* **271,** 1–19.

USDA (2001) Natural Resources Conservation Service. http://www.aphis.usda.gov/vs/ceah/cei/taf/emerginganimalhealthissues_files/anthrax.pdf (accessed December 23, 2009).

Van Ert, M. N., Easterday, W. R., Huynh, L. Y., Okinaka, R. T., Hugh-Jones, M. E., Ravel, J., Zanecki, S. R., Pearson, T., Simonson, T. S., U'ren, J. M., Kachur, S. M., Leadem-Dougherty, R. R., Rhoton, S. D., Zinser, G., Farlow, J., Coker, P. R., Smith, K. L., Wang, B., Kenefic, L. J., Fraser-Liggett, C. M., Wagner, D. M., and Keim, P. (2007a) Global genetic population structure of *Bacillus anthracis*. *PloS One* **2,** E461.

Van Ert, M. N., Easterday, W. R., Simonson, T. S., U'ren, J. M., Pearson, T., Kenefic, L. J., Busch, J. D., Huynh, L. Y., Dukerich, M., Trim, C. B., Beaudry, J., Welty-Bernard, A., Read, T., Fraser, C. M., Ravel, J., and Keim, P. (2007b) Strain-specific single-nucleotide polymorphism assays for the *Bacillus anthracis* Ames strain. *J Clin Microbiol* **45,** 47–53.

Van Ness, G. B., Plotkin, S. A., Nuffaker, R. H., and Evans, W. G. (1959) The Oklahoma-Kansas anthrax epizootic of 1957. *J Am Vet Med Assoc* **1,** 125–129.

WHO (2008) The World Health Organization Collaborating Center for Remote Sensing and Geographic Information Systems for Public Health: The World Anthrax Data Site. http://www.vetmed.lsu.edu/whocc/ (accessed December 23, 2009).

Young, J. B. (1975) Epizootic of anthrax in Falls County, Texas. *J Am Vet Med Assoc* **167,** 842–843.

Chapter 10

Population Genetics of *Campylobacter*

SAMUEL K. SHEPPARD, MARTIN C. J. MAIDEN, AND DANIEL FALUSH

10.1 INTRODUCTION

The genus *Campylobacter* comprises 18 recognized species of microaerobic ε-proteobacteria (On, 2001; Humphrey et al., 2007; Debruyne et al., 2008). Of these, *Campylobacter jejuni* and *Campylobacter coli* are the most medically important inhabiting the gastrointestinal tract of numerous animal species and causing gastroenteritis in humans. In the past two decades, these zoonoses have emerged as the most common bacterial causes of gastroenteritis worldwide (Friedman et al., 2000) causing approximately 2.5 million annual cases in the United States and 340,000 cases in the United Kingdom (Allos, 2001; Kessel et al., 2001), over three times the number of cases caused by *Salmonella*, *Escherichia coli* O157 : H7, and *Listeria monocytogenes* combined (FSA, 2002; CDC, 2008). The ability to infect multiple hosts (often benignly) and humans, via food and the environment, has contributed to the prevalence of these pathogens. However, questions remain about the genetic basis and ecology of host specificity and niche adaptation in *Campylobacter* and how this relates to the emergence of human infection.

Campylobacter is part of the commensal gut microbiota of many species of farm and wild animals and birds, and contamination of human food chains can occur at any point from the farm to the consumer. Potential sources of human infection include contaminated meat, poultry, water, milk, and contact with animals (Kapperud et al., 2003; Friedman et al., 2004). Analytical epidemiology, such as risk assessment and case-control studies, provides indirect evidence for the origin of disease (Friedman et al., 2000, 2004; Neimann et al., 2003; Ethelberg et al., 2005; Mylius et al., 2007; Nauta et al., 2007; Stafford et al., 2007), but the relative contribution of different infection reservoirs has been difficult to establish.

In contrast to gastroenteritis caused by *Escherichia coli*, most human *Campylobacter* infections are sporadic, with very few recognized outbreaks that might indicate a common infection source (Pebody et al., 1997; Friedman et al., 2000; Frost et al., 2002; Adak et al., 2005). This has made interventions for control of transmission difficult because of

Bacterial Population Genetics in Infectious Disease, Edited by D. Ashley Robinson, Daniel Falush, and Edward J. Feil

uncertainty over source attribution and has led to effort being put into the development of molecular typing techniques to link disease to source. Genetic typing of bacteria has enhanced epidemiological investigation of outbreaks of foodborne pathogens including *E. coli* O157:H7 (Bender et al., 1997), *Salmonella enterica* (Bender et al., 2001), and *L. monocytogenes* (Olsen et al., 2005). Microbial typing schemes such as serotyping, pulsed-field gel electrophoresis (PFGE), and *flaA* typing (Harrington et al., 1997; Wassenaar et al., 1998; Steinbrueckner et al., 2001) have all been attempted for *Campylobacter*, but the utility of these schemes has been hampered by poor concordance between methods as well as often by poor correlations between genotype and phenotypes such as host specificity, which contrasts with the strong associations found between *Salmonella* serotypes and other genetic and phenotypic traits (Morales et al., 2005; Tankouo-Sandjong et al., 2007). These analyses suggest that the high recombination rates in *Campylobacter* may help to explain the absence of strong correlations. Multilocus sequence typing (MLST) has the advantage of complete portability between laboratories and also facilitates explicit evolutionary analysis of the factors responsible for high diversity and lack of strong correlation among traits.

10.2 HUMAN INFECTION

Reported cases of human *Campylobacter* infection in high-income countries are characterized by the onset of gastroenteritis within 2–5 days with symptoms including diarrhoea, fever, and abdominal pain and, less commonly, vomiting and bloody diarrhoea (Gillespie et al., 2006). Rare extraintestinal complications, including reactive arthritis and the neurological conditions of Guillain–Barré and Miller–Fisher (Servan et al., 1995) syndromes, can occur in the weeks following infection (Tam et al., 2006; Pope et al., 2007). The majority of cases of campylobacteriosis (85–97%) in the United Kingdom and in the United States remain unreported (Mead et al., 1999; Wheeler et al., 1999), and asymptomatic infection may be relatively common with symptomatic disease incidence estimated to be below 1% in high-income countries (Blaser et al., 1983; Wheeler et al., 1999; Friedman et al., 2000).

The epidemiology of campylobacteriosis in low-income countries appears to be different. There is some evidence that the symptoms are less severe (Coker et al., 2002) and that both symptomatic and asymptomatic infection are over 100 times more common in young children than in high-income countries (Oberhelman and Taylor, 2000). Serological and serial culture studies have shown that as many as 10% of asymptomatic children under 1 year old are infected with *C. jejuni* and that by the age of two, most children have been infected on multiple occasions (Richardson et al., 1983; Calva et al., 1988; Figueroa et al., 1989; Pazzaglia et al., 1991). It is not known if, under these circumstances, humans act as a reservoir for *Campylobacter* infection. In most cases, it is difficult to distinguish secondary transmission from shared primary infection, and while there is evidence for person-to-person transmission, from infected food handlers (Olsen et al., 2001) and among homosexual men (Gaudreau and Michaud, 2003), the role of humans as long-term hosts in low-income countries needs further investigation. If humans are an infection reservoir for particular *Campylobacter* strains associated with asymptomatic infection, then these potentially represent ancestral human-associated lineages. Disease isolate data sets from industrialized countries—including Australia, the United States, and England—have relatively similar genetic composition and F_{ST} values, calculated from concatenated nucleotide sequences of the MLST loci, suggesting a shared gene pool. Whereas isolates from these countries are 10% differentiated from disease isolates in the Dutch West Indies (Curaçao) on the basis of F_{ST} (Dingle et al., 2008). An understanding of the global epidemiology of human campylobacteriosis will be enhanced by the use of MLST as a common typing method, but the efficacy of this technique is dependent upon the availability of samples from diverse sources including clinical isolates from low-income countries.

10.3 GENETIC STRUCTURE

Campylobacter is a diverse organism and this makes it difficult to adequately catalog the genetic structure, even for the "core genome," which comprises the genic and nongenic regions that are shared by the great majority of strains. The problem is compounded by the high rates of genetic exchange among these bacteria. The high levels of recombination in *C. jejuni* can be demonstrated in various ways. First, in simple terms, a tree based on concatenating multiple genetic regions from each strain (e.g., MLST data) has little evidence of deep genetic structure that would indicate long periods of independent evolution of different groups. Second, the continued discovery of new genotypes, when the discovery of new alleles has reached an asymptote, suggests that the majority of genetic variation is generated by the reassortment of existing alleles, not the generation of new ones. Finally, *C. jejuni* strains that are distantly related on the global tree often share the same allele at individual MLST loci, which provides strong evidence of genetic exchange. The high rate of sharing of specific alleles provides evidence that the average size of imported fragments is larger than the average size of MLST fragments, so that new alleles are broken up relatively infrequently, and most of the variation within *C. coli* and *C. jejuni* genotypes is the result of reassortment of known alleles (Sheppard et al., 2008). Estimates of the frequency of recombination and the size of imported fragments vary, and several model-based approaches have been used to describe recombination. Using an approximate likelihood method for analyzing polymorphic sites, recombination of fragments with short tract lengths (225–750 bp) was found to be of a similar rate and magnitude to mutation in *Campylobacter* (Fearnhead et al., 2005). Other studies have estimated the average size of recombination fragments to be higher, approximately 3.3 kbp (Schouls et al., 2003), and an approach using a combined population genetics microevolutionary model suggests that recombination has a fundamental role in *Campylobacter* evolution, generating twice as much diversity as de novo mutation (Wilson et al., 2009).

Unlike the human gastric pathogen *Helicobacter pylori* (Suerbaum et al., 1998), recombination in *C. jejuni* has not been sufficient to abolish the signals of clonal structure, and some seven-locus MLST types have been isolated multiple times in different places, including different host species. In addition to identical genotypes, there are many instances of pairs of strains differing at one or two of the seven loci. As in other bacterial species, this sharing of a majority of alleles is strong evidence of recent clonal descent. However, clonal relationships cannot be fully resolved with seven-locus genotypes. Both old relationships and very young ones are problematic, the former because frequent recombination removes the evidence of clonal relatedness, the latter because of an absence of events within the seven loci. Clonal complexes are units that summarize the information that MLST provides about intermediate-level relationships. They are best thought of as pragmatic designations of relatedness; each complex is unlikely to strictly represent an evolutionary lineage since strains may easily be excluded because of recent recombination events at a few loci. Nevertheless, their utility in epidemiological analysis does principally

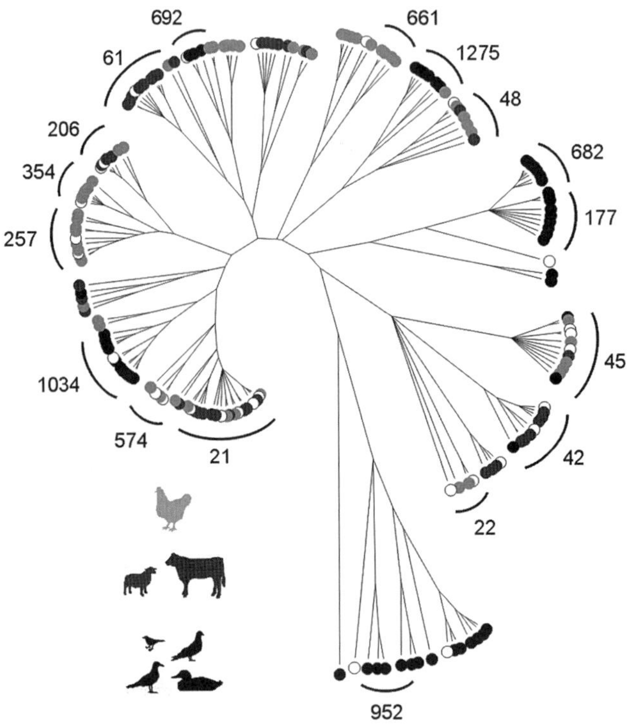

Figure 10.1 *C. jejuni* and *C. coli* host-associated genetic lineages from concatenated MLST loci. Clonal complex designations were used to annotate the phylogeny generated by ClonalFrame. The terminal branch nodes are color coded according to isolate source: black, wild bird; dark gray, ruminant; light gray, chickens and chicken meat; white, multiple sources. Data from McCarthy et al. (2007), Colles et al. (2008, 2009), and Sheppard et al. (2009a).

come from the information they contain on genealogical relationships; clonal complexes are most likely to correlate with important phenotypes if they correlate strongly with units of descent.

The relationships among *C. jejuni* isolates collected from both agricultural sources and wild birds can be expressed as a genealogical tree (Fig. 10.1). The extent to which host species harbor distinct *C. jejuni* and *C. coli* genotypes is revealed, and both host-associated sequence types (STs) and clonal complexes can be identified, but often with substantial overlap. For example, the ST-257 and ST-61 complexes are associated with chickens and ruminants, respectively. However, this division is not complete and there are clonal complexes, including the ST-21 and ST-45 complexes, that are found in multiple hosts. One possible explanation for this is that some lineages may be adapted to more than one host, but increasing the resolution of genealogical reconstructions by typing additional loci (Suerbaum et al., 2001; Manning et al., 2003) provides evidence that niche-specific genetic adaptation has occurred but that the signal may not be contained within the standard seven MLST loci. Clonal complex can be a rather poor predictor of host association (McCarthy et al., 2007), and better prediction of host, based on genotype, is possible by using methods that estimate the composition of host-specific gene pools (McCarthy et al., 2007; Wilson et al., 2008). The models in these analyses assume that strains in a chicken, for example, will import DNA from other strains in chickens and acquire a host signature

Table 10.1 Pairwise F_{ST} of Concatenated MLST Profiles among Isolates from Different Bird Host Species

	Host group 1				
Host group 2	Ducks and geese	Chicken	Gulls	Passerines	Pigeons
Ducks and geese	0	—	—		—
Chicken	0.14	0	—	—	—
Gulls	0.24	0.14	0	—	—
Passerines	0.35	0.25	0.16	0	—
Pigeons	0.25	0.12	–0.03	0.102	0

Source: Data from Sheppard et al., 2009a.

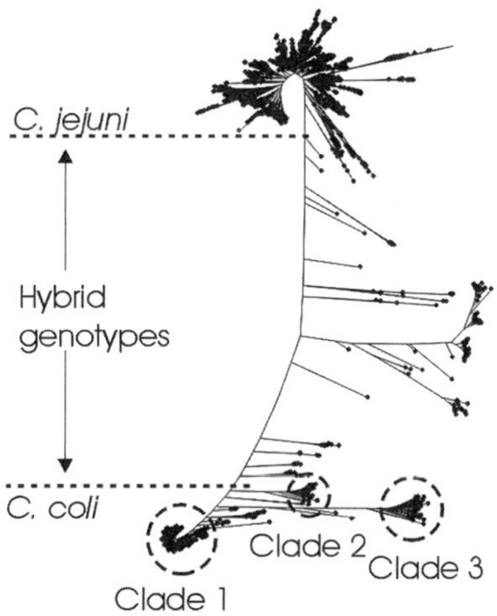

Figure 10.2 Genetic relatedness of 3705 *C. jejuni* and *C. coli* genotypes (concatenated MLST alleles) from the pubMLST database. Three *C. coli* clades and intermediate genotypes are visible on the neighbor-joining tree.

that is independent of their clonal background. Despite the evidence for differentiated gene pools, there appears to be sufficient genetic exchange within the global *C. jejuni* population to prevent progressive divergence and eventual speciation. For example, F_{ST} between the bird host species (Table 10.1). All of the values are <0.5, indicating substantial genetic exchange that is responsible for most of polymorphism within each one.

C. coli is the closest relative of *C. jejuni* but has a different population structure. First, rather than being a single rapidly recombining population, there are three clades (Fig. 10.2). For the majority of strains within each clade, most loci cluster with others from that clade. Furthermore, nucleotide-based F_{ST} values between *C. jejuni* and *C. coli* clades indicate low levels of gene flow (Table 10.2). These clades are not clonal complexes because they have more genetic diversity, differing at up to seven loci. These three clades, therefore, appear to be in the early stages of speciation on the basis of genetic isolation and are entirely different from clonal complexes in *C. jejuni*. Despite this qualitative difference in the population structure of the two species, the absolute level of nucleotide diversity within each clade is lower than between *C. jejuni* strains, and the amount of divergence between them is also very modest, such that the overall diversity within the two species is comparable.

Table 10.2 Population Pairwise F_{ST} of Concatenated MLST Profiles among *C. jejuni* and *C. coli* Clades 1–3 from Multiple Host Species

	Population 1			
Population 2	*C. jejuni*	*C. coli* (clade 1)	*C. coli* (clade 2)	*C. coli* (clade 3)
C. jejuni	0	—	—	—
C. coli (clade 1)	0.89	0	—	—
C. coli (clade 2)	0.87	0.92	0	—
C. coli (clade 3)	0.87	0.95	0.84	0

Source: Data from Sheppard et al., 2008.

The great majority of the *C. coli* strains that have been genotyped to date belong to clade 1. This numerical dominance may be a consequence of sampling and potentially reflects the dominance of this clade in agricultural sources. Indeed, evolutionary analysis by ClonalFrame (Sheppard et al., 2008) suggests that clades 2 and 3 are more diverse, indicative of a higher historical population size. Those clade 2 and clade 3 isolates that have been recovered are typically from environmental waters. The ecology of the strains from these clades is unknown, but the absence of gene flow suggests strong barriers to recombination, for example, because of highly distinct niches.

The relatively simple dichotomy between the genetic structuring within the two species is complicated by evidence of a very high recent rate of exchange between *C. jejuni* and *C. coli* clade 1. This gene flow is two way. There are 27 (2.2%) *C. coli* alleles that have been imported by at least one *C. jejuni* strain, and 60 (20.1%) *C. jejuni* alleles that have been imported by *C. coli* strains. The rate of exchange seems to be approximately the same for all seven MLST loci. Many more *C. jejuni* isolates have been genotyped than *C. coli* so that the average amount of DNA that has been imported from the other species per strain is much greater for *C. coli*.

One possible explanation is that this high rate of gene flow is the result of the co-colonization by these species of an agricultural niche. This is supported by two observations. First, almost all of the introgressed alleles are also found in the donor species (Sheppard et al., 2008), indicating that the imports occurred recently enough to have not accumulated mutations. Second, the introgressed alleles found within *C. coli* were typical of alleles found in *C. jejuni* from farm sources. In evolutionary terms, agriculture is a very new niche. Therefore, the coinfection of chicken and ruminants by similar strains, despite the differences in the biology of their digestive tracts, suggests that this niche has acquired specifically adapted lineages of bacteria rather than sharing a common gene pool with a preexisting natural reservoir. Animals in agricultural environments are highly unusual in terms of diet, genetic and age structure, density, and many other details of habitation. This novel niche has been colonized by both species, and the sharing of the niche seems to have led to a high rate of genetic exchange genome-wide. This high rate of exchange seems to hold for nonhomologous as well as homologous DNA. Some *C. jejuni* strains are more similar to particular strains of *C. coli* in gene content than they are to other strains of *C. jejuni* (Debruyne et al., 2008).

The genetic structure of other *Campylobacter* species has been less comprehensively investigated. There are many other species and they often show strong host associations (Waldenstrom et al., 2002; Miller et al., 2005; van Bergen et al., 2005; Humphrey et al.,

Table 10.3 Example Members of the Family Campylobacteriaceae with Some of the Animal Hosts from Which They Have Been Isolated

Species	Sources
Campylobacter fetus	Cattle, sheep
Campylobacter helveticus	Cats, dogs
Campylobacter hominis	Humans
Campylobacter gracilis	Humans
Campylobacter insulaenigrae	Seals, porpoises
Campylobacter hyoilei	Pigs
Campylobacter upsaliensis	Cats, dogs
Campylobacter lari	Wild birds, chickens, environmental waters, dogs, cats
C. coli	Pigs, chickens, cattle, sheep, wild birds, cats, dogs
C. jejuni	Pigs, chickens, cattle, sheep, wild birds, cats, dogs, environmental waters

Source: Adapted from Humphrey et al., 2007.

2007) (Table 10.3). *C. coli* and *C. jejuni* species have a particularly broad host range, as is characteristic in emerging diseases (Cleaveland et al., 2001), and have been isolated from farm animals (Rosef et al., 1985; Miller et al., 2006; Sheppard et al., 2009a,b), cats and dogs (S. K. Sheppard, pers. obs.), and wild birds (Waldenstrom et al., 2002; Colles et al., 2008, 2009), as well as other wild animals(Rosef et al., 1985; Sheppard et al., 2009a). *C. jejuni* has also been isolated from environmental water samples and sand from bathing beaches (Bolton et al., 1999), but it is not clear if this niche is a genuine reservoir of genotypes or simply reflects fecal contamination from a primary host.

10.4 MODELS OF *CAMPYLOBACTER* EVOLUTION

We have described the population structure at various levels principally based upon inferences from analysis of seven housekeeping loci. Here we discuss various ideas about how the genetic structure appears as it does. Some degree of structuring of genotypes is expected under entirely neutral models of evolution. In clonal organisms like *Yersinia pestis* or *Salmonella enterica* ssp. *enterica* serovar Typhi, all of the ancestry of the DNA in the present sample is derived from a single founding cell (Achtman and Wagner, 2008). Therefore, a single bifurcating tree can be used to represent both the ancestry of the cells and also the DNA within them. However, if the bacteria undergo homologous recombination, the DNA present in the population may not have been copied from the founding cell but will, in general, trace its descent to multiple cells that were present in the population at the time that this clonal common ancestor existed (Fig. 10.3).

Under a neutral model, the shape of the tree of clonal descent of a sample is described by a stochastic model called the coalescent (Kingman, 1982, 2000). The key feature of the coalescent in a population of constant size is that most of the coalescence events will be in the recent past (Fig. 10.4). The reason is that before the first coalescence event has occurred, each lineage can potentially coalesce with $n - 1$ other lineages, where n is the sample size. As coalescences occur going backward in evolutionary time, the number of lineages progressively decreases. The last coalescence event, traced to the ancestral cell

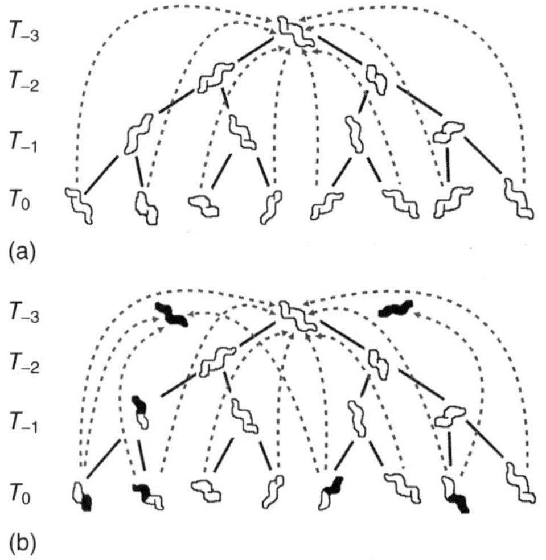

Figure 10.3 Bifurcating tree representing the proliferation of isolates over time ($T_{-3} - T_0$). The genetic relatedness of isolates sampled at T_0 to the ancestor at T_{-3} is represented by a broken gray line in (a) an entirely clonal model where all cells trace their origin directly to the founder and (b) in bacteria that undergo homologous recombination, where the DNA present in the population at T_0 may trace its descent to multiple cells in the population at the time that the clonal common ancestor existed.

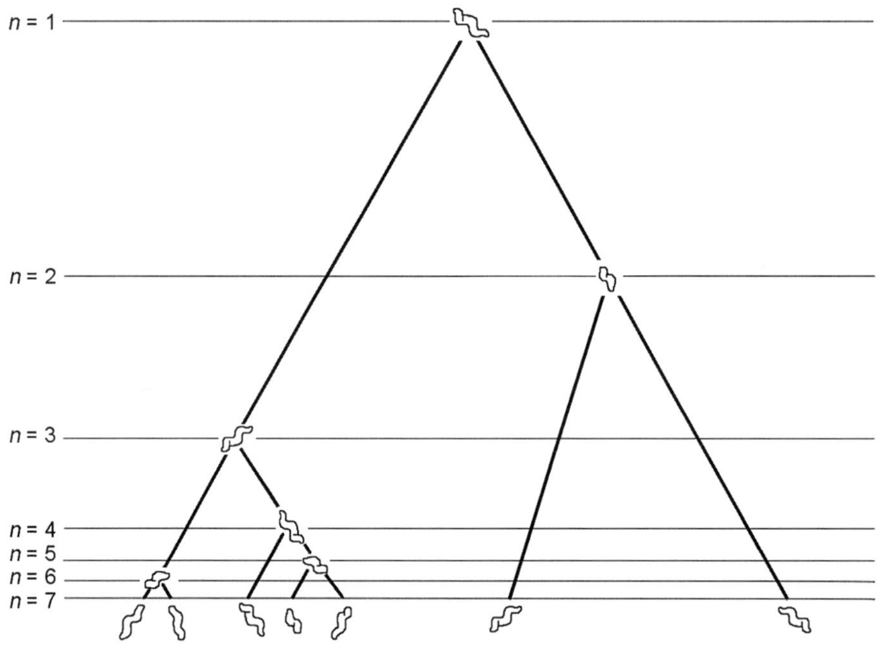

Figure 10.4 The decent of *Campylobacter* described by the coalescent. In this example, there are eight (n) samples and each lineage can potentially coalesce with $n - 1$ other lineages. As lineages coalescence going backward in evolutionary time, the average rate of coalescence decreases. The last coalescence event is traced to the ancestral cell.

of the sample, on average takes nearly half the time of the entire coalescence tree, although there is a great deal of stochastic variation from tree to tree (Hudson, 1983).

The shape of the coalescent tree describing clonal relationships has important consequences for describing patterns of variation in bacterial populations. Because the great majority of coalescence events are recent in most random trees, there will be clumps of strains that share a very recent common ancestor. Depending on the rate of mutation and homologous recombination, these clusters of strains are likely to be very similar genetically. There is therefore an entirely neutral mechanism by which the observed clonal complex structure can be explained.

The structuring of the *C. jejuni* population, with many clonal complexes associated with particular host species, highlights the potential role that natural selection also plays in determining the population structure of the species. One way of describing structuring is the ecotype. Cohan defines ecotypes as "newly divergent, ecologically distinct populations" (Cohan and Koeppel, 2008) and ascribes several strong properties to them. Ecotypes, which are founded only once, are ecologically distinct, so that they escape each other's periodic selection and drift events without concern for the degree of sexual isolation among them. Different ecotypes are irreversibly separate because they are out of range of one another's selection and drift events and because recombination is too rare to prevent their adaptive divergence. Furthermore, because they are ecologically distinct, they are able to coexist in the future.

While several aspects of the ecotype model are relevant to the evolution of *Campylobacter* species, adaptation appears to be more fluid than this model suggests, and it is difficult to delineate *Campylobacter* strains clearly into distinct ecotypes according to Cohan's proscriptive definition. First, host associations can be identified at the level of sequence type, clonal complex, clade, and species, and some groups are associated with multihost niches, for example, ruminants and agriculture. Within these groups, there are sometimes sublineages that have a more restricted host range. Further, some niches, like chickens, have been invaded by multiple distinct lineages. It might be argued that there are multiple niches within the chicken host; however, the very rapid fluctuations in the frequency of different lineages within the chicken hosts suggest that these subniches will be difficult or impossible to identify.

On the basis of these observations, a more fluid model of adaptation seems more relevant to *Campylobacter*, at least within the emerging agricultural niche. Even in stable niches, such as wild animal guts, there is a trade-off between being particularly adapted to the specific host that the bacteria is currently infecting and the ability to spread by colonizing new environments to which the bacteria will be imperfectly adapted. This trade-off could result in considerable genetic and adaptive fluidity even in stable environments. In any case, there is no compelling argument to link ecotypes with any particular level of genetic relatedness in *Campylobacter*. Because of the properties expected of neutral populations, there is also no obvious need to invoke ecotypes to explain the existence of clonal complexes. Some mechanism is required to maintain the distinct lineages of *C. coli* and the genetic isolation of *C. coli* and *C. jejuni*, but since these lineages have not historically exchanged DNA at high frequency, these entities do not fit the ecotype model, which posits that ecotypes are stable despite high gene flow.

Even though selection may not be necessary to explain the existence of clonal complexes, it remains likely that it is responsible for the presence of specific lineages in particular environments. Moreover, seven-locus genotypes, as markers of descent, are likely to correlate strongly with the adaptive loci that are responsible for adaptation to specific environments. This signal can be exploited in finding genotype–phenotype correlations, and clonal complexes may provide weak indirect inference of the adaptive nature of lineages, although this is very speculative owing to the limited information contained within seven-locus genotypes (Fig. 10.5).

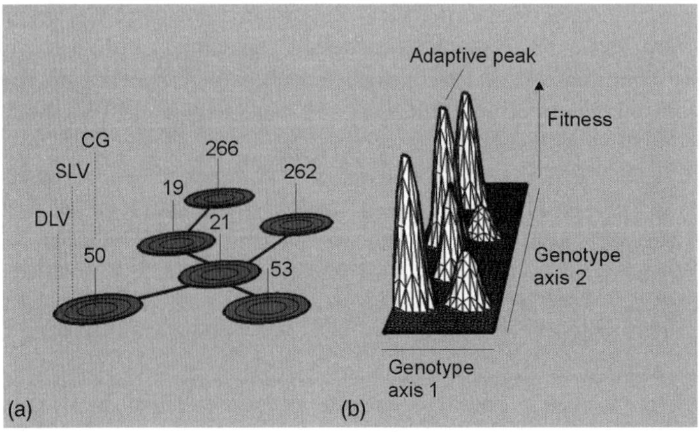

Figure 10.5 Speculation on the adaptive signal in the MLST loci. (a) Examples of *C. jejuni* allelic profiles matching at five loci belonging to the ST-21 clonal complex. The predicted primary founder, ST-21, is defined, and other central genotypes (CGs) are linked to this. Within groups, concentric circles represent single-locus variants (SLVs) and double-locus variants (DLVs). (b) Hypothetical adaptive landscape in relation to clonal complex substructure. Peaks are local fitness optima for genome-wide variation; troughs represent suboptimal genotype combinations that will be selected against impeding evolutionary transitions between peaks. See color insert.

10.5 CLADES AND SPECIES

The discovery of the three-clade structure and the potential that *C. coli* clade 1 and *C. jejuni* may be converging in an agricultural niche (Sheppard et al., 2008) has sparked considerable debate among evolutionary microbiologists (Cohan and Koeppel, 2008; Doolittle, 2008). These clades (Fig. 10.2) lead to several questions; for example, what has allowed this clade structure to be maintained? How do they relate to species? Why are no such clades present in *C. jejuni* despite the overall higher diversity and the higher frequency of the species in many of the environments that have been sampled?

There are several possible barriers to genetic exchange between clades. The simplest explanation is a general reduction in the overall level of recombination. We know that there is recombination within each *C. coli* clade from a version of the four-gamete test applied to the two most common MLST alleles in each clade population. We performed this test for each of the 21 combinations of pairs of fragments. If all four combinations of the two alleles are present, this provides good evidence that recombination has reassorted the alleles. Within clade 1, 20/21 combinations were present in all four combinations, reflecting the high sample size in the sample as well as a high recombination rate. In clades 2 and 3, there was evidence of reassortment in 14 and 5 of the 21 pairwise analyses, respectively. There is, therefore, frequent recombination within each clade. We also know that some proportion, at least, of *C. coli* clade 1 strains are highly recombinogenic because of the large numbers of alleles imported from *C. jejuni*. Therefore, it seems unlikely that the clades have diverged simply due to a uniform reduction in the rate of recombination but that barriers to recombination are involved.

Three broad classes of barrier can be described to recombination between clades: (i) mechanistic barriers—imposed by the homology dependence of recombination (Fraser et al., 2007) or other factors promoting DNA specificity, such as restriction/modification systems (Eggleston and West, 1997); (ii) ecological barriers—a consequence of physical separation of bacterial populations in distinct niches; and (iii) adaptive barriers—implying selection against hybrid genotypes (Zhu et al., 2001).

Currently, the relative importance of these three different classes of barrier are unclear. The recent increase in recombination between *C. coli* and *C. jejuni* is obviously of particular interest since it indicates a substantial recent change, and the fact that gene flow has occurred in both directions is consistent with physical proximity playing a role but is also consistent, for example, with a common vector, such a bacteriophage, infecting both lineages (Didelot et al., 2007). Adaptive explanations are also relevant because the two populations have invaded a similar niche and therefore may require similar genes. The fact that there appears to be an approximately even level of gene flow at the seven-MLST loci provides some evidence against a simple adaptive hypothesis, but it is also possible that the bacteria have evolved mechanisms for higher levels of genetic exchange, or lower levels of specificity in the DNA that they take up, in order to facilitate rapid adaptation to the new niche. Genome-wide studies of patterns of exchange will allow these different possibilities to be tested and will provide a good understanding of the mechanisms by which barriers to recombination arise, how they are disrupted, and the consequences of recombination for ecological adaptation.

10.6 CONCLUSION

Developments in molecular subtyping, in particular MLST, have greatly enhanced the study of bacterial population genetics. The compilation of large genotype archives and the development of associated analysis software have been particularly valuable in describing the role of horizontal genetic exchange in bacterial speciation and in shaping population structure. In *Campylobacter*, MLST has facilitated the description of genotype–host associations, disease attribution to particular lineages, and investigation of the evolutionary forces that generate and maintain the genetic structure. This, however, is only the beginning of the task required to catalogue and to understand the bewildering level of complexity in ecologically diverse genera such as *Campylobacter*. Questions remain about the nature of a genomic species and how they are maintained in the face of recombination, the microevolutionary events associated with niche/host adaptation, and the rate of adaptation between different species/lineages and the potential for variation in the adaptive strategy. Genome-wide approaches to mapping bacterial diversity have already proved effective for enhancing the understanding of bacterial evolution and have the potential to unravel the phenotypic basis of genetic diversity in *Campylobacter* and to investigate the dynamics of these complex microbial communities.

REFERENCES

Achtman, M. and Wagner, M. (2008) Microbial diversity and the genetic nature of microbial species. *Nature Reviews. Microbiology* **6**, 431–440.

Adak, G. K., Meakins, S. M., Yip, H., Lopman, B. A., and O'Brien, S. J. (2005) Disease risks from foods, England and Wales, 1996–2000. *Emerging Infectious Diseases* **11,** 365–372.

Allos, B. (2001) *Campylobacter jejuni* infections: Update on emerging issues and trends. *Clinical Infectious Diseases* **32,** 1201–1206.

Bender, J. B., Hedberg, C. W., Besser, J. M. et al. (1997) Surveillance by molecular subtype for *Escherichia coli* O157:H7 infections in Minnesota by molecular subtyping. *New England Journal of Medicine* **337,** 388–394.

Bender, J. B., Hedberg, C. W., Boxrud, D. J. et al. (2001) Use of molecular subtyping in surveillance for *Salmonella enterica* serotype Typhimurium. *New England Journal of Medicine* **344,** 189–195.

Blaser, M. J., Taylor, D. N., and Feldman, R. A. (1983) Epidemiology of *Campylobacter jejuni* infections. *Epidemiologic Reviews* **5,** 157–176.

Bolton, F. J., Surman, S. B., Martin, K., Wareing, D. R. A., and Humphrey, T. J. (1999) Presence of

Campylobacter and *Salmonellae* in sand from bathing beaches. *Epidemiology and Infection* **122,** 7–13.

Calva, J. J., Ruiz-Palacios, G. M., Lopez-Vidal, A. B., Ramos, A., and Bojalil, R. (1988) Cohort study of intestinal infection with *Campylobacter* in Mexican children. *Lancet* **1**, 503–506.

CDC (2008) *Division of Foodborne, Bacterial and Mycotic Diseases (DFBMD) Listing*. Centers for Disease Control and Prevention, Atlanta, GA.

Cleaveland, S., Laurenson, M. K., and Taylor, L. H. (2001) Diseases of humans and their domestic mammals: Pathogen characteristics, host range and the risk of emergence. *Philosophical Transactions of the Royal Society of London. Series B, Biological Sciences* **356,** 991–999.

Cohan, F. M. and Koeppel, A. F. (2008) The origins of ecological diversity in prokaryotes. *Current Biology* **18,** R1024–R1034.

Coker, A. O., Isokpehi, R. D., Thomas, B. N., Amisu, K. O., and Obi, C. L. (2002) Human campylobacteriosis in developing countries. *Emerging Infectious Diseases* **8,** 237–244.

Colles, F. M., Dingle, K. E., Cody, A. J., and Maiden, M. C. (2008) Comparison of *Campylobacter* populations in wild geese with those in starlings and free-range poultry on the same farm. *Applied and Environmental Microbiology* **74,** 3583–3590.

Colles, F. M., McCarthy, N. D., Howe, J. C. et al. (2009) Dynamics of *Campylobacter* colonization of a natural host, *Sturnus vulgaris* (European Starling). *Environmental Microbiolgy* **11**, 258–267.

Debruyne, L., Gevers, D., and Vandamme, P. (2008) Taxonomy of the family Campylobacteraceae. In *Campylobacter* (eds. I. Nachamkin, C. M. Szymanski, and M. J. Blaser), pp. 3–25. ASM Press, Washington, DC.

Didelot, X., Achtman, M., Parkhill, J., Thomson, N. R., and Falush, D. (2007) A bimodal pattern of relatedness between the *Salmonella* Paratyphi A and Typhi genomes: Convergence or divergence by homologous recombination? *Genome Research* **17,** 61–68.

Dingle, K. E., McCarthy, N. D., Cody, A. J., Peto, T. E., and Maiden, M. C. (2008) Extended sequence typing of *Campylobacter* spp., United Kingdom. *Emerging Infectious Diseases* **14,** 1620–1622.

Doolittle, W. F. (2008) Microbial evolution: Stalking the wild bacterial species. *Current Biology* **18,** R565–R567.

Eggleston, A. K. and West, S. C. (1997) Recombination initiation: Easy as A, B, C, D … chi? *Current Biology* **7,** R745–R749.

Ethelberg, S., Simonsen, J., Gerner-Smidt, P., Olsen, K. E., and Molbak, K. (2005) Spatial distribution and registry-based case-control analysis of *Campylobacter* infections in Denmark, 1991–2001. *American Journal of Epidemiology* **162,** 1008–1015.

Fearnhead, P., Smith, N. G., Barrigas, M., Fox, A., and French, N. (2005) Analysis of recombination in *Campylobacter jejuni* from MLST population data. *Journal of Molecular Evolution* **61,** 333–340.

Figueroa, G., Galeno, H., Troncoso, M., Toledo, S., and Soto, V. (1989) Prospective study of *Campylobacter jejuni* infection in Chilean infants evaluated by culture and serology. *Journal of Clinical Microbiology* **27,** 1040–1044.

Fraser, C., Hanage, W. P., and Spratt, B. G. (2007) Recombination and the nature of bacterial speciation. *Science* **315,** 476–480.

Friedman, C. J., Neiman, J., Wegener, H. C., and Tauxe, R. V. (2000) Epidemiology of *Campylobacter jejuni* infections in the United States and other industrialised nations. In *Campylobacter* (eds. I. Nachamkin and M. J. Blaser), pp. 121–138. ASM Press, Washington, DC.

Friedman, C. R., Hoekstra, R. M., Samuel, M. et al. (2004) Risk factors for sporadic *Campylobacter* infection in the United States: A case-control study in FoodNet sites. *Clinical Infectious Disases* **38**(Suppl 3), S285–S296.

Frost, J. A., Gillespie, I. A., and O'Brien, S. J. (2002) Public health implications of *Campylobacter* outbreaks in England and Wales, 1995–9: Epidemiological and microbiological investigations. *Epidemiology and Infection* **128,** 111–118.

FSA (2002) *Measuring Foodborne Illness Levels*. Food Standards Agency, London.

Gaudreau, C. and Michaud, S. (2003) Cluster of erythromycin- and ciprofloxacin-resistant *Campylobacter jejuni* subsp. *jejuni* from 1999 to 2001 in men who have sex with men, Quebec, Canada. *Clinical Infectious Diseases* **37,** 131–136.

Gillespie, I. A., O'Brien, S. J., Frost, J. A. et al. (2006) Investigating vomiting and/or bloody diarrhoea in *Campylobacter jejuni* infection. *Journal of Medical Microbiology* **55,** 741–746.

Harrington, C. S., Thomson Carter, F. M., and Carter, P. E. (1997) Evidence for recombination in the flagellin locus of *Campylobacter jejuni*: Implications for the flagellin gene typing scheme. *Journal of Clinical Microbiology* **35,** 2386–2392.

Hudson, R. R. (1983) Properties of a neutral allele model with intragenic recombination. *Theoretical Population Biology* **23,** 183–201.

Humphrey, T., O'Brien, S., and Madsen, M. (2007) Campylobacters as zoonotic pathogens: A food production perspective. *International Journal of Food Microbiology* **117,** 237–257.

Kapperud, G., Espeland, G., Wahl, E. et al. (2003) Factors associated with increased and decreased risk of *Campylobacter* infection: A prospective case-control study in Norway. *American Journal of Epidemiology* **158,** 234–242.

Kessel, A. S., Gillespie, I. A., O'Brien, S. J. et al. (2001) General outbreaks of infectious intestinal disease linked with poultry, England and Wales, 1992–1999. *Communicable Disease and Public Health* **4,** 171–177.

Kingman, J. F. (2000) Origins of the coalescent. 1974–1982. *Genetics* **156,** 1461–1463.

Kingman, J. F. C. (1982) On the genealogy of large populations. *Journal of Applied Probability* **19,** 27–43.

Manning, G., Dowson, C. G., Bagnall, M. C. et al. (2003) Multilocus sequence typing for comparison of veterinary and human isolates of *Campylobacter jejuni*. *Applied Environmental Microbiology* **69,** 6370–6379.

McCarthy, N. D., Colles, F. M., Dingle, K. E. et al. (2007) Host-associated genetic import in *Campylobacter jejuni*. *Emerging Infectious Diseases* **13,** 267–272.

Mead, P. S., Slutsker, L., Dietz, V. et al. (1999) Food-related illness and death in the United States. *Emerging Infectious Diseases* **5,** 607–625.

Miller, W. G., Englen, M. D., Kathariou, S. et al. (2006) Identification of host-associated alleles by multilocus sequence typing of *Campylobacter coli* strains from food animals. *Microbiology* **152,** 245–255.

Miller, W. G., On, S. L., Wang, G. et al. (2005) Extended multilocus sequence typing system for *Campylobacter coli, C. lari, C. upsaliensis*, and *C. helveticus*. *Journal of Clinical Microbiology* **43,** 2315–2329.

Morales, C. A., Porwollik, S., Frye, J. G. et al. (2005) Correlation of phenotype with the genotype of egg-contaminating *Salmonella enterica* serovar Enteritidis. *Applied Environmental Microbiology* **71,** 4388–4399.

Mylius, S. D., Nauta, M. J., and Havelaar, A. H. (2007) Cross-contamination during food preparation: A mechanistic model applied to chicken-borne *Campylobacter*. *Risk Analysis* **27,** 803–813.

Nauta, M. J., Jacobs-Reitsma, W. F., and Havelaar, A. H. (2007) A risk assessment model for *Campylobacter* in broiler meat. *Risk Analysis* **27,** 845–861.

Neimann, J., Engberg, J., Molbak, K., and Wegener, H. C. (2003) A case-control study of risk factors for sporadic *Campylobacter* infections in Denmark. *Epidemiology and Infection* **130,** 353–366.

Oberhelman, R. A. and Taylor, D. N. (2000) *Campylobacter* infections in developing countries. In *Campylobacter* (eds. I. Nachamkin and M. J. Blaser), pp. 139–153. ASM Press, Washington, DC.

Olsen, S. J., Hansen, G. R., Bartlett, L. et al. (2001) An outbreak of *Campylobacter jejuni* infections associated with food handler contamination: The use of pulsed-field gel electrophoresis. *Journal of Infectious Diseases* **183,** 164–167.

Olsen, S. J., Patrick, M., Hunter, S. B. et al. (2005) Multistate outbreak of *Listeria monocytogenes* infection linked to delicatessen turkey meat. *Clinical Infectious Disases* **40,** 962–967.

On, S. L. (2001) Taxonomy of *Campylobacter, Arcobacter, Helicobacter* and related bacteria: Current status, future prospects and immediate concerns. *Symposium Series (Society for Applied Microbiology)* **30,** 1S–15S.

Pazzaglia, G., Bourgeois, A. L., el Diwany, K. et al. (1991) *Campylobacter* diarrhoea and an association of recent disease with asymptomatic shedding in Egyptian children. *Epidemiology and Infection* **106,** 77–82.

Pebody, R. G., Ryan, M. J., and Wall, P. G. (1997) Outbreaks of *Campylobacter* infection: Rare events for a common pathogen. *Communicable Disease Report. CDR Review* **7,** R33–R37.

Pope, J. E., Krizova, A., Garg, A. X., Thiessen-Philbrook, H., and Ouimet, J. M. (2007) *Campylobacter* reactive arthritis: A systematic review. *Seminars in Arthritis and Rheumatism* **37,** 48–55.

Richardson, N. J., Koornhof, H. J., Bokkenheuser, V. D., Mayet, Z., and Rosen, E. U. (1983) Age related susceptibility to *Campylobacter jejuni* infection in a high prevalance population. *Archives of Disease in Childhood* **58,** 616–619.

Rosef, O., Kapperud, G., Lauwers, S., and Gondrosen, B. (1985) Serotyping of *Campylobacter jejuni, Campylobacter coli*, and *Campylobacter laridis* from domestic and wild animals. *Applied and Environmental Microbiology* **49,** 1507–1510.

Schouls, L. M., Reulen, S., Duim, B. et al. (2003) Comparative genotyping of *Campylobacter jejuni* by amplified fragment length polymorphism, multilocus sequence typing, and short repeat sequencing: Strain diversity, host range, and recombination. *Journal of Clinical Microbiology* **41,** 15–26.

Servan, J., Elghozi, D., Waisbord, P., and Duclos, H. (1995) [Miller–Fisher syndrome. Role of *Campylobacter jejuni* infection]. *Presse Médicale* **24,** 651.

Sheppard, S. K., Dallas, J. F., Macrae, M. et al. (2009a) Campylobacter genotypes from food animals, environmental sources and clinical disease in Scotland 2005/6. *International Journal of Food Microbiology* **134**, 96–103.

Sheppard, S. K., Dallas, J. F., Strachan, N. J. et al. (2009b) *Campylobacter* genotyping to determine the source of human infection. *Clinical Infectious Diseass* **48,** 1072–1078.

Sheppard, S. K., McCarthy, N. D., Falush, D., and Maiden, M. C. (2008) Convergence of *Campylobacter* species: Implications for bacterial evolution. *Science* **320,** 237–239.

Stafford, R. J., Schluter, P., Kirk, M. et al. (2007) A multi-centre prospective case-control study of *Campylobacter* infection in persons aged 5 years and older in Australia. *Epidemiology and Infection* **135,** 978–988.

Steinbrueckner, B., Ruberg, F., and Kist, M. (2001) Bacterial genetic fingerprint: A reliable factor in the study of the epidemiology of human *Campylobacter* enteritis? *Journal of Clinical Microbiology* **39,** 4155–4159.

Suerbaum, S., Lohrengel, M., Sonneveld, A., Ruberg, F., and Kist, M. (2001) Allelic diversity and recombination in *Campylobacter jejuni*. *Journal of Bacteriology* **183,** 2553–2559.

Suerbaum, S., Maynard Smith, J., Bapumia, K. et al. (1998) Free recombination within *Helicobacter pylori*. *Proceedings of the National Academy of Sciences of the United States of America* **95,** 12619–12624.

Tam, C. C., Rodrigues, L. C., Petersen, I. et al. (2006) Incidence of Guillain–Barre syndrome among patients with *Campylobacter* infection: A general practice research database study. *Journal of Infectious Diseases* **194,** 95–97.

Tankouo-Sandjong, B., Sessitsch, A., Liebana, E. et al. (2007) MLST-v, multilocus sequence typing based on virulence genes, for molecular typing of *Salmonella enterica* subsp. *enterica* serovars. *Journal of Microbiological Methods* **69,** 23–36.

van Bergen, M. A., Dingle, K. E., Maiden, M. C. et al. (2005) Clonal nature of *Campylobacter fetus* as defined by multilocus sequence typing. *Journal of Clinical Microbiology* **43,** 5888–5898.

Waldenstrom, J., Broman, T., Carlsson, I. et al. (2002) Prevalence of *Campylobacter jejuni, Campylobacter lari,* and *Campylobacter coli* in different ecological guilds and taxa of migrating birds. *Applied and Environmental Microbiology* **68,** 5911–5917.

Wassenaar, T. M., Geilhausen, B., and Newell, D. G. (1998) Evidence for genome instability in *Campylobacter jejuni* isolated from poultry. *Applied and Environmental Microbiology* **64,** 1816–1821.

Wheeler, J. G., Sethi, D., Cowden, J. M. et al. (1999) Study of infectious intestinal disease in England: Rates in the community, presenting to general practice, and reported to national surveilance. *British Medical Journal* **318,** 1046–1050.

Wilson, D. J., Gabriel, E., Leatherbarrow, A. J. et al. (2009) Rapid evolution and the importance of recombination to the gastroenteric pathogen *Campylobacter jejuni. Molecular Biology and Evolution* **26,** 385–397.

Wilson, D. J., Gabriel, E., Leatherbarrow, A. J. H. et al. (2008) Tracing the source of campylobacteriosis. *PLoS Genetics* **26,** e1000203.

Zhu, P., van der Ende, A., Falush, D. et al. (2001) Fit genotypes and escape variants of subgroup III *Neisseria meningitidis* during three pandemics of epidemic meningitis. *Proceedings of the National Academy of Sciences of the United States of America* **98**, 5234–5239.

Chapter 11

Population Genetics of *Enterococcus*

Rob J. Willems

11.1 INTRODUCTION

Enterococci are ubiquitous in nature and can be found in soil, water, food, animals, and humans. In animals, they are part of the normal microbiota (Aarestrup et al., 2002). In farm animals, the most encountered enterococcal species are *Enterococcus faecalis*, *Enterococcus faecium*, *Enterococcus hirae*, and *Enterococcus durans*. Also from cats, dogs, and horses, these species are among the most frequently isolated *Enterococcus* species. In a large variety of insects, including beetles, flies, bees, and termites, enterococci have been isolated with *E. faecalis*, *E. faecium*, and *Enterococcus casseliflavus* as the most common ones. Besides the fact that enterococci are normal inhabitants of the gastrointestinal (GI) tract of animals, they have also been reported as the causative agent of disease, mainly mastitis in cattle and diarrhea or endocarditis, septicemia, and encephalomalacia in birds (Aarestrup et al., 2002). Several enterococcal species have been found associated with plants, although it is unclear whether this represents environmental contamination. Enterococci are also used in food fermentation as nonstarter lactic acid bacteria in cheeses, especially artisan cheese and sausages, mainly produced in Southern Europe, to improve flavor, taste, and texture (Giraffa, 2003; Hugas et al., 2003).

In humans, enterococci are natural inhabitants of the GI tract and, as such, belong to the complex community of bacteria that constitute the microbiota of the large intestine. While facultative anaerobes like enterococci are minority players in the distal colon and feces, some studies indicate that enterococci can occur in high numbers in the cecum of humans (Marteau et al., 2001). In the large bowel, enterococci play a role in preventing invasion by pathogenic organisms through a mechanism known as colonization resistance. However, apart from being harmless commensals inhabiting the digestive tract, enterococci are also known for more than 100 years as being capable of causing serious infections, like endocarditis and urinary tract infections (Murray, 2000). In the last two decades, a dramatic shift has been observed in the number of enterococcal infections. Whereas in the 1980s enterococci were considered mere colonizers of the GI tract, only occasionally causing infections, they now rank after coagulase-negative staphylococci and *Staphylococcus aureus* as the third most common cause of health care-associated infections, more common than, for example, *Escherichia coli*, *Pseudomonas aeruginosa*, and *Enterobacter* species (Hidron et al., 2008). The most common enterococcal infections

Bacterial Population Genetics in Infectious Disease, Edited by D. Ashley Robinson, Daniel Falush, and Edward J. Feil

include those of the urinary tract, wounds, bloodstream, and endocardium. Recent data from the National Healthcare Safety Network rank enterococci as the second leading cause of central line-associated bloodstream infections and the third leading cause of catheter-associated urinary tract infections and surgical site infections.

The rapid increase in enterococcal infections since the 1980s of the previous century coincided with the emergence of high-level resistance to beta-lactam antibiotics followed by high-level vancomycin resistance mainly in *E. faecium* (Grayson et al., 1991; Martone, 1998). Nowadays, 80% of all pathogenic *E. faecium* isolates from health care-associated infections are vancomycin resistant, while this is only 7% for *E. faecalis* (Hidron et al., 2008). Since high-level ampicillin and vancomycin resistance is predominantly seen in *E. faecium*, the emergence of vancomycin-resistant enterococci (VREs) also increased the proportion of enterococcal infections caused by *E. faecium*. While traditionally 90% of all enterococcal infections were caused by *E. faecalis* and only 10% by *E. faecium*, the proportion of *E. faecium* gradually increased to 40% after the turn of the century, thereby partly replacing *E. faecalis* (Iwen et al., 1997; Murdoch et al., 2002; Treitman et al., 2005; Hidron et al., 2008; Top et al., 2008).

11.2 ANTIBIOTIC RESISTANCE

Enterococci, specifically most strains of *E. faecium* and *E. faecalis*, are low-level intrinsically resistant to various antibiotics from different antibiotic classes including cell wall-active agents like beta-lactam antibiotics (particular cephalosporins) aminoglycosides, clindamycin, and lincomycin. The production of penicillin-binding proteins (PBP), PBP-4, -5, or -6, with a low affinity for beta-lactam antibiotics explains, for most of the cases, the low susceptibility of enterococci for this class of antibiotics (Fontana et al., 1983; Williamson et al., 1985). In addition, the two-component system, CroRS, is also required for intrinsic beta-lactam resistance in *E. faecalis* (Comenge et al., 2003). Low-level aminoglycoside resistance is due to low uptake of these compounds because of the relatively impermeable cell wall of enterococci and/or the binding of aminoglycosides to components of the peptidoglycan (Moellering and Weinberg, 1971). Intrinsic antibiotic resistance seriously hampers efficient treatment of enterococcal infections. In the last two to three decades, this has become an even bigger challenge due to the emergence of acquired resistance, either through mutation or through the acquisition of foreign genes through horizontal gene transfer of plasmids or transposons, which makes some enterococcal clones resistant to all currently available antibiotics. This is illustrated by the increase in high-level beta-lactam and gentamicin resistance among clinical *E. faecium* isolates. While in the early 1980s less than 10% of all *E. faecium* isolates were ampicillin resistant and none were high-level gentamicin-resistant, this gradually increased to 75–80% and 25–30%, respectively, at the turn of the century (Grayson et al., 1991; Iwen et al., 1997). Nowadays, up to 90% of health care-associated *E. faecium* are ampicillin resistant (Hidron et al., 2008). Interestingly, these numbers are considerably lower for *E. faecalis*. In 2000, the prevalence of high-level gentamicin resistance was below 20%, while ampicillin resistance was virtually absent in *E. faecalis* (Murdoch et al., 2002). Also in the latest overview of the National Healthcare Safety Network, ampicillin resistance is found in less than 5% of health care-associated *E. faecalis* infections (Hidron et al., 2008).

High-level resistance to beta-lactam antibiotics is mainly due either to overproduction of the low-affinity PBP5 or to mutations in PBP5, which even lowers the affinity for beta-lactam antibiotics (Ligozzi et al., 1996; Zorzi et al., 1996; Rybkine et al., 1998; Rice et al., 2001, 2004). Due to the fact that high-level beta-lactam resistance is the result of chromosomal mutations, the emergence of ampicillin-resistant *E. faecium* is thought to be the result of clonal expansion. Although this is probably true in the majority of cases, the finding of conjugational transfer of low-affinity *pbp5* through linkage with transposable elements like *Tn5382* suggests that the spread of ampicillin resistance among clinical *E. faecium* isolates may also be due, in part, to the horizontal gene transfer of low-affinity *pbp5* (Carias et al., 1998; Dahl et al., 2000; Rice et al., 2005).

A large array of genetic elements encoding aminoglycoside-modifying enzymes exists in enterococci (Chow, 2000). However, over 90% of enterococci that are high-level resistant to gentamicin carry the *aac(6′)-Ie-aph(2″)-Ia* gene, which encodes a bifunctional enzyme with acetylating and phosphorylating activity (Ferretti et al., 1986; Chow, 2000). In addition, enterococci contain at least eight other aminoglycoside resistance genes. The most common acquired resistance mechanism in enterococci is resistance to macrolide–lincosamide–streptogramin B (MLS) antibiotics mediated by the *ermB* gene (Jensen et al., 1999). In addition, several other genes conferring MLS resistance are found in enterococci as well as genes conferring resistance to new streptogramin combinations of quinupristin and dalfopristin, as well as chloramphenicol and tetracycline (Pepper et al., 1986; Jones et al., 1998; Werner et al., 2002). Resistance to quinolones is mainly the result of chromosomal mutations, though Qnr-like pentapeptide repeat proteins, implicated in plasmid-mediated quinolone resistance, have also been identified in enterococci (Brisse et al., 1999; El Amin et al., 1999; Kanematsu et al., 1998; Leavis et al., 2006; Rodriguez-Martinez et al., 2008). Especially disturbing is the fact that also against the latest developed antibiotics, linezolid and daptomycin, resistance has been reported in enterococci (Arias et al., 2007; Kainer et al., 2007; Aksoy and Unal, 2008; Montero et al., 2008; Scheetz et al., 2008). The high amount of resistance genes often located on mobile genetic elements makes enterococci the most resistant opportunistic nosocomial pathogens with an increasing impact on patients' health care. In combination with their high capacity of genetic exchange, they represent perfect hubs for resistance genes facilitating horizontal gene transfer among bacterial species.

11.3 VANCOMYCIN RESISTANCE

Acquisition of vancomycin resistance by enterococci, first reported in 1988, boosted global attention for nosocomial enterococcal infection since antibiotic therapeutic options for treating infections with VREs were almost exhausted (Leclercq et al., 1988; Uttley et al., 1988). Following the initial isolation of VRE in Europe, the VRE epidemic took off in the United States. In 15 years, the prevalence of vancomycin resistance in *E. faecium* associated with healthcare infections increased from 0% to 80% (Murdoch et al., 2002; Treitman et al., 2005; Hidron et al., 2008). As for ampicillin and gentamicin, vancomycin resistance has hardly penetrated into *E. faecalis*. Still, only 0.5–7.0% of all health care-associated *E. faecalis* are vancomycin resistant (Murdoch et al., 2002; Treitman et al., 2005; Hidron et al., 2008). The reasons behind this are not well understood.

Since their initial recovery from patients in Europe and their subsequent emergence in U.S. hospitals, VREs have been found in many countries all over the world, including countries in Europe, Latin America, Asia, and Australia. In North America, the mean annual incidence rates of VRE in Canada were significantly lower compared with the United States. Between 1999 and 2005, the percentage of enterococci that were vancomycin resistant increased 2.8-fold, from 1.16% to 3.25% ($p < 0.001$) (Ofner-Agostini et al., 2008). In most European countries, the prevalence rates are not as high as in the United States, although over the past 6 years, vancomycin resistance in *E. faecium* causing invasive infections increased significantly in six countries, namely, Germany, Greece, Ireland, Israel, Slovenia, and Turkey (European Antimicrobial Resistance Surveillance System [EARSS], 2007). In 2007, seven countries reported more than 10% vancomycin resistance among invasive *E. faecium* isolates: Italy (11%), Germany (15%), United Kingdom (21%), Israel (24%), Portugal (29%), Ireland (33%), and Greece (37%) (European Antimicrobial Resistance Surveillance System (EARSS), 2007). In Latin America, VRE prevalence rates range between 2.9% and 6.6%, which is considerably lower than in North America. Also in the Asia Pacific region, vancomycin resistance is more predominantly found in *E. faecium* than in *E. faecalis*. Up to 13% of *E. faecium* recovered from infection sites in this region were vancomycin resistant, while this was found in only 0.4% of *E. faecalis* (Biedenbach et al., 2007; Park et al., 2007).

Today, seven types of vancomycin resistance have been described: vanA, -B, -C, -D, -E, -G, and -L (Courvalin, 2006; Boyd et al., 2008). These types have in common that the genes encoding vancomycin resistance are organized in operons encoding enzymes for the synthesis of low-affinity peptidoglycan precursors and the elimination of endogenously produced high-affinity precursors (Courvalin, 2006). The different "Van" types can be distinguished on the basis of DNA sequence, the type of low-affinity precursor produced, the level of vancomycin resistance, and whether the resistance operons are located on mobile genetic element. The finding that resistance to vancomycin can be plasmid mediated (Uttley et al., 1988) predicted that dissemination of glycopeptide resistance is not only the result of clonal expansion of VRE but also of horizontal gene transfer of vancomycin resistance genes. This became even more apparent when *vanA* and *vanB* gene clusters were found to be contained on transposable elements. VanA type of vancomycin resistance encoded by the *vanA* operon was the first type of vancomycin resistance described and nowadays is the most prevalent type of vancomycin resistance. The *vanA* operon is contained on transposon Tn*1546* and derivatives of this element (Arthur et al., 1993; Handwerger and Skoble, 1995; Jensen et al., 1998; Woodford et al., 1998; de Lencastre et al., 1999; Willems et al., 1999). Since Tn*1546* does not encode conjugative functions, lateral transfer can only occur after integration into transferable elements such as plasmids or conjugative transposons. Today, Tn*1546* and Tn*1546*-like elements are found in various conjugative plasmids or nonself-transmissible plasmids that can be mobilized, including pheromone-responsive plasmids (Handwerger et al., 1990) and broad-host-range plasmids (Flannagan et al., 2003) as well as on larger composite and conjugative transposons (Handwerger and Skoble, 1995; Arthur et al., 1997; de Lencastre et al., 1999).

All this illustrates the enormous potential of vancomycin resistance gene dissemination that has contributed to the global VRE epidemic. It also means that studies aimed at disclosing the epidemiology of vancomycin resistance must include investigations on plasmid and even gene epidemiology. Detailed analysis of *vanA* gene clusters revealed a high level of sequence conservation with only a small number of single-nucleotide polymorphism. Sources of DNA heterogeneity included, in addition to a scarce number of mutations, deletions encompassing the first two open reading frames of Tn*1546* encoding a transposase and a resolvase or the last two genes, *vanY* and *vanZ*, and insertions of different insertion sequence (IS) elements (Jensen et al., 1998; Woodford et al., 1998; Willems et al., 1999; Werner et al., 2006). The finding of polymorphisms in Tn*1546* and Tn*1546*-like elements facilitated the studies on the epidemiology of this transposon. However, the exact impact of horizontal gene transfer of the *vanA* gene cluster on the dissemination of vancomycin resistance is hampered by the lack of a common nomenclature for Tn*1546* variants. Because different transposon typing systems exist, comparison of epidemiological studies on the presence, prevalence, and spread of Tn*1546* derivatives is cumbersome. Nevertheless, from several studies, it is apparent that horizontal gene transfer of *vanA* transposons contribute significantly to the transmission of vancomycin resistance (Jensen et al., 1998; Woodford et al., 1998; Willems et al., 1999; Palazzo et al., 2006; Garcia-Migura et al., 2007; Park et al., 2007; Novais et al., 2008; Sung et al., 2008; Talebi et al., 2008).

11.4 VRE: A ZOONOSIS OR NOT?

While the VRE prevalence rates in the United States rapidly increased in hospitals in the 1990s, the prevalence rates in European hospitals remained low during that period despite the fact that VREs were increasingly found in animal husbandry and in the environment (Bonten et al., 2001; Aarestrup et al., 2002). It was proposed that the use of avoparcin, a vancomycin analogue, as an antimicrobial growth promoter (AMGP) promoted selection of VRE in farm animals (Wegener et al., 1999). In contrast to the high prevalence of VRE in farm animals in Europe, VREs were not found in nonhospital environments in the United States, most probably due to the fact that avoparcin was never licensed to be used as AMGP in the United States. VREs were not only found in high densities in farm animals in Europe but were also frequently recovered in healthy humans in the general population (Bonten et al., 2001). Moreover, molecular typing of VRE from animals and healthy humans indicated multiple events of clonal spread between animal and man (van den Bogaard et al., 1997; Aarestrup et al., 2002; Hammerum et al., 2004).

The rapid increase of VRE among hospitalized patients in the United States in combination with the emergence of a large reservoir of VRE in the community prompted Denmark, Norway, and Germany to ban the use of avoparcin as an AMGP in animal husbandry, which was soon followed by a European-wide ban in April 1997. Surveillance studies following this ban documented an important decline in the prevalence of VRE in animals and in healthy humans in several European countries, although in some countries, VRE remarkably persisted mainly in poultry (Borgen et al., 2000; van den Bogaard et al., 2000; Heuer et al., 2002; Manson et al., 2004; Lim et al., 2006; Ghidan et al., 2008). This decrease of VRE in the community coincided with an increase of VRE among hospitalized patients in Europe (see above). Furthermore, in the United States, VRE emerged in the 1990s without an apparent community reservoir. The inverse epidemiological link between community and hospital prevalence of VRE in the United States and in Europe highly questions direct clonal spread between VRE selected in animals and in hospitalized patients. Also, molecular epidemiological studies did not demonstrate a clear link between the hospital and community reservoir of VRE. Instead, amplified fragment length polymorphism of 255 VREs recovered from different human and animal sources identified host specificity of *E. faecium* and suggested the existence of a distinct genetic subpopulation of *E. faecium* among the majority of hospitalized patients, genetically different from the majority of strain isolates from animals (Willems et al., 2000). This observation triggered more profound research into the population biology of both *E. faecium* and *E. faecalis*.

11.5 POPULATION STRUCTURE AND GENETIC EVOLUTION: SIMILARITIES AND DIFFERENCES BETWEEN *E. FAECIUM* AND *E. FAECALIS*

To provide insights in the existence, distribution, and dynamics of specific enterococcal subpopulations or lineages and their evolutionary descent, multilocus sequence typing

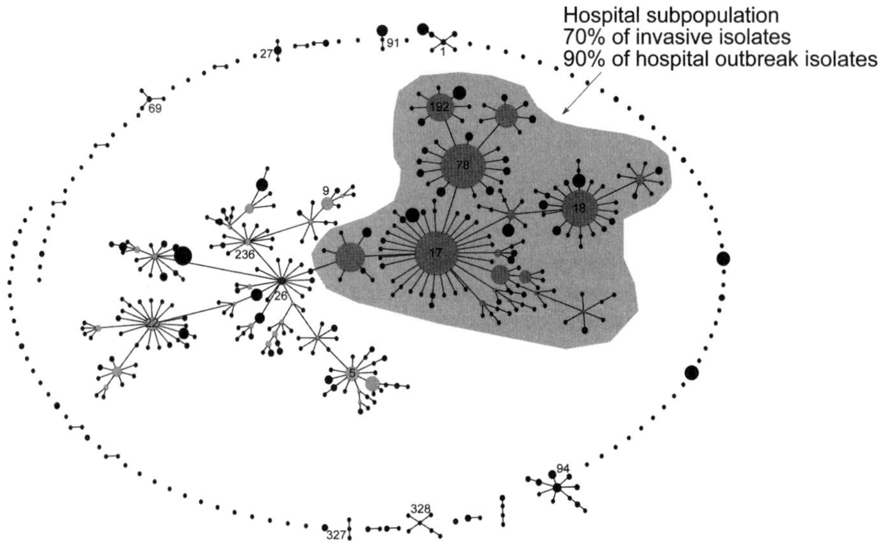

Figure 11.1 Population snapshot of 1543 *E. faecium* isolates (http://efaecium.mlst.net/) representing 452 STs based on MLST allelic profiles using the eBURST algorithm (Feil et al., 2004). This snapshot shows all clonal complexes, singletons, and patterns of evolutionary descent. The size of the circles indicates their prevalence in the MLST database. Numbers correspond to the STs of major (sub)group founders and lines connect single-locus variants, STs that differ in only one of the seven housekeeping genes. Hospital subpopulation, representing the majority of hospital outbreaks and clinical infections, is indicated.

(MLST) schemes were developed for *E. faecium* and *E. faecalis* (Homan et al., 2002; Ruiz-Garbajosa et al., 2006). eBURST (Feil et al., 2004) clustering of MLST data of a collection of more than 400 *E. faecium* isolates from hospital outbreaks, clinical infections in humans, surveillance (feces) isolates from hospitalized patients and from nonhospitalized persons, and surveillance isolates from different farm (swine, poultry, and veal calves) and pet (dogs and cats) animals revealed that the majority of outbreak-associated and clinical isolates, representing hospital-acquired *Enterococcus faecium* (HA-Efm), clustered together in a hospital subpopulation, designated complex 17, distinct from animal and human community isolates (Willems et al., 2005). Increasing the analysis to more than 1200 strains using data from the *E. faecium* MLST web-based database (http://efaecium.mlst.net/) confirms the previously identified population structure (Fig. 11.1). However, MLST also disclosed that eBURST is not able to divide the *E. faecium* population in distinct clonal complexes (CCs) but that the majority (59%) of all sequence types (STs) are part of a single large, straggly group that includes HA-Efm as well as community and animal isolates.

Thus, despite the fact that HA-Efm isolates seem to group distinct from community and animal isolates, from the eBURST clustering it is not immediately apparent that all hospital-acquired strains are closely related evolutionarily and are less related to nonhospital-acquired strains. To further evaluate genetic divergence between HA-Efm and non-HA-Efm strains contained in the *E. faecium* MLST web-based database, levels of pairwise genetic distances (F_{ST}) using Arlequin 2.0 (Schneider et al., 2000) as well as shared polymorphisms and fixed differences using DnaSP 4.0 (Rozas et al., 2003) were

Table 11.1 Genetic Variation of *E. faecium* MLST Loci between Hospital-Acquired (HA-Efm) (n = 136) and Nonhospital-Acquired (Non-HA-Efm) (n = 270) Isolates

	MLST housekeeping genes						
	atpA	ddl	gdh	purK	gyd	pstS	adk
Length of aligned nucleotide sequences[a]	556	465	530	492	395	583	437
Number of alleles	59	45	39	47	31	64	29
Total number of polymorphic sites	123	43	173	63	59	77	91
Total no. of polymorphisms	136	44	193	67	64	80	91
F_{ST}	0.04	0.14	0.17	0.11	0.14	0.00	0.00
P value F_{ST} (10,000 permutations)	0.09	0.001	0.04	0.03	0.14	0.4	0.992
Shared polymorphisms	118	21	157	9	51	50	86
Fixed differences	0	0	0	0	0	0	0

[a]The total number of polymorphic sites, total number of polymorphisms, shared polymorphisms, and fixed differences were calculated using DnaSP 4.0 (Rozas et al., 2003). F_{ST} and F_{ST} p values were calculated using Arlequin 2.0 (Schneider et al., 2000).

calculated (Table 11.1). F_{ST} indicates levels of gene flow and has a theoretical minimum of 0, indicating no differentiation in the population, thus free genetic exchange, and a maximum of 1, indicating no gene flow and fixation of alleles in different populations. However, this index rarely reaches the maximum of 1 and a F_{ST} of >0.15 already denotes a considerable differentiation (Litvintseva et al., 2006). The high number of polymorphisms shared by HA-Efm and non-HA-Efm for almost all MLST housekeeping genes suggests a high level of genetic similarity. There were also no fixed differences between the nucleotide sequences of HA-Efm and non-HA-Efm in any of the seven housekeeping genes, and the F_{ST} for four of the seven housekeeping genes were close to zero, indicating high gene flow between HA-Efm and non-HA-Efm, with sequences not genetically statistically different ($p > 0.05$). This suggests a common gene pool for these housekeeping genes with frequent gene exchange between HA-Efm and non-HA-Efm. It also indicates that HA-Efm do not constitute an isolated ecotype with a coherent self-contained gene pool, at least when four of the seven examined housekeeping genes that belong the *E. faecium* core genome are taken into account. However, three genes, *ddl*, *gdh*, and *purK*, had F_{ST} values ranging from 0.11 to 0.17 ($p < 0.05$), suggesting significant divergence of these genes between HA-Efm and non-HA-Efm. Furthermore, eBURST clustering indicates that HA-Efm and non-HA-Efm isolates are not randomly mixed but display distinct grouping, as mentioned above, despite the fact that eBURST may not infer the population structure of *E. faecium* entirely correctly (see below). Also, non-HA-Efm isolates tend to cluster according to the host they were isolated from. This becomes even more apparent when the distribution of *E. faecium* STs among sources is examined in more detail, not at the level of eBURST groups but at the level of STs (Table 11.2). Of all HA-Efm isolates, almost 50% have STs that are uniquely found among HA-Efm isolates, and 27% of the HA-Efm isolates have STs that are shared with surveillance isolates from hospitalized

Table 11.2 Distribution of *E. faecium* STs among Sources

Source	No. of isolates (%) with STs unique for source	No. of isolates (%) with STs shared with other sources								Total
		HA	HS	C	S	P	VC	D	Total shared	
Hospital-acquired (HA)	342 (48)	—	189 (27)	77 (11)	33 (5)	5 (0.7)	1 (0.1)	64 (9)	369 (52)	711
Hospital surveillance (HS)	58 (35)	64 (38)	—	18 (11)	14 (8)	2 (1)	1 (0.6)	11 (7)	110 (65)	168
Community (C)	35 (53)	4 (6)	14 (21)	—	8 (12)	4 (6)	1 (2)	0	31 (47)	66
Swine (S)	49 (69)	6 (8)	8 (11)	7 (10)	—	0	0	1 (1)	22 (31)	71
Poultry (P)	85 (86)	5 (5)	1 (1)	6 (6)	0	—	1 (1)	1 (1)	14 (14)	99
Veal calves (VC)	18 (90)	1 (5)	1 (5)	0	0	0	—	0	2 (10)	20
Dogs (D)	24 (35)	31 (46)	9 (12)	0	1 (1)	2 (3)	1 (1)	—	44 (65)	68

patients. This means that only 25% of HA-Efm isolates share STs with nonhospital isolates and less than 15% with animal isolates. It is interesting to note that 9% of HA-Efm isolates share STs with dog strains, while 46% of dog isolates share STs with HA-Efm. This suggests that dogs share isolates that are also recovered from invasive sites and are associated with outbreaks in hospitalized patients. The finding that *E. faecium* in dogs are similar to HA-Efm isolates was also reported recently (Damborg et al., 2008). In contrast to dog isolates, the majority of *E. faecium* isolates from swine, poultry, and veal calves represent STs that are unique to the source, confirming that *E. faecium* isolates are primarily host specific (Willems et al., 2000, 2005). Nonoverlapping STs between HA-Efm and community and animal isolates indicate a distinct evolutionary history of hospital and nonhospital human isolates. The observation that the genetic diversity (GD) on the ST level of 701 HA-Efm (GD = 0.93, CI_{95} = 0.92–0.94) is significantly lower than that of 66 human community isolates (GD = 0.985, CI_{95} = 0.971–0.999) suggests that the hospital environment functions as a population bottleneck restricting GD. Niche adaptation causing selective sweeps that purges ST GD in populations has been proposed previously (Fraser et al., 2009).

The difference in GD may also be indicative of difference in age of the hospital subpopulation and nonhospital *E. faecium* populations. Preliminary evidence for this comes from the fact that of all invasive (mainly blood) *E. faecium* isolates from hospitalized patients isolated that group within the eBURST-based hospital subpopulation (n = 537) (Fig. 11.1), only one isolate originates from before the 1990s. Eleven other invasive isolates from before the 1990s (92%) group are distinct from the hospital subpopulation. This may suggest that the hospital subpopulation as defined by eBURST (Fig. 11.1) is relatively young. Since the number of invasive isolates from before the 1990s is extremely low in the database (http://efaecium.mlst.net/), conclusions about age of the hospital subpopulation based on these data should be drawn with considerable care.

However, epidemiological data also point toward a relatively recent emergence of HA-Efm. One of the characteristics of contemporary HA-Efm, which will be discussed in more detail below, is the fact that these isolates are ampicillin resistant, while ampicillin resistance is almost absent in non-HA-Efm (Leavis et al., 2003; Coque et al., 2005; Willems et al., 2005). It is interesting in this respect that prevalence data from U.S. hospitals indicate that ampicillin resistance has only started to emerge from the mid-1980s (Grayson et al., 1991; Iwen et al., 1997; Murdoch et al., 2002). Whether or not the hospital subpopulation of Ha-Efm has emerged relatively recently, they represent highly successful clones capable of intra- and interhospital spread. Some HA-Efm STs have even spread globally as illustrated for two STs (Fig. 11.2).

Clustering of *E. faecalis* MLST data (http://efaecalis.mlst.net/) by eBURST reveals a different population structure compared to that of *E. faecium*. The eBURST-based population structure is not dominated by one large group but by several smaller CCs (Fig. 11.3). Despite the fact that in some CCs hospital-derived isolates seem to dominate (Nallapareddy et al., 2005; Ruiz-Garbajosa et al., 2006; Kawalec et al., 2007; McBride et al., 2007), the distinction between hospital-acquired *E. faecalis* (HA-Efs) and community isolates is not as apparent as it is in *E. faecium* (McBride et al., 2007). In a set of 211 *E. faecalis* isolates, almost two-thirds of HA-Efs isolates had STs that were shared by non-HA-Efs, and more than a third of these isolates shared STs with isolates from the community (Table 11.3). Also, the majority of isolates from the community and animals shared STs with isolates from other ecological niches, while this was not the case for *E. faecium*. This suggests less ecological isolations among this set of *E. faecalis* isolates and more disseminations of clones among different sources.

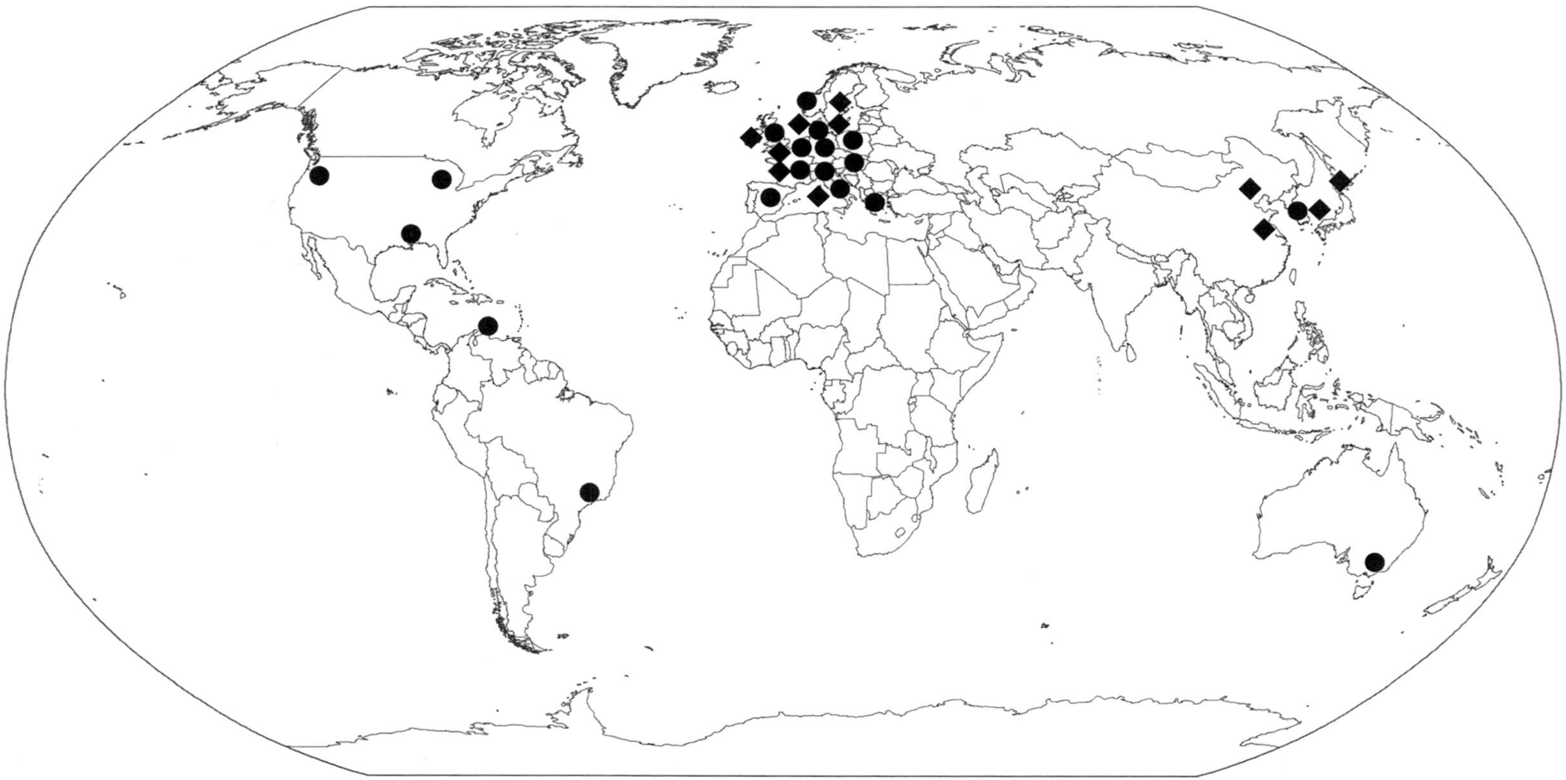

Figure 11.2 Global distribution of HA-Efm isolates with ST-17 (black circles) and ST-78 (black diamonds) isolates.

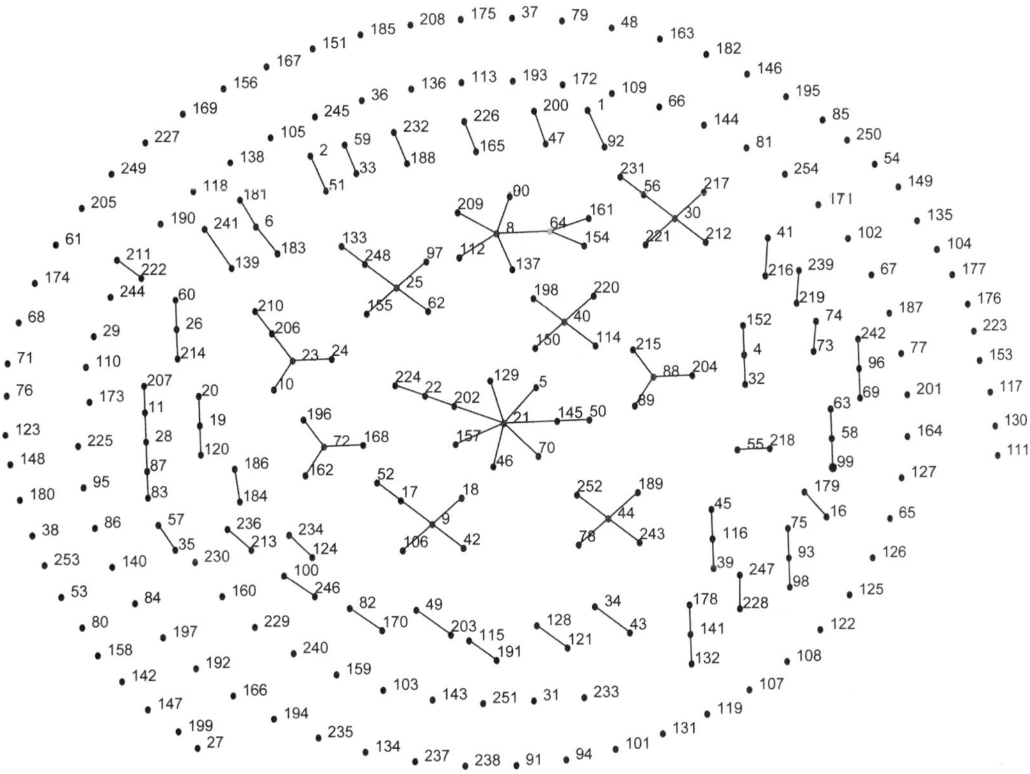

Figure 11.3 Population snapshot of 643 *E. faecalis* isolates (http://efaecalis.mlst.net/) representing 249 STs based on MLST allelic profiles using the eBURST algorithm (Feil et al., 2004). See also Fig. 11.1 for the explanation of the eBURST-based population snapshot. See color insert.

Table 11.3 Distribution of *E. faecalis* STs among Sources

Source	No of isolates (%) with STs unique for source	No. of isolates (%) with STs shared with other sources					Total
		HA	HS	C	A	Total shared	
Hospital-acquired (HA)	117 (38)	—	79 (26)	57 (18)	56 (18)	192 (62)	309
Hospital surveillance (HS)	22 (27)	28 (35)	—	17 (21)	14 (17)	59 (73)	81
Community (C)	26 (40)	15 (23)	15 (23)	—	9 (14)	39 (60)	65
Animal (A)	46 (42)	36 (33)	15 (14)	13 (12)	—	64 (58)	110

11.6 WHAT IS DRIVING GD IN *E. FAECIUM* AND *E. FAECALIS*?

The collection of a large amount of sequence information from MLST data also allowed an evaluation of the origin of genetic variability in *E. faecium* and *E. faecalis*. Gene tree comparisons of individual MLST housekeeping genes of *E. faecium* and *E. faecalis* revealed that the majority of all 42 pairwise comparisons of the seven MLST loci were

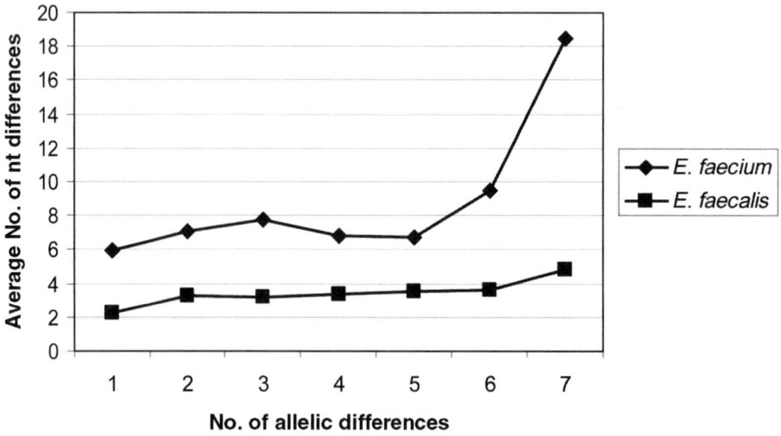

Figure 11.4 Sequence diversity versus allelic diversity. The average number of nucleotide (nt) differences in nonidentical alleles for all pairwise comparisons of 446 *E. faecium* STs and 253 *E. faecalis* STs was calculated separately for allelic profiles that differ in one to seven alleles using BLAND software (Feil et al., 2003). This computation shows no positive correlation between the number of nucleotide differences and allelic differences, which suggests that recombination has played an important role in the genetic diversification of *E. faecium* and *E. faecalis*.

incongruent. For *E. faecium*, incongruence was found for 22/42 (60%) of all pairwise comparisons, while this was 100% (42/42) for *E. faecalis* (Willems et al., 2005; Ruiz-Garbajosa et al., 2006). In *E. faecalis*, in almost all cases, the random tree was a better fit to the gene of investigation than to the most incongruent MLST gene (Ruiz-Garbajosa et al., 2006). This lack of congruence between gene tree topologies in the individual housekeeping genes indicates extensive recombination in both *E. faecalis* and *E. faecium* since in populations where recombination is rare, the genetic relationships inferred using the sequence of one housekeeping gene should be congruent with those obtained using other housekeeping genes (Feil et al., 2001). An important role for recombination in genetic diversification in *E. faecium* and *E. faecalis* was confirmed by the lack of a positive trend between the average number of nucleotide differences in nonidentical alleles for all pairwise comparisons of 446 *E. faecium* STs and 253 *E. faecalis* STs calculated using BLAND software (Feil et al., 2003) (Fig. 11.4). The finding of high average numbers of nucleotide differences in the nonidentical alleles of single-locus variants (SLVs) in *E. faecium* and *E. faecalis*, 5.9 and 2.3, respectively, also points toward frequent recombination. Furthermore, of all *E. faecalis* allelic differences between group ancestor–SLV pairs ($n = 43$), as predicted by eBURST, 28 included >1 nucleotide difference, were not unique in the database, and thus were most likely a result of recombination. Fifteen allelic differences included only a single nucleotide change, were unique within the dataset, and thus were most likely a result of mutation, leading to a recombination (r) to mutation (m) ratio (r/m) per locus of 1.9. For *E. faecium*, including group and subgroup ancestor–SLV pairs ($n = 150$), these numbers were 127 recombinational events and 23 mutations (per locus r/m = 5.5). The observations that most SLVs, 65% in *E. faecalis* and 85% in *E. faecium*, have arisen by recombination rather than by point mutations, that no positive correlation exists between the degree of allelic diversity and the number of nucleotide differences in nonidentical alleles, and that most of the comparisons of MLST gene tree topologies were incongruent strongly indicate that GD in *E. faecalis* and *E. faecium* has mainly been driven by recombination.

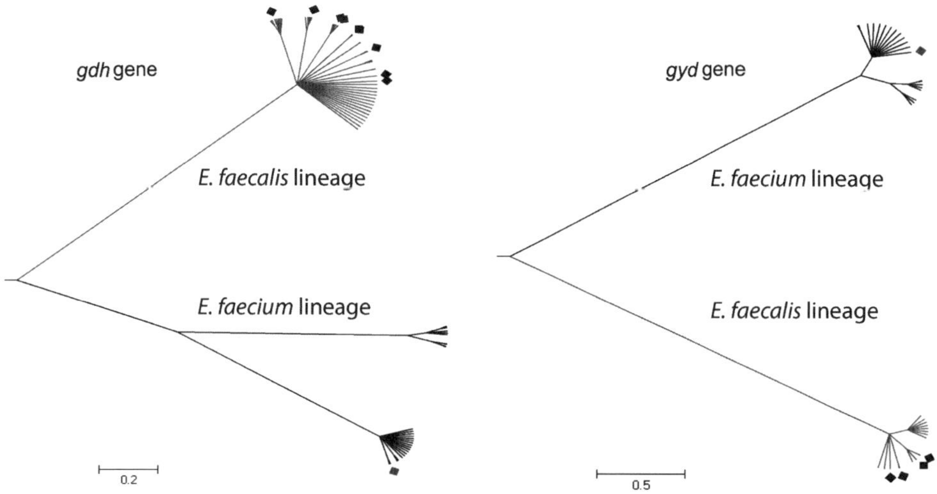

Figure 11.5 Phylogeny of the *gdh* and *gyd* genes of *E. faecium* and *E. faecalis* inferred by ClonalFrame (Didelot and Falush, 2007). The program was run for 100,000 Markov chain Monte Carlo (MCMC) iterations including 50,000 burn-in iterations. ClonalFrame produced for both *gdh* and *gyd* distinct *E. faecium* and *E. faecalis* branches. Diamonds indicate for both trees *E. faecium* isolates in the *E. faecalis* branch and vice versa, demonstrating interspecies horizontal gene transfer of these housekeeping genes.

Recombination of MLST housekeeping genes is not limited to intraspecies transfer. A phylogenetic analysis of the *gyd* and *gdh* genes present in both *E. faecium* and *E. faecalis* MLST schemes using ClonalFrame (Didelot and Falush, 2007) indicates that these genes can also be transferred between the two species (Fig. 11.5). In general, sequence identity between *gyd* and *gdh* of *E. faecium* and *E. faecalis* is around 73%, resulting in a distinct *E. faecium* and *E. faecalis* lineage for both genes. However, within the *E. faecalis gyd* and *gdh* lineages, alleles originating from *E. faecium* are found and vice versa.

The predicted high level of recombination in *E. faecium* and *E. faecalis* seems to contrast with two recent reports in which recombination levels, based on the MLST housekeeping genes in these two enterococcal species, were calculated to be only low or intermediate (Perez-Losada et al., 2006; Vos and Didelot, 2009). This apparent difference may be explained by the fact that these studies calculated mainly intragenic recombination, while most probably, the length of the recombining fragment in *E. faecium* and *E. faecalis* is larger than that of the MLST locus. Conjugational transfer of large DNA fragments has been extensively reported in enterococci (Clewell, 1990; Francois et al., 1997; Rice and Carias, 1998; Dahl and Sundsfjord, 2003).

High-level recombination as observed in *E. faecium* and *E. faecalis* may also influence the reliability of the eBURST clustering. As observed in Fig. 11.1, the population snapshot of *E. faecium* displays a large, straggly group that includes approximately 59% of all STs. The presence of a single large, straggly eBURST group is a strong indicator that STs have been erroneously linked as a result of a high-population recombination rate (ρ) and a low-population mutation rate (θ) (Turner et al., 2007). In populations with a high ρ/θ ratio, the accuracy of eBURST links, that is, the number of true links divided by all drawn links, drops to 55–64%, which means that in these populations, approximately 40% of the drawn links are not accurate (Turner et al., 2007). Using simulated populations with known ancestry, Turner and coworkers showed that in populations with extremely high ρ/θ ratios, groups or CCs that do not share recent common ancestry are inappropriately linked. As demonstrated above, high-level recombination relative to mutation is being predicted for *E. faecium*, indicating that evolutionary descent in *E. faecium* is most probably not displayed correctly by eBURST (Turner et al., 2007). This would mean that STs in the large, straggly group in Fig. 11.1 most probably do not share a recent common ancestry but should be split up into distinct CCs with STs like ST5, ST9, ST17, ST18, ST22, ST26, ST78, ST192, and ST236 as founders of distinct CCs. This would also mean that HA-Efm within the hospital subpopulation in Fig. 11.1, which previously has been designated complex 17 (Willems et al., 2005), have not been evolved from a single founder (ST17) but that different HA-Efm CCs (e.g., ST17, ST18, ST78, and ST192) have evolved independently. A neighbor-net tree built from concatenated sequences of 16 group and subgroup founders, STs 1, 5, 9, 17, 18, 22, 26, 27, 69, 78, 91, 94, 192, 236, 327, 328 (Fig. 11.1), using SplitsTree4 (Huson and Bryant, 2006), indeed suggests that HA-Efm STs 17, 18, 78, and 192 have not evolved recently from a common ancestor distinct from non-HA-Efm (Fig. 11.6).

Although recombination levels in *E. faecalis* were also predicted to be high, the eBURST population snapshot does not display a large, straggly group including the majority of STs as seen in *E. faecium*. In contrast, the population snapshot of *E. faecalis* is more dominated by small CCs and by a relatively high number of singletons. This may be explained by the fact that recombination rates in *E. faecalis* are lower than in *E. faecium* and/or that mutation rates, that is, the rate at which new alleles arise, may be higher in *E. faecalis* than in *E. faecium*. Simulations predict that in the case of a high level of recombination in combination with a high mutation rate (i.e., a high number of different alleles), but with a ρ/θ ratio still >1, eBURST-based population snapshots lead to small CCs and a high number of singletons, as seen in *E. faecalis*, while in populations with a high level of recombination in combination with low levels of mutations (i.e., a low number of different alleles), eBURST displays the large, straggly groups as seen in *E. faecium* (Hanage et al., 2006; Turner et al., 2007).

11.7 THE ACCESSORY GENOME OF *E. FAECIUM* AND *E. FAECALIS*

Although HA-Efm may not have evolved from one recent evolutionary ancestor, they share much of their auxiliary genetic repertoire. A first indication for this came from the finding that the majority of clinical and outbreak-associated *E. faecium* isolates harbor the *esp* gene encoding the enterococcal surface protein Esp contained on a putative pathogenicity island, while this gene was virtually absent in nonhospital isolates (Baldassarri et al., 2001; Willems et al., 2001; Woodford et al., 2001; Coque et al., 2002; Leavis et al., 2003, 2004). A more comprehensive insight into the gene content of hospital and nonhospital *E. faecium* isolates came from comparative genomic hybridizations (CGH) using a mixed whole genome array (Leavis et al., 2007). From these experiments the accessory genome of *E. faecium* was estimated to comprise 35% of the genome content. Using a Bayesian-based clustering based on gene content inferred from the CGH data, a distinct hospital clade was identified supported by a Bayesian posterior probability of 1.0, meaning that 100% of all phylogenies showed this branch, which largely overlap with the MLST-based hospital subpopulation as indicated in Fig. 11.1. In total, 44 of the 46 (96%) hospital clade isolates based on gene content, representing seven STs, group within the MLST-based hospital subpopulation, while 50 of the 51 (98%) nonhospital clade isolates, representing 36 STs, clustered outside this subpopulation. In total, more than 100 genes were identified, which were specifically enriched in HA-Efm isolates, representing antibiotic resistance

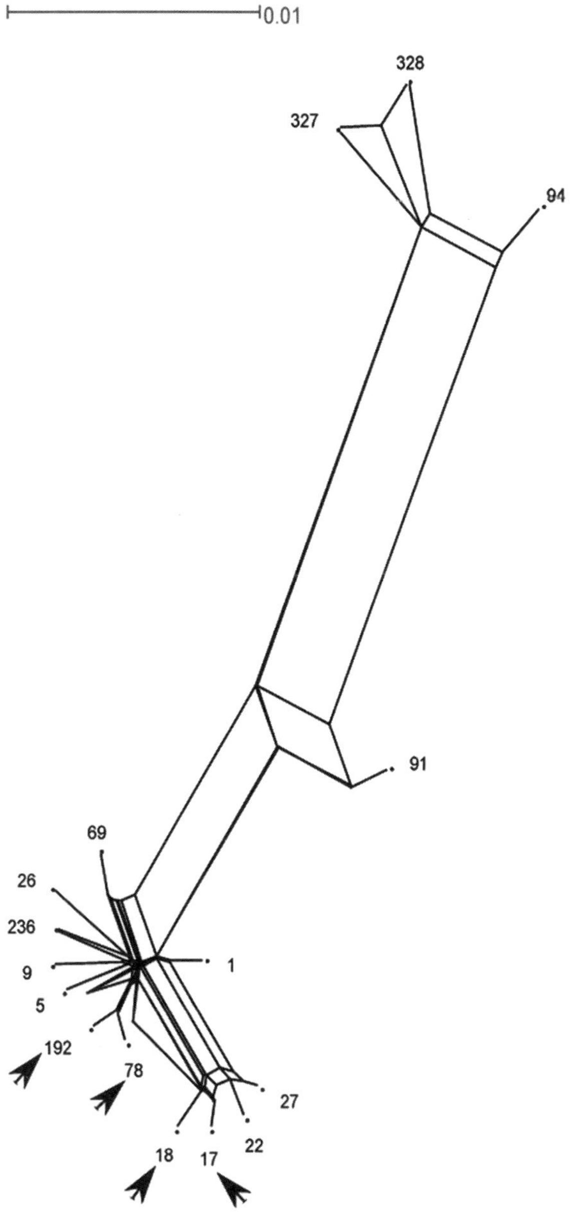

Figure 11.6 A neighbor-net tree constructed from a concatenation of the seven MLST housekeeping genes from 15 *E. faecium* STs (STs 1, 5, 9, 17, 18, 22, 26, 27, 69, 78, 91, 94, 236, 327, and 328) representing major (sub)group founders according to the eBURST clustering (see also Fig. 11.1) using SplitsTree4 (Huson and Bryant, 2006). STs that belong to the hospital subpopulation according to eBURST (STs 17, 18, 78, and 192) are indicated by arrows. The network does not conclusively support that the hospital subpopulation STs have recently evolved from a common ancestor.

genes, genes tentatively involved in carbohydrate metabolism, putative cell surface proteins, specific IS element with IS*16* as being the most discriminatory between HA-Efm and non-HA-Efm, and several genes encoding for proteins with unknown function (Leavis et al., 2007).

In follow-up studies, two genes encoding cell surface proteins were found to be enriched in HA-Efm (Hendrickx et al., 2007) as well as three pilin gene clusters (Hendrickx et al., 2007, 2008). Using a combination of a mathematical algorithm followed by PCR and sequencing, a novel genomic island was identified, highly specific for HA-Efm and tentatively encoding a metabolic pathway involved in carbohydrate transport and

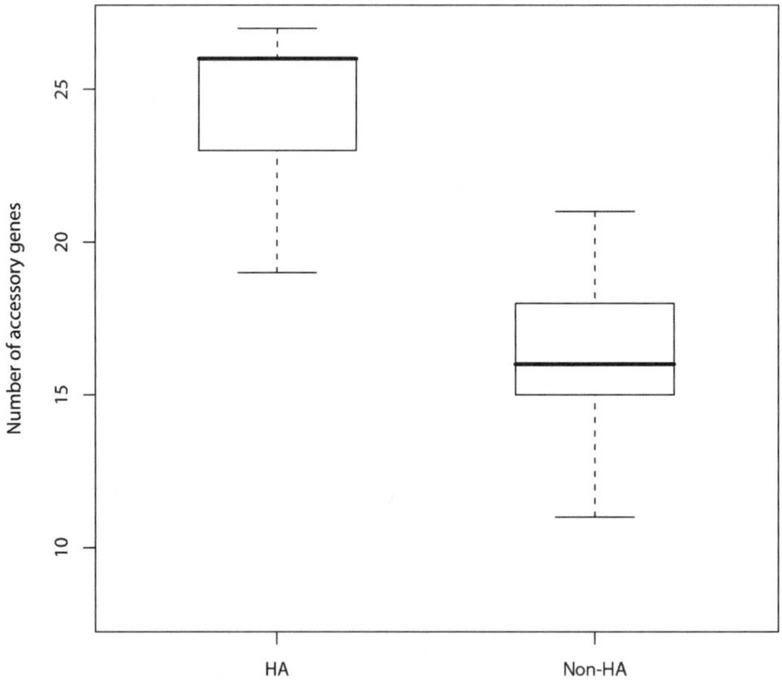

Figure 11.7 Boxplot displaying the distribution of 26 cell surface protein genes and a genomic island tentatively encoding a novel sugar uptake system among 42 HA-Efm and 88 non-HA-Efm. The box stretches from the lower hinge (defined as the twenty-fifth percentile) to the upper hinge (the seventy-fifth percentile). The median is shown as a line across the box. The ends of the vertical lines or "whiskers" indicate the minimum and maximum data values.

metabolism (Heikens et al., 2008). Also, another potential virulence determinant encoded by the *hylEfm* also predominates in clinical *E. faecium* strains (Rice et al., 2003). A comparison of accessory gene content, focusing on 26 cell surface protein genes, including the pilin gene clusters and the putative metabolic island, between 42 HA-Efm and 88 non-HA-Efm showed that HA-Efm are clearly enriched in these auxiliary elements. This may have facilitated the evolutionary development and ecological dominance of these clones in hospitalized patients (Fig. 11.7). Not only putative virulence genes or genes encoding metabolic pathways are enriched in HA-Efm but they also harbor antibiotic resistance genes that are virtually absent among non-HA-Efm isolates. Ampicillin resistance and high-level quinolone resistance are traits specifically acquired by HA-Efm (Leavis et al., 2003, 2006; Coque et al., 2005; Willems et al., 2005).

The association between the accessory genome and the *E. faecalis* genetic background is less clear. This is well illustrated by the difference in distribution of the *esp* virulence gene between *E. faecium* and *E. faecalis*. While esp_{Efm} seems to be more restricted to isolates recovered from hospitalized patients, esp_{Efs} is found much more dispersed among nonhospital and even nonhuman *E. faecalis*, like dogs, pigs, birds, and environmental sources (Eaton and Gasson, 2002; Hammerum and Jensen, 2002; Harada et al., 2005; Poeta et al., 2006; Shankar et al., 2006; Whitman et al., 2007). However, a recent study, identifying esp_{Efm} in environmental samples from the Pacific coast environment suggests that the host range of esp_{Efm} may be broader than initially thought (Layton et al., 2009). Several putative and proven virulence determinants have been described in *E. faecalis*, but most seem to be ubiquitously present in this species. A more comprehensive study that examined the presence of 18 auxiliary traits like virulence genes, antibiotic resistance genes, and polymorphisms in capsule operon in a collection of 106 strains isolated over the past 100 years revealed that virulence and antibiotic resistance genes can be found in many diverse lineages (McBride et al., 2007). However, in some lineages dominated by isolates that have caused infectious outbreaks, virulence traits are overrepresented although not as apparent as in *E. faecium*.

11.8 SUMMARY, CONCLUSIONS, AND FUTURE PERSPECTIVES

Enterococci prevail in highly diverse environmental niches and have become one of the most important nosocomial pathogens over the last two decades. They are characterized by a high level of resistance to a wide variety of antimicrobial agents and by a high propensity for intra- and interhospital spread. This makes enterococcal emergence extremely difficult to control. Of additional importance is the emergence of *E. faecium*, the enterococcal species expressing the highest levels of antimicrobial resistance and partly replacing *E. faecalis* as a cause of enterococcal infections. Population biology studies have revealed that the emergence of *E. faecium* infections is mirrored by a change in the *E. faecium* population structure with an evolutionary development of a novel *E. faecium* subpopulation of HA-Efm. This genetic subpopulation, characterized by a unique genetic repertoire, has virtually been absent before the 1990s with less than 10% of invasive and hospital outbreak-related isolates from this period belonging to this subpopulation, while nowadays, it contains more than 83% of HA-Efm. The rapid evolutionary development of the *E. faecium* hospital subpopulation, driven by frequent recombination and lateral acquisition of accessory DNA, is a clear illustration of quantum evolution rather than gradual evolutionary change. As such, HA-Efm resemble "hopeful monsters" (Sarich, 1980; Turner and Feil, 2007), a term coined to describe an event of instantaneous evolutionary saltation generating organisms that have the potential of establishing new evolutionary lineages, however with questionable fitness. Although HA-Efm probably represent a fitness peak in hospitals, their low prevalence in the community (Top et al., 2007) suggests that in the absence of the selective constraints imposed by the hospital environment, HA-Efm suffer a fitness deficit compared to the indigenous *E. faecium* population. In addition to high recombination frequencies and a highly sophisticated machinery of DNA exchange through conjugational transfer, clonal evolution is also affected by the population size, the size of the gene pool, and the genetic content of the ecological niche they are part of. Since the vast preponderance of enterococcal existence occurs as a member of the GI tract of man and animals or in the environment, enterococcal evolution has mostly occurred in these ecological niches with ample possibilities of exchanging DNA with many other organisms. Several *E. faecalis* and *E. faecium* genomes have been and are currently being sequenced, which will provide more information about the enterococcal core- and pan-genome and on whether *E. faecalis* and *E. faecium* have an open or a closed genome. This will give us more insights into the genetic potential of these organisms. Functional genomics, transcriptomics, proteomics, and metabolomics data, combined with *in vivo* studies of potential virulence determinants, clinical epidemiological data, and mathematical modeling, will provide knowledge on host–enterococcal interactions and may provide explanations for its recent success. Improved insights may also allow the development of novel strategies to counteract the emergence of this nosocomial pathogen.

REFERENCES

Aarestrup, F. M., Butaye, P., and Witte, W. (2002) Nonhuman reservoirs of enterococci. In *The Enterococci: Pathogenesis, Molecular Biology and Antibiotic Resistance* (eds. M. S. Gilmore, D. B. Clewell, P. Courvalin, G. M. Dunny, B. E. Murray, and L. B. Rice), pp. 55–99. American Society for Microbiology, Washington, DC.

Aksoy, D. Y. and Unal, S. (2008) New antimicrobial agents for the treatment of gram-positive bacterial infections. *Clin Microbiol Infect* **14,** 411–420.

Arias, C. A., Torres, H. A., Singh, K. V. et al. (2007) Failure of daptomycin monotherapy for endocarditis caused by an *Enterococcus faecium* strain with vancomycin-resistant and vancomycin-susceptible subpopulations and evidence of in vivo loss of the *vanA* gene cluster. *Clin Infect Dis* **45,** 1343–1346.

Arthur, M., Depardieu, F., Gerbaud, G. et al. (1997) The VanS sensor negatively controls VanR-mediated transcriptional activation of glycopeptide resistance genes of Tn*1546* and related elements in the absence of induction. *J Bacteriol* **179,** 97–106.

Arthur, M., Molinas, C., Depardieu, F., and Courvalin, P. (1993) Characterization of Tn*1546*, a Tn*3*-related transposon conferring glycopeptide resistance by synthesis of depsipeptide peptidoglycan precursors in *Enterococcus faecium* BM4147. *J Bacteriol* **175,** 117–127.

Baldassarri, L., Bertuccini, L., Ammendolia, M. G. et al. (2001) Variant *esp* gene in vancomycin-sensitive *Enterococcus faecium. Lancet* **357,** 1802.

Biedenbach, D. J., Bell, J. M., Sader, H. S. et al. (2007) Antimicrobial susceptibility of gram-positive bacterial isolates from the Asia-Pacific region and an in vitro evaluation of the bactericidal activity of daptomycin, vancomycin, and teicoplanin: A SENTRY Program Report (2003–2004). *Int J Antimicrob Agents* **30,** 143–149.

Bonten, M. J., Willems, R., and Weinstein, R. A. (2001) Vancomycin-resistant enterococci: Why are they here, and where do they come from? *Lancet Infect Dis* **1,** 314–325.

Borgen, K., Simonsen, G. S., Sundsfjord, A. et al. (2000) Continuing high prevalence of VanA-type vancomycin-resistant enterococci on Norwegian poultry farms three years after avoparcin was banned. *J Appl Microbiol* **89,** 478–485.

Boyd, D. A., Willey, B. M., Fawcett, D. et al. (2008) Molecular characterization of *Enterococcus faecalis* N06-0364 with low-level vancomycin resistance harboring a novel D-Ala-D-Ser gene cluster, *vanL. Antimicrob Agents Chemother* **52,** 2667–2672.

Brisse, S., Fluit, A. C., Wagner, U. et al. (1999) Association of alterations in ParC and GyrA proteins with resistance of clinical isolates of *Enterococcus faecium* to nine different fluoroquinolones. *Antimicrob Agents Chemother* **43,** 2513–2516.

Carias, L. L., Rudin, S. D., Donskey, C. J., and Rice, L. B. (1998) Genetic linkage and cotransfer of a novel, *vanB*-containing transposon (Tn*5382*) and a low-affinity penicillin-binding protein 5 gene in a clinical vancomycin-resistant *Enterococcus faecium* isolate. *J Bacteriol* **180,** 4426–4434.

Chow, J. W. (2000) Aminoglycoside resistance in enterococci. *Clin Infect Dis* **31,** 586–589.

Clewell, D. B. (1990) Movable genetic elements and antibiotic resistance in enterococci. *Eur J Clin Microbiol Infect Dis* **9,** 90–102.

Comenge, Y., Quintiliani, R. J., Li, L. et al. (2003) The CroRS two-component regulatory system is required for intrinsic beta-lactam resistance in *Enterococcus faecalis. J Bacteriol* **185,** 7184–7192.

Coque, T. M., Willems, R., Canton, R. et al. (2002) High occurrence of *esp* among ampicillin-resistant and vancomycin-susceptible *Enterococcus faecium* clones from hospitalized patients. *J Antimicrob Chemother* **50,** 1035–1038.

Coque, T. M., Willems, R. J., Fortun, J. et al. (2005) Population structure of *Enterococcus faecium* causing bacteremia in a Spanish university hospital: Setting the scene for a future increase in vancomycin resistance? *Antimicrob Agents Chemother* **49,** 2693–2700.

Courvalin, P. (2006) Vancomycin resistance in gram-positive cocci. *Clin Infect Dis* **42**(Suppl 1)**,** S25–S34.

Dahl, K. H., Lundblad, E. W., Rokenes, T. P. et al. (2000) Genetic linkage of the *vanB2* gene cluster to Tn*5382* in vancomycin-resistant enterococci and characterization of two novel insertion sequences. *Microbiology* **146,** 1469–1479.

Dahl, K. H. and Sundsfjord, A. (2003) Transferable *vanB2* Tn*5382*-containing elements in fecal streptococcal strains from veal calves. *Antimicrob Agents Chemother* **47,** 2579–2583.

Damborg, P., Sorensen, A. H., and Guardabassi, L. (2008) Monitoring of antimicrobial resistance in healthy dogs: First report of canine ampicillin-resistant *Enterococcus faecium* clonal complex 17. *Vet Microbiol* **132,** 190–196.

de Lencastre, H., Brown, A. E., Chung, M. et al. (1999) Role of transposon Tn*5482* in the epidemiology of vancomycin-resistant *Enterococcus faecium* in the pediatric oncology unit of a New York City Hospital. *Microb Drug Resist* **5,** 113–129.

Didelot, X. and Falush, D. (2007) Inference of bacterial microevolution using multilocus sequence data. *Genetics* **175,** 1251–1266.

Eaton, T. J. and Gasson, M. J. (2002) A variant enterococcal surface protein Esp(fm) in *Enterococcus faecium*; distribution among food, commensal, medical, and environmental isolates. *FEMS Microbiol Lett* **216,** 269–275.

El Amin, N. A., Jalal, S., and Wretlind, B. (1999) Alterations in GyrA and ParC associated with fluoroquinolone resistance in *Enterococcus faecium. Antimicrob Agents Chemother* **43,** 947–949.

European Antimicrobial Resistance Surveillance System (EARSS) (2007) EARSS Annual Report 2007. http://www.rivm.nl/earss/ (accessed January 29, 2009).

FEIL, E. J., COOPER, J. E., GRUNDMANN, H. et al. (2003) How clonal is *Staphylococcus aureus*? *J Bacteriol* **185,** 3307–3316.

FEIL, E. J., HOLMES, E. C., BESSEN, D. E. et al. (2001) Recombination within natural populations of pathogenic bacteria: Short-term empirical estimates and long-term phylogenetic consequences. *Proc Natl Acad Sci U S A* **98,** 182–187.

FEIL, E. J., LI, B. C., AANENSEN, D. M. et al. (2004) eBURST: Inferring patterns of evolutionary descent among clusters of related bacterial genotypes from multilocus sequence typing data. *J Bacteriol* **186,** 1518–1530.

FERRETTI, J. J., GILMORE, K. S., and COURVALIN, P. (1986) Nucleotide sequence analysis of the gene specifying the bifunctional 6′-aminoglycoside acetyltransferase 2″-aminoglycoside phosphotransferase enzyme in *Streptococcus faecalis* and identification and cloning of gene regions specifying the two activities. *J Bacteriol* **167,** 631–638.

FLANNAGAN, S. E., CHOW, J. W., DONABEDIAN, S. M. et al. (2003) Plasmid content of a vancomycin-resistant *Enterococcus faecalis* isolate from a patient also colonized by *Staphylococcus aureus* with a VanA phenotype. *Antimicrob Agents Chemother* **47,** 3954–3959.

FONTANA, R., CERINI, R., LONGONI, P. et al. (1983) Identification of a streptococcal penicillin-binding protein that reacts very slowly with penicillin. *J Bacteriol* **155,** 1343–1350.

FRANCOIS, B., CHARLES, M., and COURVALIN, P. (1997) Conjugative transfer of *tet(S)* between strains of *Enterococcus faecalis* is associated with the exchange of large fragments of chromosomal DNA. *Microbiology* **143,** 2145–2154.

FRASER, C., ALM, E. J., POLZ, M. F. et al. (2009) The bacterial species challenge: Making sense of genetic and ecological diversity. *Science* **323,** 741–746.

GARCIA-MIGURA, L., LIEBANA, E., and JENSEN, L. B. (2007) Transposon characterization of vancomycin-resistant *Enterococcus faecium* (VREF) and dissemination of resistance associated with transferable plasmids. *J Antimicrob Chemother* **60,** 263–268.

GHIDAN, A., DOBAY, O., KASZANYITZKY, E. J. et al. (2008) Vancomycin-resistant enterococci (VRE) still persist in slaughtered poultry in hungary 8 years after the ban on avoparcin. *Acta Microbiol Immunol Hung* **55,** 409–417.

GIRAFFA, G. (2003) Functionality of enterococci in dairy products. *Int J Food Microbiol* **88,** 215–222.

GRAYSON, M. L., ELIOPOULOS, G. M., WENNERSTEN, C. B. et al. (1991) Increasing resistance to beta-lactam antibiotics among clinical isolates of *Enterococcus faecium*: A 22-year review at one institution. *Antimicrob Agents Chemother* **35,** 2180–2184.

HAMMERUM, A. M. and JENSEN, L. B. (2002) Prevalence of *esp*, encoding the enterococcal surface protein, in *Enterococcus faecalis* and *Enterococcus faecium* isolates from hospital patients, poultry, and pigs in Denmark. *J Clin Microbiol* **40,** 4396.

HAMMERUM, A. M., LESTER, C. H., NEIMANN, J. et al. (2004) A vancomycin-resistant *Enterococcus faecium* isolate from a Danish healthy volunteer, detected 7 years after the ban of avoparcin, is possibly related to pig isolates. *J Antimicrob Chemother* **53,** 547–549.

HANAGE, W. P., FRASER, C., and SPRATT, B. G. (2006) The impact of homologous recombination on the generation of diversity in bacteria. *J Theor Biol* **239,** 210–219.

HANDWERGER, S., PUCCI, M. J., and KOLOKATHIS, A. (1990) Vancomycin resistance is encoded on a pheromone response plasmid in *Enterococcus faecium* 228. *Antimicrob Agents Chemother* **34,** 358–360.

HANDWERGER, S. and SKOBLE, J. (1995) Identification of chromosomal mobile element conferring high-level vancomycin resistance in *Enterococcus faecium*. *Antimicrob Agents Chemother* **39,** 2446–2453.

HARADA, T., TSUJI, N., OTSUKI, K., and MURASE, T. (2005) Detection of the *esp* gene in high-level gentamicin resistant *Enterococcus faecalis* strains from pet animals in Japan. *Vet Microbiol* **106,** 139–143.

HEIKENS, E., VAN SCHAIK, W., LEAVIS, H. L. et al. (2008) Identification of a novel genomic island specific to hospital-acquired clonal complex 17 *Enterococcus faecium* isolates. *Appl Environ Microbiol* **74,** 7094–7097.

HENDRICKX, A. P., BONTEN, M. J., VAN LUIT-ASBROEK, M. et al. (2008) Expression of two distinct types of pili by a hospital-acquired *Enterococcus faecium* isolate. *Microbiology* **154,** 3212–3223.

HENDRICKX, A. P., VAN WAMEL, W. J., POSTHUMA, G. et al. (2007) Five genes encoding surface-exposed LPXTG proteins are enriched in hospital-adapted *Enterococcus faecium* clonal complex 17 isolates. *J Bacteriol* **189,** 8321–8332.

HEUER, O. E., PEDERSEN, K., ANDERSEN, J. S., and MADSEN, M. (2002) Vancomycin-resistant enterococci (VRE) in broiler flocks 5 years after the avoparcin ban. *Microb Drug Resist* **8,** 133–138.

HIDRON, A. I., EDWARDS, J. R., PATEL, J. et al. (2008) NHSN annual update: Antimicrobial-resistant pathogens associated with healthcare-associated infections: Annual summary of data reported to the National Healthcare Safety Network at the Centers for Disease Control and Prevention, 2006–2007. *Infect Control Hosp Epidemiol* **29,** 996–1011.

HOMAN, W. L., TRIBE, D., POZNANSKI, S. et al. (2002) Multilocus sequence typing scheme for *Enterococcus faecium*. *J Clin Microbiol* **40,** 1963–1971.

HUGAS, M., GARRIGA, M., and AYMERICH, M. T. (2003) Functionality of enterococci in meat products. *Int J Food Microbiol* **88,** 223–233.

HUSON, D. H. and BRYANT, D. (2006) Application of phylogenetic networks in evolutionary studies. *Mol Biol Evol* **23,** 254–267.

IWEN, P. C., KELLY, D. M., LINDER, J. et al. (1997) Change in prevalence and antibiotic resistance of *Enterococcus* species isolated from blood cultures over an 8-year period. *Antimicrob Agents Chemother* **41,** 494–495.

JENSEN, L. B., AHRENS, P., DONS, L. et al. (1998) Molecular analysis of Tn*1546* in *Enterococcus faecium* isolated from animals and humans. *J Clin Microbiol* **36,** 437–442.

JENSEN, L. B., FRIMODT-MOLLER, N., and AARESTRUP, F. M. (1999) Presence of *erm* gene classes in gram-positive bacteria of animal and human origin in Denmark. *FEMS Microbiol Lett* **170,** 151–158.

JONES, R. N., BEACH, M. L., PFALLER, M. A., and DOERN, G. V. (1998) Antimicrobial activity of gatifloxacin tested against 1676 strains of ciprofloxacin-resistant gram-positive cocci isolated from patient infections in North and South America. *Diagn Microbiol Infect Dis* **32,** 247–252.

KAINER, M. A., DEVASIA, R. A., JONES, T. F. et al. (2007) Response to emerging infection leading to outbreak of linezolid-resistant enterococci. *Emerg Infect Dis* **13,** 1024–1030.

KANEMATSU, E., DEGUCHI, T., YASUDA, M. et al. (1998) Alterations in the GyrA subunit of DNA gyrase and the ParC subunit of DNA topoisomerase IV associated with quinolone resistance in *Enterococcus faecalis*. *Antimicrob Agents Chemother* **42,** 433–435.

KAWALEC, M., PIETRAS, Z., DANILOWICZ, E. et al. (2007) Clonal structure of *Enterococcus faecalis* isolated from Polish hospitals: Characterization of epidemic clones. *J Clin Microbiol* **45,** 147–153.

LAYTON, B. A., WALTERS, S. P., and BOEHM, A. B. (2009) Distribution and diversity of the enterococcal surface protein (*esp*) gene in animal hosts and the Pacific coast environment. *J Appl Microbiol* **106,** 1521–1531.

LEAVIS, H., TOP, J., SHANKAR, N. et al. (2004) A novel putative enterococcal pathogenicity island linked to the *esp* virulence gene of *Enterococcus faecium* and associated with epidemicity. *J Bacteriol* **186,** 672–682.

LEAVIS, H. L., WILLEMS, R. J., TOP, J., and BONTEN, M. J. (2006) High-level ciprofloxacin resistance from point mutations in *gyrA* and *parC* confined to global hospital-adapted clonal lineage CC17 of *Enterococcus faecium*. *J Clin Microbiol* **44,** 1059–1064.

LEAVIS, H. L., WILLEMS, R. J., TOP, J. et al. (2003) Epidemic and nonepidemic multidrug-resistant *Enterococcus faecium*. *Emerg Infect Dis* **9,** 1108–1115.

LEAVIS, H. L., WILLEMS, R. J., VAN WAMEL, W. J., et al. (2007) Insertion sequence-driven diversification creates a globally dispersed emerging multiresistant subspecies of *E. faecium*. *PLoS Pathog* **3,** e7.

LECLERCQ, R., DERLOT, E., DUVAL, J., and COURVALIN, P. (1988) Plasmid-mediated resistance to vancomycin and teicoplanin in *Enterococcus faecium*. *N Engl J Med* **319,** 157–161.

LIGOZZI, M., PITTALUGA, F., and FONTANA, R. (1996) Modification of penicillin-binding protein 5 associated with high-level ampicillin resistance in *Enterococcus faecium*. *Antimicrob Agents Chemother* **40,** 354–357.

LIM, S. K., KIM, T. S., LEE, H. S. et al. (2006) Persistence of vanA-type *Enterococcus faecium* in Korean livestock after ban on avoparcin. *Microb Drug Resist* **12,** 136–139.

LITVINTSEVA, A. P., THAKUR, R., VILGALYS, R., and MITCHELL, T. G. (2006) Multilocus sequence typing reveals three genetic subpopulations of *Cryptococcus neoformans* var. g*rubii* (serotype A), including a unique population in Botswana. *Genetics* **172,** 2223–2238.

MANSON, J. M., SMITH, J. M., and COOK, G. M. (2004) Persistence of vancomycin-resistant enterococci in New Zealand broilers after discontinuation of avoparcin use. *Appl Environ Microbiol* **70,** 5764–5768.

MARTEAU, P., POCHART, P., DORE, J. et al. (2001) Comparative study of bacterial groups within the human cecal and fecal microbiota. *Appl Environ Microbiol* **67,** 4939–4942.

MARTONE, W. J. (1998) Spread of vancomycin-resistant enterococci: Why did it happen in the United States? *Infect Control Hosp Epidemiol* **19,** 539–545.

MCBRIDE, S. M., FISCHETTI, V. A., LEBLANC, D. J. et al. (2007) Genetic diversity among *Enterococcus faecalis*. *PLoS One* **2,** e582.

MOELLERING, R. C. J. and WEINBERG, A. N. (1971) Studies on antibiotic syngerism against enterococci. II. Effect of various antibiotics on the uptake of 14 C-labeled streptomycin by enterococci. *J Clin Invest* **50,** 2580–2584.

MONTERO, C. I., STOCK, F., and MURRAY, P. R. (2008) Mechanisms of resistance to daptomycin in *Enterococcus faecium*. *Antimicrob Agents Chemother* **52,** 1167–1170.

MURDOCH, D. R., MIRRETT, S., HARRELL, L. J. et al. (2002) Sequential emergence of antibiotic resistance in enterococcal bloodstream isolates over 25 years. *Antimicrob Agents Chemother* **46,** 3676–3678.

MURRAY, B. E. (2000) Vancomycin-resistant enterococcal infections. *N Engl J Med* **342,** 710–721.

NALLAPAREDDY, S. R., WENXIANG, H., WEINSTOCK, G. M., and MURRAY, B. E. (2005) Molecular characterization of a widespread, pathogenic, and antibiotic resistance-receptive *Enterococcus faecalis* lineage and dissemination of its putative pathogenicity island. *J Bacteriol* **187,** 5709–5718.

NOVAIS, C., FREITAS, A. R., SOUSA, J. C. et al. (2008) Diversity of Tn*1546* and its role in the dissemination of vancomycin-resistant enterococci in Portugal. *Antimicrob Agents Chemother* **52,** 1001–1008.

OFNER-AGOSTINI, M., JOHNSTON, B. L., SIMOR, A. E. et al. (2008) Vancomycin-resistant enterococci in Canada: Results from the Canadian nosocomial infection surveillance program, 1999–2005. *Infect Control Hosp Epidemiol* **29,** 271–274.

PALAZZO, I. C., CAMARGO, I. L., ZANELLA, R. C., and DARINI, A. L. (2006) Evaluation of clonality in enterococci isolated in Brazil carrying Tn*1546*-like elements associated with *vanA* plasmids. *FEMS Microbiol Lett* **258,** 29–36.

PARK, I. J., LEE, W. G., LIM, Y. A., and CHO, S. R. (2007) Genetic rearrangements of Tn*1546*-like elements in vancomycin-resistant *Enterococcus faecium* isolates collected from hospitalized patients over a seven-year period. *J Clin Microbiol* **45,** 3903–3908.

PEPPER, K., Le BOUGUENEC, C., DE CESPEDES, G., and HORAUD, T. (1986) Dispersal of a plasmid-borne chloramphenicol resistance gene in streptococcal and enterococcal plasmids. *Plasmid* **16,** 195–203.

PEREZ-LOSADA, M., BROWNE, E. B., MADSEN, A. et al. (2006) Population genetics of microbial pathogens estimated from multilocus sequence typing (MLST) data. *Infect Genet Evol* **6,** 97–112.

Poeta, P., Costa, D., Rodrigues, J., and Torres, C. (2006) Detection of genes encoding virulence factors and bacteriocins in fecal enterococci of poultry in Portugal. *Avian Dis* **50,** 64–68.

Rice, L. B., Bellais, S., Carias, L. L. et al. (2004) Impact of specific *pbp5* mutations on expression of beta-lactam resistance in *Enterococcus faecium. Antimicrob Agents Chemother* **48,** 3028–3032.

Rice, L. B. and Carias, L. L. (1998) Transfer of Tn5385, a composite, multiresistance chromosomal element from *Enterococcus faecalis. J Bacteriol* **180,** 714–721.

Rice, L. B., Carias, L. L., Hutton-Thomas, R. et al. (2001) Penicillin-binding protein 5 and expression of ampicillin resistance in *Enterococcus faecium. Antimicrob Agents Chemother* **45,** 1480–1486.

Rice, L. B., Carias, L., Rudin, S. et al. (2003) A potential virulence gene, *hylEfm*, predominates in *Enterococcus faecium* of clinical origin. *J Infect Dis* **187,** 508–512.

Rice, L. B., Carias, L. L., Rudin, S. et al. (2005) *Enterococcus faecium* low-affinity *pbp5* is a transferable determinant. *Antimicrob Agents Chemother* **49,** 5007–5012.

Rodriguez-Martinez, J. M., Velasco, C., Briales, A. et al. (2008) Qnr-like pentapeptide repeat proteins in gram-positive bacteria. *J Antimicrob Chemother* **61,** 1240–1243.

Rozas, J., Sanchez-DelBarrio, J. C., Messeguer, X., and Rozas, R. (2003) DnaSP, DNA polymorphism analyses by the coalescent and other methods. *Bioinformatics* **19,** 2496–2497.

Ruiz-Garbajosa, P., Bonten, M. J., Robinson, D. A. et al. (2006) Multilocus sequence typing scheme for *Enterococcus faecalis* reveals hospital-adapted genetic complexes in a background of high rates of recombination. *J Clin Microbiol* **44,** 2220–2228.

Rybkine, T., Mainardi, J. L., Sougakoff, W. et al. (1998) Penicillin-binding protein 5 sequence alterations in clinical isolates of *Enterococcus faecium* with different levels of beta-lactam resistance. *J Infect Dis* **178,** 159–163.

Sarich, V. M. (1980) A macromolecular perspective on "The Material Basis of Evolution." *Experientia Suppl* **35,** 27–31.

Scheetz, M. H., Knechtel, S. A., Malczynski, M. et al. (2008) Increasing incidence of linezolid-intermediate or -resistant, vancomycin-resistant *Enterococcus faecium* strains parallels increasing linezolid consumption. *Antimicrob Agents Chemother* **52,** 2256–2259.

Schneider, S., Roessli, D., and Excoffier, L. (2000) *Arlequin Version 2.000: A Software for Population Genetic Data Analysis*. University of Geneva, Geneva.

Shankar, N., Baghdayan, A. S., Willems, R. et al. (2006) Presence of pathogenicity island genes in *Enterococcus faecalis* isolates from pigs in Denmark. *J Clin Microbiol* **44,** 4200–4203.

Sung, K., Khan, S. A., and Nawaz, M. S. (2008) Genetic diversity of Tn*1546*-like elements in clinical isolates of vancomycin-resistant enterococci. *Int J Antimicrob Agents* **31,** 549–554.

Talebi, M., Pourshafie, M. R., Katouli, M., and Mollby, R. (2008) Molecular structure and transferability of Tn*1546*-like elements in *Enterococcus faecium* isolates from clinical, sewage, and surface water samples in Iran. *Appl Environ Microbiol* **74,** 1350–1356.

Top, J., Willems, R., Blok, H. et al. (2007) Ecological replacement of *Enterococcus faecalis* by multiresistant clonal complex 17 *Enterococcus faecium. Clin Microbiol Infect* **13,** 316–319.

Top, J., Willems, R., van der Velden, S. et al. (2008) Emergence of clonal complex 17 *Enterococcus faecium* in The Netherlands. *J Clin Microbiol* **46,** 214–219.

Treitman, A. N., Yarnold, P. R., Warren, J., and Noskin, G. A. (2005) Emerging incidence of *Enterococcus faecium* among hospital isolates (1993 to 2002). *J Clin Microbiol* **43,** 462–463.

Turner, K. M. and Feil, E. J. (2007) The secret life of the multilocus sequence type. *Int J Antimicrob Agents* **29,** 129–135.

Turner, K. M., Hanage, W. P., Fraser, C. et al. (2007) Assessing the reliability of eBURST using simulated populations with known ancestry. *BMC Microbiol* **7,** 30.

Uttley, A. H., Collins, C. H., Naidoo, J., and George, R. C. (1988) Vancomycin-resistant enterococci. *Lancet* **1,** 57–58.

van den Bogaard, A. E., Bruinsma, N., and Stobberingh, E. E. (2000) The effect of banning avoparcin on VRE carriage in The Netherlands. *J Antimicrob Chemother* **46,** 146–148.

van den Bogaard, A. E., Jensen, L. B., and Stobberingh, E. E. (1997) Vancomycin-resistant enterococci in turkeys and farmers. *N Engl J Med* **337,** 1558–1559.

Vos, M. and Didelot, X. (2009) A comparison of homologous recombination rates in bacteria and archaea. *ISME J* **3,** 199–208.

Wegener, H. C., Aarestrup, F. M., Jensen, L. B. et al. (1999) Use of antimicrobial growth promoters in food animals and *Enterococcus faecium* resistance to therapeutic antimicrobial drugs in Europe. *Emerg Infect Dis* **5,** 329–335.

Werner, G., Dahl, K. H., and Willems, R. J. L. (2006) Composite elements encoding antibiotic resistance in *Enterococcus faecium* and *Enterococcus faecalis* In *Drug Resistance of Enterococci: Epidemiology and Molecular Mechanisms* (ed. N. Kobayashi), pp. 157–208. Research Signpost, Kerala, India.

Werner, G., Klare, I., and Witte, W. (2002) Molecular analysis of streptogramin resistance in enterococci. *Int J Med Microbiol* **292,** 81–94.

Whitman, R. L., Przybyla-Kelly, K., Shively, D. A., and Byappanahalli, M. N. (2007) Incidence of the enterococcal surface protein (*esp*) gene in human and animal fecal sources. *Environ Sci Technol* **41,** 6090–6095.

Willems, R. J., Homan, W., Top, J. et al. (2001) Variant *esp* gene as a marker of a distinct genetic lineage of vancomycin-resistant *Enterococcus faecium* spreading in hospitals. *Lancet* **357,** 853–855.

WILLEMS, R. J., TOP, J., VAN DEN BRAAK, N. et al. (1999) Molecular diversity and evolutionary relationships of Tn*1546*-like elements in enterococci from humans and animals. *Antimicrob Agents Chemother* **43,** 483–491.

WILLEMS, R. J., TOP, J., VAN DEN BRAAK, N. et al. (2000) Host specificity of vancomycin-resistant *Enterococcus faecium. J Infect Dis* **182,** 816–823.

WILLEMS, R. J., TOP, J., VAN SANTEN, M. et al. (2005) Global spread of vancomycin-resistant *Enterococcus faecium* from distinct nosocomial genetic complex. *Emerg Infect Dis* **11,** 821–828.

WILLIAMSON, R., le BOUGUENEC, C., GUTMANN, L., and HORAUD, T. (1985) One or two low affinity penicillin-binding proteins may be responsible for the range of susceptibility of *Enterococcus faecium* to benzylpenicillin. *J Gen Microbiol* **131,** 1933–1940.

WOODFORD, N., ADEBIYI, A. M., PALEPOU, M. F., and COOKSON, B. D. (1998) Diversity of VanA glycopeptide resistance elements in enterococci from humans and nonhuman sources. *Antimicrob Agents Chemother* **42,** 502–508.

WOODFORD, N., SOLTANI, M., and HARDY, K. J. (2001) Frequency of *esp* in *Enterococcus faecium* isolates. *Lancet* **358,** 584.

ZORZI, W., ZHOU, X. Y., DARDENNE, O. et al. (1996) Structure of the low-affinity penicillin-binding protein 5 PBP5fm in wild-type and highly penicillin-resistant strains of *Enterococcus faecium. J Bacteriol* **178,** 4948–4957.

Chapter 12

Population Biology of Lyme Borreliosis Spirochetes

Klaus Kurtenbach, Anne Gatewood Hoen, Stephen J. Bent, Stephanie A. Vollmer, Nicholas H. Ogden, and Gabriele Margos

12.1 INTRODUCTION

Two questions are central to the scientific debate in population biology and ecology: How do population fluctuations arise and how is diversity generated and maintained (Bjornstad and Grenfell, 2001; Hudson and Bjornstad, 2003)? Despite a different terminology, the same fundamental questions are at the center of contemporary infectious disease epidemiology (Grenfell and Bjornstad, 2005). While these questions have been resolved for several directly transmitted pathogens, the processes operating in the population biology of vector-borne pathogens are much less understood. Here, we use Lyme borreliosis (LB) as an example to shed light on the evolutionary ecology and population biology of vector-borne zoonotic diseases.

LB is the most prevalent vector-borne disease of humans in the temperate zone (Steere et al., 2004; Kurtenbach et al., 2006). It was named after the town Old Lyme in Southern Connecticut, United States, where it was described in 1976 (Steere et al., 1976, 1978). The disease is caused by several species of the LB group of spirochetes (also referred to as *Borrelia burgdorferi* sensu lato). The bacteria are maintained in nature by complex zoonotic transmission cycles involving hard ticks of the family Ixodidae and a vast number of small and medium-sized vertebrate host species (Kurtenbach et al., 2006). Humans are accidental hosts and are not involved in the life cycle or evolution of the spirochetes (Steere et al., 2004). Larger animals, such as deer, are not susceptible to LB spirochetes, but they serve as reproductive hosts for the ticks and are required for the maintenance of the tick populations (Wilson et al., 1985). The principal tick vectors of LB in Europe, Asia, and North America, *Ixodes ricinus*, *Ixodes persulcatus*, *Ixodes scapularis*, and *Ixodes pacificus*, have a three-stage life cycle that takes 2–5 years to complete and requires feeding on one vertebrate host per stage (Kurtenbach et al., 2002b; Randolph et al., 2002; Piesman and Gern, 2004). The two immature stages (larva and nymph) normally feed on rodents and other small- and medium-sized mammals and birds, while adult female ticks mainly feed on larger animals. Adult male ticks do not feed. Ticks of the genus *Ixodes* are capable

Bacterial Population Genetics in Infectious Disease, Edited by D. Ashley Robinson, Daniel Falush, and Edward J. Feil

of off-host dispersal of only a few meters (Falco and Fish, 1991). However, vertebrate hosts, particularly deer, but possibly also birds, are believed to drive the dispersal of these ticks (Rand et al., 1998; Scott et al., 2001; Ogden et al., 2008b).

LB spirochetes are transmitted to vertebrate hosts by infected ticks (mainly the nymphal stage) during the bloodmeal, which lasts between three and seven consecutive days (Piesman and Gern, 2004). In susceptible vertebrate hosts, the spirochetes disseminate and may persist for prolonged periods (Hanincova et al., 2008). Such infected hosts may subsequently pass on the spirochetes to noninfected ticks, thereby completing the transmission cycle of the LB spirochetes. The hosts often remain infective to the ticks for a prolonged period of time; however, the duration of infectivity depends on the genetic background of the spirochetal strain and, perhaps, also of the host species (Hanincova et al., 2008). Reciprocal horizontal transmission of the spirochetes between ticks and hosts is essential for the survival of the bacteria because the transovarial transmission of LB spirochetes to questing larvae is rare or absent; no direct transmission among ticks or vertebrate hosts occurs; and because LB spirochetes cannot survive free-living conditions (Kurtenbach et al., 2006).

At present, LB spirochetes constitute a group of 17 named species (Margos et al., 2009; Rudenko et al., 2009a,b). These species and their genotypes are not evenly distributed across the globe. For example, *Borrelia garinii*, *Borrelia valaisiana*, and *Borrelia afzelii* are widely distributed across Eurasia, but are not found in *I. scapularis* or *I. pacificus* ticks in North America. *B. burgdorferi* (previously referred to as *B. burgdorferi* sensu stricto) occurs both in the Old and New World; however, it is relatively rare in Europe and has not been recorded in Asia (Piesman and Gern, 2004). Several species are more restricted to particular regions. *Borrelia japonica*, for instance, is abundant in the Far East, mainly in Japan (Masuzawa, 2004), and *Borrelia lusitaniae* is found mainly around the Mediterranean Basin (L. Vitorino, unpublished observations). Depending on the ecological conditions, the relative abundance of these species varies. For example, in some regions of Western Europe, such as the British Isles, *B. garinii* and *B. valaisiana* are dominant (Kurtenbach et al., 2002b), whereas *B. afzelii* is probably the most abundant species of the LB group in many terrestrial habitats of continental Eurasia (Hubalek and Halouzka, 1997; Kurtenbach et al., 2006).

Although *I. ricinus*, *I. persulcatus*, and *I. scapularis* are considered generalist feeders (Piesman and Gern, 2004), most species of LB group spirochetes occupy different ecological niches as defined by different spectra of vertebrate hosts they can infect (Kurtenbach et al., 2002a,b, 2006). For example, *B. valaisiana* and most *B. garinii* strains are maintained by birds, and *B. afzelii* and *B. lusitaniae* are specialized to rodents and lizards, respectively (Kurtenbach et al., 2002a; Hanincova et al., 2003a,b; Dsouli et al., 2006; Richter and Matuschka, 2006; Taragel'ova et al., 2008). *B. burgdorferi* appears to be less specialized, as genotypes of this species were found to be infectious for several phylogenetically distant host species in the United States (Hanincova et al., 2006). Host association of LB spirochetes has been shown to be determined by interactions of a group of outer surface proteins, collectively referred to as complement regulator-acquiring surface proteins (CRASPs) (Bykowski et al., 2008), with components of the hosts' complement system (Kurtenbach et al., 1998; 2002a,b). Furthermore, frequency-dependent selection, mediated by adaptive immunity, seems to shape the population structure of *B. burgdorferi* s.l. (Qiu et al., 1997). The distributions of LB species and their intraspecific phylogeographic population structures are strongly influenced by these processes (Kurtenbach et al., 2006). Accumulating data suggest that the migration rates of bird-associated spirochetal genotypes are substantial (Gylfe et al., 2000), supported by data on spatial admixture of *B. garinii* and *B. valaisiana* genotypes across much of Europe (S. A. Vollmer et al., unpublished observations). On the other hand, *B. afzelii*, a genospecies associated with rodents (that are known to migrate much more slowly than most birds), displays a pronounced phylogeographic structure (Vollmer et al., unpublished observations). The limited geographic range of *B. lusitaniae* (De Michelis et al., 2000; Vitorino et al., 2008) and its association with reptilian hosts (Dsouli et al., 2006; Richter and Matuschka, 2006) suggest that its migration rate is also low.

In this chapter, we aim to illuminate evolutionary processes of LB group spirochetes that shape the patterns obtained by phylogeographic approaches in order to understand past and present demographic trends and to predict future epidemiological trends of these environmentally maintained pathogens in an ever-changing world. We first describe the highly unusual genome organization of the bacteria followed by a review of currently used molecular typing tools. We then explore the population biology of the spirochetes in North America and in Europe and provide phylogeographic evidence to infer evolutionary pathways and to understand adaptive radiation and speciation of LB spirochetes. Finally, we discuss possible future research avenues and the applications of new conceptual frameworks and analytical tools, in particular landscape genetics (Manel et al., 2003; Storfer et al., 2007).

12.2 GENOME ORGANIZATION OF LB SPIROCHETES

LB spirochetes possess a fragmented genome that is highly unusual for bacteria. It consists of a linear chromosome and a large number of linear and circular plasmids (Saint Girons et al., 1992; Fraser et al., 1997; Casjens et al., 2000). The linearity of genome fragments is sustained by telomeres, which are small inverted repeats with covalently closed ends. ResT, a plasmid-encoded telomere resolvase, is essential for the regeneration of the hairpin telomeres after duplication of genome fragments (Tourand et al., 2003). Whole genome sequences of the *B. burgdorferi* strains B31 (clone MI) and ZS7, *B. garinii* strain PBi (now renamed *Borrelia bavariensis* sp. nov.; Margos et al. 2009), and *B. afzelii* strain Pko have been completed (Fraser et al., 1997; Casjens et al., 2000; Glockner et al., 2006). At the time of writing, a number of additional genome projects were in progress (14 *B. burgdorferi* strains, 2 *B. garinii* strains [Far04 and PBr], *Borrelia spielmanii* strain A14S, *B. afzelii* strain ACA-1, and *B. valaisiana* strain VS116).

The sequenced *B. burgdorferi* B31 clone MI genome has been fully assembled and annotated, except for some unclonable regions adjacent to telomeres (Fraser et al., 1997; Casjens et al., 2000). Assembly of plasmid sequences proved to be difficult due to a mosaic structure, resulting in large stretches of similar sequences on different plasmids. Therefore, plasmid sequences are not completely assembled for all genomes of LB spirochetes sequenced to date (Glockner et al., 2004, 2006).

The genomic repertoire of *B. burgdorferi* clone MI consists of the linear chromosome (910 kbp) and 9 circular and 12 linear plasmids (>600 kbp) (Table 12.1). The total plasmid number of B31 may even be higher (total of 24), since some clones derived of the B31 isolate contain plasmids that are not present in MI (Casjens et al., 2000). It is known that plasmids may be lost during isolation and/or *in vitro* culture (Norris et al., 1997).

The plasmids of LB spirochetes are named according to their size and whether they are linear (lp) or circular (cp). For example, lp54 is a linear plasmid of 54 kb and cp32 a circular plasmid with a molecular size of 32 kb. The linear chromosome, cp26, and lp54 are colinear, while the other plasmids may differ significantly (Glockner et al., 2004,

Table 12.1 Comparison of Genomes of *B. burgdorferi* Strain B31 (Clone MI), *B. garinii* (*B. bavariensis*) Strain PBi, and *B. afzelii* Strain Pko

	B. burgdorferi MI	*B. afzelii* Pko	*B. garinii* PBi
Chromosome (kbp)	910	907	905
No. of open reading frame (ORF)	853	856	832
Plasmids (kbp)	610	507	372
Total no. of plasmids	21 (24)	15 (16)	11
Linear (lp)	12	6 (7)	8
Circular (cp)	9 (12)	9	3
Size range (lp)	5–56	6–60	21–59
Size range (cp)	9–32	27–30	28–31
Orthologous (lp)	lp54	lp60	lp59
Orthologous (cp)	cp28	cp27	cp26
Multiple plasmids/cell	7 × cp32	8 × cp30	cp29, cp30

2006). The size of plasmids in *B. burgdorferi* clone MI ranges from 5 to 56 kbp, in *B. afzelii* strain Pko from 6 to 60 kbp, and in *B. garinii* (*B.* bavariensis) strain PBi from 21 to 59 kbp (Casjens, 2000; Glockner et al., 2006). Homologous plasmids may vary in size in different LB species. For example, lp54, a plasmid carrying the gene encoding the outer surface protein A (*ospA*), has a molecular size of 54 kbp in *B. burgdorferi* B31 (Casjens et al., 2000), 59 kbp in *B. garinii* PBi, 60 kbp in *B. afzelii* Pko (Glockner et al., 2006), and 70 kbp in *B. lusitaniae* (Vitorino et al., 2009). In total, the plasmid fraction of isolate PBi comprises 29% of the whole genome (11 plasmids of which 2 were assembled), 36% of the genome in *B. afzelii* isolate Pko (15/16 plasmids), while the plasmid fraction of the *B. burgdorferi* B31 clone MI represents >40% of the genome (Casjens et al., 2000; Glockner et al., 2006).

The relationships among the plasmids are complex. In *B. burgdorferi* B31, many plasmids contain large stretches of DNA rearrangements, insertions, nonhomologous recombinations, and disrupted or mutationally changed coding regions (pseudogenes), suggesting a turbulent evolutionary history of these genomic fragments, the biological significance of which is unclear. For example, an almost intact copy of a cp32-like plasmid was found in lp56. Ten out of 12 linear plasmids in *B. burgdorferi* showed a high proportion of pseudogenes (i.e., lp5, lp17, lp21, lp25, lp28-1, lp28-3, lp28-4, lp36, lp38, and the non-cp32-like portion of lp56) (Casjens et al., 2000). Insertions of plasmid sequences into the chromosome have also been described, and in *B. burgdorferi*, the right arm of the chromosome can vary in length in different isolates (Casjens et al., 1997a; Huang et al., 2004).

Although the plasmid repertoire of LB spirochetes may differ considerably at the intra- and interspecific level, lp54 and cp26 of *B. burgdorferi* and their homologues in other LB species are indispensable and belong, in addition to the main chromosome, to the basic genome inventory of all LB group spirochetes (Glockner et al., 2006).

Virtually all genes encoding proteins involved in cell maintenance, often referred to as housekeeping genes, are located on the linear chromosome. The majority of outer surface proteins (Osp) are products of plasmid-located genes. A number of these N-terminally lipidated proteins (Coleman et al., 1986; Brandt et al., 1990) critically interact with the bacterial environment, that is, the tick or the vertebrate host, stressing the importance of extrachromosomal elements in the life cycle of LB spirochetes. The cp32 plasmids, a family of highly similar elements in *B. burgdorferi*, represent temperate prophage genomes, for which transduction has been shown to be a mechanism of reshuffling genetic material among spirochete strains (Eggers et al., 2002). Individual LB spirochetes can harbor multiple cp32 (Casjens et al., 1997b; Eggers et al., 2002; Stevenson and Miller, 2003). These plasmids (and their homologues) carry loci encoding the OspE-related proteins. They are sometimes collectively referred to as *erps*, although various names have previously been used, such as *ospE*, *ospF*, *elp*, *p21*, *bbk2.10*, *bbk2.11*, and *pG* (Casjens et al., 1997b; Stevenson et al., 2000; Stevenson and Miller, 2003). Their gene products include proteins (CRASPs) that have been demonstrated to bind host complement regulator factor H or factor H-like molecules, protecting LB spirochetes from host complement-mediated killing (Kraiczy et al., 2001a,b; Stevenson and Miller, 2003). As the repertoire of these gene products seems to determine the spirochetal ecotype, horizontal transfer of prophage DNA among strains might hold the key to the adaptive radiation of LB spirochetes (Cohan, 2002; Kurtenbach et al., 2002a,b; Stevenson and Miller, 2003).

The Vmp-like (*vls*) locus is located on the linear plasmid lp28-1 and encodes a 35-kDa surface-exposed and immunogenic lipoprotein, VlsE. Apart from the expression site (*vlsE*), the *vls* locus of *B. burgdorferi* contains 15 silent *vls* cassettes, which serve as sequence donors to the expression site. The recombination rates at the *vls* locus are very high, and therefore, this locus represents an interesting antigenic variation system that has been shown to facilitate immune evasion and persistent infection in the host (Coutte et al., 2009).

OspA and OspC are the best studied outer surface proteins of LB spirochetes. These proteins are differentially expressed in questing ticks and vertebrate hosts, and antigenic phase variation of these proteins is believed to play a pivotal role in the transmission cycle of LB spirochetes (Kurtenbach et al., 2002b). OspA expression is essential in the process of survival and/or dissemination from the midgut of the tick during its bloodmeal (Yang et al., 2004; Battisti et al., 2008). OspA has been the target of a transmission-blocking vaccine that was available commercially until 2002 (Abbott, 2006). In contrast, OspC is required for colonization of the tick's salivary glands (Pal et al., 2004; Fingerle et al., 2007) and/or for early infection of the vertebrate host (Stewart et al., 2006; Tilly et al., 2007). Both *ospA* and *ospC* have been frequently used as molecular markers in population genetics and in epidemiological studies of LB spirochetes.

The loci encoding the large (5S, 23S) and small subunit (16S) ribosomal RNA (rRNA) form a cluster on the linear chromosome. This cluster contains a single copy of the 16S rRNA (*rrs*) approximately 2 kbp upstream of tandemly repeated 23S-5S rRNA (*rrlA-rrfA*, *rrlB-rrfB*) loci. The tandem repeats of 23S-5S are separated by an additional intergenic spacer (IGS) of approximately 200 bp, a feature unique to LB spirochetes. The length of the 16S-23S (*rrs-rrl*) and the 23S-5S (*rrfA-rrlB*) IGS may vary in different LB species (Schwartz et al., 1992; Gazumyan et al., 1994; Ojaimi et al., 1994). These noncoding loci have also been widely used for inter- and intraspecies phylogenetic analysis of LB spirochetes (Postic et al., 1994, 1998; Liveris et al., 1995; Bunikis et al., 2004).

Several studies of *B. burgdorferi* in the United States found that *ospA*, *ospC*, and the 16S-23S IGS form a linkage group. These findings supported the hypothesis that *B. burgdorferi* in the United States is clonal (Dykhuizen et al., 1993; Wang et al., 1999; Bunikis et al., 2004; Wormser et al., 2008). However, the model of a strictly clonal evolution of LB spirochetes has repeatedly been challenged because recombination events, especially at *ospC*, and plasmid exchange within and between LB species were detected (Wang et al., 1999; Qiu et al., 2004). For *B. lusitaniae*, we have recently shown that recombination events may even occur on the chromosome. However, the overall ratio of recombination to mutation was very low, suggesting that the linear chromosome of LB spirochetes is relatively clonal (Vitorino et al., 2008). In nature, the dynamics of infection in host and vector populations will determine the opportunity for mixing of different genotypes in ways that allow the horizontal transfer of genetic material. This will be the key to the rates of genetic change in LB spirochete populations (Kurtenbach et al., 2006).

12.3 GENOTYPING OF LB SPIROCHETES AND PHYLOGENETIC TOOLS

Unambiguous genotyping systems are crucial to capture epidemiological and ecological patterns of microbial populations and to illuminate the evolutionary processes that shape such patterns in space and time. Several of the LB species known to date have been delineated using whole DNA–DNA hybridization (Baranton et al., 1992; Kawabata et al., 1993; Masuzawa et al., 2001). While DNA–DNA hybridization served for many years as the standard for bacterial species delineation (Wayne et al., 1987), it is a method that requires a specialized laboratory, and questions exist about interpretation and reproducibility (Stackebrandt and Ebers, 2006). Most ecological, population genetics, or epidemiological studies have been performed using analyses of single loci, such as IGS regions, the 16S rRNA locus, the genes encoding the decorin-binding protein A (*dbpA*) or the flagellin B (*flaB*), as well as *ospA* and *ospC* (Wilske et al., 1993, 1996; Postic et al., 1994; Marconi et al., 1995; Will et al., 1995; Fukunaga et al., 1996; Dykhuizen and Baranton, 2001; Michel et al., 2004; Schulte-Spechtel et al., 2006). Although some of these single loci, or particular combinations thereof (Bunikis et al., 2004; Richter et al., 2006; Rudenko et al., 2009), have been convenient for species assignment of strains or to address particular epidemiological questions, they may be unsuitable to resolve evolutionary relationships among LB species because it is not always possible to define outgroups in phylogenetic trees. For example, both the 5S-23S IGS and *ospA* are present in LB spirochete genomes but not in relapsing fever spirochetes or other bacteria (Schwartz et al., 1992; Postic et al., 1994).

Multilocus sequence typing (MLST) and multilocus sequence analysis (MLSA) are extremely powerful and practical molecular tools for population genetics studies and assignment of large numbers of strains to bacterial species (Maiden et al., 1998; Feil and Spratt, 2001; Gevers et al., 2006; Bishop et al., 2009). Coupled with phenotypic and ecological information, MLSA is currently replacing whole DNA–DNA hybridization for species delineation of LB spirochetes and other bacteria, which, in many cases, shows good agreement with whole DNA–DNA hybridization, but with the advantages that cumulative databases can be generated to assign strains to species using the Internet (Richter et al., 2006; Postic et al., 2007; Chu et al., 2008; Bishop et al., 2009). MLST/MLSA schemes have been applied to many directly transmitted pathogens (http://www.mlst.net/) but, so far, only to very few vector-borne microbial populations, such as *Yersinia pestis*, the agent of plague, or LB spirochetes (Achtman et al., 1999; Richter et al., 2006; Postic et al., 2007; Vitorino et al., 2007, 2008; Chu et al., 2008; Margos et al., 2008; Rudenko et al., 2009).

Several previous studies have used sequence information of multiple combined loci to characterize LB spirochetes (Bunikis et al., 2004; Qiu et al., 2004; Richter et al., 2006;

Attie et al., 2007; Postic et al., 2007). However, these typing approaches deviate from typical MLST/MLSA schemes developed for other microbial pathogens (Maiden et al., 1998; Enright and Spratt, 1999; Gevers et al., 2005) in that different evolutionary classes of loci were combined, such as hypervariable genes encoding outer surface proteins, conserved housekeeping genes, or noncoding loci. Analyses of loci with different evolutionary rates can allow researchers to assess population structure over a wider range of temporal scales, but combining loci that evolve at different rates and under different tree structures poses problems in inferring phylogenetic trees (Matsen et al., 2008), and sequence data from genes that are subject to selective forces must be interpreted in the context of that selection.

Most MLST/A schemes are based on nucleotide sequences of internal fragments of multiple housekeeping genes, which are evolving nearly neutrally. Because of the combination of multiple housekeeping genes, MLSA is, in most cases, highly discriminatory while retaining signatures of longer-term evolutionary relationships or clonal stability (Enright and Spratt, 1999). An MLST/A scheme that avoids the previously described problems by using solely housekeeping genes was developed for LB spirochetes (Margos et al., 2008, 2009; Vitorino et al., 2008) (http://www.mlst.net). Briefly, eight housekeeping genes were amplified by nested polymerase chain reaction (PCR) directly from DNA extracted from ticks followed by a determination of their nucleotide sequences. For each locus, sequences that differ in one or more nucleotides are assigned to different alleles. The combination of all loci gives rise to the allelic profile or sequence type, which unambiguously identifies a strain. Based on the analyses of MLSA allelic profile data using the eBURST algorithm, relationships among LB spirochetes can be inferred. This bioinformatics tool visualizes clustering of closely related genotypes and can predict the founding genotype of each clonal complex (Feil et al., 2004) (see also Section 2.1, Chapter 2, in this book). In another approach to infer evolutionary pathways among LB spirochetes and other bacteria, the sequences of the housekeeping genes are concatenated for each strain into a single string of sequence, followed by generating phylogenetic trees, for example, using Bayesian phylogenetic inferences, the maximum likelihood algorithm, or the neighbor-joining method (Gevers et al., 2005; Margos et al., 2008).

The MLSA scheme described here is not only a powerful tool to demarcate sequence clusters or species among the bacteria but is also able to establish phylogenetic relationships within and among LB species, because gene trees can be rooted with housekeeping genes of relapsing fever spirochetes (Margos *et al.*, 2009). Bayesian mixture modeling techniques revealed that these housekeeping genes, but not *ospC*, belong to the same evolutionary class (E. Loza et al., unpublished observations). This renders the data amenable to detailed evolutionary studies of LB group spirochaetes. Since MLSA can be applied directly to infected ticks, LB spirochetes do not need to be cultured for these analyses, which is a decisive advantage for epidemiological and population genetics studies using large numbers of samples (Margos *et al.*, 2008).

Several culturable bacterial species have been studied using population genomics, driven by the advent of new technologies, such as the 454 Life Sciences sequencing approach (Roche), that make draft genome sequencing faster and cheaper (Hall, 2007; Holt et al., 2008). However, whole genome sequencing is still not a feasible approach to study the population biology and epidemiology of LB spirochetes because the presently available technologies require pure DNA from isolated bacterial strains. Culturing of LB spirochetes is time-consuming and introduces sampling biases (Norris et al., 1997). Therefore, it is likely that population studies of LB spirochetes will rely on targeted PCR amplification and sequencing of partial genomes for quite some time. In our view, MLSA based on housekeeping genes provides enough power to characterize LB group spirochetes at the different phylogenetic levels required for evolutionary, epidemiological, and landscape genetics studies.

12.4 POPULATION BIOLOGY AND EVOLUTION OF LB SPIROCHETES

12.4.1 Dispersal of Ticks and LB Spirochetes

The mechanisms whereby LB spirochetes disperse are surprisingly poorly studied. These bacteria can only migrate within the range of vector competent tick species. For populations of LB spirochetes to extend their distributional range, dispersal of tick vectors is clearly crucial because, outside the range of the vectors, the bacteria would have no means of being transmitted. Ticks cannot fly, so they are unable to move actively over long distances on their own (Falco and Fish, 1991) or by being carried on the wind, as are many insect vectors (Purse et al., 2005). Therefore, ticks of the genus *Ixodes* migrate passively when attached to hosts, which may occur on a number of spatial scales, being predominantly driven by mammalian hosts at a shorter range (Madhav et al., 2004) and likely by migratory birds over longer distances (Ogden et al., 2008b). Migration of adult ticks on large hosts, such as deer, is likely to be the most effective mechanism of tick dispersal. Immature ticks dispersed by hosts are less likely to establish new tick populations due to high mortality.

While the presence of ticks is essential for LB spirochetes to thrive, ticks are less important for migration of the bacteria than vertebrate hosts. Except for some unusual circumstances (Ogden et al., 1997), adult ticks play little or no role in the transmission cycles of LB spirochetes, which means that infected nymphal ticks transported by hosts (and which will molt into adult ticks) will rarely introduce the spirochetes into a newly established tick vector population. Larval ticks can be dispersed on hosts only for relatively short distances compared to nymphs (because they attach for shorter periods to their hosts), and they are nearly always uninfected prior to feeding. Thus, for larval ticks to contribute to spirochete migration, they must be carried by infective vertebrate hosts, which infect the larvae while they take a bloodmeal. For this reason, dispersal of larval ticks also appears to be a relatively inefficient mode of long-distance migration of the bacteria. It is most likely that LB spirochetes migrate mainly via persistently infected vertebrate hosts, which can then establish new infections in local questing nymphal ticks (i.e., the effective vector stage) by virtue of infecting local larval ticks. Consequently, we hypothesize that the migration rates of LB spirochetes match those of their main reservoir hosts. This idea is consistent with findings of phylogeographic studies, showing that LB spirochetes associated with highly mobile hosts, such as birds, are much less structured spatially than those associated with rodent or reptilian hosts (Gylfe et al., 2000; Vitorino et al., 2008; Vollmer et al., unpublished observations).

12.4.2 Transatlantic Diversification and Global Origin of *B. burgdorferi* (Sensu Stricto)

In the United States, LB has spread from two major focal regions to affect large areas of the Northeast and Midwest over the past few decades, with more than 27,000 cases reported from these regions in 2007 (CDC, 2009). The spread of LB into its current distribution in the United States is primarily due to the recent range expansion of the blacklegged tick *I. scapularis*, the principal vector of *B. burgdorferi* to humans (Chen et al., 2005). It is likely that the wide host range of *B. burgdorferi* has been facilitating its epidemic dispersal in the Northeastern United States (Hanincova et al., 2006; Kurtenbach et al., 2006). However, it remains unclear whether the different strains of *B. burgdorferi* in North America are specialized to different vertebrate host species or subpopulations (Brisson and Dykhuizen, 2004; Hanincova et al., 2006).

Using MLST based on housekeeping genes, we have demonstrated that North American and European *B. burgdorferi* populations correspond to distinct lineages (Margos et al., 2008). The finding that the populations of *B. burgdorferi* sampled in North America and Europe differ genetically suggests limited present-day migration of the strains between the regions due to geographic barriers, such as the Atlantic Ocean, the Great Plains, or the Rocky Mountains. This is corroborated by the distinct distributional ranges of *I. ricinus*, *I. scapularis*, and *I. pacificus* ticks, the principal vectors of LB in Europe, in Eastern North America, and in the Pacific region of North America, respectively (Piesman and Gern, 2004).

Importantly, there is strong evidence that *B. burgdorferi* originated in Europe and not in North America as proposed previously based on diversity patterns of *ospC* (Marti Ras et al., 1997; Dykhuizen and Baranton, 2001; Margos et al., 2008). The conflicting scenarios of speciation and origin of *B. burgdorferi* are likely to be related to different evolutionary pathways of the housekeeping genes and *ospC*. This is supported by the finding that the topologies of the MLST tree and *ospC* tree differ (Margos et al., 2008). Recombination at the housekeeping genes was found to be very rare relative to mutation, indicating that chromosomal loci are relatively clonal (Vitorino et al., 2008). Signals of recombination, however, have been found for *ospC* (Dykhuizen and Baranton, 2001; Vitorino et al., 2008), which may explain the different tree topology of this gene. In addition, the *ospC* tree is characterized by deeper branching than the MLST and IGS trees, suggesting that selective forces have affected nucleotide substitution rates among the *ospC* major groups.

Balancing selection could have maintained ancient polymorphisms of *ospC* in the wake of past population bottlenecks, as proposed previously (Wang et al., 1999; Dykhuizen and Baranton, 2001; Qiu et al., 2002, 2004). Balancing selection is a form of frequency-dependent selection and should result in a dN/dS ratio of >1 for genes under immune selection. However, we and others have found an overall dN/dS ratio of <1 for *ospC*, a gene encoding an immunodominant outer surface lipoprotein of *B. burgdorferi* (Wang et al., 1999). Sliding window analysis and a "sitewise likelihood ratio" method (Massingham and Goldman, 2005; Margos et al., 2008) showed that different parts of *ospC* display different dN/dS ratios. This indicates that some regions of the gene are under positive immune selection, while others are more conserved probably due to functional constraints. Consistent with this are recent findings that OspC is essential to colonize the tick's salivary glands (Pal et al., 2004; Fingerle et al., 2007) and/or for early infection of the vertebrate host (Stewart et al., 2006; Tilly et al., 2007). Balancing selection is also likely to homogenize the spatial frequency distribution of *ospC* alleles, even if geographic population structure is detected at other loci. In fact, for the Northeastern United States, geographic uniformity was observed at *ospC* (Qiu et al., 2002). In addition, European and North American *B. burgdorferi* populations were not consistently distinguished when using *ospC* as a marker, whereas MLST clearly demonstrated that European and North American populations of *B. burgdorferi* constitute distinct lineages (Margos et al., 2008).

12.4.3 Phylogeographic Population Structure of *B. burgdorferi* in Eastern North America

The range expansion of *I. scapularis* into its current distribution in the Northeast and Midwest of the United States has been associated with the reintroduction of deer following the reforestation of much of the Eastern United States since the mid-twentieth century (Spielman et al., 1985). Previous to this, much of the deciduous forest cover in the eastern part of the country was cleared for farming and for use in manufacturing during the early agricultural and industrial development of the United States (Halls, 1984). Forest clearing along with unregulated hunting led to the elimination of deer populations and, as a result, of *I. scapularis* populations throughout much of the region where both had presumably been present and widely distributed during precolonial times. Isolated deer herds were, however, continuously present on Long Island, New York (Cronon, 1983) and on smaller islands offshore of Massachusetts (Halls, 1984) as well as in remote areas of Wisconsin and possibly in other Midwestern states (Christensen, 1959). These refugial herds also appear to have supported *I. scapularis* populations that allowed for local continuous maintenance of *B. burgdorferi* (Collins et al., 1949; Persing et al., 1990).

Although the expansion of deer populations and the subsequent spread of *I. scapularis* is well documented as the major cause of LB emergence in both the Northeast and Midwest, these two foci of LB endemicity appear to be discontinuous and spreading independently of one another (Callister et al., 1988; Lastavica et al., 1989; Pinger et al., 1996; Cortinas et al., 2002; Diuk-Wasser et al., 2006; Hamer et al., 2007). However, our empirical knowledge of the origins and movement of *B. burgdorferi* in North America is known only from patchy entomological records of vector presence and incomplete case reports of human disease.

Using MLST based on housekeeping genes, we have analyzed the population structure of *B. burgdorferi* in host-seeking ticks that were collected from the vegetation between May and September in 2004 and 2005 in a systematic study across much of Eastern United States (Diuk-Wasser et al., 2006; Hoen et al., 2009). The *B. burgdorferi* populations from the Midwest and Northeast were found to be genetically divergent, and in no instance were two samples collected in both regions with identical sequence types. However, the two regional populations of *B. burgdorferi* were found to be closely related, strongly suggesting that they have a shared evolutionary past and that they once constituted an admixed population. In subsequent MLST analyses of a large number of strains from the Northeast, however, in two cases, samples with identical sequence types were found in both regions (S. J. Bent et al., unpublished observations). This finding does not change the interpretation of these data and provides further support either for a very low (but nonzero) migration rate or for the presence of strains that have simply not changed since the vicariance of the two regional populations.

Distributions of sequence mismatch frequencies revealed signatures of population expansion in both the Northeast and Midwest United States and allowed for estimates of the timescale of ancient expansion events of *B. burgdorferi* (Harpending, 1994). The number of mutational steps since expansion, τ, was ~20 for both populations under all supported expansion scenarios. This parameter is related to time, t, since expansion, according to the formula $\tau = 2\mu t$, where μ is the mutation rate per nucleotide per year. Considering the time *B. burgdorferi* spends during its life cycle reproducing in both its tick and vertebrate hosts, approximations of its doubling time observed in the laboratory (De Silva and Fikrig, 1995), and typical rates of spontaneous mutation per generation in bacteria (Drake et al., 1998), we estimated the mutation rate for *B. burgdorferi* to be on the order of 10^{-8} to 10^{-9} substitutions per site per year. Even considering a wide range of mutation rates spanning several orders of magnitude (10^{-10} – 10^{-6}), the time since expansion of *B. burgdorferi* in North America indicated by the mismatch distribution parameters is still at least several thousand years and is probably closer to a million years. This indicates prehistoric population growth and spread of *B. burgdorferi* populations within each of the two regions long before the emergence of modern LB (Hoen et al., 2009). These signatures of ancient demographic processes for *B. burgdorferi* exhibited in the housekeeping genes, in particular population and spatial expansions occurring on the order of several thousands to millions of years ago, suggest that the recent near-simultaneous *B. burgdorferi* expansions out of separate relict foci in the Northeastern and Midwestern United States over the last several decades are independent events. Thus, LB spirochetes were likely prevalent in North America long before the arrival of humans.

The eBURST analysis revealed important clues about the origins of *B. burgdorferi* in North America (Hoen et al., 2009). Four rooted clonal complexes out of 37 sequence types were identified. Interestingly, all four of the clonal complexes had founding sequence types with distributions restricted to a few sites in coastal New England and in Southern New York State (Fig. 12.1). The findings suggest that this pattern is a consequence of an ancient spread of *B. burgdorferi* in an east-to-west direction. This evidence for an ancient east-to-west dispersal of *B. burgdorferi* in the United States is of particular interest in light of a previous study using the same multilocus markers that found a European origin for *B. burgdorferi* strains circulating in North America (Margos et al., 2008). Our finding that all four rooted clonal complexes of *B. burgdorferi* had ancestral genotypes found exclusively in coastal sites of the Northeastern United States is consistent with the finding that *B. burgdorferi* originated in Europe.

Populations of *I. scapularis* and *B. burgdorferi* are currently emerging in Southern Quebec, Canada, but in many of the identified tick populations, there is no evidence of local *B. burgdorferi* transmission yet (Ogden et al., 2008c). In locations with evidence of local transmission, prevalence is still very low in rodent hosts and ticks (C. Bouchard et al., unpublished observations), suggesting pioneer colonization of *B. burgdorferi* in the region. An MLST analysis of LB spirochetes in *I. scapularis* ticks collected in passive surveillance (Ogden et al., 2006) as well as in those collected in field collection efforts (Bouchard et al., unpublished observations) supports the notion that *B. burgdorferi* in Southern Quebec stems from sites in Northeastern United States and that occasional pioneer colonization events are associated with random selection of certain MLST types (N. H. Ogden et al., unpublished observations). Further studies are needed to confirm this. Such studies of emerging areas of *B. burgdorferi* endemicity may be fruitful in understanding the selection processes and the rates of dispersal of *B. burgdorferi* (distance per unit time) relative to mutation rates (substitutions per year), which may be used to test hypotheses and to validate our current phylogeographic models of LB spirochetes.

12.4.4 Phylogeographic Population Structures of LB Spirochetes in Europe

For most of the eight recorded European species of LB group spirochetes, it is established that they are specialized to groups of vertebrate host species (Kurtenbach et al., 1998, 2002b, 2006; Richter et al., 2004; Margos et al., 2009). Host specialization is likely to affect the evolution, population structure and epidemiology of LB spirochetes

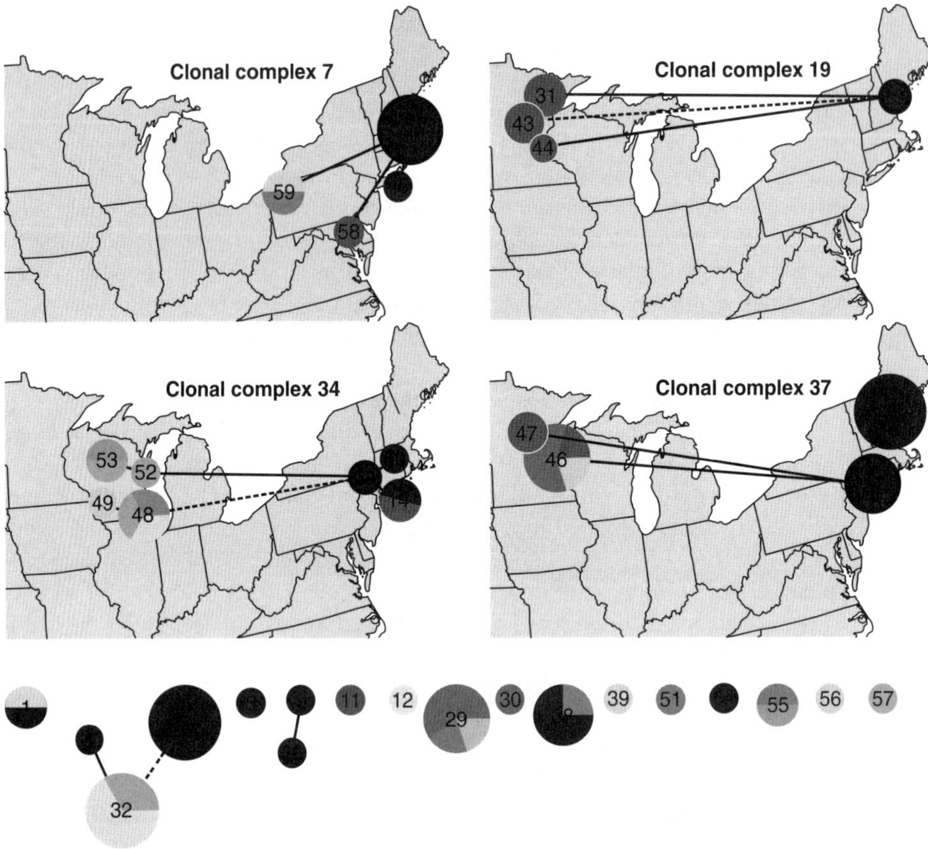

Figure 12.1 Geographic distributions of clonal complexes. Clonal complexes are based on multilocus allelic profiles defined by the eBURST algorithm. Each circle represents a unique sequence type. Circle size corresponds to sample size. Sequence types (STs) connected by a solid line are single-locus variants and STs connected by a dotted line are double locus variants. Inferred founders of clonal complexes are outlined in black. For illustration purposes, the four clonal complexes with an inferred founding ST are plotted on maps of the study area; STs are plotted at approximately the centroid of all of the sampling locations where the ST was found, with adjustments made to avoid completely overlapping circles. The bottom row contains singletons and complexes with no inferred founder. Clonal complexes with inferred founders are named for their founding ST numerical assignment (original figure from Hoen et al. [2009], doi:10.1073/pnas.0903810106, reproduced with permission from *Proceedings of the National Academy of Sciences of the United States of America*).

because migration of vertebrate hosts seems to determine the migration of the bacteria. In order to test the hypothesis that the phylogeographic structures of European LB species are shaped by host association, we have extended an MLST scheme developed for *B. burgdorferi* (Margos et al., 2008) (http://www.mlst.net/) to MLSA, being able to analyze all the European LB species (Margos et al., 2009). Like the original MLST scheme, this MLSA scheme is based solely on single-copy housekeeping genes located on the linear chromosome of LB group spirochetes. The phylogenetic MLSA trees were rooted with sequences of orthologous housekeeping genes of the relapsing fever spirochetes *Borrelia duttonii*, *Borrelia hermsii*, and *Borrelia turicatae* as an outgroup (Fig. 12.2).

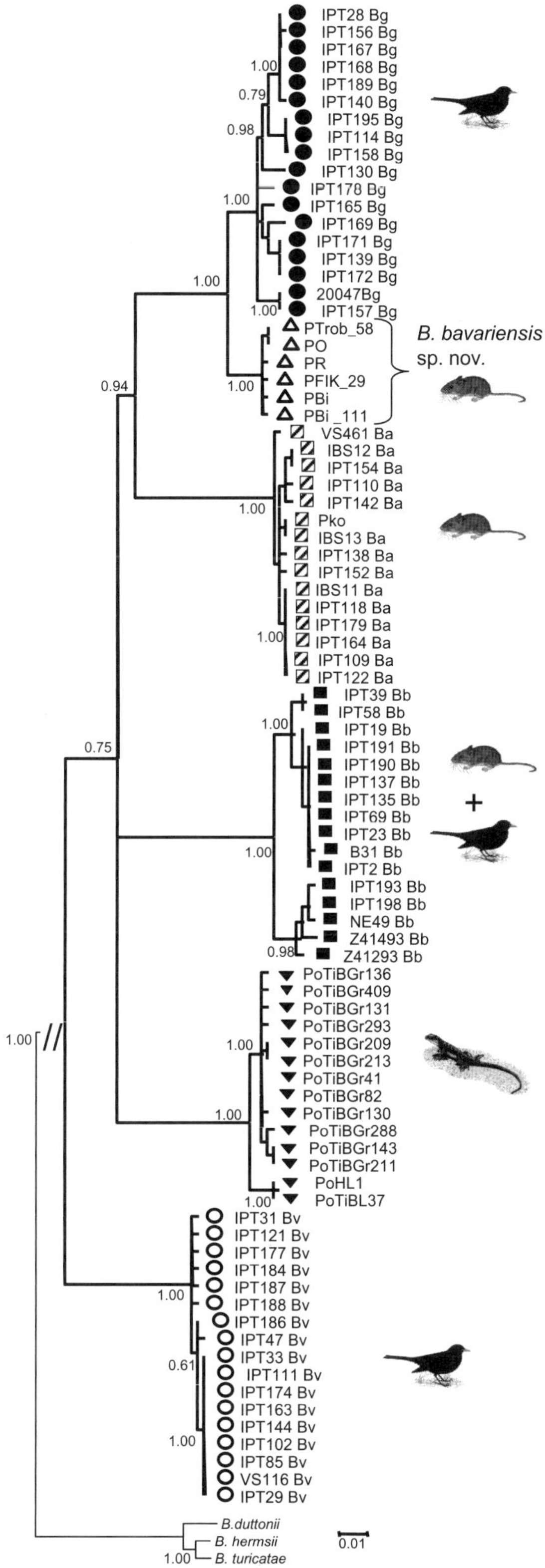

Figure 12.2 Rooted Bayesian phylogenetic inference of concatenated housekeeping gene sequences of LB group spirochetes. Posterior probability values of clades are provided. The samples have been assigned to LB species labeled as follows: *B. burgdorferi*, ■; *B. afzelii*, ▨; *B. garinii*, ●; *B. bavariensis*, △; *B. valaisiana*, ○; and *B. lusitaniae*, ▼. The tree was rooted with sequences of the relapsing fever spirochetes *B. duttonii*, *B. hermsii*, and *B. turicatae*. The branch length of the outgroup is not according to scale as indicated by slashes. Scale bar =1% divergence.

The samples analyzed in this study comprised DNA derived from questing ticks collected in sites in Latvia, England, and Portugal, as well as from cultured tick isolates obtained in France and Portugal. First, the MLSA scheme unambiguously assigned all the tested European strains to five previously defined LB species, with pronounced genetic gaps between the species clusters (Margos et al., 2009; Vollmer et al., unpublished observations) (Fig. 12.2). This is an important finding because the entire regional gene pool of LB spirochetes should be present in questing ticks, and the nested PCR primers were designed to pick up any genotype of LB spirochetes from a tick sample. Thus, the lack of a genetic continuum means that the species clustering found in our study was not due to sampling bias.

Second, while the strains of *B. garinii* and *B. valaisiana* were found to be spatially admixed across Europe (i.e., they are associated with birds, including migratory birds) (Fig. 12.3), pronounced phylogeographic structuring was observed for *B. afzelii* being associated with rodent populations, which disperse only 200–300 m/year (Fig. 12.4). Furthermore, for *B. afzelii*, but not for *B. garinii* and *B. valaisiana*, the English Channel appears to be an efficient barrier to migration, as the populations of *B. afzelii* from continental Europe and Great Britain were found to constitute genetically distinct lineages (Vollmer et al., unpublished observations). For *B. lusitaniae* in Portugal, a pronounced fine-scale phylogeographic population structure over a short geographic distance was observed using MLSA (Vitorino et al., 2008). Lizards of the family Lacertidae have been identified as reservoir hosts of *B. lusitaniae* (Dsouli et al., 2006; Richter and Matuschka, 2006). The distribution of these and other Mediterranean lizard populations is known to be highly parapatric (Paulo et al., 2008), which is likely to be the cause of the geographic structuring of *B. lusitaniae*.

The commonly used molecular markers *ospA* and *ospC* seemed unsuitable for phylogeographic analyses of European LB spirochetes at a smaller geographic scale (Vitorino et al., 2008; Vollmer et al., unpublished observations). The reasons for the lack of clear geographic signals contained in the *ospA* sequences remain unknown, since no recombination events were detected for this gene. As discussed earlier for *ospC* of *B. burgdorferi*, recombination and balancing selection are possible processes that homogenize the spatial frequency distribution of *ospC* alleles of European LB spirochetes, and either of these processes or both may generate a uniform geographic structure (Qiu et al., 2002; Vitorino et al., 2008).

In contrast to *B. burgdorferi*, which occurs in the Old and New World, *B. garinii*, *B. valaisiana*, *B. afzelii*, and *B. lusitaniae* are confined to Eurasia. The geographic and phylogenetic patterns indicate that all the European species of LB spirochetes, not only *B. burgdorferi* (Margos et al., 2008, 2009), evolved in the Old World. Although no fossil records of LB spirochetes exist, and a precise molecular clock has not yet been established, patterns of distributions of sequence mismatch frequencies suggest a time estimate of divergence of the LB species in the order of several millions of years. It is possible that the divergence of LB species coincided with the expansion and radiation of birds and mammals and with the emergence of *Ixodes*-like hard ticks ~70 million years ago (De La Fuente, 2003).

Taken together, the phylogeographic studies of European LB spirochetes discussed in this section show that the different migration patterns within the LB group of spirochetes are directly linked with those of their vertebrate hosts, strongly supporting the idea that levels and patterns of host specialization of tick-borne microparasites affect their evolution and geographic spread.

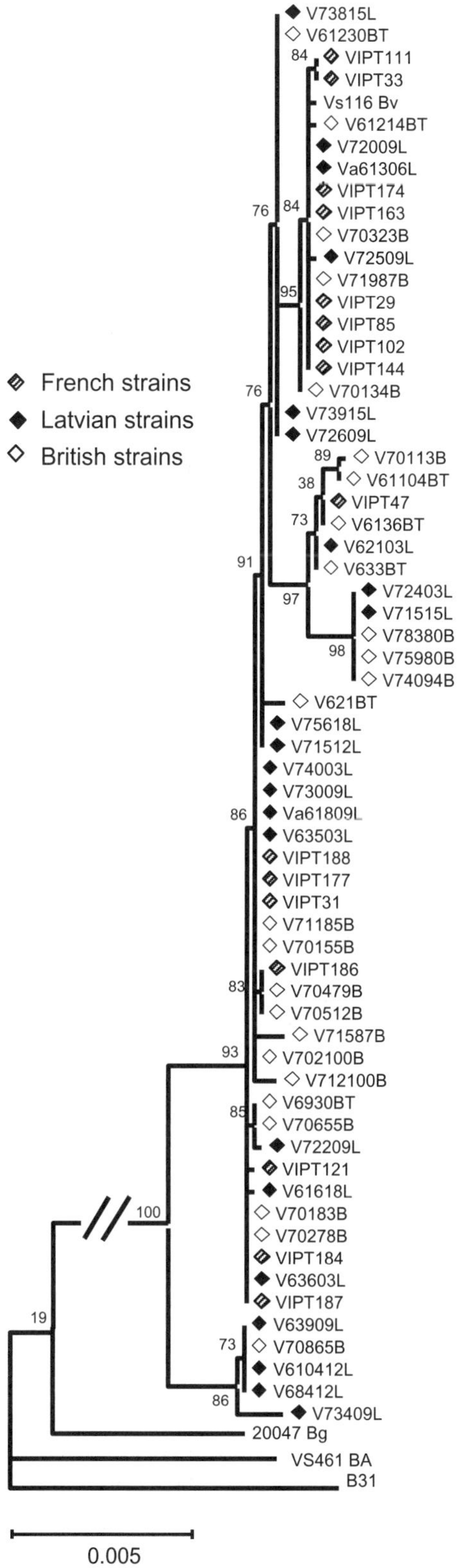

Figure 12.3 Rooted PHYML tree of concatenated housekeeping gene sequences of *B.* valais*iana* strains from Britain, Latvia, and France. An estimation of approximate likelihood branch support is provided for each clade. The tree is rooted using *B. afzelii* (VS461)-, *B. garinii* (20047)-, and *B. burgdorferi* (B31)-type strains. The branch length of the outgroup is not according to scale as indicated by slashes. Scale bar =0.5% divergence.

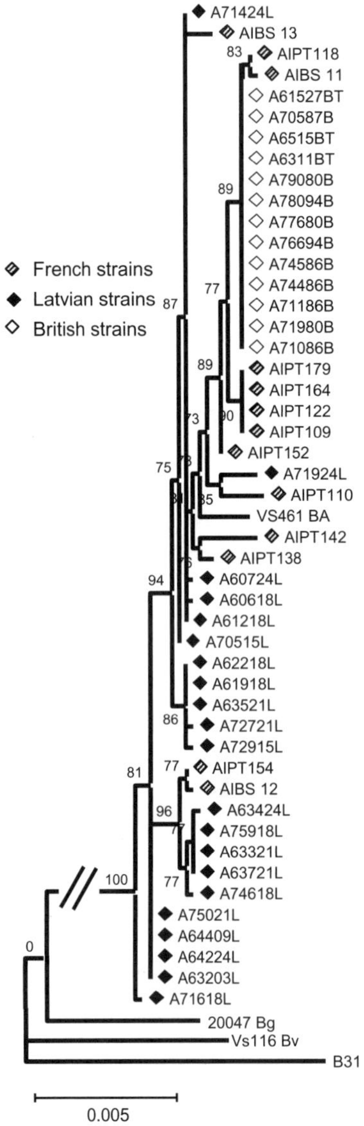

Figure 12.4 Rooted PHYML tree of concatenated housekeeping gene sequences of *B. afzelii* strains from Britain, Latvia, and France. An estimation of approximate likelihood branch support is provided for each clade. The tree was rooted using *B. valaisiana* (VS116)-, *B. garinii* (20047)-, and *B. burgdorferi* (B31)-type strains. The branch length of the outgroup is not according to scale as indicated by slashes. Scale bar =0.5% divergence.

12.4.5 The Basic Reproduction Number (R_0) and LB Epidemiology

The singular measure of pathogen fitness is the basic reproduction number R_0, giving the number of new infections that arise directly from one infected host introduced into a totally naive host population (Anderson and May, 1992). It is an artificial measure (except perhaps in zones of expansion of the range of infectious agents), but based on its theoretical attributes, we can develop indices for R_0 that allow for comparisons of the fitness of different strains of LB spirochetes in the laboratory, in the field and *in silico*. The key index is the number of infected ticks produced by an individual infected reservoir host, which determines how many new infected hosts will acquire infection, via ticks, from the index case. In simple terms, this index is the sum of (i) the proportion of feeding ticks that acquire infection from the infected host (i.e., the host-to-tick transmission coefficient, β_{v-t}) and (ii) the duration of host infectivity for ticks. These qualities are functions of the pathogenesis of infection, that is, how generally disseminated the tick-borne bacterium is throughout the host's body (and particularly those tissues such as skin, tissue fluid, and blood that feeding ticks come into contact with) and how abundant the bacterium is in tissues and fluids that feeding ticks have access to. How well the bacteria can evade the host innate and acquired immune responses determines both the abundance of the bacteria and the duration of infection. The diversity and generality of mechanisms whereby vector-borne pathogens evade the host immune response stands as testament to the importance of persistent host infections in the ecology of vector-borne diseases (Kurtenbach et al., 2006).

The classical dogma is the transmission–virulence trade-off hypothesis: higher abundance of pathogens in the host increases transmission but also virulence at the same time (Ewald, 1983; Massad, 1987). For some tick-borne pathogens, there is evidence that increased pathogen multiplication or abundance in the host is accompanied by greater host-to-tick transmission efficiency and increased pathogenicity (Young et al., 1996; Ogden et al., 2002, 2003). This can be evidenced by variation in severity of disease or higher mortality among individuals with different levels of parasitemia (Dolan et al., 1984) or by a more general association of disease severity with peaks in parasitemia during the course of infection (Brodie et al., 1986). Host mortality effectively curtails the duration of infectivity of infected hosts, while morbidity reduces host movement and contact with ticks, thereby indirectly reducing the number of infected ticks produced from an infected host. *B. burgdorferi* is often considered nonpathogenic for natural hosts, such as the white-footed mouse in North America (Wright and Nielsen, 1990). However, some strains may increase host mortality (Moody et al., 1994) through indirect effects on survival (such as increased predation) that can be difficult to detect in nature (Johnson et al., 2006) yet are potentially important drivers of bacterial evolution (Alizon, 2008). Indeed, the paradigm that *B. burgdorferi* and other LB group spirochetes are not pathogenic for natural hosts may suffer from the fact that we are "looking under the lamppost": for example, North American variants of *B. burgdorferi* that are highly adapted to white-footed mice are likely to be particularly abundant in the Northeastern United States, and perhaps, therefore, those variants may have been used in studies on pathogenicity. Further studies are needed to understand under what circumstances *Borrelia* infections may influence natural host survival and the trade-offs between virulence and transmission. We have hypothesized, however, that the evolutionary trajectories of tick-borne pathogens to maximize transmission and persistence of infection, while minimizing mortality and morbidity, would drive increasing host specialization and multiple niche polymorphism (Kurtenbach et al., 2006).

The intrinsic dynamics of evolutionary processes are driven by the relationships between vertebrate host and bacterium. These and their consequences for transmission and mortality or morbidity can be influenced by extrinsic factors. Recent modeling studies have highlighted seasonality of tick vectors as a potentially important influence on fitness of tick-borne pathogens (Ogden et al., 2007, 2008a). Hosts infected by nymphal ticks must remain infected and infective until they encounter larval ticks for the transmission cycle to be perpetuated. In some circumstances, for example, the life cycle of *I. scapularis* in Northeastern North America (Wilson and Spielman, 1985; Fish, 1993; Gatewood et al., 2009), there can be considerable temporal separation in the seasonal activity periods for nymphal and larval ticks. Under such circumstances, infected hosts must remain alive and infective in the period between infection by nymphs and transmission of infection to larvae, so we have suggested that this seasonality in tick activity will increase the fitness of strains that are less pathogenic and longer-lived and, therefore, are possibly more adapted to their host. In other words, this seasonality pattern may drive host specialization and multiple niche polymorphism (Kurtenbach et al., 2006; Ogden et al., 2007). However, in other regions of North America, seasonality of nymphal and larval *I. scapularis* may be more synchronous, and mathematical models predict that more generalist variants, that is, any that are pathogenic for hosts and have short-lived infections, can survive better (Ogden et al., 2008a). The seasonality of the ticks and the host-seeking behavior are determined to a considerable degree by temperature-mediated effects on tick development and questing height in the vegetation (Randolph and Storey, 1999; Ogden et al., 2004, 2005), and, therefore, it is likely that climate shapes the allele frequency distribution of LB spirochetes (Ogden et al., 2008a).

An analysis of the relationships between climate and the seasonal activity of *I. scapularis* and *B. burgdorferi* genotype frequency in 30 geographically diverse sites in the Northeastern and Midwestern United States provided empirical evidence for the effects of climate on the allele frequency distribution of LB spirochetes (Gatewood et al., 2009). It showed that variation in summer and winter temperature cycle extremes is associated with variation in the seasonal synchrony of *I. scapularis* larval and nymphal host-seeking activity in North America. This is primarily driven by differences in the seasonality of larvae, which exhibit a pattern of host seeking earlier in the season in the Midwestern United States, resulting in greater seasonal overlap between nymphs and larvae in these areas as compared with the Northeast (Fig. 12.5). *B. burgdorferi* strains found in host-seeking *I. scapularis* nymphs collected in these sites were broadly typed at the 16S-23S IGS by determining the restriction fragment length polymorphism sequence types 1–3 (RSTs; Wormser et al., 1999). RST 1 strains were more prevalent than the other RSTs in sites where immature tick activity is less synchronous. This observation is consistent with the hypothesis that a temporal gap between nymphal and larval feeding confers a higher relative fitness to these strains compared with other strains. This is of particular interest in light of two recent studies that compared the transmission dynamics of two North American strains of *B. burgdorferi*, an invasive RST 1 strain (BL206) isolated from human blood (Wang et al., 2001), and a slowly disseminating RST 3 strain (B348) isolated from an erythema migrans lesion (Wang et al., 2002). A population simulation study (Ogden et al., 2007) found that BL206 was weakly favored when immature tick feeding was synchronous but was more strongly favored when the temporal gap between peak nymphal and larval feeding was long. Another recent study experimentally compared the fitness of BL206 and B348 by measuring the duration of infection in mice with each strain by their ability to infect larval ticks (Hanincova et al., 2008). The authors found that the duration of BL206 infection in the white-footed mouse *Peromyscus leucopus* lasted for at least 79 days, while B348 infection declined dramatically within 40 days. This confirmed the result of an earlier study, which found that B348 infection dropped sharply by 21 days and approached zero by day 42 (Derdakova et al., 2004).

Previous studies paint a complex and incomplete picture of strain-specific variation in dissemination and persistence for *B. burgdorferi*. In the absence of phenotypic data from other isolates or an understanding of the genetic basis of persistence, we cannot assume that all RST 1 strains share the high relative persistence of infectiousness observed in BL206. Nevertheless, these studies form the foundation for the hypothesis that less invasive and persistent strains would have a relative fitness disadvantage in regions with low seasonal overlap of subadult tick activity, thereby allowing the more persistent strains to reach a higher prevalence than they would under synchronous conditions. The finding

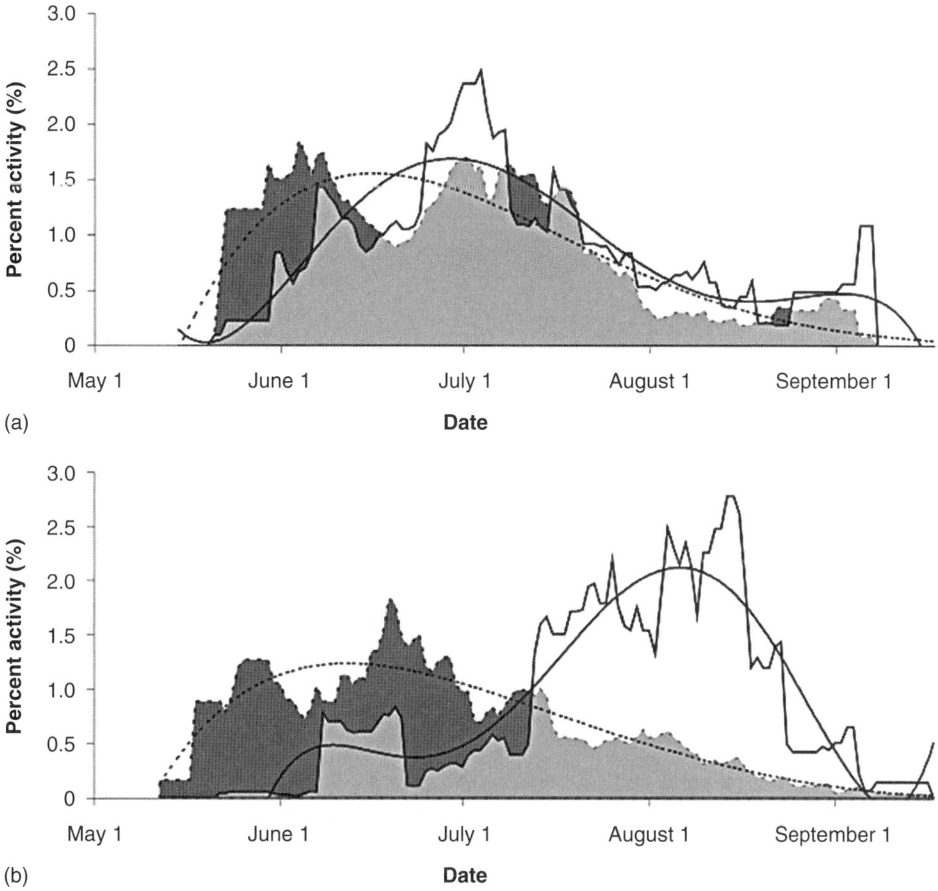

Figure 12.5 Seasonal activity of larvae and nymphs. Seasonal activity curves of immature tick host seeking based on 2-week moving averages of larval (solid lines/no shading) and nymphal (dashed lines/dark gray shading) percent activity in sites with (a) synchronous and (b) asynchronous peaks in immature host seeking. Observations of seasonal activity were pooled into two groups divided at the median of the seasonal synchrony index. Overlapping areas under both larval and nymphal seasonality curves are shaded in light gray. Fifth-order polynomial trend lines were fitted for illustration purposes and are shown in black (original figure from Gatewood et al. [2009], doi:10.1128/AEM.02622-08, reproduced with permission from the American Society for Microbiology).

that RST 1 strains are more prevalent in areas with low seasonal synchrony of immature tick host-seeking activity is consistent with the hypothesis that seasonality-related drivers of selection for persistent strains in part underlie the uneven geographic frequency distribution of different *B. burgdorferi* genotypes in North America.

Undoubtedly, diverse evolutionary forces determine the allele frequencies of LB spirochetes that we observe in host-seeking ticks. With this in mind, we propose (i) that environmental features, such as climate, can modify tick phenology and host-seeking behavior to select for certain strains of LB spirochetes over others; (ii) that these selective pressures are moderated by the opposing forces of balancing selection that drive the maintenance of diversity; and (iii) that the relative strengths of these and other evolutionary forces in a given environment determine the resulting frequencies of different strains of

LB spirochetes. Future studies that elucidate the relationships between climate, tick phenology, host-seeking behavior, and the microevolution within the context of the diversity of factors, including host community composition, regulating this complex disease system are needed.

12.5 DO LB SPECIES EXIST?

The increase in numbers of completed prokaryotic genome sequences of closely related bacteria has sparked a debate about the nature of bacterial species, emphasizing the need for a theory-based prokaryotic species concept. It has been suggested repeatedly that genetic clustering of bacterial populations is shaped by cohesive forces, such as recombination or selection (Lan and Reeves, 2001; Cohan, 2002; Konstantinidis and Tiedje, 2005; Achtman and Wagner, 2008), and that such clusters may have the quintessential properties of species. However, horizontal gene transfer of auxillary, plasmid-located genes may occasionally occur between unrelated prokaryotes and can lead to "evolutionary jumps" (e.g., resistance to antibiotics or host switches), which may enable new strains to emerge and to occupy a different ecological niche. Such ecotypic changes may not be observed immediately at chromosomal housekeeping genes, stressing the importance to take ecological data into account when delineating novel bacterial species (Cohan, 2002; Gevers et al., 2005; Achtman and Wagner, 2008).

According to the biological species concept developed by Ernst Mayr (Mayr, 1942), species are groups of organisms whose diversity is purged by sex (de Queiroz, 2005). Due to a different mode of reproduction compared to most eukaryotic organisms, the biological species concept cannot be applied to bacteria, raising the question what, if they exist, bacterial species are. A consensus about the nature of bacterial species has not been reached in bacteriology. As a consequence, most bacteriologists define bacterial species through arbitrary genetic or phenotypic cutoffs (Goris et al., 2007). In an attempt toward the development of a more general concept, the cohesion species concept has been proposed (Meglitsch, 1954; Cohan, 2002). According to this concept, species are maintained as distinct groups by cohesive forces. This is sex in most eukaryotic species and selection in asexually reproducing organisms, such as bacteria. Cohan has demonstrated for several bacteria that ecotypes are remarkably congruent with sequence clusters as revealed by MLST/A and has suggested that such lineages could be regarded as species (Cohan, 2002; Koeppel et al., 2008).

LB spirochetes comprise distinct ecotypes that are broadly defined by their spectrum of vertebrate hosts (Kurtenbach et al., 2002a, 2006). Ecotypes of LB group spirochetes can, therefore, more easily be determined than those of free-living bacteria. In most cases published so far, different ecotypes of LB spirochetes correspond to different defined species, while the reverse is not always the case. Importantly, within the wider *B. garinii* clade, MLSA detected a cluster of strains, referred to as OspA serotype 4 strains, known to be maintained by rodents in nature (Margos et al., 2009) (Fig. 12.2). The genetic distance between bird-associated and rodent-associated *B. garinii* was greater than 0.0170 (the genetic species threshold determined for the MLSA scheme developed by Margos and colleagues [2009]). We have, therefore, suggested to raise OspA serotype 4 strains to species status and to name it *B. bavariensis* sp. nov. because of its discovery in Bavaria, Germany. These findings, furthermore, indicate that MLSA has the power to detect and to demarcate ecotypes of LB group spirochetes a priori of their ecological characterization.

The Eurasian species *B. valaisiana* and *B. garinii* form distinct sequence clusters, although they are, with the exception of OspA serotype 4 strains, very similar ecologically in that they occur sympatrically and utilize the same spectrum of tick vectors and avian hosts (Hanincova et al., 2003b; Kurtenbach et al., 2006; Taragel'ova et al., 2008). It is likely that *B. valaisiana* and *B. garinii* evolved allopatrically, which may explain their pronounced genetic distance despite their shared ecotype. Strains assigned to *B. garinii* by methods other than MLSA, on the other hand, represent at least two ecotypes (rodent associated vs. bird associated), which are congruent with distinct clusters revealed by MLSA. Although a previous study suggested that OspA serotype 4 strains represent a recently emerged clonal lineage within *B. garinii* (Marconi et al., 1999), such strains were found to be at the base of the *B. garinii* clade in the MLSA tree in our study (Margos et al., 2009). This suggests that specialization of OspA serotype 4 strains to rodents is a more ancient trait and that genetic elements of *B. garinii* to survive in birds were acquired more recently. A more recent adaptation of *B. garinii* to birds as reservoir hosts could also explain the present-day sympatric distribution of *B. valaisiana* and *B. garinii*. Furthermore, it suggests that adaptation to avian hosts evolved at least twice independently in the LB group of spirochetes (i.e., in *B. valaisiana* and *B. garinii*). Phylogenetic studies of other bird-associated LB species, such as *Borrelia turdi*, may provide more information on the evolution of host specialization.

Within *B. burgdorferi*, there exists substantial diversity in the form of the different genotypes, as classified using either single-locus sequences (*ospC*, *ospA*, IGS) or multiple loci (MLST). These genotypes exhibit a low level of immunological cross reactivity and therefore have been postulated to assort more or less independently in nature with these immunological properties driving negative frequency-dependent selection (Qiu et al., 1997; Wang et al., 1999; J. E. Brown et al., unpublished observations). While there is some evidence that different *B. burgdorferi* genotypes may have different fitness in different hosts (Brisson and Dykhuizen, 2004), these differences are not well-supported and do not appear to constitute clearly different ecotypes (Hanincova et al., 2006). Therefore, and because the intraspecific genetic distances among *B. burgdorferi* strains are below the threshold for species delineation, the different lineages of *B. burgdorferi* can be regarded as conspecific (Postic et al., 2007; Margos et al., 2008).

Selection and isolation driven by host specialization appear to constitute the cohesive forces that purge the diversity of clusters of LB group spirochetes. From a pragmatic point of view, raising sequence clusters that correspond to ecotypes of LB group spirochetes to bacterial species status would have the advantage of being ecologically, epidemiologically, and clinically predictive. In view of the cohesion species concept, we conclude that the observed sequence clusters and divergent ecotypes within the LB group of spirochetes are more than operational taxonomic units in that these groups are likely to be irreversibly separated through their evolutionary history. Bacterial groups with such properties deserve to be regarded as species.

12.6 FUTURE RESEARCH AVENUES

In this chapter, we have presented phylogeographic patterns of LB spirochetes at intermediate spatial scales and have discussed possible evolutionary and ecological processes that shape these patterns. Future studies of the population biology of LB spirochetes and other tick-borne pathogens are likely to investigate the population structures of the pathogens at much finer spatial and temporal scales. On small spatial scales, the ecological processes associated with migration tend to homogenize the distribution of genotypes, and the mutation rate may be sufficiently low as to allow the same sequence types to appear across the region. Understanding the fine-scale genetic patterns of the bacteria within a region can be facilitated by the analysis of allele frequencies. Any observed differences may be due to selective or stochastic forces.

A highly promising future research avenue in LB research is the application of landscape genetics. Landscape genetics is a new scientific area that combines genetic data of populations (preferably in the form of allele frequencies) with spatial statistics and geographic information systems (GIS) (Manel et al., 2003; Storfer et al., 2007). Using this approach, we will be in the position to search for landscape characteristics that shape the genetic patterns of LB spirochetes at all spatial and, perhaps, temporal scales. Landscape genetics is particularly powerful if multiple neutral genetic markers of the populations of interest are used.

For an emerging pathogen with expanding zones of endemicity, such as *B. burgdorferi* in North America, it would be useful to identify landscape features that allow for or discourage its spread into new areas, empirically and beyond simple habitat suitability mapping. Several methodologies have been developed for identifying barriers to migration manifested as zones of lower-than-expected gene flow, or higher-than-average genetic distance relative to geographic distance in a spatially referenced data set of neutral genotype frequencies. Wombling is a procedure whereby allele frequencies are interpolated to form a continuous surface, and the partial derivative is computed at each point on the surface to generate a topology of genetic distance (Womble, 1951; Barbujani et al., 1989). Zones with large slopes represent possible barriers to gene flow. The direction of the slope at adjacent samples is compared to distinguish between actual patterns and noise. Another method, known as Monmonier's algorithm, connects all adjacent samples by lines on a map to form a network of genetic distances (Monmonier, 1973). The segment of the network with the largest associated genetic distance is selected and used to begin creating a boundary that is extended along the path of greatest genetic distance, which is considered the most likely barrier to gene flow in the network. The putative barriers to gene flow identified by such algorithms can be compared to GIS maps of satellite-derived remotely sensed landscape features, digital elevation models, and climate data. These comparisons may take the form of more natural or anthropogenic landscape features that serve as barriers to the spread of LB spirochetes.

The PATHMATRIX algorithm is another method for identifying physical boundaries to gene flow (Ray, 2005). In this method, maps identifying landscape features that are putative boundaries to gene flow, known as friction maps, are inputted along with point locations of samples. The estimated effective distance between each pair of points, which takes into account the estimated permeability of the barrier, is calculated for comparison with genetic distance. Unlike wombling and Monmonier's algorithm, when using PATHMATRIX, potential barriers must be identified a priori. This could be done using digital elevation models and other remotely sensed data. Landscape features that could be investigated using this method include urban corridors, lakes, major rivers, and large expanses of treeless regions such as farmland. Large areas that have been suggested to represent ideal tick habitats such as coastlines, areas high in forest edges, or large suburbanized areas will be tested for above-average gene flow that could implicate them as corridors of the spread of LB spirochetes.

Among other drivers, we have identified climate, via effects on tick seasonality, as a possibly significant future or even current influence on evolutionary processes of LB spirochetes. If this is the case, then we can expect climate change to impact directly on evolution of LB spirochetes, in terms of the relative fitness of generalist and specialist variants. In the near future, these effects may be predictable because the most immediate effects of climate warming may be on tick seasonality rather than more complex and less easily predictable effects of climate change on vertebrate host community structure (Ogden et al., 2008a).

As we develop landscape genetics models that shed light on the patterns and processes governing the spread and genetic structure of LB spirochetes, we can begin to consider the impacts of anthropogenic landscape change and global climate change on their distribution and evolution.

ACKNOWLEDGMENTS

We thank Durland Fish, Liliana R. Vitorino, Margarida Collares-Pereira, Edward J. Feil, David M. Aanensen, Brian G. Spratt, Ira Schwartz, Klara Hanincova, Maria Duik-Wasser, Michael Donaghy, Muriel Cornet, Martine Garnier, Antra Bormane, Volker Fingerle, Bettina Wilske, Sarah E. Randolph, Andrias Hoigaard, and Joseph Piesman for valuable discussions and for providing ticks or bacterial strains. The research was funded by the Wellcome Trust, United Kingdom, the Biotechnology and Biological Sciences Research Council, United Kingdom, the National Institutes of Health, United States of America, the Centers for Disease Control and Prevention, United States of America, and the Public Health Agency of Canada.

REFERENCES

Abbott, A. (2006) Lyme disease: Uphill struggle. *Nature* **439,** 524–525.

Achtman, M. and Wagner, M. (2008) Microbial diversity and the genetic nature of microbial species. *Nat Rev Microbiol* **6,** 431–440.

Achtman, M., Zurth, K., Morelli, G. et al. (1999) *Yersinia pestis*, the cause of plague, is a recently emerged clone of *Yersinia pseudotuberculosis*. *Proc Natl Acad Sci U S A* **96,** 14043–14048.

Alizon, S. (2008) Transmission-recovery trade-offs to study parasite evolution. *Am Nat* **172,** E113–E121.

Anderson, R. M. and May, R. M. (1992) *Infectious Diseases of Humans: Dynamics and Control* Oxford University Press, Oxford.

Attie, O., Bruno, J. F., Xu, Y. et al. (2007) Co-evolution of the outer surface protein C gene (*ospC*) and intraspecific lineages of *Borrelia burgdorferi* sensu stricto in the Northeastern United States. *Infect Genet Evol* **7,** 1–12.

Baranton, G., Postic, D., Saint Girons, I. et al. (1992) Delineation of *Borrelia burgdorferi* sensu stricto, *Borrelia garinii* sp. nov., and group VS461 associated with Lyme borreliosis. *Int J Syst Bacteriol* **42,** 378–383.

Barbujani, G., Oden, N. L., and Sokal, R. R. (1989) Detecting regions of abrupt change in maps of biological variables. *Syst Zool* **38,** 376–389.

Battisti, J. M., Bono, J. L., Rosa P. A. et al. (2008) Outer surface protein A protects Lyme disease spirochetes from acquired host immunity in the tick vector. *Infect Immun* **76,** 5228–5237.

Bishop, C. J., Aanensen, D. M., Jordan, G. E. et al. (2009) Assigning strains to bacterial species via the internet. *BMC Biology* **7,** 3.

Bjornstad, O. N. and Grenfell, B. T. (2001) Noisy clockwork: Time series analysis of population fluctuations in animals. *Science* **293,** 638–643.

Brandt, M. E., Riley, B. S., Radolf, J. D., and Norgard, M. V. (1990) Immunogenic integral membrane proteins of *Borrelia burgdorferi* are lipoproteins. *Infect Immun* **58,** 983–991.

Brisson, D. and Dykhuizen, D. E. (2004) *ospC* diversity in *Borrelia burgdorferi*: Different hosts are different niches. *Genetics* **168,** 713–722.

Brodie, T. A., Holmes, P. H., and Urquhart, G. M. (1986) Some aspects of tick-borne diseases of British sheep. *Vet Rec* **118,** 415–418.

Bunikis, J., Garpmo, U., Tsao, J. et al. (2004) Sequence typing reveals extensive strain diversity of the Lyme borreliosis agents *Borrelia burgdorferi* in North America and *Borrelia afzelii* in Europe. *Microbiology* **150,** 1741–1755.

Bykowski, T., Woodman, M. E., Cooley, A. E. et al. (2008) *Borrelia burgdorferi* complement regulator-acquiring surface proteins (BbCRASPs): Expression patterns during the mammal-tick infection cycle. *Int J Med Microbiol* **298**(Suppl 1), 249–256.

Callister, S. M., Agger, W. A., Schell, R. F., and Ellingson, J. L. (1988) *Borrelia burgdorferi* infection surrounding La Crosse, Wis. *J Clin Microbiol* **26,** 2632–2636.

CASJENS, S. (2000) *Borrelia* genomes in the year 2000. *J Mol Microbiol Biotechnol* **2,** 401–410.

CASJENS, S., MURPHY, M., DELANGE, M. et al. (1997a) Telomeres of the linear chromosomes of Lyme disease spirochaetes: Nucleotide sequence and possible exchange with linear plasmid telomeres. *Mol Microbiol* **26,** 581–596.

CASJENS, S., VAN VUGT, R., TILLY, K. et al. (1997b) Homology throughout the multiple 32-kilobase circular plasmids present in Lyme disease spirochetes. *J Bacteriol* **179,** 217–227.

CASJENS, S., PALMER, N., VAN VUGT, R. et al. (2000) A bacterial genome in flux: The twelve linear and nine circular extrachromosomal DNAs in an infectious isolate of the Lyme disease spirochete *Borrelia burgdorferi*. *Mol Microbiol* **35,** 490–516.

CDC (2009) Summary of notifiable diseases, United States, 2007. *MMWR Morb Mort Wkly Rep* **56**(53)**,** 39.

CHEN, H., WHITE, D. J., CARACO, T. B., and STRATTON, H. H. (2005) Epidemic and spatial dynamics of Lyme disease in New york State, 1990-2000. *J Med Entomol* **42,** 899–908.

CHRISTENSEN, E. M. (1959). A historical view of the ranges of the white-tailed deer in Northern Wisconsin forests. *Am Midl Nat* **61,** 230–238.

CHU, C. Y., LIU, W., JIANG, B. G. et al. (2008) Novel genospecies of *Borrelia burgdorferi* sensu lato from rodents and ticks in Southwestern China. *J Clin Microbiol* **46,** 3130–3133.

COHAN, F. M. (2002). What are bacterial species? *Annu Rev Microbiol* **56,** 457–487.

COLEMAN, J. L., BENACH, J. L., BECK, G., and HABICHT, G. S. (1986) Isolation of the outer envelope from *Borrelia burgdorferi*. *Zentralbl Bakteriol Mikrobiol Hyg A* **263,** 123–126.

COLLINS, D. L., NARDY, R. V., and GLASGOW, R. D. (1949) Further notes on the host relationships of ticks on Long Island. *J Econ Entomol* **42,** 159.

CORTINAS, M. R., GUERRA, M. A., JONES, C. J., and KITRON, U. (2002) Detection, characterization, and prediction of tick-borne disease foci. *Int J Med Microbiol* **291**(Suppl 33)**,** 11-20.

COUTTE, L., BOTKIN, D. J., GAO, L., and NORRIS, S. J. (2009) Detailed analysis of sequence changes occurring during vlsE antigenic variation in the mouse model of *Borrelia burgdorferi* infection. *PLoS Pathog* **5,** e1000293.

CRONON, W. (1983) *Changes in the land: Indians, colonists, and the ecology of New England.* Hill and Wang, New York.

DE LA FUENTE, J. (2003) The fossil record and the origin of ticks (Acari: Parasitiformes: Ixodida). *Exp Appl Acarol* **29,** 331–344.

DE MICHELIS, S., SEWELL, H. S., COLLARES-PEREIRA, M. et al. (2000) Genetic diversity of *Borrelia burgdorferi* sensu lato in ticks from mainland Portugal. *J Clin Microbiol* **38,** 2128–2133.

DE QUEIROZ, K. (2005) Ernst Mayr and the modern concept of species. *Proc Natl Acad Sci U S A* **102,** 6600–6607.

DERDAKOVA, M., DUDIOAK, V., BREI, B. et al. (2004) Interaction and transmission of two *Borrelia burgdorferi* sensu stricto strains in a tick-rodent maintenance system. *Appl Environ Microbiol* **70,** 6783–6788.

DE SILVA, A. M. and FIKRIG, E. (1995) Growth and migration of *Borrelia burgdorferi* in *Ixodes* ticks during blood feeding. *Am J Trop Med Hyg* **53,** 397–404.

DIUK-WASSER, M. A., GATEWOOD, A. G., CORTINAS, M. R. et al. (2006) Spatiotemporal patterns of host-seeking *Ixodes scapularis* nymphs (Acari: Ixodidae) in the United States. *J Med Entomol* **43,** 166–176.

DOLAN, T. T., YOUNG, A. S., LOSOS, G. J. et al. (1984) Dose dependent responses of cattle to *Theileria parva* stabilate. *Int J Parasitol* **14,** 89–95.

DRAKE, J. W., CHARLESWORTH, B., CHARLESWORTH, D., and CROW, J. F. (1998) Rates of spontaneous mutation. *Genetics* **148,** 1667–1686.

DSOULI, N., YOUNSI-KABACHII, H., POSTIC, D. et al. (2006) Reservoir role of lizard *Psammodromus algirus* in transmission cycle of *Borrelia burgdorferi* sensu lato (Spirochaetaceae) in Tunisia. *J Med Entomol* **43,** 737–742.

DYKHUIZEN, D. E. and BARANTON, G. (2001) The implications of a low rate of horizontal transfer in *Borrelia*. *Trends Microbiol* **9,** 344–350.

DYKHUIZEN, D. E., POLIN, D. S., DUNN, J. J. et al. (1993) *Borrelia burgdorferi* is clonal: Implications for taxonomy and vaccine development. *Proc Natl Acad Sci U S A* **90,** 10163–10167.

EGGERS, C. H., CAIMANO, M. J., CLAWSON, M. L. et al. (2002) Identification of loci critical for replication and compatibility of a *Borrelia burgdorferi* cp32 plasmid and use of a cp32-based shuttle vector for the expression of fluorescent reporters in the Lyme disease spirochaete. *Mol Microbiol* **43,** 281–295.

ENRIGHT, M. C. and SPRATT, B. G. (1999) Multilocus sequence typing. *Trends Microbiol* **7,** 482–487.

EWALD, P. W. (1983) Host-parasite relations, vectors, and the evolution of disease severity. *Annu Rev Ecol Syst* **14,** 465–485.

FALCO, R. C. and FISH, D. (1991) Horizontal movement of adult *Ixodes dammini* (Acari: Ixodidae) attracted to CO_2-baited traps. *J Med Entomol* **28,** 726–729.

FEIL, E. J. and SPRATT, B. G. (2001) Recombination and the population structures of bacterial pathogens. *Annu Rev Microbiol* **55,** 561–590.

FEIL, E. J., LI, B. C., AANENSEN, D. M. et al. (2004) eBURST: Inferring patterns of evolutionary descent among clusters of related bacterial genotypes from multilocus sequence typing data. *J Bacteriol* **186,** 1518–1530.

FINGERLE, V., GOETTNER, G., GERN, L. et al. (2007) Complementation of a *Borrelia afzelii* OspC mutant highlights the crucial role of OspC for dissemination of *Borrelia afzelii* in *Ixodes ricinus*. *Int J Med Microbiol* **297,** 97–107.

FISH, D. (1993) Population ecology of *Ixodes dammini*. In *Ecology and Environmental Management of Lyme Disease* (ed. H. S. Ginsberg), pp. 25–42. Rutgers University Press, New Brunswick, NJ.

FRASER, C. M., CASJENS, S., HUANG, W. M. et al. (1997) Genomic sequence of a Lyme disease spirochaete, *Borrelia burgdorferi*. *Nature* **390,** 580–586.

FUKUNAGA, M., OKADA, K., NAKAO, M. et al. (1996) Phylogenetic analysis of *Borrelia* species based on flagellin gene sequences and its application for molecular typing of Lyme disease borreliae. *Int J Syst Bacteriol* **46,** 898–905.

GATEWOOD, A. G., LIEBMAN, K. A., VOURC'H, G. et al. (2009) Climate and tick seasonality predict *Borrelia burgdorferi* genotype distribution. *Appl Environ Microbiol* **75,** 2476–2483.

GAZUMYAN, A., SCHWARTZ, J. J., LIVERIS, D., and SCHWARTZ, I. (1994) Sequence analysis of the ribosomal RNA operon of the Lyme disease spirochete, *Borrelia burgdorferi*. *Gene* **146,** 57–65.

GEVERS, D., COHAN, F. M., LAWRENCE, J. G. et al. (2005) Opinion: Re-evaluating prokaryotic species. *Nat Rev Microbiol* **3,** 733–739.

GEVERS, D., DAWYNDT, P., VANDAMME, P. et al. (2006) Stepping stones towards a new prokaryotic taxonomy. *Philos Trans R Soc Lond B Biol Sci* **361,** 1911–1916.

GLOCKNER, G., LEHMANN, R., ROMUALDI, A. et al. (2004) Comparative analysis of the *Borrelia garinii* genome. *Nucleic Acids Res* **32,** 6038–6046.

GLOCKNER, G., SCHULTE-SPECHTEL, U., SCHILHABEL, M. et al. (2006) Comparative genome analysis: Selection pressure on the *Borrelia* vls cassettes is essential for infectivity. *BMC Genomics* **7,** 211.

GORIS, J., KONSTANTINIDIS, K. T., KLAPPENBACH, J. A. et al. (2007) DNA-DNA hybridization values and their relationship to whole-genome sequence similarities. *Int J Syst Evol Microbiol* **57,** 81–91.

GRENFELL, B. and BJORNSTAD, O. (2005) Sexually transmitted diseases: Epidemic cycling and immunity. *Nature* **433,** 366–367.

GYLFE, A., BERGSTROM, S., LUNDSTROM, J., and OLSEN, B. (2000) Reactivation of *Borrelia* infection in birds. *Nature* **403,** 724–725.

HALL, N. (2007) Advanced sequencing technologies and their wider impact in microbiology. *J Exp Biol* **210,** 1518–1525.

HALLS, L. K. (1984) *White-Tailed Deer: Ecology and Management*. Stackpole Books, Harrisburg, PA.

HAMER, S. A., ROY, P. L., HICKLING, G. J. et al. (2007) Zoonotic pathogens in *Ixodes scapularis*, Michigan. *Emerg Infect Dis* **13,** 1131–1133.

HANINCOVA, K., KURTENBACH, K., DIUK-WASSER, M. et al. (2006) Epidemic spread of Lyme borreliosis, Northeastern United States. *Emerg Infect Dis* **12,** 604–611.

HANINCOVA, K., OGDEN, N. H., DIUK-WASSER, M. et al. (2008) Fitness variation of *Borrelia burgdorferi* sensu stricto strains in mice. *Appl Environ Microbiol* **74,** 153–157.

HANINCOVA, K., SCHAFER, S. M., ETTI, S. et al. (2003a) Association of *Borrelia afzelii* with rodents in Europe. *Parasitology* **126,** 11–20.

HANINCOVA, K., TARAGELOVA, V., KOCI, J. et al. (2003b) Association of *Borrelia garinii* and *B. valaisiana* with songbirds in Slovakia. *Appl Environ Microbiol* **69,** 2825–2830.

HARPENDING, H. C. (1994) Signature of ancient population growth in a low-resolution mitochondrial DNA mismatch distribution. *Hum Biol* **66,** 591–600.

HOEN, A. G., MARGOS, G., BENT, S. J. et al. (2009) Phylogeography of *Borrelia burgdorferi* in the eastern United States reveals multiple independent Lyme disease emergence events. *Proc Natl Acad Sci U S A.* **106**(35), 15013–15018.

HOLT, K. E., PARKHILL, J., MAZZONI, C. J. et al. (2008) High-throughput sequencing provides insights into genome variation and evolution in *Salmonella Typhi*. *Nat Genet* **40,** 987–993.

HUANG, W. M., ROBERTSON, M., ARON, J., and CASJENS, S. (2004) Telomere exchange between linear replicons of *Borrelia burgdorferi*. *J Bacteriol* **186,** 4134–4141.

HUBALEK, Z. and HALOUZKA, J. (1997) Distribution of *Borrelia burgdorferi* sensu lato genomic groups in Europe, a review. *Eur J Epidemiol* **13,** 951–957.

HUDSON, P. J. and BJORNSTAD, O. N. (2003) Ecology. Vole stranglers and lemming cycles. *Science* **302,** 797–798.

JOHNSON, P. T., STANTON, D. E., PREU, E. R. et al. (2006) Dining on disease: How interactions between infection and environment affect predation risk. *Ecology* **87,** 1973–1980.

KAWABATA, H., MASUZAWA, T., and YANAGIHARA, Y. (1993) Genomic analysis of *Borrelia japonica* sp. nov. isolated from *Ixodes ovatus* in Japan. *Microbiol Immunol* **37,** 843–848.

KOEPPEL, A., PERRY, E. B., SIKORSKI, J. et al. (2008) Identifying the fundamental units of bacterial diversity: A paradigm shift to incorporate ecology into bacterial systematics. *Proc Natl Acad Sci U S A* **105,** 2504–2509.

KONSTANTINIDIS, K. T. and TIEDJE, J. M. (2005) Genomic insights that advance the species definition for prokaryotes. *Proc Natl Acad Sci U S A* **102,** 2567–2572.

KRAICZY, P., SKERKA, C., BRADE, V., and ZIPFEL, P. F. (2001a) Further characterization of complement regulator-acquiring surface proteins of *Borrelia burgdorferi*. *Infect Immun* **69,** 7800–7809.

KRAICZY, P., SKERKA, C., KIRSCHFINK, M. et al. (2001b) Mechanism of complement resistance of pathogenic *Borrelia burgdorferi* isolates. *Int Immunopharmacol* **1,** 393–401.

KURTENBACH, K., HANINCOVA, K., TSAO, J. I. et al. (2006) Fundamental processes in the evolutionary ecology of Lyme borreliosis. *Nat Rev Microbiol* **4,** 660–669.

KURTENBACH, K., PEACEY, M., RIJPKEMA, S. G. et al. (1998) Differential transmission of the genospecies of *Borrelia burgdorferi* sensu lato by game birds and small rodents in England. *Appl Environ Microbiol* **64,** 1169–1174.

KURTENBACH, K., DE MICHELIS, S., ETTI, S. et al. (2002a) Host association of *Borrelia burgdorferi* sensu lato-the key role of host complement. *Trends Microbiol* **10,** 74–79.

KURTENBACH, K., SCHAEFER, S. M., DE MICHELIS, S. et al. (2002b) *Borrelia burgdorferi* s.l. in the vertebrate host.

In *Lyme Borreliosis: Biology of the Infectious Agents and Epidemiology of Disease* (eds. J. S. Gray, O. Kahl, R. S. Lane, and G. Stanek), pp. 117–148. CABI Publishing, Wallingford.

LAN, R. and REEVES, P. R. (2001) When does a clone deserve a name? A perspective on bacterial species based on population genetics. *Trends Microbiol* **9,** 419–424.

LASTAVICA, C. C., WILSON, M. L., BERARDI, V. P. et al. (1989) Rapid emergence of a focal epidemic of Lyme disease in coastal Massachusetts. *N Engl J Med* **320,** 133–137.

LIVERIS, D., GAZUMYAN, A., and SCHWARTZ, I. (1995) Molecular typing of *Borrelia burgdorferi* sensu lato by PCR-restriction fragment length polymorphism analysis. *J Clin Microbiol* **33,** 589–595.

MADHAV, N. K., BROWNSTEIN, J. S., TSAO, J. I., and FISH, D. (2004) A dispersal model for the range expansion of blacklegged tick (Acari: Ixodidae). *J Med Entomol* **41,** 842–852.

MAIDEN, M. C., BYGRAVES, J. A., FEIL, E. et al. (1998) Multilocus sequence typing: A portable approach to the identification of clones within populations of pathogenic microorganisms. *Proc Natl Acad Sci U S A* **95,** 3140–3145.

MANEL, S., SCHWARTZ M. K., LUIKART, G., and RABERLET, P. (2003) Landscape genetics: Combining landscape ecology and populations genetics. *Trends Ecol Evol* **18,** 189–197.

MARCONI, R. T., HOHENBERGER, S., JAURIS-HEIPKE, S. et al. (1999) Genetic analysis of *Borrelia garinii* OspA serotype 4 strains associated with neuroborreliosis: Evidence for extensive genetic homogeneity. *J Clin Microbiol* **37,** 3965-3970.

MARCONI, R. T., LIVERIS, D., and SCHWARTZ, I. (1995) Identification of novel insertion elements, restriction fragment length polymorphism patterns, and discontinuous 23S rRNA in Lyme disease spirochetes: Phylogenetic analyses of rRNA genes and their intergenic spacers in *Borrelia japonica* sp. nov. and genomic group 21038 (*Borrelia andersonii* sp. nov.) isolates. *J Clin Microbiol* **33,** 2427–2434.

MARGOS, G., GATEWOOD, A. G., AANENSEN, D. M. et al. (2008) MLST of housekeeping genes captures geographic population structure and suggests a European origin of *Borrelia burgdorferi. Proc Natl Acad Sci U S A* **105,** 8730–8735.

MARGOS, G., VOLLMER, S. A., CORNET, M. et al. (2009) A new *Borrelia* species defined by multilocus sequence analysis of housekeeping genes. *Appl Environ Microbiol* **75,** 5410–5416.

MARTI RAS, N., POSTIC, D., FORETZ, M., and BARANTON, G. (1997) *Borrelia burgdorferi* sensu stricto, a bacterial species "made in the U.S.A."? *Int J Syst Bacteriol* **47,** 1112–1117.

MASSAD, E. (1987) Transmission rates and the evolution of pathogenicity. *Evolution* **41,** 1127–1130.

MASSINGHAM, T. and GOLDMAN, N. (2005) Detecting amino acid sites under positive selection and purifying selection. *Genetics* **169,** 1753–1762.

MASUZAWA, T. (2004) Terrestrial distribution of the Lyme borreliosis agent *Borrelia burgdorferi* sensu lato in East Asia. *Jpn J Infect Dis* **57,** 229–235.

MASUZAWA, T., TAKADA, N., KUDEKEN, M. et al. (2001) *Borrelia sinica* sp. nov., a Lyme disease-related *Borrelia* species isolated in China. *Int J Syst Evol Microbiol* **51,** 1817–1824.

MATSEN, F. A., MOSSEL, E., and STEEL, M. (2008) Mixed-up trees: The structure of phylogenetic mixtures. *Bull Math Biol.* **70**(4)**,** 1115–1139.

MAYR, E. (1942) *Systematics and the Origin of Species, from the Viewpoint of a Zoologist.* Harvard University Press, Cambridge, MA.

MEGLITSCH, P. A. (1954) On the nature of species. *Syst Zool* **3,** 491–503.

MICHEL, H., WILSKE, B., HETTCHE, G. et al. (2004) An *ospA*-polymerase chain reaction/restriction fragment length polymorphism-based method for sensitive detection and reliable differentiation of all European *Borrelia burgdorferi* sensu lato species and OspA types. *Med Microbiol Immunol* **193,** 219–226.

MONMONIER, M. (1973) Maximum-difference barriers—Alternative numerical regionalization method. *Geogr Anal* **5,** 245–261.

MOODY, K. D., TERWILLIGER, G. A., HANSEN, G. M., and BARTHOLD, S. W. (1994) Experimental *Borrelia burgdorferi* infection in *Peromyscus leucopus. J Wildl Dis* **30,** 155–161.

NORRIS, D. E., JOHNSON, B. J., PIESMAN, J. et al. (1997) Culturing selects for specific genotypes of *Borrelia burgdorferi* in an enzootic cycle in Colorado. *J Clin Microbiol* **35,** 2359–2364.

OGDEN, N. H., BIGRAS-POULIN, M., HANINCOVA, K. et al. (2008a) Projected effects of climate change on tick phenology and fitness of pathogens transmitted by the North American tick *Ixodes scapularis. J Theor Biol* **254,** 621–632.

OGDEN, N. H., LINDSAY, L. R., HANINCOVA, K. et al. (2008b) Role of migratory birds in introduction and range expansion of *Ixodes scapularis* ticks and of *Borrelia burgdorferi* and *Anaplasma phagocytophilum* in Canada. *Appl Environ Microbiol* **74,** 1780–1790.

OGDEN, N. H., ST-ONGE, L., BARKER, I. K. et al. (2008c) Risk maps for range expansion of the Lyme disease vector, *Ixodes scapularis*, in Canada now and with climate change. *Int J Health Geogr* **7,** 24.

OGDEN, N. H., BIGRAS-POULIN, M., O'CALLAGHAN, C. J. et al. (2005) A dynamic population model to investigate effects of climate on geographic range and seasonality of the tick *Ixodes scapularis. Int J Parasitol* **35,** 375–389.

OGDEN, N. H., BIGRAS-POULIN, M., O'CALLAGHAN, C. J. et al. (2007) Vector seasonality, host infection dynamics and fitness of pathogens transmitted by the tick *Ixodes scapularis. Parasitology* **134,** 209–227.

OGDEN, N. H., CASEY, A. N., FRENCH, N. P. et al. (2002) Natural *Ehrlichia phagocytophila* transmission coefficients from sheep "carriers" to *Ixodes ricinus* ticks vary with the numbers of feeding ticks. *Parasitology* **124,** 127–136.

Ogden, N. H., Casey, A. N., Woldehiwet, Z., and French, N. P. (2003) Transmission of *Anaplasma phagocytophilum* to *Ixodes ricinus* ticks from sheep in the acute and post-acute phases of infection. *Infect Immun* **71,** 2071–2078.

Ogden, N. H., Lindsay, L. R., Beauchamp, G. et al. (2004) Investigation of relationships between temperature and developmental rates of tick *Ixodes scapularis* (Acari: Ixodidae) in the laboratory and field. *J Med Entomol* **41,** 622–633.

Ogden, N. H., Nuttall, P. A., and Randolph, S. E. (1997) Natural Lyme disease cycles maintained via sheep by co-feeding ticks. *Parasitology* **115**(Pt 6)**,** 591–599.

Ogden, N. H., Trudel, L., Artsob, H. et al. (2006) *Ixodes scapularis* ticks collected by passive surveillance in Canada: Analysis of geographic distribution and infection with Lyme borreliosis agent *Borrelia burgdorferi*. *J Med Entomol* **43,** 600–609.

Ojaimi, C., Davidson, B. E., Saint Girons, I., and Old, I. G. (1994) Conservation of gene arrangement and an unusual organization of rRNA genes in the linear chromosomes of the Lyme disease spirochaetes *Borrelia burgdorferi, B. garinii* and *B. afzelii*. *Microbiology* **140**(Pt 11)**,** 2931–2940.

Pal, U., Yang, X., Chen, M. et al. (2004) OspC facilitates *Borrelia burgdorferi* invasion of *Ixodes scapularis* salivary glands. *J Clin Invest* **113,** 220–230.

Paulo, O. S., Pinheiro, J., Miraldo, A. et al. (2008) The role of vicariance vs. dispersal in shaping genetic patterns in ocellated lizard species in the western Mediterranean. *Mol Ecol* **17,** 1535–1551.

Persing, D. H., Telford, S. R. III, Rys, P. N. et al. (1990) Detection of *Borrelia burgdorferi* DNA in museum specimens of *Ixodes dammini* ticks. *Science* **249,** 1420–1423.

Piesman, J. and Gern, L. (2004) Lyme borreliosis in Europe and North America. *Parasitology* **129**(Suppl), S191–S220.

Pinger, R. R., Timmons, L., and Karris, K. (1996) Spread of *Ixodes scapularis* (Acari: Ixodidae) in Indiana: Collections of adults in 1991-1994 and description of a *Borrelia burgdorferi*-infected population. *J Med Entomol* **33,** 852–855.

Postic, D., Assous, M. V., Grimont, P. A., and Baranton, G. (1994) Diversity of *Borrelia burgdorferi* sensu lato evidenced by restriction fragment length polymorphism of *rrf* (5S)-*rrl* (23S) intergenic spacer amplicons. *Int J Syst Bacteriol* **44,** 743–752.

Postic, D., Ras, N. M., Lane, R. S. et al. (1998) Expanded diversity among Californian borrelia isolates and description of *Borrelia bissettii* sp. nov. (formerly *Borrelia* group DN127). *J Clin Microbiol* **36,** 3497–3504.

Postic, D., Garnier, M., and Baranton, G. (2007) Multilocus sequence analysis of atypical *Borrelia burgdorferi* sensu lato isolates—Description of *Borrelia californiensis* sp. nov., and genomospecies 1 and 2. *Int J Med Microbiol* **297,** 263–271.

Purse, B. V., Mellor, P. S., Rogers, D. J. et al. (2005) Climate change and the recent emergence of bluetongue in Europe. *Nat Rev Microbiol* **3,** 171–181.

Qiu, W. G., Bosler, E. M., Campbell, J. R. et al. (1997) A population genetic study of *Borrelia burgdorferi* sensu stricto from eastern Long Island, New York, suggested frequency-dependent selection, gene flow and host adaptation. *Hereditas* **127,** 203–216.

Qiu, W. G., Dykhuizen, D. E., Acosta, M. S., and Luft, B. J. (2002) Geographic uniformity of the Lyme disease spirochete (*Borrelia burgdorferi*) and its shared history with tick vector (*Ixodes scapularis*) in the Northeastern United States. *Genetics* **160,** 833–849.

Qiu, W. G., Schutzer, S. E., Bruno, J. F. et al. (2004) Genetic exchange and plasmid transfers in *Borrelia burgdorferi* sensu stricto revealed by three-way genome comparisons and multilocus sequence typing. *Proc Natl Acad Sci U S A* **101,** 14150–14155.

Rand, P. W., Lacombe, E. H., Smith, R. P. Jr., and Ficker, J. (1998) Participation of birds (Aves) in the emergence of Lyme disease in Southern Maine. *J Med Entomol* **35,** 270–276.

Randolph, S. E., Green, R. M., Hoodless, A. N., and Peacey, M. F. (2002) An empirical quantitative framework for the seasonal population dynamics of the tick *Ixodes ricinus*. *Int J Parasitol* **32,** 979–989.

Randolph, S. E. and Storey, K. (1999) Impact of microclimate on immature tick-rodent host interactions (Acari: Ixodidae): Implications for parasite transmission. *J Med Entomol* **36,** 741–748.

Ray, N. (2005) PATHMATRIX: A geographical information system tool to compute effective distances among samples. *Mol Ecol Notes* **5,** 177–180.

Richter, D. and Matuschka, F. R. (2006) Perpetuation of the Lyme disease spirochete *Borrelia lusitaniae* by lizards. *Appl Environ Microbiol* **72,** 4627–4632.

Richter, D., Postic, D., Sertour, N. et al. (2006) Delineation of *Borrelia burgdorferi* sensu lato species by multilocus sequence analysis and confirmation of the delineation of *Borrelia spielmanii* sp. nov. *Int J Syst Evol Microbiol* **56,** 873–881.

Richter, D., Schlee, D. B., Allgower, R., and Matuschka, F. R. (2004) Relationships of a novel Lyme disease spirochete, *Borrelia spielmanii* sp. nov., with its hosts in Central Europe. *Appl Environ Microbiol* **70,** 6414–6419.

Rudenko, N., Golovchenko, M., Grubhoffer, L., and Oliver, J. H. Jr. (2009a) *Borrelia carolinensis* sp.nov.-—A new (14th) member of *Borrelia burgdorferi* sensu lato complex from the Southeastern United States. *J Clin Microbiol* **47,** 134–141.

Rudenko, N., Golovchenko, M., Lin, T., Gao, L., Grubhoffer, L., and Oliver, J.H. Jr. (2009b) Delineation of a new species of the *Borrelia burgdorferi* sensu lato complex, *Borrelia americana* sp. nov. *J Clin Microbiol* **47**(12)**,** 3875–3880.

Saint Girons, I., Norris, S. J., Gobel, U. et al. (1992) Genome structure of spirochetes. *Res Microbiol* **143,** 615–621.

Schulte-Spechtel, U., Fingerle, V., Goettner, G. et al. (2006) Molecular analysis of decorin-binding protein A (DbpA) reveals five major groups among European *Borrelia burgdorferi* sensu lato strains with impact for the development of serological assays and indicates lateral gene transfer of the dbpA gene. *Int J Med Microbiol* **296**(Suppl 40)**,** 250–266.

Schwartz, J. J., Gazumyan, A., and Schwartz, I. (1992) rRNA gene organization in the Lyme disease spirochete, *Borrelia burgdorferi. J Bacteriol* **174,** 3757–3765.

Scott, J. D., Fernando, K., Banerjee, S. N. et al. (2001) Birds disperse ixodid (Acari: Ixodidae) and *Borrelia burgdorferi*-infected ticks in Canada. *J Med Entomol* **38,** 493–500.

Spielman, A., Wilson, M. L., Levine, J. F., and Piesman, J. (1985) Ecology of *Ixodes dammini*-borne human babesiosis and Lyme disease. *Annu Rev Entomol* **30,** 439–460.

Stackebrandt, E. and Ebers, J. (2006) Taxonomic parameters revisited: Tarnished gold standards. *Microbiol Today* **33,** 152–155.

Steere, A. C., Coburn, J., and Glickstein, L. (2004) The emergence of Lyme disease. *J Clin Invest* **113,** 1093–1101.

Steere, A. C., Hardin, J. A., and Malawista, S. E. (1978) Lyme arthritis: A new clinical entity. *Hosp Pract* **13,** 143–158.

Steere, A. C., Malawista, S. E., Snydman, D. R., and Andiman, W. A. (1976) Cluster of arthritis in children and adults in Lyme, Connecticut. *Arthritis Rheum* **19,** 824.

Stevenson, B. and Miller, J. C. (2003) Intra- and interbacterial genetic exchange of Lyme disease spirochete *erp* genes generates sequence identity amidst diversity. *J Mol Evol* **57,** 309–324.

Stevenson, B., Zuckert, W. R., and Akins, D. R. (2000) Repetition, conservation, and variation: The multiple cp32 plasmids of *Borrelia* species. *J Mol Microbiol Biotechnol* **2,** 411–422.

Stewart, P. E., Wang, X., Bueschel, D. M. et al. (2006) Delineating the requirement for the *Borrelia burgdorferi* virulence factor OspC in the mammalian host. *Infect Immun* **74,** 3547–3553.

Storfer, A., Murphy, M. A., Evans, J. S. et al. (2007) Putting the "landscape" in landscape genetics. *Heredity* **98,** 128–142.

Taragel'ova, V., Koci, J., Hanincova, K. et al. (2008) Blackbirds and song thrushes constitute a key reservoir of *Borrelia garinii*, the causative agent of borreliosis in Central Europe. *Appl Environ Microbiol* **74,** 1289–1293.

Tilly, K., Bestor, A., Jewett, M. W., and Rosa, P. (2007) Rapid clearance of Lyme disease spirochetes lacking OspC from skin. *Infect Immun* **75,** 1517–1519.

Tourand, Y., Kobryn, K., and Chaconas, G. (2003) Sequence-specific recognition but position-dependent cleavage of two distinct telomeres by the *Borrelia burgdorferi* telomere resolvase, ResT. *Mol Microbiol* **48,** 901–911.

Vitorino, L., Chelo, I. M., Bacellar, F., and Ze-Ze, L. (2007) Rickettsiae phylogeny: A multigenic approach. *Microbiology* **153,** 160–168.

Vitorino, L. R., Margos, G., Feil, E. J. et al. (2008) Fine-scale phylogeographic structure of *Borrelia lusitaniae* revealed by multilocus sequence typing. *PLoS One* **3,** e4002.

Wang, G., Ojaimi, C., Iyer, R. et al. (2001) Impact of genotypic variation of *Borrelia burgdorferi* sensu stricto on kinetics of dissemination and severity of disease in C3H/HeJ mice. *Infect Immun* **69,** 4303–4312.

Wang, G., Ojaimi, C., Wu, H. et al. (2002). Disease severity in a murine model of Lyme borreliosis is associated with the genotype of the infecting *Borrelia burgdorferi* sensu stricto strain. *J Infect Dis* **186,** 782–791.

Wang, G., van Dam, A. P., and Dankert, J. (1999) Evidence for frequent OspC gene transfer between *Borrelia valaisiana* sp. nov. and other Lyme disease spirochetes. *FEMS Microbiol Lett* **177,** 289–296.

Wang, I. N., Dykhuizen, D. E., Qiu, W. et al. (1999) Genetic diversity of *ospC* in a local population of *Borrelia burgdorferi* sensu stricto. *Genetics* **151,** 15–30.

Wayne, L. G., Brenner, D. J., Colwell, R. R. et al. (1987) Report of the ad hoc committee on reconciliation of approaches to bacterial systematics. *Int J Syst Bacteriol* **37,** 463–464.

Will, G., Jauris-Heipke, S., Schwab, E. et al. (1995) Sequence analysis of *ospA* genes shows homogeneity within *Borrelia burgdorferi* sensu stricto and *Borrelia afzelii* strains but reveals major subgroups within the *Borrelia garinii* species. *Med Microbiol Immunol* **184,** 73–80.

Wilske, B., Busch, U., Eiffert, H. et al. (1996) Diversity of OspA and OspC among cerebrospinal fluid isolates of *Borrelia burgdorferi* sensu lato from patients with neuroborreliosis in Germany. *Med Microbiol Immunol* **184,** 195–201.

Wilske, B., Preac-Mursic, V., Gobel, U. B. et al. (1993) An OspA serotyping system for *Borrelia burgdorferi* based on reactivity with monoclonal antibodies and *ospA* sequence analysis. *J Clin Microbiol* **31,** 340–350.

Wilson, M. L., Adler, G. H., and Spielman, A. (1985) Correlation between abundance of deer and that of the deer tick, *Ixodes dammini* (Acari, Ixodidae). *Ann Entomol Soc Am* **78,** 172–176.

Wilson, M. L. and Spielman, A. (1985) Seasonal activity of immature *Ixodes dammini* (Acari: Ixodidae). *J Med Entomol* **22,** 408–414.

Womble, W. H. (1951) Differential systematics. *Science* **114,** 315–322.

Wormser, G. P., Brisson, D., Liveris, D. et al. (2008) *Borrelia burgdorferi* genotype predicts the capacity for hematogenous dissemination during early Lyme disease. *J Infect Dis* **198,** 1358–1364.

WORMSER, G. P., LIVERIS, D., NOWAKOWSKI, J. et al. (1999) Association of specific subtypes of *Borrelia burgdorferi* with hematogenous dissemination in early Lyme disease. *J Infect Dis* **180,** 720–725.

WRIGHT, S. D. and NIELSEN, S. W. (1990) Experimental infection of the white-footed mouse with *Borrelia burgdorferi*. *Am J Vet Res* **51,** 1980–1987.

YANG, X. F., PAL, U., ALANI, S. M. et al. (2004) Essential role for OspA/B in the life cycle of the Lyme disease spirochete. *J Exp Med* **199,** 641–648.

YOUNG, A. S., DOLAN, T. T., MORZARIA, S. P. et al. (1996) Factors influencing infections in *Rhipicephalus appendiculatus* ticks fed on cattle infected with *Theileria parva*. *Parasitology* **113**(Pt 3), 255–266.

Chapter 13

Population Genetics of *Neisseria meningitidis*

ULRICH VOGEL, CHRISTOPH SCHOEN, AND JOHANNES ELIAS

13.1 INTRODUCTION

The gram-negative species *Neisseria meningitidis* belongs to the genus *Neisseria* within the beta-subgroup of proteobacteria. Many of the prominent neisserial species are restricted to the human host, such as *N. meningitidis* itself, but also *Neisseria gonorrhoeae*, *Neisseria lactamica*, *Neisseria sicca*, and *Neisseria flavescens*. The scientific interest in Neisseriae is certainly catalyzed by the unquestionable medical importance of two species, that is, *N. meningitidis*, which causes sepsis and meningitis in toddlers, infants, and adolescents (Rosenstein et al., 2001), and *N. gonorrhoeae*, the agent of gonorrhea (Sarafian and Knapp, 1989). *N. meningitidis* has been attracting molecular epidemiologists and population geneticists alike for its property of maintaining lineage structure in spite of its constant reinvention by horizontal gene transfer (HGT). Furthermore, a fascinating and not fully understood dichotomy of asymptomatic carriage and invasive, frequently fatal, disease, led to the fact that the meningococcus (the trivial name for *N. meningitidis*) is one of the model organisms for the study of population structures of pathogenic bacteria. Notwithstanding theoretical considerations and implications of the study of meningococcal population structures, it is noteworthy that knowledge of diversity, transmission patterns, and clonal stability affects future predictions on how the population structure will react upon vaccine pressure; thus, knowledge of the population biology influences directly vaccine development and application (Rappuoli, 2007).

13.2 A BRIEF HISTORY OF TYPING OF MENINGOCOCCI

Serological typing of meningococci by microprecipitation and later on slide agglutination laid the grounds to distinguish between variants of meningococci, first on the basis of the capsular serogroup (Slaterus, 1961). Work in the 1970s and 1980s extended these early approaches to subcapsular antigens, the porins and the lipopolysaccharides, by the use of antisera and monoclonal antibodies (Frasch and Chapman, 1972; Frasch et al., 1985). The typing schemes composed of the serogroup (the capsule), the serotype (porin B protein),

Bacterial Population Genetics in Infectious Disease, Edited by D. Ashley Robinson, Daniel Falush, and Edward J. Feil

the serosubtype (porin A protein), and the immunotype lipopolysaccharide (LPS) enabled the first uniform nomenclature and a variety of downstream applications such as outbreak investigations (Cartwright et al., 1986) and outer membrane vesicle vaccine design (Bjune et al., 1991). Nevertheless, the immunological typing approach had weaknesses, mainly resulting from the higher rate of emergence of new antigen variants compared to the generation of specific reagents identifying them. The adoption of DNA sequence typing of antigens or their variable regions consecutively provided an open approach of high accuracy, interlaboratory reproducibility, harmonization of nomenclature, and portability (Feavers et al., 1992b; Zapata et al., 1992; Jolley and Maiden, 2006; Jolley et al., 2007).

In parallel, the application of typing methods investigating neutrally evolving genes significantly broadened the view on meningococcal populations by analyzing housekeeping genes either by multilocus enzyme electrophoresis (MLEE) (Caugant et al., 1986b) or by multilocus sequence typing (MLST) (Maiden et al., 1998). An MLEE of meningococci convincingly demonstrated that there is some kind of lineage structure of meningococci, with successful lineages being observed on a global scale. Based on the principles of MLEE, the development of MLST was a major methodological step forward to characterize population structures not only of meningococci but also of a large number of other bacterial species (Maiden, 2006). The beauty of the meningococcal MLST scheme is made perfect by the fact that this is one of the few schemes designed to serve several species of the genus (Bennett et al., 2005, 2007). Not surprisingly, the neisserial MLST database is the largest MLST database with >7000 profiles and >12,000 strain data sets at the time of writing.

13.3 SPECIES SEPARATION

There is convincing evidence that *N. meningitidis*, *N. gonorrhoeae*, and *N. lactamica*, to name the best studied species, are separated (Vazquez et al., 1993), despite the facts (i) that many neisserial species share the same natural habitat or ecological niche (the human nasopharynx), (ii) that cross-reactive antigens are expressed (Kim et al. 1989; Derrick et al. 1999; Kremastinou et al. 1999), and (iii) that there are reports on HGT across species barriers and evolutionary relatedness of genes (Lujan et al., 1991; Zhou and Spratt, 1992; Bowler et al., 1994; Feil et al., 1995; Vazquez et al., 1995; Zhou et al., 1997; Smith et al., 1999; Zhu et al., 2001; Bennett et al., 2008, 2009) (see also Table 13.1). Thus, according to the concept, classical biochemical traits are reliably used in clinical laboratories to distinguish, for example, between *N. meningitidis* (positive for the gamma-glutamyltransferase [Riou et al., 1982]) and *N. lactamica* (positive for the beta-galactosidase [Hollis et al., 1969]). DNA–DNA hybridization studies (Hoke and Vedros, 1982) showed separation of the three species as did partial 16S rDNA sequencing (Harmsen et al., 2001). Comparing *N. meningitidis*, *N. lactamica*, and *N. gonorrhoeae* by MLST, which provides a fragmentary but fairly representative view on the genome sequence, dissected at least these three species, although there were problems with others (Hanage et al., 2005). Neisserian MLST therefore serves as a fairly good example for the application of pan-genus multilocus sequence analysis to species concepts as proposed by Gevers et al. (2005). The discussion of species separation of *Streptococcus pneumoniae* and *Streptococcus oralis* presented recently (Fraser et al., 2007) might also fit for the neisserial species discussed herein: Despite being a sexual species and despite possibilities for recombination, the overall sequence diversity is so high that intergenic recombination rates are reduced so much that species barriers are not abrogated.

Table 13.1 Nonexhaustive List of Meningococcal Genes That Were Either Altered or Acquired by Horizontal Gene Transfer (HGT) from the Same or Related Species

Gene	Species involved	Type of HGT	Reference
A) Housekeeping genes from the meningococcal core genome where homologous intragenic recombination resulted in allelic conversion			
16S rDNA	*N. meningitidis* and other commensal *Neisseria* spp.	Interspecies	Smith et al. (1999)
abcZ	*N. meningitidis*	Intraspecies	Holmes et al. (1999) and Jolley et al. (2005)
adk	*N. meningitidis* and other commensal *Neisseria* spp.	Intra- and interspecies	Feil et al. (1995, 1996, 2001)
argF	*N. meningitidis, Neisseria cinerea*, and other commensal *Neisseria* spp.	Interspecies	Smith et al. (1999) and Zhou and Spratt (1992)
aroE	*N. meningitidis* and other commensal *Neisseria* spp.	Interspecies	Feil et al. (2001), Holmes et al. (1999), Jolley et al. (2005), and Zhou et al. (1997)
fumC	*N. meningitidis*	Intraspecies	Jolley et al. (2005)
gdh	*N. meningitidis*	Intraspecies	Feil et al. (2001) and Jolley et al. (2005)
glnA	*N. meningitidis* and other commensal *Neisseria* spp.	Interspecies	Zhou et al. (1997)
gyrA	*N. meningitidis* and *N. lactamica*	Interspecies	Wu et al. (2009)
iga	*N. meningitidis* and *N. gonorrhoeae*	Intra- and interspecies	Lomholt et al. (1992, 1995) and Morelli et al. (1997)
opa	*N. meningitidis* and *N. gonorrhoeae*	Intra- and interspecies	Hobbs et al. (1994, 1998) and Morelli et al. (1997)
opc	*N. meningitidis*	Intraspecies	Seiler et al. (1996)
pdhC	*N. meningitidis*	Intraspecies	Feilet al. (2001), Holmes et al. (1999), and Jolley et al. (2005)
penA	*N. meningitidis, N. cinerea*, and *N. flavescens*	Interspecies	Bowler et al. (1994) and Spratt et al. (1992)
pgm	*N. meningitidis*	Intraspecies	Feil et al. (2001) and Jolley et al. (2005)
porA	*N. meningitidis, N. gonorrhoeae, Neisseria polysaccharea*, and *N. lactamica*	Intra- and interspecies	Bygraves et al. (1999), Derrick et al. (1999), Feavers et al. (1992a), and Suker et al. (1994)
porB	*N. meningitidis* and *N. gonorrhoeae*	Intra- and interspecies	Bygraves et al. (1999), Derrick et al. (1999), and Vazquez et al. (1995)
recA	*N. meningitidis* and other commensal *Neisseria* spp.	Interspecies	Smith et al. (1999)
rho	*N. meningitidis* and other commensal *Neisseria* spp.	Interspecies	Smith et al. (1999)
tbpB	*N. meningitidis*	Intraspecies	Linz et al. (2000) and Zhu et al. (2001)

(*Continued*)

Table 13.1 (*Continued*)

Gene	Species involved	Type of HGT	Reference
B) Meningococcal genes that were acquired via homologous intergenic recombination[a]			
synX-D	*N. meningitidis* serogroup B and C strains	Intraspecies	Swartley et al. (1997)
opcA	*N. meningitidis* and unidentified source species	Interspecies	Zhu et al. (1999)
tbpB	*N. meningitidis, N. lactamica*, and other commensal *Neisseria* spp.	Interspecies	Linz et al. (2000)
hmbR	*N. meningitidis* and other commensal *Neisseria* spp.	Interspecies	Kahler et al. (2001)
C) Meningococcal genes that were acquired via nonhomologous recombination			
sodC, bio gene cluster	*N. meningitidis* and *Haemophilus* spp.	Interspecies	Kroll et al. (1998)
lav	*N. meningitidis* and *H. influenzae*	Interspecies	Davis et al. (2001)

[a]For a more comprehensive list of genes forming minimal mobile elements, the reader is referred to Snyder et al. (2007).

From a variety of methodological perspectives, the three neisserial species under discussion thus appear to be well-defined, but nevertheless communicating, recombining taxonomic entities. The evolutionary origin of meningococci, however, remains an interesting and not fully resolved issue. Dating the age of the species has not been successfully achieved, although the absence of historical reports on meningococcal disease suggests that meningococcal disease emerged not long before the early nineteenth century (Cartwright, 2006). Genomic analysis suggests that the acquisition of the insertion sequence IS*1655* was a turning point coinciding with speciation of meningococci (Schoen et al., 2008). In contrast, the early view that meningococci necessarily are encapsulated by polysaccharides or at least carry genetic material associated to polysaccharide synthesis was disproven by the identification of capsule null locus meningococci, which, instead of harboring the genomic island for capsule synthesis and transport, displayed a genomic organization at this locus resembling that of *N. lactamica* and *N. gonorrhoeae* (Claus et al., 2002; Dolan-Livengood et al., 2003). This finding suggested that capsule null locus meningococci might represent contemporary descendants of unencapsulated ancient meningococci (Vogel and Claus, 2004). Genome-based phylogeny including one capsule null locus isolate supported this hypothesis, because it placed the capsule null locus strain closer to *N. gonorrhoeae* and *N.* lactamica than it did to other meningococcal strains (Schoen et al., 2008).

Given that *N. meningitidis* can be clearly separated from other neisserial species, its probably complicated evolutionary history as well as its variability at the capsule locus points to the fact that this species is far from being genetically monomorphic, is highly recombinogenic, and hence remains fuzzy at its edges (Hanage et al., 2005). Consequently, *N. meningitidis* has been summarized as a "metapopulation that consists of numerous, semi-discrete lineages that are linked by recombination" (Achtman and Wagner, 2008). The composition of this metalineage and forces shaping them will be discussed below. However, the next paragraph will first describe the biological material that is available for population studies.

13.4 SAMPLING STRATEGIES

For meningococci, there are abundant strain collections available representing isolates from invasive disease sampled for medical or epidemiological purposes. Isolates are available on a global scale over several decades with an emphasis on the later half of the twentieth century. Theoretical conclusions made on the basis of invasive strain collections must take into account that invasive isolates represent only the tip of the iceberg, as many strains and lineages never or only rarely cause disease (Yazdankhah et al., 2004). Furthermore, cross-sectional studies, especially when they are regionally confined, are biased because they reflect only a snapshot of lineages circulating over time, as the life span of many clonal complexes within a region is limited to only a few years (Buckee et al., 2008). It should also be highlighted that enhanced surveillance of invasive isolates installed to accompany outbreak management might be biased due to the overrepresentation of the outbreak strain (Krizova and Musilek, 1995; Baker et al., 2001; Aguilera et al., 2002). Likewise, there are tremendous differences of serogroups and lineages circulating in countries and continents, to mention only the persistent anchorage of serogroup A meningococci in the African meningitis belt paralleled by their current absence in Western Europe and in North America.

The analysis of the meningococcal population biology would be incomplete without samples from asymptomatic carriers, mostly obtained within the frame of cross-sectional studies (Gold et al., 1978; Block et al., 1999; Maiden and Stuart, 2002; Yazdankhah et al., 2004; Claus et al., 2005; Findlow et al., 2007; Yaro et al., 2007; Maiden et al., 2008; Mueller et al., 2008). Carrier strains usually are not collected as part of infinite surveillance tasks but are sampled on the basis of defined projects over a limited time period. The Czech carriage collections with repeated sampling efforts over three decades (Buckee et al., 2008), or massive swabbing campaigns such as the one accompanying the U.K. serogroup C conjugate vaccine initiative (Maiden et al., 2008), therefore represent invaluable resources.

13.5 THE CLONAL COMPLEXES OF MENINGOCOCCI

In the 1980s, the pioneering work of D. A. Caugant demonstrated that the species *N. meningitidis* is composed of a variety of clonal complexes or lineages that differ in their electrophoretic mobility patterns of neutral housekeeping enzymes as elucidated by MLEE (Caugant et al., 1986a,b, 1987, 1988; Selander et al., 1986). The lineages were tagged with designations that are still in use, such as "cluster A4" or "lineage 3." Nowadays, these historical names are applied in conjunction with multilocus sequence type designations (e.g., ST-8 complex/cluster A4).

The above cited work by Caugant et al. highlighted important principles still capable of bearing nowadays: (i) Meningococci can be divided into complexes of genetically related clones; (ii) the clonal complexes besides their MLEE patterns share other common traits such as capsular polysaccharide and subcapsular antigens; (iii) successful lineages are found on a global scale; and (iv) the clonal complexes found among healthy carriers of meningococci differ substantially from those recovered from invasive disease.

It was consecutively shown that meningococci tend to exchange DNA and to recombine, thereby distorting single-locus trees and vertical ancestry (Lujan et al., 1991; Zhou and Spratt, 1992; Smith et al., 1993; Bowler et al., 1994; Feil et al., 1995; Vazquez et al., 1995; Zhou et al., 1997). Using DNA sequence data and well-defined strain collection, Jolley et al. quantified the size of horizontally transferred DNA fragments to about 1.1 kb; they showed more than 10% of sites of housekeeping genes were segregating; furthermore, recombination was about 10 times more important for the generation of diversity than mutation (Jolley et al., 2005). The fragment length deduced from housekeeping genes by Jolley et al. was lower than previously measured for the *tbpB* region encoding the transferring-binding protein, for which a median size of 5.1 kb was estimated (Linz et al., 2000).

In general, the interpretation of results of MLEE studies was very much facilitated by the accumulation of large amounts of DNA sequence data, either as MLST data, as antigen sequencing data, or as whole genome data (Maiden, 2008). MLST data are now available for more than 7000 profiles in the public *Neisseria* database; antigen sequence data are available, for example, for PorA (Russell et al., 2004), PorB (Russell et al., 2004; Urwin et al., 2004), FetA (Thompson et al., 2003; Bennett et al., 2009), Opa (Callaghan et al., 2008), PenA (Taha et al., 2007), and the factor H-binding protein (Fletcher et al., 2004). Furthermore, several complete genome sequences are available (Parkhill et al., 2000; Tettelin et al., 2000; Bentley et al., 2007; Peng et al., 2008; Schoen et al., 2008), and many more are apparently under investigation at the time of writing.

The discriminatory power of MLST and antigen sequence typing (in the following referred to as "finetypes") is highlighted here by data from Germany, which were obtained by the Bavarian meningococcal carriage study (Claus et al., 2005) and for strains from invasive diseases (Elias et al., 2006; Reinhardt et al., 2008). Figure 13.1 illustrates that there is an only incomplete overlap of clonal lineages obtained from invasive disease (Germany, 2002 through 2008, partial data set) and carriage (Bavaria, 1999 and 2000). Furthermore, some lineages (e.g., the ST-11 complex/ET-37 complex), despite causing a major share of the invasive disease burden, are underrepresented in carriage. The overall diversity of carrier isolates apparently is higher than that of invasive isolates. These findings are in concordance with what has been shown for the Czech data set (Yazdankhah et al., 2004). The discriminatory index of antigen sequence typing (serogroup: PorA variable region 1, PorA variable region 2: FetA) in its currently proposed form (Jolley et al., 2007) has been reported for the German data set to be 0.963 (95% confidence interval [CI] 0.959–0.968). This demonstrates the tremendous diversity of antigens of invasive isolates even in a geographically confined region. The data for the extended data set (2002–2008) are highlighted in Fig. 13.2. Some finetypes regularly occurred over the whole period of time at high frequency, whereas others were observed only occasionally or even only once. Local emergence of variants and their ready extinction might be one reason why the intersection of finetypes identified in a geographically and temporally confined carriage study (Oppermann et al., 2006) with the finetypes of invasive strains recovered from the whole of Germany from 2002 to 2008 is less than 50% (Fig. 13.3). Another factor contributing to this divergence is the indubitable differences of virulence attributes of clones with comparable success regarding transmission resulting in high carriage rates. In the local swabbing campaign described above (Oppermann et al., 2006), 14 of 74 strains belonged to the finetype cnl:P1.18,25-1:F5-5, which was never observed as the causative agent of invasive disease in the whole country between 2002 and 2008. Cnl is the abbreviation for the above-mentioned capsule null locus (Claus et al., 2002; Weber et al., 2006). These meningococci lack all genes for capsule synthesis and transport. Thus, the only

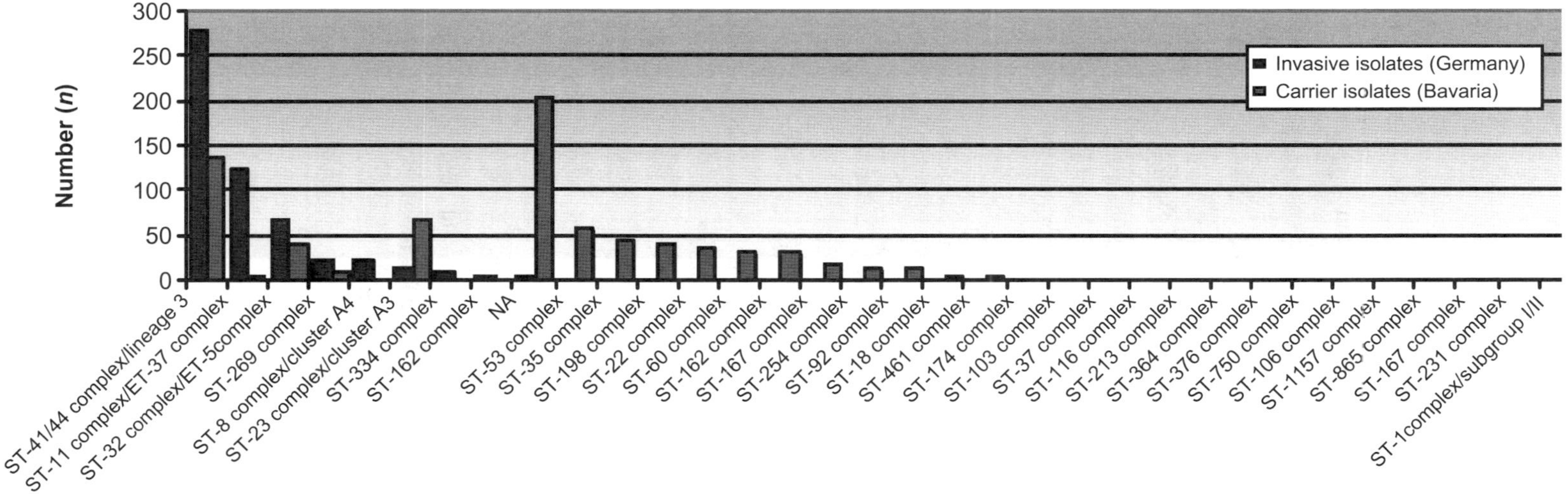

Figure 13.1 Comparison of the distribution of clonal complexes between invasive isolates and carrier isolates. Invasive isolates were collected from 2002 to 2008 by the German reference laboratory for meningococci, where MLST is performed on a selection of isolates. Carrier isolates were obtained during the carriage study in Bavaria in 1999 and 2000 (Claus et al., 2005). Note that among the invasive isolates, there is a bias toward the ST-41/44 complex strains due to enhanced surveillance of this particular lineage. NA: clonal complex not assigned.

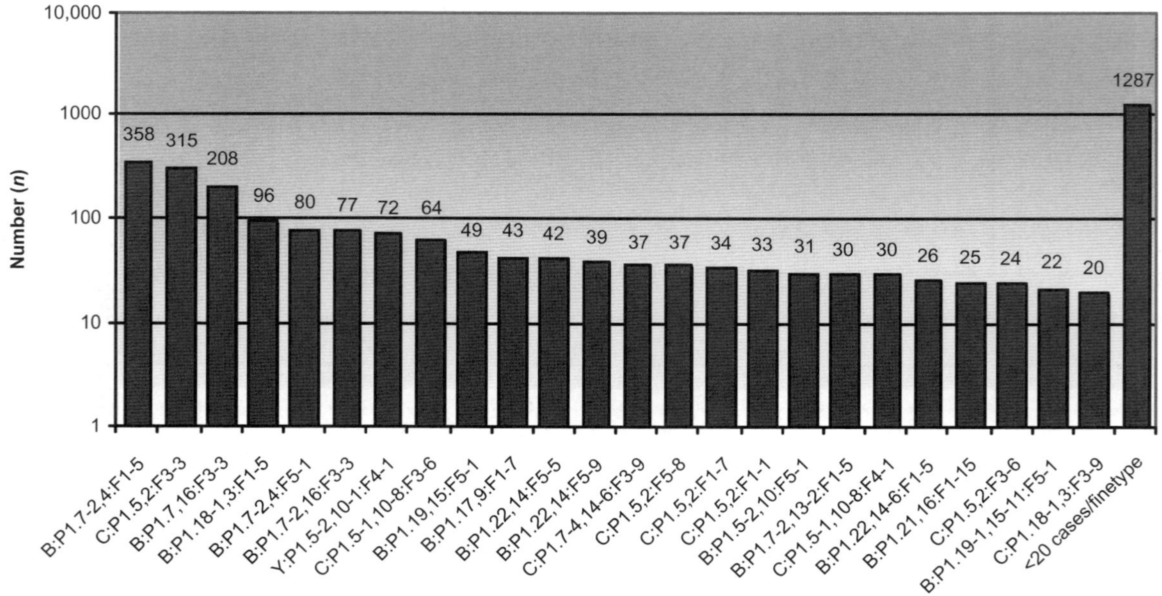

Figure 13.2 Finetype distribution among the collection of invasive meningococcal strains collected between 2002 and 2008 by the German Reference Laboratory for Meningococci. Number of finetypes: $N = 657$; number of cases: $N = 3079$.

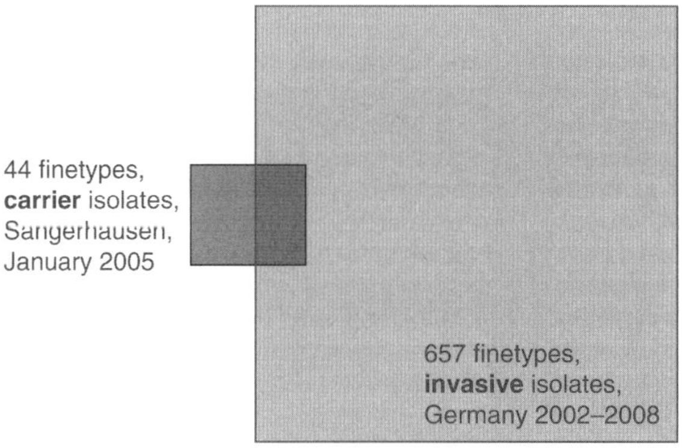

Figure 13.3 Intersection of finetypes (serogroup: PorA variable region 1, PorA variable region 2: FetA) identified in a geographically and temporally refined carriage study (Oppermann et al., 2006) with finetypes recovered from Germany, 2002–2008.

virulence factor that appears to be indispensible in meningococci (Vogel et al., 1996; Geoffroy et al., 2003) is not expressed by these variants, which provides an explanation for their almost exclusive commensal behavior.

Within-host evolution and consecutive extinction of less fit variants could also be observed by detailed analysis of carrier isolates. It is obvious that encapsulated meningococci accumulate mutations of their capsule synthesis during circulation. Some of those mutations are reversible (Hammerschmidt et al., 1996a,b); however, many of those are irreversible, and the variant strains obviously are readily lost due to a reduction of their transmission rates (Weber et al., 2006). Phenotypic changes during a microepidemic spread have also been demonstrated for porins and capsule antigens (Swartley et al., 1997; Vogel et al., 2000; van Der Ende et al., 2003). An antigenic shift of several outer membrane proteins might even trigger the prominent changes in disease epidemiology over time (Harrison et al., 2006).

Meningococcal finetypes and multilocus sequence types do not evolve independently from each other. Antigen sequence types are structured with regard to their combinations, and the number of combinations is lower than would be expected if they were assorted at random (Urwin et al., 2004). This has implications for molecular epidemiology: Antigen sequence types may serve as a more or less reliable marker for clonal lineages. Furthermore, their number is limited in nature, despite the huge number of possible antigen variants (at the time of writing: 12 serogroups × 523 PorA VR1 peptides × 188 PorA VR2 peptides × 290 FetA peptides = 342,167,520 possible variants), which allows the design of protein-based vaccines including a limited number of variants, for example, of the porin A peptides (Urwin et al., 2004; van den Dobbelsteen et al., 2007).

Finally, from an epidemiological point of view, the establishment of comprehensive epidemiological databases comprising molecular typing data provides the opportunity to objectively assess clusters and outbreaks (Elias et al., 2006) and, furthermore, to visualize data in geographic information systems including spatiotemporal information (Reinhardt et al., 2008). From these exercises it becomes clear that finetypes differ in their spatiotemporal migration patterns (Fig. 13.4). Whereas B:P1.7-2,4:F1-5 strains, which belong to the ST-41/44 complex/lineage 3, apparently tend to circulate in confined regions for longer periods of time, others like C:P1.5,2:F3-3 are much more mobile. One might speculate that B:P1.7-2,4:F1-5 elicits a less effective herd immunity during circulation, as the P1.7-2,4 antigen has been reported to be a weak immunization candidate in outer membrane vesicle vaccines (Luijkx et al., 2003). This implies that immune selection on this particular porin antigen is poor, resulting in extended circulation times of this lineage in a transmission

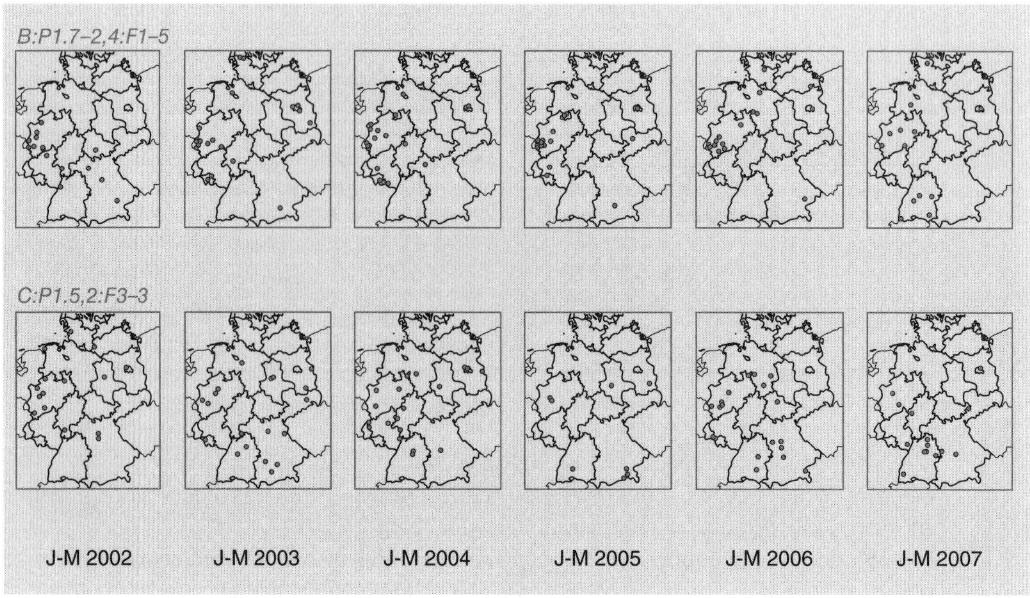

Figure 13.4 Geographic spread of two distinct meningococcal finetypes between 2002 and 2008 from Germany. The months January to March are depicted for each year. The finetype B:P1.7-2,4:F1-5 mostly belongs to the ST-41/44 complex; most strains of the finetype C:P1.5,2:F3-3 are ST-11 complex. B:P1.7-2,4:F1-5 persists in Western Germany, whereas C:P1.5,2:F3-3 appears to occur with a more or less random distribution. The maps were generated with EpiScanGIS (http://www.episcangis.org/) (Reinhardt et al., 2008). See color insert.

network. Extended seroprevalence and carriage studies are needed to confirm this hypothesis; however, the duration of the New Zealand meningococcus B epidemic caused by this lineage, to give an example, has been remarkable (Baker et al., 2001).

13.6 FORCES SHAPING THE MENINGOCOCCAL METALINEAGE

As already mentioned in the introduction to this chapter, the fact that *Neisseria* sp. are naturally transformable species, being competent for the uptake of DNA throughout their entire life cycle (Koomey, 1998), which permeates all aspects of meningococcal population biology; genetic diversity of meningococci is constantly driven more by recombination than by mutation (Feil et al., 1999, 2001). Co-colonization of the nasopharynx with other meningococcal variants (Caugant et al., 2007) or with commensal Neisseriae (Linz et al., 2000) provides a common gene pool serving as a reservoir of genetic diversity (Maiden et al., 1996). HGT between different meningococcal strains as well as different neisserial species can consecutively result in gene content changes in the recipient's genome as well as allelic exchanges in, for example, housekeeping genes. Regions of putatively horizontally transferred DNA in the meningococcal genome comprise so-called minimal mobile elements (Snyder et al., 2007), islands of horizontally transferred DNA (Tettelin et al., 2000), canonical genomic islands, and (defective) prophages. This diversity of mobile genetic elements was shown to substantially contribute to the highly flexible meningococcal gene pool (Hotopp et al., 2006). In fact, in the seven meningococcal genomes sequenced to date, only 67% of the genes in each genome belong to the so-called core genome, which thus accounts for only about 40% of the meningococcal pan-genome (Schoen et al., 2009). This flexibility in gene content is paralleled also by a large number of genes from the meningococcal core genome that show signs of intragenic recombination (Table 13.1). This highly flexible genome facilitates adaptation of the colonizing meningococcal cell to the randomly fluctuating environment of the human nasopharynx with the residential microflora and local immune defenses (Schoen et al., 2007).

At the population genetic level, this diversity provides the raw material for the local and temporarily bounded accumulation of genetic variants over time and the generation of so-called genoclouds (Achtman et al., 2001). It has been shown that a neutral microepidemic model of evolution of *N. meningitidis*, in which neutral genetic drift combined with local microepidemic spread, shapes a theoretical recombinogenic population according to what is observed for meningococci (Fraser et al., 2005). Nevertheless, the emerging descendant variants are also the adaptive consequence of rising herd immunity within the human population. Their expansion is regularly limited as an effect of reduced fitness of descendants and their periodic selection, and by bottlenecks, for example, as a consequence of transregional spread and introduction of only a few clones into a new transmission network (Zhu et al., 2001). Meningococci display signs of a highly recombinogenic population with purifying events and consecutive clonal expansion of fit variants. Their population structure has therefore been categorized as "epidemic" (Smith et al., 1993). Clone dominance in such an epidemic structure might be very strong, for example, for serogroup A disease in Africa, which shows properties of clonal disease despite HGT (Bart et al., 2001; Leimkugel et al., 2007).

Invasive disease can be considered as a dead end for meningococcal transmission. Therefore, only selective forces acting on carriage of strains and their transmission are active in the transformation of the meningococcal population. The duration of carriage and the number of effectively infected contacts of a carrier determine the spread of bacteria within a population. Subtle fitness differences between strains affect clonal expansion and consequently also modulate invasive disease rates (Buckee et al., 2008). The major force acting on carried meningococci is immune selection. Meningococcal carriage gives rise to antibodies against meningococci, which consecutively terminate carriage of a defined clone and block its transmission. A major immunogenic antigen of meningococci is porin A, which is used as a vaccine antigen in outer membrane vesicles (Bjune et al., 1991). Mathematical modeling has suggested that immune selection results into a strain structure with nonoverlapping hypervariable epitopes of PorA (Gupta et al., 1996; Gupta and Maiden, 2001). The Opa proteins of meningococci, which bind to receptors on the host cell surface, are present in three to four gene copies with two hypervaribale regions per locus. Structuring of the variants was also shown for this protein family, both in carriage and disease isolates (Callaghan et al., 2008). Mathematical modeling again was consistent with strong immune selection being the driving force for structuring allelic variants.

In a "multilocus model of pathogen population structure" published recently (Buckee et al., 2008), the role of competition between lineages was investigated to explore whether competition would shape the meningococcal population structure to a species of several coexisting lineages, which display only a minor fraction of possible allele combinations. The model assumed minor differences of transmissibility between sequence types, which compete with each other, and it assumed that coinfection is possible. Increasing competition eliminated less fit lineages up to a level, where only very few lineages coexist. When immune selection was chosen as competitive force and lineages were defined both by housekeeping allele combinations and two antigenic variants, increasing immune selection associated lineages with antigen variants either stably or—with higher selective pressure—

oscillated over time. The model fits reality as most sequence types in Czech strains collected within 27 years were short-lived, and there was some oscillation of antigen association with sequence types. The high clonality of African serogroup A meningococci (Leimkugel et al., 2007), with a very limited genetic diversity, might result from very high levels of competition under the particular social, climate, and ethnic conditions present in the African meningitis belt.

Besides the forces described above, other influences stabilizing the lineage structure might be taken into account. The extent of DNA transformation might be affected by restriction enzymes. A variety of differentially distributed restriction modification systems in meningococci have been identified (Bart et al., 2000; Claus et al., 2000a,b). The type II R.NmeDI cleaves double-stranded DNA in proximity to the recognition sequence RCCGGY and generates a 25-bp fragment (Kwiatek and Piekarowicz, 2007). The respective restriction modification (RM) system was found to be located between the genes *pheS* and *pheT* and was restricted almost exclusively to strains belonging to the ST-11 complex/ET-37 complex and ST-8 complex/cluster A4. Interestingly, these two lineages share alleles at several multilocus sequence type loci; many of the strains are serogroup C and share related PorA and PorB peptides. *NmeBI* occupied the same genomic locus as *nmeDI*, but R.NmeDI was suggested to be an isoschizomer to *Hga*I. *NmeBI* was present in ST-32 complex/ET-5 complex and in ST-41/44 complex/lineage 3 isolates, both pathogenic lineages frequently expressing the serogroup B polysaccharide and related PorA and PorB antigens. The presence of R.NmeBI recognition sites in donor DNA from a M.NmeBI-free donor reduced the transformation rate by 3.8-fold, if an R.NmeDI-positive recipient was used (Claus et al., 2000a). This experimental finding suggested that RM systems at least partially direct the flow of DNA within a meningococcal population thereby promoting some kind of ecological separation.

13.7 VIRULENCE, A MYSTERIOUS TRAIT

The chapter was introduced by the statement that the disease burden of meningococci and gonococci has catalyzed the interest in their population biology. For meningococci, we are now in the peculiar situation that probably more is known about the population structure, genetic diversity, and genome architecture than about the virulence itself. Wonderful work has been conducted dissecting the role of various components in pathogenicity, but still it is unclear just what determines a virulent clone. Much of the lack of knowledge may be attributed to inadequate animal models for this pathogen, which is so adapted to the human host (Vogel and Frosch, 1999), but with the advent of more and more sophisticated animal models, this situation might be improved (Alonso et al., 2003; Johansson et al., 2003; Zarantonelli et al., 2007). Certainly, only 5 out of 12 capsular polysaccharides have been regularly associated with invasive meningococcal disease (Frosch and Vogel, 2006), but the emergence of serogroup X disease in Africa should be a reminder about the unpredictable behavior of this organism (Djibo et al., 2003; Leimkugel et al., 2007).

In an attempt to identify genomic islands specific to invasive disease, genomic comparisons revealed the presence of a filamentous phage within the meningococcal disease associated (MDA) island, which was associated with strains invasive in adolescents (Bille et al., 2005, 2008). Interestingly, the association with disease was not visible in children. One might therefore speculate that the presence of the phage somehow provides a benefit for high transmission rates or invasive properties in a cohort displaying natural immunity to the pathogen (Bille et al., 2008). One should bear in mind that immunity to the meningococcus increases with age (Goldschneider et al., 1969). Transmission patterns among adolescents presumably are very different from those found in infants and in toddlers, who acquire bacteria from parents, from and within the family, and from caretakers. Transmission among teenagers, on the other hand, is very much dependent on behavior and is mostly horizontal (Maclennan et al., 2006). A selection for MDA-positive strains among adolescents might therefore be due to their increased immunity and specific transmission patterns. The role MDA plays is still speculative. Recent genome analysis paved the grounds for the speculation that MDA excision and insertion in a dynamic genome might result in oscillations of gene expression profiles without changing gene content itself (Schoen et al., 2008). This scenario might trigger an increase of attack rates for short periods of time. Experimental data proving this hypothesis are still lacking.

In parallel to the epidemiological approaches described above, attempts have also been undertaken to model virulence in the meningococcal population. Meningococcal disease occurs sporadically and as clusters or outbreaks. Transmissions leading to invasive disease are rare even in potentially pathogenic strains, giving rise to a stochastic disease distribution and to unpredictable outbreaks. A stochastic mathematical model employing two distinct meningococcal variants, one being apathogenic, the other causing disease occasionally, was able to mimic small-sized outbreaks only if genetic diversity was present (Stollenwerk et al., 2004). In the aforementioned work by Buckee et al. (2008), virulence was regarded as a kind of luxury component of the bacteria that could only be afforded if there was some kind of compensation for the shortening of the infectious period. Without competition, virulence attributes according to this model could be acquired by all circulating sequence types (STs). With increasing competition, the situation changed and only highly transmissible strains were able to tolerate the loss of infectious contacts by severely affecting the host. This finding implies that besides virulence factors such as the polysaccharide capsule, effective transmissibility of strains is essential for causing disease. As this especially holds true for situations where competition is high, one might speculate that it is increased competition in adolescents compared to infants/toddlers that requires the presence of the MDA region to cause disease in this age segment.

Epidemiological models explicitly taking into account the within-host infection dynamics of *N. meningitidis* further suggest that the virulence of this bacterium might actually be an inadvertent consequence of short-sighted within-host evolution being exasperated by the increased mutation rates associated with phase shifting, where rapid phase shifting evolves as an adaptation for colonization of diverse hosts (Meyers et al., 2003). Phase shifting is a mutational process that turns specific genes on and off, in particular, contingency loci that code for virulence determinants such as pili, lipopolysaccharides, capsular polysaccharides, and outer membrane proteins (Moxon et al., 2006). Computational analyses identified over 80 potentially phase-variable genes in the meningococcal genomes (Saunders et al., 2000; Snyder et al., 2001) with experimental evidence of phase variation currently for 15 genes. Taking into account the results from the epidemiological studies as well as the predictions of the modeling approaches, *N. meningitidis* should consequently be viewed best not as an obligate but more as an accidental pathogen (Moxon and Jansen, 2005).

13.8 POPULATION EFFECT OF MENINGOCOCCAL VACCINES

Vaccines that elicit a mucosal immunity counteracting carriage might affect the contemporary regional meningococcal lineage composition. The best studied example for this effect is the meningococcal C conjugate (MCC) vaccine. This vaccine delivered by three manufacturers was first introduced in the United Kingdom in 1999 as part of a massive vaccination campaign. The extraordinary scientific companionship of this vaccine campaign, which also included carriage studies, very early registered effects of the vaccine on serogroup C meningococcal carriage rates (Maiden and Stuart, 2002; Maiden et al., 2008). Furthermore, herd immunity effects were identified (Ramsay et al., 2003). Clearly, the success of the campaign regarding its outcome measure, the number of invasive serogroup C cases, was at least partially due to herd immunity effects, which has also been investigated by mathematical modeling (Trotter et al., 2005).

A major anthropogenic manipulation of a bacterial population might not be without consequences considering that lineages compete with each other. Furthermore, it was speculated that immune escape variants arise from the increasing immune pressure in the host population (Maiden and Spratt, 1999). Fortunately, serogroup switching under vaccination of ST-11 complex/ET-37 complex was not observed (Gray et al., 2006). Whether this is mostly attributable to the vigorous concept and execution of the campaign remains to be determined. The British experience is most important for all upcoming vaccine licensures, that is, the meningococcus A conjugate vaccine project (LaForce et al., 2007) and the serogroup B protein vaccines, which are currently under investigation (Giuliani et al., 2006; McNeil et al., 2009). The induction of mucosal immunity and herd immunity effects should be monitored carefully during the introduction of the vaccines.

13.9 ANTIBIOTIC RESISTANCE AND MENINGOCOCCAL LINEAGES

For many pathogens, especially those causing nosocomial infections, the rise of antibiotic resistance represents another human-made factor shaping bacterial population structures. An example for this is *Staphylococcus aureus*, where lineage composition differs between methicillin-resistant, mostly hospital-associated, and methicillin-sensitive isolates (Feil and Spratt, 2001), although admittedly, also for *S. aureus*, MLST is not a fully reliable tool to distinguish between resistant and sensitive strains (Turner and Feil, 2007). Meningococcal disease is treated with penicillin or with cephalosporins. Close contacts of cases might receive prophylactic treatment with rifampicin, ciprofloxacin, or ceftriaxone. During asymptomatic carriage, the bacteria might furthermore be exposed to antibiotic treatment for reasons unrelated to meningococci.

The genetic mechanisms behind resistance to penicillin, rifampicin, and ciprofloxacin have been unraveled (Vazquez et al., 2007). Of note, reduced penicillin susceptibility has been reported for *N. lactamica*, a potential source of DNA for meningococci (Arreaza et al., 2002). Resistance to ciprofloxacin has been identified in different parts of the world (Anonymous, 2008; Nair et al., 2008; Skoczynska et al., 2008; Wu et al., 2009), and resistance to rifampicin might cause problems in outbreak management (Almog et al., 1994). True resistance to penicillin is rarely reported (Nair et al., 2008), but reduced susceptibility to penicillin due to mutations of the transpeptidase domain of the penicillin binding protein PBP2 needs careful monitoring (Antignac et al., 2001). Interestingly, the emergence of penicillin resistance in the pneumococcus, which is also living in the nasopharynx, is alarming in some countries with up to 50% of strains exhibiting minimal inhibitory concentrations above 1–2 μg/mL (Jefferson et al., 2006). It may be assumed that pneumococci acquire resistance genotypes from other oral streptococci (Chi et al., 2007). In contrast, a European survey showed that among 1644 meningococcal isolates, none showed a minimal inhibitory concentration of >1 μg/mL, with most isolates displaying even ≤0.125 μg/mL (Taha et al., 2007). The reasons for the difference between meningococci and pneumococci are yet unexplored. Our own unpublished work suggests that the consequences of uptake of variant PBP2 alleles from *N. lactamica* are moderate, and other factors contribute to the much higher minimal inhibitory concentrations in this species (H. Claus et al., unpublished data). Of note, there might be differences between serogroups and lineages with regard to penicillin susceptibility. Sixty-four of 99 serogroup W135 strains and 96 of 97 ST-8/cluster A4 isolates harbored variant PBP2 alleles (Taha et al., 2007). This might imply that there are differences between lineages with regard to the emergence of antibiotic resistance; however, confounding effects of sampling site and time of isolation have not been ruled out.

13.10 CONCLUDING REMARKS

The meningococcus has become a model organism for the study of pathogen population biology. The journey is ongoing, with more and more neisserial genomes becoming available through novel high-throughput sequencing technology, with more and even better-designed meningococcal carriage studies, and with an ever-increasing amount of typing data available for bioinformatic analysis and modeling approaches. It is a fascinating feature of the neisserial research community that knowledge of population structure has had and will have a direct effect on the molecular epidemiology and public health management of the disease. This especially holds true for the implementation of novel vaccines that promise to have the potential to significantly reduce mortality caused by the disease, if knowledge of the population dynamics of the organism will be carefully considered.

REFERENCES

Achtman, M., van Der Ende, A., Zhu, P., Koroleva, I. S., Kusecek, B., Morelli, G., Schuurman, I. G. A., Brieske, N., Zurth, K., Kostyukova, N. N., and Platonov, A.E. (2001) Molecular epidemiology of serogroup A meningitis in Moscow, 1969 to 1997. *Emerg Infect Dis* **7,** 420–427.

Achtman, M. and Wagner, M. (2008) Microbial diversity and the genetic nature of microbial species. *Nat Rev Microbiol* **6,** 431–440.

Aguilera, J. F., Perrocheau, A., Meffre, C., and Hahne, S. (2002) Outbreak of serogroup W135 meningococcal disease after the Hajj pilgrimage, Europe, 2000. *Emerg Infect Dis* **8,** 761–767.

Almog, R., Block, C., Gdalevich, M., Lev, B., Wiener, M., and Ashkenazi, S. (1994) First recorded outbreaks of meningococcal disease in the Israel Defence Force: Three clusters due to serogroup C and the emergence of resistance to rifampicin. *Infection* **22,** 69–71.

Alonso, J. M., Guiyoule, A., Zarantonelli, M. L., Ramisse, F., Pires, R., Antignac, A., Deghmane, A. E., Huerre, M., van Der, W. S., and Taha, M. K. (2003) A model of meningococcal bacteremia after respiratory superinfection in influenza A virus-infected mice. *FEMS Microbiol Lett* **222,** 99–106.

Anonymous (2008) Emergence of fluoroquinolone-resistant *Neisseria meningitidis*–Minnesota and North Dakota, 2007–2008. *MMWR Morb Mortal Wkly Rep* **57,** 173–175.

Antignac, A., Kriz, P., Tzanakaki, G., Alonso, J. M., and Taha, M. K. (2001) Polymorphism of *Neisseria meningitidis* penA gene associated with reduced susceptibility to penicillin. *J Antimicrob Chemother* **47,** 285–296.

Arreaza, L., Salcedo, C., Alcala, B., and Vazquez, J. A. (2002) What about antibiotic resistance in *Neisseria lactamica*? *J Antimicrob Chemother* **49,** 545–547.

Baker, M. G., Martin, D. R., Kieft, C. E., and Lennon, D. (2001) A 10-year serogroup B meningococcal disease epidemic in New Zealand: Descriptive epidemiology, 1991–2000. *J Paediatr Child Health* **37,** S13–S19.

Bart, A., Barnabe, C., Achtman, M., Dankert, J., van Der, E. A., and Tibayrenc, M. (2001) The population structure of *Neisseria meningitidis* serogroup A fits the predictions for clonality. *Infect Genet Evol* **1,** 117–122.

Bart, A., Dankert, J., and van Der Ende, A. (2000) Representational difference analysis of *Neisseria meningitidis* identifies sequences that are specific for the hypervirulent lineage III clone. *FEMS Microbiol Lett* **188,** 111–114.

Bennett, J. S., Callaghan, M. J., Derrick, J. P., and Maiden, M. C. (2008) Variation in the *Neisseria lactamica* porin, and its relationship to meningococcal PorB. *Microbiology* **154,** 1525–1534.

Bennett, J. S., Griffiths, D. T., McCarthy, N. D., Sleeman, K. L., Jolley, K. A., Crook, D. W., and Maiden, M. C. (2005) Genetic diversity and carriage dynamics of *Neisseria lactamica* in infants. *Infect Immun* **73,** 2424–2432.

Bennett, J. S., Jolley, K. A., Sparling, P. F., Saunders, N. J., Hart, C. A., Feavers, I. M., and Maiden, M. C. (2007) Species status of *Neisseria gonorrhoeae*: Evolutionary and epidemiological inferences from multilocus sequence typing. *BMC Biol* **5,** 35.

Bennett, J. S., Thompson, E. A., Kriz, P., Jolley, K. A., and Maiden, M.C. (2009) A common gene pool for the *Neisseria* FetA antigen. *Int J Med Microbiol* **299,** 133–139.

Bentley, S. D., Vernikos, G. S., Snyder, L. A., Churcher, C., Arrowsmith, C., Chillingworth, T., Cronin, A., Davis, P. H., Holroyd, N. E., Jagels, K., Maddison, M., Moule, S., Rabbinowitsch, E., Sharp, S., Unwin, L., Whitehead, S., Quail, M. A., Achtman, M., Barrell, B., Saunders, N. J., and Parkhill, J. (2007) Meningococcal genetic variation mechanisms viewed through comparative analysis of serogroup C strain FAM18. *PLoS Genet* **3,** e23.

Bille, E., Ure, R., Gray, S. J., Kaczmarski, E. B., McCarthy, N. D., Nassif, X., Maiden, M. C., and Tinsley, C. R. (2008) Association of a bacteriophage with meningococcal disease in young adults. *PLoS One* **3,** e3885.

Bille, E., Zahar, J. R., Perrin, A., Morelle, S., Kriz, P., Jolley, K. A., Maiden, M. C., Dervin, C., Nassif, X., and Tinsley, C. R. (2005) A chromosomally integrated bacteriophage in invasive meningococci. *J Exp Med* **201,** 1905–1913.

Bjune, G., Hoiby, E. A., Gronnesby, J. K., Arnesen, O., Fredriksen, J. H., Halstensen, A., Holten, E., Lindbak, A. K., Nokleby, H., Rosenqvist, E., Solberg, L. K., Closs, O., Eng, J., Frøholm, L. O., Lystad, A., Bakketeig, L. S., Hareide, B., Halstensen, A., Holten, E., Eng, J. (1991) Effect of outer membrane vesicle vaccine against group B meningococcal disease in Norway. *Lancet* **338**, 1093–1096.

Block, C., Gdalevich, M., Buber, R., Ashkenazi, I., Ashkenazi, S., and Keller, N. (1999) Factors associated with pharyngeal carriage of *Neisseria meningitidis* among Israel Defense Force personnel at the end of their compulsory service. *Epidemiol Infect* **122,** 51–57.

Bowler, L. D., Zhang, Q. Y., Riou, J. Y., and Spratt, B. G. (1994) Interspecies recombination between the penA genes of *Neisseria meningitidis* and commensal *Neisseria* species during the emergence of penicillin resistance in *N. meningitidis*: Natural events and laboratory simulation. *J Bacteriol* **176,** 333–337.

Buckee, C. O., Jolley, K. A., Recker, M., Penman, B., Kriz, P., Gupta, S., and Maiden, M. C. (2008) Role of selection in the emergence of lineages and the evolution of virulence in *Neisseria meningitidis*. *Proc Natl Acad Sci U S A* **105,** 15082–15087.

Bygraves, J.A., Urwin, R., Fox, A.J., Gray, S.J., Russell, J.E., Feavers, I.M. and Maiden, M.C. (1999) Population genetic and evolutionary approaches to analysis of *Neisseria meningitidis* isolates belonging to the ET-5 complex. *J Bacteriol* **181,** 5551–5556.

Callaghan, M. J., Buckee, C. O., Jolley, K. A., Kriz, P., Maiden, M. C., and Gupta, S. (2008) The effect of immune selection on the structure of the meningococcal opa protein repertoire. *PLoS Pathog* **4,** e1000020.

Cartwright, K. A. (2006) Historical aspects. In *Handbook of Meningococcal Disease* (eds. M. Frosch and M. C. J. Maiden) pp. 1–13. Wiley-VCH, Weinheim, Germany.

Cartwright, K. A., Stuart, J. M., and Noah, N. D. (1986) An outbreak of meningococcal disease in Gloucestershire. *Lancet* **2,** 558–561.

Caugant, D. A., Bovre, K., Gaustad, P., Bryn, K., Holten, E., Hoiby, E. A., and Froholm, L. O. (1986a) Multilocus genotypes determined by enzyme electrophoresis of *Neisseria meningitidis* isolated from patients with systemic disease and from healthy carriers. *J Gen Microbiol* **132,** 641–652.

Caugant, D. A., Froholm, L. O., Bovre, K., Holten, E., Frasch, C. E., Mocca, L. F., Zollinger, W. D., and Selander, R. K. (1986b) Intercontinental spread of a genetically distinctive complex of clones of *Neisseria meningitidis* causing epidemic disease. *Proc Natl Acad Sci U S A* **83,** 4927–4931.

Caugant, D. A., Kristiansen, B. E., Froholm, L. O., Bovre, K., and Selander, R. K. (1988) Clonal diversity of *Neisseria meningitidis* from a population of asymptomatic carriers. *Infect Immun* **56,** 2060–2068.

Caugant, D. A., Mocca, L. F., Frasch, C. E., Froholm, L. O., Zollinger, W. D., and Selander, R. K. (1987) Genetic structure of *Neisseria meningitidis* populations in relation to serogroup, serotype, and outer membrane protein pattern. *J Bacteriol* **169,** 2781–2792.

Caugant, D. A., Tzanakaki, G., and Kriz, P. (2007) Lessons from meningococcal carriage studies. *FEMS Microbiol Rev* **31,** 52–63.

Chi, F., Nolte, O., Bergmann, C., Ip, M., and Hakenbeck, R. (2007) Crossing the barrier: Evolution and spread of a major class of mosaic pbp2x in *Streptococcus pneumoniae, S. mitis* and *S. oralis*. *Int J Med Microbiol* **297,** 503–512.

Claus, H., Friedrich, A., Frosch, M., and Vogel, U. (2000a) Differential distribution of two novel restriction-modification systems in clonal lineages of *Neisseria meningitidis*. *J Bacteriol* **182,** 1296–1303.

Claus, H., Stoevesandt, J., Frosch, M., and Vogel, U. (2000b) Genetic isolation of meningococci of the electrophoretic type 37 complex. *J Bacteriol* **183,** 2570–2575.

Claus, H., Maiden, M. C., Maag, R., Frosch, M., and Vogel, U. (2002) Many carried meningococci lack the genes required for capsule synthesis and transport. *Microbiology* **148,** 1813–1819.

Claus, H., Maiden, M. C., Wilson, D. J., McCarthy, N. D., Jolley, K. A., Urwin, R., Hessler, F., Frosch, M., and Vogel, U. (2005) Genetic analysis of meningococci carried by children and young adults. *J Infect Dis* **191,** 1263–1271.

Davis, J., Smith, A. L., Hughes, W. R., and Golomb, M. (2001) Evolution of an autotransporter: Domain shuffling and lateral transfer from pathogenic *Haemophilus* to *Neisseria*. *J Bacteriol* **183,** 4626–4635.

Derrick, J. P., Urwin, R., Suker, J., Feavers, I. M., and Maiden, M. C. (1999) Structural and evolutionary inference from molecular variation in *Neisseria porins*. *Infect Immun* **67,** 2406–2413.

Djibo, S., Nicolas, P., Alonso, J. M., Djibo, A., Couret, D., Riou, J. Y., and Chippaux, J. P. (2003) Outbreaks of serogroup X meningococcal meningitis in Niger 1995–2000. *Trop Med Int Health* **8,** 1118–1123.

Dolan-Livengood, J. M., Miller, Y. K., Martin, L. E., Urwin, R., and Stephens, D. S. (2003) Genetic basis for nongroupable *Neisseria meningitidis*. *J Infect Dis* **187,** 1616–1628.

Elias, J., Harmsen, D., Claus, H., Hellenbrand, W., Frosch, M., and Vogel, U. (2006) Spatiotemporal analysis of invasive meningococcal disease, Germany. *Emerg Infect Dis* **12,** 1689–1695.

Feavers, I. M., Heath, A. B., Bygraves, J. A., and Maiden, M. C. (1992a) Role of horizontal genetic exchange in the antigenic variation of the class 1 outer membrane protein of *Neisseria meningitidis*. *Mol Microbiol* **6,** 489–495.

Feavers, I. M., Suker, J., McKenna, A. J., Heath, A. B., and Maiden, M. C. (1992b) Molecular analysis of the serotyping antigens of *Neisseria meningitidis*. *Infect Immun* **60,** 3620–3629.

Feil, E., Carpenter, G., and Spratt, B. G. (1995) Electrophoretic variation in adenylate kinase of *Neisseria meningitidis* is due to inter- and intraspecies recombination. *Proc Natl Acad Sci U S A* **92,** 10535–10539.

Feil, E., Zhou, J., Maynard Smith, J., and Spratt, B. G. (1996) A comparison of the nucleotide sequences of thc adk and recA genes of pathogenic and commensal *Neisseria* species: Evidence for extensive interspecies recombination within adk. *J Mol Evol* **43,** 631–640.

Feil, E. J., Holmes, E.C., Bessen, D. E., Chan, M. S., Day, N. P., Enright, M. C., Goldstein, R., Hood, D. W., Kalia, A., Moore, C. E., Zhou, J., and Spratt, B. G. (2001) Recombination within natural populations of pathogenic bacteria: Short-term empirical estimates and long-term phylogenetic consequences. *Proc Natl Acad Sci U S A* **98,** 182–187.

Feil, E. J., Maiden, M. C., Achtman, M., and Spratt, B. G. (1999) The relative contributions of recombination and mutation to the divergence of clones of *Neisseria meningitidis*. *Mol Biol Evol* **16,** 1496–1502.

Feil, E. J. and Spratt, B. G. (2001) Recombination and the population structures of bacterial pathogens. *Annu Rev Microbiol* **55,** 561–590.

Findlow, H., Vogel, U., Mueller, J. E., Curry, A., Njanpop-Lafourcade, B. M., Claus, H., Gray, S. J., Yaro, S., Traore, Y., Sangare, L., Nicolas, P., Gessner, B. D., and Borrow, R. (2007) Three cases of invasive meningococcal disease caused by a capsule null locus strain circulating among healthy carriers in Burkina Faso. *J Infect Dis* **195,** 1071–1077.

Fletcher, L. D., Bernfield, L., Barniak, V., Farley, J. E., Howell, A., Knauf, M., Ooi, P., Smith, R. P., Weise, P., Wetherell, M., Xie, X., Zagursky, R., Zhang, Y., and Zlotnick, G. W. (2004) Vaccine potential of the *Neisseria meningitidis* 2086 lipoprotein. *Infect Immun* **72,** 2088–2100.

Frasch, C. E. and Chapman, S. S. (1972) Classification of *Neisseria meningitidis* group B into distinct serotypes. I. Serological typing by a microbactericidal method. *Infect Immun* **5,** 98–102.

Frasch, C. E., Zollinger, W. D., and Poolman, J. T. (1985) Serotype antigens of *Neisseria meningitidis* and a proposed scheme for designation of serotypes. *Rev Infect Dis* **7,** 504–510.

Fraser, C., Hanage, W. P., and Spratt, B. G. (2005) Neutral microepidemic evolution of bacterial pathogens. *Proc Natl Acad Sci U S A* **102,** 1968–1973.

Fraser, C., Hanage, W. P., and Spratt, B. G. (2007) Recombination and the nature of bacterial speciation. *Science* **315,** 476–480.

Frosch, M. and Vogel, U. (2006) Structure and genetics of the meningococcal capsule. In *Handbook of Meningococcal Disease* (eds. M. Frosch, and M. C. J. Maiden), pp. 145–162. Wiley-VCH, Weinheim.

Geoffroy, M. C., Floquet, S., Metais, A., Nassif, X., and Pelicic, V. (2003) Large-scale analysis of the meningococcus genome by gene disruption: Resistance to complement-mediated lysis. *Genome Res* **13,** 391–398.

Gevers, D., Cohan, F. M., Lawrence, J. G., Spratt, B. G., Coenye, T., Feil, E. J., Stackebrandt, E., Van de, P. Y., Vandamme, P., Thompson, F. L., and Swings, J. (2005) Opinion: Re-evaluating prokaryotic species. *Nat Rev Microbiol* **3,** 733–739.

Giuliani, M. M., du-Bobie, J., Comanducci, M., Arico, B., Savino, S., Santini, L., Brunelli, B., Bambini, S., Biolchi, A., Capecchi, B., Cartocci, E., Ciucchi, L., Di, M. F., Ferlicca, F., Galli, B., Luzzi, E., Masignani, V., Serruto, D., Veggi, D., Contorni, M., Morandi, M., Bartalesi, A., Cinotti, V., Mannucci, D., Titta, F., Ovidi, E., Welsch, J. A., Granoff, D., Rappuoli, R., and Pizza, M. (2006) A universal vaccine for serogroup B meningococcus. *Proc Natl Acad Sci U S A* **103,** 10834–10839.

Gold, R., Goldschneider, I., Lepow, M. L., Draper, T. F., and Randolph, M. (1978) Carriage of *Neisseria meningitidis* and *Neisseria lactamica* in infants and children. *J Infect Dis* **137,** 112–121.

Goldschneider, I., Gotschlich, E. C., and Artenstein, M. S. (1969) Human immunity to the meningococcus. II. Development of natural immunity. *J Exp Med* **129,** 1327–1348.

Gray, S. J., Trotter, C. L., Ramsay, M. E., Guiver, M., Fox, A. J., Borrow, R., Mallard, R. H., and Kaczmarski, E. B. (2006) Epidemiology of meningococcal disease in England and Wales 1993/94 to 2003/04: Contribution and experiences of the Meningococcal Reference Unit. *J Med Microbiol* **55,** 887–896.

Gupta, S. and Maiden, M. C. (2001) Exploring the evolution of diversity in pathogen populations. *Trends Microbiol* **9,** 181–185.

Gupta, S., Maiden, M. C., Feavers, I. M., Nee, S., May, R. M., and Anderson, R. M. (1996) The maintenance of strain structure in populations of recombining infectious agents [see comments]. *Nat Med* **2,** 437–442.

Hammerschmidt, S., Hilse, R., van Putten, J. P., Gerardy Schahn, R., Unkmeir, A., and Frosch, M. (1996a) Modulation of cell surface sialic acid expression in *Neisseria meningitidis* via a transposable genetic element. *EMBO J* **15,** 192–198.

Hammerschmidt, S., Muller, A., Sillmann, H., Mühlenhoff, M., Borrow, R., Fox, A., van Putten, J., Zollinger, W. D., Gerardy-Schahn, R., and Frosch, M. (1996b) Capsule phase variation in *Neisseria meningitidis* serogroup B by slipped strand mispairing in the polysialyltransferase gene (siaD): Correlation with bacterial invasion and the outbreak of meningococcal disease. *Mol Microbiol* **20,** 1211–1220.

Hanage, W. P., Fraser, C., and Spratt, B. G. (2005) Fuzzy species among recombinogenic bacteria. *BMC Biol* **3,** 6.

Harmsen, D., Singer, C., Rothganger, J., Tonjum, T., de Hoog, G. S., Shah, H., Albert, J., and Frosch, M. (2001) Diagnostics of Neisseriaceae and Moraxellaceae by ribosomal DNA sequencing: Ribosomal differentiation of medical microorganisms. *J Clin Microbiol* **39,** 936–942.

Harrison, L. H., Jolley, K. A., Shutt, K. A., Marsh, J. W., O'Leary, M., Sanza, L. T., and Maiden, M. C. (2006) Antigenic shift and increased incidence of meningococcal disease. *J Infect Dis* **193,** 1266–1274.

Hobbs, M. M., Malorny, B., Prasad, P., Morelli, G., Kusecek, B., Heckels, J. E., Cannon, J. G., and Achtman, M. (1998) Recombinational reassortment among opa genes from ET-37 complex *Neisseria meningitidis* isolates of diverse geographical origins. *Microbiology* **144**(Pt 1)**,** 157–166.

Hobbs, M. M., Seiler, A., Achtman, M., and Cannon, J. G. (1994) Microevolution within a clonal population of pathogenic bacteria: Recombination, gene duplication and horizontal genetic exchange in the opa gene family of *Neisseria meningitidis*. *Mol Microbiol* **12,** 171–180.

Hoke, C. and Vedros, N. A. (1982) Taxonomy of the Neisseriae: Deoxyribonucleic acid base composition, interspecific transformation, and deoxyribonucleic acid hybridization. *Int J Syst Bacteriol* **32,** 57–66.

Hollis, D. G., Wiggins, G. L., and Weaver, R. E. (1969) *Neisseria lactamicus* sp. n., a lactose-fermenting species resembling *Neisseria meningitidis*. *Appl Microbiol* **17,** 71–77.

Holmes, E. C., Urwin, R., and Maiden, M. C. (1999) The influence of recombination on the population structure and evolution of the human pathogen *Neisseria meningitidis*. *Mol Biol Evol* **16,** 741–749.

Hotopp, J. C., Grifantini, R., Kumar, N., Tzeng, Y. L., Fouts, D., Frigimelica, E., Draghi, M., Giuliani, M. M., Rappuoli, R., Stephens, D. S., Grandi, G., and Tettelin, H. (2006) Comparative genomics of *Neisseria meningitidis*: Core genome, islands of horizontal transfer and pathogen-specific genes. *Microbiology* **152,** 3733–3749.

Jefferson, T., Ferroni, E., Curtale, F., Giorgi, R. P., and Borgia, P. (2006) *Streptococcus pneumoniae* in Western Europe: Serotype distribution and incidence in children less than 2 years old. *Lancet Infect Dis* **6,** 405–410.

Johansson, L., Rytkonen, A., Bergman, P., Albiger, B., Kallstrom, H., Hokfelt, T., Agerberth, B., Cattaneo, R., and Jonsson, A. B. (2003) CD46 in meningococcal disease. *Science* **301,** 373–375.

Jolley, K. A., Brehony, C., and Maiden, M. C. (2007) Molecular typing of meningococci: Recommendations for target choice and nomenclature. *FEMS Microbiol Rev* **31,** 89–96.

Jolley, K. A. and Maiden, M. C. (2006) AgdbNet—Antigen sequence database software for bacterial typing. *BMC Bioinformatics* **7,** 314.

Jolley, K. A., Wilson, D. J., Kriz, P., McVean, G., and Maiden, M. C. (2005) The influence of mutation, recombination, population history, and selection on patterns of genetic diversity in *Neisseria meningitidis*. *Mol Biol Evol* **22,** 562–569.

Kahler, C. M., Blum, E., Miller, Y. K., Ryan, D., Popovic, T., and Stephens, D. S. (2001) exl, an exchangeable genetic island in *Neisseria meningitidis*. *Infect Immun* **69,** 1687–1696.

Kim, J. J., Mandrell, R. E., and Griffiss, J. M. (1989) *Neisseria lactamica* and *Neisseria meningitidis* share lipooligosaccharide epitopes but lack common capsular and class 1, 2, and 3 protein epitopes. *Infect Immun* **57,** 602–608.

Koomey, M. (1998) Competence for natural transformation in *Neisseria gonorrhoeae*: A model system for studies of horizontal gene transfer. *APMIS Suppl* **84,** 56–61.

Kremastinou, J., Tzanakaki, G., Pagalis, A., Theodondou, M., Weir, D. M., and Blackwell, C. C. (1999) Detection of IgG and IgM to meningococcal outer membrane proteins in relation to carriage of *Neisseria meningitidis* or *Neisseria lactamica*. *FEMS Immunol Med Microbiol* **24,** 73–78.

Krizova, P. and Musilek, M. (1995) Changing epidemiology of meningococcal invasive disease in the Czech republic caused by new clone *Neisseria meningitidis* C:2a: P1.2(P1.5), ET-15/37. *Cent Eur J Public Health* **3,** 189–194.

Kroll, J. S., Wilks, K. E., Farrant, J. L., and Langford, P. R. (1998) Natural genetic exchange between *Haemophilus* and *Neisseria*: Intergeneric transfer of chromosomal genes between major human pathogens. *Proc Natl Acad Sci U S A* **95,** 12381–12385.

Kwiatek, A. and Piekarowicz, A. (2007) The restriction endonuclease R.NmeDI from *Neisseria meningitidis* that recognizes a palindromic sequence and cuts the DNA on both sides of the recognition sequence. *Nucleic Acids Res* **35,** 6539–6546.

LaForce, F. M., Konde, K., Viviani, S., and Preziosi, M. P. (2007) The Meningitis Vaccine Project. *Vaccine* **25**(Suppl 1)**,** A97–A100.

Leimkugel, J., Hodgson, A., Forgor, A. A., Pfluger, V., Dangy, J. P., Smith, T., Achtman, M., Gagneux, S., and Pluschke, G. (2007) Clonal waves of *Neisseria* colonisation and disease in the African meningitis belt: Eight-year longitudinal study in northern Ghana. *PLoS Med* **4,** e101.

LINZ, B., SCHENKER, M., ZHU, P., and ACHTMAN, M. (2000) Frequent interspecific genetic exchange between commensal Neisseriae and *Neisseria meningitidis*. *Mol Microbiol* **36,** 1049–1058.

LOMHOLT, H., POULSEN, K., CAUGANT, D. A. and KILIAN, M. (1992) Molecular polymorphism and epidemiology of *Neisseria meningitidis* immunoglobulin A1 proteases. *Proc Natl Acad Sci U S A* **89,** 2120–2124.

LOMHOLT, H., POULSEN, K., and KILIAN, M. (1995) Comparative characterization of the iga gene encoding IgA1 protease in *Neisseria meningitidis, Neisseria gonorrhoeae* and *Haemophilus influenzae*. *Mol Microbiol* **15,** 495–506.

LUIJKX, T. A., VAN, D. H., HAMSTRA, H. J., KUIPERS, B., VAN DER, L. P., VAN, A. L., and VAN DEN, D. G. (2003) Relative immunogenicity of PorA subtypes in a multivalent *Neisseria meningitidis* vaccine is not dependent on presentation form. *Infect Immun* **71,** 6367–6371.

LUJAN, R., ZHANG, Q. Y., SAEZ NIETO, J. A., JONES, D. M., and SPRATT, B. G. (1991) Penicillin-resistant isolates of *Neisseria lactamica* produce altered forms of penicillin-binding protein 2 that arose by interspecies horizontal gene transfer. *Antimicrob Agents Chemother* **35,** 300–304.

MACLENNAN, J., KAFATOS, G., NEAL, K., ANDREWS, N., CAMERON, J. C., ROBERTS, R., EVANS, M. R., CANN, K., BAXTER, D. N., MAIDEN, M. C., and STUART, J. M. (2006) Social behavior and meningococcal carriage in British teenagers. *Emerg Infect Dis* **12,** 950–957.

MAIDEN, M. C. (2006) Multilocus sequence typing of bacteria. *Annu Rev Microbiol* **60,** 561–588.

MAIDEN, M. C. (2008) Population genomics: Diversity and virulence in the *Neisseria*. *Curr Opin Microbiol* **11,** 467–471.

MAIDEN, M. C., BYGRAVES, J. A., FEIL, E., MORELLI, G., RUSSELL, J. E., URWIN, R., ZHANG, Q., ZHOU, J., ZURTH, K., CAUGANT, D. A., FEAVERS, I. M., ACHTMAN, M., and SPRATT, B. G. (1998) Multilocus sequence typing: A portable approach to the identification of clones within populations of pathogenic microorganisms. *Proc Natl Acad Sci U S A* **95,** 3140–3145.

MAIDEN, M. C., IBARZ-PAVON, A. B., URWIN, R., GRAY, S. J., ANDREWS, N. J., CLARKE, S. C., WALKER, A. M., EVANS, M. R., KROLL, J. S., NEAL, K. R., ALA'ALDEEN, D. A., CROOK, D. W., CANN, K., HARRISON, S., CUNNINGHAM, R., BAXTER, D., KACZMARSKI, E., MACLENNAN, J., CAMERON, J. C., and STUART, J. M. (2008) Impact of meningococcal serogroup C conjugate vaccines on carriage and herd immunity. *J Infect Dis* **197,** 737–743.

MAIDEN, M. C., MALORNY, B., and ACHTMAN, M. (1996) A global gene pool in the Neisseriae [letter]. *Mol Microbiol* **21,** 1297–1298.

MAIDEN, M. C. and SPRATT, B. G. (1999) Meningococcal conjugate vaccines: New opportunities and new challenges. *Lancet* **354,** 615–616.

MAIDEN, M. C. and STUART, J. M. (2002) Carriage of serogroup C meningococci 1 year after meningococcal C conjugate polysaccharide vaccination. *Lancet* **359,** 1829–1831.

MCNEIL, L. K., MURPHY, E., ZHAO, X. J., GUTTMANN, S., HARRIS, S., SCOTT, A., TAN, C., MACK, M., DASILVA, I., ALEXANDER, K., JIANG, H. Q., ZHU, D., MININNI, T., ZLOTNICK, G. W., HOISETH, S. K., JONES, T. R., PRIDE, M., JANSEN, K. U., and ANDERSON, A. (2009) Detection of LP2086 on the cell surface of *Neisseria meningitidis* and its accessibility in the presence of serogroup B capsular polysaccharide. *Vaccine* **27,** 3417–3421.

MEYERS, L. A., LEVIN, B. R., RICHARDSON, A. R., and STOJILJKOVIC, I. (2003) Epidemiology, hypermutation, within-host evolution and the virulence of *Neisseria meningitidis*. *Proc Biol Sci* **270,** 1667–1677.

MORELLI, G., MALORNY, B., MULLER, K., SEILER, A., WANG, J. F., del VALLE, J., and ACHTMAN, M. (1997) Clonal descent and microevolution of *Neisseria meningitidis* during 30 years of epidemic spread. *Mol Microbiol* **25,** 1047–1064.

MOXON, E. R. and JANSEN, V. A. (2005) Phage variation: Understanding the behaviour of an accidental pathogen. *Trends Microbiol* **13,** 563–565.

MOXON, R., BAYLISS, C., and HOOD, D. (2006) Bacterial contingency loci: The role of simple sequence DNA repeats in bacterial adaptation. *Annu Rev Genet* **40,** 307–333.

MUELLER, J. E., YARO, S., MADEC, Y., SOMDA, P. K., IDOHOU, R. S., LAFOURCADE, B. M., DRABO, A., TARNAGDA, Z., SANGARE, L., TRAORE, Y., FONTANET, A., and GESSNER, B. D. (2008) Association of respiratory tract infection symptoms and air humidity with meningococcal carriage in Burkina Faso. *Trop Med Int Health* **13,** 1543–1552.

NAIR, D., DAWAR, R., DEB, M., CAPOOR, M. R., SINGAL, S., UPADHAYAY, D. J., AGGARWAL, P., DAS, B., and SAMANTARAY, J. C. (2008) Outbreak of meningococcal disease in and around New Delhi, India, 2005–2006: A report from a tertiary care hospital. *Epidemiol Infect* **137,** 1–7.

OPPERMANN, H., THRIENE, B., IRMSCHER, H. M., GRAFE, L., BORRMANN, M., BELLSTEDT, D., KAYNAK, S., HELLENBRAND, W., and VOGEL, U. (2006) Meningokokken-Trägerstatus von Gymnasiasten und mögliche Risikofaktoren. *Gesundheitswesen* **68,** 633–637.

PARKHILL, J., ACHTMAN, M., JAMES, K. D., BENTLEY, S. D., CHURCHER, C., KLEE, S. R., MORELLI, G., BASHAM, D., BROWN, D., CHILLINGWORTH, T., DAVIES, R. M., DAVIS, P., DEVLIN, K., FELTWELL, T., HAMLIN, N., HOLROYD, S., JAGELS, K., LEATHER, S., MOULE, S., MUNGALL, K., QUAIL, M. A., RAJANDREAM, M. A., RUTHERFORD, K. M., SIMMONDS, M., SKELTON, J., WHITEHEAD, S., SPRATT, B. G., BARRELL, B. G. (2000) TI Complete DNA sequence of a serogroup A strain of *Neisseria meningitidis* Z2491. *Nature* **404,** 502–506.

PENG, J., YANG, L., YANG, F., YANG, J., YAN, Y., NIE, H., ZHANG, X., XIONG, Z., JIANG, Y., CHENG, F., XU, X., CHEN, S., SUN, L., LI, W., SHEN, Y., SHAO, Z., LIANG, X., XU, J., and JIN, Q. (2008) Characterization of ST-4821 complex, a unique *Neisseria meningitidis* clone. *Genomics* **91,** 78–87.

Ramsay, M. E., Andrews, N. J., Trotter, C. L., Kaczmarski, E. B., and Miller, E. (2003) Herd immunity from meningococcal serogroup C conjugate vaccination in England: Database analysis. *BMJ* **326,** 365–366.

Rappuoli, R. (2007) Bridging the knowledge gaps in vaccine design. *Nat Biotechnol* **25,** 1361–1366.

Reinhardt, M., Elias, J., Albert, J., Frosch, M., Harmsen, D., and Vogel, U. (2008) EpiScanGIS: An online geographic surveillance system for meningococcal disease. *Int J Health Geogr* **7,** 33.

Riou, J. Y., Buissiere, J., Richard, C., and Guibourdenche, M. (1982) Intérêt de la recherche de la gamma-glutamyl transferase chez les *Neisseriaceae*. *Ann Microbiol (Paris)* **133,** 387–392.

Rosenstein, N. E., Perkins, B. A., Stephens, D. S., Popovic, T., and Hughes, J. M. (2001) Meningococcal disease. *N Engl J Med* **344,** 1378–1388.

Russell, J. E., Jolley, K. A., Feavers, I. M., Maiden, M. C., and Suker, J. (2004) PorA variable regions of *Neisseria meningitidis*. *Emerg Infect Dis* **10,** 674–678.

Sarafian, S. K. and Knapp, J. S. (1989) Molecular epidemiology of gonorrhea. *Clin Microbiol Rev* **2**(Suppl), S49–S55.

Saunders, N. J., Jeffries, A. C., Peden, J. F., Hood, D. W., Tettelin, H., Rappuoli, R., and Moxon, E. R. (2000) Repeat-associated phase variable genes in the complete genome sequence of *Neisseria meningitidis* strain MC58. *Mol Microbiol* **37,** 207–215.

Schoen, C., Blom, J., Claus, H., Schramm-Gluck, A., Brandt, P., Muller, T., Goesmann, A., Joseph, B., Konietzny, S., Kurzai, O., Schmitt, C., Friedrich, T., Linke, B., Vogel, U., and Frosch, M. (2008) Whole-genome comparison of disease and carriage strains provides insights into virulence evolution in *Neisseria meningitidis*. *Proc Natl Acad Sci U S A* **105,** 3473–3478.

Schoen, C., Joseph, B., Claus, H., Vogel, U., and Frosch, M. (2007) Living in a changing environment: Insights into host adaptation in *Neisseria meningitidis* from comparative genomics. *Int J Med Microbiol* **297,** 601–613.

Schoen, C., Tettelin, H., Parkhill, J., and Frosch, M. (2009) Genome flexibility in *Neisseria meningitidis*. **27** (Suppl 2)**,** B103–11.

Seiler, A., Reinhardt, R., Sarkari, J., Caugant, D. A., and Achtman, M. (1996) Allelic polymorphism and site-specific recombination in the opc locus of *Neisseria meningitidis*. *Mol Microbiol* **19,** 841–856.

Selander, R. K., Caugant, D. A., Ochman, H., Musser, J. M., Gilmour, M. N., and Whittam, T. S. (1986) Methods of multilocus enzyme electrophoresis for bacterial population genetics and systematics. *Appl Environ Microbiol* **51,** 873–884.

Skoczynska, A., Alonso, J. M. and Taha, M. K. (2008) Ciprofloxacin resistance in *Neisseria meningitidis*, France. *Emerg Infect Dis* **14,** 1322–1323.

Slaterus, K. W. (1961) Serological typing of meningococci by means of micro-precipitation. *Antonie Van Leeuwenhoek* **27,** 305–315.

Smith, J. M., Smith, N. H., O'Rourke, M., and Spratt, B. G. (1993) How clonal are bacteria? *Proc Natl Acad Sci U S A* **90,** 4384–4388.

Smith, N. H., Holmes, E. C., Donovan, G. M., Carpenter, G. A., and Spratt, B. G. (1999) Networks and groups within the genus *Neisseria*: Analysis of argF, recA, rho, and 16S rRNA sequences from human *Neisseria* species. *Mol Biol Evol* **16,** 773–783.

Snyder, L. A., Butcher, S. A., and Saunders, N. J. (2001) Comparative whole-genome analyses reveal over 100 putative phase-variable genes in the pathogenic *Neisseria* spp. *Microbiology* **147,** 2321–2332.

Snyder, L. A., McGowan, S., Rogers, M., Duro, E., O'Farrell, E., and Saunders, N. J. (2007) The repertoire of minimal mobile elements in the *Neisseria* species and evidence that these are involved in horizontal gene transfer in other bacteria. *Mol Biol Evol* **24,** 2802–2815.

Spratt, B. G., Bowler, L. D., Zhang, Q. Y., Zhou, J., and Smith, J. M. (1992) Role of interspecies transfer of chromosomal genes in the evolution of penicillin resistance in pathogenic and commensal *Neisseria* species. *J Mol Evol* **34,** 115–125.

Stollenwerk, N., Maiden, M. C., and Jansen, V. A. (2004) Diversity in pathogenicity can cause outbreaks of meningococcal disease. *Proc Natl Acad Sci U S A* **101,** 10229–10234.

Suker, J., Feavers, I. M., Achtman, M., Morelli, G., Wang, J. F., and Maiden, M. C. (1994) The porA gene in serogroup A meningococci: Evolutionary stability and mechanism of genetic variation. *Mol Microbiol* **12,** 253–265.

Swartley, J. S., Marfin, A. A., Edupuganti, S., Liu, L. J., Cieslak, P., Perkins, B., Wenger, J. D., and Stephens, D. S. (1997) Capsule switching of *Neisseria meningitidis*. *Proc Natl Acad Sci U S A* **94,** 271–276.

Taha, M. K., Vazquez, J. A., Hong, E., Bennett, D. E., Bertrand, S., Bukovski, S., Cafferkey, M. T., Carion, F., Christensen, J. J., Diggle, M., Edwards, G., Enriquez, R., Fazio, C., Frosch, M., Heuberger, S., Hoffmann, S., Jolley, K. A., Kadlubowski, M., Kechrid, A., Kesanopoulos, K., Kriz, P., Lambertsen, L., Levenet, I., Musilek, M., Paragi, M., Saguer, A., Skoczynska, A., Stefanelli, P., Thulin, S., Tzanakaki, G., Unemo, M., Vogel, U., and Zarantonelli, M. L. (2007) Target gene sequencing to characterize the penicillin G susceptibility of *Neisseria meningitidis*. *Antimicrob Agents Chemother* **51,** 2784–2792.

Tettelin, H., Saunders, N. J., Heidelberg, J., Jeffries, A. C., Nelson, K. E., Eisen, J. A., Ketchum, K. A., Hood, D. W., Peden, J. F., Dodson, R. J., Nelson, W. C., Gwinn, M. L., DeBoy, R., Peterson, J. D., Hickey, E. K., Haft, D. H., Salzberg, S. L., White, O., Fleischmann, R. D., Dougherty, B. A., Mason, T., Ciecko, A., Parksey, D. S., Blair, E., Cittone, H., Clark, E. B., Cotton, M. D., Utterback, T.R., Khouri, H., Qin, H., Vamathevan, J., Gill, J., Scarlato, V., Masignani, V., Pizza, M., Grandi, G.,

SUN, L., SMITH, H. O., FRASER, C. M., MOXON, E. R., RAPPUOLI, R., and VENTER, J. C. (2000) Complete genome sequence of *Neisseria meningitidis* serogroup B strain MC58. *Science* **287**, 1809–1815.

THOMPSON, E. A., FEAVERS, I. M., and MAIDEN, M. C. (2003) Antigenic diversity of meningococcal enterobactin receptor FetA, a vaccine component. *Microbiology* **149,** 1849–1858.

TROTTER, C. L., GAY, N. J., and EDMUNDS, W. J. (2005) Dynamic models of meningococcal carriage, disease, and the impact of serogroup C conjugate vaccination. *Am J Epidemiol* **162,** 89–100.

TURNER, K. M. and FEIL, E. J. (2007) The secret life of the multilocus sequence type. *Int J Antimicrob Agents* **29,** 129–135.

URWIN, R., RUSSELL, J. E., THOMPSON, E. A., HOLMES, E. C., FEAVERS, I. M., and MAIDEN, M. C. (2004) Distribution of surface protein variants among hyperinvasive meningococci: Implications for vaccine design. *Infect Immun* **72,** 5955–5962.

VAN DEN DOBBELSTEEN, G. P., VAN DIJKEN, H. H., PILLAI, S., and VAN, A. L. (2007) Immunogenicity of a combination vaccine containing pneumococcal conjugates and meningococcal PorA OMVs. *Vaccine* **25,** 2491–2496.

VAN DER ENDE, A., HOPMAN, C. T., KEIJZERS, W. C., SPANJAARD, L., LODDER, E. B., VAN KEULEN, P. H., and DANKERT, J. (2003) Outbreak of meningococcal disease caused by PorA-deficient meningococci. *J Infect Dis* **187,** 869–871.

VAZQUEZ, J. A., BERRON, S., O'ROURKE, M., CARPENTER, G., FEIL, E., SMITH, N. H., and SPRATT, B. G. (1995) Interspecies recombination in nature: A meningococcus that has acquired a gonococcal PIB porin. *Mol Microbiol* **15,** 1001–1007.

VAZQUEZ, J. A., DE LA, F. L., BERRON, S., O'ROURKE, M., SMITH, N. H., ZHOU, J., and SPRATT, B. G. (1993) Ecological separation and genetic isolation of *Neisseria gonorrhoeae* and *Neisseria meningitidis*. *Curr Biol* **3,** 567–572.

VAZQUEZ, J. A., ENRIQUEZ, R., ABAD, R., ALCALA, B., SALCEDO, C., and ARREAZA, L. (2007) Antibiotic resistant meningococci in Europe: Any need to act? *FEMS Microbiol Rev* **31,** 64–70.

VOGEL, U. and CLAUS, H. (2004) Genetic lineages and their traits in Neisseria meningitidis. *Int J Med Microbiol* **294,** 75–82.

VOGEL, U., CLAUS, H., and FROSCH, M. (2000) Rapid serogroup switching in *Neisseria meningitidis*. *N Engl J Med* **342,** 219–220.

VOGEL, U. and FROSCH, M. (1999) Infant rat model of acute meningitis. In *Handbook of Animal Models of Infection* (eds. O. Zak, and M. Sande), pp. 619–626. Academic Press, London.

VOGEL, U., HAMMERSCHMIDT, S., and FROSCH, M. (1996) Sialic acids of both the capsule and the sialylated lipooligosaccharide of *Neisseria meningitidis* serogroup B are prerequisites for virulence of meningococci in the infant rat. *Med Microbiol Immunol* **185,** 81–87.

WEBER, M. V., CLAUS, H., MAIDEN, M. C., FROSCH, M., and VOGEL, U. (2006) Genetic mechanisms for loss of encapsulation in polysialyltransferase-gene-positive meningococci isolated from healthy carriers. *Int J Med Microbiol* **296,** 475–484.

WU, H. M., HARCOURT, B. H., HATCHER, C. P., WEI, S. C., NOVAK, R. T., WANG, X., JUNI, B. A., GLENNEN, A., BOXRUD, D. J., RAINBOW, J., SCHMINK, S., MAIR, R. D., THEODORE, M. J., SANDER, M. A., MILLER, T. K., KRUGER, K., COHN, A. C., CLARK, T. A., MESSONNIER, N. E., MAYER, L. W., and LYNFIELD, R. (2009) Emergence of ciprofloxacin-resistant *Neisseria meningitidis* in North America. *N Engl J Med* **360,** 886–892.

YARO, S., TRAORE, Y., TARNAGDA, Z., SANGARE, L., NJANPOPLAFOURCADE, B. M., DRABO, A., FINDLOW, H., BORROW, R., NICOLAS, P., GESSNER, B. D., and MUELLER, J. E. (2007) Meningococcal carriage and immunity in western Burkina Faso, 2003. *Vaccine* **25S1,** A42–A46.

YAZDANKHAH, S. P., KRIZ, P., TZANAKAKI, G., KREMASTINOU, J., KALMUSOVA, J., MUSILEK, M., ALVESTAD, T., JOLLEY, K. A., WILSON, D. J., MCCARTHY, N. D., CAUGANT, D. A., and MAIDEN, M. C. (2004) Distribution of serogroups and genotypes among disease-associated and carried isolates of *Neisseria meningitidis* from the Czech Republic, Greece, and Norway. *J Clin Microbiol* **42,** 5146–5153.

ZAPATA, G. A., VANN, W. F., RUBINSTEIN, Y., and FRASCH, C. E. (1992) Identification of variable region differences in *Neisseria meningitidis* class 3 protein sequences among five group B serotypes. *Mol Microbiol* **6,** 3493–3499.

ZARANTONELLI, M. L., SZATANIK, M., GIORGINI, D., HONG, E., HUERRE, M., GUILLOU, F., ALONSO, J. M., and TAHA, M. K. (2007) Transgenic mice expressing human transferrin as a model for meningococcal infection. *Infect Immun* **75,** 5609–5614.

ZHOU, J., BOWLER, L. D., and SPRATT, B. G. (1997) Interspecies recombination, and phylogenetic distortions, within the glutamine synthetase and shikimate dehydrogenase genes of *Neisseria meningitidis* and commensal *Neisseria* species. *Mol Microbiol* **23,** 799–812.

ZHOU, J. and SPRATT, B. G. (1992) Sequence diversity within the argF, fbp and recA genes of natural isolates of *Neisseria meningitidis*: Interspecies recombination within the argF gene. *Mol Microbiol* **6,** 2135–2146.

ZHU, P., MORELLI, G., and ACHTMAN, M. (1999) The opcA and (psi)opcB regions in *Neisseria*: Genes, pseudogenes, deletions, insertion elements and DNA islands. *Mol Microbiol* **33,** 635–650.

ZHU, P., VAN DER, E. A., FALUSH, D., BRIESKE, N., MORELLI, G., LINZ, B., POPOVIC, T., SCHUURMAN, I. G., ADEGBOLA, R. A., ZURTH, K., GAGNEUX, S., PLATONOV, A. E., RIOU, J. Y., CAUGANT, D. A., NICOLAS, P., and ACHTMAN, M. (2001) Fit genotypes and escape variants of subgroup III *Neisseria meningitidis* during three pandemics of epidemic meningitis. *Proc Natl Acad Sci U S A* **98,** 5234–5239.

Chapter 14

Population Genetics of Pathogenic *Escherichia coli*

ERICK DENAMUR, BERTRAND PICARD, AND OLIVIER TENAILLON

14.1 INTRODUCTION

Escherichia coli is one of the best-studied model organisms. A considerable amount of knowledge in terms of genetics, molecular biology, physiology, and biochemistry has been accumulated on K-12, a unique human commensal strain isolated in 1922 in Palo Alto. However, K-12 and derivative strains are far from representing the huge diversity of the species encountered in the wild (Hobman et al., 2007).

In nature, the total *E.coli* population has been estimated to 10^{20} (Whitman et al., 1998). *E. coli* strains alternate between their primary habitat, the gut of vertebrates, and their secondary habitat, water and sediments, where they are excreted from the primary habitat (Savageau, 1983). But *E. coli* is also a devastating versatile pathogen being involved in a large range of pathologies, from intestinal to extraintestinal diseases (Kaper et al., 2004). As such, it is a major cause of human morbidity and mortality around the world, entering the inglorious top five of the most damaging infectious agents. Each year, *E. coli* causes more than two million deaths due to infant diarrheas and extraintestinal infections (mainly septicemia derived from urinary tract infection [UTI]) (Kosek et al., 2003; Russo and Johnson, 2003). Moreover, each year, *E. coli* is also responsible for approximately 150 million cases of uncomplicated cystitis (Russo and Johnson, 2003). It is acting as an opportunistic, facultative pathogen both in humans and in domesticated animals, but also as a human-specific obligate pathogen (enteroinvasive *E. coli* [EIEC] and *Shigella*). As we will see, *Shigella*, which have been elevated to the genus order with four species (*dysenteriae*, *flexneri*, *boydii*, and *sonnei*) based on their capacity to generate a specific mucosal invasive diarrhea and their biochemical characteristics, in fact belong to the *E. coli* species. Other pathotypes that result in diarrheal diseases are enteropathogenic *E. coli* (EPEC), enterohemorrhagic *E. coli* (EHEC), enterotoxinogenic *E. coli* (ETEC), enteroaggregative *E. coli* (EAEC), and diffusely adherent *E. coli* (DAEC) (Kaper et al., 2004). On the other hand, extraintestinal *E. coli* (ExPEC) (Russo and Johnson, 2000) are responsible for UTIs, sepsis, and newborn meningitis. At the cellular level, this dichotomy between facultative and obligate pathogens is reflected by the fact that facultative

Bacterial Population Genetics in Infectious Disease, Edited by D. Ashley Robinson, Daniel Falush, and Edward J. Feil

pathogenic *E. coli* strains remain extracellular, whereas *Shigella* and EIEC are true intracellular pathogens that can replicate within epithelial cells and macrophages (Sansonetti et al., 1999).

Due to its diversity of niches, large population size, and various lifestyles (pathogen and commensal), *E. coli* has always been used as a model in population genetics studies, benefiting from each conceptual and/or technical advance. Population genetics studies of this versatile species are of high interest as they allow the understanding of the emergence of virulence at a population scale. They are complementary of the studies that decipher the molecular mechanisms of virulence.

14.2 *E. COLI* POPULATION GENETICS: CLONAL OR NOT CLONAL?

14.2.1 The Extent of Recombination

The genetic structure of the *E. coli* population has now been scrutinized for several decades. The early analyses using serotypes or multilocus enzyme electrophoresis (MLEE) have revealed a clear clonal structure with strong linkage disequilibrium among the different protein or antigenic loci studied, but no geographical structure. Yet, with the first DNA sequence data of individual genes in the 1980s, proofs of recombination at the molecular level accumulated (Guttman and Dykhuizen, 1994). As early as in 1983, Milkman and Crawford (1983) pointed to clustered base substitutions in the *trp* gene operon that they interpreted as possible recombination events. Incongruence between trees reconstructed from individual genes and the species phylogeny evaluated as a consensus tree of the genes or the MLEE tree (DuBose et al., 1988; Bisercic et al., 1991; Dykhuizen and Green, 1991; Milkman and Bridges, 1993) appeared as further proof of recombination at the gene level.

How can multiloci analysis reveal a clear clonal structure while traces of recombination are found in most genes? Indeed the answer lies both in the ratio of recombination to mutation and in the length of the DNA recombined. Experimental studies suggested that even if the DNA is entering the cell in large fragments, due to restriction, only short fragments will be recombined. Moreover, a superimposition of replacements of a quite large size generates a mosaic of shorter fragments (Milkman and Bridges, 1993). In agreement with those experiments, coalescent simulation coupled with approximate Bayesian calculation applied to more than 1900 conserved genes in 20 genomes revealed that recombination was twice more frequent than mutation but involved very short fragments (Touchon et al., 2009). Further analysis revealed that even if a base is 100 times more likely to be involved in a recombination event than to be mutated, an appropriate phylogenic structure of the population can be recovered through classical phylogenetic analysis, provided enough loci are sampled. The conclusion is that, despite a higher recombination rate than a mutation rate, the mode of recombination (involving short fragments) is compatible with an apparent clonal population structure and a clear phylogenetic signal (Touchon et al., 2009).

14.2.2 The Impact of Recombination on Population Genetics Inference

If the quasi-clonal structure of the species is now better framed, the existence of recombination still has a strong impact, such that methods applicable to asexual species cannot be applied directly. Indeed, recombination involving short fragments (Schierup and Hein, 2000) has a very large impact on branch length, so that any quantitative modeling made on phylogeny-derived branch length should be avoided. The impact of recombination on branch length is multiple: Whenever a fragment is transferred between two distant clones, it decreases their genetic distance and simultaneously increases the distance between the recipient strain and its closely related strains. As a result, terminal branches are much longer than expected and internal branches are much shorter. Such a pattern is frequently analyzed as a proof of population expansion (Wirth et al., 2006); however, population expansion is linked to an excess of rare alleles (negative Tajima's *D*), while recombination is not. When looking at the *E. coli* species phylogeny, one found this exact same pattern: extreme terminal branch length and short internal branch length, and this occurred whether all sites or only synonymous sites were analyzed. Yet, there was no excess of rare alleles on synonymous sites (negative Tajima's *D*), suggesting that recombination is responsible for this pattern. Hence, we doubt that any signal of population expansion can be inferred from *E. coli* sequence data, without serious corrections to take into account the effect of recombination. Moreover, as old polymorphisms are more likely to have been involved in some recombination, we doubt that subtracting presumably recombined sites from the analysis will generate appropriate branch length; the informative sites supporting internal branches will be much more filtered than recent sites generating the same pattern of long external branches and short internal ones.

Similarly, the quasi-clonal structure of the population suggests that pure population genetic methods of analysis, assuming strong recombination and no linkage between sites, should be taken with care, as shown in *Salmonella* (Falush et al., 2006). For instance, the identification of subgroups is highly sensitive to sampling and the topology of the species tree. Hence, when population structure software was used to assign *E. coli* strains into four groups according to their MLST sequences, many strains appeared as recombinant, expressing alleles of several groups (Wirth et al., 2006). Indeed, more recent studies using additional data have shown that, in agreement with the phylogenetic analysis, more than four groups were indeed required to analyze the species (Gordon et al., 2008; Jaureguy et al., 2008). The new groups (C, E, and F; see below) (Jaureguy et al., 2008) as well as the *Shigella* groups (Pupo et al., 2000; Escobar-Paramo et al., 2003) carried most of the strains characterized as recombinant in the previous analysis (Wirth et al., 2006), suggesting that this recombinant status is model dependent. We therefore are quite skeptical about the presumed link between recombination and virulence in *E. coli* that was suggested using those methods (Wirth et al., 2006). More studies are required, with unbiased sampling and analysis taking care of *E. coli* quasi-clonal structure, before strains can be assessed to be recombinant based on population genetics analysis.

14.2.3 Genome Organization and Recombination

Already in pre-genomic era it appeared that recombination varied among genes, some genes showing no trace of recombination as *gapA* (Nelson et al., 1991), *celC*, *crr*, and *gutB* (Hall and Sharp, 1992), others being highly recombined as *gnd* (Bisercic et al., 1991). This high level of recombination of *gnd* is in fact related to its close proximity with the *rfb* locus, which is under diversifying selection, and results in a "bastion of polymorphism" (Milkman et al., 2003). Another bastion of polymorphism also under diversifying selection is located at the *hsd* locus coding for the type I restriction and modification systems that enable bacteria to distinguish "foreign" DNA from their own (Barcus et al., 1995). As summarized by Milkman, the recombined parts of genes correspond to the "clonal segments" (Milkman and Stoltzfus, 1988), whereas the species phylogeny can be assimilated to the "clonal frame" (Milkman and Bridges, 1990).

In agreement with the previous gene analyses, a genome-wide analysis of recombination revealed that variations occurred along the genome. A large-scale pattern revealed that recombination was lower around the terminus (Ter) macrodomain, centered on the terminus of replication (Touchon et al., 2009). This presumably results from the fewer copies of the Ter macrodomain within the cell and from its aggregation mediated by the MatP/*matS* association (Mercier et al., 2008). At a smaller scale, using phylogenetic analysis, the two major bastions of polymorphism were found back: the *rfb* locus and the region in which both restriction system and mannose-sensitive adhesion system (Fim) are found (Touchon et al., 2009). Some other less important hot spots of recombination were identified and could be for most of them associated with locus under diversifying selection: outer membrane proteins, mutation rate control, and phage resistance systems (Touchon et al., 2009). This suggests that at those loci, it is the association between recombination and selection that leaves some traces at the genomic scale. The more intense the selection on the new allele transferred, the larger the genomic region that will be affected. A highly selected recombinant will invade the population before any further recombination will alter the signal of the initial recombination event.

14.2.4 Genome Plasticity

Genome scale analysis revealed another interesting feature of *E. coli* species: the plasticity of its genome. Logically, the first *E. coli* genome to have been sequenced in 1997 was from the commensal-derived laboratory strain K-12 (Blattner et al., 1997). Up to now, 31 *E. coli/Shigella* genomes are available. As soon as the second *E. coli* genome from an enterohemorrhagic isolate was deciphered in 2001, it appears clearly that over 30% of the genes in the pathogenic strain were unique to that organism, compared to the K-12 strain (Perna et al., 2001), pointing to the importance of another form of recombination: acquisition and loss of genes. The pattern of diversity emerging from these genomic data is an individual *E. coli* strain that has an average of 4721 genes (Hendrickson, 2009), with a core genome (i.e., genes present in all the strains) of approximately 2000 genes with high homology, and part of the pan-genome (i.e., the collection of all genes found in the *E. coli* strains) made approximately of 18,000 genes, including insertion sequence- and prophage-like elements (Rasko et al., 2008; Touchon et al., 2009). Interestingly, the resulting high gene flux made of gain and loss events occurs at precise positions in the genome, the hot spots of integration (Fig. 14.1) (Touchon et al., 2009). Some of these hot spots correspond to tRNA or phage integration hot spots, as described for a long time, but the majority of them have no specific molecular signature to date (Touchon et al., 2009). An example is the genomic island contents at the *pheV* tRNA in 12 *E. coli* strains (Fig. 14.2), illustrating the high plasticity that can be observed at a single hot spot of integration. Finally, a link was found between some hot spots of integration and some hot spots of recombination. It appears that once a new cluster of gene is incorporated within the genome of one strain, if the locus provides any advantage, it can spread through the species thanks to homologous recombination of the conserved flanking parts (Schubert et al., 2009; Touchon et al., 2009).

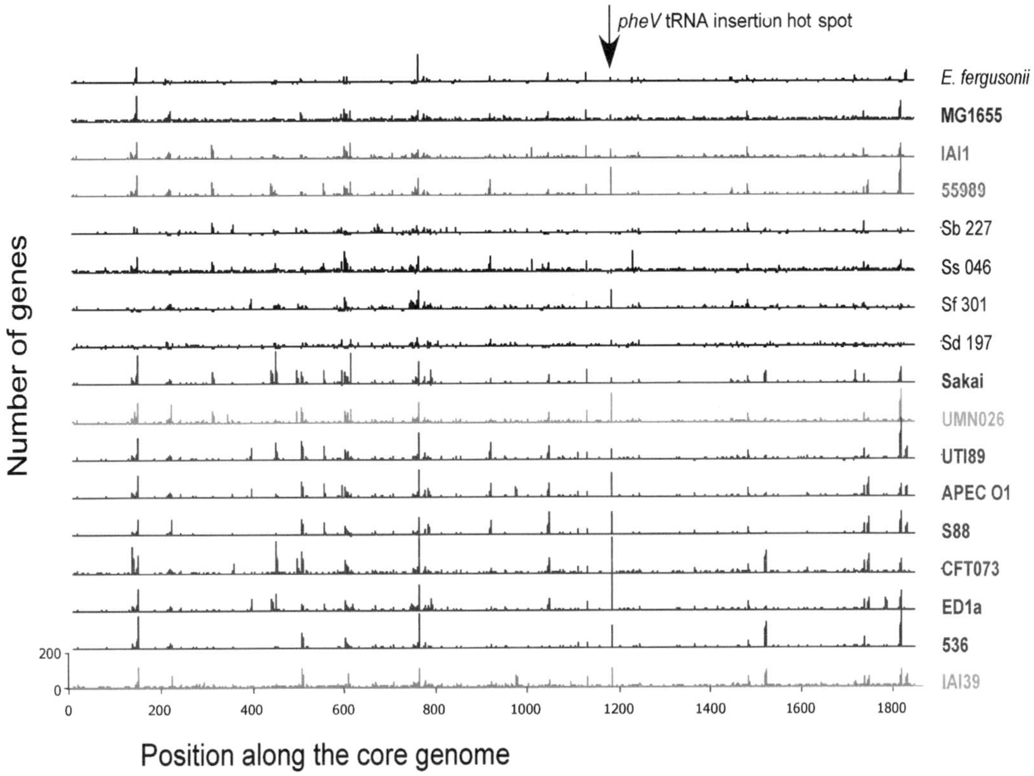

Figure 14.1 Global view of insertion/deletion hot spots on the chromosome of 16 *Escherichia coli/Shigella* and *Escherichia fergusonii*. Number of genes (ranging from 0 to 200) in indels along the genomes according to the ancestral gene order of the core genome (Touchon et al., 2009). The numbers on the *x*-axis represent the order of genes in the core genome, as in *E. coli* K-12 MG1655. Sb 227: *Shigella boydii* 4, Ss 046: *Shigella sonnei*, Sf 301: *Shigella flexneri* 2a, Sd 197: *Shigella dysenteriae* 1. See color insert.

14.3 THE *E. COLI* PHYLOGENETIC STRUCTURE

As we have mentioned before, a clear structure of the species emerges in multilocus analyses; let us now describe it. An MLEE-based phenogram using 38 enzymes including 4 esterases (Selander et al., 1986; Goullet and Picard, 1989) individualized four main groups, A, B1, B2, and D, and two accessory groups, C and E (Selander et al., 1987; Herzer et al., 1990), within the species. The concatenation of individual gene sequences obtained by MLST retrieved these groups whatever the MLST scheme used (Lecointre et al., 1998; Reid et al., 2000; Escobar-Paramo et al., 2004c; Johnson et al., 2006b; Wirth et al., 2006) both by phylogenetic (with or without removal of recombination events) and population genetic (unsupervised population assignment algorithms) approaches. In addition, it allowed a true phylogeny of the species by rooting the tree on an outgroup.

A meaningful and robust tree topology, in agreement with previous MLST studies (Reid et al., 2000; Escobar-Paramo et al., 2004a; Hershberg et al., 2007), was obtained using the 1878 genes of the *Escherichia* core genome and the 2.6 million nucleotides of the *E. coli* chromosomal backbone (Touchon et al., 2009). The use of *Escherichia*

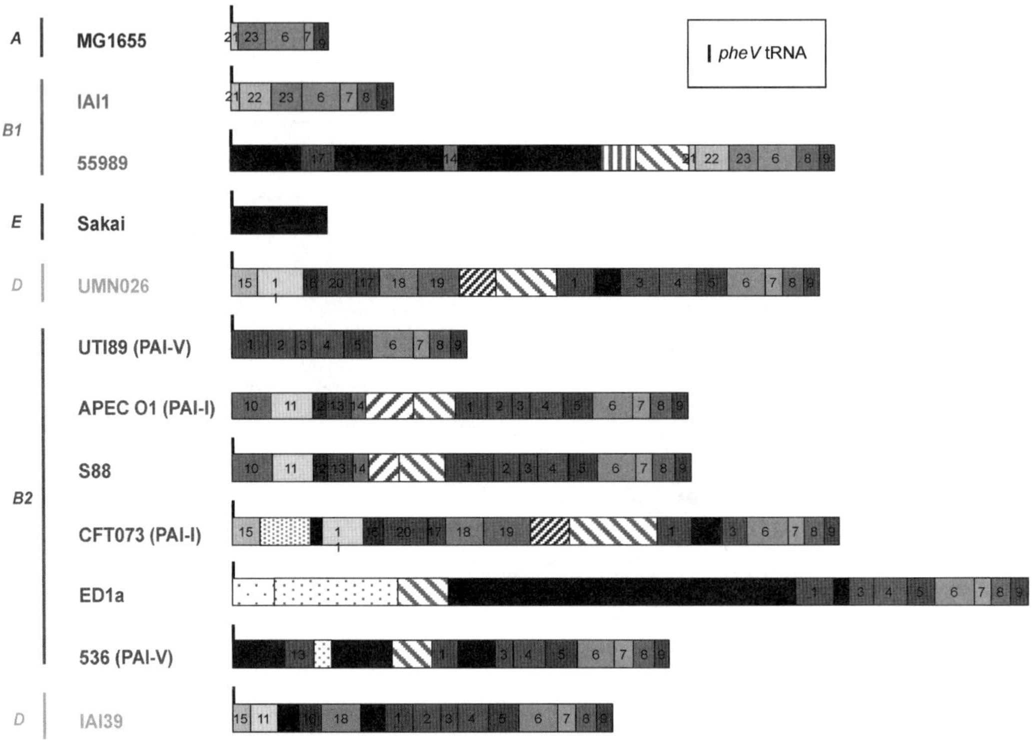

Figure 14.2 The genomic island at the *pheV* tRNA insertion hot spot in 12 different *Escherichia coli* strains. The figure provides a synthetic view of the *pheV* tRNA insertion hot spot in the different *E. coli* strains (Touchon et al., 2009). In strain APEC O1, the *pheV* tRNA gene is absent. This large genomic island has been divided into subregions (or modules), which are found in only a subset of the compared *E. coli* strains. Homologous modules have the same color code and identifying number throughout. A total of 23 homologous modules were defined. Black modules are strain specific. Modules with hatched patterns correspond to repeated regions. Modules with gray dotted patterns are found in other strains but at another genomic location. The pathogenicity island PAI-V in UTI89 and 536 or PAI-I in APEC O1 and CFT073 ends just before module number 6. See color insert.

fergusonii, which is the closest relative of *E. coli* (Lawrence et al., 1991), instead of *Salmonella* as the outgroup avoided the long-branch attraction artifact (Felsenstein, 1978). The first split in the *E. coli* phylogenetic history leads on one hand to the strains of group B2 and a subgroup within D that we called F (Jaureguy et al., 2008), and on the other hand, to the rest of the species (Touchon et al., 2009). The remaining strains of group D then emerge, followed by those of group E. Finally, the A and B1 groups are sister groups (Jaureguy et al., 2008; Touchon et al., 2009) (Fig. 14.3).

When comparing the level of diversity within these phylogenetic groups, both at the nucleotide and gene content levels, the strains of group B2 exhibit the higher level (Touchon et al., 2009). It could correspond to a subspecies (Lescat et al., 2009b) with its own genetic structure, as at least nine phylogenetic subgroups have been observed (Le Gall et al., 2007).

In the 2000s, a rapid, simple, and robust PCR triplex method based on the presence of three genes (*chuA*, *yjaA*, and a gene coding for a putative lipase) allowing the A, B1,

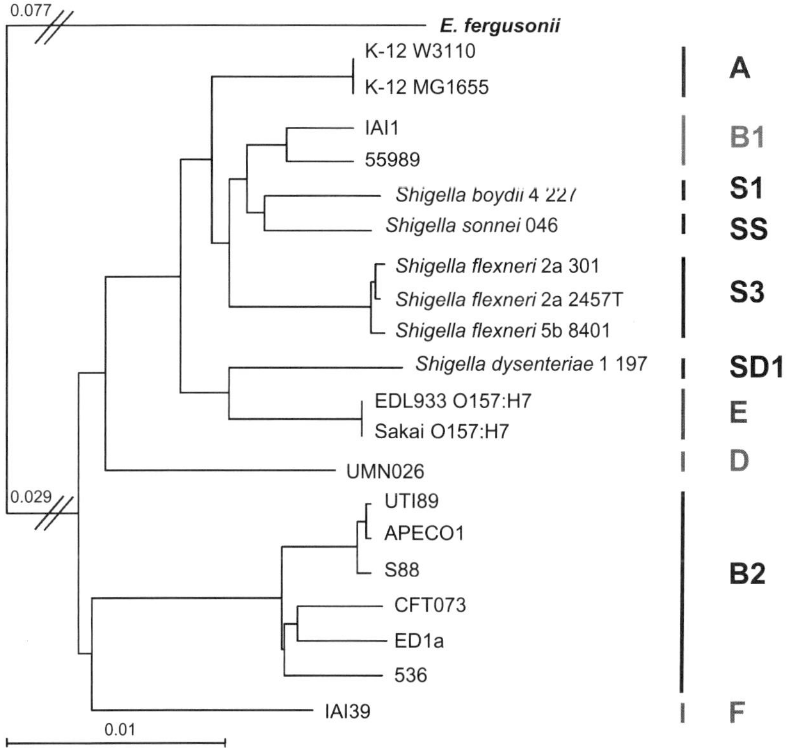

Figure 14.3 Maximum likelihood phylogenetic tree of the 20 *Escherichia coli* and *Shigella* strains as reconstructed from the sequences of the 1878 genes of the *Escherichia* core genome. The earliest diverging species, *E.fergusonii*, was chosen to root the tree (Touchon et al., 2009). The branch length separating *E.fergusonii* from the *E. coli* strains is not to scale; the numbers above the branch indicate its length. Phylogenetic group membership of the strains is indicated with bars at the right of the figure (A, B1, B2, D, E, F and S1, S3, SS, SD1 for *E. coli* and *Shigella*, respectively). SD1 = *S. dysenteriae* serotype 1; SS = *Shigella sonnei*. See color insert.

B2, and D phylogroup determination has become very popular (Clermont et al., 2000). Eighty to eighty-five percent of the phylogroup memberships assigned using this method are correct when compared to MLST data. However, the accuracy with which strains are assigned to the correct phylogroup depends on their genotype. For example, strains yielding a genotype consistent with phylogroups B1 and B2 are assigned correctly 95% of the time, whereas strains failing to yield any PCR products are not well assigned (Gordon et al., 2008). Likewise, strains of group E appear D. However, despite these small drawbacks, this method has allowed a large number of isolates all over the world to be typed by various investigators.

Now that we know that the structure of the *E. coli* species is roughly clonal, with a level of recombination allowing building a robust phylogeny, and that we have efficient molecular tools for *E. coli* typing, we can reconstruct the scenario of the evolution of the virulence within the species. We will keep the dichotomy of obligate and facultative pathogens, as they correspond to distinct pathophysiologies, but also to distinct evolutionary strategies.

14.4 THE EVOLUTIONARY HISTORY OF A HOST-SPECIFIC OBLIGATE PATHOGEN: THE *SHIGELLA* AND EIEC CASE STUDY

Shigella and EIEC strains have a characteristic form of pathogenesis involving invasion of mucosal epithelium cells of the large intestine (Sansonetti et al., 1999). The genes for this invasive property reside on a plasmid called the virulence plasmid (VP) (Buchrieser et al., 2000; Le Gall et al., 2005b). But *Shigella* and EIEC are also characterized by phenotypic properties encoded by chromosomal genes such as the lack of catabolic pathways and mobility.

14.4.1 The Origin of *Shigella* and EIEC and the Acquisition of the VP

MLEE (Ochman et al., 1983; Goullet and Picard, 1987) and *rrn* Restriction Fragment Length Polymorphism (RFLP) (Rolland et al., 1998) studies have clearly shown that, except *Shigella boydii* serotype 13, *Shigella* fall within the *E. coli* species, intermixed with *E. coli* strains. MLST analyses confirmed these data (Pupo et al., 2000; Escobar-Paramo et al., 2003; Wirth et al., 2006; Choi et al., 2007). Three main phylogenetic groups (S1, S2, and S3) and four outliers, containing exclusively *Shigella* strains, were identified. The S1 group encompasses strains of three nomen species: *S. boydii* (serotypes 1, 2, 3, 4, 6, 8, 10, 14, and 18), *Shigella dysenteriae* (serotypes 3, 4, 5, 6, 7, 9, 11, 12, and 13), and *Shigella flexneri* (serotypes 6 and 6a); the S2 group encompasses *S. dysenteriae* serotype 2 and *S. boydii* (serotypes 5, 7, 9, 11, 15, 16, and 17) strains, whereas the S3 group is made almost of *S. flexneri* (serotypes 1a, 1b, 2a, 2b, 3a, 3b, 3c, 4a, 4b, 5, X, and Y) strains, with *S. boydii* serotype 12 strains. The four outliners are composed of *Shigella sonnei* and *S. dysenteriae* serotypes 1, 8, and 10 strains, respectively (Pupo et al., 2000; Escobar-Paramo et al., 2003). Several clusters of EIEC were also identified (Escobar-Paramo et al., 2003; Lan et al., 2004). MLST data (Escobar-Paramo et al., 2003, 2004a), reinforced by phylogenetic analysis of the core genomes (Touchon et al., 2009), showed that *Shigella* and EIEC emerged within the *E. coli* evolutionary history after the split D group/rest of the species (Fig. 14.3).

The most plausible evolutionary scenario for the emergence of *Shigella* is the multiple independent origins of clones (Pupo et al., 2000; Yang et al., 2007). However, when several authors have studied the phylogenetic history of the VP genes, a striking correlation between the phylogenetic groups obtained from the VP genes and the chromosomal genes was observed, despite some horizontal gene transfer events (Lan et al., 2001, 2004; Escobar-Paramo et al., 2003; Yang et al., 2007). This indicates that, even though the origin of *Shigella* and EIEC occurred multiple times with several VP acquisitions, the VPs arrived early in *Shigella*/EIEC evolution, followed by a stable VP/chromosome coevolution. These data have lead some authors to propose an alternative scenario in which the acquisition of the VP in an ancestral *E. coli* strain preceded the diversification by radiation of all *Shigella* and EIEC groups (Escobar-Paramo et al., 2003).

14.4.2 Convergence of Characters: Drift and Adaptive Evolution

Besides the VP, *Shigella* and EIEC exhibit numerous negative characteristics due to gene inactivation or loss. These characters are often the result of convergent evolution, that is, the presence of different molecular defects in the distinct *Shigella* and EIEC lineages

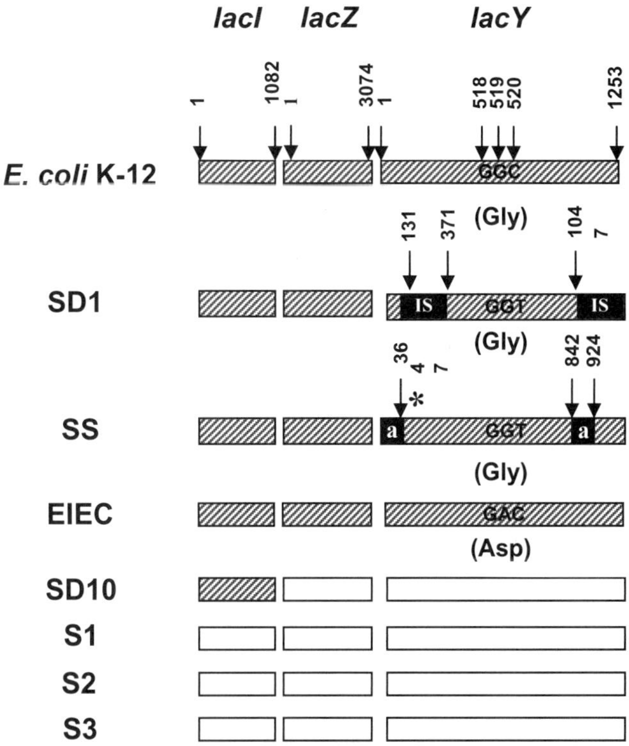

Figure 14.4 Example of convergent evolution leading to the inactivation of the lactose operon in *Shigella* and enteroinvasive *Escherichia coli* strains. Two genes of the *lac* operon necessary for lactose metabolism (*lacZ* and *lacY*) and the *lacI* gene of *E. coli* K-12 are represented by dashed boxes (not to scale) on the first line, and numbers indicate the coordinates of the corresponding nucleotides in each gene. Dashed and empty boxes indicate that the corresponding gene is present or not, respectively. Black boxes correspond to exogenous sequences that have replaced the corresponding *lacY* sequence (IS = *IS*911; a = sequence homologous to a portion of *ynjC* of *E. coli* K-12). The asterisk at position 47 of the *lacY* gene in strains of group SS represents the insertion of nine nucleotides (GAAAACGCA). Nucleotides 518–520 (GGC/GGT and GAC) code for wild type residue Gly-173 and mutated residue Asp-173 of the *lacY* gene product, respectively (Escobar-Paramo et al., 2003). SD1 = *S. dysenteriae* serotype 1; SS = *Shigella sonnei*; SD10 = *Shigella dysenteriae* serotype 10; S1, S2, S3 = main *Shigella* phylogenetic groups; EIEC = enteroinvasive *E. coli*.

(Ito et al., 1991; Escobar-Paramo et al., 2003) (Fig. 14.4). Two evolutionary scenarios, nonexclusive, can lead to this pattern. The very peculiar *Shigella*/EIEC intracellular lifestyle results in a reduced effective population size and in a less efficient selection (Hershberg et al., 2007). In this context, gene loss can be seen as the result of independent mutation accumulation. However, experimental (Maurelli et al., 1998; Fernandez et al., 2001; Prunier et al., 2007) and *in silico* genomic analyses (Touchon et al., 2009) suggest a footprint of selection through an antagonistic pleiotropy mechanism of adaptation (Cooper and Lenski, 2000). Probably, both processes are at work, with some of the inactivations being adaptive.

Interestingly, convergence has also been observed in *Shigella* and EIEC for the level of transcripts. A subset of genes of the core genome coding for transporters and binding proteins is overexpressed in *Shigella* and EIEC when compared to other *E. coli*, revealing the role of selection in shaping transcriptome polymorphism (Le Gall et al., 2005a).

14.5 WHAT MAKES YOU AN OPPORTUNISTIC PATHOGEN?

All the other *E. coli* pathotypes can be considered as facultative, opportunistic pathogens. Contrary to *Shigella*, they can be encountered in a wide spectrum of hosts. Even the highly pathogenic O157:H7 EHEC strain in humans is found as commensal in cattle. The passage of the frontier between commensalism and pathogenicity results both from the bacteria that acquired specific virulence genes and from the status of the host (different species, different ages within a single host [newborns, elderly], anatomical modifications, immunocompromised hosts). In this context, it is mandatory to first study the population genetics of commensal strains to understand the evolution of virulence. We will then study in this commensal framework the population genetics of virulent strains.

14.5.1 The Population Genetics of Commensal Strains

In animals, the major environmental force shaping the genetic structure of *E. coli* gut population is the domestication status of their host (Escobar-Paramo et al., 2006). Domesticated animals have a decreased proportion of B2 strains (from 30% in wild animals to 14% and 11% in farm and zoo animals, respectively) and an increase in A strains (14% to 27% and 26 %, respectively) (data compiled from 254 animals [Ochman and Selander, 1984; Gordon and Cowling, 2003; Escobar-Paramo et al., 2006; Baldy-Chudzik et al., 2008]) compared to their wild counterparts.

Similarly, large changes in *E. coli* group prevalence are found among different human populations. According to their *E. coli* group prevalence, human populations can roughly be split in two groups. Commensal strains isolated from Europe (France, Croatia) in the 1980s, Africa (Mali, Benin), Asia (Pakistan), and South America (French Guyana, Colombia, Bolivia) belong mainly to A (55%) and B1 (21%) phylogenetic groups, whereas strains from D (14%) and especially B2 (10%) phylogenetic groups are uncommon (data compiled from 550 subjects [Duriez et al., 2001; Escobar-Paramo et al., 2004b; Pallecchi et al., 2007; Nowrouzian et al., 2009]). Conversely, strains isolated from Europe (France, Sweden) in the 2000s, North America (United States), Japan, and Australia belong mainly to B2 (43%), followed by A (24%), D (21%), and B1 (12%) phylogenetic groups (data compiled from 567 subjects [Obata-Yasuoka et al., 2002; Zhang et al., 2002; Nowrouzian et al., 2003; Watt et al., 2003; Gordon et al., 2005]). Socioeconomic factors, such as dietary habits and level of hygiene, more than the host's genetics, are presumably the main factors accounting for this phylogenetic group distribution. This is indicated by the dramatic shift in the proportion of B2 (10–30%) and A (60–30%) strains during the last 20 years in France and by the modification of the *E. coli* microbiota during controlled human migration between metropolitan France and French Guyana (Skurnik et al., 2008). Furthermore, the morphological, physiological, and dietary differences that occur among human individuals of different sex or age influence the distribution of the *E. coli* genotypes (Gordon et al., 2005).

14.5.2 The Population Genetics of Opportunistic Pathogens

Facultative pathogens can be classified according to the pathophysiology of the disease they cause as extraintestinal and intestinal pathogens. A clear link between extraintestinal virulence and B2 (and at a lesser extent D) phylogenetic group strains has been reported for a long time. By studying almost 200 strains from acute extraintestinal infections (mainly UTIs but also septicemia and miscellaneous infections), Goullet and Picard (1986) showed that 40% of the strains belonged to the B2 phylogenetic group, as compared to 7% of strains obtained from stools of healthy individuals. This percentage increases to 50%–65%, 75%, and 84% when strains from septicemia (Johnson et al., 1991; Jaureguy et al., 2008), newborn meningitis, and urosepsis in newborns free of major urinary tract abnormalities (Bidet et al., 2007) are studied, respectively. The most frequently phylogenetic group observed in extraintestinal infections, after the B2 group, is the D group. Thus, strains belonging to the D phylogenetic group are isolated in 23% and 16% of septicemia (Jaureguy et al., 2008) and newborn meningitis (Bidet et al., 2007), respectively. Furthermore, within the B2 group, the majority of extraintestinal isolates belong to two major lineages, named subgroups II and IX (Le Gall et al., 2007), clonal complexes 1 and 4 (Jaureguy et al., 2008), ST73 and 95 (Wirth et al., 2006), or ST27 and 29 (Bidet et al., 2007). Among the strains isolated from bacteremia, those two lineages were more frequently associated with urosepsis (Jaureguy et al., 2008). At the opposite, the strains of subgroup VIII exhibiting an O81 type appear to be almost always human commensals (Clermont et al., 2008). Specialization can go further as it has been shown that within subgroup IX, O18:K1:H7 strains cause neonatal meningitis but appear to be almost unable to cause urosepsis, suggesting a gut translocation in the pathophysiology of the disease. In contrast, O45:K1:H7 strains are able to cause both meningitis and urosepsis in infants, suggesting a digestive or urinary portal of entry for the meningitis (Bidet et al., 2007). These epidemiological data have been corroborated by animal studies using a mouse model of extraintestinal virulence that tests the intrinsic virulence of the strains (Picard et al., 1999). It appears that the B2 group strains have overall the greatest virulence (Johnson et al., 2006a), although some B2 strains can be avirulent (Le Gall et al., 2007; Clermont et al., 2008).

A correlation between the phylogenetic history of the strains and the host characteristics and the pathophysiology of the disease is clearly observed in septicemia. B2 strains are more frequently observed in men without underlying diseases developing a community acquired septicemia of urinary tract origin, whereas non-B2 strains are more frequently associated with immunocompromised women, nosocomial origin, and nonurinary tract source of septicemia (Picard and Goullet, 1988; Jaureguy et al., 2007).

The picture for the EPEC strains is somewhat different. The more health-threatening ETEC and EHEC strains belong to other than B2 and D phylogenetic groups (Escobar-Paramo et al., 2004a). Two main groups of EHEC strains have been identified, the EHEC group 1 encompassing the O157:H7 strains and belonging to the E phylogenetic group and the EHEC group 2 encompassing the O111:H8 and O26:H11 clones and belonging to the B1 phylogenetic group (Reid et al., 2000). It has been proposed that, within the E phylogenetic group, O157:H7 strains emerged from the O55:H7 EPEC strains by stepwise evolution (Feng et al., 2007). Most typical EPEC fall into one of four clonal lineages, three belonging to the B2 phylogenetic groups (EPEC 1 with O55:H6 and O127:H6 clones, EPEC 3 with O86:H34 clone, and EPEC 4 with O119:H6 clone) and one to the B1 phylogenetic group, EPEC 2 with O128:H2 and O111:H2 clones (Escobar-Paramo et al., 2004a; Lacher et al., 2007; Newton et al., 2009). EAEC, DAEC, and atypical EPEC (strains with the locus of enterocyte effacement [LEE] but without the EPEC adherence factor [EAF] VP) for which the pathogenicity is less clear (Nataro and Kaper, 1998), are widespread all over the species genetic diversity (Escobar-Paramo et al., 2004a; Afset et al., 2008).

14.5.3 The Arrival of the Virulence Genes, the Pathoadaptative Mutations, and the Genetic Background of the Strain

Epidemiological and animal studies, based either on character association or direct inactivation of the gene, have implicated numerous genes, linked physically on pathogenicity island (Hacker and Kaper, 2000) or not, as "virulence genes." Studies comparing the phylogenetic history of the strains to the phylogenetic history of the virulence genes have clearly shown that both histories are incongruent (i.e., the obtained trees do not group the strains in the same way). This indicates multiple parallel acquisitions, with phenotypic convergence (Reid et al., 2000; Escobar-Paramo et al., 2004a; Lacher et al., 2007). The arrival of some of these virulence genes in only some specific *E. coli* clones, as observed for ExPEC, ETEC, and EHEC strains, is a strong argument for a role of the genetic background of the strain in the acquisition and expression of virulence (Escobar-Paramo et al., 2004a). This could reflect a molecular fine-tuning between the chromosomal backbone and the new arriving gene, resulting from epistatic interactions.

Allelic variation as observed in the FimH fimbrial protein, called pathoadaptive mutation, has also been implicated in virulence (Sokurenko et al., 1998). Likewise, the role of the genetic background has been noted. The A27V mutation occurred several times in the B2 strains, where it has been shown to be adaptive (Hommais et al., 2003).

14.5.4 The Coincidental Hypothesis for the "Virulence Factors"

Theoretical and empirical studies have created a convincing conceptual framework regarding the evolution of virulence for obligate pathogens. However, the evolution of virulence in facultative pathogens is poorly understood. The selective advantages associated with colonization of blood in bacteremia or of cerebrospinal fluid in meningitis are not clear, as such infections often lead to rapid host death and poor transmission to new hosts. As we have seen, the ability to initiate such infections requires that the cell possesses many elaborate traits usually termed virulence factors. The apparent contradiction between the presence and maintenance of many virulence factors and the presumed poor selective advantage accruing to the cell from extraintestinal infection has led to the idea that the virulence genes evolved and are maintained by selection for other roles that they play in the ecology of the bacteria, especially in commensalism (Levin, 1996; Le Gall et al., 2007). This view leads to the hypothesis that infections caused by facultative pathogens occur by accident, the virulence being a by-product of commensalism.

Extraintestinal virulence genes, coding for adhesins, iron capture systems, toxins, and protectins, correlate with successful gut colonization in humans (Wold et al., 1992; Nowrouzian et al., 2006; Moreno et al., 2009), in dogs (Johnson et al., 2008), and in piglets (Schierack et al., 2008), whereas the adhesin intimin involved in intraintestinal pathogenicity is essential for the colonization of bovine rectal mucosa (Sheng et al., 2006). Likewise, lipopolysaccharide (Alsam et al., 2006), Shiga toxins (Steinberg and Levin, 2007), and extraintestinal virulence genes (Diard et al., 2007) enhance survival by providing protection against predation caused by protozoa (amoeba [Alsam et al., 2006] and *Tetrahymena* [Steinberg and Levin, 2007]) or nematodes (Diard et al., 2007).

The prevalence of these genes is variable among commensal populations. At a global scale, the human microbiota is characterized by a higher prevalence of virulence genes than the nonhuman ones (Escobar-Paramo et al., 2006). In animals, the presence of virulence genes increases with the increase in body mass, a reflection of the gut complexity (Escobar-Paramo et al., 2006). Hence, virulence factors and their change in prevalence among hosts may reflect some local adaptation to commensal habitats rather than virulence per se. An additional argument for the coincidental hypothesis is the fact that, within the more basal group of the species, the B2 phylogenetic group, extraintestinal virulence is a widespread ancestral character; virulence evolved by gene losses (Le Gall et al., 2007).

14.6 THE VIRULENCE RESISTANCE TRADE-OFF

A trade-off between virulence and resistance to various classes of antibiotics (beta-lactams, aminoglycosides, sulfonamides, quinolones) has been reported as B2 strains expressing virulence factors and exhibiting the classically urovirulence-associated serotypes were less resistant than the non-B2 strains lacking these determinants (Picard and Goullet, 1989; Johnson et al., 1991, 1994; Jaureguy et al., 2007). This trade-off is particularly clear for quinolone resistance as it has been retrieved whatever the criteria to quantify the virulence: clinical data (Velasco et al., 2001), virulence genes (Vila et al., 2002; Branger et al., 2005; Horcajada et al., 2005), pathogenicity islands (Houdouin et al., 2006), and phylogenetic groups (Johnson et al., 2002; Branger et al., 2005). Interestingly, integrons, which are molecular tools allowing the capture, integration, and expression of gene cassettes encoding antibiotic resistance (Mazel, 2006), have also been less frequently observed in B2 than in non-B2 isolates (Skurnik et al., 2005).

A molecular mechanism for this trade-off has been suggested for quinolones as, *in vitro*, quinolones induce partial or total loss of pathogenicity islands in uropathogenic strains by SOS-dependent or -independent pathways (Soto et al., 2006). More generally, this trade-off could be the result of the alternative insertion of genomic modules bearing virulence genes or pathogenicity islands, as recently reported at the bastion of polymorphism close to the *fim* operon (Lescat et al., 2009a).

14.7 CONCLUDING REMARKS

There is a need to study the species as a whole to decipher how a clone becomes virulent and/or resistant to antibiotics. We should understand the selective pressures acting on the species in its different primary or secondary habitat, as well as in pathological conditions, in the context of a highly dynamic genome evolving in a quasi-clonal population structure. New technological developments (MacLean et al., 2009) will allow the generation of numerous complete genomes, rendering possible the population genomics era. However, only ecologically well-characterized collections of strains will be useful to pinpoint how harmless and necessary bacteria can become a fearsome killer.

ACKNOWLEDGMENTS

Our laboratory is funded by the Institut National de la Santé et de la Recherche Médicale, the Université Paris 7, the Fondation pour la Recherche Médicale, and the Agence Nationale de la Recherche.

REFERENCES

AFSET, J. E., ANDERSSEN, E., BRUANT, G., HAREL, J., WIELER, L., and BERGH, K. (2008) Phylogenetic backgrounds and virulence profiles of atypical enteropathogenic *Escherichia coli* strains from a case-control study using multilocus sequence typing and DNA microarray analysis. *J Clin Microbiol* **46,** 2280–2290.

ALSAM, S., JEONG, S. R., SISSONS, J., DUDLEY, R., KIM, K. S., and KHAN, N. A. (2006) *Escherichia coli* interactions with acanthamoeba: A symbiosis with environmental and clinical implications. *J Med Microbiol* **55,** 689–694.

BALDY-CHUDZIK, K., MACKIEWICZ, P., and STOSIK, M. (2008) Phylogenetic background, virulence gene profiles, and genomic diversity in commensal *Escherichia coli* isolated from ten mammal species living in one zoo. *Vet Microbiol* **131,** 173–184.

BARCUS, V. A., TITHERADGE, A. J., and MURRAY, N. E. (1995) The diversity of alleles at the hsd locus in natural populations of *Escherichia coli. Genetics* **140,** 1187–1197.

BIDET, P., MAHJOUB-MESSAI, F., BLANCO, J., DEHEM, M., AUJARD, Y., BINGEN, E., and BONACORSI, S. (2007) Combined multilocus sequence typing and O serogrouping distinguishes *Escherichia coli* subtypes associated with infant urosepsis and/or meningitis. *J Infect Dis* **196,** 297–303.

BISERCIC, M., FEUTRIER, J. Y., and REEVES, P. R. (1991) Nucleotide sequences of the *gnd* genes from nine natural isolates of *Escherichia coli*: Evidence of intragenic recombination as a contributing factor in the evolution of the polymorphic *gnd* locus. *J Bacteriol* **173,** 3894–3900.

BLATTNER, F. R., PLUNKETT, G. III, BLOCH, C. A., PERNA, N. T., BURLAND, V., RILEY, M., COLLADO-VIDES, J., GLASNER, J. D., RODE, C. K., MAYHEW, G. F., GREGOR, J., DAVIS, N. W., KIRKPATRICK, H. A., GOEDEN, M. A., ROSE, D. J., MAU, B., and SHAO, Y. (1997) The complete genome sequence of *Escherichia coli* K-12. *Science* **277,** 1453–1474.

BRANGER, C., ZAMFIR, O., GEOFFROY, S., LAURANS, G., ARLET, G., THIEN, H. V., GOURIOU, S., PICARD, B., and DENAMUR, E. (2005) Genetic background of *Escherichia coli* and extended-spectrum beta-lactamase type. *Emerg Infect Dis* **11,** 54–61.

BUCHRIESER, C., GLASER, P., RUSNIOK, C., NEDJARI, H., D'HAUTEVILLE, H., KUNST, F., SANSONETTI, P., and PARSOT, C. (2000) The virulence plasmid pWR100 and the repertoire of proteins secreted by the type III secretion apparatus of *Shigella flexneri. Mol Microbiol* **38,** 760–771.

CHOI, S. Y., JEON, Y. S., LEE, J. H., CHOI, B., MOON, S. H., VON SEIDLEIN, L., CLEMENS, J. D., DOUGAN, G., WAIN, J., YU, J., LEE, J. C., SEOL, S. Y., LEE, B. K., SONG, J. H., SONG, M., CZERKINSKY, C., CHUN, J., and KIM, D. W. (2007) Multilocus sequence typing analysis of *Shigella flexneri* isolates collected in Asian countries. *J Med Microbiol* **56,** 1460–1466.

CLERMONT, O., BONACORSI, S., and BINGEN, E. (2000) Rapid and simple determination of the *Escherichia coli* phylogenetic group. *Appl Environ Microbiol* **66,** 4555–4558.

CLERMONT, O., LESCAT, M., O'BRIEN, C. L., GORDON, D. M., TENAILLON, O., and DENAMUR, E. (2008) Evidence for a human-specific *Escherichia coli* clone. *Environ Microbiol* **10,** 1000–1006.

COOPER, V. S. and LENSKI, R. E. (2000) The population genetics of ecological specialization in evolving *Escherichia coli* populations. *Nature* **407,** 736–739.

DIARD, M., BAERISWYL, S., CLERMONT, O., GOURIOU, S., PICARD, B., TADDEI, F., DENAMUR, E., and MATIC, I. (2007) *Caenorhabditis elegans* as a simple model to study phenotypic and genetic virulence determinants of extraintestinal pathogenic *Escherichia coli. Microbes Infect* **9,** 214–223.

DUBOSE, R. F., DYKHUIZEN, D. E., and HARTL, D. L. (1988) Genetic exchange among natural isolates of bacteria: Recombination within the *phoA* gene of *Escherichia coli. Proc Natl Acad Sci U S A* **85,** 7036–7040.

DURIEZ, P., CLERMONT, O., BONACORSI, S., BINGEN, E., CHAVENTRE, A., ELION, J., PICARD, B., and DENAMUR, E. (2001) Commensal *Escherichia coli* isolates are phylogenetically distributed among geographically distinct human populations. *Microbiology* **147,** 1671–1676.

DYKHUIZEN, D. E. and GREEN, L. (1991) Recombination in *Escherichia coli* and the definition of biological species. *J Bacteriol* **173,** 7257–7268.

ESCOBAR-PARAMO, P., CLERMONT, O., BLANC-POTARD, A. B., BUI, H., LE BOUGUENEC, C., and DENAMUR, E. (2004a) A specific genetic background is required for acquisition and expression of virulence factors in *Escherichia coli. Mol Biol Evol* **21,** 1085–1094.

ESCOBAR-PARAMO, P., GRENET, K., LE MENAC'H, A., RODE, L., SALGADO, E., AMORIN, C., GOURIOU, S., PICARD, B., RAHIMY, M. C., ANDREMONT, A., DENAMUR, E., and RUIMY, R. (2004b) Large-scale population structure of human commensal *Escherichia coli* isolates. *Appl Environ Microbiol* **70,** 5698–700.

ESCOBAR-PARAMO, P., GIUDICELLI, C., PARSOT, C., and DENAMUR, E. (2003) The evolutionary history of *Shigella* and enteroinvasive *Escherichia coli* revised. *J Mol Evol* **57,** 140–148.

ESCOBAR-PARAMO, P., LE MENAC'H, A., LE GALL, T., AMORIN, C., GOURIOU, S., PICARD, B., SKURNIK, D., and DENAMUR, E. (2006) Identification of forces shaping the commensal *Escherichia coli* genetic structure by comparing animal and human isolates. *Environ Microbiol* **8,** 1975–1984.

ESCOBAR-PARAMO, P., SABBAGH, A., DARLU, P., PRADILLON, O., VAURY, C., DENAMUR, E., and LECOINTRE, G. (2004c) Decreasing the effects of horizontal gene transfer on bacterial phylogeny: The *Escherichia coli* case study. *Mol Phylogenet Evol* **30,** 243–250.

FALUSH, D., TORPDAHL, M., DIDELOT, X., CONRAD, D. F., WILSON, D. J., and ACHTMAN, M. (2006) Mismatch induced speciation in *Salmonella*: Model and data. *Philos Trans R Soc Lond B Biol Sci* **361,** 2045–2053.

Felsenstein, J. (1978) Cases in which parsimony or compatibility methods will be positively misleading. *Syst Zool* **27,** 401–410.

Feng, P. C., Monday, S. R., Lacher, D. W., Allison, L., Siitonen, A., Keys, C., Eklund, M., Nagano, H., Karch, H., Keen, J., and Whittam, T. S. (2007) Genetic diversity among clonal lineages within *Escherichia coli* O157:H7 stepwise evolutionary model. *Emerg Infect Dis* **13,** 1701–1706.

Fernandez, I. M., Silva, M., Schuch, R., Walker, W. A., Siber, A. M., Maurelli, A. T., and Mccormick, B. A. (2001) Cadaverine prevents the escape of *Shigella flexneri* from the phagolysosome: A connection between bacterial dissemination and neutrophil transepithelial signaling. *J Infect Dis* **184,** 743–753.

Gordon, D. M., Clermont, O., Tolley, H., and Denamur, E. (2008) Assigning *Escherichia coli* strains to phylogenetic groups: Multi-locus sequence typing versus the PCR triplex method. *Environ Microbiol* **10,** 2484–2496.

Gordon, D. M. and Cowling, A. (2003) The distribution and genetic structure of *Escherichia coli* in Australian vertebrates: Host and geographic effects. *Microbiology* **149,** 3575–3586.

Gordon, D. M., Stern, S. E., and Collignon, P. J. (2005) Influence of the age and sex of human hosts on the distribution of *Escherichia coli* ECOR groups and virulence traits. *Microbiology* **151,** 15–23.

Goullet, P. and Picard, B. (1986) Highly pathogenic strains of *Escherichia coli* revealed by the distinct electrophoretic patterns of carboxylesterase B. *J Gen Microbiol* **132,** 1853–1858.

Goullet, P. and Picard, B. (1987) Differentiation of *Shigella* by esterase electrophoretic polymorphism. *J Gen Microbiol* **133,** 1005–1017.

Goullet, P. and Picard, B. (1989) Comparative electrophoretic polymorphism of esterases and other enzymes in *Escherichia coli*. *J Gen Microbiol* **135,** 135–143.

Guttman, D. S. and Dykhuizen, D. E. (1994) Clonal divergence in *Escherichia coli* as a result of recombination, not mutation. *Science* **266,** 1380–1383.

Hacker, J. and Kaper, J. B. (2000) Pathogenicity islands and the evolution of microbes. *Annu Rev Microbiol* **54,** 641–679.

Hall, B. G. and Sharp, P. M. (1992) Molecular population genetics of *Escherichia coli*: DNA sequence diversity at the *celC*, *crr*, and *gutB* loci of natural isolates. *Mol Biol Evol* **9,** 654–665.

Hendrickson, H. (2009) Order and disorder during *Escherichia coli* divergence. *PLoS Genet* **5,** E1000335.

Hershberg, R., Tang, H., and Petrov, D. A. (2007) Reduced selection leads to accelerated gene loss in *Shigella*. *Genome Biol* **8,** R164.

Herzer, P. J., Inouye, S., Inouye, M., and Whittam, T. S. (1990) Phylogenetic distribution of branched RNA-linked multicopy single-stranded DNA among natural isolates of *Escherichia coli*. *J Bacteriol* **172,** 6175–6181.

Hobman, J. L., Penn, C. W., and Pallen, M. J. (2007) Laboratory strains of *Escherichia coli*: Model citizens or deceitful delinquents growing old disgracefully? *Mol Microbiol* **64,** 881–885.

Hommais, F., Gouriou, S., Amorin, C., Bui, H., Rahimy, M. C., Picard, B., and Denamur, E. (2003) The FimH A27V mutation is pathoadaptive for urovirulence in *Escherichia coli* B2 phylogenetic group isolates. *Infect Immun* **71,** 3619–3622.

Horcajada, J. P., Soto, S., Gajewski, A., Smithson, A., Jimenez De Anta, M. T., Mensa, J., Vila, J., and Johnson, J. R. (2005) Quinolone-resistant uropathogenic *Escherichia coli* strains from phylogenetic group B2 have fewer virulence factors than their susceptible counterparts. *J Clin Microbiol* **43,** 2962–2964.

Houdouin, V., Bonacorsi, S., Bidet, P., Bingen-Bidois, M., Barraud, D. and Bingen, E. (2006) Phylogenetic background and carriage of pathogenicity island-like domains in relation to antibiotic resistance profiles among *Escherichia coli* urosepsis isolates. *J Antimicrob Chemother* **58,** 748–751.

Ito, H., Kido, N., Arakawa, Y., Ohta, M., Sugiyama, T., and Kato, N. (1991) Possible mechanisms underlying the slow lactose fermentation phenotype in *Shigella* spp. *Appl Environ Microbiol* **57,** 2912–2917.

Jaureguy, F., Carbonnelle, E., Bonacorsi, S., Clec'h, C., Casassus, P., Bingen, E., Picard, B., Nassif, X., and Lortholary, O. (2007) Host and bacterial determinants of initial severity and outcome of *Escherichia coli* sepsis. *Clin Microbiol Infect* **13,** 854–862.

Jaureguy, F., Landreau, L., Passet, V., Diancourt, L., Frapy, E., Guigon, G., Carbonnelle, E., Lortholary, O., Clermont, O., Denamur, E., Picard, B., Nassif, X., and Brisse, S. (2008) Phylogenetic and genomic diversity of human bacteremic *Escherichia coli* strains. *BMC Genomics* **9,** 560.

Johnson, J. R., Clabots, C., and Kuskowski, M. A. (2008) Multiple-host sharing, long-term persistence, and virulence of *Escherichia coli* clones from human and animal household members. *J Clin Microbiol* **46,** 4078–4082.

Johnson, J. R., Clermont, O., Menard, M., Kuskowski, M. A., Picard, B., and Denamur, E. (2006a) Experimental mouse lethality of *Escherichia coli* isolates, in relation to accessory traits, phylogenetic group, and ecological source. *J Infect Dis* **194,** 1141–1150.

Johnson, J. R., Owens, K. L., Clabots, C. R., Weissman, S. J., and Cannon, S. B. (2006b) Phylogenetic relationships among clonal groups of extraintestinal pathogenic *Escherichia coli* as assessed by multi-locus sequence analysis. *Microbes Infect* **8,** 1702–1713.

Johnson, J. R., Goullet, P., Picard, B., Moseley, S. L., Roberts, P. L., and Stamm, W. E. (1991) Association of carboxylesterase B electrophoretic pattern with presence and expression of urovirulence factor determinants and antimicrobial resistance among strains of *Escherichia coli* that cause urosepsis. *Infect Immun* **59,** 2311–2315.

Johnson, J. R., Orskov, I., Orskov, F., Goullet, P., Picard, B., Moseley, S. L., Roberts, P. L., and Stamm, W. E. (1994) O, K, and H antigens predict virulence factors, carboxylesterase B pattern, antimicrobial resistance, and host compromise among *Escherichia coli* strains causing urosepsis. *J Infect Dis* **169,** 119–126.

Johnson, J. R., Van Der Schee, C., Kuskowski, M. A., Goessens, W., and Van Belkum, A. (2002) Phylogenetic background and virulence profiles of fluoroquinolone-resistant clinical *Escherichia coli* isolates from The Netherlands. *J Infect Dis* **186,** 1852–1856.

Kaper, J. B., Nataro, J. P., and Mobley, H. L. (2004) Pathogenic *Escherichia coli. Nat Rev Microbiol* **2,** 123–140.

Kosek, M., Bern, C., and Guerrant, R. L. (2003) The global burden of diarrhoeal disease, as estimated from studies published between 1992 and 2000. *Bull World Health Organ* **81,** 197–204.

Lacher, D. W., Steinsland, H., Blank, T. E., Donnenberg, M. S., and Whittam, T. S. (2007) Molecular evolution of typical enteropathogenic *Escherichia coli*: Clonal analysis by multilocus sequence typing and virulence gene allelic profiling. *J Bacteriol* **189,** 342–350.

Lan, R., Alles, M. C., Donohoe, K., Martinez, M. B., and Reeves, P. R. (2004) Molecular evolutionary relationships of enteroinvasive *Escherichia coli* and *Shigella* spp. *Infect Immun* **72,** 5080–5088.

Lan, R., Lumb, B., Ryan, D., and Reeves, P. R. (2001) Molecular evolution of large virulence plasmid in *Shigella* clones and enteroinvasive *Escherichia coli. Infect Immun* **69,** 6303–6309.

Lawrence, J. G., Ochman, H., and Hartl, D. L. (1991) Molecular and evolutionary relationships among enteric bacteria. *J Gen Microbiol* **137,** 1911–1921.

Lecointre, G., Rachdi, L., Darlu, P., and Denamur, E. (1998) *Escherichia coli* molecular phylogeny using the incongruence length difference test. *Mol Biol Evol* **15,** 1685–1695.

Le Gall, T., Clermont, O., Gouriou, S., Picard, B., Nassif, X., Denamur, E., and Tenaillon, O. (2007) Extraintestinal virulence is a coincidental by-product of commensalism in B2 phylogenetic group *Escherichia coli* strains. *Mol Biol Evol* **24,** 2373–2384.

Le Gall, T., Darlu, P., Escobar-Paramo, P., Picard, B., and Denamur, E. (2005a) Selection-driven transcriptome polymorphism in *Escherichia coli/Shigella* species. *Genome Res* **15,** 260–268.

Le Gall, T., Mavris, M., Martino, M. C., Bernardini, M. L., Denamur, E., and Parsot, C. (2005b) Analysis of virulence plasmid gene expression defines three classes of effectors in the type III secretion system of *Shigella flexneri. Microbiology* **151,** 951–962.

Lescat, M., Calteau, A., Hoede, C., Barbe, V., Touchon, M., Rocha, E., Tenaillon, O., Medigue, C., Johnson, J. R., and Denamur, E. (2009a) A module located at a chromosomal integration hot spot is responsible for the multidrug resistance of a reference strain from *Escherichia coli* clonal group A. *Antimicrob Agents Chemother* **53,** 2283–2288.

Lescat, M., Hoede, C., Clermont, O., Garry, L., Darlu, P., Tuffery, P., Denamur, E., and Picard, B. (2009b) *aes*, the gene encoding the esterase B in *Escherichia coli*, is a powerful phylogenetic marker of the species. *BMC Microbiol*, in press.

Levin, B. R. (1996) The evolution and maintenance of virulence in microparasites. *Emerg Infect Dis* **2,** 93–102.

MacLean, D., Jones, J. D., and Studholme, D. J. (2009) Application of 'next-generation' sequencing technologies to microbial genetics. *Nat Rev Microbiol* **7,** 287–296.

Maurelli, A. T., Fernandez, R. E., Bloch, C. A., Rode, C. K., and Fasano, A. (1998) "Black holes" and bacterial pathogenicity: A large genomic deletion that enhances the virulence of *Shigella* spp. and enteroinvasive *Escherichia coli. Proc Natl Acad Sci U S A* **95,** 3943–3948.

Mazel, D. (2006) Integrons: Agents of bacterial evolution. *Nat Rev Microbiol* **4,** 608–620.

Mercier, R., Petit, M. A., Schbath, S., Robin, S., El Karoui, M., Boccard, F., and Espeli, O. (2008) The MatP/*matS* site-specific system organizes the terminus region of the *E. coli* chromosome into a macrodomain. *Cell* **135,** 475–485.

Milkman, R. and Bridges, M. M. (1990) Molecular evolution of the *Escherichia coli* Chromosome. III. Clonal frames. *Genetics* **126,** 505–517.

Milkman, R. and Bridges, M. M. (1993) Molecular evolution of the *Escherichia coli* chromosome. IV. Sequence comparisons. *Genetics* **133,** 455–468.

Milkman, R. and Crawford, I. P. (1983) Clustered third-base substitutions among wild strains of *Escherichia coli. Science* **221,** 378–380.

Milkman, R., Jaeger, E., and McBride, R. D. (2003) Molecular evolution of the *Escherichia coli* chromosome. VI. Two regions of high effective recombination. *Genetics* **163,** 475–483.

Milkman, R. and Stoltzfus, A. (1988) Molecular evolution of the *Escherichia coli* chromosome. II. Clonal segments. *Genetics* **120,** 359–366.

Moreno, E., Johnson, J. R., Perez, T., Prats, G., Kuskowski, M. A., and Andreu, A. (2009) Structure and urovirulence characteristics of the fecal *Escherichia coli* population among healthy women. *Microbes Infect* **11,** 274–280.

Nataro, J. P. and Kaper, J. B. (1998) Diarrheagenic *Escherichia coli. Clin Microbiol Rev* **11,** 142–201.

Nelson, K., Whittam, T. S., and Selander, R. K. (1991) Nucleotide polymorphism and evolution in the glyceraldehyde-3-phosphate dehydrogenase gene (*gapA*) in natural populations of *Salmonella* and *Escherichia coli. Proc Natl Acad Sci U S A* **88,** 6667–6671.

Newton, H. J., Sloan, J., Bulach, D. M., Seemann, T., Allison, C. C., Tauschek, M., Robins-Browne, R. M., Paton, J. C., Whittam, T. S., Paton, A. W., and Hartland, E. L. (2009) Shiga toxin-producing *Escherichia coli* strains negative for locus of enterocyte effacement. *Emerg Infect Dis* **15,** 372–80.

Nowrouzian, F., Hesselmar, B., Saalman, R., Strannegard, I. L., Aberg, N., Wold, A. E., and Adlerberth, I. (2003) *Escherichia coli* in infants' intestinal microflora: Colonization rate, strain turnover, and virulence gene carriage. *Pediatr Res* **54,** 8–14.

NOWROUZIAN, F. L., ADLERBERTH, I., and WOLD, A. E. (2006) Enhanced persistence in the colonic microbiota of *Escherichia coli* strains belonging to phylogenetic group B2: Role of virulence factors and adherence to colonic cells. *Microbes Infect* **8,** 834–840.

NOWROUZIAN, F. L., OSTBLOM, A. E., WOLD, A. E., and ADLERBERTH, I. (2009) Phylogenetic group B2 *Escherichia coli* strains from the bowel microbiota of Pakistani infants carry few virulence genes and lack the capacity for long-term persistence. *Clin Microbiol Infect* **15,** 466–472.

OBATA-YASUOKA, M., BA-THEIN, W., TSUKAMOTO, T., YOSHIKAWA, H., and HAYASHI, H. (2002) Vaginal *Escherichia coli* share common virulence factor profiles, serotypes and phylogeny with other extraintestinal *E. coli*. *Microbiology* **148,** 2745–2752.

OCHMAN, H. and SELANDER, R. K. (1984) Standard reference strains of *Escherichia coli* from natural populations. *J Bacteriol* **157,** 690–693.

OCHMAN, H., WHITTAM, T. S., CAUGANT, D. A., and SELANDER, R. K. (1983) Enzyme polymorphism and genetic population structure in *Escherichia coli* and *Shigella*. *J Gen Microbiol* **129,** 2715–2726.

PALLECCHI, L., LUCCHETTI, C., BARTOLONI, A., BARTALESI, F., MANTELLA, A., GAMBOA, H., CARATTOLI, A., PARADISI, F., and ROSSOLINI, G. M. (2007) Population structure and resistance genes in antibiotic-resistant bacteria from a remote community with minimal antibiotic exposure. *Antimicrob Agents Chemother* **51,** 1179–1184.

PERNA, N. T., PLUNKETT, G. III, BURLAND, V., MAU, B., GLASNER, J. D., ROSE, D. J., MAYHEW, G. F., EVANS, P. S., GREGOR, J., KIRKPATRICK, H. A., POSFAI, G., HACKETT, J., KLINK, S., BOUTIN, A., SHAO, Y., MILLER, L., GROTBECK, E. J., DAVIS, N. W., LIM, A., DIMALANTA, E. T., POTAMOUSIS, K. D., APODACA, J., ANANTHARAMAN, T. S., LIN, J., YEN, G., SCHWARTZ, D. C., WELCH, R. A., and BLATTNER, F. R. (2001) Genome sequence of enterohaemorrhagic *Escherichia coli* O157:H7. *Nature* **409,** 529–533.

PICARD, B., GARCIA, J. S., GOURIOU, S., DURIEZ, P., BRAHIMI, N., BINGEN, E., ELION, J., and DENAMUR, E. (1999) The link between phylogeny and virulence in *Escherichia coli* extraintestinal infection. *Infect Immun* **67,** 546–553.

PICARD, B. and GOULLET, P. (1988) Correlation between electrophoretic types B1 And B2 of carboxylesterase B and host-dependent factors in *Escherichia coli* septicaemia. *Epidemiol Infect* **100,** 51–61.

PICARD, B. and GOULLET, P. (1989) Correlation between electrophoretic types B1 And B2 of carboxylesterase B and sex of patients in *Escherichia coli* urinary tract infections. *Epidemiol Infect* **103,** 97–103.

PRUNIER, A. L., SCHUCH, R., FERNANDEZ, R. E., MUMY, K. L., KOHLER, H., MCCORMICK, B. A., and MAURELLI, A. T. (2007) *nadA* and *nadB* of *Shigella flexneri* 5a are antivirulence loci responsible for the synthesis of quinolinate, a small molecule inhibitor of *Shigella* pathogenicity. *Microbiology* **153,** 2363–2372.

PUPO, G. M., LAN, R., and REEVES, P. R. (2000) Multiple independent origins of *Shigella* clones of *Escherichia coli* and convergent evolution of many of their characteristics. *Proc Natl Acad Sci U S A* **97,** 10567–10572.

RASKO, D. A., ROSOVITZ, M. J., MYERS, G. S., MONGODIN, E. F., FRICKE, W. F., GAJER, P., CRABTREE, J., SEBAIHIA, M., THOMSON, N. R., CHAUDHURI, R., HENDERSON, I. R., SPERANDIO, V., and RAVEL, J. (2008) The pangenome structure of *Escherichia coli*: Comparative genomic analysis of *E. coli* commensal and pathogenic isolates. *J Bacteriol* **190,** 6881–6893.

REID, S. D., HERBELIN, C. J., BUMBAUGH, A. C., SELANDER, R. K., and WHITTAM, T. S. (2000) Parallel evolution of virulence in pathogenic *Escherichia coli*. *Nature* **406,** 64–67.

ROLLAND, K., LAMBERT-ZECHOVSKY, N., PICARD, B., and DENAMUR, E. (1998) *Shigella* and enteroinvasive *Escherichia coli* strains are derived from distinct ancestral strains of *E. coli*. *Microbiology* **144**(Pt 9), 2667–2672.

RUSSO, T. A. and JOHNSON, J. R. (2000) Proposal for a new inclusive designation for extraintestinal pathogenic isolates of *Escherichia coli*: ExPEC. *J Infect Dis* **181,** 1753–1754.

RUSSO, T. A. and JOHNSON, J. R. (2003) Medical and economic impact of extraintestinal infections due to *Escherichia coli*: Focus on an increasingly important endemic problem. *Microbes Infect* **5,** 449–456.

SANSONETTI, P. J., TRAN VAN NHIEU, G., and EGILE, C. (1999) Rupture of the intestinal epithelial barrier and mucosal invasion by *Shigella flexneri*. *Clin Infect Dis* **28,** 466–475.

SAVAGEAU, M. A. (1983) *Escherichia coli* habitats, cell types, and moleculat mechanisms of gene control. *Am Nat* **122,** 732–744.

SCHIERACK, P., WALK, N., EWERS, C., WILKING, H., STEINRUCK, H., FILTER, M., and WIELER, L. H. (2008) ExPEC-typical virulence-associated genes correlate with successful colonization by intestinal *E. coli* in a small piglet group. *Environ Microbiol* **10,** 1742–1751.

SCHIERUP, M. H. and HEIN, J. (2000) Consequences of recombination on traditional phylogenetic analysis. *Genetics* **156,** 879–891.

SCHUBERT, S., DARLU, P., CLERMONT, O., WIESER, A., MAGISTRO, G., HOFFMANN, C., WEINERT, K., TENAILLON, O., MATIC, I., and DENAMUR, E. (2009) Role of intraspecies recombination in the spread of pathogenicity islands within the *Escherichia coli* species. *PLoS Pathog* **5,** E1000257.

SELANDER, R. K., CAUGANT, D. A., OCHMAN, H., MUSSER, J. M., GILMOUR, M. N., and WHITTAM, T. S. (1986) Methods of multilocus enzyme electrophoresis for bacterial population genetics and systematics. *Appl Environ Microbiol* **51,** 873–884.

SELANDER, R. K., CAUGANT, D. A., and WHITTAM, T. S. (1987) Genetic structure and variation in natural populations of *Escherichia coli*. In *Escherichia coli and Salmonella Typhimurium. Cellular and Molecular Biology* (ed. F. C. Neidhardt), pp. 1625–1648. American Society for Microbiology, Washington, DC.

SHENG, H., LIM, J. Y., KNECHT, H. J., LI, J., and HOVDE, C. J. (2006) Role of *Escherichia coli* O157:H7 virulence factors in colonization at the bovine terminal rectal mucosa. *Infect Immun* **74,** 4685–4693.

SKURNIK, D., BONNET, D., BERNEDE-BAUDUIN, C., MICHEL, R., GUETTE, C., BECKER, J. M., BALAIRE, C., CHAU, F., MOHLER, J., JARLIER, V., BOUTIN, J. P., MOREAU, B., GUILLEMOT, D., DENAMUR, E., ANDREMONT, A., and RUIMY, R. (2008) Characteristics of human intestinal *Escherichia coli* with changing environments. *Environ Microbiol* **10,** 2132–2137.

SKURNIK, D., LE MENAC'H, A., ZURAKOWSKI, D., MAZEL, D., COURVALIN, P., DENAMUR, E., ANDREMONT, A., and RUIMY, R. (2005) Integron-associated antibiotic resistance and phylogenetic grouping of *Escherichia coli* isolates from healthy subjects free of recent antibiotic exposure. *Antimicrob Agents Chemother* **49,** 3062–3065.

SOKURENKO, E. V., CHESNOKOVA, V., DYKHUIZEN, D. E., OFEK, I., WU, X. R., KROGFELT, K. A., STRUVE, C., SCHEMBRI, M. A., and HASTY, D. L. (1998) Pathogenic adaptation of *Escherichia coli* by natural variation of the FimH adhesin. *Proc Natl Acad Sci U S A* **95,** 8922–8926.

SOTO, S. M., JIMENEZ DE ANTA, M. T., and VILA, J. (2006) Quinolones induce partial or total loss of pathogenicity islands in uropathogenic *Escherichia coli* by SOS-dependent or -independent pathways, respectively. *Antimicrob Agents Chemother* **50,** 649–653.

STEINBERG, K. M. and LEVIN, B. R. (2007) Grazing protozoa and the evolution of the *Escherichia coli* O157:H7 *Shiga* toxin-encoding prophage. *Proc Biol Sci* **274,** 1921–1929.

TOUCHON, M., HOEDE, C., TENAILLON, O., BARBE, V., BAERISWYL, S., BIDET, P., BINGEN, E., BONACORSI, S., BOUCHIER, C., BOUVET, O., CALTEAU, A., CHIAPELLO, H., CLERMONT, O., CRUVEILLER, S., DANCHIN, A., DIARD, M., DOSSAT, C., KAROUI, M. E., FRAPY, E., GARRY, L., GHIGO, J. M., GILLES, A. M., JOHNSON, J., LE BOUGUENEC, C., LESCAT, M., MANGENOT, S., MARTINEZ-JEHANNE, V., MATIC, I., NASSIF, X., OZTAS, S., PETIT, M. A., PICHON, C., ROUY, Z., RUF, C. S., SCHNEIDER, D., TOURRET, J., VACHERIE, B., VALLENET, D., MEDIGUE, C., ROCHA, E. P., and DENAMUR, E. (2009) Organised genome dynamics in the *Escherichia coli* species results in highly diverse adaptive paths. *PLoS Genet* **5,** E1000344.

VELASCO, M., HORCAJADA, J. P., MENSA, J., MORENO-MARTINEZ, A., VILA, J., MARTINEZ, J. A., RUIZ, J., BARRANCO, M., ROIG, G., and SORIANO, E. (2001) Decreased invasive capacity of quinolone-resistant *Escherichia coli* in patients with urinary tract infections. *Clin Infect Dis* **33,** 1682–1686.

VILA, J., SIMON, K., RUIZ, J., HORCAJADA, J. P., VELASCO, M., BARRANCO, M., MORENO, A., and MENSA, J. (2002) Are quinolone-resistant uropathogenic *Escherichia coli* less virulent? *J Infect Dis* **186,** 1039–1042.

WATT, S., LANOTTE, P., MEREGHETTI, L., MOULIN-SCHOULEUR, M., PICARD, B., and QUENTIN, R. (2003) *Escherichia coli* strains from pregnant women and neonates: Intraspecies genetic distribution and prevalence of virulence factors. *J Clin Microbiol* **41,** 1929–1935.

WHITMAN, W. B., COLEMAN, D. C., and WIEBE, W. J. (1998) Prokaryotes: The unseen majority. *Proc Natl Acad Sci U S A* **95,** 6578–6583.

WIRTH, T., FALUSH, D., LAN, R., COLLES, F., MENSA, P., WIELER, L. H., KARCH, H., REEVES, P. R., MAIDEN, M. C., OCHMAN, H., and ACHTMAN, M. (2006) Sex and virulence in *Escherichia coli*: An evolutionary perspective. *Mol Microbiol* **60,** 1136–1151.

WOLD, A. E., CAUGANT, D. A., LIDIN-JANSON, G., DE MAN, P., and SVANBORG, C. (1992) Resident colonic *Escherichia coli* strains frequently display uropathogenic characteristics. *J Infect Dis* **165,** 46–52.

YANG, J., NIE, H., CHEN, L., ZHANG, X., YANG, F., XU, X., ZHU, Y., YU, J., and JIN, Q. (2007) Revisiting the molecular evolutionary history of *Shigella* spp. *J Mol Evol* **64,** 71–79.

ZHANG, L., FOXMAN, B., and MARRS, C. (2002) Both urinary and rectal *Escherichia coli* isolates are dominated by strains of phylogenetic group B2. *J Clin Microbiol* **40,** 3951–3955.

Chapter 15

Population Genetics of *Salmonella*: Selection for Antigenic Diversity

KRISTEN BUTELA AND JEFFREY LAWRENCE

15.1 INTRODUCTION

Salmonellae are gram-negative, rod-shaped γ-proteobacteria. As members of the Enterobacteriaceae, they are related to the animal pathogens *Escherichia coli*, *Shigella dysenteriae*, *Klebsiella pneumoniae*, and *Yersinia pestis* and to the plant pathogen *Erwinia carotovora*. Salmonellae are pathogens of many mammalian species, commonly associated with both foodborne illness and enteric fever. In humans, *Salmonella* causes typhoid and paratyphoid fevers and is one of the primary causes of foodborne bacterial illness, resulting in at least 1.5 million cases each year (Mead et al., 1999). Salmonellosis typically results in abdominal cramps, diarrhea, nausea, vomiting, and fever. Symptoms occur within 6–72h after ingestion of an infectious dose, and while most infected individuals recover after 5–7 days, those with severe diarrhea may develop life-threatening dehydration or may experience spread of the *Salmonella* to the blood and other body tissues (Hohmann, 2001). In rare cases, salmonellosis can lead to Reiter's syndrome, a condition characterized by chronic inflammation and pain in the joints, eyes, and urethra (Dworkin et al., 2001). The financial costs of foodborne illness in humans in the United States caused by the six most common bacterial pathogens were estimated to be $2.9 to $6.7 billion per year a decade ago (Buzby et al., 1996), a figure that continues to rise. The high prevalence and significant financial impact of *Salmonella* infection led Voetsch et al. (2004) to conclude that *Salmonella* "presents a major ongoing burden to public health." Despite being pathogenic to many hosts, *Salmonella* often adopts a commensal lifestyle in the intestines of some reptiles (Briones et al., 2004; Chambers and Hulse, 2006; Hahn et al., 2007), birds (Hubalek et al., 1995; Refsum et al., 2002), and small mammals (Thigpen et al., 1975; Kourany et al., 1976; Handeland et al., 2002).

Two species are currently recognized within the genus *Salmonella*: *Salmonella bongori* and *Salmonella enterica*; *S. enterica* is further divided into six subspecies (Table 15.1). Warm-blooded animals are the primary hosts for subspecies *enterica* and *salmae*, whereas cold-blooded animals and the environment are reservoirs for all other subspecies (Grimont and Weill, 2007). The genus *Salmonella* is further divided into 2579

Bacterial Population Genetics in Infectious Disease, Edited by D. Ashley Robinson, Daniel Falush, and Edward J. Feil

Table 15.1 Classification of the Genus *Salmonella* (Tindall et al., 2005; Grimont and Weill, 2007)

Species	Subspecies	Serotypes
S. bongori	—	22
S. enterica	*enterica* (I)	1531
S. enterica	*salmae* (II)	505
S. enterica	*arizonae* (IIIa)	99
S. enterica	*diarizonae* (IIIb)	336
S. enterica	*houtenae* (IV)	73
S. enterica	*indica* (VI)	13

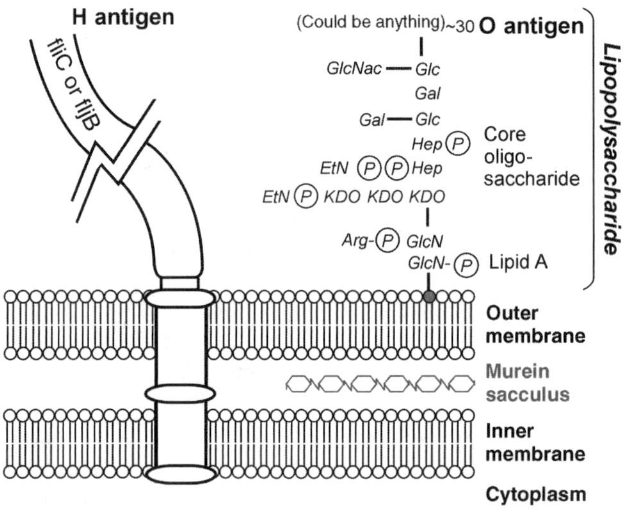

Figure 15.1 Schematic diagram of the *Salmonella enterica* outer surface depicting the O and H antigens.

antigenically distinct strains, or serovars, 2557 serovars in *S. enterica* and 22 serovars in *S. bongori* (Grimont and Weill, 2007). Subspecies *enterica* (group I) includes the strains responsible for the majority of cases of mammalian illness; this well-studied group contains 1531 serovars, although this relative overrepresentation may reflect our collective intense interest in mammalian disease more than the distribution of serovars among natural isolates. Serovars are defined by two major highly diverse surface molecules, the O and H antigens.

The O antigen is the outermost portion of *Salmonella*'s lipopolysaccharide (LPS) layer (Fig. 15.1); it is a constitutively expressed repeating polysaccharide unit and is the most abundant molecule on the cell surface (Samuel and Reeves, 2003). The variable component of the *Salmonella* O-antigen polysaccharide is produced by the *Salmonella rfb* genes, which encode various sugar synthases and transferases that assemble the repeating polysaccharide units of the O antigen (Samuel and Reeves, 2003). These units are assembled into long chains by the Rfc O-antigen polymerase (Samuel and Reeves, 2003). The O-antigen polysaccharide is then linked by the WaaL O-antigen ligase to lipid A, located on the outer membrane side of LPS, via a core oligosaccharide that is highly conserved across enteric bacteria (Schnaitman and Klena, 1993; Heinrichs et al., 1998; Samuel and Reeves, 2003). While the O antigen may be modified, such as in the case of the acetylation of the *Salmonella* LT2 O antigen by the unlinked *oafA* gene (Slauch et al., 1996), variability in saccharide composition and linkage is conferred by variable gene content at the *rfb* locus.

The H antigen is conferred by the filament of *Salmonella*'s peritrichous flagellae, which are used for swimming. Unlike the O antigen, the H antigen is only expressed under certain environmental conditions (Chilcott and Hughes, 2000). In most serovars, phase switching occurs between one of two phases of the H antigen, which are encoded by separate genes (Lederberg and Edwards, 1953; Lederberg and Iino, 1956; Chilcott and Hughes, 2000). Serotypes also differ at other antigenic proteins not included in the classification scheme, including the major outer membrane porins OmpC (Singh et al., 1995), OmpD (Singh et al., 1996; Santiviago et al., 2003), and PhoE (Singh et al., 1992; Spierings et al., 1992). Porins are transmembrane proteins that permit the diffusion of large molecules through the cell's outer membrane; they are typically highly conserved at transmembrane regions but are highly diverse at exposed, external loops (Nikaido and Vaara, 1985; Nikaido, 2003). This chapter will focus on diversity at the *Salmonella* O antigen. Below, we will discuss how genetic variability at the O-antigen-encoding *rfb* operon cannot be explained by conventional models, and will develop a framework for the maintenance of antigenetic variability in *Salmonella* populations.

15.1.1 Frequency-Dependent Selection and Antigenic Diversity

An interesting property of loci encoding antigenic determinants is that many are hypervariable; that is, given the expectations of diversity afforded by neutral variation (Kimura, 1983), chromosomal loci encoding antigens (or loci linked to those that do) are far more variable than one would expect. This is true for *Salmonella*'s H-antigen-encoding *fliC* gene (Smith and Selander, 1990; Smith et al., 1990; McQuiston et al., 2004; Jacob Sonne-Hansen, 2005), and the O-antigen-encoding *rfb* operon (Verma et al., 1988; Verma and Reeves, 1989; Brown et al., 1991, 1992; Jiang et al., 1991; Liu et al., 1991, 1993, 1995; Lee et al., 1992a,b; Wang et al., 1992; Xiang et al., 1993, 1994). For loci in many bacteria, antigenic diversity can be explained by frequency-dependent selection (Fig. 15.2a; Levin, 1988). Here, the fitness contribution of any given antigen-encoding allele depends not only on the function of its product but also on its frequency in the entire population. There are two general models one can consider where rarity is advantageous.

First, rare antigens can confer a high degree of fitness on a spatial scale (Fig. 15.2b). For example, cells with commonly found antigens are more likely to encounter non-naive hosts, whereas cells with rare antigens are more likely to encounter and successfully infect hosts that are naive to its antigen. As these cells multiply, their antigenic profile becomes more common in the bacterial population, leading to a lower population of susceptible hosts (Fig. 15.2b). Therefore, frequency-dependent selection favors pathogens that possess mechanisms to maintain population-level antigenic diversity over those that are unable to maintain such antigenic diversity.

Alternatively, rare antigens can provide an advantage on a temporal scale. Once it has successfully infected a host, a pathogen can continue to evade the host's immune system to prolong the infection and can increase reproductive capability only if it alters its antigenic profile. Pathogens that are able to switch their antigenic profile every few generations

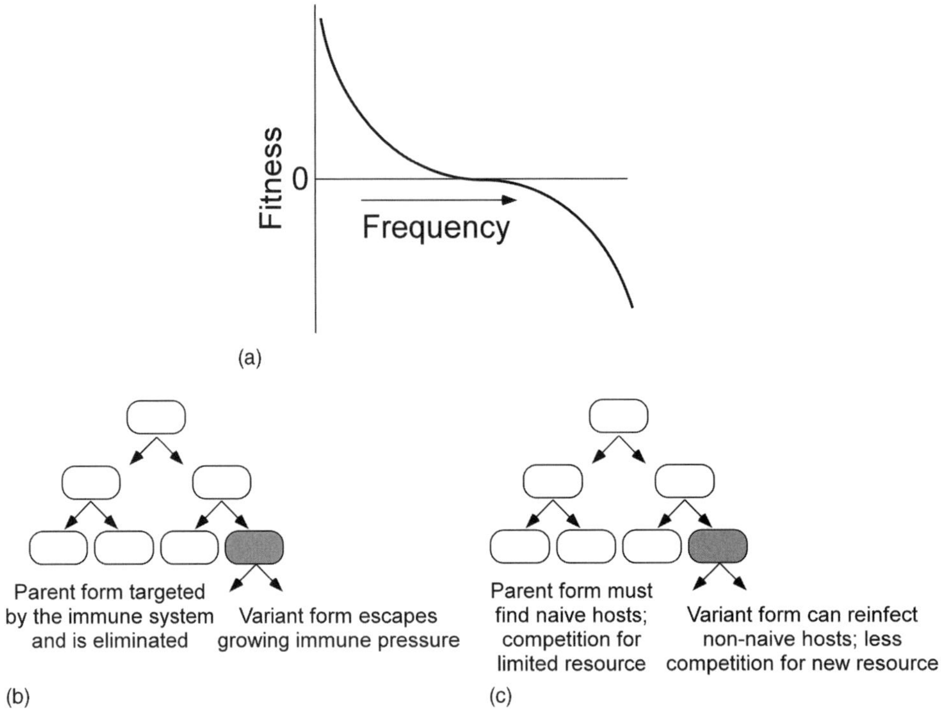

Figure 15.2 (a) An overview of frequency-dependent selection. (b) Spatial heterogeneity model for frequency-dependent selection; here, variant daughter cells gain an advantage by being able to infect hosts that have been exposed to their parents' antigens. (c) Temporal heterogeneity model for frequency-dependent selection; here, variant daughter cells persist in a single host that has begun to respond to its parent cells' antigens.

have a distinct advantage in prolonging infection, engaging in an "arms race" with the immune system of the host organism (Fig. 15.2c). Continual presentation of the same antigenic profile would allow the host adaptive immune system to mount defenses against the invading organism, whereas organisms that have the capacity to switch antigenic profiles have much better chances at evading the host immune system, reproducing, and maintaining the infection.

Because environments are dynamic, no one antigenic profile will have a sustained high fitness level over both space and time (Levin, 1988). Thus, traditional host–pathogen dynamics generally favor selection of organisms that generate heritable, random, and reversible antigenic variation. As discussed below, many pathogens possess molecular mechanisms that produce generation timescale diversity, whereby daughter cells are often antigenically distinct from their parents. As a result, the population as a whole will be diverse at antigen-encoding loci as a result of constant change at short timescales.

15.1.2 The Conventional Wisdom Fails for *Salmonella* Diversity

Yet, neither of the two models for host–pathogen interaction appears to apply to highly variable strains of *S. enterica*. Here, organisms are not infected by recently unencountered serotypes, and *Salmonella* does not alter its antigenic profile during the course of infection.

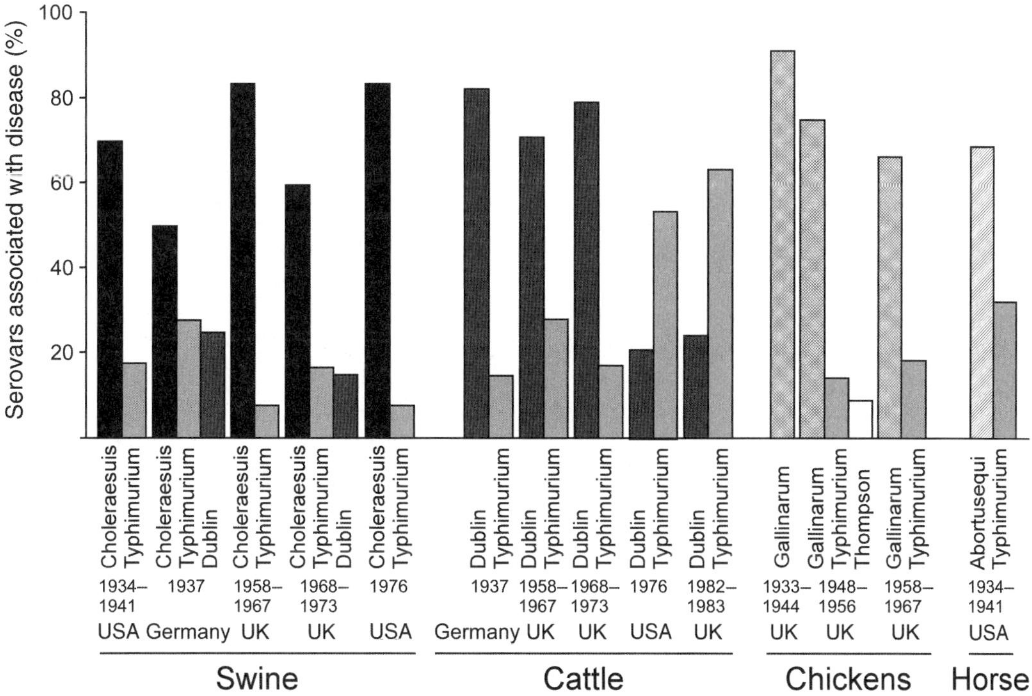

Figure 15.3 Host–serovar specificity in *Salmonella*. The proportion of different serovars is shown for outbreaks in different mammalian hosts. Data from Rabsch et al. (2002).

Rather, each host of *Salmonella* is infected by only a small subset of *Salmonella* serovars (Rabsch et al., 2002), the composition of which is specific to each host (Fig. 15.3). For example, swine are commonly infected with serovar Choleraesuis, cattle with serovar Dublin, poultry with serovar Gallinarium, sheep by serovar Abortus-ovis, horses by serovar Abortus-equi, and so on (Fig. 15.3). Beyond the sample of one outbreak in horses, this pattern is consistent across both time (several decades) and space (several countries). This poorly understood pattern of infection is referred to as host–serovar specificity. Both of the antigens used to classify serovars have been implicated in *Salmonella* virulence (Grossman et al., 1987; Parker and Guard-Petter, 2001; Schmitt et al., 2001; Thomsen et al., 2003; Bergman et al., 2005a; Cummings et al., 2005; Murray et al., 2006; Keestra et al., 2008), and efforts to link *Salmonella* antigenic diversity to host–serovar specificity have mainly focused on the relationship between *Salmonella* and the host immune system (Bolton et al., 1999; Uzzau et al., 2001; Meyerholz and Stabel, 2003). Whereas diversity at the H-antigen-encoding loci and other minor antigenic loci has been linked to immune system evasion (Baumler et al., 1998; Norris and Baumler, 1999; Humphries et al., 2001, 2005; Ikeda et al., 2001; Parker and Guard-Petter, 2001; Schmitt et al., 2001; Althouse et al., 2003; Cummings et al., 2005; Secundino et al., 2006; Keestra et al., 2008), the exact role of the O antigen in determining host–serovar specificity remains unclear.

Variability at any antigenic locus reflects the sum of selective pressures and stochastic processes acting on it. As we discuss below, the nature of the variability at the *Salmonella rfb* locus reflects a suite of selective pressures experienced by this pathogen that differs from those influencing antigenic loci in other species. We propose an alternative mechanism explaining the maintenance of *Salmonella* O-antigen diversity that focuses on

Salmonella's lifestyle in the broader intestinal ecosystem. Here, the benefits of generation timescale constancy—that is, avoiding variability—outweigh any benefits of generation timescale diversity. As a result, frequency-dependent selection cannot be invoked to explain antigenic diversity in *Salmonella*. Rather, diversity at the population level must be maintained by diversifying selection acting on the different antigenic types. To outline this model, we begin by contrasting antigenic diversity in a number of pathogens to highlight both the differences between the selective regimes faced by *Salmonella* and other pathogens and the differences in the mechanisms by which diversity is generated and maintained.

15.2 GENERATION TIMESCALE DIVERSIFICATION

For many pathogens, population-level diversity reflects the production of variant antigenic types on a generation timescale. That is, diversity at the species level results from the collection of highly variable cells that are produced at the cellular level. Critically, these organisms possess clear mechanisms that produce variant daughter cells. Three of these organisms, each possessing different mechanisms that reflect three different selective regimes, are discussed below. The lack of selective sweeps purging the variability reflects the difference in the timescales over which the two processes (creating and purging variability) act.

15.2.1 *Haemophilus*: Extending Persistence Times

Pathogens may continually change surface antigens by stochastic activation or inactivation of constituent genes. This process is mediated by DNA polymerase slippage on repeated nucleotide tracts termed contingency loci (Moxon et al., 1994, 2006). This frequent slipped-strand mispairing may occur during DNA replication and repair, resulting in the insertion or deletion of a single nucleotide or a repeated nucleotide group (Moxon et al., 2006; Wisniewski-Dye and Vial, 2008). Thus, antigenically diverse daughter cells are continually produced. If located within the coding sequence of a gene, these contingency loci can cause changes in gene expression at the translational level by changing the gene's reading frame and the location of the stop codon (Levinson and Gutman, 1987). If located within the promoter of a gene, they may alter the RNA polymerase binding sites or may facilitate premature transcription termination (Levinson and Gutman, 1987; Wisniewski-Dye and Vial, 2008). Contingency loci permit heritable, reversible, and random genotypic changes to antigen-encoding genes, changing the expression of these genes nearly every generation (Bayliss et al., 2001).

Notable contingency loci in the human commensal *Haemophilus influenzae* include the *lic1*, *lic2*, and *lic3* genes responsible for the major surface antigen LPS biosynthesis (Weiser et al., 1989, 1990; High et al., 1993, 1996; Roche et al., 1994; Roche and Moxon, 1995; Hood et al., 1996; Hosking et al., 1999). Expression of different types of LPS forms in *Haemophilus* is regulated by polymerase slippage at the 5′ region of these genes, which contain a variable number of CAAT repeats (Weiser et al., 1989, 1990). Polymerase slippage at these sequence repeats alters the translation of *lic* genes by placing the genes in or out of the proper reading frame (Weiser et al., 1989). LPS variation in *Haemophilus* is critical for avoiding attack by the host adaptive immune system and for maintenance of colonization within the restricted niche of the upper respiratory tract in individual hosts (Weiser and Pan, 1998). *Haemophilus* can also switch from commensal to opportunistic pathogen, causing meningitis and septicemia in infected hosts (Michaels and Norden, 1977; Moxon, 1985; Zwahlen et al., 1986). The lifestyle switch from commensal to pathogen is dependent on LPS variability at both intrastrain (Tolan et al., 1986) and interstrain (Inzana 1983, 1987) levels.

The ability of *Haemophilus* to continually vary its antigenic profile through polymerase slippage gives it the ability to avoid recognition by the host immune system, prolonging colonization in one particular host. Frequency-dependent selection arising from selective pressure from the host immune system explains the maintenance of antigenic diversity in the *Haemophilus* population, as no one surface antigen confers a distinct temporal advantage over the others. When surface antigens are recognized, immune defenses are produced by the host organism, which can eliminate *Haemophilus* from that host. *Haemophilus* has evolved contingency loci to vary surface antigens randomly at a high frequency, which decreases the probability of an effective immune system defense and prolongs the *Haemophilus* within-host life cycle (Fig. 15.2c). Thus, the mechanism generating phenotypic variability provides insight into the selective forces acting on this species.

15.2.2 *Neisseria*: Infecting Non-Naive Hosts

Like *Haemophilus*, *Neisseria meningitidis* is a commensal of the human upper respiratory tract and has the capacity to adopt a pathogenic lifestyle, causing meningitis and septicemia (Caugant, 2008). Contingency loci have been associated with the ability of *Neisseria* to maintain its commensal lifestyle, regulating many genes involved in the biosynthesis of antigens (Tettelin et al., 2000; Snyder et al., 2001), such as LPS (Kurzai et al., 2005), pili (Tinsley and Heckels, 1986), opacity proteins (Stern and Meyer, 1987), capsular polysaccharides (Hammerschmidt et al., 1996; Lavitola et al., 1999), and the PorA outer membrane protein (van der Ende et al., 2000). Expression of the PorA protein, which is used to identify the serosubtype of *Neisseria* strains, is regulated by variation in the number of nucleotide repeats both in the promoter region and within the coding sequence of the gene. Insertions or deletions caused by slipped-strand mispairing regulates PorA expression at the transcriptional level in the homopolymeric G-repeat region located in the promoter (van der Ende et al., 1995) and at the translational level in the homopolymeric A-repeat region located within the coding sequence (van der Ende et al., 2000). In addition, expression of the *lgtABE* genes responsible for LPS biosynthesis is regulated by slipped-strand mispairing at a homopolymeric tract of G repeats upstream of *lgtA*, the first gene of the locus (Jennings et al., 1995, 1999).

Antigenic diversity in the *Neisseria meningitidis* type IV pili, a major target of the host immune system (Caugant, 2008), is generated by gene conversion. The antigenic portion of the pilus is PilE, which consists of a conserved N-terminus and a variable C-terminus (Potts and Saunders, 1988). Variation at the PilE C-terminus arises from RecA-dependent nonreciprocal recombination between the expressed *pilE* gene and several silent *pilS* loci found up to hundreds of bases away from the *pilE* gene (Perry et al. 1987, 1988; Potts and Saunders, 1988). Donation from a *pilS* gene to *pilE* is based on short sequence homology and occurs through several RecA-mediated crossover events between genes (Koomey et al., 1987; Sechman et al., 2006). Type IV pili are involved in attachment and colonization of *N. meningitidis* to mucosal membranes (McGee et al., 1983), and variations in pili have been linked to changes in antibiotic resistance (Manning et al., 1991) and adhesion to host surfaces (Stephens et al., 1982; Greenblatt et al., 1988; Rytkonen et al.,

2004). Gene conversion at the hypervariable region of *pilE* can be explained by frequency-dependent selection, as generation of novel antigenic variants is the key to the ability of *Neisseria* to colonize non-naive hosts (Andrews and Gojobori, 2004).

High rates of antigenic phase switching in *Neisseria*, as in the case with PorA and LPS, most likely evolved as a response to the selection pressure to establish a commensal relationship with non-naive hosts (Meyers et al., 2003; Schoen et al., 2008). *Neisseria* is typically cleared from the host in a few days to several months after initial colonization (Caugant et al., 2007), during which time the host immune system builds up a defensive response to *Neisseria* based on the recognition of bacterial surface antigens. Phase variation of surface antigens allows *Neisseria* to reestablish a commensal relationship with non-naive hosts that have built up adaptive immune responses from previous colonizations (Fig. 15.2b). Frequency-dependent selection at the spatial level explains the high degree of antigenic variation maintained in the *Neisseria* population, as strains with more common antigenic profiles will encounter a greater number of non-naive hosts. Antigenic phase variation allows *Neisseria* to continually evade host immune systems to colonize new hosts, regardless of the prior colonization status of the host. This is especially important, given that *Neisseria* colonizes approximately 10% of the population at any given time in industrialized countries (Fontanals et al., 1996).

Interestingly, *Neisseria* virulence could be viewed as a rare consequence of phase variation in which a commensal switches its antigenic profile to a pathogenic form, allowing tissue invasion and migration of bacteria into the host bloodstream (Meyers et al. 2003; Caugant et al., 2007). Pathogenic *Neisseria* are rarely transmitted between hosts; rather, pathogenicity arises from within a commensal population. Therefore, the selection pressure to establish commensal colonization of non-naive hosts likely drives selection for antigenic phase variation rather than pressure on pathogenic forms to infect non-naive hosts. As with *Haemophilus*, the mechanism for creating phenotypic diversity sheds light on selective forces acting on this species.

15.2.3 *Bacteroides*: Avoiding Innate Immune Responses

Bacteroides fragilis, a major gram-negative bacterial inhabitant of the human intestine (Ley et al., 2006; Xu et al., 2007), synthesizes a large number of phase-variable surface antigens using site-specific inversion (Kuwahara et al., 2004; Cerdeno-Tarraga et al., 2005). In this mechanism, short, inverted DNA repeats flank the invertible element, which typically contains a promoter for adjacent antigen-encoding genes (van de Putte and Goosen, 1992). These repeated sequences are recognized and brought together in a synapse by a DNA invertase, which cleaves DNA through strand exchange, resulting in reciprocal recombination and inversion of the DNA segment flanked by the repeated sequences (van de Putte and Goosen, 1992; Wisniewski-Dye and Vial, 2008). DNA invertases belong to one of two classes based on the mechanism by which they cleave and ligate DNA: the serine site-specific recombinases (Ssr) or the tyrosine (or lambda) site-specific recombinases (Tsr) (Gopaul and Van Duyne, 1999; Smith and Thorpe, 2002). Inversion results in changes in orientation of promoters for various genes, which in turn affects gene expression. Like contingency loci, invertible DNA regions produce random, heritable, and reversible changes in antigenic genotypes.

B. fragilis is able to produce eight distinct capsular polysaccharides determined by the expression of the PSA, PSB, PSC, PSD, PSE, PSF, PSG and PSH loci (Kuwahara et al., 2004; Cerdeno-Tarraga et al., 2005; Coyne et al., 2008). Expression of each capsular polysaccharide locus, with the exception of PSC, is regulated by specific inversions of DNA termed *fin* regions (Patrick et al., 2003). Promoters for the capsular polysaccharide-encoding loci (except for PSC) are located in the *fin* regions immediately upstream of each locus, with the transcription of each locus dependent on the orientation of its promoter (Coyne et al., 2003; Patrick et al., 2003; Kuwahara et al., 2004; Cerdeno-Tarraga et al., 2005). Inversion of *fin* sites mediated by the Ssr Mpi recombinase can switch genes on or off at random (Coyne et al., 2003), and the transcriptional status of each promoter is independent of the expression of other capsule-encoding loci. The only exception is the PSC locus, which produces a default capsular polysaccharide that is thought to act as a "fail-safe" in the event all seven other loci are turned off (Krinos et al., 2001; Coyne et al., 2003). Each *B. fragilis* cell has the capacity to express any suite of capsular polysaccharides simply based on the inversion status of *fin* regions, resulting in local host-level population antigenic diversity.

Production of multiple surface polysaccharides has been demonstrated for the successful long-term colonization of the intestine by *B. fragilis* (Coyne et al., 2003; Liu et al., 2008). Because the human intestinal ecosystem is a very dynamic and competitive environment, many factors could be responsible for maintaining diverse surface antigens in *B. fragilis*. Avoidance of bacteriophages, adhesion to changing intestinal surfaces, or competition with other commensals or pathogenic bacteria could also be involved in the maintenance of mechanisms that permit *Bacteroides* to vary its surface antigenic profile. *B. fragilis* expression of PSA has also been shown to actively protect against intestinal colitis caused by *Helicobacter hepaticus* in an animal model (Mazmanian et al., 2008), further highlighting the complex role antigenic phase variation plays in the lifestyle of *B. fragilis*. Underlying these complex interactions is the close association *B. fragilis* forms with the intestinal mucosa. Because *Bacteroides* species form a majority of the cells in the intestinal microbiota (Savage, 1977; Backhed et al., 2005; Gill et al., 2006), it is a likely target for sampling by dendritic cells, which would result in IgA excretion targeting overrepresented O-antigen epitopes (Macpherson and Uhr, 2004; Macpherson, 2006; Macpherson and Slack, 2007). Therefore, continual variation of surface antigens likely protects *B. fragilis* from attack by the host immune system (Kuwahara et al., 2004). Within-host frequency-dependent selection may favor a diverse array of capsular polysaccharide production by *B. fragilis*, as the continual presence of a predominant polysaccharide antigen may result in the host immune system mounting defenses against *B. fragilis* and clearing it from the intestine.

Many site-specific invertible regions found in bacteria, especially those of the Ssr family, have been imported from bacteriophages (Smith and Thorpe, 2002). Bacteriophage P1 contains the *cin* recombinase that controls the phase-variable expression of tail fiber genes, altering the host range of P1 depending on which tail fibers are expressed (Iida et al., 1982; Hiestand-Nauer and Iida, 1983; Iida, 1984). Host range of the temperate coliphage Mu is also dependent on a site-specific recombinase, encoded by the *gin* gene (Kamp et al., 1978; van de Putte et al., 1980). *E. coli* encodes the Pin recombinase, which is involved in flagellar variation; it is similar to the Gin and Cin proteins and rescues Mu *gin* mutants (Enomoto et al., 1983; Plasterk et al., 1983). DNA invertases can control a wide variety of phenotypes in bacteria and bacteriophages that are under frequency-dependent selection, including host range and antigenic variation. Thus, this mechanism of phase variability, used by *Bacteroides* to provide generation timescale diversity, has the potential to be widely distributed among bacteria.

15.2.4 Mechanisms for Generating Diversity Reflect an Organism's Selective Regime

In all of the cases discussed above, it is beneficial for cells of one antigenic type to yield an offspring of a different antigenic type. The continual switching of surface antigens presents a host–pathogen arms race, in which antigen switching occurs in response to selective pressure from the host adaptive and innate immune systems. *Neisseria*, *Haemophilus*, and *Bacteroides* contact the adaptive and/or innate components of the host immune system, so it is advantageous for these organisms to maintain molecular mechanisms that permit continual switching of surface antigen profiles to evade immune defenses within and among hosts. The maintenance of molecular mechanisms that permit frequent antigenic phase switching can allow microorganisms to prolong infection or colonization within an individual host and can increase the likelihood of infecting non-naive hosts. Frequency-dependent selection explains the maintenance of antigenic diversity for such organisms, as host immune responses prevent any one antigenic profile from dominating a population of infectious or commensal microorganisms for more than a brief period of time.

When a particular surface antigen becomes common in a population, the chances that host immune systems mount defenses against that antigen increase. Once an immune response is mounted against a particular surface antigen, cells expressing that surface antigen are more likely to be eliminated by the host immune system, whereas cells that have switched antigens can continue to evade host defenses, prolonging infection within a host or spreading to non-naive hosts. Under conditions that favor frequency-dependent selection mediated by host immune systems, populations of microorganisms that possess the capacity to generate an offspring with different surface antigens than parent cells have a greater survival advantage over microorganisms that are unable to vary cell surface antigens.

Perhaps nowhere is this phenomenon better illustrated than with the HIV, where excess phenotypic diversity is one of the main obstacles to creating successful treatment methods and potential vaccines for HIV (Rambaut et al., 2004). Here, the mechanism generating diversity is intrinsic to viral reproduction, but the concept is the same. HIV and other retroviruses are especially prone to mutation, as the molecular mechanisms used for retroviral reproduction (lack of proofreading and short replication time) are in themselves highly mutagenic (Galetto and Negroni, 2005; Ramirez et al., 2008). In addition to its high mutation rate, HIV undergoes approximately three recombination events per genome per replication, one of the highest recombination rates of all organisms (Zhuang et al., 2002). In HIV, recombination typically occurs between two coinfecting virus particles through the actions of the viral reverse transcriptase, which can switch between the strands of the copackaged viruses (Ramirez et al., 2008) and can occur between viruses of the same subtype, different subtype, or different groups (Fang et al., 2004; Kalish et al., 2004; McCutchan et al., 2005; Iwabu et al., 2008). High rates of mutation and recombination create a virus population with high, continually changing antigenic diversity, which allows HIV to rapidly adapt surface antigens in response to selective pressure from host immune systems. HIV is able to prolong within-host infection through rapid changes of surface antigens and other properties, which enables it to evade host immune system defenses.

15.3 ANTIGENIC DIVERSITY IN *SALMONELLA*

Salmonella serovars are classified by their two highly variable antigens, the O-antigen polysaccharide and the H-antigen flagellar filament. As we discuss here, the mechanisms producing this diversity vary greatly from the mechanisms discussed above. Therefore, the selective regime responsible for maintenance of that diversity must be different than selective pressure from the host immune system driving antigenic diversity in *Neisseria*, *Haemophilus*, and *Bacteroides*.

15.3.1 Salmonella H-Antigen Diversity

The *Salmonella* H antigen is conferred by flagellin, the major filament of its peritrichous flagellae. Unlike many constitutively expressed antigens, flagellae are expressed only under certain environmental conditions. Expression of flagellin is regulated by the flagellar master regulator genes *flhCD* in response to starvation conditions when locomotion is advantageous (Kutsukake, 1997; Yanagihara et al., 1999). The FlhCD master regulatory proteins upregulate numerous genes for the synthesis of the flagellum, motor proteins for flagellar rotation, and the chemotaxis signal-transduction system, which controls the direction of rotation. The flagellar filament is the final, outermost portion of the bacterial flagellum to be synthesized and assembled; because they form the exposed portion of the flagellum, *Salmonella* flagellins are targets of both the innate and adaptive host immune systems (Salazar-Gonzalez and McSorley, 2005; Sanders et al., 2006, 2008, 2009; Nempont et al., 2008). Flagellin binds Toll-like receptor 5 (TLR5), activating a proinflammatory response by the innate immune system (Andersen-Nissen et al., 2005; Feuillet et al., 2006). Flagellin is also recognized by memory $CD4^+$ T cells, which are involved in the clearance of *Salmonella* from infected phagocytes (Bergman et al., 2005a,b; Cummings et al., 2005).

The most common form of flagellin found in *Salmonella* is encoded by the *fliC* gene, which is embedded within the major flagellar gene locus. The *fliC* alleles found in different serovars of *Salmonella* are quite diverse (Fig. 15.4a); in *E. coli* (where they have been studied more intensively), it has been proposed that strong frequency-dependent selection leads to this diversity (Wang et al., 2003). In addition, most *Salmonella* serovars have the capacity to produce one of two possible forms of antigenic flagellin at any one time, with the alternate (H2 antigen) flagellin-encoding genes found in the unlinked *fljBA* operon (Kalir et al., 2001; Aldridge et al., 2006; Yamamoto and Kutsukake, 2006). Like the *fliC* gene, alleles of the *fljB* flagellin gene are also hypervariable (McQuiston et al., 2004). In general, different *fljB* and *fliC* alleles are well conserved at the 5′ and 3′ ends with hypervariable regions in the middle of the genes (Fig. 15.4a; Smith and Selander, 1990; Wang et al., 2003; McQuiston et al., 2004). These regions correspond to the functionally constrained N- and C-termini and the antigenically exposed middle domain of flagellin (McQuiston et al., 2004), respectively. The diversity of flagellin-encoding genes accounts for 114 different serotype combinations made of 99 antigenically distinct H-antigen factors (Grimont and Weill, 2007). In a minority of serovars, plasmid-encoded elements create production of a third flagellar phase or influence the H1 or H2 serotypes (Smith and Selander, 1991; Baker et al., 2007).

Phase-variable expression of the two flagellins is controlled by site-specific inversion of *hin*, a region of DNA most likely arising from the integration of a Mu-like phage (van de Putte and Goosen, 1992); the *hin* region bears structural similarity to the *fin* regions of *B. fragilis* and the *cin* region of bacteriophage P1 (Patrick et al., 2003). Aside from the gene for the *cis*-acting invertase, the *hin* region contains the promoter for the *fljBA* operon (Fig. 15.4b,c). Under appropriate environmental conditions, expression of flagellae is turned on by the actions of the flagellar transcriptional regulatory factors FlhCD (Kutsukake, 1997). To begin, the type III secretory system that comprises the transmembrane core of

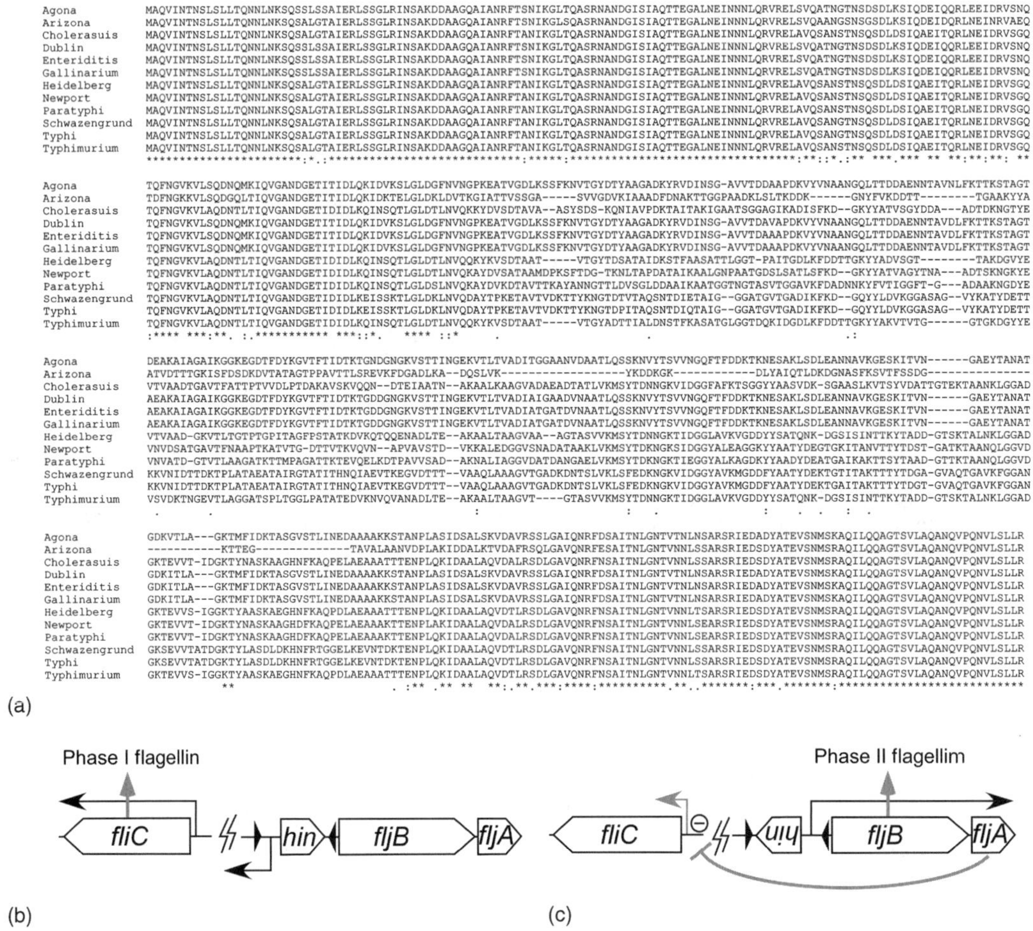

Figure 15.4 (a) Diversity at the *Salmonella fliC* locus. (b) Mechanism of phase switching at the *Salmonella fljBA* locus. Phase I flagellin is expressed from the *fliC* gene. (c) When the *hin* region inverts between the two *hix* sites, phase II flagellin is expressed from the *fljB* gene, while the FljA repressor prevents *fliC* expression.

the flagellum is synthesized (Chilcott and Hughes, 2000). After this is complete, a flagellin gene is expressed, and export of the flagellin results in filament synthesis outside the cell (Chilcott and Hughes, 2000). When the *hin* gene is in one orientation (Fig. 15.4b), the *fljBA* promoter is in the incorrect orientation for *fljBA* expression; the *fljBA* operon remains silent and the *fliC* gene is transcribed, thus producing the H1 antigen (Aldridge et al., 2006). Stochastic expression of the *hin* gene results in site-specific recombination at the two *hix* sites that flank the *hin* gene (Zieg and Simon, 1980). In the opposite orientation, the *fljBA* operon acquires an indirectly FlhCD-responsive promoter. FljA represses both the transcription and translation of the flagellin-encoding *fliC* gene (Aldridge et al., 2006; Yamamoto and Kutsukake, 2006), while FljB encodes an alternate flagellin, presenting the H2 antigen (Fig. 15.4c). Hence, switching between the H1 and H2 antigens is reversible, heritable, and occurs at random on a generation timescale.

While phase switching allows two H antigens to be variably expressed, the variable H-antigen-encoding genes found in the *Salmonella* population are not variably expressed on generation timescales. That is, variable alleles are found in the population, but *Salmonella* cells produce daughter cells with the same two H-antigen flagellins as their parents. Moreover, the conservation of the central, flagellin-variable domain (Smith and Selander, 1990; Wang et al., 2003; McQuiston et al., 2004) suggests that mutations that alter the sequence of the flagellin are counter-selected rather than placed under positive selection. That is, if frequency-dependent selection acted on these proteins, one would expect an excess of nonsynonymous substitution in the variable domain of the flagellin gene; however, this is not observed. Therefore, *Salmonella* H-antigenic diversity has a generation timescale component in its phase switching, but population-level selection must act to maintain excess diversity at the constituent *fliC* and *fljB* loci.

15.3.2 *Salmonella* Fimbrial Diversity

Salmonella has several different types of fimbriae, which are involved in attachment of *Salmonella* to intestinal epithelia (Humphries et al., 2001; Althouse et al., 2003). Some of these adhesion factors undergo phase variation, although the molecular mechanism by which this is accomplished is poorly characterized compared to that of the H antigen. The long polar fimbriae-encoding *lpf* operon undergoes generation timescale, heritable phase variation, and expression is required for *Salmonella* colonization of Peyer's patches (Humphries et al., 2005). LpfA, the major subunit of long polar fimbriae, has been shown to elicit an antigenic response in mice (Humphries et al., 2005). Although the role of fimbrial diversity present in *Salmonella* is unclear, the capacity of *Salmonella* to phase regulate the expression of the *lpf* operon demonstrates that *Salmonella* is capable of using multiple molecular mechanisms of phase variation to regulate surface antigenic diversity.

15.3.3 Salmonella O-Antigen Diversity

The O antigen, the outermost layer of the gram-negative LPS, is a repeating sugar unit found on the outside of the cell and is the most abundant cell surface molecule in *Salmonella* (Schnaitman and Klena, 1993; Samuel and Reeves, 2003). The O antigen is synthesized by various sugar synthases and transferases encoded by the *rfb* genes, a cluster of genes typically 10–35 kb in length located between the *gnd* and *galF* genes on the *Salmonella* physical map (Brahmbhatt et al., 1988; McClelland et al., 2001; Reeves and Wang, 2002). While flanking genes are little changed among *Salmonella* serovars, the *rfb* operon varies widely in gene composition among *Salmonella* serovars (Reeves, 1993; Reeves et al., 1996; Reeves and Wang, 2002; Fig. 15.5a). For example, serovars Typhi, Typhimurium, and Dublin share a common set of genes responsible for the 4, 12 and 9, 12 serotypes (the 1 epitope is conferred by a prophage and the [5] epitope is conferred by an unlinked gene). Yet, the *rfb* operon conferring the 3,10 serotype on serovar Weltevreden has gained and lost numerous genes relative to serovar Typhimurium. The Choleraesuis serovar is even more extreme, sharing absolutely no genes in common with serovar Typhimurium in the *rfb* operon (Fig. 15.5a). This variability in gene content results in varying patterns of content, linkage, and order of sugars composing the antigen (Samuel and Reeves, 2003). Variation at the *rfb* locus arose over time by a series of horizontal gene transfer events (Verma et al., 1988; Jiang et al., 1991; Liu et al., 1991; Brown et al., 1992; Wang et al., 1992, 2002, 2007; Reeves, 1993; Xiang et al., 1993, 1994; Curd et al., 1998; Li and Reeves, 2000; Kaniuk et al., 2002; Reeves and Wang, 2002), where the

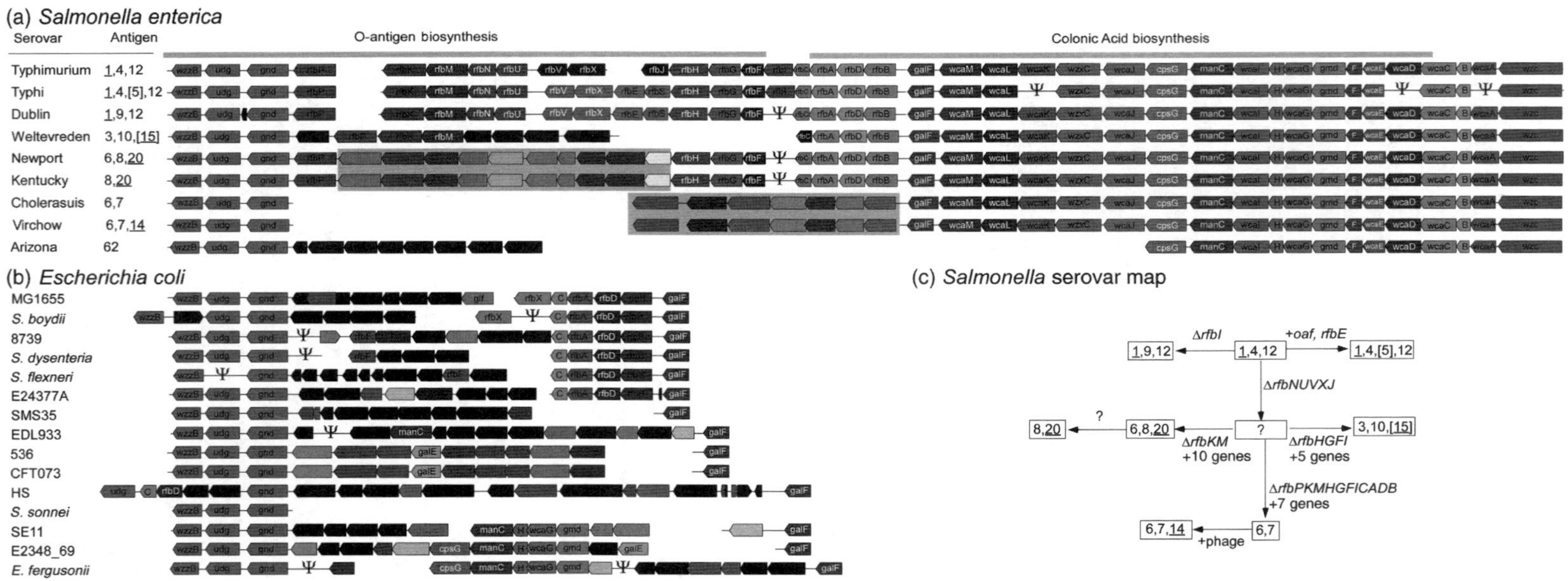

Figure 15.5 (a) Alignment of *rfb* operon regions of *Salmonella* strains. Orthologous genes are shaded with the same color. (b) Alignment of *rfb* operon regions of *E. coli* strains. Genes with *Salmonella* orthologues are shown in cognate colors. (c) Differences between *Salmonella* O antigens. A color version of this figure appears in the center of this volume. See color insert.

introduced genes encode enzymes for the synthesis and assembly of novel sugar configurations or compositions. With respect to the O antigen, the only time a cell can become antigenically distinct from its parent is through the acquisition and maintenance of a new *rfb*-like gene from another bacterium.

Mutating the *rfb* locus affects the pathogenicity of *Salmonella*, but it is not clear why. Although LPS does provide a mechanism for adhesion to eukaryotic cell surfaces, SPI-1-encoded genes are responsible for pathogenicity-specific cell adhesion. *Salmonella* also contains numerous fimbrial genes involved in many forms of adhesion (Humphries et al., 2001). Therefore, *rfb*-encoded genes likely influence the efficacy of infection via a more indirect route. Similar patterns of variability in *rfb* operon composition are seen among serovars of the closely related species *E. coli* (Fig. 15.5b). In both cases, recombination has introduced foreign genes into the operon, enabling different sugars to be synthesized and novel biochemical linkages to be created. This results in differences among O-antigenic types that reflect stable gene loss and gain (Fig. 15.5c), not mutational change, slipped-strand mispairing, gene conversion events, or site-specific recombination in invertible segments.

Although the sugar synthases and transferases encoded by the *rfb* genes are responsible for the majority of the hypervariable region of the O antigen, other genes have been shown to affect the serotype of the O antigen. In serovar Typhimurium LT2, O-factor [5] is an acetylation of the O antigen mediated by *oafA* (Slauch et al., 1996), and O-factor 27 is due to $wzy_{\alpha 1-6}$ (Wang et al., 2002), an O-antigen chain-length regulator closely linked to *rfb*. The O antigen is a crucial virulence determinant and most likely also protects *Salmonella* from harsh environmental conditions such as desiccation (Thomsen et al., 2003; Garmiri et al., 2008). It is not clear what roles are played by O-antigen modifications, and the range of modifications catalyzed—and their distributions among strains—is not well studied. For these reasons, this review focuses on variation at the *rfb* locus itself.

15.4 WHY ARE DIVERSE H AND O ANTIGENS MAINTAINED IN *SALMONELLA*?

As illustrated in Fig. 15.5, the phenotypic diversity of *S. enterica* serovars reflects the structurally distinct *rfb* operons they harbor, wherein nonhomologous genes encode enzymes for the synthesis of different sugars and their attachment into structurally distinct polysaccharides to be placed on the exterior of the cell. Aside from the notable changes in gene inventory at the *rfb* operon, the genes flanking this locus show elevated diversity as well (Fig. 15.6), again suggesting that variation-purging selective sweeps do not affect this region of the chromosome. This phenomenon was first described in the late 1980s, when alleles of the *rfb*-proximal *gnd* gene (see Fig. 15.5 for location) proved far more diverse in strains of *E. coli* than alleles of other loci (Dykhuizen and Green, 1991). Unlike genes elsewhere in the *Salmonella* or *E. coli* chromosomes, genes flanking the *rfb* locus maintain very high levels of polymorphism; this increase in variability in the *rfb* region is evident when multiple genes are assessed using complete genome sequences (Fig. 15.6). Rather than reflecting unusual selective regimes affecting these loci directly, the excess variation has been attributed to linkage to the *rfb* operon (Nelson and Selander, 1994; Milkman et al., 2003). At loci unlinked to the *rfb* operon, beneficial alleles may arise by mutational processes. Selective sweeps operate to purge diverse alleles as the beneficial allele is transferred among strains by homologous recombination (e.g., see Guttman and Dykhuizen, 1994). Such selective sweeps could not occur in the proximity of the *rfb*

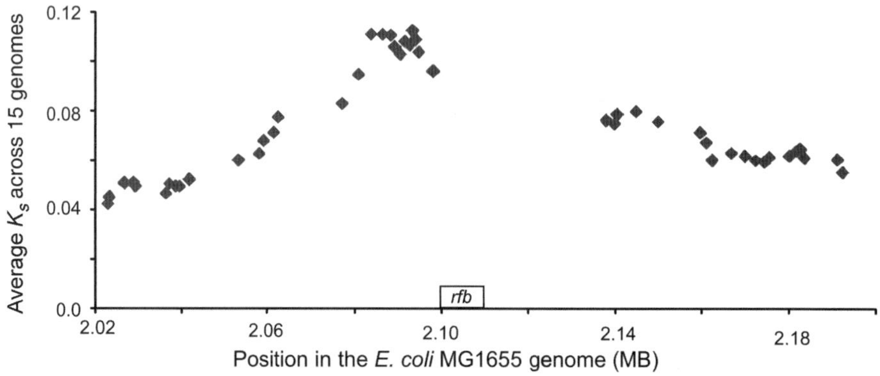

Figure 15.6 Genetic diversity of genes flanking the *E. coli rfb* operon. The average divergence of synonymous sites among pairwise comparisons of genes shared among 16 completely sequenced *E. coli* genomes is plotted according to their position in the *E. coli* K12 genome. The diversity of the set of completely sequenced *E. coli* strains enabled this genome-scale analysis; the poor sampling (as yet) of completely sequenced *Salmonella* genomes precluded a robust analysis in that taxon at this scale.

operon if variant forms have selective value because recombinants would have decreased fitness.

This excess genetic variability raises an interesting conundrum. As discussed above, many organisms have mechanisms for producing antigenic diversity at the capsular or O-antigen polysaccharide using generation timescale mechanisms that act upon otherwise statically encoded information that is similar between strains. *Salmonella* utilizes such mechanisms to generate phenotypic diversity at the H antigen and at fimbriae, where genotypically similar organisms can be phenotypically distinct. Yet, *Salmonella* does not alter the O antigen on a generation timescale. Thus, any model invoking frequency-dependent selection as a rapid response to changing environmental conditions—either to extend an infection or to infect non-naive hosts—cannot apply to the diversity being maintained at the O-antigen-encoding *rfb* operon.

This is not to say that the H and O antigens do not experience such selective pressure. Like the H antigen, the O antigen is a target of the host innate immune system and is recognized by the Toll-like receptor 4 (TLR4) (Muroi and Tanamoto, 2002; Royle et al., 2003; Vazquez-Torres et al., 2004). The O antigen in complex with LPS stimulates production of antibodies by the adaptive immune system, which has been shown to at least temporarily protect a host from reinfection (Ding et al., 1990; Robbins et al., 1992; Muthukkumar and Muthukkaruppan, 1993). Because the O- and H-antigens elicit a host immune response, frequency-dependent selective pressure from the immune system has been offered to explain *rfb* diversity. But given the lack of a mechanism to create generation timescale diversity, as well as the robust phenomenon of host/serovar specificity (Fig. 15.2), the failure of many studies attempting to link O-antigen diversity to variations in pathogenicity and immune system evasion is not surprising (Reeves, 1995; Baumler et al., 1998; Bolton et al., 1999; Kingsley and Baumler, 2000; Uzzau et al., 2001; Milkman et al., 2003).

The lack of variation-purging recombination at or near the *Salmonella rfb* locus suggests that the different diverse forms are advantageous for the strains that stably harbor them. That is, there is no *rfb* allele that is favored by all strains of *Salmonella*. Whereas many bacteria employ mechanisms for creating generation timescale diversity that ensures production of antigenically distinct progeny, *Salmonella* produces daughter cells that retain phenotypic identity with their parents at the O antigen even while differing at other antigens. This suggests *Salmonella* strains that differ at the O-antigen lead significantly different lifestyles wherein each different O-antigen form provides an advantage not realized by other strains. While this argument essentially assigns the different serovars to distinct niches, we do not consider these strains to be representatives of wholly different species. Strains of *Salmonella* do experience interstrain homologous recombination, leading to species-wide selective sweeps. These sweeps simply do not occur at the *rfb* operon.

15.4.1 Diversifying Selection in Salmonella

We suggest that the different O antigens stably expressed by different serovars of *Salmonella* confer advantages in different environments; thus, genetic diversity above that predicted by the neutral theory would reflect the action of diversifying selection rather than frequency-dependent selection. Here, no single allele can confer a benefit in all environments, precluding a selective sweep at the *rfb* operon. Moreover, genes closely linked to the *rfb* operon also fail to experience selective sweeps since recombination there would likely result in problematic introduction of less-suited genes at the *rfb* operon (see Fig. 15.6). Yet, recombination at unlinked loci would still occur, providing strains of *Salmonella* with the genotypic and phenotypic cohesion expected of a bacterial species (Dykhuizen and Green, 1991). Thus, *Salmonella* strains could be considered different strains of the same species everywhere except at the *rfb* operon, where they carry adaptations to distinctly different environments.

The phenomenon of host–serovar specificity is consistent with this model. Unlike other bacterial pathogens, specific serovars of *Salmonella* are consistently associated with disease states in different host animals (Fig. 15.2). That is, rather than requiring infecting strains present an O-antigen that is novel to the vertebrate host, *Salmonella* strains that mount successful infections consistently present the same antigen to particular hosts (Rabsch et al., 2002). These data support the hypothesis that the nature of the host environment favors particular O-antigenic types of *Salmonella*.

This model is not at odds with the observation that other intestinal bacteria—for example, species of *Bacteroides* as discussed above—gain a benefit from producing daughter cells with variant O antigens. Rather, we posit that the advantage gained by *Salmonella*'s retention of its parental O antigen outweighs any detriment this lack of variability incurs. *Bacteroides* is a major constituent of the intestinal microflora (Gill et al., 2006; Ley et al., 2006); therefore, it is likely to be heavily sampled by the host immune system resulting in targeted IgA excretion (Macpherson and Uhr, 2004). Because *Salmonella* is such a rare member of the intestinal microbiota, it would not be targeted by the immune system, and constant switching of the O antigen would not be advantageous. Instead, we propose that the particular features of its parental O antigen would provide more direct benefits.

15.4.2 Differential Distribution of Bacterial Strains

Central to this hypothesis is the supposition that *Salmonella* is differentially distributed in natural environments. Beyond the pattern of host–serovar specificity (Fig. 15.3), data from several species of intestinal bacteria indicate that genotypes are differentially distributed among host species. This phenomenon has been exploited for microbial source tracking (MST), whereby the source of fecal contamination in water is preliminarily identified by virtue of the genotypes of contaminating bacteria found in the water (Scott et al., 2002; Simpson et al., 2002; Barnes and Gordon, 2004). Common methods for MST include rep-PCR, pulsed-field gel electrophoresis, viral typing, antibiotic resistance profiles or multilocus sequence typing (Griffith et al., 2003; Harwood et al., 2003; Myoda et al., 2003); all methods are grounded in the observation that genotypes of intestinal bacteria are not randomly distributed in the enteric environments of mammalian hosts and exploit these patterns to infer the source of water contamination.

Differential distribution of bacteria begins at the species level where, for example, genera of enteric bacteria are differentially distributed among major lineages of mammals showing that mammalian intestines are not all uniform environments (Gordon and FitzGibbon, 1999). Closely related species within a genus are also differentially distributed, such as the strains of *Enterococcus* used in MST (Wheeler et al., 2002); here, water being contaminated from untreated human wastewater can be discriminated from runoff from a cattle farm by the relative abundances of *Enterococcus* strains. In addition, strains of *Bacteroides* have differential distribution among intestinal environments (Cotta et al., 2003; Huang et al., 2003; Rigottier-Gois et al., 2003). Lastly, and most importantly from our perspective, genotypically distinct strains within a single bacterial species can also be differentially distributed. For example, strains of *E. coli*, another species widely used in MST (Johnson et al., 2004; Ram et al., 2004; Stoeckel et al., 2004), are differentially distributed among mammals, whereby mammals having different diets or dwelling in different environments harbor different genotypes of *E. coli* (Gordon et al., 2002; Gordon and Cowling, 2003). The use of genotypic differences among bacteria found in different intestinal environments for MST indicates that these differences are stable, robust, and repeatable.

Why do different environments favor *Salmonella* with different O antigens? Although *Salmonella* is broadly noted for its pathogenic effects, it also dwells in intestinal environments as a harmless commensal, likely for a much greater percentage of time. For example, many captive reptiles asymptomatically shed *Salmonella* that can be transmitted to humans, especially small children or immunocompromised individuals, resulting in pathogenic infection (Cohen et al., 1980; Bergmire-Sweat et al., 2008). Interestingly, while the conditions in the reptile intestine favor a more commensal lifestyle for *Salmonella*, the introduction of these same cells into human intestines results in a switch to pathogenicity. Unlike *Neisseria* and *Haemophilus*, where the switch from commensal to pathogen is dependent on immune system evasion mediated by O-antigen phase variation, the same *Salmonella* cells harmlessly inhabiting reptiles cause gastroenteritis in humans without any change in the nature of the O antigen. Particular O antigens could contribute to differential survival in different environments independent of pathogenic behavior.

The reasons why *Salmonella* can adopt a commensal lifestyle in one organism and can cause pathogenic infection in another organism are complex. Many genes could show adaptive differences in response to abiotic variation among environments, such as differences in oxygen tension, pH, ionic strength, salinity, or the availability of nutrients. Yet, it is difficult to attribute advantages to particular O antigens in response to such differences. A particular O antigen could also provide greater competitive abilities in certain environments; for example, they may mediate more effective adhesion to some intestinal mucins, resulting in a greater chance for invasion of intestinal epithelial cells. However, adhesion to intestinal mucins is very complex and most likely involves many factors, and the role of differential adhesion to mucin mediated by different O antigens is not particularly well tested. In addition, *Salmonella* possesses several other mechanisms for attachment, including fimbriae (Chessa et al., 2009) and SPI-1 (Klein et al., 2000), making differential mucin attachment an unlikely explanation for O-antigen diversity. Lastly, different O antigens may provide defense against predation in the intestine.

15.4.3 Predation as a Selective Force

The connection between O-antigen variation and predation is clear: the O antigen is the most abundant molecule on the surface of the cell, thereby being a likely ligand for predator/prey interactions. Many organisms are potential predators of bacteria in intestinal environments, such as ciliates, bacteriophages, and amoebae. Ciliates will consume any bacteria that are sufficiently small to pass through their feeding comb, and bacteriophages are highly specialized on very few strains within any given bacterial species. In contrast, amoebae are generalist predators, recognize prey by cell–cell contact, and are abundant predators in water, soil, and intestinal environments (Rodriguez-Zaragoza, 1994; Hahn and Hofle, 2001; Ronn et al., 2002). Because phagocytotic amoebae rely on cell–cell contact to recognize their prey, then one would expect different serovars to be recognized with different efficiencies. Moreover, amoebae would be expected to be differentially distributed among vertebrate intestinal environments. As a result, amoebae could mediate diversifying selection at the O-antigen-encoding loci, allowing different serovars of *Salmonella* to gain fitness in particular environments where amoebae consume them less rapidly.

For many reasons we posit that an active response to potential predation is not likely. *Salmonella* are nonmotile in the gut, so cells do not have the capacity to swim away from predators unlike ciliates, which do exhibit a behavioral response to predation (Kusch, 1993). Bacteria do not have time to adapt behavioral responses to predation, as predation results in cell death. Because cell death is constantly occurring in the intestinal lumen, any chemical alarmone-mediated responses would be constitutively active independent of predation risk. The small size of bacteria prevents their escape by size refugia. Defensive cell wall thickening is not seen in vegetative cells, although they are a feature of persistent spores; defensive structures, such as those seen in protozoa (Kuhlmann and Heckmann, 1985; Wicklow, 1988; Fyda and Wiackowski, 1992) or *Daphnia* (Dodson, 1989), are neither evident nor thought to be effective since they would act on the molecular scale and do not impede chemical degradation in the food vacuole. Simply put, bacteria are unable to adopt many of the strategies employed by higher organisms to escape predation. The key to the bacterial prey–predator relationship is that successful bacteria outcompete other bacterial prey, which leads to a fitness advantage in the intestine. They do not avoid predation entirely; as long as a given *Salmonella* serovar is consumed by predators less efficiently than other serovars, the given serovar has a better chance of survival in that particular environment than other serovars. Therefore, predation will impact the genetic structure of species, like *Salmonella*, which is found across multiple environments, resulting in differential distribution of serovars across environments.

15.4.4 Amoebae-Mediated Diversity at the *Salmonella rfb* Operon

Facets of this model have been established with rigor. First, it is clear that amoebae consume bacteria as prey in natural environments. Not only have bactivorous amoebae been isolated from ground water and soil but also from intestinal environments (Wildschutte et al., 2004; Wildschutte and Lawrence, 2007). Amoebae within the intestinal lumen consume intestinal bacteria, thus limiting both bacterial growth yield and persistence time within the lumen of any individual host (Sharp et al., 1994). We posit that amoebae are the major general predators of bacteria found in vertebrate intestines, and that the feeding preferences of these amoebae affect the structure of intestinal bacterial populations.

For amoebae predators to influence the distribution of their prey, they cannot be randomly distributed in the environment; if they were, a prey bacterium could not persist in an environment where it could avoid all of the resident predators (since they would be constantly changing). Some survey experiments on pathogenic and commensal populations suggest that amoebae are not randomly distributed in the environment, with particular species of hosts harboring specific populations of amoebae. Differential distribution of pathogenic protozoa has been described for *Entamoeba*; *Entamoeba invadens* causes disease in reptiles (Donaldson et al., 1975), including ball pythons (Kojimoto et al., 2001), whereas *Entamoeba histolytica* causes disease in humans (Leber, 1999; Pozio, 2003). *Entamoeba suis* and *Entamoeba chattoni* infect nonhuman mammals, yet a related but distinct species preferentially infects birds (Martinez-Diaz et al., 2000). The amoeba *Vannella platypodia* was found to infect multiple fishes (Dykova et al., 1998), while *Neoparamoeba* preferentially colonizes gills (Fiala and Dykova, 2003). The microsporidian *Encephalitozoon cuniculi* is a pathogen of rabbits and dogs, whereas *E. intestinalis*, *E. hellem*, and *E. bieneusi* are opportunistic pathogens of humans (Wasson and Peper, 2000). Commensal protozoa also show differential distribution among hosts. For example, the nonpathogenic amoeba *Paravahlkampfia ustiana* was isolated multiple times from the intestines of skinks (Schuster et al., 2003). Lastly, our data show that there is differential distribution of *Naegleria*, *Hartmannella*, *Tetramitus*, and *Acanthamoeba* among amphibian, fish, and reptile intestinal tracts (Wildschutte and Lawrence, 2007). Based on this information, we conclude that the population of amoebae in a given host intestine is most likely stable and specific to that particular species of host. Therefore, when *Salmonella* cells enter an intestinal lumen, they are faced with predictable communities of amoebae that are encountered among all individuals of that host species.

Intestinal amoebae do not simply consume bacteria indiscriminately; rather, amoebae can discriminate between different bacterial strains provided as prey (Wildschutte et al., 2004). When presented with different serovars of *Salmonella* as prey, amoebae will consistently consume one serovar more quickly than another, less preferred serovar (Fig. 15.7). In this example, the fitness of any given serovar is calculated relative to the group of serovars; those serovars that are eaten faster by a particular predator have a low relative fitness, while those eaten more slowly have a higher relative fitness. This discrimination is evident even when amoebae are presented with both strains at the same time; one strain is consumed from the mixed population more quickly than its fitter competitor (Wildschutte et al., 2004). The basis for this discrimination is complex, but strains that differ only by virtue of their O antigens still experience differential predation (Wildschutte et al., 2004). Thus, we link the O antigen to *Salmonella* susceptibility to predation.

Not only do amoebae discriminate among prey bacteria based on the nature of the O antigen, but different amoebae also have different feeding preferences (Fig. 15.7). For example, while strain SARB36 was not readily consumed by the amoeba *Naegleria gruberi* strain NL, it was the most preferred strain when facing *Naegleria* strain F1-9 (Fig. 15.7). The feeding preferences of amoebae shown in Fig. 15.7 do not show any significant similarity ($R = 0.06$, $p > 0.1$), and these amoebae were isolated from different environments. This leads to the possibility that *Salmonella* serovars may experience differential survival in different environments as the result of their ability to avoid the resident amoeboid predators in that environment. That is, differential susceptibility to predation by

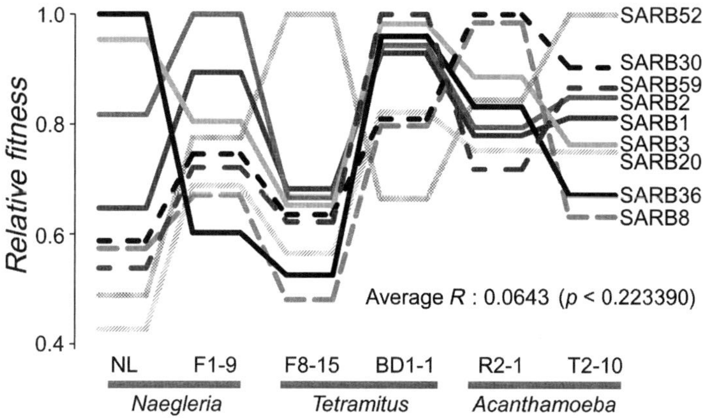

Figure 15.7 Fitness of *Salmonella* strains against protozoan predators. Each amoeba was tested against nine *Salmonella* reference collection B (SARB) strains. The feeding preferences of each predator were determined separately against nine antigenically diverse serovars of *Salmonella*. The least preferred strain was assigned a fitness of 1.0 for each predator. Data from Wildschutte and Lawrence (2007) and Wildschutte et al. (2004).

amoebae in a given host could influence which serovars from the entire *Salmonella* population may be best suited to survive there. Because the intestinal lumen of each host species harbors different populations of predators with different feeding preferences, no one form of the *Salmonella* O antigen can confer enhanced resistance to predation in all environments.

One prediction of this model is that unrelated predators that inhabit a single environment must share feeding preferences, which would allow a single serovar of *Salmonella* to escape predation by the suite of predators it would face in its preferred environment. When predators were isolated from the intestinal tracts of fish, their feeding preferences were significantly more similar than one would expect (Wildschutte and Lawrence, 2007). In the example shown here (Fig. 15.8a; $p < 10^{-5}$), even unrelated amoebae from two families isolated from the same host species shared a common set of feeding preferences. The similar fitness values of the five tested serovars against the 16 predators isolated from the same environment in Fig. 15.8a stands in stark contrast to their varying fitness values against six predators from a different environment shown in Fig. 15.7 (Wildschutte and Lawrence, 2007). Overall, predators isolated from the same environment (including the intestinal tracts of goldfish, tadpole, or turtles) shared feeding preferences (closed markers in Fig. 15.8b), whereas those from different environments do not (open markers in Fig. 15.8b). Unlike the dynamic interactions between pathogens and host immune systems that are observed with *Neisseria* and *Haemophilus*, *Salmonella* serovars face stable selective pressure from predation each time they encounter an individual of a particular species of host. The expression of an O antigen that confers enhanced resistance to predation in that particular environment affords that serovar with a greater chance of survival and replication within that host intestine. In contrast, other serovars are more readily consumed by the resident predators, eliminating those serovars from the environment in a manner akin to clearance from a host of *Neisseria* or *Haemophilus* strains having O antigens that are quickly recognized by a non-naive immune system.

The mechanistic basis for this congruence in feeding preference among unrelated amoebae isolated from the same host is not clear. One model posits that intestinal amoebae recognize host mucins as attachment sites, and unrelated amoebae share an inclination to

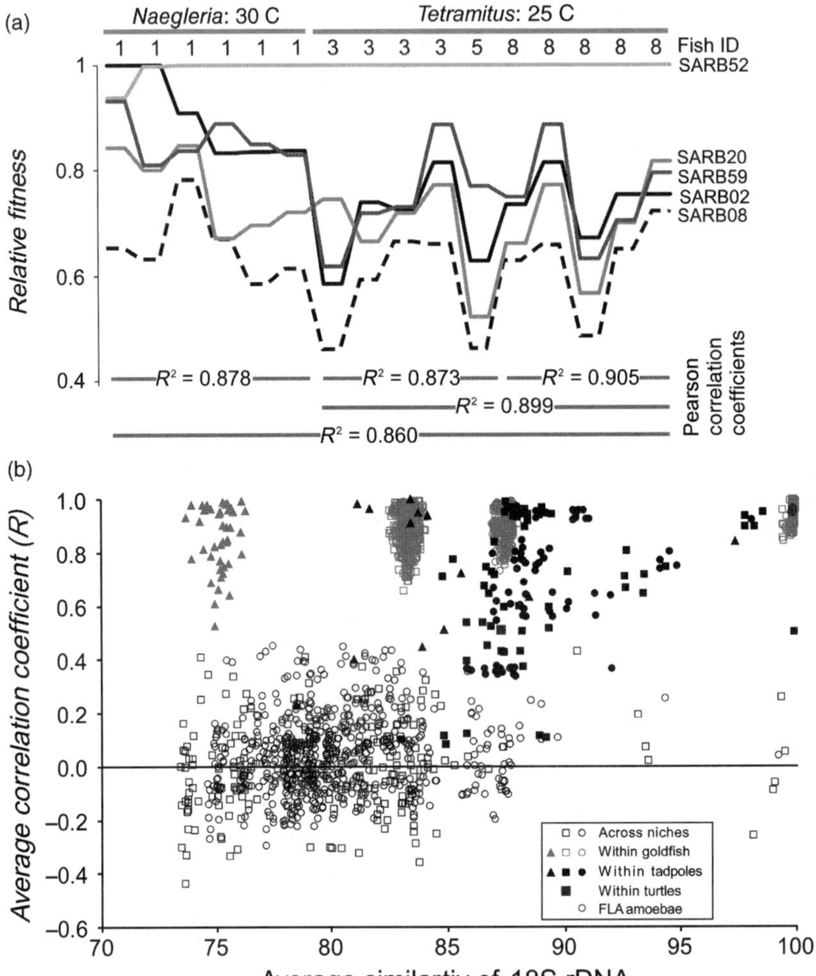

Figure 15.8 (a) Fitness of *Salmonella* against 16 predator amoebae isolated from four separate goldfish. Amoebae are numbered according to the fish from which it was isolated (fish number 1, 3, 5, or 8). (b) Relationship between feeding preference and environment. Average correlation coefficients are calculated for feeding preferences of two, three, and four different predators; these values are plotted against the average similarity of their 18S rDNA loci. Data for amoebae isolated from different environments are shown in open gray markers, and data from amoebae from the same environment are shown in black and/or closed markers. Adapted from Wildschutte and Lawrence (2007).

bind to the intestinal wall without trying to consume it (Wildschutte and Lawrence, 2007). Amoebae encounter two different sets of carbohydrates in their environment. The intestinal epithelium is covered in mucins, and binding to these carbohydrates allows the protozoa to remain in the lumen and avoid expulsion. The bacteria they consume are covered in LPS carbohydrates, and our experiments have demonstrated that they use the O antigens to discriminate among prey (Wildschutte et al., 2004; Wildschutte and Lawrence, 2007). This model proposes that amoebae bind differently to food than they do to intestinal mucins. If a bacterial O antigen resembles the intestinal mucins of its host, it could act as molecular camouflage. This similarly would confound intestinal amoebae attempting to

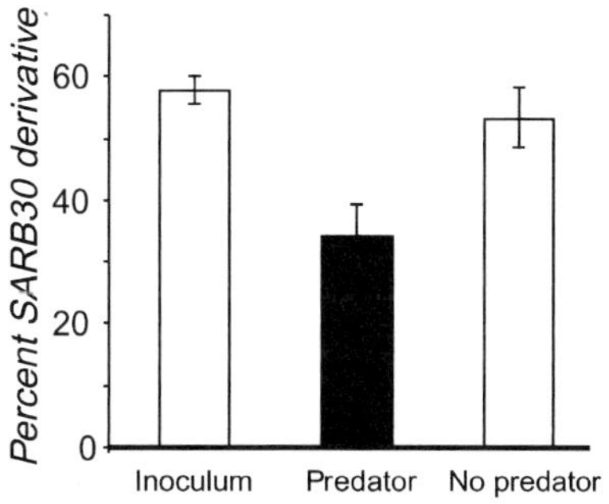

Figure 15.9 Predator-mediated survival of *Salmonella* within fish. The derivative of two antigenically distinct serovars of *Salmonella* was introduced into goldfish. The proportion of the SARB30 derivative was assessed in the food inocula and in the intestinal contents of metronidazole-treated and untreated goldfish after 24- to 48-h incubation.

discriminate between food and housing. Such camouflaged bacteria would benefit from increased resistance to predation over serovars with O antigens that are different from the mucins of that host. In effect, bacteria may better escape predation by molecular mimicry of host mucins.

By our model, one would predict that *Salmonella* would experience differential survival within intestinal environments by virtue of differential susceptibility to predation by intestinal amoebae. This was tested directly by introducing derivatives of two different *Salmonella* serovars into the intestinal tracts of goldfish by oral inoculation (Fig. 15.9; K. Butela and J. G. Lawrence, unpublished data). After 24–48 h, intestinal contents were isolated and the proportion of the two strains was assessed. Relative to the inocula, the proportion of the two *Salmonella* strains was unchanged in goldfish that had been treated with the antiprotozoal drug metronidazole. In contrast, the derivative of SARB30 (O-serotype 6,7) did much more poorly in fish where intestinal amoebae were present than the derivative of SARB20 (O-serotype 8,20). These data suggest that amoebae preferred to consume SARB30 over SARB20, thus mediating differential survivorship *in vivo*.

If predators mediate the differential survival of *Salmonella* in host intestinal tracts, any serovar bearing the O antigen that confers the greatest relative resistance to predation in that environment would have a high fitness in that particular gut environment relative to isogenic strains displaying other serotypes. In the example above, one could predict that any serovar with an 8,20 O antigen would have a higher fitness than any serovar with a 6,7 O antigen. Such a phenomenon has been observed in chickens, a natural host of *Salmonella*. For many years, the predominant illness-causing strains residing in chickens were members of serovar Gallinarium (Rabsch et al., 2002), which became the target of intense efforts to eradicate *Salmonella* from commercial chickens (Rabsch et al., 2000). While efforts to create Gallinarium-free chickens were successful, this niche was then filled by serovar Enteriditis (Rabsch et al., 2000); strikingly, these serovars share a common O-antigen serotype, 1,9,12. Based on our model, the 1,9,12 O antigen would confer increased resistance to predation from amoebae within the chicken intestine, and the selective elimination of Gallinarium from chickens simply provided the opportunity for another serovar with the 1,9,12 O antigen to occupy this niche. That is, we propose that the 1,9,12 O antigen provides a benefit to competitors to fill this niche in shielding the strains from predation; therefore, it was not surprising that serovar Enteriditis filled the niche vacated by serovar Gallinarium.

15.4.5 Diversifying Selection and the Nature of Bacterial Species

Antigenic diversity at the genes encoding the H and O antigens in *S. enterica* presents a conundrum of sorts. *Salmonella* has no mechanism for creating diverse O antigens; rather, O antigens are consistent between parent and daughter cells, and the same O-antigenic types cause disease in the same host species (Fig. 15.2). Yet, other intestinal organisms, such as *Bacteroides*, have mechanisms to generate O-antigen diversity on short timescales, presumably to escape the pressures exerted by the innate immune system. In general, we posit that *Salmonella* strains experience a greater advantage in retaining their O-antigen phenotype than they would reap if their O antigens could change. In the model detailed above, this advantage is manifested in their ability to mimic host mucins and to avoid predation by intestinal amoebae, all of which share feeding preferences when found in the same intestinal environment.

Yet, this model somewhat invalidates the question we are asking: Why are strains of the same species genetically diverse at antigenic loci? If different strains of *Salmonella* have advantages in different environments, why are they assigned to the same species? That is, different organisms persisting in different environments fit the description of different bacterial species, and it is not surprising that different bacterial species are genetically distinct. However, simply placing different serovars of *Salmonella* into different species does not solve this problem. *S. enterica* is described as a single species because recombination occurs between the serovars at many loci (Selander et al., 1996). As a result, selective sweeps distribute beneficial alleles at many loci across different *Salmonella* serovars. Therefore, they share a common gene pool and conform to plausible definitions of bacterial species (Dykhuizen and Green, 1991).

We can reconcile these viewpoints by proposing that serovars of *S. enterica* belong to the same species when assessed by many genes but belong to different species when assessed by the *rfb* operon, which adapts different serovars to different environments, each with different complements of intestinal predators. This model proposes that bacterial lineages can undergo speciation, but that species boundaries are not as well defined as in obligately sexual, eukaryotic lineages (Lawrence, 2002; Retchless and Lawrence, 2007). Gene flow in typical eukaryotes entails the formation of diploid zygotes via syngamy of haploid parental genomes; therefore, gene exchange requires the production of viable, fertile offspring. The lack of gene flow between species imposes genetic isolation to all genes in the genome. In bacteria, gene exchange involves the unidirectional transfer of small regions of the genome. As a result, two serovars of *Salmonella* can exchange genes readily at many loci—thereby placing them in the same species at those genes—but fail to exchange genes at other loci. We propose that genetic isolation is imposed at the *rfb* operon because particular O antigens adapt each serovar to particular intestinal environments by increasing their fitness against communities of intestinal predators. Since recombinants would have increased susceptibility to predation, they would be less fit, and cells recombining at the *rfb* locus would be counter-selected. Therefore, serovars of *Salmonella* can be considered genetically isolated at this locus or, in effect, at different species. Consistent with this model, the rfb operon appears to have been genetically isolated very early in the diversification of the nascent *E. coli* and *Salmonella* genomes (Retchless and Lawrence, 2007). While gene exchange was not occurring there, genetic exchange continued elsewhere in the genome for an additional 100 million years.

15.5 CONCLUSIONS

Like many pathogens, strains of *S. enterica* show excess genetic diversity at loci encoding antigenic determinants. Frequency-dependent selection has often been invoked to explain high levels of diversity in bacterial populations, and many bacteria harbor molecular mechanisms that promote the creation of generation timescale phenotypic diversity that is expected under these models. *Salmonella* differs in that diverse alleles are found within the population, but individuals are phenotypically stable. Here we presented a model whereby diversifying selection acts across the intestinal lumens of vertebrate hosts to favor different serovars of *Salmonella* in different environments. The interaction of *Salmonella* with differentially distributed amoebae predators is consistent with a role for predators in the differential survival of different serovars in different environments. In this way, the stable ecological niche of a bacterial strain, rather than its occasional participation in pathogenic behaviors, would provide the primary evolutionary force shaping its genetic diversity at these loci.

REFERENCES

Aldridge, P., Gnerer, J., Karlinsey, J. E., and Hughes, K. T. (2006) Transcriptional and translational control of the *Salmonella fliC* gene. *Journal of Bacteriology* **188,** 4487–4496.

Althouse, C., Patterson, S., Fedorka-Cray, P., and Isaacson, R. E. (2003) Type 1 fimbriae of *Salmonella enterica* serovar Typhimurium bind to enterocytes and contribute to colonization of swine *in vivo*. *Infection and Immunity* **71,** 6446–6452.

Andersen-Nissen, E., Smith, K. D., Strobe, K. L. et al. (2005) Evasion of Toll-like receptor 5 by flagellated bacteria. *Proceedings of the National Academy of Sciences of the United States of America* **102,** 9247–9252.

Andrews, T. D. and Gojobori, T. (2004) Strong positive selection and recombination drive the antigenic variation of the PilE protein of the human pathogen *Neisseria meningitidis*. *Genetics* **166,** 25–32.

Backhed, F., Ley, R. E., Sonnenburg, J. L. et al. (2005) Host-bacterial mutualism in the human intestine. *Science* **307,** 1915–1920.

Baker, S., Hardy, J., Sanderson, K. E. et al. (2007) A novel linear plasmid mediates flagellar variation in *Salmonella* Typhi. *PLoS Pathogens* **3,** e59.

Barnes, B. and Gordon, D. M. (2004) Coliform dynamics and the implications for source tracking. *Environmental Microbiology* **6,** 501–509.

Baumler, A. J., Tsolis, R. M., Ficht, T. A., and Adams, L. G. (1998) Evolution of host adaptation in *Salmonella enterica*. *Infection and Immunity* **66,** 4579–4587.

Bayliss, C. D., Field, D., and Moxon, E. R. (2001) The simple sequence contingency loci of *Haemophilus influenzae* and *Neisseria meningitidis*. *Journal of Clinical Investigation* **107,** 657–662.

Bergman, M. A., Cummings, L. A., Alaniz, R.C. et al. (2005a) CD4+-T-cell responses generated during murine *Salmonella enterica* serovar Typhimurium infection are directed towards multiple epitopes within the natural antigen FliC. *Infection and Immunity* **73,** 7226–7235.

Bergman, M. A., Cummings, L. A., Barrett, S. L. R. et al. (2005b) CD4+ T cells and Toll-like receptors recognize *Salmonella* antigens expressed in bacterial surface organelles. *Infection and Immunity* **73,** 1350–1356.

Bergmire-Sweat, D., Schlegel, J., Marin, C. et al. (2008) Multistate outbreak of human *Salmonella* infections associated with exposure to turtles–United States, 2007–2008. *Morbidity and Mortality Weekly Report* **57,** 69–72.

Bolton, A. J., Osborne, M. P., Wallis, T. S., and Stephen, J. (1999) Interaction of *Salmonella choleraesuis, Salmonella dublin* and *Salmonella typhimurium* with porcine and bovine terminal ileum *in vivo*. *Microbiology* **145**(Pt 9), 2431–2441.

Brahmbhatt, H. N., Wyk, P., Quigley, N. B., and Reeves, P. R. (1988) Complete physical map of the *rfb* gene cluster encoding biosynthetic enzymes for the O-antigen of *Salmonella typhimurium* LT2. *Journal of Bacteriology* **170,** 98–102.

Briones, V., Tellez, S., Goyache, J. et al. (2004) *Salmonella* diversity associated with wild reptiles and amphibians in Spain. *Environmental Microbiology* **6,** 868–871.

Brown, P. K., Romana, L. K., and Reeves, P. R. (1991) Cloning of the *rfb* gene cluster of a group C2 *Salmonella* strain: Comparison with the *rfb* regions of groups B and D. *Molecular Microbiology* **5,** 1873–1881.

Brown, P. K., Romana, L. K., and Reeves, P. R. (1992) Molecular analysis of the *rfb* gene cluster of *Salmonella* serovar Muenchen (strain M67): The genetic basis of the polymorphism between groups C2 and B. *Molecular Microbiology* **6,** 1385–1394.

Buzby, J. C., Roberts, T., Lin, C.-T. J. and MacDonald, J. M. (1996) *Bacterial Foodborne Disease: Medical Costs and Productivity Losses*, p. 93. United States Department of Agriculture, Washington, DC.

Caugant, D. A. (2008) Genetics and evolution of *Neisseria meningitidis*: Importance for the epidemiology of meningococcal disease. *Infection, Genetics and Evolution* **8,** 558–565.

Caugant, D. A., Tzanakaki, G., and Kriz, P. (2007) Lessons from meningococcal carriage studies. *FEMS Microbiology Reviews* **31,** 52–63.

Cerdeno-Tarraga, A. M., Patrick, S., Crossman, L. C. et al. (2005) Extensive DNA inversions in the *B. fragilis* genome control variable gene expression. *Science* **307,** 1463–1465.

Chambers, D. L. and Hulse, A. C. (2006) *Salmonella* serovars in the herpetofauna of Indiana County, Pennsylvania. *Applied Environmental Microbiology* **72,** 3771–3773.

Chessa, D., Winter, M. G., Jakomin, M., and Baumler, A. J. (2009) *Salmonella enterica* serotype Typhimurium Std fimbriae bind terminal alpha(1,2)fucose residues in the cecal mucosa. *Molecular Microbiology* **71,** 864–875.

Chilcott, G. S. and Hughes, K. T. (2000) Coupling of flagellar gene expression to flagellar assembly in *Salmonella enterica* serovar Typhimurium and *Escherichia coli. Microbiology and Molecular Biology Reviews* **64,** 694–708.

Cohen, M. L., Potter, M., Pollard, R., and Feldman, R. A. (1980) Turtle-associated salmonellosis in the United States. Effect of public health action, 1970 to 1976. *Journal of the American Medical Association* **243,** 1247–1249.

Cotta, M. A., Whitehead, T. R., and Zeltwanger, R. L. (2003) Isolation, characterization and comparison of bacteria from swine faeces and manure storage pits. *Environmental Microbiology* **5,** 737–745.

Coyne, M. J., Chatzidaki-Livanis, M., Paoletti, L. C., and Comstock, L. E. (2008) Role of glycan synthesis in colonization of the mammalian gut by the bacterial symbiont *Bacteroides fragilis. Proceedings of the National Academy of Sciences of the United States of America* **105,** 13099–13104.

Coyne, M. J., Weinacht, K. G., Krinos, C. M., and Comstock, L. E. (2003) Mpi recombinase globally modulates the surface architecture of a human commensal bacterium. *Proceedings of the National Academy of Sciences of the United States of America* **100,** 10446–10451.

Cummings, L. A., Barrett, S. L., Wilkerson, W. D. et al. (2005) FliC-specific CD4+ T cell responses are restricted by bacterial regulation of antigen expression. *Journal of Immunology* **174,** 7929–7938.

Curd, H., Liu, D., and Reeves, P.R. (1998) Relationships among the O-antigen gene clusters of Salmonella enterica groups B, D1, D2, and D3. *Journal of Bacteriology* **180,** 1002–1027.

Ding, H. F., Nakoneczna, I., and Hsu, H. S. (1990) Protective immunity induced in mice by detoxified *Salmonella* lipopolysaccharide. *Journal of Medical Microbiology* **31,** 95–102.

Dodson, S. I. (1989) The ecological role of chemical stimuli for the zooplankton: Predator-induced morphology in *Daphnia. Oecologia* **78,** 361–367.

Donaldson, M., Heyneman, D., Dempster, R., and Garcia, L. (1975) Epizootic of fatal amoebiasis among exhibited snakes: Epidemiologic, pathologic, and chemotherapeutic considerations. *American Journal of Veterinary Research* **36,** 807–817.

Dworkin, M. S., Shoemaker, P. C., Goldoft, M. J., and Kobayashi, J. M. (2001) Reactive arthritis and Reiter's syndrome following an outbreak of gastroenteritis caused by *Salmonella enteritidis. Clinical Infectious Diseases* **33,** 1010–1014.

Dykhuizen, D. E. and Green, L. (1991) Recombination in *Escherichia coli* and the definition of biological species. *Journal of Bacteriology* **173,** 7257–7268.

Dykova, I., Lom, J., Machackova, B., and Peckova, H. (1998) *Vexillifera expectata* sp. n. and other non-encysting amoebae isolated from organs of freshwater fish. *Folia Parasitology* **45,** 17–26.

Enomoto, M., Oosawa, K., and Momota, H. (1983) Mapping of the *pin* locus coding for a site-specific recombinase that causes flagellar-phase variation in *Escherichia coli* K-12. *Journal of Bacteriology* **156,** 663–668.

Fang, G., Weiser, B., Kuiken, C. et al. (2004) Recombination following superinfection by HIV-1. *AIDS* **18,** 153–159.

Feuillet, V., Medjane, S., Mondor, I. et al. (2006) Involvement of Toll-like receptor 5 in the recognition of flagellated bacteria. *Proceedings of the National Academy of Sciences of the United States of America* **103,** 12487–12492.

Fiala, I. and Dykova, I. (2003) Molecular characterisation of *Neoparamoeba* strains isolated from gills of *Scophthalmus maximus. Diseases of Aquatic Organisms* **55,** 11–16.

Fontanals, D., Van Esso, D., Pons, I. et al. (1996) Asymptomatic carriage of *Neisseria meningitidis* in a randomly sampled population. Serogroup, serotype and subtype distribution and associated risk factors. *Clinical Microbiology and Infection* **2,** 145–146.

Fyda, J. and Wiackowski, K. (1992) Predator-induced morphological defences in the ciliate *Colprium. European Journal of Protistology* **28,** 341.

Galetto, R. and Negroni, M. (2005) Mechanistic features of recombination in HIV. *AIDS Reviews* **7,** 92–102.

Garmiri, P., Coles, K. E., Humphrey, T. J., and Cogan, T. A. (2008) Role of outer membrane lipopolysaccharides in the protection of *Salmonella enterica* serovar Typhimurium from desiccation damage. *FEMS Microbiology Letters* **281,** 155–159.

Gill, S. R., Pop, M., Deboy, R. T. et al. (2006) Metagenomic analysis of the human distal gut microbiome. *Science* **312,** 1355–1359.

Gopaul, D. N. and Van Duyne, G. D. (1999) Structure and mechanism in site-specific recombination. *Current Opinion in Structural Biology* **9,** 14–20.

Gordon, D. M., Bauer, S., and Johnson, J. R. (2002) The genetic structure of *Escherichia coli* populations in primary and secondary habitats. *Microbiology* **148,** 1513–1522.

GORDON, D. M. and COWLING, A. (2003) The distribution and genetic structure of *Escherichia coli* in Australian vertebrates: Host and geographic effects. *Microbiology* **149,** 3575–3586.

GORDON, D. M. and FITZGIBBON, F. (1999) The distribution of enteric bacteria from Australian mammals: Host and geographical effects. *Microbiology* **145,** 2663–2671.

GREENBLATT, J. J., FLOYD, K., PHILIPPS, M. E., and FRASCH, C. E. (1988) Morphological differences in *Neisseria meningitidis* pili. *Infection and Immunity* **56,** 2356–2362.

GRIFFITH, J. F., WEISBERG, S. B., and MCGEE, C. D. (2003) Evaluation of microbial source tracking methods using mixed fecal sources in aqueous test samples. *Journal of Water and Health* **1,** 141–151.

GRIMONT, P. A. D. and WEILL, F.-X. (2007) *Antigenic Formulae of the Salmonella Serovars*. WHO Collaborating Centre for Reference and Research on *Salmonella*, Institut Pasteur, Paris.

GROSSMAN, N., SCHMETZ, M. A., FOULDS, J. et al. (1987) Lipopolysaccharide size and distribution determine serum resistance in *Salmonella montevideo*. *Journal of Bacteriology* **169,** 856–863.

GUTTMAN, D. S. and DYKHUIZEN, D. E. (1994) Detecting selective sweeps in naturally occurring *Escherichia coli*. *Genetics* **138,** 993–1003.

HAHN, D., GAERTNER, J., FORSTNER, M. R., and ROSE, F. L. (2007) High-resolution analysis of salmonellae from turtles within a headwater spring ecosystem. *FEMS Microbiology Ecology* **60,** 148–155.

HAHN, M. W. and HOFLE, M. G. (2001) Grazing of protozoa and its effect on populations of aquatic bacteria. *FEMS Microbiology Ecology* **35,** 113–121.

HAMMERSCHMIDT, S., MÜLLER, A., SILLMANN, H. et al. (1996) Capsule phase variation in *Neisseria meningitidis* serogroup B by slipped-strand mispairing in the polysialyltransferase gene (*siaD*): Correlation with bacterial invasion and the outbreak of meningococcal disease. *Molecular Microbiology* **20,** 1211–1220.

HANDELAND, K., REFSUM, T., JOHANSEN, B. S. et al. (2002) Prevalence of *Salmonella typhimurium* infection in Norwegian hedgehog populations associated with two human disease outbreaks. *Epidemiology and Infection* **128,** 523–527.

HARWOOD, V. J., WIGGINS, B., HAGEDORN, C. et al. (2003) Phenotypic library-based microbial source tracking methods: Efficacy in the California collaborative study. *Journal of Water and Health* **1,** 153–166.

HEINRICHS, D. E., YETHON, J. A., and WHITFIELD, C. (1998) Molecular basis for structural diversity in the core regions of the lipopolysaccharides of *Escherichia coli* and *Salmonella enterica*. *Molecular Microbiology* **30,** 221–232.

HIESTAND-NAUER, R. and IIDA, S. (1983) Sequence of the site-specific recombinase gene *cin* and of its substrates serving in the inversion of the C segment of bacteriophage P1. *EMBO Journal* **2,** 1733–1740.

HIGH, N. J., DEADMAN, M. E., and MOXON, E. R. (1993) The role of a repetitive DNA motif (5′-CAAT-3′) in the variable expression of the *Haemophilus influenzae* lipopolysaccharide epitope alpha Gal(1-4)beta Gal. *Molecular Microbiology* **9,** 1275–1282.

HIGH, N. J., JENNINGS, M. P., and MOXON, E. R. (1996) Tandem repeats of the tetramer 5′-CAAT-3′ present in *lic2A* are required for phase variation but not lipopolysaccharide biosynthesis in *Haemophilus influenzae*. *Molecular Microbiology* **20,** 165–174.

HOHMANN, E. L. (2001) Nontyphoidal salmonellosis. *Clinical Infectious Diseases* **32,** 263–269.

HOOD, D. W., DEADMAN, M. E., JENNINGS, M. P. et al. (1996) DNA repeats identify novel virulence genes in *Haemophilus influenzae*. *Proceedings of the National Academy of Sciences of the United States of America* **93,** 11121–11125.

HOSKING, S. L., CRAIG, J. E., and HIGH, N. J. (1999) Phase variation of *lic1A, lic2A* and *lic3A* in colonization of the nasopharynx, bloodstream and cerebrospinal fluid by *Haemophilus influenzae* type b. *Microbiology* **145**(Pt 11), 3005–3011.

HUANG, Y., UMEDA, M., TAKEUCHI, Y. et al. (2003) Distribution of *Bacteroides forsythus* genotypes in a Japanese periodontitis population. *Oral Microbiology and Immunology* **18,** 208–214.

HUBALEK, Z., SIXL, W., MIKULASKOVA, M. et al. (1995) Salmonellae in gulls and other free-living birds in the Czech Republic. *Cenralt European Journal of Public Health* **3,** 21–24.

HUMPHRIES, A., DERIDDER, S., and BAUMLER, A. J. (2005) *Salmonella enterica* serotype Typhimurium fimbrial proteins serve as antigens during infection of mice. *Infection and Immunity* **73,** 5329–5338.

HUMPHRIES, A. D., TOWNSEND, S. M., KINGSLEY, R. A. et al. (2001) Role of fimbriae as antigens and intestinal colonization factors of *Salmonella* serovars. *FEMS Microbiology Letters* **201,** 121–125.

IIDA, S. (1984) Bacteriophage P1 carries two related sets of genes determining its host range in the invertible C segment of its genome. *Virology* **134,** 421–434.

IIDA, S., MEYER, J., KENNEDY, K. E., and ARBER, W. (1982) A site-specific, conservative recombination system carried by bacteriophage P1. Mapping the recombinase gene *cin* and the cross-over sites cix for the inversion of the C segment. *EMBO Journal* **1,** 1445–1453.

IKEDA, J. S., SCHMITT, C. K., DARNELL, S. C. et al. (2001) Flagellar phase variation of *Salmonella enterica* serovar Typhimurium contributes to virulence in the murine typhoid infection model but does not influence *Salmonella*-induced enteropathogenesis. *Infection and Immunity* **69,** 3021–3030.

INZANA, T. J. (1983) Electrophoretic heterogeneity and interstrain variation of the lipopolysaccharide of *Haemophilus influenzae*. *Journal of Infectious Diseases* **148,** 492–499.

INZANA, T. J. (1987) Lipopolysaccharide gel profiles of *Haemophilus influenzae* type b for epidemiologic analysis. *Journal of Clinical Microbiology* **25,** 2252.

Iwabu, Y., Mizuta, H., Kawase, M. et al. (2008) Superinfection of defective human immunodeficiency virus type 1 with different subtypes of wild-type virus efficiently produces infectious variants with the initial viral phenotypes by complementation followed by recombination. *Microbes and Infection* **10,** 504–513.

Jacob Sonne-Hansen, S. M. J. (2005) Molecular serotyping of *Salmonella*: Identification of the phase 1 H antigen based on partial sequencing of the *fliC* gene. *Acta Pathologica, Microbiologica et Immunologica Scandinavica* **113,** 340–348.

Jennings, M. P., Hood, D. W., Peak, I. R. et al. (1995) Molecular analysis of a locus for the biosynthesis and phase-variable expression of the lacto-N-neotetraose terminal lipopolysaccharide structure in *Neisseria meningitidis*. *Molecular Microbiology* **18,** 729–740.

Jennings, M. P., Srikhanta, Y. N., Moxon, E. R. et al. (1999) The genetic basis of the phase variation repertoire of lipopolysaccharide immunotypes in *Neisseria meningitidis*. *Microbiology* **145**(Pt 11)**,** 3013–3021.

Jiang, X. M., Neal, B., Santiago, F. et al. (1991) Structure and sequence of the *rfb* (O antigen) gene cluster of *Salmonella* serovar Typhimurium (strain LT2). *Molecular Microbiology* **5,** 695–713.

Johnson, L. K., Brown, M. B., Carruthers, E. A. et al. (2004) Sample size, library composition, and genotypic diversity among natural populations of *Escherichia coli* from different animals influence accuracy of determining sources of fecal pollution. *Applied and Environmental Microbiology* **70,** 4478–4485.

Kalir, S., McClure, J., Pabbaraju, K. et al. (2001) Ordering genes in a flagella pathway by analysis of expression kinetics from living bacteria. *Science* **292,** 2080–2083.

Kalish, M. L., Robbins, K. E., Pieniazek, D. et al. (2004) Recombinant viruses and early global HIV-1 epidemic. *Emerging Infectious Diseases* **10,** 1227–1234.

Kamp, D., Kahmann, R., Zipser, D. et al. (1978) Inversion of the G DNA segment of phage Mu controls phage infectivity. *Nature* **271,** 577–580.

Kaniuk, N. A., Monteiro, M. A., Parker, C. T., and Whitfield, C. (2002) Molecular diversity of the genetic loci responsible for lipopolysaccharide core oligosaccharide assembly within the genus *Salmonella*. *Molecular Microbiology* **46,** 1305–1318.

Keestra, A. M., de Zoete, M. R., van Aubel, R. A. M. H., and van Putten, J. P. M. (2008) Functional characterization of chicken TLR5 reveals species-specific recognition of flagellin. *Molecular Immunology* **45,** 1298–1307.

Kimura, M. (1983) *The Neutral Theory of Molecular Evolution*. Cambridge University Press, Cambridge.

Kingsley, R. A. and Baumler, A. J. (2000) Host adaptation and the emergence of infectious disease: The *Salmonella* paradigm. *Molecular Microbiology* **36,** 1006–1014.

Klein, J. R., Fahlen, T. F., and Jones, B. D. (2000) Transcriptional organization and function of invasion genes within *Salmonella enterica* serovar Typhimurium pathogenicity island 1, including the *prgH, prgI, prgJ, prgK, orgA, orgB*, and *orgC* genes. *Infection and Immunity* **68,** 3368–3376.

Kojimoto, A., Uchida, K., Horii, Y. et al. (2001) Amebiasis in four ball pythons, *Python reginus*. *Journal of Veternary Medical Sciences* **63,** 1365–1368.

Koomey, M., Gotschlich, E. C., Robbins, K. et al. (1987) Effects of *recA* mutations on pilus antigenic variation and phase transitions in *Neisseria gonorrhoeae*. *Genetics* **117,** 391–398.

Kourany, M., Bowdre, L., and Herrer, A. (1976) Panamanian forest mammals as carriers of *Salmonella*. *American Journal of Tropical Medicine and Hygiene* **25,** 449–455.

Krinos, C. M., Coyne, M. J., Weinacht, K. G. et al. (2001) Extensive surface diversity of a commensal microorganism by multiple DNA inversions. *Nature* **414,** 555–558.

Kuhlmann, H.-W. and Heckmann, K. (1985) Interspecies morphogens regulating prey-predator relationships in protozoa. *Science* **227,** 1347–1349.

Kurzai, O., Schmitt, C., Claus, H. et al. (2005) Carbohydrate composition of meningococcal lipopolysaccharide modulates the interaction of *Neisseria meningitidis* with human dendritic cells. *Cellular Microbiology* **7,** 1319–1334.

Kusch, J. (1993) Behavioural and morphological changes in ciliates induced by the predator *Amoeba proteus*. *Oecologia* **96,** 354–359.

Kutsukake, K. (1997) Autogenous and global control of the flagellar master operon, *flhD*, in *Salmonella typhimurium*. *Molecular and General Genetics* **254,** 440–448.

Kuwahara, T., Yamashita, A., Hirakawa, H. et al. (2004) Genomic analysis of *Bacteroides fragilis* reveals extensive DNA inversions regulating cell surface adaptation. *Proceedings of the National Academy of Sciences of the United States of America* **101,** 14919–14924.

Lavitola, A., Bucci, C., Salvatore, P. et al. (1999) Intracistronic transcription termination in polysialyltransferase gene (*siaD*) affects phase variation in *Neisseria meningitidis*. *Molecular Microbiology* **33,** 119–127.

Lawrence, J. G. (2002) Gene transfer in bacteria: Speciation without species? *Theoretical Population Biology* **61,** 449–460.

Leber, A. L. (1999) Intestinal amebae. *Clinics in Laboratory Medicine* **19,** 601–619.

Lederberg, J. and Edwards, P. R. (1953) Sero-typic recombination in *Salmonella*. *Journal of Immunology* **71,** 232–240.

Lederberg, J. and Iino, T. (1956) Phase variation in *Salmonella*. *Genetics* **41,** 743–757.

Lee, S. J., Romana, L. K., and Reeves, P. R. (1992a) Cloning and structure of group C1 O antigen (*rfb* gene cluster) from *Salmonella* enterica serovar Montevideo. *Journal of General Microbiology* **138,** 305–312.

Lee, S. J., Romana, L. K., and Reeves, P. R. (1992b) Sequence and structural analysis of the *rfb* (O antigen) gene cluster from a group C1 *Salmonella enterica* strain. *Journal of General Microbiology* **138,** 1843–1855.

LEVIN, B. R. (1988) Frequency-dependent selection in bacterial populations. *Philosophical Transaction of the Royal Society of London. Series B, Biological Sciences* **319,** 459–72.

LEVINSON, G. and GUTMAN, G. A. (1987) Slipped-strand mispairing: A major mechanism for DNA sequence evolution. *Molecular Biology and Evolution* **4,** 203–221.

LEY, R. E., PETERSON, D. A., and GORDON, J. I. (2006) Ecological and evolutionary forces shaping microbial diversity in the human intestine. *Cell* **124,** 837–848.

LI, Q. and REEVES, P. R. (2000) Genetic variation of dTDP-L-rhamnose pathway genes in *Salmonella enterica. Microbiology* **146**(Pt 9), 2291–2307.

LIU, C. H., LEE, S. M., VANLARE, J. M. et al. (2008) Regulation of surface architecture by symbiotic bacteria mediates host colonization. *Proceedings of the National Academy of Sciences of the United States of America* **105,** 3951–3956.

LIU, D., HAASE, A. M., LINDQVIST, L. et al. (1993) Glycosyl transferases of O-antigen biosynthesis in *Salmonella enterica*: Identification and characterization of transferase genes of groups B, C2, and E1. *Journal of Bacteriology* **175,** 3408–3413.

LIU, D., LINDQVIST, L., and REEVES, P. R. (1995) Transferases of O-antigen biosynthesis in *Salmonella enterica*: Dideoxyhexosyltransferases of groups B and C2 and acetyltransferase of group C2. *Journal of Bacteriology* **177,** 4084–4088.

LIU, D., VERMA, N. K., ROMANA, L. K., and REEVES, P. R. (1991) Relationships among the *rfb* regions of *Salmonella* serovars A, B, and D. *Journal of Bacteriology* **173,** 4814–4819.

MACPHERSON, A. J. (2006) IgA adaptation to the presence of commensal bacteria in the intestine. *Current Topics in Microbiology and Immunology* **308,** 117–136.

MACPHERSON, A. J. and SLACK, E. (2007) The functional interactions of commensal bacteria with intestinal secretory IgA. *Current Opinion in Gastroenterology* **23,** 673–678.

MACPHERSON, A. J. and UHR, T. (2004) Induction of protective IgA by intestinal dendritic cells carrying commensal bacteria. *Science* **303,** 1662–1665.

MANNING, P. A., KAUFMANN, A., ROLL, U. et al. (1991) L-pilin variants of *Neisseria gonorrhoeae* MS11. *Molecular Microbiology* **5,** 917–926.

MARTINEZ-DIAZ, R. A., HERRERA, S., CASTRO, A., and PONCE, F. (2000) *Entamoeba* sp. (Sarcomastigophora: Endamoebidae) from ostriches (*Struthio camelus*) (Aves: Struthionidae). *Veterinary Parasitology* **92,** 173–179.

MAZMANIAN, S. K., ROUND, J. L., and KASPER, D. L. (2008) A microbial symbiosis factor prevents intestinal inflammatory disease. *Nature* **453,** 620–625.

MCCLELLAND, M., SANDERSON, K. E., SPIETH, J. et al. (2001) Complete genome sequence of *Salmonella enterica* serovar Typhimurium LT2. *Nature* **413,** 852–856.

MCCUTCHAN, F. E., HOELSCHER, M., TOVANABUTRA, S. et al. (2005) In-depth analysis of a heterosexually acquired human immunodeficiency virus type 1 superinfection: Evolution, temporal fluctuation, and intercompartment dynamics from the seronegative window period through 30 months postinfection. *Journal of Virology* **79,** 11693–11704.

MCGEE, Z. A., STEPHENS, D. S., HOFFMAN, L. H. et al. (1983) Mechanisms of mucosal invasion by pathogenic *Neisseria. Review of Infectious Diseases* **5**(Suppl 4), S708–S714.

MCQUISTON, J. R., PARRENAS, R., ORTIZ-RIVERA, M. et al. (2004) Sequencing and comparative analysis of flagellin genes *fliC, fljB*, and *flpA* from *Salmonella. Journal of Clinical Microbiology* **42,** 1923–1932.

MEAD, P. S., SLUTSKER, L., DIETZ, V. et al. (1999) Food-related illness and death in the United States. *Emerging Infectious Diseases* **5,** 607–625.

MEYERHOLZ, D. K. and STABEL, T. J. (2003) Comparison of early ileal invasion by *Salmonella enterica* serovars Choleraesuis and Typhimurium. *Veterinary Pathology* **40,** 371–375.

MEYERS, L. A., LEVIN, B. R., RICHARDSON, A. R., and STOJILJKOVIC, I. (2003) Epidemiology, hypermutation, within-host evolution and the virulence of *Neisseria meningitidis. Proceedings of the Royal Society of London, Series B, Biological Sciences* **270,** 1667–1677.

MICHAELS, R. H. and NORDEN, C. W. (1977) Pharyngeal colonization with *Haemophilus influenzae* type b: A longitudinal study of families with a child with meningitis or epiglottitis due to *H. influenzae* type b. *Journal of Infectious Diseases* **136,** 222–228.

MILKMAN, R., JAEGER, E., and MCBRIDE, R. D. (2003) Molecular evolution of the *Escherichia coli* chromosome. VI. Two regions of high effective recombination. *Genetics* **163,** 475–483.

MOXON, E. R. (1985) The molecular basis of *Haemophilus influenzae* virulence. *Journal of the Royal College of Physicians of London* **19,** 174–178.

MOXON, E. R., RAINEY, P. B., NOWAK, M. A., and LENSKI, R. E. (1994) Adaptive evolution of highly mutable loci in pathogenic bacteria. *Current Biology* **4,** 24–33.

MOXON, R., BAYLISS, C., and HOOD, D. (2006) Bacterial contingency loci: The role of simple sequence DNA repeats in bacterial adaptation. *Annual Review of Genetics* **40,** 307–333.

MUROI, M. and TANAMOTO, K. (2002) The polysaccharide portion plays an indispensable role in *Salmonella* lipopolysaccharide-induced activation of NF-kappaB through human Toll-like receptor 4. *Infection and Immunity* **70,** 6043–6047.

MURRAY, G. L., ATTRIDGE, S. R., and MORONA, R. (2006) Altering the length of the lipopolysaccharide O antigen has an impact on the interaction of *Salmonella enterica* serovar Typhimurium with macrophages and complement. *Journal of Bacteriology* **188,** 2735–2739.

MUTHUKKUMAR, S. and MUTHUKKARUPPAN, V. R. (1993) Mechanism of protective immunity induced by porin-lipopolysaccharide against murine salmonellosis. *Infection and Immunity* **61,** 3017–3025.

MYODA, S. P., CARSON, C. A., FUHRMANN, J. J. et al. (2003) Comparison of genotypic-based microbial source tracking methods requiring a host origin database. *Journal of Water and Health* **1,** 167–180.

Nelson, K. and Selander, R. K. (1994) Intergeneric transfer and recombination of the 6-phosphogluconate dehydrogenase gene (*gnd*) in enteric bacteria. *Proceedings of the National Academy of Sciences of the United States of America* **91,** 10227–10231.

Nempont, C., Cayet, D., Rumbo, M. et al. (2008) Deletion of flagellin's hypervariable region abrogates antibody-mediated neutralization and systemic activation of TLR5-dependent immunity. *Journal of Immunology* **181,** 2036–2043.

Nikaido, H. (2003) Molecular basis of bacterial outer membrane permeability revisited. *Microbiology and Molecular Biology Reviews* **67,** 593–656.

Nikaido, H. and Vaara, M. (1985) Molecular basis of bacterial outer membrane permeability. *Microbiological Reviews* **49,** 1–32.

Norris, T. L. and Baumler, A. J. (1999) Phase variation of the *lpf* operon is a mechanism to evade cross-immunity between *Salmonella* serotypes. *Proceedings of the National Academy of Sciences of the United States of America* **96,** 13393–13398.

Parker, C. T. and Guard-Petter, J. (2001) Contribution of flagella and invasion proteins to pathogenesis of *Salmonella enterica* serovar Enteritidis in chicks. *FEMS Microbiology Letters* **204,** 287–291.

Patrick, S., Parkhill, J., McCoy, L. J. et al. (2003) Multiple inverted DNA repeats of *Bacteroides fragilis* that control polysaccharide antigenic variation are similar to the hin region inverted repeats of *Salmonella typhimurium*. *Microbiology* **149,** 915–924.

Perry, A. C., Hart, C. A., Nicolson, I. J. et al. (1987) Inter-strain homology of pilin gene sequences in *Neisseria meningitidis* isolates that express markedly different antigenic pilus types. *Journal of General Microbiology* **133,** 1409–1418.

Perry, A. C., Nicolson, I. J., and Saunders, J. R. (1988) *Neisseria meningitidis* C114 contains silent, truncated pilin genes that are homologous to *Neisseria gonorrhoeae pil* sequences. *Journal of Bacteriology* **170,** 1691–1697.

Plasterk, R. H., Brinkman, A., and van de Putte, P. (1983) DNA inversions in the chromosome of *Escherichia coli* and in bacteriophage Mu: Relationship to other site-specific recombination systems. *Proceedings of the National Academy of Sciences of the United States of America* **80,** 5355–5358.

Potts, W. J. and Saunders, J. R. (1988) Nucleotide sequence of the structural gene for class I pilin from *Neisseria meningitidis*: Homologies with the *pilE* locus of *Neisseria gonorrhoeae*. *Molecular Microbiology* **2,** 647–653.

Pozio, E. (2003) Foodborne and waterborne parasites. *Acta Microbiologica Polonica* **52,** 83–96.

Rabsch, W., Andrews, H., Kingsley, R. A. et al. (2002) *Salmonella enterica* serotype Typhimurium and its host-adapted variants. *Infection and Immunity* **70,** 2249–2255.

Rabsch, W., Hargis, B. M., Tsolis, R. M. et al. (2000) Competitive exclusion of *Salmonella enteritidis* by *Salmonella gallinarum* in poultry. *Emerging Infectious Diseases* **6,** 443–448.

Ram, J. L., Ritchie, R. P., Fang, J. et al. (2004) Sequence-based source tracking of *Escherichia coli* based on genetic diversity of beta-glucuronidase. *Journal of Environmental Quality* **33,** 1024–1032.

Rambaut, A., Posada, D., Crandall, K. A., and Holmes, E. C. (2004) The causes and consequences of HIV evolution. *Nature Reviews. Genetics* **5,** 52–61.

Ramirez, B. C., Simon-Loriere, E., Galetto, R., and Negroni, M. (2008) Implications of recombination for HIV diversity. *Virus Research* **134,** 64–73.

Reeves, P. (1993) Evolution of *Salmonella* O antigen variation by interspecific gene transfer on a large scale. *Trends in Genetics* **9,** 17–22.

Reeves, P. (1995) Role of O-antigen variation in the immune response. *Trends in Microbiology* **3,** 381–386.

Reeves, P. P. and Wang, L. (2002) Genomic organization of LPS-specific loci. *Current Topics in Microbiology and Immunology* **264,** 109–135.

Reeves, P. R., Hobbs, M., Valvano, M. A. et al. (1996) Bacterial polysaccharide synthesis and gene nomenclature. *Trends in Microbiology* **4,** 495–503.

Refsum, T., Heir, E., Kapperud, G. et al. (2002) Molecular epidemiology of *Salmonella enterica* serovar Typhimurium isolates determined by pulsed-field gel electrophoresis: Comparison of isolates from avian wildlife, domestic animals, and the environment in Norway. *Applied and Environmental Microbiology* **68,** 5600–5606.

Retchless, A. C. and Lawrence, J. G. (2007) Temporal fragmentation of speciation in bacteria. *Science* **317,** 1093–1096.

Rigottier-Gois, L., Rochet, V., Garrec, N. et al. (2003) Enumeration of *Bacteroides* species in human faeces by fluorescent *in situ* hybridisation combined with flow cytometry using 16S rRNA probes. *Systematic and Applied Microbiology* **26,** 110–118.

Robbins, J. B., Chu, C., and Schneerson, R. (1992) Hypothesis for vaccine development: Protective immunity to enteric diseases caused by nontyphoidal salmonellae and shigellae may be conferred by serum IgG antibodies to the O-specific polysaccharide of their lipopolysaccharides. *Clinical Infectious Diseases* **15,** 346–361.

Roche, R. J., High, N. J., and Moxon, E. R. (1994) Phase variation of *Haemophilus influenzae* lipopolysaccharide: Characterization of lipopolysaccharide from individual colonies. *FEMS Microbiology Letters* **120,** 279–283.

Roche, R. J. and Moxon, E. R. (1995) Phenotypic variation of carbohydrate surface antigens and the pathogenesis of *Haemophilus influenzae* infections. *Trends in Microbiology* **3,** 304–309.

Rodriguez-Zaragoza, S. (1994) Ecology of free-living amoebae. *Critical Reviews in Microbiology* **20,** 225–241.

Ronn, R., McCaig, A. E., Griffiths, B. S., and Prosser, J. I. (2002) Impact of protozoan grazing on bacterial community structure in soil microcosms. *Applied and Environmental Microbiology* **68,** 6094–6105.

Royle, M. C., Totemeyer, S., Alldridge, L. C. et al. (2003) Stimulation of Toll-like receptor 4 by lipopolysaccharide during cellular invasion by live *Salmonella typhimurium* is a critical but not exclusive event leading to macrophage responses. *Journal of Immunology* **170,** 5445–5454.

Rytkonen, A., Albiger, B., Hansson-Palo, P. et al. (2004) *Neisseria meningitidis* undergoes PilC phase variation and PilE sequence variation during invasive disease. *Journal of Infectious Diseases* **189,** 402–409.

Salazar-Gonzalez, R. M. and McSorley, S. J. (2005) *Salmonella* flagellin, a microbial target of the innate and adaptive immune system. *Immunology Letters* **101,** 117–122.

Samuel, G. and Reeves, P. (2003) Biosynthesis of O-antigens: Genes and pathways involved in nucleotide sugar precursor synthesis and O-antigen assembly. *Carbohydrate Research* **338,** 2503–2519.

Sanders, C. J., Franchi, L., Yarovinsky, F. et al. (2009) Induction of adaptive immunity by flagellin does not require robust activation of innate immunity. *European Journal of Immunology* **39,** 359–371.

Sanders, C. J., Moore, D. A. III, Williams, I. R., and Gewirtz, A. T. (2008) Both radioresistant and hemopoietic cells promote innate and adaptive immune responses to flagellin. *Journal of Immunology* **180,** 7184–7192.

Sanders, C. J., Yu, Y., Moore, D. A. III et al. (2006) Humoral immune response to flagellin requires T cells and activation of innate immunity. *Journal of Immunology* **177,** 2810–2818.

Santiviago, C. A., Toro, C. S., Hidalgo, A. A. et al. (2003) Global regulation of the *Salmonella enterica* serovar Typhimurium major porin, OmpD. *Journal of Bacteriology* **185,** 5901–5905.

Savage, D. C. (1977) Microbial ecology of the gastrointestinal tract. *Annual Review of Microbiology* **31,** 107–133.

Schmitt, C. K., Ikeda, J. S., Darnell, S. C. et al. (2001) Absence of all components of the flagellar export and synthesis machinery differentially alters virulence of *Salmonella enterica* serovar Typhimurium in models of typhoid fever, survival in macrophages, tissue culture invasiveness, and calf enterocolitis. *Infection and Immunity* **69,** 5619–5625.

Schnaitman, C. A. and Klena, J. D. (1993) Genetics of lipopolysaccharide biosynthesis in enteric bacteria. *Microbiological Reviews* **57,** 655–682.

Schoen, C., Blom, J., Claus, H. et al. (2008) Whole-genome comparison of disease and carriage strains provides insights into virulence evolution in *Neisseria meningitidis*. *Proceedings of the National Academy of Sciences of the United States of America* **105,** 3473–3478.

Schuster, F. L., De Jonckheere, J. F., Moura, H. et al. (2003) Isolation of a thermotolerant *Paravahlkampfia* sp. from lizard intestine: Biology and molecular identification. *Journal of Eukaryotic Microbiology* **50,** 373–378.

Scott, T. M., Rose, J. B., Jenkins, T. M. et al. (2002) Microbial source tracking: Current methodology and future directions. *Applied and Environmental Microbiology* **68,** 5796–5803.

Sechman, E. V., Kline, K. A., and Seifert, H. S. (2006) Loss of both Holliday junction processing pathways is synthetically lethal in the presence of gonococcal pilin antigenic variation. *Molecular Microbiology* **61,** 185–193.

Secundino, I., López-Macías, C., Cervantes-Barragán, L. et al. (2006) *Salmonella* porins induce a sustained, lifelong specific bactericidal antibody memory response. *Immunology* **117,** 59–70.

Selander, R. K., Li, J., and Nelson, K. (1996) Evolutionary genetics of *Salmonella enterica*. In *Escherichia coli and Salmonella typhimurium: Cellular and Molecular Biology*, 2nd ed. (eds. F. C. Neidhardt, R. Curtiss III, J. L. Ingraham et al.), pp. 2691–2707 American Society for Microbiology, Washington, DC.

Sharp, R., Hazlewood, G. P., Gilbert, H. J., and O'Donnell, A.G. (1994) Unmodified and recombinant strains of *Lactobacillus plantarum* are rapidly lost from the rumen by protozoal predation. *Journal of Applied Bacteriology* **76,** 110–117.

Simpson, J. M., Santo Domingo, J. W., and Reasoner, D. J. (2002) Microbial source tracking: State of the science. *Environmental Science & Technology* **36,** 5279–5288.

Singh, S. P., Miller, S., Williams, Y. U. et al. (1996) Immunochemical structure of the OmpD porin from *Salmonella typhimurium*. *Microbiology* **142**(Pt 11), 3201–3210.

Singh, S. P., Singh, S. R., Williams, Y. U. et al. (1995) Antigenic determinants of the OmpC porin from *Salmonella typhimurium*. *Infection and Immunity* **63,** 4600–4605.

Singh, S. P., Upshaw, Y., Abdullah, T. et al. (1992) Structural relatedness of enteric bacterial porins assessed with monoclonal antibodies to *Salmonella typhimurium* OmpD and OmpC. *Journal of Bacteriology* **174,** 1965–1973.

Slauch, J. M., Lee, A. A., Mahan, M. J., and Mekalanos, J. J. (1996) Molecular characterization of the *oafA* locus responsible for acetylation of *Salmonella typhimurium* O-antigen: *oafA* is a member of a family of integral membrane trans-acylases. *Journal of Bacteriology* **178,** 5904–5909.

Smith, M. C. M. and Thorpe, H. M. (2002) Diversity in the serine recombinases. *Molecular Microbiology* **44,** 299–307.

Smith, N. H., Beltran, P., and Selander, R. K. (1990) Recombination of *Salmonella* phase 1 flagellin genes generates new serovars. *Journal of Bacteriology* **172,** 2209–2216.

Smith, N. H. and Selander, R. K. (1990) Sequence invariance of the antigen-coding central region of the phase 1 flagellar filament gene (*fliC*) among strains of

Salmonella typhimurium. Journal of Bacteriology **172,** 603–609.

Smith, N. H. and Selander, R. K. (1991) Molecular genetic basis for complex flagellar antigen expression in a triphasic serovar of *Salmonella. Proceedings of the National Academy of Sciences of the United States of America* **88,** 956–960.

Snyder, L. A., Butcher, S. A., and Saunders, N. J. (2001) Comparative whole-genome analyses reveal over 100 putative phase-variable genes in the pathogenic *Neisseria* spp. *Microbiology* **147,** 2321–32.

Spierings, G., Elders, R., van Lith, B. et al. (1992) Characterization of the *Salmonella typhimurium phoE* gene and development of *Salmonella*-specific DNA probes. *Gene* **122,** 45–52.

Stephens, D. S., McGee, Z. A., Melly, M. A. et al. (1982) Attachment of pathogenic *Neisseria* to human mucosal surfaces: Role in pathogenesis. *Infection* **10,** 192–195.

Stern, A. and Meyer, T. F. (1987) Common mechanism controlling phase and antigenic variation in pathogenic neisseriae. *Molecular Microbiology* **1,** 5–12.

Stoeckel, D. M., Mathes, M. V., Hyer, K. E. et al. (2004) Comparison of seven protocols to identify fecal contamination sources using *Escherichia coli. Environmental Science & Technology* **38,** 6109–6117.

Tettelin, H., Saunders, N. J., Heidelberg, J. et al. (2000) Complete genome sequence of *Neisseria meningitidis* serogroup B strain MC58. *Science* **287,** 1809–1815.

Thigpen, J. E., Moore, J. A., Gupta, B. N., and Feldman, D. B. (1975) Opossums as a reservoir for Salmonellae. *Journal of the American Veterinary Medical Association* **167,** 590–592.

Thomsen, L. E., Chadfield, M. S., Bispham, J. et al. (2003) Reduced amounts of LPS affect both stress tolerance and virulence of *Salmonella enterica* serovar Dublin. *FEMS Microbiology Letters* **228,** 225–231.

Tindall, B. J., Grimont, P. A., Garrity, G. M., and Euzeby, J. P. (2005) Nomenclature and taxonomy of the genus *Salmonella. International Journal of Systematic and Evolutionary Microbiology* **55,** 521–524.

Tinsley, C. R. and Heckels, J. E. (1986) Variation in the expression of pili and outer membrane protein by *Neisseria meningitidis* during the course of meningococcal infection. *Journal of General Microbiology* **132,** 2483–2490.

Tolan, R. W. Jr., Munson, R. S. Jr., and Granoff, D. M. (1986) Lipopolysaccharide gel profiles of *Haemophilus influenzae* type b are not stable epidemiologic markers. *Journal of Clinical Microbiology* **24,** 223–227.

Uzzau, S., Leori, G. S., Petruzzi, V. et al. (2001) *Salmonella enterica* serovar-host specificity does not correlate with the magnitude of intestinal invasion in sheep. *Infection and Immunity* **69,** 3092–3099.

van de Putte, P., Cramer, S., and Giphart-Gassler, M. (1980) Invertible DNA determines host specificity of bacteriophage Mu. *Nature* **286,** 218–222.

van de Putte, P. and Goosen, N. (1992) DNA inversions in phages and bacteria. *Trends in Genetics* **8,** 457–462.

van der Ende, A., Hopman, C. T., Zaat, S. et al. (1995) Variable expression of class 1 outer membrane protein in *Neisseria meningitidis* is caused by variation in the spacing between the -10 and -35 regions of the promoter. *Journal of Bacteriology* **177,** 2475–2480.

van der Ende, A., Hopman, C. T. P., and Dankert, J. (2000) Multiple mechanisms of phase variation of PorA in *Neisseria meningitidis. Infection and Immunity* **68,** 6685–6690.

Vazquez-Torres, A., Vallance, B. A., Bergman, M. A. et al. (2004) Toll-like receptor 4 dependence of innate and adaptive immunity to *Salmonella*: Importance of the Kupffer cell network. *Journal of Immunology* **172,** 6202–6208.

Verma, N. and Reeves, P. (1989) Identification and sequence of *rfbS* and *rfbE,* which determine antigenic specificity of group A and group D salmonellae. *Journal of Bacteriology* **171,** 5694–5701.

Verma, N. K., Quigley, N. B., and Reeves, P. R. (1988) O-antigen variation in *Salmonella* spp.: *rfb* gene clusters of three strains. *Journal of Bacteriology* **170,** 103–107.

Voetsch, A. C., Van Gilder, T. J., Angulo, F. J. et al. (2004) FoodNet estimate of the burden of illness caused by nontyphoidal *Salmonella* infections in the United States. *Clinical Infectious Diseases* **38,** S127–S134.

Wang, L., Andrianopoulos, K., Liu, D. et al. (2002) Extensive variation in the O-antigen gene cluster within one *Salmonella enterica* serogroup reveals an unexpected complex history. *Journal of Bacteriology* **184,** 1669–1677.

Wang, L., Romana, L. K., and Reeves, P. R. (1992) Molecular analysis of a *Salmonella enterica* group E1 *rfb* gene cluster: O antigen and the genetic basis of the major polymorphism. *Genetics* **130,** 429–443.

Wang, L., Rothemund, D., Curd, H., and Reeves, P. R. (2003) Species-wide variation in the *Escherichia coli* flagellin (H-antigen) gene. *Journal of Bacteriology* **185,** 2936–2943.

Wang, W., Perepelov, A. V., Feng, L. et al. (2007) A group of *Escherichia coli* and *Salmonella enterica* O antigens sharing a common backbone structure. *Microbiology* **153,** 2159–2167.

Wasson, K. and Peper, R. L. (2000) Mammalian microsporidiosis. *Veterinary Pathology* **37,** 113–128.

Weiser, J. N., Love, J. M., and Moxon, E. R. (1989) The molecular mechanism of phase variation of *H. influenzae* lipopolysaccharide. *Cell* **59,** 657–665.

Weiser, J. N., Maskell, D. J., Butler, P. D. et al. (1990) Characterization of repetitive sequences controlling phase variation of *Haemophilus influenzae* lipopolysaccharide. *Journal of Bacteriology* **172,** 3304–3309.

Weiser, J. N. and Pan, N. (1998) Adaptation of *Haemophilus influenzae* to acquired and innate humoral immunity based on phase variation of lipopolysaccharide. *Molecular Microbiology* **30,** 767–775.

Wheeler, A. L., Hartel, P. G., Godfrey, D. G. et al. (2002) Potential of *Enterococcus faecalis* as a human fecal indicator for microbial source tracking. *Journal of Environmental Quality* **31,** 1286–1293.

WICKLOW, B. J. (1988) Developmental polymorphism induced by introspecific predation in the ciliated protozooan *Onychodromus quadricornutus*. *Journal of Protozoology* **35,** 137–141.

WILDSCHUTTE, H. and LAWRENCE, J. G. (2007) Differential *Salmonella* survival against communities of intestinal amoebae. *Microbiology* **153,** 1781–1789.

WILDSCHUTTE, H., WOLFE, D. M., TAMEWITZ, A., and LAWRENCE, J.G. (2004) Protozoan predation, diversifying selection, and the evolution of antigenic diversity in *Salmonella*. *Proceedings of the National Academy of Sciences of the United States of America* **101,** 10644–10649.

WISNIEWSKI-DYE, F. and VIAL, L. (2008) Phase and antigenic variation mediated by genome modifications. *Antonie Van Leeuwenhoek* **94,** 493–515.

XIANG, S. H., HAASE, A. M., and REEVES, P. R. (1993) Variation of the *rfb* gene clusters in *Salmonella enterica*. *Journal of Bacteriology* **175,** 4877–4884.

XIANG, S. H., HOBBS, M., and REEVES, P. R. (1994) Molecular analysis of the *rfb* gene cluster of a group D2 *Salmonella enterica* strain: Evidence for its origin from an insertion sequence-mediated recombination event between group E and D1 strains. *Journal of Bacteriology* **176,** 4357–4365.

XU, J., MAHOWALD, M. A., LEY, R. E. et al. (2007) Evolution of symbiotic bacteria in the distal human intestine. *PLoS Biology* **5,** e156.

YAMAMOTO, S. and KUTSUKAKE, K. (2006) FljA-mediated posttranscriptional control of phase 1 flagellin expression in flagellar phase variation of *Salmonella enterica* serovar Typhimurium. *Journal of Bacteriology* **188,** 958–967

YANAGIHARA, S., IYODA, S., OHNISHI, K. et al. (1999) Structure and transcriptional control of the flagellar master operon of *Salmonella typhimurium*. *Genes & Genetic Systems* **74,** 105–111.

ZHUANG, J., JETZT, A. E., SUN, G. et al. (2002) Human immunodeficiency virus type 1 recombination: Rate, fidelity, and putative hot spots. *Journal of Virology* **76,** 11273–11282.

ZIEG, J. and SIMON, M. (1980) Analysis of the nucleotide sequence of an invertible controlling element. *Proceedings of the National Academy of Sciences of the United States of America* **77,** 4196–4200.

ZWAHLEN, A., RUBIN, L. G., and MOXON, E. R. (1986) Contribution of lipopolysaccharide to pathogenicity of *Haemophilus influenzae*: Comparative virulence of genetically-related strains in rats. *Microbial Pathogenesis* **1,** 465–473.

Chapter 16

Population Genetics of *Staphylococcus*

DAVIDA S. SMYTH AND D. ASHLEY ROBINSON

16.1 INTRODUCTION

The genus *Staphylococcus* includes 40 species of gram-positive, nonmotile, facultatively anaerobic cocci (Murray et al., 2009). *Staphylococcus aureus* is the most aggressive human pathogen of the genus, though it temporarily colonizes 30–50% of the human population in an asymptomatic fashion (Kuehnert et al., 2006). Colonization of the anterior nares readily allows *S. aureus* transmission to other skin surfaces and fomites. *S. aureus* is unique among staphylococci in its production of numerous toxins and degradative enzymes that contribute to its virulence. Specific diseases caused by specific toxins are recognized (e.g., scalded skin syndrome caused by exfoliative toxins), but the relative contribution of specific virulence factors to the most common diseases is poorly understood. *S. aureus* globally causes 39% of skin and soft tissue infections, 23% of ventilator-associated pneumonia, and 22% of bloodstream infections (Diekema et al., 2001). In addition, *S. aureus* can cause economically important diseases in food animals such as cattle, chickens, goat, sheep, and rabbits. Transmission can occur between humans and animal hosts, including companion animals such as cats and dogs (Manian, 2003). Contamination of food products with *S. aureus* toxins is also an important cause of food poisoning, and the staphylococcal enterotoxin B (SEB) toxin, in particular, has been classified as a category B biothreat by the U.S. government.

S. aureus can be distinguished from most other human-colonizing staphylococci by the presence of the coagulase protein, which serves as a cofactor for fibrinogen activation. Approximately 17 species of so-called coagulase-negative staphylococci (CNS) are known to cause disease in humans (Kloos and Bannerman, 1994). These species have been classified into as few as 6 and as many as 12 species groups depending on the molecular tools used for analysis (Kloos, 1990; Takahashi et al., 1999; Ghebremedhin et al., 2008); a single staphylococcal phylogeny has yet to be established that could resolve this issue. CNS require host assistance, such as immunosuppression or access to normally sterile tissues provided by indwelling medical devices, to cause human disease. CNS are among the most frequently isolated microorganisms in clinical labs (Kloos and Bannerman, 1994), even though it is often the case that their isolation from clinical specimens represents

Bacterial Population Genetics in Infectious Disease, Edited by D. Ashley Robinson, Daniel Falush, and Edward J. Feil

contamination rather than infection. Nonetheless, some interesting observations have been made with respect to CNS colonization preferences. Examples include *Staphylococcus capitis* colonization of the scalp and *Staphylococcus auricularis* colonization of the outer ear (Kloos and Bannerman, 1994).

S. aureus and the most abundant CNS species of humans, *Staphylococcus epidermidis*, together account for 59% of bacteremia isolates in the United States and in Europe (Marshall et al., 1998; Richards et al., 1999). In addition to their abundance, they have acquired or have otherwise developed resistance to most classes of antibiotics. The dissemination of methicillin-resistant staphylococci, which cannot be effectively treated with beta-lactams, provides a global public health challenge (Grundmann et al., 2006). In the United States, approximately 60% of *S. aureus* isolates and 89% of *S. epidermidis* isolates from bacteremia are methicillin-resistant (Marshall et al., 1998; Richards et al., 1999). Even countries with historically low prevalences of resistance must continually guard against resistant staphylococci (Bartels et al., 2007). Consequently, considerable effort has been invested in the development and application of strain typing techniques that reveal transmission events at local levels and that enable surveillance of problematic strains at national and international levels. Some of these typing techniques have also been central to the study of staphylococcal population genetics.

This chapter focuses on the population genetics of *S. aureus* and *S. epidermidis*. We discuss the tools used to study these pathogens and the natural groups that have been identified and subsequently studied from an infectious disease perspective. We also discuss some genetic and population processes that have been implicated in creating and maintaining genetic variation in these species.

16.2 OVERVIEW OF THE STAPHYLOCOCCAL POPULATION STRUCTURE

16.2.1 Some Tools of the Trade

The first strain typing technique used to study bacterial population genetics was multilocus enzyme electrophoresis (MLEE) (Milkman, 1973). This technique detects genetic variation indirectly through the electrophoretic mobility of expressed, water-soluble enzymes. The amino acid sequences of the enzymes determine their electrostatic charges and, hence, their rates of migration during electrophoresis. Different electromorphs are equated with different alleles at the genetic loci that encode the enzymes. Even though MLEE fails to detect all of the underlying nucleotide sequence variation at the enzyme loci, the ability to identify individual alleles allows a straightforward population genetics analysis.

In 1976, both Zimmerman and Kloos (1976) and Schleifer et al. (1976) used esterase allelic variation to distinguish different staphylococcal species. In 1987, Branger and Goullet published a seminal study that compared esterase allelic variation among 105 isolates of methicillin-resistant *Staphylococcus aureus* (MRSA) and methicillin-susceptible *Staphylococcus aureus* (MSSA) from international sources. Not only did their study provide one of the first quantitative population genetics analyses of *S. aureus*, by using a sampling bias-corrected measure of genetic diversity (see Appendix 1), but they also were among the first to show that methicillin resistance occurred in genetically diverse strains. This conclusion was subsequently supported by Musser and Kapur in 1992 using 15 enzyme loci and a sample of 254 isolates of MRSA.

Since the 1990s, many techniques that detect nucleotide sequence variation indirectly, through characteristics of restriction enzyme sites, primer-binding sites, or combinations of the two types of sites, have been applied to staphylococci. Some of the popular techniques include amplified fragment length polymorphism (AFLP), multiple locus variable number tandem repeat analysis (MLVA), pulsed-field gel electrophoresis (PFGE), randomly amplified polymorphic DNA (RAPD), and repetitive element PCR, all of which use electrophoretic approaches to resolve complex patterns of DNA bands (e.g., de Sousa et al., 1992; Jarraud et al., 2002; Sabat et al., 2003). As a result, such techniques are generally referred to as band-based, image-based, or molecular fingerprinting techniques. With these techniques, it is often not possible to identify individual alleles at loci with confidence because bands of similar size might not represent homologous DNA that is the basis for evolutionary comparisons.

With staphylococci, a molecular epidemiology emphasis has resulted in the use of band-based techniques that detect high levels of genetic variation, but little consideration has been given to their suitability for population genetics. For example, criteria for interpreting PFGE banding patterns were developed for the expressed purpose of local outbreak investigations (Tenover et al., 1995), yet they are routinely applied to much broader populations of staphylococci. In general, the population genetics role of band-based techniques has been limited to clustering of strains based on gross similarity in their banding patterns. However, even strain similarity may not be accurately reflected because of deficiencies in study design and analysis. Detailed work with *Escherichia coli* has shown that the choice and number of restriction enzymes used with PFGE greatly affect the accuracy of the technique (Singer et al., 2004). Whereas six or more enzymes would be needed to provide reasonable estimates of strain similarity for *E. coli* (Davis et al., 2003), PFGE protocols for *S. aureus* and *S. epidermidis* use single restriction enzymes (e.g., Chung et al. 2000; McDougal et al., 2003; Murchan et al., 2003). We note that equations are available to estimate nucleotide sequence similiarity, rather than gross similarity in banding patterns, from both restriction fragment (Upholt, 1977) and restriction site (Nei and Li, 1979) data. In addition, more than strain similarity can be studied with band-based techniques; correlations between genetic distances that are calculated from different band-based techniques can provide an indication of linkage disequilibrium (LD) and relative clonality in the population (Tibayrenc, 1995).

DNA sequence-based typing techniques, namely, multilocus sequence typing (MLST), have proven invaluable for the population genetics analysis of staphylococci. MLST represents a technological progression of MLEE into an era of high-throughput nucleotide sequencing. It involves direct sequencing of portions of housekeeping genes, which can be loosely defined as loci that are essential to normal cellular function and that presumably accumulate genetic variation free of diversifying selection. Each unique nucleotide sequence at a locus is given a numeric allele designation, and the unique combination of alleles across all loci is given a numeric sequence type (ST) designation. A mature MLST scheme based on seven housekeeping loci is available for *S. aureus* (Enright et al., 2000), and its publicly available database (http://saureus.mlst.net/) contains records of 1391 STs at present. With *S. epidermidis*, three different MLST schemes were initially proposed (Wisplinghoff et al., 2003; Kozitskaya et al., 2005; Wang et al., 2003), but a consensus scheme making use of more variable loci from these different schemes has prevailed (Thomas et al., 2007). The publicly available database (http://sepidermidis.mlst.net/) for the consensus MLST scheme of *S. epidermidis* contains records of 211 STs at present.

16.2.2 Natural Groups: From STs to Subspecies

One important function of bacterial population genetics is to elucidate the phylogeny of strains within a named species. Phylogenies are "gold standards" for studying evolution; they provide hypotheses for the genealogy of strains, detailing their patterns of descent and amounts of genetic variation accumulated over time. As such, phylogenies will reveal natural groups that are often important analysis units for molecular epidemiology. Natural groups can be defined as monophyletic groups, which are assemblages of bacteria with a unique common ancestor. We now discuss some natural groups of staphylococci, from fine to coarse hierarchical levels.

With both *S. aureus* and *S. epidermidis*, the multilocus ST is thought to be a natural group. STs are equated with bacterial clones, though it is recognized that variation affecting infectious disease can occur within STs. Early work identified single-nucleotide polymorphisms (SNPs) as well as insertion–deletion polymorphisms within *S. aureus* STs, using genes that encode proven or putative surface proteins (Robinson and Enright, 2003; Gomes et al., 2005; Kuhn et al., 2006). *S. aureus* ST8 was shown to be represented by several different clones that each have different gene expression patterns that affect virulence properties (Li et al., 2009). SNPs are the most abundant variations in genomes (Morin et al., 2004) and can even occur within *S. aureus* clones that are defined by a combination of ST, PFGE type, and antibiotic resistance gene type (Kennedy et al., 2008). Since MLST merely samples genetic variation from genomes, the ST or any other genotypically defined clone not based on complete genome sequences will likely represent a group of closely related clones. Recent work has found that *S. aureus* ST5 and ST239 each have SNPs that identify natural groups within the confines of the ST (i.e., there is population structure within an ST) (Nübel et al., 2008; Smyth et al., 2009b).

The clustering of STs by the eBURST algorithm has been a popular approach in the field. eBURST makes use of the numeric allele designations of STs without regard to their underlying nucleotide differences (Feil et al., 2004). With a user-defined level of allele sharing, eBURST identifies groups of closely related STs called clonal complexes (CCs). Putative founders of CCs are assigned through a frequency criterion that has a sound theoretical basis (Crandall and Templeton, 1993). However, the relationships within CCs are worked out with an algorithm of unproven accuracy. The space of all possible relationships within CCs is not explored by eBURST; only a single topology is returned and it is not measured against similar topologies. The algorithm might be aptly classified as pseudo-parsimony. A nonparametric bootstrapping procedure is used to assess the reliability of the founder assignments. However, it is important to emphasize that the CCs identified by eBURST, as well as the founder assignments and the relationships within CCs, are dependent on the sample used for analysis. To decrease the risk of identifying artificial groups, eBURST analysis in a given study should probably make use of the entire public MLST database for clustering their STs into CCs.

S. aureus and *S. epidermidis* present different population structures as inferred from eBURST. *S. aureus* shows 40 major CCs, 17 minor CCs with no candidate founders, and 210 singletons not affiliated with CCs (Fig. 16.1a). Validation of ST memberships in various *S. aureus* CCs has been done previously with conventional phylogenetic analyses (Feil et al., 2003; Robinson and Enright, 2003; Sakwinska et al., 2009). eBURST analysis of *S. epidermidis* shows 8 major CCs, 6 minor CCs, and 46 singletons (Fig. 16.2a). Sixty-one percent of the *S. epidermidis* STs are clustered within a single CC, previously called CC2 (Miragaia et al., 2007). A population structure of this sort has been noted by Turner et al. (2007) to contain potentially inaccurate relationships. Consistent with this warning,

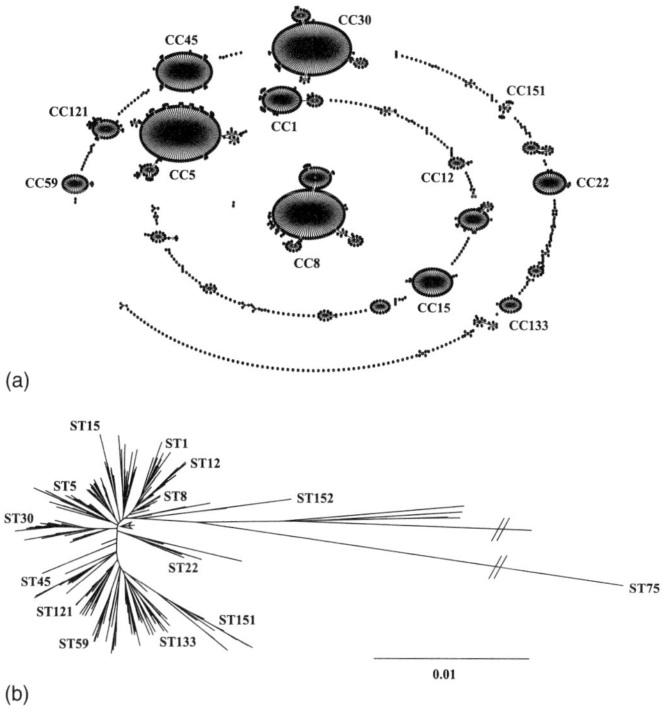

Figure 16.1 Overview of population structure in *S. aureus*. (a) eBURST analysis of all 1391 STs in the MLST database. Each circle represents a unique ST. Lines connect STs that differ at a single locus, though not all such connections are depicted. Names of various CCs are indicated. (b) Neighbor-joining phylogenetic tree based on concatenated MLST sequences. The tree shows relationships between 265 diverse STs that are representative of each CC and the singleton STs that do not cluster into CCs; two STs with insertion–deletion polymorphisms were dropped from the analysis. The arrow indicates the branch that subdivides the left half of the tree into two groups (above and below the branch) that contain most of the STs in the database. The branch leading to CC75 is much longer than is depicted. Scale is in substitutions per site. See color insert.

we found that an eBURST analysis of the first 74 STs in the *S. epidermidis* MLST database produced different results from that obtained with a larger database of 182 STs (Wong et al., 2009) and still different results from the present database of 211 STs. For example, predicted founder assignments have changed from ST2 to ST6 to ST5 with the growth of the MLST database. These STs possibly represent separate clusters of *S. epidermidis* (Fig. 16.2b), though more sequence data and more detailed analyses are required to test this hypothesis. Interestingly, the assignment of ST6 as a founder for CC2 would have remained consistent throughout the development of the MLST database, if founder assignments had been based on the lowest average number of pairwise locus differences to all other STs rather than the default frequency criterion (data not shown). Nucleotide diversity (i.e., average number of pairwise nucleotide differences per site) based on concatenated MLST sequences is comparable for *S. aureus* and *S. epidermidis*; both species average nucleotide diversities of 0.001 within the major and minor CCs and 0.01 when comparing single representatives from each of these CCs. Differences in the mechanisms that shape genetic variation in these species will be examined later in this chapter.

How are *S. aureus* CCs related to each other? MLST data from Feil et al. (2003) hinted that *S. aureus* might be subdivided into two subspecies groups that each contain

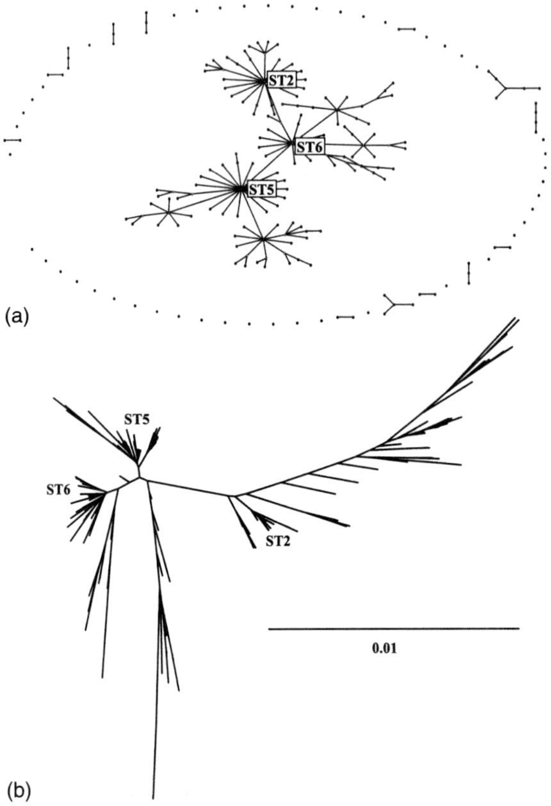

Figure 16.2 Overview of population structure in *S. epidermidis*. (a) eBURST analysis of all 211 STs in the MLST database. Each circle represents a unique ST. Lines connect STs that differ at a single locus, though not all such connections are depicted. Names of various STs within the large, "straggly" CC are indicated. (b) Neighbor-joining phylogenetic tree based on concatenated MLST sequences. The tree shows relationships between 210 STs; one ST with insertion–deletion polymorphisms was dropped from the analysis. Scale is in substitutions per site. See color insert.

multiple CCs. Further investigation by Robinson et al. (2005b), using both the seven MLST loci as well as seven surface protein-encoding loci, confirmed the notion of two groups. Cooper and Feil (2006) sequenced 33 gene fragments and resolved additional phylogenetic structure within the two groups. A neighbor-joining tree based on concatenated MLST sequences reflects the relationships between CCs and the two groups (Fig. 16.1b). Both groups include widely disseminated and clinically important CCs. The appearance of longer branches within group 1 has been noted (Robinson et al., 2005b; Cooper and Feil, 2006), which might indicate that different processes have shaped the two groups over time, but these hypotheses require testing. The phylogenetic position of some CCs (e.g., CC22) is unclear because they cluster in different groups, sometimes with strong statistical support, depending on the sample used for analysis (Robinson and Enright, 2003; Sakwinska et al., 2009). Nonetheless, the signature of the two groups is detectable at a wide variety of loci, including within the allelic groups defined at the accessory gene regulator (*agr*) (Robinson et al., 2005b), capsular biosynthesis locus (unpublished data), coagulase locus (Watanabe et al., 2009), and even in microarray data sets of gene content (Lindsay et al., 2006; Monecke et al., 2008).

Interestingly, recent studies have revealed additional *S. aureus* CCs, including CC152 and CC75, that fall outside of the two groups (McDonald et al., 2006; Ruimy et al., 2008; Fig. 16.1b). The divergence of these two CCs appears to be due to the accumulation of mutations in MLST loci over time; eight times as many polymorphisms were noted in comparisons between CC75 and groups 1 and 2 versus comparisons between groups 1 and

2 alone (Ng et al., 2009). CC75 is the most basal member of *S. aureus* identified at present. Both CC152 and CC75 are known from colonization specimens and from community-acquired MRSA infections; no phenotype has yet been found that distinguishes them from other *S. aureus* (Giffard et al., 2009).

An alternative view of natural groups within *S. aureus* was summarized by Wright et al. (2005). Their hypothesis is grounded on the important biological role of the *agr*. *agr* encodes a two-component signal transduction system that is also a quorum-sensing system. All staphylococcal species examined contain a homologous *agr* locus (Dufour et al., 2002). In *S. aureus*, *agr* functions as a master regulator of virulence gene expression (Novick, 2003). A hypervariable region defines four allelic groups that Wright et al. (2005) consider to be a fundamental basis for subdividing the species; however, each *agr* group occurs in multiple CCs, and there is no evidence that the different CCs of a given *agr* group are monophyletic (Robinson et al., 2005b). Over short time frames, a stable relationship exists between *agr* groups and CCs, but, over long time frames, *agr* groups have been shuffled between CCs. Statistically supported signatures of recombination at the *agr* locus have been presented, and a case for ancestral polymorphism at the locus has been made (Robinson et al., 2005b). Evidence of *agr* recombination in the canine pathogen *Staphylococcus intermedius* has also been presented (Bannhoer et al., 2007).

16.3 STAPHYLOCOCCAL POPULATION STRUCTURE IN SPECIFIC DISEASE CONTEXTS

16.3.1 Colonization and Disease

With facultative (opportunistic) pathogens such as *S. aureus* and *S. epidermidis*, an ability to colonize hosts asymptomatically is often considered a characteristic of ecological success. However, Sakwinska et al. (2009) observed that the most abundant *S. aureus* genotypes in a Swiss population were not necessarily the most abundant in nasal samples in terms of colony-forming units (Sakwinska et al., 2009). CC30 and CC45 from *S. aureus* subspecies group 1 are common colonizers in both Europe and North America (Melles et al., 2008). In addition, CC59 and CC121 from subspecies group 1 were unexpectedly abundant in carriage samples from Southwest China (Fan et al., 2009). On the other hand, the top four CCs from carriage samples in Mali, Africa included the divergent CC152 and CC5, CC8, and CC15 from subspecies group 2. These results could be interpreted to mean that virtually any *S. aureus* genotype can colonize humans and that host susceptibility or the geographic distribution of *S. aureus* diversity will determine the array of colonizing CCs in any given locale. Several host susceptibility factors for colonization have been identified (van Belkum et al., 2007, 2009a,b), but their role in colonization of specific *S. aureus* genotypes is not currently known.

The relative ability of different *S. aureus* genotypes to cause infection and to specialize in different types of infection is still unclear. Booth et al. (2001) studied a collection of 405 clinical *S. aureus* strains and 55 colonization strains. Two PFGE types were significantly overrepresented from particular infection sites, one type from respiratory tract infections and the other type from blood cultures. Feil et al. (2003) found that strains from colonization and invasive disease were evenly distributed among CCs. Wertheim et al. (2005) confirmed this observation, but they also observed that infection with a strain from a CC was associated with a significantly higher risk of mortality than infection with a strain not affiliated with a CC. Peacock et al. (2002) showed an intriguing correlation between the number of virulence factors identified by PCR and the frequency with which strains were isolated from invasive disease versus colonization, and an association between specific virulence factors and CCs. However, a microarray-based analysis by Lindsay et al. (2006) did not demonstrate an association between specific virulence factors and invasiveness. With the resolution of current typing techniques, it appears that virtually any *S. aureus* genotype can cause an invasive infection if given the opportunity (van Belkum et al., 2009b).

S. epidermidis is ubiquitious on human skin (Kloos and Musselwhite, 1975; Carr and Kloos, 1977). However, associations between certain *S. epidermidis* genotypes and disease have been recorded. Kozitskaya et al. (2005) found that one multilocus ST was overrepresented from clinical versus nonclinical strains, and it frequently contained genes that encode for biofilm and methicillin resistance traits. Moreover, Monk et al. (2008) recently found that *S. epidermidis* from prosthetic valve endocarditis and native valve endocarditis may represent genetically distinct subpopulations of different virulence potentials. These studies are important because they provide a genetic clarity to the view that CNS are an emerging source of infectious disease (Chu et al., 2008).

16.3.2 MRSA Infections

In 2005, MRSA infections were estimated to cause approximately 19,000 deaths of hospitalized patients in the U.S., which was more than the number of deaths caused by AIDS (Klevens et al., 2007). This finding emphasized the need for continued surveillance and study of this pathogen. Fortunately, much progress has been made in understanding both the genetics and the population genetics of resistance. The methicillin resistance phenotype is encoded by the *mecA* operon, which is located on a mobile genetic element called staphylococcal chromosomal cassette *mec* (SCC*mec*) (Ito et al., 2001). SCC*mec* elements can encode resistance to a variety of antimicrobials due to their carriage of integrated plasmids and transposons. PCR-based typing and DNA sequencing have identified at least seven major structural variants of SCC*mec* and numerous minor variants. SCC*mec* types I–III are commonly found from hospital-associated (HA)-MRSA, whereas SCC*mec* types IV and V are commonly found from both community-associated (CA-) and HA-MRSA (Deurenberg and Stobberingh, 2008). Do MRSAs derive from a single, special clone, or do they arise independently under the selective pressure of antimicrobials?

The single clone theory of MRSA evolution proposed that all MRSAs had evolved from a common susceptible (MSSA) ancestor by a single acquisition of *mecA* (Kreiswirth et al., 1993). In contrast, the multiple clone theory proposed a more dynamic scenario (Branger and Goullet, 1987). Data from MLEE (Musser and Kapur, 1992), microarray analysis (Fitzgerald et al., 2001), and MLST (Enright et al., 2002) indicated that resistance had arisen within genetically diverse lineages of *S. aureus*, supporting the theory of multiple MRSA origins. Enright et al. (2002) traced the origin of globally predominant HA-MRSA to five CCs, CC5, CC8, CC22, CC30, and CC45, and further noted that different strains of the same ST could carry different SCC*mec* elements. A subsequent study using 14 housekeeping and surface protein-encoding loci led to the conclusion that SCC*mec* was acquired by MSSA a minimum of 20 times (Robinson and Enright, 2003). A recent study using SNP discovery at 108 loci to construct a phylogeny for ST5 concluded that this single ST had acquired SCC*mec* at least 23 times (Nübel et al., 2008). Thus, there are likely hundreds of independent MRSA clones circulating in various locales. It has been proposed that CNS species might serve as the reservoir of SCC*mec* elements that occasionally transfer to *S. aureus* (Archer et al., 1994). It is currently unknown how often interspecies transfers of SCC*mec* occur and how often local MRSA clones disseminate globally.

CA-MRSAs were reported as a cause of skin and soft tissue infections in otherwise healthy aboriginal patients in Australia in 1993 (Udo et al., 1993) and in Native American populations in Wisconsin, USA in the early 1990s (Shukla et al., 2004). These reports were of interest because they indicated that MRSA was no longer restricted to nosocomial environments. CA-MRSAs are noted for their carriage of the Panton-Valentine leukocidin (PVL) (Robinson et al., 2005a; Voyich et al., 2006; Labandeira-Rey et al., 2007), though evidence has accumulated that other loci have an important role in their virulence (Bubeck et al., 2007; Wang et al., 2007). Several different lineages of CA-MRSA have been identified in different geographical locations, including ST1 in Europe, in Asia, and in the United States; ST8 in Europe and in the United States (i.e., the USA300 clone); ST30 in Australia, in Europe, and in South America; ST59 in Asia and in the United States; and ST80 in Asia, in Europe, and in the Middle East (Diep et al., 2006; Tristan et al., 2006). In recent times, CA-MRSA clones have been observed to replace HA-MRSA clones in hospital settings in the United States and in Taiwan and are increasing in prevalence in countries that have maintained a low prevalence of MRSA, namely, the Netherlands, Denmark, and Norway (Bartels et al., 2007; Stam-Bolink et al., 2007; Fang et al., 2008). At present, the distinction between HA-MRSA and CA-MRSA may be fading (Popovich et al., 2008).

16.3.3 Animal Infections

Several staphylococcal species are noted for their ability to cause disease in animals. *Staphylococcus pseudintermedius* commonly causes canine pyoderma; *Staphylococcus hyicus* is responsible for exudative epidermitis and arthritis in pigs; *Staphylococcus schleiferi* ssp. *coagulans* causes canine otitis externa; and *Staphylococcus delphini* has been reported as a cause of suppurative skin lesions in dolphins (Sato et al., 1990; Devriese et al., 2005; Yamashita et al., 2005). While studies of the population structure of these species are rare (Bannoehr et al., 2007), much work has focused on *S. aureus*, which can cause disease in a range of animals. *S. aureus* ssp. *anaerobius* is a cause of lymphadenitis in sheep (de la Fuente and Suarez, 1985), and it is capable of causing disease in human farm workers (Peake et al., 2006). The phylogenetic position of this taxon within *S. aureus* is currently unknown.

Early studies of *S. aureus* ssp. *aureus* investigated variation in numerous phenotypic traits and revealed the existence of human-, bovine-, ovine-, and poultry-specific biotypes (Devriese and Oeding, 1976; Devriese, 1984; Hébert et al., 1988). MLEE contributed to knowledge of the population structure of bovine-associated *S. aureus*. For example, 357 geographically diverse strains from bovine milk samples were shown to belong to 39 electrophoretic types (ETs), including the predominant ET3 that had achieved an international distribution (Kapur et al., 1995). Subsequent work involving a variety of band-based techniques confirmed the hypothesis that bovine-associated *S. aureus* was clonal and exhibited a host specialization (Fitzgerald et al., 1997; Zadoks et al., 2002; Jørgensen et al., 2005; van Leeuwen et al., 2005). Recent microarray studies have revealed that bovine-associated lineages encode variants of several genes, including *fnbA*, *fnbB*, and *coa*. Mobile genetic element-encoded genes such as *chp*, *scn*, and *sak* were less commonly observed than in their human counterparts (Guinane et al., 2008; Sung et al., 2008). Recently, whole genome sequencing of a representative strain of ET3 provided evidence of extensive loss-of-function mutations, particularly among the genes that encode surface proteins (Herron et al., 2002, 2007). In addition, 47 gene sequences were identified as unique to bovine strains (Herron et al., 2007). Similarly, other studies have revealed abundant polymorphisms in surface protein loci from sheep and from other ruminants (Smyth et al., 2009a; Vautor et al., 2009).

MLST has more precisely defined several animal-associated lineages, including CC97, CC133, and CC151 (Smith et al., 2005; Ben Zakour et al., 2008; Smyth et al., 2009a). ST151 (i.e., ET3) and ST133 have yet to be isolated from humans; interestingly, these ruminant-associated STs cluster together on the species tree (Fig. 16.1b). MLST has also aided the study of the transmission and population structure of emerging animal-associated MRSA. For example, ST398 was refractory to typing by PFGE using standard protocols owing to a novel DNA methylation enzyme. Using MLST, ST398 has been identified in horses, poultry, pigs, dogs, cats, and human veterinary workers and healthcare staff (Witte et al., 2007; Nemati et al., 2008; Wulf et al., 2008). Other notable animal-associated STs include ST121, associated with rabbits (Vancraeynest et al., 2006), and ST5, associated with chickens (Smyth et al., 2009a).

16.3.4 Linking Natural Groups with Specific Traits

Equally important to improved typing techniques are improved methods of analysis for addressing whether bacterial populations are structured according to a trait of interest. One approach is the G_{ST}-like differentiation statistics noted in Appendix 1. Another approach is the Slatkin–Maddison (SM) test (Slatkin and Maddison, 1989), which was originally developed to estimate gene flow between subpopulations that are resolvable on a phylogenetic tree. The SM test requires a reliable phylogeny, which can be challenging to obtain for within-ST subpopulations because of the large amount of sequence data required (Nübel et al., 2008; Smyth et al., 2009b). The SM test has been used successfully to study both within-host and between-host viral populations and has proven to be among the most powerful for detecting population structure differences (Zárate et al., 2007). Essentially, any trait (e.g., geography, virulence, antibiotic resistance) can be mapped onto a tree using the parsimony approach of Fitch (1971) (implemented by MacClade software). Even moderately recombinant species might be amenable to the SM test if the signal of recombination could be removed from the data used for phylogeny reconstruction or if linkage groups could be identified and treated separately. We have recently used the SM test in conjunction with a differentiation statistic (K_{ST}) to study geographic, temporal, and methicillin resistance traits of *S. aureus* ST239 (Smyth et al., 2009b). The two approaches are complementary, but should be used on clone-corrected data (i.e., one example strain per genotype per character state) to remove artifactual associations that will arise because of overrepresented clones. An example application of the SM test is presented in Fig. 16.3. These and other statistical approaches can greatly contribute to progress in understanding how staphylococcal natural groups are linked to traits of infectious disease relevance.

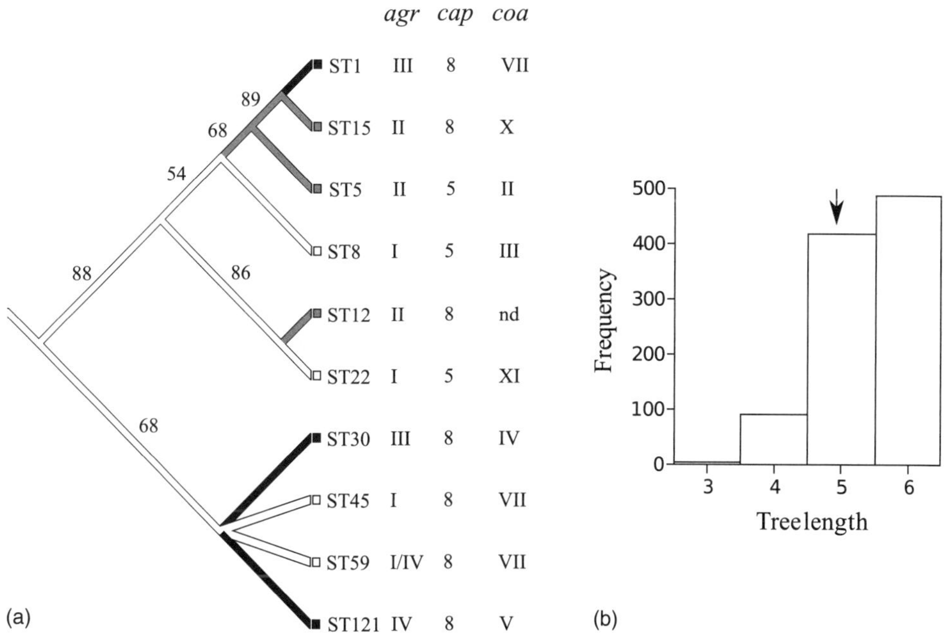

Figure 16.3 Phylogenetic distribution of the allelic groups at several loci. (a) Neighbor-joining phylogenetic tree based on concatenated MLST sequences of 10 diverse STs; numbers on branches are bootstrap proportions. The tree was rooted with CC75, which was subsequently pruned from the tree. Note that ST22 has moved in comparison to its position in Fig. 16.1b. The table adjacent to the tree shows allelic groups at the *agr*, *cap*, and *coa* loci (data from Robinson et al., 2005b, Watanabe et al., 2009, and unpublished data). The allelic groups at these loci cannot be accounted for by a single origin on the MLST tree, most likely because of recombination events between STs. Tree shading indicates the most parsimonious reconstruction of *agr* groups, calculated with MacClade software. (b) Slatkin–Maddison test that *agr* groups are linked to the phylogenetic structure depicted by the MLST tree. The null distribution of tree lengths is under a hypothesis of free migration of *agr* groups between STs. The arrow shows the minimum number of *agr* migration events required by the tree in panel a. Since the tree contained a polytomy (i.e., lack of resolution for the branching arrangements in subspecies group 1), the observed tree length was averaged over all possible dichotomous resolutions of the polytomy. We fail to reject the null hypothesis of free migration of *agr* groups between STs ($p = 0.513$). We conclude that *agr* allelic groups are not linked to the phylogenetic structure depicted by MLST.

16.4 ORIGIN AND MAINTENANCE OF STAPHYLOCOCCAL GENETIC VARIATION

16.4.1 Mutation and Recombination

Nonrandom associations between different genetic variations describe a condition known as LD. A clonal evolutionary process, where drift and selection determine the fate of mutations, will naturally produce LD. Conversely, genetic recombination is often inferred when LD cannot be detected, though rapid population growth (Slatkin, 1994) and under-powered data can produce effects similar to recombination. The LD approach to studying bacterial population structure has a long tradition (Whittam et al., 1983) and has led to some important breakthroughs in understanding (Smith, 1993).

For both *S. aureus* and *S. epidermidis*, LD is implied by the moderately good congruence that occurs between genotypes defined by different typing techniques. For example, de Lencastre and colleagues made quantitative comparisons between PFGE and MLST data for 198 diverse *S. aureus* strains (Faria et al., 2008) and 138 diverse *S. epidermidis* strains (Miragaia et al., 2008). When PFGE types were defined on a UPGMA dendrogram with a similarity threshold of 79–80%, their congruence with STs was comparable as reflected by an adjusted Rand index of 0.34 for *S. aureus* and 0.31 for *S. epidermidis*. While the adjusted Rand index accounts for chance matching of types, and values closer to one indicate better congruence, lower values could reflect technical and diversity differences in the typing techniques rather than inherent conflicts at the loci that are assessed by the techniques.

When individual alleles can be identified by the typing techniques, a variety of association indices can be used to measure LD more precisely. The index of association, I_A, has been one of the most popular indices used with bacterial data. I_A is based on the variance in allelic mismatches (from the mismatch distribution) and is significantly greater than zero when LD is present (see Haubold and Hudson, 2000). Our analysis of the MLST data of Feil et al. (2003), which included a sample of colonization and disease *S. aureus* strains from Oxford, England, provides evidence of significant LD when the 334 *S. aureus* strains are considered ($I_A^S = 0.54$) and when the subset of 74 *S. aureus* STs is considered ($I_A^S = 0.33$). Similarly, Miragaia et al. (2007) found significant LD among MLST loci in a geographically diverse sample of 217 *S. epidermidis* strains ($I_A^S = 0.29$) and in the subset of 74 *S. epidermidis* STs ($I_A^S = 0.17$). Maynard Smith (1994) found that recombination would need to be 20 times more frequent than mutation for simulated data to appear recombinant by I_A. LD measures do not provide a full description of the processes that influence staphylococcal population structure; however, the observed LD does indicate that neither of the two species are freely recombining across much of their diversity. Alternatively, the observed LD could be explained by the existence of natural groups within these species, where recombination is frequent within but not between groups (Souza et al., 1992).

An approach described by Guttman and Dykhuizen (1994) and detailed by Feil et al. (2000) allows more direct estimates of the relative roles of mutation and recombination on allelic variation. This approach first involves inference of closely related ancestor–descendant pairs of STs (e.g., using eBURST), followed by investigation of the types of nucleotide changes that occur between them. Mutations are attributed to single base pair differences that are unique and are not found elsewhere in the gene pool (i.e., the MLST database), whereas recombinations are attributed to multiple base pair differences and to alleles that occur in unrelated CCs. Using this approach, it was found that *S. aureus* MLST alleles change 15 times more often by mutation than by recombination (Feil et al., 2003). A fourfold predominance of mutation over recombination events was also seen with *S. aureus* colonization strains from Mali, Africa (Ruimy et al., 2008). In contrast, Kozitskaya et al. (2005) and Miragaia et al. (2007) used two different *S. epidermidis* MLST schemes and found that recombination was 4.0 and 2.5 times more frequent than mutation, respectively, for generating allelic variation.

These estimates of mutation and recombination events are probably very crude; we know essentially nothing about their sampling properties. However, do they inform our understanding of staphylococcal population structure? It is the case that recombination events in *S. aureus* are rare enough that they call attention to themselves. Robinson and Enright (2004) found that the recombination of a large, contiguous region of the chromosome between ST8- and ST30-like parent strains led to the origin of the ST239 pandemic MRSA clonal group. The ST34 MSSA clonal group was also found to be recombinant in origin, with ST10/ST145- and ST30-like parents. It was concluded that hundreds of kilobases of DNA had been transferred en bloc to form these *S. aureus* hybrids (Robinson and Enright, 2004). These observations challenge the notion that chromosomal recombination events in natural populations of bacteria are limited to short, localized replacements (Smith et al., 1991). Moreover, the ST239 story shows that mixing large portions of the chromosomes of unrelated CCs can produce a new pathogen with enhanced virulence potential (Amaral et al., 2005; Edgeworth et al., 2007) rather than an unfit amalgamation destined for extinction. In fact, we think that rare, large chromosomal recombinations between *S. aureus* CCs might even provide a mechanism for the birth of new CCs; CCs that cluster with some statistical support frequently appear to have shuffled their alleles at loci that encode important phenotypes (e.g., *agr*, *cap*, *coa*) (Fig. 16.3). While recombination in *S. aureus* is rare, it can be highly relevant to infectious disease (Hughes and Friedman, 2005).

S. epidermidis probably undergoes more frequent recombination than does *S. aureus*, but just how much more is unclear. While the eBURST structure, standardized I_A values, and r:m estimates described above support this hypothesis, one piece of data does not. Pérez-Losada et al. (2006) used a coalescent-based estimator of recombination and mutation rates, which are scaled to account for differences in effective population size. Their results suggested that *S. epidermidis* allelic variation is, on average, more influenced by mutation relative to recombination than is the case with *S. aureus* allelic variation. Two caveats might explain this discrepancy. First, the *S. epidermidis* MLST database was poorly developed at the time of their analysis; data on a small number of strains of low diversity were available. Second, their analysis made use of a gene conversion model of recombination that requires recombination breakpoints to fall within the investigated sequences in order to be detected; if recombinations tend to mobilize sequences longer than MLST alleles (e.g., >450 bp), they might be missed in the analysis. Clark and Zheng (2008) found that patterns of LD in simulated bacterial data were greatly influenced by the type of recombination model employed. Thus, it will be important to determine the most appropriate model of recombination for staphylococcal species.

16.4.2 Drift and Selection

Multiple models that attempt to explain standing genetic variation in bacterial populations are available (Levin, 1981; Smith et al., 1993; Majewski and Cohan, 1999; Buckee et al., 2008; Fraser et al., 2009). These models differ in the relative importance attributed to drift versus selection, competition (or clonal interference; Park and Krug, 2007), and in the role of ecological adaptation. All of these models have elements that are relevant to staphylococcal populations, but, at present, it is not possible to reject any of them as a single theoretical framework.

Both random genetic drift and natural selection probably have important roles in shaping genetic variation in staphylococcal populations, at different time frames. Consider again the *S. aureus* MLST data of Feil et al. (2003), where the ratio of synonymous (d_S) to nonsynonymous (d_N) mutations in MLST loci averages 3.1 within the major and minor CCs and 7.6 when comparing single representatives from each of these CCs. Likewise, for the *S. epidermidis* MLST data of Miragaia et al. (2007), this ratio averages 5.3 within the major and minor CCs and 9.2 when comparing single representatives from each of these CCs. Numerous nonsynonymous mutations in housekeeping genes are evident in both species, but their numbers decrease with increasing genetic distance. Why? One could hypothesize that all of these mutations are effectively neutral (Fraser et al., 2009).

Alternatively, one could hypothesize that some of these mutations confer more or less efficient housekeeping functions and are therefore subject to selection (Buckee et al., 2008). We suggest that purifying (negative) selection may become more efficient at removing the slightly deleterious nonsynonymous mutations over the time frame represented by the birth of new CCs.

Over short time frames, the relative roles of drift and selection in shaping genetic variation are well known. When a neutral mutation arises in a population of N individuals, its fixation probability is $1/N$ and its extinction probability is $1 - 1/N$ (Kimura, 1962). These simple equations show that a neutral mutation has a much greater chance of going extinct than becoming fixed and that fixation is more probable in smaller populations than in larger populations. s represents the selection coefficient, which is the percentage change in fitness conferred by a mutation. It is also known that mutations behave as if they were neutral if $s < 1/N$ (Kimura and Takahata, 1983). This equation shows that a given mutation can be neutral in smaller populations and not in larger populations. In other words, for determining the fate of genetic variation, drift is more important in smaller populations and selection is more important in larger populations. Because mutations are inherited en bloc, the fate of a mutation also depends on the selection operating at linked sites (Smith and Haigh, 1974). This linkage can temper the rare beneficial mutations and can accentuate the more abundant class of slightly deleterious mutations; beneficial mutations can go extinct and slightly deleterious mutations can become fixed. Genetic recombination provides a way to break up linkage, and it might help to reduce the load of slightly deleterious mutations (Hill and Robertson, 1966). Thus, population size and recombination rates should be important parameters for determining the fate of genetic variation in staphylococcal populations. The reader is referred to Chapters 1, 2, 4, and 6 from this book for more information on the challenges associated with estimating effective population size and recombination rates in bacteria.

We note that, over long time frames, population size has no effect on the fate of a neutral mutation. If μ represents the mutation rate per generation, then the expected number of mutations per generation is $N\mu$. The rate at which new mutations become fixed per generation, also called the substitution rate, is then $1/N \times N\mu = \mu$ (Kimura, 1968). This equation shows that, for strictly neutral mutations, the substitution rate is simply the mutation rate. Substitution rate is independent of population size because smaller populations have fewer mutations with higher fixation probabilities, whereas larger populations have more mutations with lower fixation probabilities. These relationships allow for a long time frame molecular clock, which can be of practical use for studying waves of epidemics that unfold over centuries (Feng et al., 2008). However, when studying even more recent events, the full range of nucleotide changes can provide useful information. Using 32 variable loci and strains of known isolation dates, an absolute mutation rate for the *S. aureus* ST239 clonal group was recently estimated at $\sim 10^{-6}$ nucleotide changes per site per year (Smyth et al., 2009b). The method used overshoots the substitution rate because it includes all nucleotide changes, neutral and otherwise. However, it is much closer to the $\sim 10^{-5}$ mutation rate estimates recently reported from *Campylobacter jejuni* (Wilson et al., 2009), *Helicobacter pylori* (Falush et al., 2001), and *Neisseria gonorrhoeae* (Pérez-Losada et al., 2007) than to the standard $\sim 10^{-8}$ mutation rate estimate from *E. coli* (Achtman et al., 1999). The *E. coli* estimate was used recently to study the *S. aureus* ST5 clonal group (Nübel et al., 2008). Direct estimates of mutation rates from different *S. aureus* clonal groups should provide powerful new tools for studying their evolution and epidemiology.

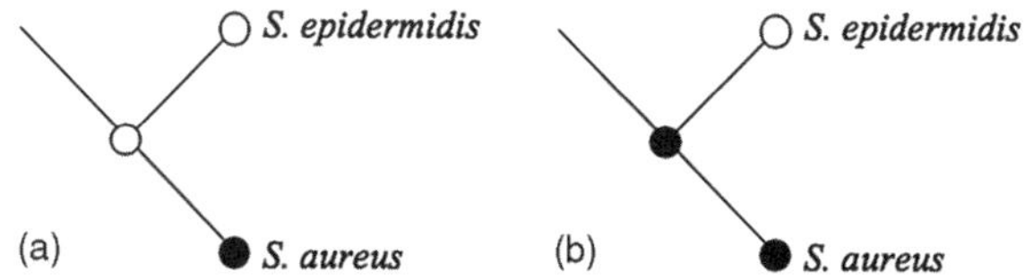

Figure 16.4 Two competing models for the origin of different population structures in *S. aureus* and in *S. epidermidis*. (a) The common ancestor of these species had a recombinant population structure, which requires an explanation for the changes in the *S. aureus* population structure. (b) The common ancestor of these species had a relatively clonal population structure, which requires an explanation for the changes in the *S. epidermidis* population structure.

16.5 MACROEVOLUTIONARY CONSIDERATIONS AND CONCLUDING REMARKS

The different population structures and processes at work within *S. aureus* and *S. epidermidis* have interesting implications for understanding the origin of these species. We can construct two competing speciation models that differ in the recombination status of the most recent common ancestor (MRCA) of these species. Under a model with a recombinant MRCA (Fig. 16.4a), a recombinant sister group combined with evidence for early shuffling of traits in *S. aureus* would suggest a lowered recombination rate for this pathogen. We note that the population-scaled recombination rate is proportional to the effective population size and the recombination rate per generation, $C = 2N_e r$. Either a decrease in population size (e.g., due to niche specialization for the anterior nares) or a decrease in recombination rate (e.g., due to the appearance of barriers to horizontal gene transfer) would provide a feasible explanation for a lowered recombination rate. Emmett and Kloos (1979) had pointed to arginine auxotrophy as a trait that could lock some staphylococci into specialized niches. This explanation might predict that other niche-specialized species (e.g., *S. capitis*) would have a population structure more similar to *S. aureus* than to *S. epidermidis*. As for the other explanation, a role for the *S. aureus* Sau1 restriction–modification system in destroying DNA imported by conjugation, transduction, and transformation has been demonstrated (Waldron and Lindsay, 2006), but a recent study suggested that this system is strain dependent and may not be the sole predictor of gene transfer pathways (Veiga and Pinho, 2009).

Under a model with a clonal MRCA (Fig. 16.4b), the data would suggest that *S. epidermidis* had experienced an increased recombination rate. *S. epidermidis* is predominant and is more persistent than *S. aureus* in the colonization of skin surfaces (Kloos and Musselwhite, 1975; Carr and Kloos, 1977); it reaches higher population sizes everywhere except an *S. aureus*-colonized anterior nares. Can population size differences alone account for the different population structures of these species? This explanation might predict that species with intermediately sized populations (e.g., *Staphylococcus hominis*) would show population structures intermediate between the relative clonality in *S. aureus* and the higher rates of recombination in *S. epidermidis*. To test these hypotheses about staphylococcal speciation, we need to know more about divergent *S. aureus* groups such as CC75 and CC152, and we need much more genetic information for *S. epidermidis* and other CNS.

Perhaps the most progress has been made in identifying coarse natural groups within *S. aureus*. Indeed, the natural groups of *S. aureus* colonization are considered by some to be largely solved (van Belkum et al., 2009a). Identifying natural groups at increasingly finer scales (e.g., within STs) should continue to yield new information. However, taxonomy cannot be population genetics' only contribution to staphylococcal biology. Almost 20 years ago, Kloos (1990) said, "Although considerable progress has been made in describing the variety of species, subspecies, and strains of staphylococci found in natural populations, there is currently little information available on their population or community interactions, thus offering a timely challenge to those interested in staphylococcal ecology and population genetics." With the public health threats posed by untreatable *S. aureus* infections (Sievert et al., 2008) and by emerging infections caused by CNS (Chu et al., 2008), the challenge remains timely. The extent of large chromosomal recombinations, their mechanisms, and the best way to model these events are just beginning to be explored. Finally, we have a poor understanding of the relative roles of different population processes in determining the fate of staphylococcal genetic variation, and we have no more than a taxonomic catalog, little mechanistic understanding, for how natural groups relate to infectious disease. With new high-throughput, high-coverage techniques for the detection of sequence variation and the wider application of statistical methods of analysis, all of these questions can be expected to yield answers.

REFERENCES

Achtman, M., Zurth, K., Morelli, G. et al. (1999) *Yersinia pestis*, the cause of plague, is a recently emerged clone of *Yersinia pseudotuberculosis*. *Proceedings of the National Academy of Sciences of the United States of America* **96,** 14043–14048.

Amaral, M. M., Coelho, R., Flores, R. P. et al. (2005) The predominant variant of the Brazilian epidemic clonal complex of methicillin-resistant *Staphylococcus aureus* has an enhanced ability to produce biofilm and to adhere to and invade airway epithelial cells. *Journal of Infectious Diseases* **192,** 801–810.

Archer, G. L., Niemeyer, D. M., Thanassi, J. A. et al. (1994) Dissemination among staphylococci of DNA sequences associated with methicillin resistance. *Antimicrobial Agents and Chemotherapy* **38,** 447–454.

Bannoehr, J., Ben Zakour, N. L., Waller, A. S. et al. (2007) Population genetic structure of the *Staphylococcus intermedius* group: Insights into *agr* diversification and the emergence of methicillin-resistant strains. *Journal of Bacteriology* **189,** 8685–8692.

Bartels, M. D., Boye, K., Rhod Larsen, A. et al. (2007) Rapid increase of genetically diverse methicillin-resistant *Staphylococcus aureus*, Copenhagen, Denmark. *Emerging Infectious Diseases* **13,** 1533–1540.

Ben Zakour, N. L., Sturdevant, D. E., Even, S. et al. (2008) Genome-wide analysis of ruminant *Staphylococcus aureus* reveals diversification of the core genome. *Journal of Bacteriology* **190,** 6302–6317.

Booth, M. C., Pence, L. M., Mahasreshti, P. et al. (2001) Clonal associations among *Staphylococcus aureus* isolates from various sites of infection. *Infection and Immunity* **69,** 345–352.

Branger, C. and Goullet, P. (1987) Esterase electrophoretic polymorphism of methicillin-sensitive and methicillin-resistant strains of *Staphylococcus aureus*. *Journal of Medical Microbiology* **24,** 275–281.

Bubeck Wardenburg, J., Bae, T., Otto, M. et al. (2007) Poring over pores: Alpha-hemolysin and Panton-Valentine leukocidin in *Staphylococcus aureus* pneumonia. *Nature Medicine* **13,** 1405–1406.

Buckee, C. O., Jolley, K. A., Recker, M. et al. (2008) Role of selection in the emergence of lineages and the evolution of virulence in *Neisseria meningitidis*. *Proceedings of the National Academy of Sciences of the United States of America* **105,** 15082–15087.

Carr, D. L. and Kloos, W. E. (1977) Temporal study of the staphylococci and micrococci of normal infant skin. *Applied and Environmental Microbiology* **34,** 673–680.

Chu, V. H., Woods, C. W., Miro, J. M. et al. (2008) Emergence of coagulase-negative staphylococci as a cause of native valve endocarditis. *Clinical Infectious Diseases* **15,** 232–242.

Chung, M., de Lencastre, H., Matthews, P. et al. (2000). Molecular typing of methicillin-resistant *Staphylococcus aureus* by pulsed-field gel electrophoresis: Comparison of results obtained in a multilaboratory effort using identical protocols and MRSA strains. *Microbial Drug Resistance* **6,** 189–198.

Clark, A. G. and Zheng, Y. (2008) Dynamics of linkage disequilibrium in bacterial genomes undergoing transformation and/or conjugation. *Journal of Evolutionary Biology* **10,** 663–676.

Cooper, J. E. and Feil, E. J. (2006) The phylogeny of *Staphylococcus aureus*—Which genes make the best intra-species markers? *Microbiology* **152,** 1297–1305.

Crandall, K. A. and Templeton, A. R. (1993) Empirical tests of some predictions from coalescent theory with applications to intraspecific phylogeny reconstruction. *Genetics* **134,** 959–969.

Davis, M. A., Hancock, D. D., Besser, T. E. et al. (2003) Evaluation of pulsed-field gel electrophoresis as a tool for determining the degree of genetic relatedness between strains of *Escherichia coli* O157:H7. *Journal of Clinical Microbiology* **41,** 1843–1849.

de la Fuente, R. and Suarez, G. (1985) Respiratory deficient *Staphylococcus aureus* as the aetiological agent of "abscess disease." *Journal of Veterinary Medicine. Series B* **32,** 397–406.

de Sousa, M. A., Sanches, I. S., van Belkum, A. et al. (1992) Characterization of methicillin-resistant *Staphylococcus aureus* isolates from Portuguese hospitals by multiple genotyping methods. *Microbial Drug Resistance* **2,** 331–341.

Deurenberg, R. H. and Stobberingh, E. E. (2008) The evolution of *Staphylococcus aureus*. *Infection, Genetics and Evolution* **8,** 747–763.

Devriese, L. A. (1984) A simplified system for biotyping *Staphylococcus aureus* strains isolated from animal species. *Journal of Applied Bacteriology* **56,** 215–220.

Devriese, L. A. and Oeding, P. (1976) Characteristics of *Staphylococcus aureus* strains isolated from different animal species. *Research in Veterinary Science* **21,** 284–291.

Devriese, L. A., Vancanneyt, M., Baele, M. et al. (2005) *Staphylococcus pseudintermedius* sp. nov., a coagulase-positive species from animals. *International Journal of Systematic Evolutionary Microbiology* **55,** 1569–1573.

Diekema, D. J., Pfaller, M. A., Schmitz, F. J. et al. (2001) Survey of infections due to *Staphylococcus* species: Frequency of occurrence and antimicrobial susceptibility of isolates collected in the United States, Canada, Latin America, Europe, and the Western Pacific region for the SENTRY Antimicrobial Surveillance Program, 1997–1999. *Clinical Infectious Diseases* **32,** S114–S132.

Diep, B. A., Gill, S. R., Chang, R. F. et al. (2006) Complete genome sequence of USA300, an epidemic clone of community-acquired methicillin-resistant *Staphylococcus aureus*. *Lancet* **367,** 731–739.

Dufour, P., Jarraud, S., Vandenesch, F. et al. (2002) High genetic variability of the *agr* locus in *Staphylococcus* species. *Journal of Bacteriology* **184,** 1180–1186.

Edgeworth, J. D., Yadegarfar, G., Pathak, S. et al. (2007) An outbreak in an intensive care unit of a strain of methicillin-resistant *Staphylococcus aureus* sequence type 239 associated with an increased rate of vascular access device-related bacteremia. *Clinical Infectious Diseases* **44,** 493–501.

Emmett, M. and Kloos, W. E. (1979) The nature of arginine auxotrophy in cutaneous populations of staphylococci. *Journal of General Microbiology* **110,** 305–314.

Enright, M. C., Day, N. P., Davies, C. E. et al. (2000) Multilocus sequence typing for characterization of methicillin-resistant and methicillin-susceptible clones of *Staphylococcus aureus*. *Journal of Clinical Microbiology* **38,** 1008–1015.

Enright, M. C., Robinson, D. A., Randle, G. et al. (2002) The evolutionary history of methicillin-resistant *Staphylococcus aureus* (MRSA). *Proceedings of the National Academy of Sciences of the United States of America* **99,** 7687–7692.

Falush, D., Kraft, C., Taylor, N. S. et al. (2001) Recombination and mutation during long-term gastric colonization by *Helicobacter pylori*: Estimates of clock rates, recombination size, and minimal age. *Proceedings of the National Academy of Sciences of the United States of America* **98,** 15056–15061.

Fan, J., Shu, M., Zhang, G. et al. (2009) Biogeography and virulence of *Staphylococcus aureus*. *PloS One* **4,** e6216.

Fang, H., Hedin, G., Li, G. et al. (2008) Genetic diversity of community-associated methicillin-resistant *Staphylococcus aureus* in southern Stockholm, 2000–2005. *Clinical Microbiology and Infection* **14,** 370–376.

Faria, N. A., Carriço, J. A., Oliveira, D. C. et al. (2008) Analysis of typing methods for epidemiological surveillance of both methicillin-resistant and methicillin-susceptible *Staphylococcus aureus* strains. *Journal of Clinical Microbiology* **46,** 136–144.

Feil, E. J., Cooper, J. E., Grundmann, H. et al. (2003) How clonal is *Staphylococcus aureus*? *Journal of Bacteriology* **185,** 3307–3316.

Feil, E. J., Enright, M. C., and Spratt, B. G. (2000) Estimating the relative contributions of mutation and recombination to clonal diversification: A comparison between *Neisseria meningitidis* and *Streptococcus pneumoniae*. *Research in Microbiology* **151,** 465–469.

Feil, E. J., Li, B. C., Aanensen, D. M. et al. (2004) eBURST: Inferring patterns of evolutionary descent among clusters of related bacterial genotypes from multilocus sequence typing data. *Journal of Bacteriology* **186,** 1518–1530.

Feng, L., Reeves, P. R., Lan, R. et al. (2008) A recalibrated molecular clock and independent origins for the cholera pandemic clones. *PloS One* **3,** e4053.

Fitch, W. M. (1971). Toward defining the course of evolution: Minimal change for a specific tree topology. *Systematic Zoolology* **20,** 406–416.

Fitzgerald, J. R., Meaney, W. J., Hartigan, P. J. et al. (1997) Fine-structure molecular epidemiological analysis of *Staphylococcus aureus* recovered from cows. *Epidemiology and Infection* **119,** 261–269.

Fitzgerald, J. R., Sturdevant, D. E., Mackie, S. M. et al. (2001) Evolutionary genomics of *Staphylococcus aureus*: Insights into the origin of methicillin-resistant strains and the toxic shock syndrome epidemic. *Proceedings of the National Academy of Sciences of the United States of America* **98,** 8821–8826.

Fraser, C., Alm, E. J., Polz, M. F. et al. (2009) The bacterial species challenge: Making sense of genetic and ecological diversity. *Science* **323,** 741–746.

Ghebremedhin, B., Layer, F., König, W. et al. (2008) Genetic classification and distinguishing of *Staphylococcus* species based on different partial gap, 16S rRNA, *hsp60, rpoB, sodA*, and *tuf* gene sequences. *Journal of Clinical Microbiology* **46,** 1019–1025.

Gomes, A. R., Vinga, S., Zavolan, M. et al. (2005) Analysis of the genetic variability of virulence-related loci in epidemic clones of methicillin-resistant *Staphylococcus aureus. Antimicrobial Agents and Chemotherapy* **49,** 366–379.

Grundmann, H., de Sousa, M. A., Boyce, J. et al. (2006) Emergence and resurgence of methicillin-resistant *Staphylococcus aureus* as a public-health threat. *Lancet* **368,** 874–885.

Grundmann, H., Hori, S., and Tanner, G. (2001) Determining confidence intervals when measuring genetic diversity and the discriminatory abilities of typing methods for microorganisms. *Journal of Clinical Microbiology* **39,** 4190–4192.

Guinane, C. M., Sturdevant, D. E., Herron-Olson, L. et al. (2008) Pathogenomic analysis of the common bovine *Staphylococcus aureus* clone (ET3): Emergence of a virulent subtype with potential risk to public health. *Journal of Infectious Diseases* **197,** 205–213.

Guttman, D. S. and Dykhuizen, D. E. (1994) Clonal divergence in *Escherichia coli* as a result of recombination, not mutation. *Science* **266,** 1380–1383.

Haubold, B. and Hudson, R. R. (2000) LIAN 3.0: Detecting linkage disequilibrium in multilocus data. Linkage analysis. *Bioinformatics* **16,** 847–848.

Hébert, G. A., Cooksey, R. C., Clark, N. C. et al. (1998) Biotyping coagulase-negative staphylococci. *Journal of Clinical Microbiology* **26,** 1950–1956.

Hedrick, P. W. (2005) A standardized genetic differentiation measure. *Evolution* **59,** 1633–1638.

Herron, L. L., Chakravarty, R., Dwan, C. et al. (2002) Genome sequence survey identifies unique sequences and key virulence genes with unusual rates of amino acid substitution in bovine *Staphylococcus aureus. Infection and Immunity* **70,** 3978–3981.

Herron-Olson, L., Fitzgerald, J. R., Musser, J. M. et al. (2007) Molecular correlates of host specialization in *Staphylococcus aureus. PloS One* **2,** e1120.

Hill, W. G. and Robertson, A. (1966) The effect of linkage on limits to artificial selection. *Genetics Research* **8,** 269–294.

Holsinger, K. E. and Weir, B. S. (2009) Genetics in geographically structured populations: Defining, estimating, and interpreting F_{ST}. *Nature Reviews. Genetics* **10,** 639–650.

Hughes, A. L. and Friedman, R. (2005) Nucleotide substitution and recombination at orthologous loci in *Staphylococcus aureus. Journal of Bacteriology* **187,** 2698–2704.

Hunter, P. R. and Gaston, M. A. (1988) Numerical index of the discriminatory ability of typing systems: An application of Simpson's index of diversity. *Journal of Clinical Microbiology* **26,** 2465–2466.

Ito, T., Katayama, Y., Asada, K. et al. (2001) Structural comparison of three types of staphylococcal cassette chromosome *mec* integrated in the chromosome in methicillin-resistant *Staphylococcus aureus. Antimicrobial Agents and Chemotherapy* **45,** 1323–1336.

Jarraud, S., Mougel, C., Thioulouse, J. et al. (2002) Relationships between *Staphylococcus aureus* genetic background, virulence factors, *agr* groups (alleles), and human disease. *Infection and Immunity* **70,** 631–641.

Jørgensen, H. J., Mørk, T., Caugant, D. A. et al. (2005). Genetic variation among *Staphylococcus aureus* strains from Norwegian bulk milk. *Applied Environmental Microbiology* **71,** 8352–8361.

Jost L. (2008) G(ST) and its relatives do not measure differentiation. *Molecular Ecology* **17,** 4015–4026.

Kapur, V., Sischo, W. M., Greer, R. S. et al. (1995) Molecular population genetic analysis of *Staphylococcus aureus* recovered from cows. *Journal of Clinical Microbiology* **33,** 376–380.

Kennedy, A. D., Otto, M., Braughton, K. R. et al. (2008) Epidemic community-associated methicillin-resistant *Staphylococcus* aureus: Recent clonal expansion and diversification. *Proceedings of the National Academy of Sciences of the United States of America* **105,** 1327–1332.

Kimura, M. (1962) On the probability of fixation of mutant genes in a population. *Genetics* **47,** 713–719.

Kimura, M. (1968) Evolutionary rate at the molecular level. *Nature* **217,** 624–626.

Kimura, M. and Crow, J. F. (1964) The number of alleles that can be maintained in a finite population. *Genetics* **49,** 725–738.

Kimura, M. and Takahata, N. (1983) Selective constraint in protein polymorphism: Study of the effectively neutral mutation model by using an improved pseudosampling method. *Proceedings of the National Academy of Sciences of the United States of America* **80,** 1048–1052.

Klevens, R. M., Morrison, M. A., Nadle, J. et al. (2007) Invasive methicillin-resistant *Staphylococcus aureus* infections in the United States. *Journal of the American Medical Society* **298,** 1763–1771.

Kloos, W. E. (1990) Systematics and the natural history of staphylococci 1. *Society for Applied Bacteriology symposium series* **19,** 25S–37S.

Kloos, W. E. and Bannerman, T. L. (1994) Update on clinical significance of coagulase-negative staphylococci. *Clinical Microbiology Reviews* **7,** 117–140.

Kloos, W. E. and Musselwhite, M. S. (1975) Distribution and persistence of *Staphylococcus* and *Micrococcus* species and other aerobic bacteria on human skin. *Applied Microbiology* **30,** 381–385.

Kozitskaya, S., Olson, M. E., Fey, P. D. et al. (2005) Clonal analysis of *Staphylococcus epidermidis* isolates carrying or lacking biofilm-mediating genes by multilocus sequence typing. *Journal of Clinical Microbiology* **43,** 4751–4757.

KREISWIRTH, B., KORNBLUM, J., ARBEIT, R. D. et al. (1993) Evidence for a clonal origin of methicillin resistance in *Staphylococcus aureus*. *Science* **259,** 227–230.

KUEHNERT, M. J., KRUSZON-MORAN, D., HILL, H. A. et al. (2006) Prevalence of *Staphylococcus aureus* nasal colonization in the United States, 2001–2002. *Journal of Infectious Diseases* **193,** 172–179.

KUHN, G., FRANCIOLI, P., and BLANC, D. S. (2006) Evidence for clonal evolution among highly polymorphic genes in methicillin-resistant *Staphylococcus aureus*. *Journal of Bacteriology* **188,** 169–178.

LABANDEIRA-REY, M., COUZON, F., BOISSET, S. et al. (2007) *Staphylococcus aureus* Panton-Valentine leukocidin causes necrotizing pneumonia. *Science* **315,** 1130–1133.

LEVIN, B. R. (1981) Periodic selection, infectious gene exchange and the genetic structure of *E. coli* populations. *Genetics* **99,** 1–23.

LI, M., DIEP, B. A., VILLARUZ, A. E. et al. (2009) Evolution of virulence in epidemic community-associated methicillin-resistant *Staphylococcus aureus*. *Proceedings of the National Academy of Sciences of the United States of America* **106,** 5883–5888.

LINDSAY, J. A., MOORE, C. E., DAY, N. P. et al. (2006) Microarrays reveal that each of the ten dominant lineages of *Staphylococcus aureus* has a unique combination of surface-associated and regulatory genes. *Journal of Bacteriology* **188,** 669–676.

MAJEWSKI, J. and COHAN, F. M. (1999) Adapt globally, act locally: The effect of selective sweeps on bacterial sequence diversity. *Genetics* **152,** 1459–1474.

MANIAN, F. A. (2003) Asymptomatic nasal carriage of mupirocin-resistant, methicillin-resistant *Staphylococcus aureus* (MRSA) in a pet dog associated with MRSA infection in household contacts. *Clinical Infectious Diseases* **36,** e26–e28.

MARSHALL, S. A., WILKE, W. W., PFALLER, M. A. et al. (1998) *Staphylococcus aureus* and coagulase-negative staphylococci from blood stream infections: Frequency of occurrence, antimicrobial susceptibility, and molecular (*mecA*) characterization of oxacillin resistance in the SCOPE program. *Diagnostic Microbiology Infectious Diseases* **30,** 205–214.

MCDONALD, M., DOUGALL, A., HOLT, D. et al. (2006) Use of a single-nucleotide polymorphism genotyping system to demonstrate the unique epidemiology of methicillin-resistant *Staphylococcus aureus* in remote aboriginal communities. *Journal of Clinical Microbiology* **44,** 3720–3727.

MCDOUGAL, L. K., STEWARD, C. D., KILLGORE, G. E. et al. (2003). Pulsed-field gel electrophoresis typing of oxacillin-resistant *Staphylococcus aureus* isolates from the United States: Establishing a national database. *Journal of Clinical Microbiology* **41,** 5113–5120.

MELLES, D. C., TENOVER, F. C., KUEHNERT, M. J. et al. (2008) Overlapping population structures of nasal isolates of *Staphylococcus aureus* from healthy Dutch and American individuals. *Journal of Clinical Microbiology* **46,** 235–241.

MILKMAN, R. (1973) Electrophoretic variation in *Escherichia coli* from natural sources. *Science* **182,** 1024–1026.

MIRAGAIA, M., CARRIÇO, J. A., THOMAS, J. C. et al. (2008) Comparison of molecular typing methods for characterization of *Staphylococcus epidermidis*: Proposal for clone definition. *Journal of Clinical Microbiology* **46,** 118–129.

MIRAGAIA, M., THOMAS, J. C., COUTO, I. et al. (2007) Inferring a population structure for *Staphylococcus epidermidis* from multilocus sequence typing data. *Journal of Bacteriology* **189,** 2540–2552.

MONECKE, S., SLICKERS, P., EHRICHT, R. (2008) Assignment of *Staphylococcus aureus* isolates to clonal complexes based on microarray analysis and pattern recognition. *FEMS Immunology and Medical Microbiology* **53,** 237–251.

MONK, A. B., BOUNDY, S., CHU, V. H. et al. (2008) Analysis of the genotype and virulence of *Staphylococcus epidermidis* isolates from patients with infective endocarditis. *Infection and Immunity* **76,** 5127–5132.

MORIN, P. A., LUIKART, G., WAYNE, R. K. et al. (2004) SNPs in ecology, evolution and conservation. *Trends in Ecology and Evolution* **4,** 208–216.

MURCHAN, S., KAUFMANN, M. E., DEPLANO, A. et al. (2003) Harmonization of pulsed-field gel electrophoresis protocols for epidemiological typing of strains of methicillin-resistant *Staphylococcus aureus*: A single approach developed by consensus in 10 European laboratories and its application for tracing the spread of related strains. *Journal of Clinical Microbiology* **41,** 1574–1585.

MURRAY, P. R., ROSENTHAL, K. S., and PFALLER, M. A. (2009) *Staphylococcus* and related gram-positive cocci. In *Medical Microbiology* (ed. Murray, P. R., Rosenthal, K. S., Pfaller, M. A.), pp. 209–224. Mosby-Elsevier, Philadephia, PA.

MUSSER, J. M. and KAPUR V. (1992) Clonal analysis of methicillin-resistant *Staphylococcus aureus* strains from intercontinental sources: Association of the *mec* gene with divergent phylogenetic lineages implies dissemination by horizontal transfer and recombination. *Journal of Clinical Microbiology* **30,** 2058–2063.

NEI, M. and LI, W. H. (1979) Mathematical model for studying genetic variation in terms of restriction endonucleases. *Proceedings of the National Academy of Sciences of the United States of America* **76,** 5269–5273.

NEMATI, M., HERMANS, K., LIPINSKA, U. et al. (2008) Antimicrobial resistance of old and recent *Staphylococcus aureus* isolates from poultry: First detection of livestock-associated methicillin-resistant strain ST398. *Antimicrobial Agents and Chemotherapy* **52,** 3817–3819.

NG, J. W., HOLT, D. C., LILLIEBRIDGE, R. A. et al. (2009) Phylogenetically distinct *Staphylococcus aureus* lineage prevalent among indigenous communities in northern Australia. *Journal of Clinical Microbiology* **47,** 2295–2300.

NIELSEN, R., TARPY, D. R., and REEVE, H. K. (2003) Estimating effective paternity number in social insects and the effective number of alleles in a population. *Molecular Ecology* **12,** 3157–3164.

NOVICK, R. P. (2003) Autoinduction and signal transduction in the regulation of staphylococcal virulence. *Molecular Microbiology* **48,** 1429–1449.

NÜBEL, U., ROUMAGNAC, P., FELDKAMP, M. et al. (2008) Frequent emergence and limited geographic dispersal of methicillin-resistant *Staphylococcus aureus*. *Proceedings of the National Academy of Sciences of the United States of America* **105,** 14130–14135.

PARK, S. C. and KRUG, J. (2007) Clonal interference in large populations. *Proceedings of the National Academy of Sciences of the United States of America* **104,** 18135–18140.

PEACOCK, S. J., MOORE, C. E., JUSTICE, A. et al. (2002) Virulent combinations of adhesin and toxin genes in natural populations of *Staphylococcus aureus*. *Infection and Immunity* **70,** 4987–4996.

PEAKE, S. L., PETER, J. V., CHAN, L. et al. (2006) First report of septicemia caused by an obligately anaerobic *Staphylococcus aureus* infection in a human. *Journal of Clinical Microbiology* **44,** 2311–2313.

PÉREZ-LOSADA, M., BROWNE, E. B., MADSEN, A. et al. (2006) Population genetics of microbial pathogens estimated from multilocus sequence typing (MLST) data. *Infection, Genetics and Evolution* **6,** 97–112.

PÉREZ-LOSADA, M., CRANDALL, K., ZENILMAN, J. et al. (2007) Temporal trends in gonococcal population genetics in a high prevalence urban community. *Infection, Genetics and Evolution* **7,** 271–278.

PONS, O. and PETIT, R. J. (1996) Measuring and testing genetic differentiation with ordered versus unordered alleles. *Genetics* **144,** 1237–1245.

POPOVICH, K. J., WEINSTEIN, R. A., and HOTA, B. (2008) Are community-associated methicillin-resistant *Staphylococcus aureus* (MRSA) strains replacing traditional nosocomial MRSA strains? *Clinical Infectious Diseases* **46,** 787–794.

RICHARDS, M. J., EDWARDS, J. R., CULVER, D. H. et al. (1999) Nosocomial infections in medical intensive care units in the United States. National Nosocomial Infections Surveillance System. *Critical Care Medicine* **27,** 887–892.

ROBINSON, D. A. and ENRIGHT, M. C. (2003) Evolutionary models of the emergence of methicillin-resistant *Staphylococcus aureus*. *Antimicrobial Agents and Chemotherapy* **47,** 3926–3934.

ROBINSON, D. A. and ENRIGHT, M. C. (2004) Evolution of *Staphylococcus aureus* by large chromosomal replacements. *Journal of Bacteriology* **186,** 1060–1064.

ROBINSON, D. A., KEARNS, A. M., HOLMES, A. et al. (2005a) Re-emergence of early pandemic *Staphylococcus aureus* as a community-acquired methicillin-resistant clone. *Lancet* **365,** 1256–1258.

ROBINSON, D. A., MONK, A. B., COOPER, J. E. et al. (2005b) Evolutionary genetics of the accessory gene regulator (*agr*) locus in *Staphylococcus aureus*. *Journal of Bacteriology* **187,** 8312–8321.

RUIMY, R., MAIGA, A., ARMAND-LEFEVRE, L. et al. (2008) The carriage population of *Staphylococcus aureus* from Mali is composed of a combination of pandemic clones and the divergent Panton-Valentine leukocidin-positive genotype ST152. *Journal of Bacteriology* **190,** 3962–3968.

SABAT, A., J. KRZYSZTON-RUSSJAN, W. STRZALKA, R. et al. (2003). New method for typing *Staphylococcus aureus* strains: Multiple-locus variable-number tandem repeat analysis of polymorphism and genetic relationships of clinical isolates. *Journal of Clinical Microbiology* **41,** 1801–1804.

SAKWINSKA, O., KUHN, G., BALMELLI, C. et al. (2009) Genetic diversity and ecological success of *Staphylococcus aureus* strains colonizing humans. *Applied Environmental Microbiology* **75,** 175–183.

SATO, H., TANABE, T., NAKANOWATARI, M. et al. (1990) Isolation of *Staphylococcus hyicus* subsp. *hyicus* from pigs affected with exudative epidermitis and experimental infection of piglets with isolates. *Kitasato Archives of Experimental Medicine* **63,** 119–130.

SCHLEIFER, K. H., SCHUMACHER-PERDREAU, F., GÖTZ, F. et al. (1976) Chemical and biochemical studies for the differentiation of coagulase-positive staphylococci. *Archives of Microbiology* **110,** 263–270.

SELANDER, R. K., CAUGANT, D. A., OCHMAN, H. et al. (1986) Methods of multilocus enzyme electrophoresis for bacterial population genetics and systematics. *Applied Environmental Microbiology* **51,** 873–884.

SHUKLA, S. K., STEMPER, M. E., RAMASWAMY, S. V. et al. (2004) Molecular characteristics of nosocomial and Native American community-associated methicillin-resistant *Staphylococcus aureus* clones from rural Wisconsin. *Journal of Clinical Microbiology* **42,** 3752–3757.

SIEVERT, D. M., RUDRICK, J. T., PATEL, J. B. et al. (2008) Vancomycin-resistant *Staphylococcus aureus* in the United States, 2002–2006. *Clinical Infectious Diseases* **46,** 668–674.

SIMPSON, E. H. (1949) Measurement of diversity. *Nature* **163,** 688.

SINGER, R. S., SISCHO, W. M., and CARPENTER, T. E. (2004) Exploration of biases that affect the interpretation of restriction fragment patterns produced by pulsed-field gel electrophoresis. *Journal of Clinical Microbiology* **42,** 5502–5511.

SLATKIN, M. (1994) Linkage disequilibrium in growing and stable populations. *Genetics* **137,** 331–336.

SLATKIN, M. and MADDISON, W. P. (1989) A cladistic measure of gene flow inferred from the phylogenies of alleles. *Genetics* **123,** 603–613.

SMITH, E. M., GREEN, L. E., MEDLEY, G. F. et al. (2005) Multilocus sequence typing of intercontinental bovine *Staphylococcus aureus* isolates. *Journal of Clinical Microbiology* **43,** 4737–4743.

SMITH, J. M. (1993) The role of sex in bacterial evolution. *Journal of Heredity* **84,** 326–327.

SMITH, J. M. (1994) Estimating the minimum rate of genetic transformation in bacteria. *Journal of Evolutionary Biology* **7,** 525–534.

Smith, J. M., Dowson, C. G., and Spratt, B. G. (1991) Localized sex in bacteria. *Nature* **349,** 29–31.

Smith, J. M. and Haigh J. (1974) The hitch-hiking effect of a favourable gene. *Genetics Research* **23,** 23–35.

Smith, J. M., Smith, N. H., O'Rourke, M. et al. (1993) How clonal are bacteria? *Proceedings of the National Academy of Sciences of the United States of America* **90,** 4384–4388.

Smyth, D. S., Feil, E. J., Meaney, W. J. et al. (2009a) Molecular genetic typing reveals further insights into the diversity of animal-associated *S. aureus*. *Journal of Medical Microbiology* **58,** 1343–1353.

Smyth, D. S., McDougal, L. K., Gran, F. et al. (2009b) Population structure of a hybrid clonal group of methicillin-resistant *Staphylococcus aureus*, ST239-MRSA-III. *PLoS One*, in press.

Souza, V., Nguyen, T. T., and Hudson, R. R. (1992) Hierarchical analysis of linkage disequilibrium in *Rhizobium* populations: Evidence for sex? *Proceedings of the National Academy of Sciences of the United States of America* **89,** 8389–8393.

Stam-Bolink, E. M., Mithoe, D., Baas, W. H. et al. (2007) Spread of a methicillin-resistant *Staphylococcus aureus* ST80 strain in the community of the northern Netherlands. *European Journal of Clinical Microbiology and Infectious Diseases* **26,** 723–727.

Sung, J. M., Lloyd, D. H., and Lindsay, J. A. (2008) *Staphylococcus aureus* host specificity: Comparative genomics of human versus animal isolates by multi-strain microarray. *Microbiology* **154,** 1949–1959.

Takahashi, T., Satoh, I., and Kikuchi, N. (1999) Phylogenetic relationships of 38 taxa of the genus *Staphylococcus* based on 16S rRNA gene sequence analysis. *International Journal of Systematic Bacteriology* **49,** 725–728.

Tenover, F. C., Arbeit, R. D., Goering, R. V. et al. (1995) Interpreting chromosomal DNA restriction patterns produced by pulsed-field gel electrophoresis: Criteria for bacterial strain typing. *Journal of Clinical Microbiology* **33,** 2233–2239.

Thomas, J. C., Vargas, M. R., Miragaia, M. et al. (2007) Improved multilocus sequence typing scheme for *Staphylococcus epidermidis*. *Journal of Clinical Microbiology* **45,** 616–619.

Tibayrenc, M. (1995) Population genetics and strain typing of microorganisms: How to detect departures from panmixia without individualizing alleles and loci. *Les Comptes Rendus de l'Académie des Sciences* **318,** 135–139.

Tristan, A., Bes, M., and Meugnier, H. (2006) Global distribution of Panton-Valentine leukocidin-positive methicillin-resistant *Staphylococcus aureus*, 2006. *Emerging Infectious Diseases* **13,** 594–600.

Turner, K. M., Hanage, W. P., Fraser, C. et al. (2007) Assessing the reliability of eBURST using simulated populations with known ancestry. *BMC Microbiology* **12,** 7–30.

Udo, E. E., Pearman, J. W., and Grubb, W. B. (1993) Genetic analysis of community isolates of methicillin-resistant *Staphylococcus aureus* in Western Australia. *Journal of Hospital Infection* **1993,** 97–108.

Upholt, W. B. (1977) Estimation of DNA sequence divergence from comparison of restriction endonuclease digests. *Nucleic Acids Research* **4,** 1257–1265.

van Belkum, A., Emonts, M., Wertheim, H. et al. (2007) The role of human innate immune factors in nasal colonization by *Staphylococcus aureus*. *Microbes and Infection* **9,** 1471–1477.

van Belkum, A., Melles, D. C., Nouwen, J. et al. (2009a) Co-evolutionary aspects of human colonisation and infection by *Staphylococcus aureus*. *Infection, Genetics and Evolution* **9,** 32–47.

van Belkum, A., Verkaik, N. J., de Vogel C. P. et al. (2009b) Reclassification of *Staphylococcus aureus* nasal carriage types. *Journal of Infectious Diseases* **199,** 1820–1826.

Vancraeynest, D., Haesebrouck, F., Deplano, A. et al. (2006). International dissemination of a high virulence rabbit *Staphylococcus aureus* clone. *Journal of Veterinary Medicine. B, Infectious Diseases and Veterinary Public Health* **53,** 418–422.

van Leeuwen, W. B., Melles, D. C., Alaidan, A. et al. (2005) Host- and tissue-specific pathogenic traits of *Staphylococcus aureus*. *Journal of Bacteriology* **187,** 4584–4591.

Vautor, E., Cockfield, J., Le Marechal, C. et al. (2009) Difference in virulence between *Staphylococcus aureus* isolates causing gangrenous mastitis versus subclinical mastitis in a dairy sheep flock. *Veterinary Research* **40,** 56.

Veiga, H. and Pinho, M. G. (2009) Inactivation of the SauI Type I restriction-modification system is not sufficient to generate *Staphylococcus aureus* strains capable of efficiently accepting foreign DNA. *Applied and Environmental Microbiology* **75,** 3034–3038.

Voyich, J. M., Otto, M., Mathema, B. et al. (2006) Is Panton-Valentine leukocidin the major virulence determinant in community-associated methicillin-resistant *Staphylococcus aureus* disease? *Journal of Infectious Diseases* **194,** 1761–1770.

Waldron, D. E. and Lindsay, J. A. (2006) Sau1: A novel lineage-specific type I restriction-modification system that blocks horizontal gene transfer into *Staphylococcus aureus* and between *S. aureus* isolates of different lineages. *Journal of Bacteriology* **188,** 5578–5585.

Wang, R., Braughton, K. R., Kretschmer, D. et al. (2007) Identification of novel cytolytic peptides as key virulence determinants for community-associated MRSA. *Nature Medicine* **13,** 1510–1514.

Wang, X. M., Noble, L., Kreiswirth, B. N. et al. (2003). Evaluation of a multilocus sequence typing system for *Staphylococcus epidermidis*. *Journal of Medical Microbiology* **52,** 989–998.

Watanabe, S., Ito, T., Sasaki, T. et al. (2009) Genetic diversity of staphylocoagulase genes (*coa*): Insight into the evolution of variable chromosomal virulence factors in *Staphylococcus aureus*, *PloS One* **4,** e5714.

WERTHEIM, H. F., VAN LEEUWEN, W. B., SNIJDERS, S. et al. (2005) Associations between *Staphylococcus aureus* genotype, infection, and in-hospital mortality: A nested case-control study. *Journal of Infectious Diseases* **192,** 1196–200.

WHITTAM, T. S., OCHMAN, H., and SELANDER, R. K. (1983) Multilocus genetic structure in natural populations of *Escherichia coli*. *Proceedings of the National Academy of Sciences of the United States of America* **80,** 1751–1755.

WILSON, D. J., GABRIEL, E., LEATHERBARROW, A. J. et al. (2009) Rapid evolution and the importance of recombination to the gastroenteric pathogen *Campylobacter jejuni*. *Molecular Biology and Evolution* **26,** 385–397.

WISPLINGHOFF, H., ROSATO, A. E., ENRIGHT, M. C. et al. (2003) Related clones containing SCC*mec* type IV predominate among clinically significant *Staphylococcus epidermidis* isolates. *Antimicrobial Agents and Chemotherapy* **47,** 3574–3579.

WITTE, W., STROMMENGER, B., STANEK, C. et al. (2007) Methicillin-resistant *Staphylococcus aureus* ST398 in humans and animals, Central Europe. *Emerging Infectious Diseases* **13,** 255–258.

WONG, A., REDDY, S., SMYTH, D. S. et al. (2009) Polyphyletic emergence of linezolid-resistant staphylococci in the United States. *Antimicrobial Agents and Chemotherapy*, in press.

WRIGHT, J. S. III, TRABER, K. E., CORRIGAN, R. et al. (2005) The *agr* radiation: An early event in the evolution of staphylococci. *Journal of Bacteriology* **187,** 5585–5594.

WULF, M. W., TIEMERSMA, E., KLUYTMANS, J. et al. (2008) MRSA carriage in healthcare personnel in contact with farm animals. *Journal of Hospital Infection* **70,** 186–190.

YAMASHITA, K., SHIMIZU, A., KAWANO, J. et al. (2005) Isolation and characterization of staphylococci from external auditory meatus of dogs with or without otitis externa with special reference to *Staphylococcus schleiferi* subsp. *coagulans* isolates. *Journal of Veterinary Medical Science* **67,** 263–268.

ZADOKS, R. N., VAN LEEUWEN, W. B., KREFT, D. et al. (2002) Comparison of *Staphylococcus aureus* isolates from bovine and human skin, milking equipment, and bovine milk by phage typing, pulsed-field gel electrophoresis, and binary typing. *Journal of Clinical Microbiology* **40,** 3894–3902.

ZÁRATE, S., POND, S. L., SHAPSHAK, P. et al. (2007) Comparative study of methods for detecting sequence compartmentalization in human immunodeficiency virus type 1. *Journal of Virology* **81,** 6643–6651.

ZIMMERMAN, R. J. and KLOOS W. E. (1976) Comparative zone electrophoresis of esterases of *Staphylococcus* species isolated from mammalian skin. *Canadian Journal of Microbiology* **22,** 771–779.

APPENDIX 1—DIVERSITY AND DIFFERENTIATION

Simpson's measure of diversity (Simpson, 1949) has a long tradition of use in bacterial population genetics for comparing the diversity of different loci and different strain typing techniques (Selander et al., 1986; Hunter and Gaston, 1988). A population parameter called gene diversity (or expected heterozygosity) gives the probability that two individuals selected at random from the population will be of different types:

$$D = 1 - \sum p_i^2, \tag{16.1}$$

where p_i is the proportion of the ith type in the population. Several sample estimators of gene diversity are available, which should be used with finite samples. Equation 16.1 in Chapter 6 of this book described one such estimator. Another commonly used estimator is

$$\hat{D} = 1 - \frac{\sum n_i(n_i - 1)}{N(N-1)}, \tag{16.2}$$

where n_i is the number of strains of the ith type in the sample, and N is the total number of strains in the sample. Simpson (1949) provided equations for small and large sample variances that have been used as described by Grundmann et al. (2001) to construct confidence intervals around the estimator for hypothesis testing.

Kimura and Crow (1964) introduced a measure of diversity to population genetics called the effective number of alleles. This measure, applied broadly, makes use of a population parameter that gives the number of equally frequent types that will produce the observed diversity,

$$D = \frac{1}{\sum p_i^2}, \tag{16.3}$$

where p_i is the proportion of the ith type in the population. A sampling bias-corrected estimator of the effective number of types was provided by Nielsen et al. (2003):

$$\hat{D} = \frac{(N-1)^2}{\sum p_i^2 (N+1)(N-2) + 3 - N}, \tag{16.4}$$

where p_i is the proportion of the ith type in the sample, and N is the total number of strains in the sample. In addition, Nielsen et al. (2003) provided an equation for the sample variance, which can be used to construct confidence intervals for hypothesis testing.

Jost (2008) has argued that the effective number of types provides an intuitively appealing measure of diversity and may be more valid than the gene diversity measure for partitioning diversity into within- and between-population components. For example, consider two hypothetical populations, one of which has four STs of 10 strains each, and one of which has eight STs of 10 strains each. Many would think of the second population as being twice as diverse as the first population, which would be reflected by Equation 16.3 (four vs. eight) but not by Equation 16.1 (0.75 vs. 0.875). As diversity increases, the differences in the two measures sharpen. Consider some hypothetical infection control procedure that reduces MRSA gene diversity from 0.99 to 0.95. By this measure of diversity, there would be a drop of only 4% in the two populations' probabilities that a random pair of individuals would be of different types. However, these same two populations can explain 100 and 20 equally frequent types, respectively, which is a drop in diversity of 80%!

In many cases, one may be interested in asking whether subpopulations are genetically distinct from each other (e.g., colonization strains vs. disease strains), not whether the diversity of the subpopulations is higher or lower than each other. In such cases, differentiation statistics can be used. A general equation for the level of between-subpopulation differentiation is

$$G_{ST} = \frac{D_T - D_S}{D_T} = 1 - \frac{D_S}{D_T}, \tag{16.5}$$

where D_T is the diversity of the total population and D_S is the mean within-subpopulation diversity. The concept of partitioning diversity into various hierarchical components was introduced to population genetics by Wright (Holsinger and Weir, 2009). It has been greatly expanded to incorporate the peculiarities of different typing techniques (e.g., G_{ST} for multilocus data, H_{ST} for haplotypes) and the degree of difference in types (e.g., N_{ST} for multilocus data, K_{ST} for haplotypes) (e.g., Pons and Petit, 1996). G_{ST}-like statistics are dependent on the observed levels of diversity, such that highly diverse subpopulations will produce a low value of G_{ST} regardless of whether the subpopulations are genetically distinct; note that $G_{ST} < 1 - D_S$. Thus, in order to compare differentiation using G_{ST} across studies and typing techniques, it should be standardized based on the diversity (Hedrick, 2005).

Chapter 17

Population Genetics of *Streptococcus*

Debra E. Bessen

17.1 HABITATS, TRANSMISSION, AND DISEASE

The *Streptococcus* genus contains >50 species that inhabit a broad range of hosts, including humans and domesticated animals, which they colonize as part of the normal flora, and/or cause infection. The streptococci are gram-positive bacteria that are coccoid in shape and typically grow as diplococci or in chains. The classification of streptococci has its roots in clinical microbiology; hemolytic pattern and serological differences in the Lancefield group carbohydrate present on the cell wall surface are important distinguishing features.

17.1.1 Group Carbohydrate and Taxonomy

The Lancefield method for serological grouping, based on cell wall carbohydrate and dating back to the first half of the twentieth century, provided an important foundation for classifying streptococci and related gram-positive cocci (e.g., *Enterococcus*, *Lactococcus*). Most of the recognized streptococcal species of today are either represented by a limited number of serogroups (usually one or two) or lack a serogroup carbohydrate (e.g., *Streptococcus pneumoniae*). Various biochemical tests and other phenotype characteristics (e.g., hemolytic pattern) were integrated into classification schemes that, historically, have been tailored to address the need for effective management of streptococcal diseases. The earlier taxonomies and naming of species, which eventually incorporated DNA–DNA hybridization data, have been extensively revised over the years, and the fact that streptococci undergo extensive lateral gene transfer (LGT) probably explains why. More recently, phylogenies based on 16S rRNA gene sequences have been very useful in clarifying genetic relationships among *Streptococcus* species (Facklam, 2002).

The β-hemolytic streptococci, which largely correspond to the "pyogenic division" and leave a clear hemolytic zone following growth on blood agar, include the human pathogens *Streptococcus pyogenes* and *Streptococcus agalactiae*, containing serogroups A and B carbohydrate, respectively. *Streptococcus dysgalactiae* ssp. *equisimilis* includes organisms having groups C, G, or L carbohydrate; they are often recovered as commensals, causing human disease only on occasion. Important animal pathogens in the pyogenic

Bacterial Population Genetics in Infectious Disease, Edited by D. Ashley Robinson, Daniel Falush, and Edward J. Feil

division include *Streptococcus equi* ssp. *equi* (group C, infecting equine), *Streptococcus equi* ssp. *zooepidemicus* (group C, infecting equine), *Streptococcus canis* (group G, infecting canine), as well as numerous strains of *S. agalactiae* (group B, infecting bovine). According to the 16S rRNA phylogeny, the zoonotic pathogen *Streptococcus uberis* is most closely related to the *Streptococcus equi–Streptococcus zooepidemicus* biovars.

The viridans division of streptococci is α-hemolytic (i.e., "green" hemolysis) and is often subdivided into the Mitis, Salivarius, and Mutans groups. The non-β-hemolytic streptococci include the significant human pathogen *S. pneumoniae* (Mitis group), as well as the animal pathogens *Streptococcus suis* and *Streptococcus bovis*, which on rare occasion will cause human disease. Most of this chapter on the population genetics of streptococci focuses on the human pathogens, as this is where most major research efforts have been directed.

17.1.2 The Streptococcal Pathogens

The population genetics of a bacterial pathogen can be profoundly shaped by the ecological niches it occupies, as well as by its mode(s) of transmission. The natural habitats of streptococcal pathogens, and the diseases they cause, are briefly reviewed.

S. pneumoniae (Pneumococcus)

S. pneumoniae, also known as pneumococcus, is a human-specific pathogen. It is a major cause of mortality throughout the world, primarily afflicting young children and the elderly. An estimated one million children under the age of 5 years die annually from pneumococcal disease, largely in the form of pneumonia or meningitis (http://www.who.int/). In Europe and in the United States, pneumococcal pneumonia is the most common form of community-acquired bacterial pneumonia, estimated to affect approximately 0.1% of adults each year. Pneumococci are also one of the most frequent causes of middle ear infection (otitis media), a common condition that is typically mild but, nevertheless, is responsible for many children seeking health care and extensive antibiotic usage. Sinusitis can also be attributed to pneumococcal infection in many instances. The chronic infections caused by pneumococci—otitis media and sinusitis—likely involve biofilm formation. Despite its association with both severe and mild infections, the pneumococcus is recovered most often as an inhabitant of the normal flora of the upper respiratory tract. Transmission is primarily via respiratory droplets.

S. pyogenes (Group A Streptococcus [GAS])

Humans are the sole biological hosts of *S. pyogenes*, also known as the large colony-forming β-hemolytic GAS. These organisms are responsible for a minimally estimated 616 million cases of throat infection (pharyngitis, tonsillitis) worldwide per year, and 111 million cases of skin infection (primarily nonbullous impetigo) in children of less developed countries (Carapetis et al., 2005). Streptococcal pharyngitis and impetigo are superficial, self-limiting infections that usually cause only a mild illness. These mild infections typically resolve within 2 weeks, coincident with the rise of specific host immune defenses. Most morbidity and mortality due to GAS arise from invasive and autoimmune disease, although each is somewhat rare in occurrence when compared to the highly prevalent, yet generally mild infections at the throat and skin.

Like the pneumococcus, the carrier state at the throat, whereby the organism persists in a quiescent state and does not cause clinical symptoms, is an important facet of GAS ecology. It is not unusual for throat carriage rates to exceed 20% in populations of school-age children. However, respiratory tract infection in association with throat colonization appears to occur at a much higher rate for GAS as compared with pneumococci. Less is understood about GAS acting as a commensal organism at the skin. Human-to-human transmission is usually by respiratory droplets or by direct skin contact.

*S. agalactiae (*Group B Streptococcus [GBS]*)*

S. agalactiae, also known as GBS, is a common cause of neonatal sepsis in humans and of bovine mastitis, and can infect other animals as well. In recent years in the United States, the widespread screening of pregnant women for vaginal colonization by GBS, followed by antibiotic prophylaxis, has led to a marked decrease in the incidence of early-onset neonatal disease occurring during the first week of life. Late-onset GBS disease in infants, which primarily results in meningitis, remains a continuing problem. Also, the incidence of invasive GBS disease in the elderly is steadily increasing (Phares et al., 2008).

17.1.3 Additional Human Pathogens of *Streptococcus*

S. dysgalactiae ssp. *equisimilis* bearing the C and G group carbohydrate cell surface antigens are common residents of the normal flora of the human respiratory tract, but have also been recovered in association with pharyngitis and with invasive disease. *Streptococcus mutans* is a major pathogen of dental caries in humans. *S. mutans*, along with several other viridans streptococcal species that are part of the normal flora of the oral cavity, can give rise to endocarditis following their transient entry into the bloodstream of at-risk individuals. *S. mutans* is characterized by vertical transmission, from mother to infant. Interestingly, coevolution of the bacterium and host is evidenced by the correspondence between *S. mutans* genotypes and patterns of human migration (Caufield, 2009). Other viridans streptococci found in association with native valve endocarditis include *Streptococcus sanguinis* and *Streptococcus gordonii*.

17.1.4 Additional Animal Pathogens of *Streptococcus*

The zoonotic pathogen *S. equi* ssp. *zooepidemicus* (*S. zooepidemicus*) is a common colonizer of the equine nasopharynx but occasionally causes invasive disease and can infect other mammalian hosts. A biovar of *S. zooepidemicus*, known as *S. equi* ssp. *equi* (*S. equi*), is the causative agent of strangles, an often deadly infection of horses. *S. uberis* is an important cause of bovine mastitis in dairy herds. *S. suis* is a pathogen of pigs, causing systemic infections, and has also been attributed to several recent outbreaks of meningitis in humans, particularly in China.

17.2 CLASSICAL STRAIN TYPING

Historically, serology-based typing approaches that target a variety of antigenic surface structures have proven extremely valuable for distinguishing among streptococcal isolates of the same species. Serological typing has been an essential tool for investigating the epidemiology of many streptococcal diseases and also provides an important reference point for gaining a complete understanding of the population genetic structure of a species. Both carbohydrate and protein moieties provide the basis for serological typing schemes among streptococci and include the capsular polysaccharides of pneumococci and GBS, and several surface proteins of GAS. For protein antigens, serotyping is gradually being replaced by nucleotide sequence-based approaches.

17.2.1 Capsular Polysaccharide of Pneumococci and GBS

Pneumococci possess capsular polysaccharides that provide the basis for the primary serological typing scheme. The capsule is an important virulence factor during infection of deeper tissues, and antibodies directed to the capsule can confer protective immunity. However, 91 capsular serotypes are recognized for pneumococci, thereby complicating efforts to develop a capsular vaccine having complete coverage against all virulent strains. Multiple genes are required for capsular biosynthesis, collectively known as the *cps* genes, but only a small subset of the *cps* genes is responsible for serotype specificity. The correspondence between *cps* loci and immunological serotype is generally strong, although there are some exceptions.

The capsular genes of pneumococci lie between the *dexB* and *aliA* loci and occupy a segment ranging from about 10.3 to 30.3 kb, depending on the serotype. Taken together, the unique *cps* genes for the 91 pneumococcal serotypes occupy ~1.8 Mb of genetic material, which is close to the size of a single pneumococcus genome (Bentley et al., 2006; Mavroidi et al., 2007). Thus, a large amount of genomic "real estate" is devoted to this virulence factor. Conceivably, new serotypes might be generated by shuffling genes encoding the specific glycosyltransferases, but this does not appear to be the case. Plausible explanations for the seeming nonoccurrence of this genetic event include the possibility that there is low sequence homology in the flanking regions used for crossover, or that functional constraints are imposed by the assembly of new combinations of oligosaccharide repeat units, which would probably require the exchange of genes encoding multiple biosynthetic enzymes.

At least nine capsular polysaccharide serotypes are recognized for GBS. The serotyping approach for GBS differs from that used for pneumococcus, and consequently, the genotypic and phenotype diversity may be underestimated (Slotved et al., 2007). The biosynthetic locus of serotype III consists of 16 genes occupying 15.5 Kb; a single capsular polysaccharide polymerase gene distinguishes serotypes Ia and III (Chaffin et al., 2000). As in the pneumococcus, the GBS capsule inhibits opsonophagocytosis and acts as a key virulence factor.

Like GBS and the pneumococcus, GAS also synthesizes a polysaccharide capsule that functions as an important virulence factor. The quantity of capsule produced by GAS isolates can vary markedly and is often elevated in association with increased levels of transmission via a respiratory route (Stollerman and Dale, 2008). However, the capsular material of GAS is uniformly composed of hyaluronic acid, a poor immunogen, and it is not useful for serotype determination.

17.2.2 Surface Fibrils: M Protein of GAS

During the 1920s, Dr. Rebecca Lancefield began work aimed at understanding the basis for protective immunity to GAS infection. Antiserum raised to the extractable surface antigens, known as M proteins, led to opsonophagocytosis and killing of the strain from which the M protein was derived (Lancefield, 1962). However, antiserum directed to the M protein of one organism often failed to protect against many other isolates. A serological typing scheme arose through the development of antiserum directed to M proteins of different isolates. More than 80 distinct M types were identified, and strong protective immunity was found to be M type specific.

M proteins form hairlike fibrils extending about 60 nm from the surface of the bacterial cell (Fischetti, 1989). The determinants of serological type lie at the amino termini, which correspond to the distal fibril tips. The serologically based M protein typing scheme was replaced about 12 years ago with a highly correlated nucleotide sequence-based *emm* typing scheme (Beall et al., 1996). *emm* type is defined by the 5′ end region of the *emm* gene, and genes assigned the same *emm* type have >92% nucleotide sequence identity. More than 200 *emm* types are currently listed in the online Centers for Disease Control and Prevention (CDC) database (http://www.cdc.gov/ncidod/biotech/strep/strepindex.htm). The discovery of new *emm* types has continued over recent years, suggesting that undiscovered *emm* types remain at large. Each genome of the many GAS strains examined carries genetic material for only a single *emm* type, as opposed to having a broad genetic repertoire for assembling the complete set of *emm* types.

Serum opacity factor (SOF) is a protein produced by ~50% of GAS isolates, both in a surface-bound and secreted form. This multifunctional protein binds apolipoprotein present in the serum, leading to its opacification, and also functions as an adhesin via binding to fibronectin (Gillen et al., 2008). A serological typing scheme was developed in which an SOF type-specific antiserum inhibits serum opacification activity. The *sof* gene lies ~16 kB upstream of the *emm* region.

17.2.3 The Newly Recognized Pili are Present in Several Streptococcal Species

The presence of pili, or fimbriae, on the surface of several streptococcal species is an important recent discovery (Telford et al., 2006). A third serological typing scheme for GAS, in addition to M and SOF typing, is based on the T antigen. The T antigen was originally defined by its resistance following treatment of the bacteria with trypsin. The *tee* gene of a T6-type strain was initially mapped to the highly recombinatorial fibronectin binding protein–collagen binding protein-T antigen (FCT) region of the GAS genome (Schneewind et al., 1990; Bessen and Kalia, 2002). More recently, T antigens have been characterized as forming elongated pili that function as adhesins (Kreikemeyer et al., 2005; Mora et al., 2005; Abbot et al., 2007; Manetti et al., 2007).

The FCT region of GAS, encoding pilus biosynthetic and structural genes, is positioned ~300 kb away from the *emm* region on the chromosome. There appears to be a far greater number of distinct M serotypes than T serotypes (Johnson et al., 2006). A new approach for nucleotide sequence-based typing of the genes encoding pilus structural proteins of GAS was recently introduced (Falugi et al., 2008).

Both GBS and pneumococci possess surface pili; however, these structures have not been a part of classical serotyping schemes. Originally described as the pathogenicity *rlrA* islet, the pilus-encoding genes are found in only a subset of pneumococcal strains (Hava et al., 2003; Aguiar et al., 2008). In GBS, three genetic islands of pilus gene variants have been described, whereby all strains harbor one or more islands (Margarit et al., 2009). The T serotype has been determined for numerous human isolates of group C and G streptococci and therefore, these organisms very likely express surface pili as well.

Table 17.1 MLST Findings for Pathogenic Species of Streptococci

Streptococcus species	No. of STs	No. of CCs	No. of singletons	% of STs that are singletons	% of STs in largest CC	Housekeeping gene tree topologies
S. pyogenes	458	70	202	44.1	2.8	Incongruent
S. zooepidemicus	192	34	76	39.6	4.7	Incongruent
S. pneumoniae	3758	268	822	21.9	5.3	Incongruent
S. mutans	92	13	54	58.7	8.7	n.d.
S. suis	127	11	63	49.6	15.8	n.d.
S. uberis	392	28	181	46.2	33.2	Incongruent
S. agalactiae	449	11	29	6.2	48.3	n.d.

MLST data are from online databases, as of January 2009. All schemes are based on seven housekeeping loci except *S. mutans* (eight loci).

n.d. = not detected.

17.3 MULTILOCUS SEQUENCE TYPING (MLST) BASED ON HOUSEKEEPING GENES

17.3.1 MLST Schemes for Streptococci

MLST that is based on housekeeping genes is an important tool used for understanding the population genetics of a bacterial species. MLST schemes have been developed for numerous *Streptococcus* species including pneumococci (Enright and Spratt, 1998), GAS (Enright et al., 2001), *S. suis* (King et al., 2002), GBS (Jones et al., 2003), *S. uberis* (Zadoks et al., 2005; Coffey et al., 2006), *S. mutans* (Nakano et al., 2007), and the *S. zooepidemicus* group (Webb et al., 2008). Most streptococcal MLST schemes utilize the partial sequences of seven housekeeping genes; extensive data sets are posted on the Internet at http://www.mlst.net/ and at http://www.pubmlst.org/.

Using these valuable resources of online data, the number of sequence types (STs) identified for the various streptococcal species is summarized (Table 17.1). In general, strain sampling has been more extensive for the human pathogens as compared with the zoonotic pathogens. The pneumococcus has been the most extensively characterized by MLST, whereby the number of STs presently defined for pneumococci is approaching 4000, far exceeding that uncovered for the other streptococcal species examined.

17.3.2 Contributions of Recombination and Mutation to Genetic Diversification

Recombination and mutation can be envisioned as two opposing forces in shaping the population genetic structure of a bacterial species. MLST data can provide valuable insights on the predominant mechanisms underlying genetic diversification in a bacterial population. It is generally assumed that recombination detected among housekeeping genes most often occurs via a homologous mechanism following horizontal gene transfer from a donor strain harboring a distinct allele. For streptococci, the donor organism can belong to the same species, or in some instances, to another species that is closely related. Numerous analytic methods have been applied using the MLST housekeeping gene data of various streptococcal species.

eBURST and Clonal Complexes (CCs)

The ancestor–descendant relationships between the STs, as defined by MLST data, can be inferred by eBURST analysis (Feil et al., 2004). Population snapshots were generated for three of the *Streptococcus* species using the publicly available MLST data (http://www.mlst.net/ and http://www.pubmlst.org/). Each population snapshot depicts STs as dots and CCs in which the connected STs are single-locus variants (SLVs) sharing six of the seven housekeeping alleles (Fig. 17.1).

The three species displayed—GBS, *S. suis*, and GAS—differ widely in the percent of STs present in the largest CC, measured as 48.3%, 15.8%, and 2.8%, respectively (Table 17.1). Collectively, these population structures reflect the broad range observed for simulated bacterial populations differing in the levels of recombination (ρ) and diversity generated by mutation (θ) (Turner et al., 2007). eBURST performs well in predicting ancestor–descendant relationships when 5–25% of the STs are in the largest eBURST group, yet only three of the seven streptococcal species evaluated to date fall within this intermediate range.

The eBURST population snapshot for GBS (Fig. 17.1) is suggestive of high rates of recombination with housekeeping alleles having low diversity, leading to straggly CCs with unreliable ancestor–descendant relationships. Likewise, little useful information on patterns of descent is found for GAS, but for entirely different reasons. Instead, GAS shows few linked STs, a pattern consistent with high rates of mutation generating large numbers of alleles, coupled with high rates of recombination that randomly shuffle them.

Empirical Analysis of Genetic Change in SLVs

Estimates of the relative rates of homologous recombination versus mutation can be made by evaluating the nature of the genetic changes that distinguish the descendant ST from the ancestral ST of an SLV pair. Based on an earlier and less extensive MLST data set for pneumococci, the number of alleles changed by recombination compared to mutation was estimated to be ~9 to 10 to 1 (Feil et al., 2000; Feil and Spratt, 2001). The ratio of per nucleotide site change by recombination to mutation, which incorporates the number of nucleotide changes per single recombinational event, is higher at ~50 to 1, with the average size of the recombinational replacements estimated at ~5 to 10 kb (Feil et al., 2000).

An empirical estimate of the relative ratio of recombination to mutation events has been more difficult to ascertain for GAS due to the dearth of SLVs uncovered by eBURST; this may partly be a consequence of a sampling strategy that sought to capture the full range of genetic diversity for the species rather than a focus on community outbreaks. Nonetheless, rough estimates point to a recombination-to-mutation ratio of >1 for GAS (McGregor et al., 2004b).

ClonalFrame

ClonalFrame applies MLST data to a coalescent-based approach in order to estimate the relative probabilities that a nucleotide is changed as the result of homologous recombination versus point mutation (Didelot and Falush, 2007). Estimates for the ratio of rates at which a nucleotide becomes substituted as a result of recombination or mutation for pneumococci and GAS are 23.1 (95% confidence interval [CI], 16.9–29.0) and 17.2 (95% CI, 6.8–24.4), respectively (Vos and Didelot, 2009). These data suggest very high rates of recombination for both major streptococcal pathogens. The ClonalFrame findings for

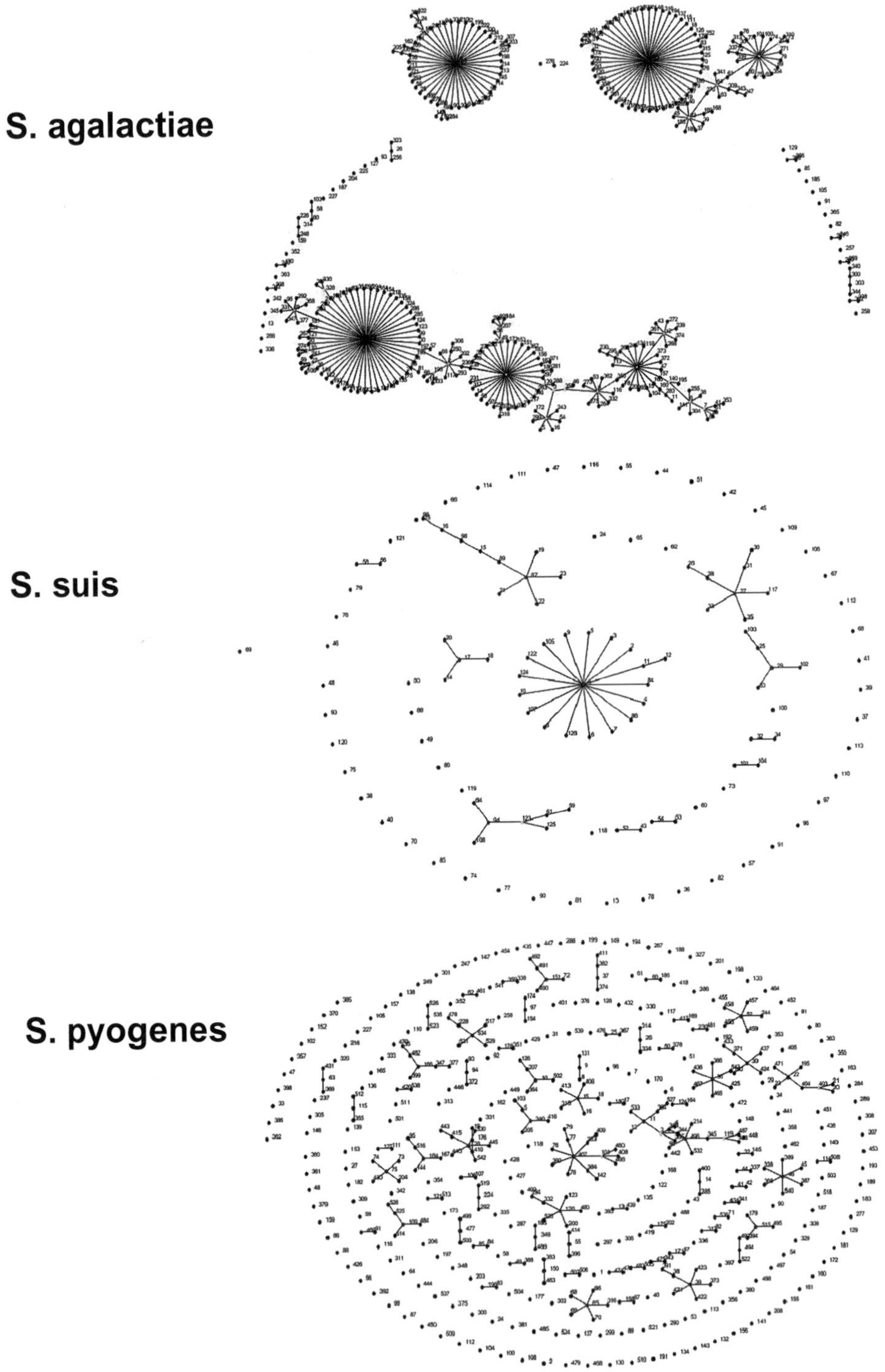

Figure 17.1 Population snapshots generated by eBURST. The population snapshots for the three streptococcal species indicated are based on the MLST data as presented in Table 17.1 and include only one strain representative of each ST. To obtain all STs in a single group, eBURST (http://www.mlst.net/) was implemented using a group definition of one identical locus (out of seven).

pneumococci and GAS are in sharp contrast to data gathered for the other low G + C Firmicutes organisms, whereby similarly calculated recombination-to-mutation ratios are ≤1 for various species assigned to the *Enterococcus*, *Staphylococcus*, or *Bacillus* genera.

Infinite Allele Model

Estimates for population recombination and mutation rates can also be made by comparisons to the neutral infinite allele model (Fraser et al., 2005). Based on MLST data for an expanded sample set of GAS isolates, high relative rates of recombination are predicted once again (Hanage et al., 2006). The value for the estimated rate of mutation (θ) is 7.1 for GAS, which is similar to that calculated for pneumococci ($\theta = 7.4$). The estimated rate of recombination (ρ) is even higher for GAS (51.2) than for pneumococci ($\rho = 29.6$). The ratios of the recombination and mutation rates show that the GAS population approaches linkage equilibrium, characteristic of extensive recombination, whereas pneumococci are in a transitional zone between a clonal and sexual population genetic structure.

Congruency of Gene Tree Topologies

Another way to evaluate the relative extent of genetic diversification by recombination versus mutation is to assess congruence via pairwise comparisons of housekeeping gene tree topologies. Representative STs corresponding to deep branches of a dendrogram constructed by clustering were identified and used to generate trees by the maximum likelihood method for each of the seven housekeeping genes of pneumococci and GAS; the tree topologies were assessed for congruency, a signature of diversification by mutation (Feil et al., 2001). Of the 42 possible pairwise gene tree comparisons, there was no evidence for statistically significant levels of congruence between gene trees topologies (Table 17.1). The lack of congruency is indicative of relatively high levels of recombination within both streptococcal species.

In the *S. zooepidemicus* group, extensive recombination is evident by the complete lack of congruency between housekeeping gene tree topologies (Webb et al., 2008) (Table 17.1). A similar lack of congruence between housekeeping gene tree topologies was found for *S. uberis* (Coffey et al., 2006). A very high ratio of recombinational to mutational events was also calculated for SLVs detected by eBURST (Tomita et al., 2008).

The GAS and pneumococci MLST data sets are noteworthy for their low levels of average pairwise sequence divergence (π), although this could be ruled out as a factor contributing to the lack of congruency among gene tree topologies via the use of model congruent data sets containing little phylogenetic signal (Feil et al., 2001). A phylogenetic tree based on the concatenated sequences of the seven housekeeping genes of GAS, for 114 strains representing 114 STs and 113 *emm* types, shows very few branches have bootstrap support >80%; the overall shape of the tree is consistent with high rates of genetic recombination within this species (Bessen, 2009).

Summary of Findings on Recombination versus Mutation

The collective MLST-based findings on the population genetic structures of pneumococci and GAS support a key role for homologous recombination in generating genetic diversity. Other *Streptococcus* species studied in depth—GBS, *S. uberis*, and *S. zooepidemicus*—also display evidence for high rates of genetic change arising from recombination, as compared to mutation.

17.3.3 Limitations of MLST Data Analysis: Mode of Sampling

The current MLST data sets for pneumococci, GAS, and GBS are the result of the combined efforts of many investigators and numerous studies. Any single study may differ from the others in its goals and therefore will have employed strategies for strain sampling that were tailored to the specific objectives.

Global sampling across wide geographic regions aims to provide a comprehensive view of the full range of diversity within the extant species. For some streptococci, isolates dating back ~80 years have been available for MLST analysis. Local sampling within a small host community over a narrow time period can often capture recent genetic changes. If recent genetic changes are rare in occurrence, collection of a very large number of bacterial isolates may be required to detect any changes. High rates of strain migration into a well-delineated host population—narrowly defined across space and time—can yield a mixture of organisms having genetic differences that may have accumulated over a very long time period, plus strain variants that arose via recent genetic change. Other sources of potential skewing or bias encountered in strain sampling can arise by overrepresentation of isolates having a clinically significant phenotype, such as an association with invasive disease or resistance to antibiotics.

The combined data in Table 17.1, although being the most comprehensive assemblage, represent numerous sampling strategies that may not be equivalent for the different streptococcal species and thus are not strictly comparable. Nonetheless, all MLST-based measures strongly point toward an important role for homologous recombination in the genetic diversification of streptococci.

17.3.4 Defining "Strain" or "Clone"

For a bacterial population that diversifies primarily by mutation, it may be (mostly) valid to equate the ST generated by MLST with strain or clone. However, at least for the pathogenic streptococcal species examined in depth, a rich history of past horizontal gene transfer events is not limited to the housekeeping alleles but also extends to the genes encoding the determinants of serotype. It has been proposed that strain is best defined in epidemiological terms by the loci that affect its transmission (Gupta et al., 1996), which for pathogenic streptococci includes the serotyping determinants that elicit protective immunity in the human host (e.g., polysaccharide capsule, surface protein). Thus, horizontal transfer of the genes giving rise to the antigenically heterogenous surface structures, to a recipient strain having a distinct genetic background as defined by MLST, can lead to the emergence of newly recognized strains or clones. As will be discussed later in this chapter, serotypic determinants of several pathogenic streptococcal species are recovered in association with distant STs and in numerous combinations.

17.4 SPECIES BOUNDARIES AND GENE FLOW

Given the high rates of recombination characterizing most *Streptococcus* species examined, combined with the numerous past revisions of the taxonomical relationships between the member species, it is unsurprising that there is evidence for a history of extensive horizontal gene transfer between species. However, with increasing amounts of sequence data available, streptococcal species boundaries are becoming better defined, even though the boundaries are not sharp for all genes.

17.4.1 Whole Genome Sequences and Pan-Genomes

MLST based on housekeeping genes has the important advantage of allowing for rapid screening of large numbers of strains. A major drawback of the MLST approach, as actually applied to streptococci, is that sequence analysis is restricted to a very small portion of the genome. Whole genome sequencing addresses this shortcoming by providing a comprehensive data set, but it also has a practical limitation in that it cannot be used to assess a large number of strains at a reasonable cost, at least not yet. Comparative genomic hybridization using microarrays designed to include many or all genes can measure gene presence or absence (or high divergence) among a large number of isolates and has proven valuable in identifying genes correlated with important phenotypes such as virulence. Resequencing arrays can detect indels and provide nucleotide sequence data, but their cost remains high and may be best suited for comparing strains having relatively few genetic differences. The application of genomics to the study of bacterial population genetics is still at an early stage of development.

Whole genome sequences have been reported, or are in progress of completion, for numerous *Streptococcus* species, including pneumococci (Tettelin et al., 2001), GAS (Ferretti et al., 2001), and GBS (Tettelin et al., 2002). More than 10 genome sequences have been determined for each of the three major human pathogen species. Additional streptococcal genome sequences are complete or in progress for *S. equi*, *S. zooepidemicus*, *S. gordonii*, *S. mutans*, *Streptococcus mitis*, *S. suis*, and *S. uberis* among several other species. The *Streptococcus* genomes range in size from ~1.7 to 2.2 Mb.

The pan-genome, or supra- or metagenome, of a bacterial species includes core genes plus the noncore accessory genes that are present in only a subset of strains and provide an important source of genetic and biological diversity. The pan-genome of GBS, as defined by six strains of five capsular serotypes, reveals that the core genes comprise ~80% of the genome of each strain (Tettelin et al., 2005). In addition, it is estimated that for every new GBS genome sequenced, an average of 33 new strain-specific genes will be identified. Similar extrapolations made for GAS at that time projected ~27 new strain-specific genes per sequenced genome. These findings support the concept of an open pan-genome and are broadly consistent with the high rates of homologous recombination resulting from lateral transfer of housekeeping genes, as estimated by population genetic analysis of MLST data.

New insights on the evolutionary history of GBS challenge the view that recombinational replacements are highly localized and involve relatively small regions of DNA. Experimentally, it can be shown that large DNA segments, up to 334 kb of the chromosome of GBS, can be transferred through conjugation from multiple initiation sites (Brochet et al., 2008b). In this study, sequence polymorphisms within eight complete GBS genome sequences were analyzed in order to determine whether there was evidence for a history of large chromosomal replacements in natural isolates. The data support a model wherein clinically important clones are derived from a single clone that evolved by exchanging large chromosomal regions with more distantly related strains.

The complete genome sequences have been published for >10 GAS strains. Exogenous genetic elements are prominent features of GAS genes, comprising about 10% of the whole genome sequence (Beres and Musser, 2007). All sequenced GAS genomes contain ≥1 prophage, and several genomes also harbor integrative and conjugative elements (ICEs), which have features of both conjugative transposons and plasmids. Together, these exogenous genetic elements account for most of the noncore genes in GAS. Several of the exogenous genetic elements present in GAS harbor virulence and/or antibiotic resistance genes. The prophage and ICEs have modular structures that are indicative of past recombinational events. Experimentally, it can be demonstrated that ICEs transfer from one streptococcal host species (e.g., *Streptococcus thermophilus*) and integrate into another recipient species (e.g., GAS) (Bellanger et al., 2009). ICEs are also major contributors to genomic diversity in GBS (Brochet et al., 2008a). Whole genome sequence findings support the view that both transduction and conjugation are important mechanisms for generating genetic diversity in GAS.

The complete genome sequence for 17 pneumococcus isolates demonstrates that slightly less than half (46%) of orthologous gene clusters are present in all 17 strains; furthermore, the supragenome of the pneumococcus is predicted to contain ~5000 genes (Hiller et al., 2007). Comparative genomic hybridization of numerous isolates of pneumococci shows that many of the noncore accessory genes are organized into several major regions of diversity (Obert et al., 2006). Pneumococci are largely commensals of the nasopharyngeal tissue or cause chronic infections in the middle ear or sinuses, where they can form biofilms that may contain multiple clones. This latter lifestyle is compatible with horizontal gene transfer, aided by the fracticidal process involving autolysis of some cells within the bacterial population and DNA uptake by other cells that are competent for transformation (Claverys et al., 2007).

In an earlier report based on only a few genome sequences, GBS was found to share 176 genes with pneumococci and 225 genes with GAS, whereas pneumococci and GAS shared only 74 genes (Tettelin et al., 2002). Conservation of gene synteny was also more pronounced between GBS and GAS (828 genes in 35 regions spanning ~1104 kb) than between GBS and pneumococci (128 genes in nine regions spanning ~131 kb). This finding is consistent with the 16S rRNA phylogeny and other clinical microbiology phenotypes placing GBS and GAS in the pyogenic division.

In an analysis of 26 genomes derived from six species of *Streptococcus*, the number of core genes in common appears to reach a plateau at about 600 (Lefebure and Stanhope, 2007). Furthermore, it is estimated that ~18% of the core genome is recombinant. Estimates on the size of the pan-genome of the *Streptococcus* genus exceed 5300 genes; however, the entire genus is far more expansive than the six species included in this study.

17.4.2 Species of the Mitis Group and Other Viridans Streptococci

The Mitis group includes pneumococci plus about a dozen other viridans species that are strict commensals and are rarely recovered in association with disease. The taxonomy of the viridans streptococci has been difficult to reconcile. Most member species of the Mitis group are competent for genetic transformation. Horizontal gene exchange followed by high rates of homologous recombination among genes with relatively low levels of sequence divergence may have contributed to the blurring of species boundaries (Fraser et al., 2007).

Descent of Commensal Species from a Pathogen

Using the concatenated sequences of four housekeeping genes, a phylogenetic analysis of Mitis group isolates reveals three major monophyletic clusters, distinguishing *S. mitis*, *Streptococcus oralis*, and *Streptococcus infantis* strains. The Mitis cluster has many deep branches, most of which correspond to individual isolates of *S. mitis*. Within the Mitis cluster, there is also a deep branch for *Streptococcus pseudopneumoniae* and another single deep branch containing all of the pneumococci isolates distributed in a tight subcluster (Kilian et al., 2008). Each deep branch of the Mitis cluster would probably represent distinct species according to current taxonomic standards.

A new model for the evolution of organisms of the Mitis cluster has been proposed (Kilian et al., 2008). The model factors in several bodies of data, including estimates of the evolutionary distance or age of the lineages as established by housekeeping gene sequences, the identical location (synteny) of remnants of virulence genes in the genomes of commensal strains (e.g., the IgA1 protease gene region), the overall pattern of genome reductions, and the differing distributions of genes unique to *S. mitis* versus *S. pneumoniae* among other streptococcal species. The favored model proposes that the entire cluster of *S. pneumoniae*, *S. pseudopneumoniae*, and *S. mitis* lineages evolved from a pneumococcus-like bacterium that was probably pathogenic to a humanlike ancestor. During adaptation to a commensal lifestyle, most of the *pneumoniae-mitis-pseudopneumoniae* cluster lineages gradually lost virulence genes and became genetically distinct due to sexual isolation in their preferred hosts. The IgA1 protease of modern-day pneumococci was probably an important facilitator of adaptation to its primary ecological niche, the human upper respiratory tract. The very small size of the early human host population probably imposed a strong bottleneck for pneumococci, as well as for other human-restricted pathogens, such as GAS.

Models for the evolution of bacterial pathogens often invoke the acquisition of virulence genes, followed by successful and rapid expansion of the pathogen population. The model for the evolution of the Mitis cluster is significant in that it more strongly supports the opposite scenario, whereby commensals evolved from a pathogen by a loss of virulence genes.

Multilocus Sequence Analysis (MLSA)

Multilocus sequence phylogenetic analysis, or MLSA, is a new internet-based tool that uses concatenates of housekeeping gene sequences to examine clustering among strains that have been assigned to closely related species by other taxonomic methods (Bishop et al., 2009) and can be found at http://www.eMLSA.net/. A universal set of primers was used to amplify seven housekeeping genes derived from multiple streptococcal species.

The streptococci examined by MLSA include many species belonging to the Mitis group, plus GAS and GBS of the pyogenic division, in addition to other viridans species such as *Streptococcus anginosus* (Anginosus group) and *Streptococcus salivarius* (Salivarius group). The seven housekeeping gene sequences used for MLSA, which differ from those employed in the MLST schemes for each of the individual streptococcal species, were concatenated and used to construct a multiple species phylogeny (Fig. 17.2). The clustering of the Mitis group species on the streptococcal MLSA reference tree (Bishop et al., 2009) is generally consistent with the previously described findings on the Mitis group (Kilian et al., 2008), although the latter approach provided greater resolution, perhaps due to different choices in housekeeping gene targets. Both of these studies show

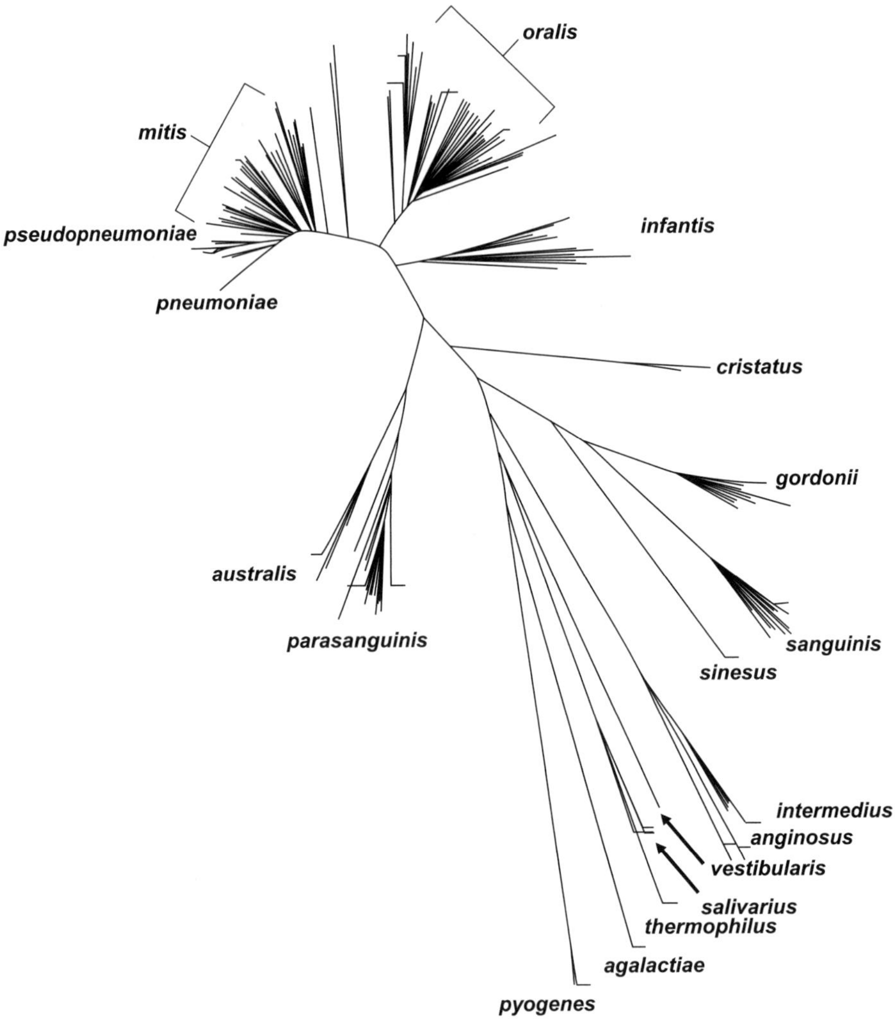

Figure 17.2 Streptococcal reference tree based on multiple housekeeping loci. The radial tree was obtained from http://www.eMLSA.net/, a portal for the electronic taxonomy of bacteria (Bishop et al., 2009). The tree is based on seven concatenated housekeeping gene sequences from 420 isolates belonging to the *Streptococcus* genus. The approximate positions of sequence clusters corresponding to numerous streptococcal species are indicated.

that recombining species that co-colonize the same niche (human upper respiratory tract and oral cavity) can be resolved using MLSA.

17.4.3 Close Genetic Relatives of GAS

Human isolates of β-hemolytic streptococci bearing group C or G cell wall carbohydrate (GCS/GGS) are often classified as *S. dysgalactiae* ssp. *equisimilis*. These organisms are common commensals of the upper respiratory tract and are occasionally found in association with disease, including invasive infections (Broyles et al., 2009). The C versus G group carbohydrate of *S. dysgalactiae* ssp. *equisimilis* isolates does not correlate with genetic clusters, based on either nucleotide sequence data for housekeeping genes or microarray-based hybridization data for virulence genes (Davies et al., 2007; Ahmad et al., 2009). The GCS/GGS of *S. dysgalactiae* ssp. *equisimilis* harbor *emm* genes that are highly homologous to those present in GAS, although the *emm* type-specific regions from GCS/GGS strains are usually distinct from the GAS-associated *emm* types.

Invasive disease isolates of *S. dysgalactiae* ssp. *equisimilis* recovered from the United States between 2002 and 2005 underwent MLST analysis (Ahmad et al., 2009). Housekeeping genes having at least some similarities to those targeted in the MLST scheme for GAS were used for GCS/GGS. The data provides evidence for genetic recombination between GCS/GGS and GAS. The MLST findings for two oral commensals of the Salivarius group may have parallels to GCS/GGS and GAS, whereby one housekeeping gene locus (*tkt*) could not be amplified in both *S. salivarius* and *S. vestibularis* using a single set of PCR primers, and alternative primers yielded highly divergent genes (Delorme et al., 2007).

Did the GCS/GGS human commensal organisms evolve from an ancestral GAS-like pathogen, as observed for the close genetic relatives of the Mitis group? There are data available that partly address this question. For example, there is strong evidence for the sharing of some virulence-related genes between GCS/GGS and GAS (Kalia and Bessen, 2004; Towers et al., 2004; Bessen et al., 2005; Davies et al., 2007). Yet, certain virulence genes appear to have originated from one species, whereas other virulence genes seem to have been transferred in the opposite direction. Furthermore, some key GAS virulence factors (e.g., the secreted cysteine proteinase SpeB) are noticeably absent from GCS/GGS. An important missing piece of the puzzle has been the complete genome sequence for one or more strains of *S. dysgalactiae* ssp. *equisimilis*, which has only recently been released. Perhaps with this added information, the evolution of species within the pyogenic division can become better elucidated.

17.4.4 Rates of Gene Gain and Loss

Recent studies have begun to address the dynamics of gene flow within and between streptococcal species. In a newly developed model-based method for using whole genome sequences to infer patterns of gene content evolution, genetic flux in GAS was found to be dominated by gain and loss of prophage genes (Didelot et al., 2009). An estimation of the rates of genetic flux indicates that, at least for some strains having multiple genome data available, prophage integration may have accelerated in recent time.

In another modeling approach using the whole genome sequences of 12 isolates originating from five *Streptococcus* species, both species-specific and intraspecies gene transfers were evaluated (Marri et al., 2006). A phylogeny based on the shared genes was used to estimate gene loss and gain, and was shown to be highest at the branch tips. The gene insertion–deletion rates were higher within a species than for branches leading to a species. Furthermore, the percentage of genes in GAS showing signatures of positive selection was much higher for those genes that were recently transferred versus ancient genes. Thus, at least some of the recently acquired genes probably have a key role in recent adaptations and may confer a survival advantage.

In an analysis of 26 genomes from six streptococcal species—pneumococci, GAS, GBS, *S. suis*, *S. thermophilus*, and *S. mutans*—the *S. suis* lineage showed the greatest amount of gene loss and gene gain (Lefebure and Stanhope, 2007). *S. suis* also showed the greatest positive selection pressure within the core genome. Positive selection of the core genome was primarily observed during species differentiation and may signify adaptation to different biological hosts.

17.5 NICHE-DRIVING GENES

For most streptococcal species, relatively little is known about environmental reservoirs, including the extent to which they even exist and their importance in the bacterial life cycle. Instead, most of our understanding of streptococci is based on their associations with biological hosts. The mammalian host species and/or tissue range restrictions and preferences are quite varied among the different *Streptococcus* species.

The emergence of new clones having unique biological properties, such as a new ecological niche, is a recurring theme among many of the streptococcal pathogens. Often this is coupled to interspecies genetic exchange. Although most species of *Streptococcus* are relatively devoid of the classical pathogenicity islands found in many other bacterial pathogens (Schmidt and Hensel, 2004), they tend to be rich in others types of genetic islands and regions of diversity. Genetic diversity introduces biological diversity, which, in turn, can shape future genetic events. This dynamic has the potential to undergo a dramatic shift when a new niche is successfully exploited by an organism, and the capacity to efficiently transmit to new hosts is retained.

17.5.1 Tissue Tropisms for Infection by GAS

GAS is a human-specific pathogen whose primary reservoirs are the epithelium of the throat and skin. It is at these two tissue sites where the organism is most successful in reproductive growth and transmission to new hosts. Decades of epidemiological studies show that some M protein serotypes of GAS have a strong tendency to cause throat infection (pharyngitis), but not superficial skin infection (impetigo), whereas other M types are often recovered from impetigo, but rarely from pharyngitis (Wannamaker, 1970; Bisno and Stevens, 2000). This observation gave rise to the concept of distinct throat and skin M types, suggesting that there exists some degree of specialization among strains of this species.

The *emm* pattern genotypes are fairly strong markers for the preferred tissue site of infection among GAS strains (Bessen et al., 2008). The *emm* pattern is based on the number and arrangement of *emm* and *emm*-like genes, and they comprise three main groups known as pattern A–C, D, and E (Hollingshead et al., 1993, 1994). Isolates of nearly all *emm* types examined to date are restricted to a single *emm* pattern grouping (McGregor et al., 2004b), and thus, the *emm* pattern can be inferred with a fair degree of accuracy based on knowledge of the *emm* type. Recent reports on population-based surveillance for GAS associated with pharyngitis or impetigo infections yield *emm* typing data on >3700 isolates collected from the six major continents (summarized in Bessen et al., 2008; Bessen, 2009). In each study cited, the designated skin isolates were from clear cases of impetigo and not from other types of skin lesions that can be difficult to distinguish in nonendemic regions, such as wound infections, which are often initiated by throat strains.

The combined population analysis for the >3700 GAS reveals a strong trend whereby isolates corresponding to the *emm* pattern A–C group have a strong predilection to cause infection at the throat, and *emm* pattern D strains have a strong tendency to cause impetigo. The relationship between the *emm* pattern and the infection site is statistically highly significant when the data for each of the surveys are combined (Bessen et al., 2008; Bessen, 2009). These data provide strong support for the division of GAS strains into three clinically and ecologically relevant groups based on *emm* pattern genotype. The *emm* pattern A–C strains are considered throat specialists, whereas *emm* pattern D strains are skin specialists. As a group, the pattern E strains are designated generalists, readily causing infections at both tissue sites.

Not only are the throat and skin specialists physically separated by their ecological niches but there is also geographic partitioning as evidenced by the prevalence of throat versus skin infection in different parts of the world, with pharyngitis predominating in temperate regions and impetigo in tropical communities (Carapetis et al., 1999). Thus, it seems likely that differential risk factors for infection are a driving force in promoting adaptation to narrow ecological niches among the specialist strains. The nonoverlapping seasonal peak incidence for pharyngitis (winter) and impetigo (summer) in some regions where both diseases occur also introduces distance.

Spatial and temporal separation might impose barriers to lateral gene exchange, raising the question of whether *emm* pattern A–C and D strains are at the beginning stages of forming new species via allopatry. However, based on MLST data, evidence for signatures of early stages of speciation within the GAS population is lacking. Numerous analyses on housekeeping alleles—including phylogenetic trees of concatenated alleles, splits graphs, fixed nucleotide differences, distribution of shared alleles—all provide strong support for extensive recombination involving housekeeping genes of the three *emm* pattern groupings (Kalia et al., 2002; Bessen, 2009). Against a background of random associations among housekeeping alleles, genes displaying linkage disequilibrium with the *emm* pattern genotype are strong candidates for conferring the tissue specificity of infection phenotype, particularly those loci mapping physically distant to the *emm* region on the genome.

Several GAS genes have been identified which display strong linkage disequilibrium with *emm* pattern grouping. Alleles at the *mga* locus comprise two discrete phylogenetic lineages of ~28% nucleotide sequence divergence; nearly all *emm* pattern A–C strains harbor an *mga-1* lineage allele, whereas pattern D and E strains have *mga-2* lineage alleles (Bessen et al., 2005). Since *mga* lies immediately upstream of the *emm* genes, strong linkage of *mga* lineages to the *emm* pattern could be the result of physical proximity, although there is also evidence for extensive genetic recombination within the *emm* region. The *mga* gene product is a regulator of gene transcription and controls the expression of *emm* genes as well as numerous genes lying outside the *emm* region (Hondorp and McIver, 2007). Other loci displaying strong linkage with the *emm* pattern and lying within the same ~55-kB region include *sof* (encoding SOF, used in strain typing), *pam* (an *emm* gene that encodes a plasminogen-binding protein), and *ska* (encoding the plasminogen activator streptokinase). Alleles of the *ska* gene comprise three distinct lineages, one of which appears to have its origins in *S. dysgalactiae* ssp. *equisimilis* (Kalia and Bessen, 2004).

Several loci within the FCT region display strong linkage disequilibrium with the *emm* pattern; the FCT and *emm* regions lie about ~0.3 Mb apart on the 1.9-Mb genome. FCT region-derived colonization factors (e.g., *cpa*, *prtF1*) may facilitate tissue-specific infection, as evidenced by the strong linkage with *emm* pattern genotype, but additional factors appear to be at play as well (Kratovac et al., 2007). Two mutually exclusive lineages of alleles having ~35% nucleotide sequence divergence correspond to the *rofA*/*nra* locus of the FCT region, which encodes global transcriptional regulators controlling the expression of pilus structural genes and non-FCT region genes as well (Kreikemeyer et al., 2003). Nearly all *emm* pattern A–C and E strains have *rofA*, whereas most *emm* pattern D strains harbor *nra* (Bessen et al., 2005). The association of *rofA* versus *nra* with the *emm* pattern group differs from the distribution of *mga* alleles in that the pattern E generalists tend to share *rofA* with pattern A–C strains, but share *mga-2* with pattern D strains. The *rofA* gene appears to have its origins in *S. dysgalactiae* ssp. *equisimilis*, providing further support for the idea that orthologous gene replacements can lead to adaptation to new ecological niches.

In summary, there is a lack of clonal congruence among strains corresponding to the three *emm* pattern-defined groups, as demonstrated with housekeeping genes. Thus, despite some niche separation driven by epidemiological trends (temporal and spatial) and innate tissue tropisms, there is no evidence for housekeeping gene sequence divergence between strains of differing *emm* patterns. This finding provides support for ecological congruence among isolates within each *emm* pattern group, whereby loci exhibiting a high degree of linkage with *emm* pattern genotypes likely contribute to the adaptations leading to throat or skin infection.

17.5.2 Bovine Origin of Human Pathogenic GBS

Unlike pneumococci and GAS, which are strictly human pathogens, GBS causes disease in humans as well as in animals, with bovine mastitis being the most intensively studied. GBS was first described as a bovine pathogen in 1887 and was reported as a rare cause of human infection in the 1930s, but it was not until many decades later (1960s) that it became recognized as a major cause of neonatal infection in humans (Bisharat et al., 2004). More recently, GBS has been responsible for invasive bacterial disease in the elderly. MLST has been used to establish the degree to which the bovine and human pathogens are two distinct populations, and to describe their relationships to human-colonizing isolates.

The CC17 strains are derived from human and bovine hosts and include isolates assigned to ST17 (human isolates) and ST61 (bovine isolates) (Bisharat et al., 2004; Jones et al., 2006; Martins et al., 2007; Bohnsack et al., 2008). Strains of the ST17 lineage display a significant association with invasive disease in neonates and thereby constitute a hyperinvasive lineage. Phylogenetic analysis indicates that the human ST-17 complex of isolates has arisen from a lineage of bovine isolates. This lineage may have been introduced fairly recently into humans from cattle but now accounts for human adult-colonizing strains as well. There are also additional CCs, aside from CC17, that are commonly found in association with human invasive GBS disease, and these lineages appear to have a different origin and may have adapted to humans over a longer time period.

Conceivably, mobile genetic elements (MGEs) might provide insights into the recent evolutionary history of GBS. Ninety-eight epidemiologically unrelated GBS isolates originating from humans (neonatal and adult disease, and colonization) or cows (udder infections) were evaluated for both MLST and the distribution of MGEs (Hery-Arnaud et al., 2007). Concatenated housekeeping gene sequences from the 98 strains were used to construct a phylogeny showing seven main divisions that closely match the CCs or subCCs identified by eBURST. The origin of human CC17 strains from bovine CC67 strains (including ST61) was confirmed by this approach. Three of the seven (sub)CCs were host specific: human CC17 and subCC19, and bovine CC67. The prevalence of MGEs among the 98 GBS isolates, specifically 10 insertion sequence elements and one group II intron, was also evaluated. Factorial analysis of correspondence with the MGEs as variables provides additional evidence for a bovine ancestor of human CC17. The findings also point to a physical barrier that reduces the exchange of genetic material between the bovine and human reservoirs.

17.5.3 Disease-Specialist Clones of GAS Arising from Lateral Transfer of GBS Genes

The R28 surface protein is expressed by some strains of GAS, including those of M type 28 (Stalhammar-Carlemalm et al., 1999). The R28 molecule is mosaic in structure, forming a chimera of three surface proteins of GBS (proteins Rib, α, and β), and suggestive of past intraspecific recombinational events (Stalhammar-Carlemalm et al., 1999). Taken together, the data support the idea that the gene encoding R28 arose in GBS and underwent horizontal transfer to GAS.

The R28-expressing GAS strains display a strong epidemiological association with puerperal sepsis, also known as childbed fever and which occurs during or shortly following childbirth. It can be experimentally shown that the R28 protein mediates adherence of GAS to cultured epithelial cells derived from human cervical tissue. GBS naturally colonize the vaginal epithelium. Thus, puerperal fever caused by GAS and neonatal sepsis caused by GBS are probably both initiated by bacteria colonizing the vaginal epithelium. In fact, the complete genome sequence of an M28 GAS strain contains the R28 gene plus six genes encoding other putative extracellular proteins, located on a 37-kB genomic island that is shared with GBS (Green et al., 2005). Therefore, the acquisition of foreign genes appears to be a key step in generating this disease-specialist clone of GAS.

The complete genome sequence of a serotype M-type 2 (M2) strain of GAS reveals a 35-kD ICE-like element harboring an R28 gene and displaying high overall homology with the 37-kD exogenous element found in the M28 strain genome (Beres and Musser, 2007). Interestingly, a portion of the pilus-associated FCT region of the serotype M2 strain is more closely related to genomic islands present in several sequenced GBS strains than to the other GAS strains. Thus, the serotype M2 strain contains two genomic regions that are closely related to genetic elements in GBS, and conceivably, both regions may contribute to puerperal sepsis that is occasionally caused by M2 strains.

17.5.4 Gene Loss and Gene Gain among Host-Restricted *S. equi*

S. equi falls within the pyogenic division and consists of two subspecies (or biovars) having distinct pathogenic properties: *S. equi* ssp. *equi* and *S. equi* ssp. *zooepidemicus*. These organisms are often referred to as *S. equi* and *S. zooepidemicus*, respectively, whereby *S. equi* is host restricted to horses and *S. zooepidemicus* is found in association with a variety of mammalian hosts, usually as a commensal but occasionally as a pathogen. MLST was performed for isolates of both biovars (Webb et al., 2008). Of the 24 *S. equi* isolates, 23 belonged to the same ST (ST179), whereas the 253 isolates of *S. zooepidemicus* were considerably more diverse (128 STs) and displayed few CCs. Reclassification of these organisms to *S. zooepidemicus* ssp. *zooepidemicus* and *S. zooepidemicus* ssp. *equi* is supported by the MLST findings and may be forthcoming in the near future.

According to a phylogenetic analysis based on the MLST data, the *S. equi* CC is clustered with nine of the *S. zooepidemicus* STs, providing strong evidence that *S. equi* evolved from an ancestral *S. zooepidemicus* strain (Webb et al., 2008). The complete genome sequence has been reported for two strains of *S. zooepidemicus* and for one strain of *S. equi* (Beres et al., 2008; Holden et al., 2009). Genetic steps leading to the emergence of the important horse pathogen *S. equi* were examined.

Comparative genome analysis points to gene loss leading to a reduction in ancestral capabilities. Relative to *S. zooepidemicus*, all 26 *S. equi* isolates exhibited a loss of genes involved in the metabolism of several carbohydrates (Holden et al., 2009). Similar types of gene loss are typical of other host-restricted bacteria that evolved from more versatile ancestors. Nutritional flexibility is also observed for the opportunistic pathogen *S. uberis* and may allow this organism to successfully occupy a discrete ecological niche (Ward et al., 2009).

There is also evidence that gene gain, leading to the introduction of novel functions, may have enabled *S. equi* to exploit a new niche (Holden et al., 2009). All 26 isolates of *S. equi* examined and about 30% of the *S. zooepidemicus* strains harbor a bacteriophage encoding a phospholipase A2 toxin sharing very high sequence homology with a gene present in some strains of *S. pyogenes*. In *S. pyogenes*, phospholipase A2 contributes to virulence in an animal model for disease (Sitkiewicz et al., 2006). Another important finding surrounds a nonribosomal peptide synthetase system for iron acquisition, encoded by a novel ICE that is conserved in all *S. equi* isolates but is absent from all *S. zooepidemicus* isolates. Combined with the MLST data, the findings on the iron acquisition-associated ICE are suggestive of it playing a pivotal role in the transition from a (mostly) commensal relationship to a pathogenic lifestyle (Heather et al., 2008; Holden et al., 2009).

17.6 BACTERIAL POPULATION DYNAMICS AND SELECTION

There is a vast body of literature on the molecular epidemiology of streptococcal pathogens in host populations. Population-based surveillance studies conducted over different time periods and/or in different host communities can address important clinical and biological questions, especially when several analyses are combined. Shifts in bacterial genetic diversity over geographic space and time, in concert with host selection pressures such as protective immunity and antibiotic use, shape the bacterial population genetic structure. Understanding these processes may provide critical insights that can aid in the development of effective infectious disease prevention and control strategies.

17.6.1 Diversity of GAS Populations and Strain Migration

Surveillance studies conducted on small spatial scales provide evidence for ample migration of GAS. In a remote tropical island community of aboriginal Australians, periodic surveillance was conducted over 2 years by collecting specimens from the throat and impetigo skin lesions (Bessen et al., 2000; McGregor et al., 2004a). Of ~500 patient visits involving 224 individual subjects, only 16 GASs were recovered from colonization of the throat, all in the absence of disease, whereas 121 isolates were obtained from skin lesions. Together, the 137 GAS isolates represent 32 *emm* types, 35 STs, and 35 unique *emm* type–ST combinations. Using the *emm* type–ST combination to define a clone, the Simpson's diversity index (Grundmann et al., 2001), *D*, is measured as 0.959 (95% CI, 0.951, 0.967), indicative of a highly diverse bacterial population. Also, there were only two CCs, with the remaining STs being singletons. The GAS strains afflicting this remote island community most likely migrated in from elsewhere. At least 90% of the 32 *emm* types had been recovered from places outside of Australia (McGregor et al., 2004a).

Similarly, high levels of GAS strain diversity are observed in other remote human populations sampled over relatively short time periods. The combined data for three remote Australian aboriginal communities on the mainland yielded 350 GAS isolates from 49 households and 1173 individuals, periodically screened over 1 or 2 years (McDonald et al., 2007a,b); 43 *emm* types were present, yet long-term throat carriage of the same *emm* type was uncommon. In far-western Nepal, 120 GAS isolates were collected from a single screening of 60 children in each of eight villages, yielding 45 *emm* types and 51 STs (Sakota et al., 2006). Thus, despite what appears to be rather limited outside contacts, many distinct strains of GAS circulate among remote human populations within a short period of time.

The three studies cited above were dominated by cases of impetigo, with few or no instances of pharyngitis, although asymptomatic throat carriage was noted at a fairly low prevalence. A large multiple site surveillance study of streptococcal pharyngitis in children was recently conducted for several urban and suburban areas in the United States and in Canada, yielding ~1900 isolates over 2 years (Shulman, 2004). In years 1 and 2, the number of distinct *emm* types recovered was 29 and 31, respectively, with 23 *emm* types shared between both years. Thus, a highly diverse set of GAS strains is also responsible for oropharyngeal disease. Using *emm* type to define a clone, the calculated Simpson's diversity index ($D = 0.891$; 95% CI, 0.885, 0.897) is somewhat less than that for the remote aboriginal Australian island population (adjusted $D = 0.938$, when clone is defined by *emm* type only; 95% CI, 0.930, 0.946) (Bessen et al., 2000). Possibly, the lower diversity for pharyngitis isolates is due to differences in sampling and/or throat versus skin specialist strains.

The pharyngitis survey also reveals striking differences between the multiple study sites in terms of the distribution of predominant *emm* types; however, the predominant *emm* types fluctuate only slightly from year to year (Shulman, 2004). Yet, a comparison of the M type distribution among pharyngitis isolates obtained from pediatric patients in Chicago during the 1960s versus 40 years later shows marked changes in the predominating M types (Shulman et al., 2006). The factors underlying these trends are probably complex and may include longer-term cyclical changes in herd immunity. Age-related changes in *emm* type distribution are observed in childhood cases of pharyngitis, with older children showing increased infection by uncommon *emm* types and a corresponding decrease in common *emm* types (Jaggi et al., 2005). Conceivably, these shifts are due to the acquisition of protective immunity following exposure to highly prevalent strains earlier in life.

17.6.2 M Serotypes of GAS as Targets of Host Immune Selection

For many GAS strains examined, protective immunity is *emm* type specific, whereby M type-specific antibodies mediate opsonization and overcome the antiphagocytic property of M protein (Lancefield, 1962; Beachey et al., 1981). Because they elicit a highly protective host immune response, the M type determinants are the targets of a promising GAS vaccine currently under development (Bisno et al., 2005; Dale et al., 2005; McNeil et al., 2005). Given the large number of *emm* types present throughout the world, the practicality of an M type-specific vaccine for the prevention of GAS disease has been the subject of much debate. The initial 26 M types that were chosen for vaccine development are present in GAS strains that are the most common causes of pharyngitis and invasive disease in the United States, yet these M types are rare or absent in many communities in other parts of the world where GAS disease leads to high levels of morbidity and mortality, often due to autoimmune sequelae such as rheumatic fever.

The CDC website currently lists ~217 *emm* types identified among GAS, represented by ~883 partial *emm* alleles (i.e., subtypes) as defined by the 150 nucleotides encoding the 50 amino-terminal residues of the mature M protein (http://www.cdc.gov/ncidod/biotech/strep/strepindex.htm). Immune selection pressures imposed by the host may drive diversification of the *emm* type-specific region, resulting in immune escape. Experimental findings provide direct evidence for amino acid changes in the M type-specific region that allow for immune escape (Jones et al., 1988; Eriksson et al., 2001; Beres et al., 2006), and such changes may explain the recent emergence of an important M3-type clone.

Calculations were made for the Ka (nonsynonymous substitutions per nonsynonymous site) and Ks (synonymous substitutions per synonymous site) values for 105 alignments of partial *emm* alleles (i.e., subtypes) corresponding to 105 *emm* types, derived from a global collection of >500 *S. pyogenes* isolates (Bessen et al., 2008). The ratio of the mean Ka to mean Ks value for the 105 alignments was 1.96, indicative of positive diversifying selection acting on the type-specific region. However, for the *emm* pattern A–C group of throat specialists, this ratio value was ~3- to 4-fold higher than for skin specialists and generalists, suggestive of stronger immune selection pressures acting on the M type-specific region in the throat specialist group of strains.

Despite the large number of distinct *emm* types and *emm* allele subtypes that exist, there also appear to be functional constraints that put the brakes on genetic diversification. The M type-specific regions of most GAS strains bind the complement regulator C4b-binding protein (C4BP). C4BP binding is achieved in the absence of a shared amino acid sequence motif, and even though substitutions can introduce antigenic change without altering C4BP binding activity (Persson et al., 2006), it seems likely that there are functional constraints on sequence variation.

Immune escape might arise following a change in *emm* type mediated by recombinational replacement of all or part of the *emm* gene. Among a genetically diverse set of nearly 600 GAS isolates, a small proportion (5.4%) of the STs examined were found in association with a substantially larger fraction (>20%) of the total *emm* types evaluated (Bessen et al., 2008). Since the ST is (by definition) unchanged, it seems likely that *emm* type replacements are recent genetic events. If strains bearing the STs of the recipient cells are highly prevalent and have high fitness, they may continue to profoundly shape GAS evolution well into the future.

There are also numerous instances of the same *emm* type (~50% of the total studied) recovered in association with distant STs differing at ≥5 of the seven housekeeping alleles (Bessen et al., 2008). Thus, for many clones, *emm* type alone is a rather poor marker. The *emm* types that are recovered on distant genetic backgrounds are disproportionately represented by the *emm* pattern D and E skin specialists and generalists, respectively. This finding is consistent with the Ka and Ks values, indicating that pattern D and E *emm* types are under less intense immune selection pressure. Interestingly, the *emm* types that bind C4BP (Persson et al., 2006) are largely represented by the pattern D and E groupings. When combined with the Ka-to-Ks ratio data (Bessen et al., 2008), the findings are suggestive of higher levels of purifying (negative) selection acting on pattern D and E *emm* types in order to preserve the C4BP binding activity. Furthermore, nearly all *emm* types recovered on distant genetic backgrounds are confined to the same *emm* pattern grouping

(McGregor et al., 2004b), indicating that recombination involving disparate *emm* pattern groups is less likely because of insufficient homology between flanking sequences and/or negative selection acting to prevent the survival of clones with mismatched *emm* types and genetic backgrounds.

A given M or *emm* type can be found in association with multiple SOF types (or *sof* alleles) and T types, as summarized in an analysis of >40,000 isolates collected worldwide over a 50-year period (Johnson et al., 2006). These findings provide further evidence that *emm* type undergoes horizontal exchange with other GAS strains. There are numerous examples of *emm* types associated with >1 SOF or *sof* type, suggestive of independent lateral transfer events involving the *sof* locus, which is positioned ~16 kb upstream of *emm*. A 450-nt region at the 5′ end of the *sof* gene was used to define the partial *sof* allele, and it strongly correlates with SOF serological types (Beall et al., 2000). The number of GAS strains or clones—as defined by unique combinations of *emm* type, *sof* type, T or pilus gene type, and ST—remains to be established. In addition to *emm*, *sof*, and *tee* (i.e., FCT region) genes, there may be additional immunodominant surface antigens that shape the population structure of the GAS species.

17.6.3 Twenty-First Century Pneumococcal Vaccines and the Impact on Population Structure

Natural exposure to the pneumococcus can lead to the development of protective immunity. A serotype-specific host immune response directed to capsular polysaccharide provides high levels of protection against invasive pneumococcal disease (IPD). However, the pneumococcus most often colonizes and acts as a commensal, a lifestyle that generally fails to elicit a potent immune response. And even though pneumococci are common causes of otitis media and sinusitis, which are true infections, it is unlikely for any one individual human host to develop strong protective immunity against more than a few of the >90 serotypes known to exist.

In general terms, vaccination can have a profound impact on herd immunity by reducing the number of susceptible hosts in a population. The 23-valent pneumococcal polysaccharide vaccine (PPV23) was licensed in the United States in 1983 and is based on 23 serotypes, which accounted for nearly 88% of pneumococcal bloodstream infections (bacteremia) at that time. However, the efficacy of PPV23 in preventing IPD is insufficient for the host populations most in need of protection from pneumococcal disease: infants and the elderly. The feeble immune response that typifies these cohorts provided the impetus to develop a vaccine conferring a more robust and longer-lasting protective immune response.

A capsular polysaccharide–protein conjugate vaccine (PCV7) was recently developed and contains seven serotypes (4, 9V, 14, 19F, 23F, 18C, 6B) conjugated to the nontoxic diphtheria toxoid; the carrier protein has a limited capacity for conjugation and explains why only seven serotypes are included. The PCV7 vaccine is designed to target serotypes that accounted for most cases of IPD among children in the United States between 1978 and 1994. Included are serotypes associated with increased resistance to penicillin, which is mediated via altered penicillin-binding proteins (PBPs) that play a key role in cell wall biosynthesis. Any effect of the PCV7 vaccine on lowering the incidence of IPD caused by the seven serotypes may be because immunized individuals are directly protected against infection, or herd immunity lowers the prevalence of the seven serotypes in the host population and thereby reduces the risk of infection in unvaccinated individuals as well.

Widespread use of the PCV7 vaccine has the potential to alter the ecology of the pneumococcus–human host interaction. Serotype replacement, whereby the vaccine strains are replaced with strains of other antigenic types, is a general concern for bacterial populations having numerous serotypes (Lipsitch, 1997). For the pneumococcus, strains bearing capsular serotypes that are excluded from the vaccine formulation might be better able to compete with microbial flora lacking the vaccine-associated serotypes. Thus, nonvaccine serotypes may fill the vacuum and achieve higher rates of colonization, and may ultimately account for a higher fraction of IPD. Alternatively, recombinational replacement of *cps* genes encoding the capsular serotype of a vaccine target strain may lead to the emergence and spread of immune escape mutants having new capsule types on genetic backgrounds that are already sufficiently fit for colonization and/or infection.

Recombinational replacement of *cps* genes, leading to a switch in capsular serotype, has been observed among natural isolates of pneumococcus. In a classic study, serotype 19F variants of the major Spanish multiresistant serotype 23F clone were shown to have large recombinational replacements at the *cps* loci; among the eight 19F isolates examined, four distinct recombinational events were discernable (Coffey et al., 1998). Experimentally, it can be shown, through the use of isogenic mutants and a mouse model for infection, that capsular serotype can have a profound effect on virulence, and this effect is influenced by the genetic background of the strain (Kelly et al., 1994).

Vaccination with PCV7, which began in the year 2000, has led to dramatic decreases in the incidence of IPD caused by the targeted serotypes, leading to many lives saved (Huang et al., 2005; Hicks et al., 2007). However, there has also been statistically significant, though less dramatic, incremental increases in IPD incidence caused by nonvaccine serotypes. The findings from a randomized clinical trial, in which children received either PCV7 or another vaccine (meningococcal), were used to characterize the relationships between genetic variation in pneumococci and PCV7 vaccination (Lipsitch et al., 2007; O'Brien et al., 2007). PCV7 reduced both the risk of acquisition and the colonization density of the vaccine serotypes, but increased the risk of acquiring nonvaccine serotypes among vaccinees and their household contacts. Serotype replacement could be largely attributed to the expansion of nonvaccine serotype clones, which were resident to the host population having received the meninogoccal vaccine (i.e., nonvaccinated with PCV7). However, possible capsular switching that generated novel ST associations with nonvaccine serotypes occurred only once, and therefore, capsular switching made only a minor contribution to serotype replacement in this host population.

Serotype 19A was a common colonizer before the introduction of PCV7; however, 19A was not included in the PCV7 vaccine because it was incorrectly anticipated that serotype 19F, which is often associated with high levels of antibiotic resistance, would afford some cross protection. Strikingly, the incidence of nonvaccine serotype invasive disease has increased since PCV7 was introduced, and a large contributor to that increase is serotype 19A. This is partly due to the expansion of a previously circulating 19A clone (ST199). However, another key factor is the recent emergence of a vaccine escape mutant arising via recombinational events. This newly discovered 19A clone has an evolutionary history that indicates replacement of the serotype 4 *cps* region of the recipient strain (ST695) with the 19A *cps* region of a donor strain (Brueggemann et al., 2007; Moore et al., 2008). Thus, this capsular switch involves replacement of a vaccine serotype with a nonvaccine serotype.

The capsular switch can be ascribed to a single genetic event involving a 39-kB segment that contains the *cps* region (Brueggemann et al., 2007). The 39-kB segment involved in this single genetic event also contains two genes flanking the *cps* region that encode PBPs, which confer increased resistance to the drug. Thus, not only is the newly emerged serotype 19A clone an immune escape mutant but it has also acquired increased resistance to a widely used antibiotic, providing an additional selective advantage. The lack of PCV7 protection against 19A, and increasing resistance within this serotype, has resulted in the expansion of drug-resistant serotype 19A; the spread of serotype 19A clones is increasing worldwide and includes countries that have not received PCV7 (Dagan and Klugman, 2008).

Over time, serotype replacement may gradually diminish the success of PCV7, and ongoing efforts for new pneumococcal vaccines having expanded coverage are deemed necessary. There are also differences in predominant serogroups in different regions of the world, driving the need for alternative vaccine formulations.

17.6.4 Future Prospects for Other Streptococcal Vaccines

Like pneumococcus, GBS has a polysaccharide capsule, although there are only nine recognized serotypes among GBS, far fewer than the 91 capsular serotypes of the pneumococcus. Vaccines based on GBS capsular polysaccharide conjugated to carrier proteins induce serotype-specific antibody, which is protective; different vaccine formulations may be necessary in parts of the world where the strain prevalence varies (Johri et al., 2006). Capsule switching following genetic recombination also occurs among GBS (Luan et al., 2005).

There are ongoing initiatives to develop a protein-based vaccine against GBS disease, and part of the impetus is the increasing number of isolates that lack a capsular serotype. Three pilus gene islands have been identified within the GBS population, and their protein products elicit a protective immune response (Margarit et al., 2009). Also, pili are expressed on the surface of 94% of GBS strains examined. Thus, a pilus-based vaccine shows promise for broad protection against GBS infection. Pilus-based vaccines are also being considered for GAS disease, since the antigenic hetereogeneity among T antigens, which are now known to be synonymous with pili, is much lower than that found for the M protein surface fibrils, and therefore, broad coverage may be more readily feasible (Falugi et al., 2008).

The potential for "pilus switching" is currently an underdeveloped area of investigation. An M2 strain of GAS harbors pili genes that have higher sequence homology to GBS pili genes than to GAS pili genes, suggestive of lateral gene exchange between species (Beres and Musser, 2007). The largely commensal species *S. dysgalactiae* ssp. *equisimilis* also possesses T antigens and may conceivably be a source for new antigenic pilus forms that enter into the gene pool of the more pathogenic GAS species.

17.6.5 Antibiotic Resistance and Selection

Decreased susceptibility to β-lactams, such as penicillin, was first recognized in the pneumococcus about 40 years ago and continues to be a growing problem in the treatment of pneumococcal disease. Increased resistance to the drug can be ascribed to PBPs having a lower binding affinity for penicillin, and thereby high concentrations of the drug are required to block bacterial growth. The *pbp* genes that lie adjacent to the *cps* region show evidence for genetic cotransfer, and thus, selection for the newly emerged clones may be driven by resistance to the antibiotic or by escape from the host immune response, or a combination of both. High levels of resistance to penicillin among pneumococci can lead to treatment failure.

GAS does not display the penicillin resistance problem characteristic of pneumococci (Horn et al., 1998), perhaps because mutations in PBPs that lead to decreased binding affinity exact a biological cost in GAS, such as a defective peptidoglycan cell wall. Thus, the direct impact of penicillin on the population structure of GAS appears to be minimal. However, GAS can cause a clinically inapparent infection at the throat, distinct from throat carriage, in that the patient lacks pharyngitis symptoms, yet still elicits a strong immune response to GAS antigens. The ability of GAS to cause a "symptom-free" infection may be an evolved trait that provides a selective advantage to the bacterium because the patient fails to seek antibiotic treatment.

Although GAS infection can usually be treated with penicillin, the organism is occasionally resistant to other antibiotics, which can adversely affect the clinical course of the disease and can shape the population biology of the species. Macrolides and lincosamides are the primary treatment for GAS infections in patients with β-lactam hypersensitivity or chronic, recurrent pharyngitis due to prior treatment failure; clindamycin (a lincosamide) is the first choice drug for patients with life-threatening soft tissue infections, such as necrotizing fasciitis, because it halts exotoxin production. Studies in Japan, Finland, and elsewhere show a strong correlation between macrolide consumption and resistance to the drug in GAS (Fujita et al., 1994; Seppala et al., 1997). In a 3-year longitudinal surveillance study in Pittsburgh, 100% of GAS isolates recovered from children were macrolide sensitive until the third year of study, when macrolide resistance was present in 48% of all isolates; all resistant isolates were M type 6 (Martin et al., 2002). Thus, a macrolide-resistant clone can have a significant survival advantage under certain conditions.

At least three genes confer resistance to macrolides in GAS: *erm(A)* and *erm(B)*, leading to ribosomal modification, and *mef(A)*, which promotes drug efflux. These genes also confer macrolide resistance in GBS (Domelier et al., 2008). In a survey of 212 macrolide-resistant GAS isolates obtained from throughout the world, at least 49 independent acquisitions of macrolide resistance were estimated, based on unique combinations of *emm* type, ST, and the resistance gene type (Robinson et al., 2006). Among this set of macrolide-resistant GAS, 22 CCs or STs were recovered from >1 continent, and together, the 22 intercontinental clones and complexes accounted for 79% of the resistant isolates. Thus, in addition to numerous independent acquisitions of resistance genes by GAS, the resistant clones and their descendants are widely spread throughout the world.

Some genetic elements carry determinants for resistance to both macrolides and tetracyclines, the latter often being conferred by *tet(M)* or *tet(O)*. Resistance to tetracyclines by GAS is estimated to have been acquired via ≥80 independent horizontal gene transfer events (Ayer et al., 2007). An evaluation of a global collection of a genetically diverse set of GAS strains, whose resistance profile was not initially known, indicated that tetracycline-resistant strains outnumbered macrolide-resistant strains by about 15-fold. Therefore, it is plausible that tetracycline usage drives the acquisition of macrolide resistance to some extent, via genetic elements harboring both resistance genes.

Numerous genetic elements harboring genes encoding resistance to erythromycin, tetracycline, and/or additional antibiotics have been characterized for several streptococcal species, including the major human pathogens. They include conjugative and nonconjugative elements that consist of plasmids, prophage, transposons, and/or ICEs (Varaldo et al., 2009). Conjugative transposons such as the Tn*916*-like elements are highly evolved for broad host range transfer. When these or ICE and prophage-like elements incorporate antibiotic resistance genes into their composite structures, they can play a major role in disseminating multiple drug resistance across numerous streptococcal species, which, in turn, can impact the genetic structure of the bacterial population.

17.7 MACHINERY OF GENETIC CHANGE, REVISITED

As mentioned throughout this chapter, streptococci are rich in MGEs and, furthermore, all (or nearly all) species exhibit relatively high levels of homologous recombination involving core genes. The genetic structure of a bacterial population can be profoundly shaped by the internal machinery that generates genetic change. The mutator phenotype, leading to elevated rates of mutation, is associated with both oxidative stress and fluoroquinolone resistance in pneumococci (Pericone et al., 2002; Gould et al., 2007), and with prophage integration at the chromosomal *mutL* locus of GAS (Scott et al., 2008). Natural competence for genetic transformation is exhibited by numerous streptococcal species, such as *S. pneumoniae*, *S. mitis*, *S. mutans*, and *S. gordonii*, whereas the major human pathogens GAS and GBS seem to lack the complete set of genetic machinery required for transformation (Claverys et al., 2007). However, some GAS strains have homologues (*sil* locus) that may be involved in DNA transfer (Hidalgo-Grass et al., 2002). ICEs harboring genes that encode type II restriction–modification cassettes may prevent the successful uptake of foreign DNA originating from certain sources (Euler et al., 2007). Clustered regularly interspaced short palindromic repeat (CRISPR) regions are present within the genomes of many strains of the numerous streptococcal species (Barrangou et al., 2007). Importantly, CRISPR regions may selectively block infection of the bacterial cell by new bacteriophage via an RNA interference-like mechanism.

In summary, the intrinsic machinery of a given bacterial cell helps to define the range of possibilities for genetic change, upon which natural selection may act. For *Streptococcus*, the diversity in genetic machinery is enormous and adds another layer of complexity to the population genetics of the member species.

REFERENCES

Abbot, E. L., Smith, W. D., Siou, G. P., Chiriboga, C., Smith, R. J., Wilson, J. A., Hirst, B. H., and Kehoe, M. A. (2007) Pili mediate specific adhesion of *Streptococcus pyogenes* to human tonsil and skin. *Cellular Microbiology* **9,** 1822–1833.

Aguiar, S. I., Serrano, I., Pinto, F. R., Melo-Cristino, J., and Ramirez, M. (2008) The presence of the pilus locus is a clonal property among pneumococcal invasive isolates. *BMC Microbiology* **8,** 41.

Ahmad, Y., Gertz, R. E. Jr., Li, Z., Sakota, V., Broyles, L. N., Van Beneden, C., Facklam, R., Shewmaker, P. L., Reingold, A., Farley, M. M., and Beall, B. W. (2009) Genetic relationships deduced from emm and multilocus sequence typing of invasive *S. dysgalactiae* subsp. *equisimilis* and *S. canis* recovered in the United States. *Journal of Clinical Microbiology* **47,** 2046–2054.

Ayer, V., Tewodros, W., Manoharan, A., Skariah, S., Luo, F., and Bessen, D. E. (2007) Tetracycline resistance in group a streptococci: Emergence on a global scale and influence on multiple-drug resistance. *Antimicrobial Agents and Chemotherapy* **51,** 1865–1868.

Barrangou, R., Fremaux, C., Deveau, H., Richards, M., Boyaval, P., Moineau, S., Romero, D. A., and Horvath, P. (2007) CRISPR provides acquired resistance against viruses in prokaryotes. *Science* **315,** 1709–1712.

Beachey, E. H., Seyer, J. M., Dale, J. B., Simpson, W. A. and Kang, A. H. (1981) Type-specific protective immunity evoked by synthetic peptide of *Streptococcus pyogenes* M protein. *Nature* **292,** 457–459.

Beall, B., Facklam, R., and Thompson, T. (1996) Sequencing *emm*-specific PCR products for routine and accurate typing of group A streptococci. *Journal of Clinical Microbiology* **34,** 953–958.

Beall, B., Gherardi, G., Lovgren, M., Forwick, B., Facklam, R. and Tyrrell, G. (2000) *Emm* and *sof* gene sequence variation in relation to serological typing of opacity factor positive group A streptococci. *Microbiology* **146,** 1195–1209.

Bellanger, X., Roberts, A. P., Morel, C., Choulet, F., Pavlovic, G., Mullany, P., Decaris, B., and Guedon, G. (2009) Conjugative transfer of the integrative conjugative elements ICESt1 and ICESt3 from

Streptococcus thermophilus. *Journal of Bacteriology* **191,** 2764–2775.

Bentley, S. D., Aanensen, D. M., Mavroidi, A., Saunders, D., Rabbinowitsch, E., Collins, M., Donohoe, K., Harris, D., Murphy, L., Quail, M. A., Samuel, G., Skovsted, I. C., Kaltoft, M. S., Barrell, B., Reeves, P. R., Parkhill, J., and Spratt, B. G. (2006) Genetic analysis of the capsular biosynthetic locus from all 90 pneumococcal serotypes. *PLoS Genetics* **2,** e31.

Beres, S. B. and Musser, J. M. (2007) Contribution of exogenous genetic elements to the group A *Streptococcus* metagenome. *PLoS One* **2,** e800.

Beres, S. B., Richter, E. W., Nagiec, M. J., Sumby, P., Porcella, S. F., DeLeo, F. R., and Musser, J. M. (2006) Molecular genetic anatomy of inter- and intraserotype variation in the human bacterial pathogen group A *Streptococcus*. *Proceedings of the National Academy of Sciences of the United States of America* **103,** 7059–7064.

Beres, S. B., Sesso, R., Pinto, S. W., Hoe, N. P., Porcella, S. F., DeLeo, F. R., and Musser, J. M. (2008) Genome sequence of a Lancefield group C *Streptococcus zooepidemicus* strain causing epidemic nephritis: New information about an old disease. *PLoS One* **3,** e3026.

Bessen, D. E., (2009) Population biology of the human restricted pathogen, *Streptococcus pyogenes*. *Infection, Genetics and Evolution* **9,** 581–593.

Bessen, D. E., Carapetis, J. R., Beall, B., Katz, R., Hibble, M., Currie, B. J., Collingridge, T., Izzo, M. W., Scaramuzzino, D. A., and Sriprakash, K. S. (2000) Contrasting molecular epidemiology of group A streptococci causing tropical and non-tropical infections of the skin and throat. *Journal of Infectious Diseases* **182,** 1109–1116.

Bessen, D. E. and Kalia, A. (2002) Genomic localization of a T-serotype locus to a recombinatorial zone encoding extracellular matrix-binding proteins in *Streptococcus pyogenes*. *Infection and Immunity* **70,** 1159–1167.

Bessen, D. E., Manoharan, A., Luo, F., Wertz, J. E., and Robinson, D. A. (2005) Evolution of transcription regulatory genes is linked to niche specialization in the bacterial pathogen *Streptococcus pyogenes*. *Journal of Bacteriology* **187,** 4163–4172.

Bessen, D. E., McGregor, K. F., and Whatmore, A. M. (2008) Relationships between *emm* and multilocus sequence types within a global collection of *Streptococcus pyogenes*. *BMC Microbiology* **8,** 59.

Bisharat, N., Crook, D. W., Leigh, J., Harding, R. M., Ward, P. N., Coffey, T. J., Maiden, M. C., Peto T., and Jones, N. (2004) Hyperinvasive neonatal group B streptococcus has arisen from a bovine ancestor. *Journal of Clinical Microbiology* **42,** 2161–2167.

Bishop, C. J., Aanensen, D. M., Jordan, G. E., Kilian, M., Hanage, W. P., and Spratt, B. G. (2009) Assigning strains to bacterial species via the internet. *BMC Biology* **7,** 3.

Bisno, A. L., Rubin, F. A., Cleary, P. P., and Dale, J. B. (2005) Prospects for a group A streptococcal vaccine: Rationale, feasibility, and obstacles—Report of a National Institute of Allergy and Infectious Diseases workshop. *Clinical Infectious Diseases* **41,** 1150–1156.

Bisno, A. L. and Stevens, D. L. (2000) Streptococcus pyogenes. In *Principles and Practice of Infectious Diseases*. (eds. G. L. Mandell, R. G. Douglas, and R. Dolin), pp. 2101–2117. Churchill Livingstone, Philadelphia, PA.

Bohnsack, J. F., Whiting, A., Gottschalk, M., Dunn, D. M., Weiss, R., Azimi, P. H., Philips, J. B. III, Weisman, L. E., Rhoads, G. G., and Lin, F. Y. (2008) Population structure of invasive and colonizing strains of *Streptococcus agalactiae* from neonates of six U.S. Academic Centers from 1995 to 1999. *Journal of Clinical Microbiology* **46,** 1285–1291.

Brochet, M., Couve, E., Glaser, P., Guedon, G., and Payot, S. (2008a) Integrative conjugative elements and related elements are major contributors to the genome diversity of *Streptococcus agalactiae*. *Journal of Bacteriology* **190,** 6913–6917.

Brochet, M., Rusniok,, C., Couve, E., Dramsi, S., Poyart, C., Trieu-Cuot, P., Kunst, F., and Glaser P. (2008b) Shaping a bacterial genome by large chromosomal replacements, the evolutionary history of *Streptococcus agalactiae*. *Proceedings of the National Academy of Sciences of the United States of America* **105,** 15961–15966.

Broyles, L. N., Van Beneden, C., Beall, B., Facklam, R., Shewmaker, P. L., Malpiedi, P., Daily, P., Reingold, A., and Farley, M. M. (2009) Population-based study of invasive disease due to beta-hemolytic streptococci of groups other than A and B. *Clinical Infectious Diseases* **48,** 706–712.

Brueggemann, A. B., Pai, R., Crook, D. W., and Beall, B. (2007) Vaccine escape recombinants emerge after pneumococcal vaccination in the United States. *PLoS Pathogens* **3,** e168.

Carapetis, J., Currie, B., and Kaplan, E. (1999) Epidemiology and prevention of group A streptococcal infections: Acute respiratory tract infections, skin infections, and their sequelae at the close of the twentieth century. *Clinical Infectious Diseases* **28,** 205–210.

Carapetis, J. R., Steer, A. C., Mulholland, E. K., and Weber, M. (2005) The global burden of group A streptococcal diseases [Review]. *Lancet Infectious Diseases* **5,** 685–694.

Caufield, P. W. (2009) Tracking human migration patterns through the oral bacterial flora. *Clinical Microbiology and Infection* **15**(Suppl 1)**,** 37–39.

Chaffin, D. O., Beres, S. B., Yim, H. H., and Rubens, C. E. (2000) The serotype of type Ia and III group B streptococci is determined by the polymerase gene within the polycistronic capsule operon. *Journal of Bacteriology* **182,** 4466–4477.

Claverys, J. P., Martin, B., and Havarstein, L. S. (2007) Competence-induced fratricide in streptococci. *Molecular Microbiology* **64,** 1423–1433.

Coffey, T. J., Enright, M. C., Daniels, M., Morona, J. K., Morona, R., Hryniewicz, W., Paton, J. C., and Spratt, B. G. (1998) Recombinational exchanges at the capsular polysaccharide biosynthetic locus lead to frequent serotype changes among natural isolates of *Streptococcus pneumoniae*. *Molecular Microbiology* **27,** 73–83.

Coffey, T. J., Pullinger, G. D., Urwin, R., Jolley, K. A., Wilson, S. M., Maiden, M. C., and Leigh, J. A. (2006) First insights into the evolution of *Streptococcus uberis*: A multilocus sequence typing scheme that enables investigation of its population biology. *Applied and Environmental Microbiology* **72,** 1420–1428.

Dagan, R. and Klugman, K. P. (2008) Impact of conjugate pneumococcal vaccines on antibiotic resistance. *Lancet of Infectious Diseases* **8,** 785–795.

Dale, J. B., Penfound, T., Chiang, E. Y., Long, V., Shulman, S. T., and Beall, B. (2005) Multivalent group A streptococcal vaccine elicits bactericidal antibodies against variant M subtypes. *Clinical and Diagnostic Laboratory Immunology* **12,** 833–836.

Davies, M. R., McMillan, D. J., Beiko, R. G., Barroso, V., Geffers, R., Sriprakash, K. S., and Chhatwal, G. S. (2007) Virulence profiling of *Streptococcus dysgalactiae* subspecies *equisimilis* isolated from infected humans reveals two distinct genetic lineages which do not segregate with their phenotypes or propensity to cause diseases. *Clinical Infectious Diseases*, **44,** 1442–1454.

Delorme, C., Poyart, C., Ehrlich, S. D., and Renault, P. (2007) Extent of horizontal gene transfer in evolution of streptococci of the *salivarius* group. *Journal of Bacteriology* **189,** 1330–1341.

Didelot, X., Darling, A., and Falush, D. (2009) Inferring genomic flux in bacteria. *Genome Research* **19,** 306–317.

Didelot, X. and Falush, D. (2007) Inference of bacterial microevolution using multilocus sequence data. *Genetics* **175,** 1251–1266.

Domelier, A. S., van der Mee-Marquet, N., Arnault, L., Mereghetti, L., Lanotte, P., Rosenau, A., Lartigue, M. F., and Quentin, R. (2008) Molecular characterization of erythromycin-resistant *Streptococcus agalactiae* strains. *Journal of Antimicrobial Chemotherapy* **62,** 1227–1233.

Enright, M. C. and Spratt, B. G. (1998) A multilocus sequence typing scheme for *Streptococcus pneumoniae*: Identification of clones associated with invasive disease. *Microbiology* **144,** 3049–3060.

Enright, M. C., Spratt, B. G., Kalia, A., Cross, J. H., and Bessen, D. E. (2001) Multilocus sequence typing of *Streptococcus pyogenes* and the relationship between *emm*-type and clone. *Infection and Immunity* **69,** 2416–2427.

Eriksson, B. K. G., Villasenor-Sierra, A., Norgren, M., and Stevens, D. L. (2001) Opsonization of T1M1 group A *Streptococcus*: Dynamics of antibody production and strain specificity. *Clinical Infectious Diseases* **32,** E24–E30.

Euler, C. W., Ryan, P. A., Martin, J. M., and Fischetti, V. A. (2007) M.SpyI, a DNA methyltransferase encoded on a mefA chimeric element, modifies the genome of *Streptococcus pyogenes*. *Journal of Bacteriology* **189,** 1044–1054.

Facklam, R., (2002) What happened to the streptococci: Overview of taxonomic and nomenclature changes [Review]. *Clinical Microbiology Reviews* **15,** 613–630.

Falugi, F., Zingaretti, C., Pinto, V., Mariani, M., Amodeo, L., Manetti, A. G., Capo, S., Musser, J. M., Orefici, G., Margarit, I., Telford, J. L., Grandi, G., and Mora, M. (2008) Sequence variation in group A *Streptococcus* pili and association of pilus backbone types with lancefield T serotypes. *Journal of Infectious Diseases* **198,** 1834–1841.

Feil, E. J., Holmes, E. C., Bessen, D. E., Chan, M.-S., Day, N. P. J., Enright, M. C., Goldstein, R., Hood, D., Kalia, A., Moore, C. E., Zhou, J., and Spratt, B. G. (2001) Recombination within natural populations of pathogenic bacteria: Short-term empirical estimates and long-term phylogenetic consequences. *Proceedings of the National Academy of Sciences of the United States of America* **98,** 182–187.

Feil, E. J., Li, B. C., Aanensen, D. M., Hanage, W. P., and Spratt, B. G. (2004) eBURST: Inferring patterns of evolutionary descent among clusters of related bacterial genotypes from multilocus sequence typing data. *Journal of Bacteriology* **186,** 1518–1530.

Feil, E. J., Smith, J. M., Enright, M. C., and Spratt, B. G. (2000) Estimating recombinational parameters in *Streptococcus pneumoniae* from multilocus sequence typing data. *Genetics* **154,** 1439–1450.

Feil, E. J. and Spratt, B. G. (2001) Recombination and the population structures of bacterial pathogens. *Annual Review of Microbiology* **55,** 561–590.

Ferretti, J. J., McShan, W. M., Ajdic, D., Savic, D. J., Savic, G., Lyon, K., Primeaux, C., Sezate, S., Suvorov, A. N., Kenton, S., Lai, H. S., Lin, S. P., Qian, Y., Jia, H. G., Najar, F. Z., Ren, Q., Zhu, H., Song, L., White, J., Yuan, X., Clifton, S. W., Roe, B. A., and McLaughlin, R. (2001) Complete genome sequence of an M1 strain of *Streptococcus pyogenes*. *Proceedings of the National Academy of Sciences of the United States of America* **98,** 4658–4663.

Fischetti, V. A. (1989) Streptococcal M protein: Molecular design and biological behavior. *Clinical Microbiology Reviews* **2,** 285–314.

Fraser, C., Hanage, W. P., and Spratt, B. G. (2005) Neutral microepidemic evolution of bacterial pathogens. *Proceedings of the National Academy of Sciences of the United States of America* **102,** 1968–1973.

Fraser, C., Hanage, W. P., and Spratt, B. G. (2007) Recombination and the nature of bacterial speciation. *Science* **315,** 476–480.

Fujita, K., Murono, K., Yoshikawa, M., and Murai, T. (1994) Decline of erythromycin resistance of group A streptococci in Japan. *Pediatric Infectious Disease Journal* **13,** 1075–1078.

Gillen, C. M., Courtney, H. S., Schulze, K., Rohde, M., Wilson, M. R., Timmer, A. M., Guzman, C. A., Nizet, V., Chhatwal, G. S., and Walker, M. J. (2008) Opacity factor activity and epithelial cell binding by the serum opacity factor protein of

Streptococcus pyogenes are functionally discrete. *Journal of Biological Chemistry* **283,** 6359–6366.

Gould, C. V., Sniegowski, P. D., Shchepetov, M., Metlay, J. P., and Weiser, J. N. (2007) Identifying mutator phenotypes among fluoroquinolone-resistant strains of *Streptococcus pneumoniae* using fluctuation analysis. *Antimicrobial Agents and Chemotherapy* **51,** 3225–3229.

Green, N. M., Zhang, S., Porcella, S. F., Nagiec, M. J., Barbian, K. D., Beres, S. B., LeFebvre, R. B., and Musser, J. M. (2005) Genome sequence of a serotype M28 strain of group a streptococcus: Potential new insights into puerperal sepsis and bacterial disease specificity. *Journal of Infectious Diseases* **192,** 760–770.

Grundmann, H., Hori, S., and Tanner, G. (2001) Determining confidence intervals when measuring genetic diversity and the discriminatory abilities of typing methods for microorganisms. *Journal of Clinical Microbiology* **39,** 4190–4192.

Gupta, S., Maiden, M. C. J., Feavers, I. M., Nee, S., May, R. M., and Anderson, R. M. (1996) The maintenance of strain structure in populations of recombining infectious agents. *Nature Medicine* **2,** 437–442.

Hanage, W. P., Fraser C., and Spratt, B. G. (2006) The impact of homologous recombination on the generation of diversity in bacteria. *Journal of Theoretical Biology* **239,** 210–219.

Hava, D. L., Hemsley, C. J., and Camilli, A. (2003) Transcriptional regulation in the *Streptococcus pneumoniae* rlrA pathogenicity islet by RlrA. *Journal of Bacteriology* **185,** 413–421.

Heather, Z., Holden, M. T., Steward, K. F., Parkhill, J., Song, L., Challis, G. L., Robinson, C., Davis-Poynter, N., and Waller, A. S. (2008) A novel streptococcal integrative conjugative element involved in iron acquisition. *Molecular Microbiology* **70,** 1274–1292.

Hery-Arnaud, G., Bruant, G., Lanotte, P., Brun, S., Picard, B., Rosenau, A., van der Mee-Marquet, N., Rainard, P., Quentin, R., and Mereghetti, L. (2007) Mobile genetic elements provide evidence for a bovine origin of clonal complex 17 of *Streptococcus agalactiae*. *Applied and Environmental Microbiology* **73,** 4668–4672.

Hicks, L. A., Harrison, L. H., Flannery, B., Hadler, J. L., Schaffner, W., Craig, A. S., Jackson, D., Thomas, A., Beall, B., Lynfield, R., Reingold, A., Farley, M. M., and Whitney, C. G. (2007) Incidence of pneumococcal disease due to non-pneumococcal conjugate vaccine (PCV7) serotypes in the United States during the era of widespread PCV7 vaccination, 1998-2004. *Journal of Infectious Diseases* **196,** 1346–1354.

Hidalgo-Grass, C., Ravins, M., Dan-Goor, M., Jaffe, J., Moses, A. E., and Hanski, E. (2002) A locus of group A *Streptococcus* involved in invasive disease and DNA transfer. *Molecular Microbiology* **46,** 87–99.

Hiller, N. L., Janto, B., Hogg, J. S., Boissy, R., Yu, S., Powell, E., Keefe, R., Ehrlich, N. E., Shen, K., Hayes, J., Barbadora, K., Klimke, W., Dernovoy, D., Tatusova, T., Parkhill, J., Bentley, S. D., Post, J. C., Ehrlich, G. D., and Hu, F. Z. (2007) Comparative genomic analyses of seventeen *Streptococcus pneumoniae* strains: Insights into the pneumococcal supragenome. *Journal of Bacteriology* **189,** 8186–8195.

Holden, M. T., Heather, Z., Paillot, R., Steward, K. F., Webb, K., Ainslie, F., Jourdan, T., Bason, N. C., Holroyd, N. E., Mungall, K., Quail, M. A., Sanders, M., Simmonds, M., Willey, D., Brooks, K., Aanensen, D. M., Spratt, B. G., Jolley, K. A., Maiden, M. C., Kehoe, M., Chanter, N., Bentley, S. D., Robinson, C., Maskell, D. J., Parkhill, J., and Waller, A. S. (2009) Genomic evidence for the evolution of *Streptococcus equi*: Host restriction, increased virulence, and genetic exchange with human pathogens. *PLoS Pathogens* **5,** e1000346.

Hollingshead, S. K., Readdy, T., Arnold, J., and Bessen, D. E. (1994) Molecular evolution of a multi-gene family in group A streptococci. *Molecular Biology and Evolution* **11,** 208–219.

Hollingshead, S. K., Readdy, T. L., Yung, D. L., and Bessen, D. E. (1993) Structural heterogeneity of the *emm* gene cluster in group A streptococci. *Molecular Microbiology* **8,** 707–717.

Hondorp, E. R. and McIver, K. S. (2007) The Mga virulence regulon: Infection where the grass is greener. *Molecular Microbiology* **66,** 1056–1065.

Horn, D., Zabriskie, J., Austrian, R., Cleary, P., Ferretti, J., Fischetti, V., Gotschlich, E., Kaplan, E., McCarty, M., Opal, S., Roberts, R., Tomasz, A., and Wachtfogel, Y. (1998) Why have group A streptococci remained susceptible to penicillin? Report on a symposium. *Clinical Infectious Diseases* **26,** 1341–1345.

Huang, S. S., Platt, R., Rifas-Shiman, S. L., Pelton, S. I., Goldmann, D., and Finkelsteins J. A. (2005) Post-PCV7 changes in colonizing pneumococcal serotypes in 16 Massachusetts communities, 2001 and 2004. *Pediatrics* **116,** e408–e413.

Jaggi, P., Tanz, R. R., Beall, B., and Shulman, S. T. (2005) Age influences the emm type distribution of pediatric group A streptococcal pharyngeal isolates. *Pediatric Infectious Disease Journal* **24,** 1089–1092.

Johnson, D. R., Kaplan, E. L., VanGheem, A., Facklam, R. R., and Beall, B. (2006) Characterization of group A streptococci (*Streptococcus pyogenes*): Correlation of M-protein and emm-gene type with T-protein agglutination pattern and serum opacity factor. *Journal of Medical Microbiology* **55,** 157–164.

Johri, A. K., Paoletti, L. C., Glaser, P., Dua, M., Sharma, P. K., Grandi, G., and Rappuoli, R. (2006) Group B *Streptococcus*: Global incidence and vaccine development. *Nature Reviews. Microbiology* **4,** 932–942.

Jones, K. F., Hollingshead, S. K., Scott, J. R., and Fischetti, V. A. (1988) Spontaneous M6 protein size mutants of group A streptococci display variation in antigenic and opsonogenic epitopes. *Proceedings of the National Academy of Sciences of the United States of America* **85,** 8271–8275.

Jones, N., Bohnsack, J. F., Takahashi, S., Oliver, K. A., Chan, M. S., Kunst, F., Glaser, P., Rusniok, C.,

Crook, D. W., Harding, R. M., Bisharat, N., and Spratt, B. G. (2003) Multilocus sequence typing system for group B *Streptococcus*. *Journal of Clinical Microbiology* **41,** 2530–2536.

Jones, N., Oliver, K. A., Barry, J., Harding, R. M., Bisharat, N., Spratt, B. G., Peto, T., and Crook, D. W. (2006) Enhanced invasiveness of bovine-derived neonatal sequence type 17 group B *Streptococcus* is independent of capsular serotype. *Clinical Infectious Diseases* **42,** 915–924.

Kalia, A. and Bessen, D. E. (2004) Natural selection and evolution of streptococcal virulence genes involved in tissue-specific adaptations. *Journal of Bacteriology* **186,** 110–121.

Kalia, A., Spratt, B. G., Enright, M. C., and Bessen, D. E. (2002) Influence of recombination and niche separation on the population genetic structure of the pathogen *Streptococcus pyogenes. Infection and Immunity* **70,** 1971–1983.

Kelly, T., Dillard, J. P., and Yother, J. (1994) Effect of genetic switching of capsular type on virulence of *Streptococcus pneumoniae. Infection and Immunity* **62,** 1813–1819.

Kilian, M., Poulsen, K., Blomqvist, T., Havarstein, L. S., Bek-Thomsen, M., Tettelin, H., and Sorensen, U. B. (2008) Evolution of *Streptococcus pneumoniae* and its close commensal relatives. *PLoS One* **3,** e2683.

King, S. J., Leigh, J. A., Heath, P. J., Luque, I., Tarradas, C., Dowson, C. G., and Whatmore, A. M. (2002) Development of a multilocus sequence typing scheme for the pig pathogen *Streptococcus suis*: Identification of virulent clones and potential capsular serotype exchange. *Journal of Clinical Microbiology* **40,** 3671–3680.

Kratovac, Z., Manoharan, A., Luo, F., Lizano, S., and Bessen, D. E. (2007) Population genetics and linkage analysis of loci within the FCT region of *Streptococcus pyogenes. Journal of Bacteriology* **189,** 1299–1310.

Kreikemeyer, B., McIver, K. S., and Podbielski, A. (2003) Virulence factor regulation and regulatory networks in *Streptococcus pyogenes* and their impact on pathogen-host interactions. *Trends in Microbiology* **11,** 224–232.

Kreikemeyer, B., Nakata, M., Oehmcke, S., Gschwendtner, C., Normann J., and Podbielski, A. (2005) *Streptococcus pyogenes* collagen type I-binding Cpa surface protein—Expression profile, binding characteristics, biological functions, and potential clinical impact. *Journal of Biological Chemistry* **280,** 33228–33239.

Lancefield, R. C. (1962) Current knowledge of the type specific M antigens of group A streptococci. *Journal of Immunology* **89,** 307–313.

Lefebure, T. and Stanhope, M. J. (2007) Evolution of the core and pan-genome of *Streptococcus*: Positive selection, recombination, and genome composition. *Genome Biology* **8,** R71.

Lipsitch, M. (1997) Vaccination against colonizing bacteria with multiple serotypes. *Proceedings of the National Academy of Sciences of the United States of America* **94,** 6571–6576.

Lipsitch, M., O'Neill, K., Cordy, D., Bugalter, B., Trzcinski, K., Thompson, C. M., Goldstein, R., Pelton, S., Huot, H., Bouchet, V., Reid, R., Santosham, M., and O'Brien, K. L. (2007) Strain characteristics of *Streptococcus pneumoniae* carriage and invasive disease isolates during a cluster-randomized clinical trial of the 7-valent pneumococcal conjugate vaccine. *Journal of Infectious Diseases* **196,** 1221–1227.

Luan, S. L., Granlund, M., Sellin, M., Lagergard, T., Spratt, B. G., and Norgren, M. (2005) Multilocus sequence typing of Swedish invasive group B *Streptococcus* isolates indicates a neonatally associated genetic lineage and capsule switching. *Journal of Clinical Microbiology* **43,** 3727–3733.

Manetti, A. G., Zingaretti, C., Falugi, F., Capo, S., Bombaci, M., Bagnoli, F., Gambellini, G., Bensi, G., Mora, M., Edwards, A. M., Musser, J. M., Graviss, E. A., Telford, J. L., Grandi, G., and Margarit, I. (2007) *Streptococcus pyogenes* pili promote pharyngeal cell adhesion and biofilm formation. *Molecular Microbiology* **64,** 968–983.

Margarit, I., Rinaudo, C. D., Galeotti, C. L., Maione, D., Ghezzo, C., Buttazzoni, E., Rosini, R., Runci, Y., Mora, M., Buccato, S., Pagani, M., Tresoldi, E., Berardi, A., Creti, R., Baker, C. J., Telford, J. L., and Grandi, G. (2009) Preventing bacterial infections with pilus-based vaccines: The group B *Streptococcus* paradigm. *Journal of Infectious Diseases* **199,** 108–115.

Marri, P. R., Hao, W., and Golding, G. B. (2006) Gene gain and gene loss in streptococcus: Is it driven by habitat? *Molecular Biology and Evolution* **23,** 2379–2391.

Martin, J. M., Green, M., Barbadora, K. A., and Wald, E. R. (2002) Erythromycin-resistant group A streptococci in schoolchildren in Pittsburgh. *New England Journal of Medicine* **346,** 1200–1206.

Martins, E. R., Pessanha, M. A., Ramirez, M., and Melo-Cristino, J. (2007) Analysis of group B streptococcal isolates from infants and pregnant women in Portugal revealing two lineages with enhanced invasiveness. *Journal of Clinical Microbiology* **45,** 3224–3229.

Mavroidi, A., Aanensen, D. M., Godoy, D., Skovsted, I. C., Kaltoft, M. S., Reeves, P. R., Bentley, S. D., and Spratt, B. G. (2007) Genetic relatedness of the *Streptococcus pneumoniae* capsular biosynthetic loci. *Journal of Bacteriology* **189,** 7841–7855.

McDonald, M. I., Towers, R. J., Andrews, R., Benger, N., Fagan, P., Currie, B. J., and Carapetis, J. R. (2007a) The dynamic nature of group A streptococcal epidemiology in tropical communities with high rates of rheumatic heart disease. *Epidemiology and Infection* **00,** 1–11.

McDonald, M. I., Towers, R. J., Fagan, P., Carapetis, J. R., and Currie, B. J. (2007b) Molecular typing of

Streptococcus pyogenes from remote aboriginal communities where rheumatic fever is common and pyoderma is the predominant streptococcal infection. *Epidemiology and Infection* **135,** 1398–1405.

McGregor, K., Bilek, N., Bennett, A., Kalia, A., Beall, B., Carapetis, J., Currie, B., Sriprakash, K., Spratt, B., and Bessen, D. (2004a) Group A streptococci from a remote community have novel multilocus genotypes but share emm-types and housekeeping alleles. *Journal of Infectious Diseases* **189,** 717–723.

McGregor, K. F., Spratt, B. G., Kalia, A., Bennett, A., Bilek, N., Beall, B., and Bessen, D. E. (2004b) Multilocus sequence typing of *Streptococcus pyogenes* representing most known *emm*-types and distinctions among sub-population genetic structures. *Journal of Bacteriology* **186,** 4285–4294.

McNeil, S. A., Halperin, S. A., Langley, J. M., Smith, B., Warren, A., Sharratt, G. P., Baxendale, D. M., Reddish, M. A., Hu, M. C., Stroop, S. D., Linden, J., Fries, L. F., Vink, P. E., and Dale, J. B. (2005) Safety and immunogenicity of 26-valent group A *Streptococcus* vaccine in healthy adult volunteers. *Clinical Infectious Diseases* **41,** 1114–1122.

Moore, M. R., Gertz, R. E. Jr., Woodbury, R. L., Barkocy-Gallagher, G. A., Schaffner, W., Lexau, C., Gershman, K., Reingold, A., Farley, M., Harrison, L. H., Hadler, J. L., Bennett, N. M., Thomas, A. R., McGee, L., Pilishvili, T., Brueggemann, A. B., Whitney, C. G., Jorgensen, J. H., and Beall, B. (2008) Population snapshot of emergent *Streptococcus pneumoniae* serotype 19A in the United States, 2005. *Journal of Infectious Diseases* **197,** 1016–1027.

Mora, M., Bensi, G., Capo, S., Falugi, F., Zingaretti, C., Manetti, A. G. O., Maggi, T., Taddei, A. R., Grandi, G., and Telford, J. L. (2005) Group A *Streptococcus* produce pilus-like structures containing protective antigens and Lancefield T antigens. *Proceedings of the National Academy of Sciences of the United States of America* **102,** 15641–15646.

Nakano, K., Lapirattanakul, J., Nomura, R., Nemoto, H., Alaluusua, S., Gronroos, L., Vaara, M., Hamada, S., Ooshima, T., and Nakagawa, I. (2007) *Streptococcus mutans* clonal variation revealed by multilocus sequence typing. *Journal of Clinical Microbiology* **45,** 2616–2625.

Obert, C., Sublett, J., Kaushal, D., Hinojosa, E., Barton, T., Tuomanen, E. I., and Orihuela, C. J. (2006) Identification of a candidate *Streptococcus pneumoniae* core genome and regions of diversity correlated with invasive pneumococcal disease. *Infection and Immunity* **74,** 4766–4777.

O'Brien, K. L., Millar, E. V., Zell, E. R., Bronsdon, M., Weatherholtz, R., Reid, R., Becenti, J., Kvamme, S., Whitney, C. G., and Santosham, M. (2007) Effect of pneumococcal conjugate vaccine on nasopharyngeal colonization among immunized and unimmunized children in a community-randomized trial. *Journal of Infectious Diseases* **196,** 1211–1220.

Pericone, C. D., Bae, D., Shchepetov, M., McCool, T., and Weiser, J. N. (2002) Short-sequence tandem and nontandem DNA repeats and endogenous hydrogen peroxide production contribute to genetic instability of *Streptococcus pneumoniae*. *Journal of Bacteriology* **184,** 4392–4399.

Persson, J., Beall, B., Linse, S., and Lindahl, G. (2006) Extreme sequence divergence but conserved ligand-binding specificity in *Streptococcus pyogenes* M protein. *PLoS Pathogens* **2,** e47.

Phares, C. R., Lynfield, R., Farley, M. M., Mohle-Boetani, J., Harrison, L. H., Petit, S., Craig, A. S., Schaffner, W., Zansky, S. M., Gershman, K., Stefonek, K. R., Albanese, B. A., Zell, E. R., Schuchat, A., and Schrag, S. J. (2008) Epidemiology of invasive group B streptococcal disease in the United States, 1999-2005. *JAMA* **299,** 2056–2065.

Robinson, D., Sutcliffe, J., Tewodros, W., Manoharan, A., and Bessen, D. (2006) Evolution and global dissemination of macrolide resistant group A streptococci. *Antimicrobial Agents and Chemotherapy* **50,** 2903–2911.

Sakota, V., Fry, A. M., Lietman, T. M., Facklam, R. R., Li, Z. Y., and Beall, B. (2006) Genetically diverse group A streptococci from children in Far-Western Nepal share high genetic relatedness with isolates from other countries. *Journal of Clinical Microbiology* **44,** 2160–2166.

Schmidt, H. and Hensel, M. (2004) Pathogenicity islands in bacterial pathogenesis. *Clinical Microbiology Reviews* **17,** 14–56.

Schneewind, O., Jones, K. F., and Fischetti, V. A. (1990) Sequence and structural characterization of the trypsin-resistant T6 surface protein of group A streptococci. *Journal of Bacteriology* **172,** 3310–3317.

Scott, J., Thompson-Mayberry, P., Lahmamsi, S., King, C. J., and McShan, W. M. (2008) Phage-associated mutator phenotype in group A streptococcus. *Journal of Bacteriology* **190,** 6290–6301.

Seppala, H., Klaukka, T., Vuopio-Varkila, J., Muotiala, A., Helenius, H., Lager, K., and Huovinen, P. (1997) The effect of changes in the consumption of macrolide antibiotics on erythromycin resistance in group A streptococci in Finland. Finnish Study Group for Antimicrobial Resistance. *New England Journal of Medicine* **337,** 441–446.

Shulman, S. (2004) Group A streptococcal pharyngitis serotype surveillance in North America, 2000-2002. *Clinical Infectious Diseases* **39,** 325–332.

Shulman, S. T., Stollerman, G., Beall, B., Dale, J. B., and Tanz, R. R. (2006) Temporal changes in streptococcal M protein types and the near-disappearance of acute rheumatic fever in the United States. *Clinical Infectious Diseases* **42,** 441–447.

Sitkiewicz, I., Nagiec, M. J., Sumby, P., Butler, S. D., Cywes-Bentley, C., and Musser, J. M. (2006) Emergence of a bacterial clone with enhanced virulence by acquisition of a phage encoding a secreted phospholipase A2. *Proceedings of the National Academy of Sciences of the United States of America* **103,** 16009–16014.

SLOTVED, H. C., KONG, F., LAMBERTSEN, L., SAUER, S., and GILBERT, G. L. (2007) Serotype IX, a proposed new *Streptococcus agalactiae* serotype. *Journal of Clinical Microbiology* **45,** 2929–2936.

STALHAMMAR-CARLEMALM, M., ARESCHOUG, T., LARSSON C., and LINDAHL, G. (1999) The R28 protein of *Streptococcus pyogenes* is related to several group B streptococcal surface proteins, confers protective immunity and promotes binding to human epithelial cells. *Molecular Microbiology* **33,** 208–219.

STOLLERMAN, G. H. and DALE, J. B. (2008) The importance of the group a streptococcus capsule in the pathogenesis of human infections: A historical perspective. *Clinical Infectious Diseases* **46,** 1038–1045.

TELFORD, J., BAROCCHI, M., MARGARIT, I., RAPPUOLI, R., and GRANDI, G., (2006) Pili in gram-positive pathogens. *Nature Reviews. Microbiology* **4,** 509–519.

TETTELIN, H., MASIGNANI V., CIESLEWICZ, M. J., DONATI, C., MEDINI, D., WARD, N. L., ANGIUOLI, S. V., CRABTREE, J., JONES, A. L., DURKIN, A. S., DEBOY, R. T., DAVIDSEN, T. M., MORA, M., SCARSELLI, M., MARGARIT Y ROS, I., PETERSON, J. D., HAUSER, C. R., SUNDARAM, J. P., NELSON, W. C., MADUPU, R., BRINKAC, L. M., DODSON, R. J., ROSOVITZ, M. J., SULLIVAN, S. A., DAUGHERTY, S. C., HAFT, D. H., SELENGUT, J., GWINN, M. L., ZHOU, L., ZAFAR, N., KHOURI, H., RADUNE, D., DIMITROV, G., WATKINS, K., O'CONNOR, K. J., SMITH, S., UTTERBACK, T. R., WHITE, O., RUBENS, C. E., GRANDI, G., MADOFF, L. C., KASPER, D. L., TELFORD, J. L., WESSELS, M. R., RAPPUOLI, R., and FRASER, C. M. (2005) Genome analysis of multiple pathogenic isolates of *Streptococcus agalactiae*: Implications for the microbial "pan-genome." *Proceedings of the National Academy of Sciences of the United States of America* **102,** 13950–13955.

TETTELIN, H., MASIGNANI, V., CIESLEWICZ, M. J., EISEN, J. A., PETERSON, S., WESSELS, M. R., PAULSEN, I. T., NELSON, K. E., MARGARIT, I., READ, T. D., MADOFF, L. C., WOLF, A. M., BEANAN, M. J., BRINKAC, L. M., DAUGHERTY, S. C., DEBOY, R. T., DURKIN, A. S., KOLONAY, J. F., MADUPU, R., LEWIS, M. R., RADUNE, D., FEDOROVA, N. B., SCANLAN, D., KHOURI, H., MULLIGAN, S., CARTY, H. A., CLINE, R. T., VAN AKEN, S. E., GILL, J., SCARSELLI, M., MORA, M., IACOBINI, E. T., BRETTONI, C., GALLI, G., MARIANI, M., VEGNI, F., MAIONE, D., RINAUDO, D., RAPPUOLI, R., TELFORD, J. L., KASPER, D. L., GRANDI, G., and FRASER, C. M. (2002) Complete genome sequence and comparative genomic analysis of an emerging human pathogen, serotype V Streptococcus agalactiae. *Proceedings of the National Academy of Sciences of the United States of America* **99,** 12391–12396.

TETTELIN, H., NELSON, K. E., PAULSEN, I. T., EISEN, J. A., READ, T. D., PETERSON, S., HEIDELBERG, J., DEBOY, R. T., HAFT, D. H., DODSON, R. J., DURKIN, A. S., GWINN, M., KOLONAY, J. F., NELSON, W. C., PETERSON, J. D., UMAYAM, L. A., WHITE, O., SALZBERG, S. L., LEWIS, M. R., RADUNE, D., HOLTZAPPLE, E., KHOURI, H., WOLF, A. M., UTTERBACK, T. R., HANSEN, C. L., MCDONALD, L. A., FELDBLYUM, T. V., ANGIUOLI, S., DICKINSON, T., HICKEY, E. K., HOLT, I. E., LOFTUS, B. J., YANG, F., SMITH, H. O., VENTER, J. C., DOUGHERTY, B. A., MORRISON, D. A., HOLLINGSHEAD, S. K., and FRASER, C. M. (2001) Complete genome sequence of a virulent isolate of *Streptococcus pneumoniae. Science* **293,** 498–506.

TOMITA, T., MEEHAN, B., WONGKATTIYA, N., MALMO, J., PULLINGER, G., LEIGH, J., and DEIGHTON, M. (2008) Identification of *Streptococcus uberis* multilocus sequence types highly associated with mastitis. *Applied and Environmental Microbiology* **74,** 114–124.

TOWERS, R. J., GAL, D., MCMILLAN, D., SRIPRAKASH, K. S., CURRIE, B. J., WALKER, M. J., CHHATWAL, G. S., and FAGAN, P. K. (2004) Fibronectin-binding protein gene recombination and horizontal transfer between group A and G streptococci. *Journal of Clinical Microbiology* **42,** 5357–5361.

TURNER, K. M., HANAGE, W. P., FRASER, C., CONNOR, T. R., and SPRATT, B. G. (2007) Assessing the reliability of eBURST using simulated populations with known ancestry. *BMC Microbiology* **7,** 30.

VARALDO, P. E., MONTANARI, M. P., and GIOVANETTI, E. (2009) Genetic elements responsible for erythromycin resistance in streptococci. *Antimicrobial Agents and Chemotherapy* **53,** 343–353.

VOS, M. and DIDELOT, X. (2009) A comparison of homologous recombination rates in bacteria and archaea. *ISME Journal* **3,** 199–208.

WANNAMAKER, L. W. (1970) Differences between streptococcal infections of the throat and of the skin. *New England Journal of Medicine* **282,** 23–31.

WARD, P. N., HOLDEN, M. T., LEIGH, J. A., LENNARD, N., BIGNELL, A., BARRON, A., CLARK, L., QUAIL, M. A., WOODWARD, J., BARRELL, B. G., EGAN, S. A., FIELD, T. R., MASKELL, D., KEHOE, M., DOWSON, C. G., CHANTER, N., WHATMORE, A. M., BENTLEY, S. D., and PARKHILL, J. (2009) Evidence for niche adaptation in the genome of the bovine pathogen *Streptococcus uberis*. *BMC Genomics* **10,** 54.

WEBB, K., JOLLEY, K. A., MITCHELL, Z., ROBINSON, C., NEWTON, J. R., MAIDEN, M. C., and WALLER, A. (2008) Development of an unambiguous and discriminatory multilocus sequence typing scheme for the *Streptococcus zooepidemicus* group. *Microbiology* **154,** 3016–3024.

ZADOKS, R. N., SCHUKKEN, Y. H., and WIEDMANN, M. (2005) Multilocus sequence typing of *Streptococcus uberis* provides sensitive and epidemiologically relevant subtype information and reveals positive selection in the virulence gene pauA. *Journal of Clinical Microbiology* **43,** 2407–2417.

Chapter 18

Population Genetics of Vibrios

NAIEL BISHARAT

18.1 INTRODUCTION

Vibrios are naturally occurring, gram-negative, inhabitants of marine, estuarine, and coastal waters worldwide. There are more than 60 species of vibrios, showing a very wide niche specialization, from colonization of fish and marine invertebrates, or attachment to plankton and algae (Farmer and Hickman-Brenner, 1992; Rosenberg and Ben-Haim, 2002; Thompson et al., 2004). Of all vibrios, *Vibrio cholerae*, *Vibrio parahaemolyticus*, and *Vibrio vulnificus* are responsible for the vast majority of human infections. *V. cholerae*, which is the causative agent of cholera, a disease that has been feared for centuries due to its high mortality rates and social disruption, is the most extensively studied and is considered one of the major problems for public health in the developing countries of Asia and Africa. Infections by vibrios are generally acquired either through ingestion of foods and water contaminated with human feces or sewage, raw fish, and seafood, or they are associated with exposure of skin lesions, such as cuts, open wounds, and abrasions, to aquatic environments and marine animals (Sack et al., 2004; Yeung and Boor, 2004; Oliver, 2006).

The past three decades have witnessed a growing amount of studies describing the population genetics of bacteria affecting humans and animals. The contribution of these studies to the understanding of the evolution and pathogenesis of bacterial pathogens has proved to be extremely important. Molecular genetic studies of vibrios provided fascinating observations. To date, the genome sequence of six vibrios, *Vibrio cholerae* N16961, *Vibrio parahaemolyticus* RIMD 2210633, *Vibrio vulnificus* YJ016 and CMCP6, *Vibrio fischeri* ES114, and *Vibrio harveyi* ATCC BAA-1116, has been completed, and several others are under way (Heidelberg et al., 2000; Chen et al., 2003; Makino et al., 2003; Ruby et al., 2005). These studies provided important insights for understanding the ecology, environmental adaptation, evolution, and pathogenicity of vibrios. In this chapter, I will review recent data on the population genetics of the three most important human-pathogenic vibrios: *V. cholerae*, *V. parahaemolyticus*, and *V. vulnificus*.

Bacterial Population Genetics in Infectious Disease, Edited by D. Ashley Robinson, Daniel Falush, and Edward J. Feil

18.1.1 Genetic Structure and Evolution

The completion of the whole genome sequencing of the three main vibrios has shed light on the unique genetic structure of vibrios and has provided interesting hints to its evolution (Heidelberg et al., 2000; Chen et al., 2003; Makino et al., 2003). Perhaps one of the most interesting genetic features of vibrios is the fact that they possess two circular chromosomes, one large and one small. To date, no vibrios with single chromosomes were found (Okada et al., 2005). It is unclear why vibrios have two chromosomes and what advantages it provides. Some have speculated that the split into two chromosomes is advantageous for DNA replication (Yamaichi et al., 1999). Some have suggested that the small chromosome might have a role in evolutionary selective pressure against integration (Heidelberg et al., 2000), while others proposed that the possession of two chromosomes might be advantageous for environmental adaptation to different lifestyles (Colwell, 1996; Schoolnik and Yildiz, 2000). Genome sequence comparison of the three main vibrios showed that the gene content and position in the large vibrio chromosomes are better conserved. The small vibrio chromosomes are more divergent in size and gene content between *V. vulnificus* and *V. cholerae*. A large fraction of genes required for growth are located on the large chromosome, while the small chromosome contains more genes required for environmental adaptation, virulence, and cell adherence (Heidelberg et al., 2000; Chen et al., 2003; Makino et al., 2003).

Different scenarios have been proposed for the origin of the small chromosome. Some have suggested that the small chromosome has arisen by excision from a large ancestral genome (Waldor and RayChaudhuri, 2000), while others have suggested that the small chromosome is a megaplasmid that has been acquired by an ancestral vibrio (Heidelberg et al., 2000). A recent study supported the later hypothesis; it showed that the origin of replication of the large chromosome of *V. cholerae* ($oriCI_{vc}$) shared features with the origin of replication of *Escherichia coli*, while the origin of replication of the small chromosome ($oriCII_{vc}$) exhibited features that are unusual for a bacterial chromosome. This suggested that the small chromosome was originally a plasmid that increased in size by acquiring more genes by horizontal gene transfer (Egan and Waldor, 2003). Evolutionary scenarios for the genome evolution of vibrios based on rates of gene loss, gene genesis (introduction of new gene), and expansion of an existing gene suggested that *V. parahaemolyticus* and *V. vulnificus* increased their gene content with acquisition and gene genesis, while the *V. cholerae* genome acquired less genes and experienced more gene decay or deletion (Gevers and Van de Peer, 2006).

18.1.2 Horizontal Gene Transfer

With the recent insights obtained from complete genome sequencing and multilocus sequence analysis of the human-pathogenic vibrios, lateral or horizontal gene transfer is more readily recognized as a central force in the microevolution of vibrios. Conjugation and transduction are considered the major modes of DNA transfer among vibrios (Boucher and Stokes, 2006). Genetic elements involved in horizontal gene transfer in vibrios include genomic islands (GIs), plasmids and conjugative elements, phages, integrons, and gene cassette arrays.

GIs are large chromosomal regions acquired by horizontal gene transfer and are generally characterized by their unique structure (e.g., G + C content, presence of mobility genes such as integrases and transposases, and flanking direct repeats) differing them from the host genome (Dobrindt et al., 2004). Most GIs that have been identified so far in vibrios contain clusters of genes related to virulence and are often called pathogenicity islands, potentially providing a previously nonpathogenic organism with pathogenic potential or enhancing the environmental fitness by providing the host genome with diverse metabolic capabilities. The first GI to be discovered among vibrios was termed vibrio pathogenicity island (VPI) (Karaolis et al., 1998); it contained a gene cluster that encoded the cholera toxin (CT) phage receptor and an essential colonization factor. The finding that this pathogenicity island was discovered among nonepidemic as well as epidemic *V. cholerae* strains suggested that these GIs are transferable within *V. cholerae* populations (Karaolis et al., 1998). Afterwards, two other GIs were identified only among seventh-pandemic strains; these were termed vibrio seventh pandemic island-I (VSP-I) and vibrio seventh pandemic island-II (VSP-II) (Dziejman et al., 2002). Another VPI termed VPI-2 was identified among toxigenic *V. cholerae* (Jermyn and Boyd, 2002). GIs were also identified among other human-pathogenic vibrios. An analysis of the complete genome sequence of *V. parahaemolyticus* RIMD 2210633 strain revealed an 80-kb region thought to be a GI (Makino et al., 2003), and a comparative analysis of complete genome sequences of two *V. vulnificus* strains (YJ016 and CMCP6) identified 14 regions that had the characteristics of GIs (Quirke et al., 2006).

Plasmids and integrative and conjugative elements (ICEs) are another important mode for horizontal gene transfer in vibrios (Boucher and Stokes, 2006). To date, more than a dozen of vibrio plasmids have been sequenced (Hazen et al., 2007). Plasmids are extremely common and diverse among vibrios, existing in different sizes and numbers even in a single strain (Boucher and Stokes, 2006). ICEs excise from the chromosomes of their hosts, transfer to a new host through conjugation, and then integrate into the chromosome again. The SXT element is a *V. cholerae*-derived ICE that was originally isolated in 1993 from a *V. cholerae* O139 clinical isolate (SXT^{MO10}) (Waldor et al., 1996). The ~100-kbp SXT element confers resistance to sulfamethoxazole, trimethoprim, chloramphenicol, and streptomycin. Since then, *V. cholerae* isolates from the Indian subcontinent and South Africa have also contained the SXT element (Dalsgaard et al., 2001; Amita et al., 2003). Given the genetic diversity, abundance, and size of these mobile genetic elements (plasmids and ICEs), a considerable amount of genetic material has the potential to be horizontally transferred within vibrio populations, thus giving rise to the emergence of new clones with unique metabolic and pathogenic profiles.

Vibriophages (phages specific to vibrios) are abundant in marine environments (Wommack and Colwell, 2000) and are considered one of the main routes for introducing genetic material into vibrios from other coexistent species (Boucher and Stokes, 2006). Perhaps the most explicit example of the impact of phages on the evolution and pathogenicity of vibrios is the CTXΦ phage of *V. cholerae*. This phage, which bears the genes *ctxA* and *ctxB* coding for CT (Waldor and Mekalanos, 1996), provided *V. cholerae* with the most critical virulence factor. Other vibrio phages associated with gene transfer among *V. cholerae* include CP-T1, RS1, and KSF-1Φ (Faruque and Mekalanos, 2003). The *V. parahaemolyticus* f237 phage is widespread among *V. parahaemolyticus* populations being identified among more than half of the clinical strains in one study (Iida et al., 2001). Studies suggested that the O3:K6 clone of *V. parahaemolyticus*, the pandemic clonal complex, might have emerged as the result of a horizontally acquired filamentous phage f237 (Nasu et al., 2000).

Integrons are genetic elements specialized in the acquisition, rearrangement, and expression of genes, notably those encoding antibiotic resistance (Hall, 2002). Integrons have an *intI* gene encoding an IntI site-specific recombinase responsible for capturing small mobile elements known as gene cassettes, an *attI* site into which cassettes are inserted, and a promoter that drives expression of the cassette-associated genes. Integrons can be found on plasmids, transposons, and chromosomes. Studies have shown that members of the Vibrionaceae family harbor mobile and chromosomal integrons (Rowe-Magnus et al., 2006). Superintegrons (SIs) are large integron islands containing a large array of gene cassettes. The *V. cholerae* SI located on chromosome 2 contains all copies of the *V. cholerae* repeat (VCR) and genes encoding products involved in drug resistance (cholarmphenicol, acetyltransferase, and fosfomycin) (Heidelberg et al., 2000). Taken together, vibrios utilize a complex system for the acquisition of exogenous genes for environmental adaptation, pathogenicity, virulence, and antibiotic resistance. GIs and phages play an important role in the spread of virulence genes and pathogenicity islands, while integrons seem to play a crucial role in providing the bacterial host with an adaptive advantage.

18.2 *V. CHOLERAE*

V. cholerae is the etiologic agent of cholera, a disease that remains endemic in the developing countries of Asia and Africa. There are more than 200 serotypes of *V. cholerae*, two of which, O1 and O139, are the only serotypes that have been identified as causing epidemics in humans; all other "non-O1/non-O139" vibrios are associated with sporadic cases of disease. Historically, *V. cholerae* serotype O1 strains, which exist in two biotypes, classical and El Tor, were responsible for all major epidemics including seven pandemics (Kaper et al., 1995). The first six cholera pandemics occurred between 1817 and 1923 and were caused by the classical biotype. The current (seventh) pandemic was caused by the El Tor biotype, which has now involved almost the entire world. Gradually, the El Tor biotype has globally replaced the classical biotype. In 1992, an epidemic clone of *V. cholerae* serotype O139 (Bengal) emerged in the Indian subcontinent (Albert et al., 1993; Ramamurthy et al., 1993; Swerdlow and Ries, 1993). *V. cholerae* serotype O1 continued to cause epidemics, reemerging in South and Central America (Ries et al., 1992; Tauxe and Blake, 1992). The two serogroups, O1 and O139, continue to cause large outbreaks of cholera in India, Bangladesh, and Central Africa. Cholera epidemics have largely occurred in coastal waters, and it has been speculated that the recent emergence of cholera in South America was caused by sewage or ballast water discharged by ships arriving from Asia (Tauxe et al, 1994).

Genetic structure analysis revealed the presence of two important genetic elements that distinguish a pathogenic *V. cholerae* from a nonpathogenic one. These are the CTX genetic element, which is the genome of a lysogenic bacteriophage designated CTXΦ that carries the genes encoding the CT, and the VPI, which carries genes for the toxin-coregulated pilus (TCP). The last 20 years of *V. cholerae* research has witnessed a great deal of knowledge and insights into the biology and genetic evolution of this unique species. *V. cholerae* is one of the best examples that nature has provided to illustrate how horizontal gene transfer has immensely impacted its evolution and has consequently affected the lives of millions around the world. This is perhaps best exemplified by the three following observations: First, the transmissible CTX element, which includes the structural genes (*ctxA* and *ctxB*) for the subunits of the CT, is the integrated genome of a filamentous bacteriophage CTXΦ (Waldor and Mekalanos, 1996; Rubin et al., 1998). Second, the bacterial receptor for CTXΦ, TCP, is encoded by an operon (*tcp*) that is a part of a transmissible pathogenicity island (Kovach et al., 1996; Karaolis et al., 1998). And lastly, the origin of the O139 clone involved a complex rearrangement of the *rfb* region in a strain of O1 El Tor, which included the deletion of genes responsible for the biosynthesis and assembly of the side chains of the O1 cell surface lipopolysaccharide (LPS) and the insertion of exogenous DNA mediating synthesis of the O139 LPS core and a capsule (Bik et al., 1995; Stroeher et al., 1995).

18.2.1 Genetic Diversity and Population Structure

Numerous studies and various molecular genetic tools have been applied over the past 25 years to study the genetic population structure of *V. cholerae*. The vast majority of these studies addressed the genetic diversity of *V. cholerae* populations isolated from infected humans during outbreaks and investigated the phylogenetic relationships of pandemic *V. cholerae* (O1 and O139), with nonpandemic clinical strains and environmental populations.

Toxigenic V. cholerae *(Serogroups O1 and O139)*

The earliest observations of the population structure of *V. cholerae* indicated that *V. cholerae* populations are genetically diverse with multiple and independent pathogenic clones emerging in different continents around the world (Kaper et al., 1982; Goldberg and Murphy, 1983; Koblavi et al., 1990; Chen et al., 1991; Popovic et al., 1993). Since then, other studies using modern genotyping tools confirmed these initial observations (see below). Wachsmuth et al. (1993) used restriction fragment length polymorphism (RFLP) of rRNA and *ctxA* genes, DNA sequence of CT B subunit gene *ctxB*, and multilocus enzyme electrophoresis (MLEE) to characterize the genetic diversity and clonal relationships of 197 globally diverse isolates. They concluded that there are at least four globally distinct toxigenic El Tor *V. cholerae* O1 clones: the seventh pandemic (Eastern Hemisphere), U.S. Gulf Coast, Australian, and the Latin American. The Latin American clone probably representing an extension of the seventh pandemic into the Western Hemisphere, while the U.S. Gulf Coast clone most likely evolved separately. These observations were confirmed by another study that analyzed nucleotide sequence diversity in a housekeeping gene, *asd*, from 45 *V. cholerae* clinical (O1), environmental nontoxigenic O1, and non-O1 isolates; it also suggested that the three pathogenic clones, the sixth pandemic clone-O1 classical biotype, the seventh-pandemic clone-O1 El Tor biotype, and the U.S. Gulf coast clone are three clones that have evolved independently from different lineages of environmental, non-O1 *V. cholerae* isolates (Karaolis et al., 1995). Cameron et al. (1994) used pulsed-field gel electrophoresis (PFGE) on 180 isolates of *V. cholerae* O1 representing six different MLEE types and 27 rRNA RFLP types (ribotypes). They found that all *V. cholerae* O1 isolates tested from the Latin American epidemic were indistinguishable by their MLEE, ribotype, or PFGE patterns, suggestive of the recent emergence of a distinct clone. The seventh-pandemic, U.S. Gulf Coast, and Australian clones were genetically diverse, displaying multiple PFGE patterns. However, in another analysis of emerging isolates from the Latin American epidemic in the early 1990s, another clone was identified (Evins et al., 1995). This group tested 447 isolates of *V. cholerae* O1 from the Western Hemisphere by MLEE. They identified two electrophoretic types (ETs) among toxigenic isolates from Latin America: 323 were ET 4; the original ET associated with the Latin American epidemic; and 29 displayed a different ET designated ET 3; 23 of these ET 3 isolates had a distinctive antimicrobial resistance pattern and an identical ribotype and nearly identical PFGE patterns.

The work by Beltran et al. (1999) provided additional insights into the population structure of *V. cholerae*. This group applied MLEE on 397 *V. cholerae* isolates (sampled over more than 60 years, from 1932 to 1995), including 143 serogroup reference strains and 244 strains from Guatemala and Mexico. They identified 279 ETs with a mean number of alleles per locus of 9.5; the mean genetic diversity per locus (*H*) was 0.4 among the O1 El Tor strains, whereas the total *V. cholerae* population showed an *H* value of 0.5. Phylogenetic analysis showed that *V. cholerae* populations are structured into two major distinct divisions (I and II) and two other distinct lineages (X and Y). The vast majority of strains resolved into division I. All epidemic *V. cholerae* O1 and O139 strains resolved into four ETs that formed a tight cluster within division I. ET 1 marked the Australian toxigenic clone; ET 2 marked a toxigenic clone that is endemic to the Gulf Coast of Mexico and the United States; ET 3 is the seventh-pandemic type, a clone that was first identified in Mexico in 1991; and ET 4 marked the O1 Inaba El Tor strains, which represent the original Latin American epidemic clone. Comparisons of the observed and expected variances of the mismatch distributions for ETs suggested high levels of recombination and limited evidence for linkage disequilibrium among *V. cholerae* populations, as also suggested by Karaolis et al. (1994). This was further supported by two other studies: The first showed lack of congruence between the phylogenetic trees deduced from four housekeeping genes (Byun et al., 1999) (Fig. 18.1), and the second by nucleotide sequence analysis

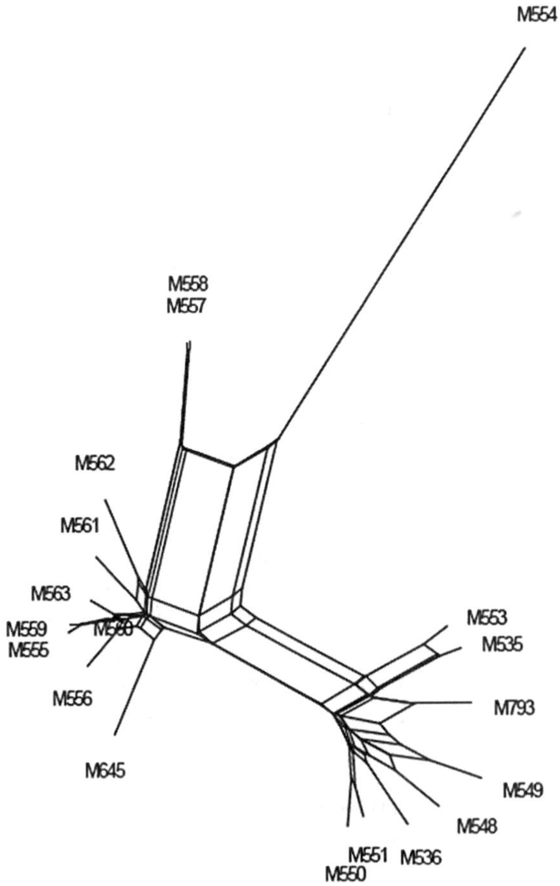

Figure 18.1 Split decomposition analysis using concatenated sequences from two housekeeping genes of *V. cholerae*, *mdh* (malate dehydrogenase) and *hlyA* (hemolysin A) (submitted to GenBank by Byun et al., 1999). Split decomposition analysis depicts a network-like structure between the sequences because of conflicting phylogenetic signals due to recombination. M645 = clinical *V. cholerae* O1 from pre-seventh-pandemic outbreaks; M793 = clinical *V. cholerae* from the seventh-pandemic outbreak; M535 and M536 = environmental O1 nontoxigenic isolates; M548, M549, M550, M551, M553, M554, M555, M556, M557, M558, M559, M560, M561, M562, M563 are all environmental, nontoxigenic, non-O1, and non-O139 isolates.

of a housekeeping gene from 45 clinical and environmental isolates, showing high levels of recombination and low levels of clonality (Karaolis et al., 1995). However, the clear structuring of *V. cholerae* populations into two major divisions was not supported by another study by Farfan et al. (2000), who also applied MLEE on 107 clinical and environmental *V. cholerae* isolates; they found that genetic diversity per locus is higher than reported by Beltran et al. (1999) and that the entire population is genetically diverse. They could not identify any significant clustering of isolates in the dendrogram with respect to serogroup, biotype, or country of isolation. Nevertheless, the results also confirmed previous work that the O139 and O1 El Tor isolates are genetically more closely related to each other than to all other subpopulations of *V. cholerae*. Stine et al. (2000) tested nucleotide sequence diversity among a single housekeeping gene, *recA*; they found that toxigenic O1/O139 El Tor strains formed a phylogenetic lineage distinctive from other populations. Others have yielded similar results (Kotetishvili et al., 2003; O'Shea et al., 2004; Reen and Boyd, 2005; Mohapatra et al., 2009).

Taken together, these studies showed that there is broad genetic diversity in *V. cholerae* populations with clear clustering of toxigenic-epidemic (O1 and O139) strains from other nonepidemic and nontoxigenic strains. Noteworthy to mention, in this regard, is the work by Thompson et al. (2003), who applied fluorescent amplified fragment length polymorphism (FAFLP) to analyze 106 *V. cholerae* O1 and non-O1 and non-O139 strains isolated from clinical specimens and the environment in Brazil between 1991 and 2001. They found that *V. cholerae* populations are genetically diverse with no clear clustering of clinical O1 strains from clinical non-O1/non-O139 strains or from environmental strains.

Prior to 1992 all reported epidemics were caused by *V. cholerae* O1. In 1992, *V. cholerae* Bengal O139 emerged in the Indian subcontinent and displaced *V. cholerae* O1 as the main cause of cholera in the region (Albert et al., 1993). The emergence of *V. cholerae* O139 prompted the question as to the evolutionary relationships of this clone with the other existing pathogenic clones. Initial observations suggested that the O139 strain is genetically very similar to strains of the seventh pandemic (Popovic et al., 1993). The work of Karaolis et al. (1994) provided additional insights into the evolution of O139 isolates. They used ribotyping to study the evolutionary relationships of seventh-pandemic clone isolates with other populations of *V. cholerae*; they studied 58 clinical strains that were isolated from patients in different countries over 62 years (1931–1993), including seventh cholera pandemic isolates, sixth pandemic isolates, strains isolated from sporadic El Tor outbreaks prior to the seventh-pandemic, U.S. Gulf Coast, and O139 Bengal isolates. The analysis indicated that the O139 isolates appear to have evolved from early seventh-pandemic isolates. This was recently confirmed by the work of Labbate et al. (2007), who used chromosomal integron assays to investigate the phylogenetic relationships within *V. cholerae* pandemic strains. They showed that an analysis of mobile gene cassette composition was able to differentiate closely related O1 El Tor and O139 strains than phylogeny based on a single locus, *recA*. It divided the El Tor group in two, placing one part closer to the O139; this has more accurately resolved the emergence of O139 from O1 El Tor. It also showed that the O1 classical strains were well separated from the O1 El Tor and O139 strains.

Two recent studies investigated the population structure of O139 Bengal strains. The first study used multilocus sequence typing (MLST) data obtained from six housekeeping genes from a collection of 29 *V. cholerae* O139 Bengal strains (Farfan et al., 2002). The study identified three distinct lineages within O139 strains; the analysis also indicated that there was little evidence for recombination within this specific population, suggestive of a recently evolving clonal population and consistent with the epidemiology of *V. cholerae* O139. This statement, however, was not supported by another study which applied DNA sequencing on 96 *V. cholerae* clinical isolates from one location, Calcuta, India (Garg et al., 2003). The authors used nine loci, including seven housekeeping genes. The 96 isolates resolved into 51 unique sequence types (STs); the number of alleles at each locus ranged from 2 to 20, and recombination was three times more likely than mutation to produce nucleotide changes, clearly displaying a genetically diverse population. The reason for the differences between the two studies (Farfan et al., 2002; Garg et al., 2003) could be explained by the fact that the first study (Farfan et al., 2002) used a small number of strains isolated during the first 2 years of O139 eruption, while the other used a larger pool that was sampled over an 8-year period, possibly better reflecting the evolving population.

A group of researchers from Bangladesh who studied the molecular epidemiology of *V. cholerae* populations isolated during epidemics in Bangladesh between 1961 and 1996 revealed clonal diversity among strains isolated during different epidemics (Faruque et al., 1998). Their work demonstrated the transient appearance and disappearance of more than six ribotypes of classical vibrios, several ribotypes of El Tor vibrios, and three different ribotypes of *V. cholerae* O139. This suggested that there is continual emergence of new clones and replacement of existing clones.

Environmental Populations

While several studies addressed the population structure and genetic diversity of *V. cholerae* populations (Chen et al., 1991; Popovic et al., 1993; Cameron et al., 1994; Choudhury et al., 1994; Karaolis et al., 1994; Colombo et al., 1997; Calia et al., 1998), only few studies have solely addressed the environmental populations, and these were studied in the context of their relationship to the epidemic serogroups of *V. cholerae*, O1 and O139 (Choudhury et al., 1994; Karaolis et al., 1995; Sharma et al., 1998).

Ecological studies have shown that non-O1 and non-O139 strains are more frequently isolated from rivers and estuarine areas than O1 and O139 (Colwell and Spira, 1992; Faruque et al., 2004), and the vast majority of environmental non-O1/non-O139 strains are nontoxigenic (Yamai et al., 1997). Furthermore, most environmental O1 strains are nontoxigenic (Colwell and Spira, 1992).

Given the low frequency with which toxigenic *V. cholerae* are found among environmental populations, how could large-scale epidemics have erupted? It has been suggested that environmental strains that acquire the relevant virulence genes, namely, the lysogenic bacteriophages CTXΦ and VPI Φ, in the aquatic ecosystem may be enriched during passage in the intestinal environment of aquatic animals or more likely in humans, thus significantly increasing the environmental load of toxigenic strains during cholera outbreaks (Faruque et al., 1998). Experiments using animal models showed that hypotoxigenic mutants of *V. cholerae* were unstable in the rabbit intestinal loop model and that during passage in the intestinal environment, they produced toxigenic revertants that eventually displaced the mutant strains *in vivo* (Mekalanos et al., 1978; Baselski et al., 1979; Mekalanos, 1983).

Studies that addressed the genetic diversity and population structure of environmental *V. cholerae* non-O1/non-O139 strains showed a highly diverse genetic structure (Dalsgaard et al., 1998; Beltran et al., 1999). Jiang et al. (2000) used amplified fragment length polymorphism (AFLP) to characterize temporal and spatial genetic diversity of 67 non-O1 *V. cholerae* strains isolated from Chesapeake Bay, USA, during 4 months, April–July 1998, at different sampling sites. The analyses identified three genetic clusters reflecting the time of the year when the strains were isolated, April and May cluster, June cluster, and July cluster. No correlation was found between genetic similarity among isolates and geographic source of isolation. The authors speculated that the population structure of *V. cholerae* undergoes a shift in genotype that is linked to changes in environmental conditions. One may extrapolate these observations to the cholera-endemic areas where *V. cholerae* populations may be transported by surface currents, or, more likely, similar environmental conditions may be selected for a specific genotype. In agreement with that, a recent study by Keymer et al. (2007), who studied the genomic and metabolic profiles of 41 non-O1/non-O139 environmental isolates from central California coastal waters and 4 clinical isolates, found a correlation between gene content and metabolic pathways. Core genes were almost universally present in strains with widely different niches, suggesting that these genes are essential for persistence in the aquatic environment, while the dispensable genes are likely to provide increased fitness for certain niche environments (e.g., cold waters, oxidative stress). Zo et al. (2002) studied the genetic structure of clinical and environmental isolates of *V. cholerae* O1 in a cholera-endemic area in Bangladesh; they found that the composition of environmental populations of toxigenic *V. cholerae* is identical to that of *V. cholerae* causing endemic cholera. The authors suggested that *V. cholerae* populations in the two distinctive habitats, humans and environment, achieve a dynamic equilibrium by rapid transfer between habitats or panmictic gene flow by the active intermingling between clinical and environmental isolates.

18.2.2 Evolution of Toxigenic *V. cholerae*

Different genotyping methods have clearly showed that the sixth- and the seventh-pandemic clones are genetically distinct clones (Koblavi et al., 1990; Cameron et al., 1994; Karaolis et al., 1994) that have evolved from nontoxigenic environmental populations (Karaolis et al., 1995). With the advent of molecular genetic tools capable of elucidating evolutionary relationships from genomic data, the scenarios for the evolution of toxigenic *V. cholerae* are nowadays more readily accepted. Faruque and Mekalanos (2003) suggested possible evolutionary pathways for the emergence of toxigenic *V. cholerae*. It is widely accepted that the acquisition of the two main virulence factors in *V. cholerae*, the TCP and CTX element, was sequential. First, *V. cholerae* O1 isolates acquired TCP, via the VPI-I (Karaolis et al., 1998), which is an essential colonization factor and the receptor for CTXΦ (Taylor et al., 1987; Waldor and Mekalanos, 1996), and then they acquired CT via CTXΦ (Waldor and Mekalanos, 1996). They speculated that in the natural ecological settings and by the influence of some environmental factors, such as water temperature, sunlight, and osmotic conditions, there is an induction of lysogenic CTXΦ in toxigenic *V. cholerae*, which in turn release extracellular CTXΦ particles into the aquatic environment (Faruque et al., 1998). These cell-free phages contribute to the emergence of novel toxigenic strains through interactions with nontoxigenic strains that exist in the aquatic environment and in the intestines of humans who consume environmental waters.

To date, several additional genomic regions have been identified mostly among epidemic O1 and O139 serogroups; these include RS1Φ (Faruque et al., 2002), VSP-I and VSP-II (Dziejman et al., 2002), and VPI-2 (Jermyn and Boyd, 2002). A comparative genomic analysis of *V. cholerae* using microarray technology was carried out in search of unique genetic profiles of the El Tor pandemic strains (Dziejman et al., 2002). The authors constructed a genomic microarray on the basis of the sequenced O1 El Tor strain N16961 from the seventh pandemic of cholera (Heidelberg et al., 2000). This array was used to compare the gene content of classical, prepandemic El Tor, pandemic El Tor, and two nontoxigenic strains to that of strain N16961. They identified four groups of genes: The first group included genes present in all El Tor strains but not in classical biotype strains; the second group included genes present only in pandemic strains; the third group included genes present only in the seventh-pandemic El Tor O1 strains; and the fourth group included genes uniquely absent from individual strains. The most notable findings from this study were that only few genes were found in group I, uniquely defining El Tor biotype from the classical biotype, and in group III genes, two chromosomal islands (VSP-I and VSP-II) were identified among seventh-pandemic strains that were lacking from the others. The authors speculated that the classical biotype strains might have evolved from a primordial environmental strain that was more El Tor-like than previously thought. They suggested that the specific array of genes found in strain N16961 could potentially encode key properties that have led to the global success of seventh-pandemic strains as agents of endemic and pandemic cholera.

One other interesting feature is the finding that although the strains tested varied in serotype, biotype, site, and year of isolation, as a group they lacked only ~1% of N16961 genes. The close genetic relatedness of the test strains to the pandemic El Tor strain prompted the question as to whether any given nonpathogenic environmental isolate has the capacity to become an epidemic or pandemic pathogen simply by acquiring the TCP island, CTXΦ, and probably few other genes. This important issue was resolved to a great extent by the work of Faruque et al. (2004). They analyzed diverse strains of *V. cholerae* isolated from environmental waters in an endemic area (Bangladesh) by direct enrichment in the intestines of adult rabbits and by conventional laboratory culture. Strains isolated by conventional culture were mostly (99.2%) negative for the major virulence gene clusters encoding TCP and CT, were nonpathogenic in animal models, and were genetically diverse. In contrast, all strains selected in rabbits were competent for colonizing infant mice, and 56.8% of these strains carried genes encoding TCP alone or both TCP and CT. They found that ribotypes of toxigenic O1 and O139 strains from the environment were similar to pandemic strains, whereas ribotypes of non-O1 non-O139 strains and TCP(–) nontoxigenic O1 strains diverged widely from the seventh-pandemic O1 and the O139 strains. The study clearly showed that most environmental *V. cholerae* strains are unlikely to attain a pandemic potential simply by the acquisition of TCP and CT genes alone and that other, yet unidentified, factors are needed to convert a nonpathogenic environmental isolate into a toxigenic strain with an epidemic or pandemic potential.

Evidence from molecular genetic studies suggests that genetic recombination plays an important role in generating genetic diversity among *V. cholerae* populations (Wachsmuth et al., 1993; Evins et al., 1995; Karaolis et al., 1995; Beltran et al., 1999). In addition, these highly recombining populations interact with other species and donors such as cell-free phages within the aquatic ecosystem. The acquisition of virulence genes by *V. cholerae* and the adaptation to the human intestinal environment has provided these strains with an evolutionary advantage over other nontoxigenic strains. The continual emergence of new strains of toxigenic *V. cholerae* and their selective enrichment during cholera outbreaks constitute an essential part of the ecosystem for the survival and evolution of *V. cholerae* and the genetic elements that mediate the transfer of virulence genes.

Besides the emergence of pathogenic strains by the acquisition of virulence gene clusters, *V. cholerae* populations seem to undergo temporal genetic and phenotypic changes leading to the emergence of diverse epidemic strains. The emergence of *V. cholerae* O139 is perhaps one of the most notable examples to illustrate that. The transformation of *V. cholerae* O1 El Tor strains into O139 involved a complex rearrangement in the genes responsible for the biosynthesis and assembly of the side chains of the O1 cell surface LPS (Berche et al., 1994; Bik et al., 1995). The observed genetic exchange in the O139 strain, which possesses all the virulence genes of the toxigenic O1 strains, seems to reflect a greater evolutionary fitness and a natural selection process associated with evading a developing immunity within an endemically infected population (Faruque and Mekalanos, 2003; Faruque and Nair, 2006).

18.3 *V. PARAHAEMOLYTICUS*

V. parahaemolyticus is a major cause of gastroenteritis, mainly in areas with high consumption of seafood (Pan et al., 1997; Daniels et al., 2000). The organism is a natural inhabitant of marine and estuarine environments and is highly adaptable to different lifestyles, a planktonic, free-swimming state; a sessile existence attached to shellfish in a commensal relation, to the bottoms of boats, or to other surfaces in the ocean, or in the host (McCarter, 1999). Strains of *V. parahaemolyticus* are serotyped on the basis of O and K antigens. Until 1995, the vast majority of *V. parahaemolyticus* infections were caused by different serotypes. However, in 1996, a new serotype emerged, O3:K6. Since then, this serotype and other genetically related serotypes, which formed the O3:K6 clone of *V. parahaemolyticus*, has spread worldwide and has become the main cause of seafood-borne bacterial associated gastroenteritis (Okuda et al., 1997; Chowdhury et al., 2000; DePaola et al., 2000; Matsumoto et al., 2000; Gonzalez-Escalona et al., 2005; Quilici et al., 2005).

Though a comprehensive understanding of the organism's ability to cause disease remains elusive, several individual factors have been identified as clearly correlated with the ability of the bacterium to cause disease in humans. Among these factors, a thermostable direct hemolysin (TDH) and the TDH-related hemolysin (TRH) have been correlated with the pathogenicity to humans (Joseph et al., 1982; Yoh et al., 1992). In general, most clinical strains posses the *tdh* gene, whereas few environmental strains do so. Although *trh*-positive strains are occasionally isolated from the environment, they are also almost exclusively pathogenic (Iida et al., 2006). A set of genes for the type III secretion system (TTSS) (Hueck, 1998), which were identified in the pathogenicity island on chromosome 2, has also been strongly suggested as being related to pathogenicity in humans (Makino et al., 2003).

18.3.1 Population Structure

Since the emergence of the pandemic clone O3:K6 as the main cause of *V. parahaemolyticus* disease worldwide, researchers in the field have focused their work on elucidating the genetic profile of the pandemic strains. An initial investigation of O3:K6 strains isolated in Calcutta, India, in 1996 revealed that all 61 strains examined shared identical traits (*tdh* positive, *trh* negative, and urease negative), with only two strains having an antibiogram different from those of the other strains (Okuda et al., 1997). In addition, the representative O3:K6 strains were shown to be genetically indistinguishable by arbitrarily primed PCR analysis and were therefore considered to be a single clone. In another study (Bag et al., 1999) from the same area of disease eruption, Calcutta, India, and using 30 clinical isolates analyzed by RFLP, *tdh* genotyping, and PFGE, the authors reached similar conclusions regarding the clonality of the O3:K6 strains. Wong et al. (2000) carried out a more comprehensive work and applied PFGE on 139 isolates of O3:K6 *V. parahaemolyticus* strains recently isolated in Taiwan. Some of these strains were isolated from travelers originating in several other Asian countries. They included O3:K6 strains isolated before and after 1996. They found that all the O3:K6 strains were grouped into two unrelated groups. The recently isolated O3:K6 strains were all in one group, consisting of eight closely related patterns, with I1 (81%) and I5 (13%) being the most frequent patterns. Pattern I1 was the major one for strains from Japan, Korea, and Taiwan. All recently isolated O3:K6 strains carried the *tdh* gene. The results in this study confirmed that the recently isolated O3:K6 strains of *V. parahaemolyticus* are genetically close to each other than to the "old" strains. These observations were further confirmed by other studies (Chowdhury et al., 2000; Matsumoto et al., 2000; Okura et al., 2003). Altogether, these studies showed that all the post-1995 O3:K6 strains were genetically identical by arbitrarily primed PCR, ribotyping, and PFGE, but they were significantly different from the genetically variable pre-1995 O3:K6 strains. Furthermore, serodiversity studies showed that genetically identical pandemic strains expressed multiple serotypes, suggesting that strains within a clonal complex may acquire previously identified serotypes by lateral gene transfer (Laohaprertthisan et al., 2003).

Two recent studies that investigated the population structure of *V. parahaemolyticus* using the MLST approach were in agreement with previous observations. The first used 81 pandemic strains of *V. parahaemolyticus* collected between 1983 and 2000 and examined four housekeeping genes all located on chromosome I (Chowdhury et al., 2004); the analyses reaffirmed the highly clonal nature of the pandemic clonal complex. The second study used a globally diverse collection of isolates and examined seven genes from both chromosomes (Gonzalez-Escalona et al., 2008); it showed that among 100 isolates, 62 different allelic combinations and an average of 33 alleles per locus were identified. Three major clonal complexes were identified; separate clonal complexes were observed for isolates originating from the Pacific and Gulf coasts of the United States, while a third complex consisted of strains belonging to the pandemic clonal complex with worldwide distribution. Phylogenetic analysis based on concatenated sequences from all seven housekeeping genes did not show any distinct clustering within *V. parahaemolyticus* strains (Fig. 18.2). Few STs (ST-1, ST-2, and ST-62) diverged from the major group. The pandemic clonal complex (designated as clonal complex 3-CC3 by Gonzalez-Escalona et al., 2008) consisted of 25 isolates that resolved into four STs (22 isolates resolved into ST 3, and three isolates resolved into three STs: ST27, ST42, and ST51).

Estimates of recombination and mutation events showed that the per allele recombination/mutation (r/m) parameter was 2.5 : 1.0, and the per site r/m parameter was 8.8 : 1.0, suggesting that genetic diversity in *V. parahaemolyticus* is primarily driven by recombination rather than mutation. Overall, the population structure of *V. parahaemolyticus* follows a model where from a background of high recombination rates, "successful" clones emerged and persisted as illustrated by the emergence of the pandemic clonal complex, O3:K6 clone of *V. parahaemolyticus* (Nair et al., 2007).

18.3.2 Pandemic O3:K6 Clone

To date, several studies support the likelihood of horizontal gene transfer among *V. parahaemolyticus* populations. The transfer of GIs, vibriophages, and integrons among *V. parahaemolyticus* populations has likely contributed to the continuing emergence of new and virulent clones. For instance, *tdh*, which encodes a virulence-associated hemolysin, has been demonstrated to exist on plasmid DNA, chromosomal DNA, and in other vibrios (Nishibuchi and Kaper, 1990; Baba et al., 1991; Yoh, Miwatani and Honda, 1992).

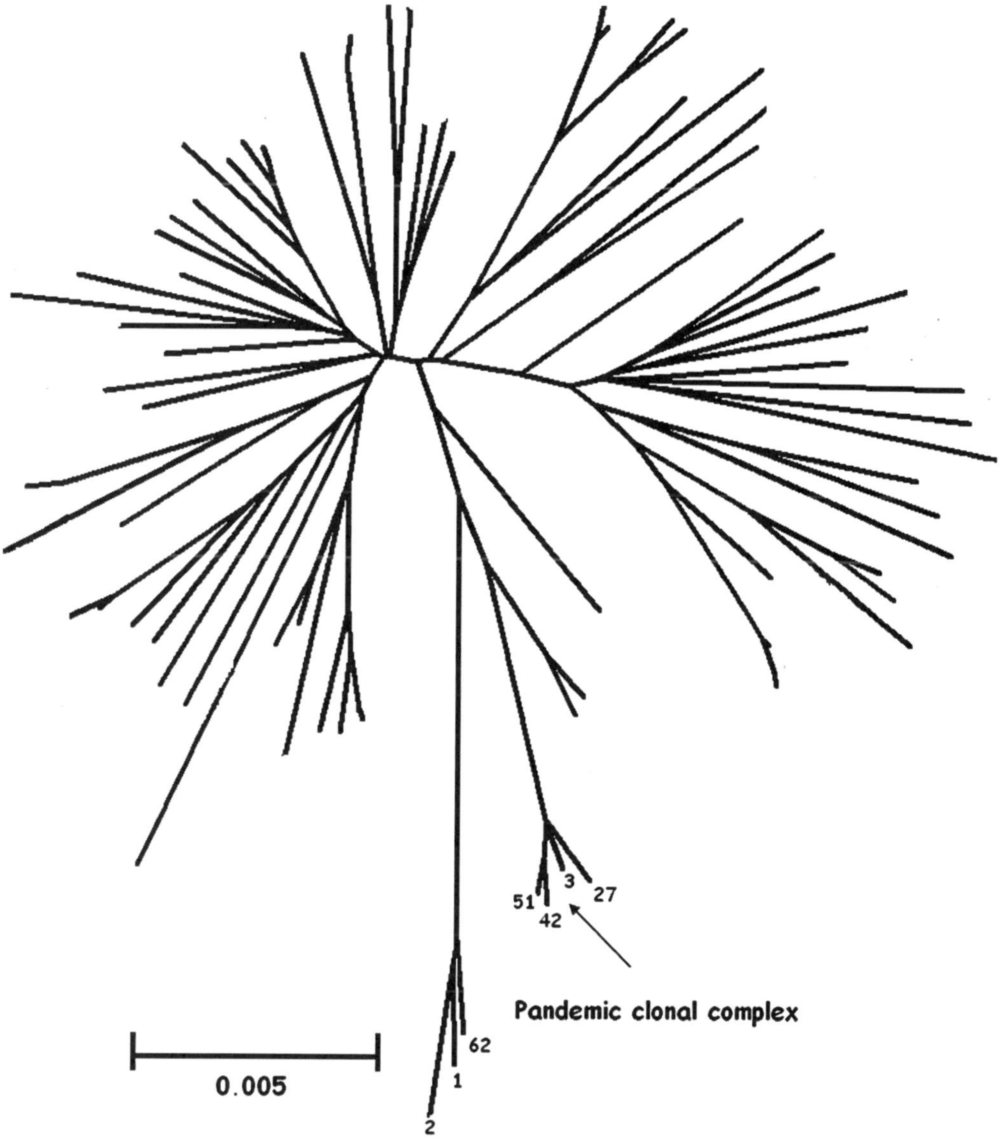

Figure 18.2 Unrooted neighbor-joining bootstrap consensus tree of concatenated sequences of seven housekeeping genes of *V. parahaemolyticus* available from http://pubmlst.org/vparahaemolyticus/ (Gonzalez-Escalona et al., 2008). The numbers at the branches indicate the sequence type (allelic profile).

Vibriophages and GIs are perhaps the more important mechanisms responsible for the evolution of new virulent clones.

Phages are abundant among *V. parahaemolyticus* populations (Baross et al., 1978; Yeung and Boor, 2004). Studies suggested that the O3:K6 clone of *V. parahaemolyticus*, the pandemic clonal complex, might have emerged as the result of a horizontally acquired filamentous phage, f237 (Nasu et al., 2000). The distribution of this phage, which bears genomic similarities to another filamentous phage, CTXΦ phage of *V. cholerae*, was examined in 96 clinical isolates of *V. parahaemolyticus* (Iida et al., 2001). They found that this phage, which possesses a unique open reading frame, ORF8, was identified among approximately 55% of the isolates, suggesting that this phage is associated not only with O3:K6 serovar but also with other recently emerging serovars of *V. parahaemolyticus*. The authors rightly concluded that such a high prevalence of the phage f237 in the *V. parahaemolyticus* strains showing pandemic spread suggests that the phage might confer some unknown advantages to the bacterium. Chan et al. (2002) described a novel filamentous phage (described as a deleted form of phage f237) from a pandemic *V. parahaemolyticus* O4:K68 strain; they speculated that these phages might protect the host bacterium against selective pressure in a certain environment before infecting humans.

A recent bioinformatics and molecular analysis of *V. parahaemolyticus* O3:K6 strain revealed seven GIs (Hurley et al., 2006). The GIs ranged in size from 10 to 81 kb and had the typical elements of GIs such as aberrant base composition (compared to the core genome) and the presence of phage-like integrases flanked by direct repeats. Molecular analysis of the distribution of these GIs among pre- and post-1995 pandemic isolates found that they were mainly present in the new O3:K6 pandemic strains (Hurley et al., 2006; Reen et al., 2006). This suggested that the new pandemic *V. parahaemolyticus* O3:K6 clone might have acquired increased pathogenic potential to humans by these GIs. A more recent analysis that applied comparative genomics using *V. parahaemolyticus* RIMD2210633, an O3:K6 strain isolated in Japan in 1996, and *V. parahaemolyticus* AQ3810 strain, an O3:K6 isolate recovered in 1983 (Boyd et al., 2008), showed that the AQ3810 did not encode 8 of the 24 regions unique to RIMD, including a T6SS (type VI secretion system) and seven GIs (*V. parahaemolyticus* [VPaI-1 to VPI-7]). The authors speculated that the most likely scenario for the evolution of the new highly virulent O3:K6 clone suggests that a pre-1995 O3:K6 strain obtained regions VPaI-1 to VPI-7, and a T6SS along with T3SS-2 (two sets of TTSS).

18.4 *V. VULNIFICUS*

V. vulnificus is considered the major cause of shellfish-associated deaths in the United States (Oliver, 1989). *V. vulnificus* causes severe systemic infections with a high mortality rate especially among immunocompromised hosts. People who are most susceptible to *V. vulnificus* infection usually suffer from a chronic disease that affects liver function, primarily cirrhosis or alcoholic liver disease. Also included are diseases associated with iron overload such as hemochromatosis, thalassemia major, and conditions directly affecting the immune function (Oliver, 2006). Strains of *V. vulnificus* are biochemically classified into three different biotypes. Human infections are almost entirely caused by strains of biotype 1, while biotype 2 strains have been reported to cause disease mainly among eels and rarely infecting humans (Tison et al., 1982). A third biotype has been recently described in Israel and has not been identified, to date, anywhere else in the world (Bisharat et al., 1999).

Isolation of *V. vulnificus* from environmental sources showed that there is a considerable strain variation and that human disease is caused by only few strains within a heterogeneous population (Jackson et al., 1997). Luckily, and despite the abundance of *V. vulnificus* in marine and estuarine waters (Oliver et al., 1982), infections with *V. vulnificus* are relatively uncommon (Centers for Disease Control and Prevention, 1993; Levine and Griffin, 1993). The U.S. Food and Drug Administration (FDA) estimates that approximately 30 fatal *V. vulnificus* cases occur each year in the United States despite estimates that 20 million Americans consume about 75 million to 80 million servings of raw oysters annually, and that 12 million to 30 million persons have one or more of the known risk factors for *V. vulnificus* infection (Oliver, 2006). These observations are clearly suggestive that not all *V. vulnificus* populations are virulent to humans.

Numerous molecular techniques have been applied in attempts to establish the genetic population structure of *V. vulnificus* (Hayat et al., 1993; Tamplin et al., 1996; Jackson et al., 1997; Arias et al., 1998; Hoi et al., 1998; Warner and Oliver, 1999; Vickery et al., 2000; Gutacker et al., 2003; Lin et al., 2003; Lin and Schwarz, 2003; Nilsson et al., 2003); these studies showed that *V. vulnificus* populations are considerably diverse. However, neither the relationship of the pathogenic clones with the environmental strains nor the identification of any distinct clones that are clearly associated with human disease could be resolved.

Several studies aimed to characterize the mechanisms utilized by *V. vulnificus* to cause severe, potentially fatal infections; however, until now, all attempts to associate genotypic characteristics of *V. vulnificus* with strain virulence have been largely unsuccessful. The only exception to that is the recent finding that disease occurrence in Israel was caused exclusively by one specific genotype (Bisharat et al., 2005). To date, no other genotypes have been implicated in human infections in Israel (N. Bisharat, unpublished data).

18.4.1 Population Structure and Genetic Diversity

Until the advent of MLST (Maiden et al., 1998), the greatest insights into the population structure of *V. vulnificus* have been obtained using MLEE (Gutacker et al., 2003). This study showed that populations of *V. vulnificus* consist of two main clusters and an eel-pathogenic clone. The application of MLST to *V. vulnificus* confirmed MLEE results and provided other interesting observations.

We have recently developed an MLST scheme for *V. vulnificus* using 10 housekeeping genes (5 from each chromosome) on a diverse global collection of isolates, which included strains from both infected humans and environmental sources (Bisharat et al., 2005). Analysis of genetic heterogeneity showed that *V. vulnificus* populations are genetically heterogeneous; strains of biotype 1 resolved into 66 STs, while biotype 2 strains resolved into 4 STs. In contrast, all 61 biotype 3 clinical isolates resolved into the same genotype, ST 8. Recent studies from Israel confirmed the genetic distinctiveness of this biotype from the other existing biotypes; yet, by using simple sequence repeat (SSR) typing and PFGE, they detected interstrain genetic genetic diversity within this homogeneous biotype (Broza et al., 2007; Zaidenstein et al., 2008).

Phylogenetic analysis showed that populations of *V. vulnificus* are structured into two main clusters with overrepresentation of environmental isolates in one cluster (cluster I) and overrepresentation of human disease isolates in the other (cluster II).

This unique division has been validated by other molecular tools (Nilsson et al., 2003; Rosche et al., 2005). Biotype 1 strains were represented in both groups, while biotype 2 strains were represented in one group dominated by environmental strains (Bisharat et al., 2005, 2007). The genetically homogeneous biotype 3 strains were placed in an intermediate position between the two main clusters (Fig. 18.3). This genotype was found to be a mosaic recombinant, with some genes originating from the first population, some from the second population, and with other genes deriving from both (Bisharat et al., 2005).

An F_{ST} value of 0.58 between the two clusters illustrates the magnitude of the significant population structure (Bisharat et al., 2007). The F_{ST} between clusters and nucleotide diversity were greater on the small chromosome compared with the large.

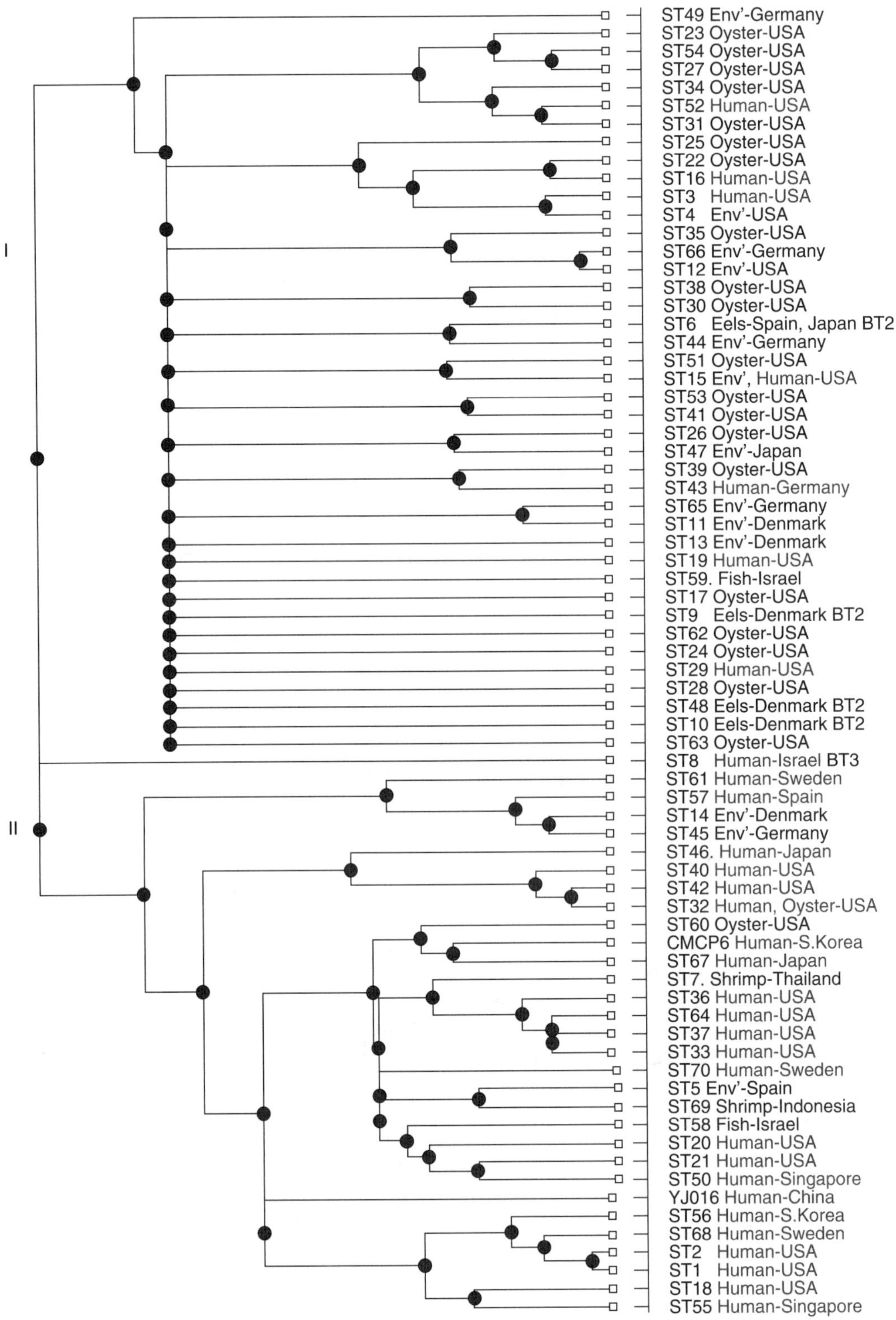

Figure 18.3 Majority-rule consensus tree based on the posterior distribution of genealogies inferred by ClonalFrame. The biotype identity is shown for some isolates; the rest of the isolates belong to biotype 1. Env' = environment. See color insert.

Nucleotide diversity was also greater for cluster II when the chromosomes were evaluated separately. On the large chromosome, F_{ST} was 0.51 and nucleotide diversity was 0.011 within cluster I, and 0.015 within cluster II. On the small chromosome, F_{ST} was 0.64 and nucleotide diversity was 0.011 within cluster I, and 0.018 within cluster II.

The majority-rule consensus tree (Fig. 18.3) shows that there is more hierarchical clustering in cluster II than in cluster I. Apparently, cluster II strains, which are associated with human disease, more strongly conform to a model of clonal evolution compared with strains identified mainly from the environment. This clearly supports the expectation of clonal relatedness in association with epidemic human disease spread. Furthermore, sampling from the environment provides scant evidence of clonal hierarchies.

18.4.2 Recombination and Mutation

Estimation of intra- and interlocus recombination, within and between clusters, and over both chromosomes, from MLST data analysis provided interesting insights into the impact of recombination and mutation in generating genetic diversity (Bisharat et al., 2007). For the large chromosome, the averages over loci within clusters showed that the population recombination rates are comparable between clusters I and II (Table 18.1), reflecting similarities in both recombination rate and effective population size. The recombination rates were marginally higher on the small chromosome, and this slight difference was greater in cluster II strains. Maximum likelihood estimates for population recombination rates over each chromosome are much lower than rates estimated within loci; the population recombination rate was lower on the small chromosome compared with the large. For the large chromosome, the maximum likelihood estimate over 6301 bp for ρ (recombination rate) was 5 ($\rho = 0.0008$ per bp); and for the small chromosome, the maximum likelihood estimate over 6025 bp for ρ is 2 ($\rho = 0.0003$ per bp). The small chromosome showed

Table 18.1 Characteristics of Sequence Diversity and Estimates of Recombination and Mutation Rates for *V. vulnificus*

Chromosome	Locus	Size of sequenced fragment (bp)	No. of SNPs (no. of alleles)	d_N/d_S ratio	Cluster I estimates $\varepsilon = \rho/\theta$	Cluster II estimates $\varepsilon = \rho/\theta$
Large	*glp*	480	46 (38)	0.011	6.0	1.9
Large	*gyrB*	459	34 (31)	0.022	3.8	0.3
Large	*mdh*	489	30 (29)	0.009	0.4	1.4
Large	*metG*	429	37 (31)	0.029	0.2	0.7
Large	*purM*	444	39 (28)	0.03	0.4	1.2
Small	*dtdS*	417	56 (46)	0.04	1.0	1.5
Small	*lysA*	465	78 (41)	0.07	1.2	0.7
Small	*pntA*	396	35 (32)	0.004	1.5	1.4
Small	*pyrC*	423	50 (35)	0.02	1.2	2.6
Small	*tnaA*	324	42 (32)	0.08	2.6	1.0

d_N/d_S = nonsynonymous/synonymous substitutions; ρ = recombination rate; θ = mutation rate; $\varepsilon = \rho/\theta$: recombination to mutation ratio.

a lower recombination rate and a shorter tract length for these recombination events compared with the large chromosome. Estimates for the relative impact of recombination and mutation on genetic diversity suggested that any single nucleotide in the *V. vulnificus* genome is one to two times more likely to change due to intralocus recombination with another strain from the same population than it is to mutation. And if the contribution of interlocus recombination to genetic diversity is taken into account, recombination in *V. vulnificus* is on the order of 10 times as important as mutation for generating novel MLST strains (Bisharat et al., 2007).

18.4.3 Possible Evolutionary Scenarios

Our previous observations (Bisharat et al., 2005, 2007) highlighted the importance of recombination in generating genetic diversity within *V. vulnificus*, yet interestingly, the genetic divergence between two clusters is being maintained across the genome and is systematic over both chromosomes. Against a background of so much potential recombination, how could that explain the distinct genetic divergence into two clusters? Two possible scenarios are suggested: In the first, the divergence may be due to import of sequence from another related species. If so, in addition to the more typical horizontal transfer of short tracts of sequence, interspecific hybrids may arise as rare events. Such hybrids may be the ancestor of cluster II strains; the subsequent clonal expansion of such hybrids would also increase the availability of divergent sequence for typical recombination events. In the second, if habitat preferences within the species vary, ecological filtering and selective loss of the most recombinant hybrids may cause the genetic divergence. Like other vibrios, *V. vulnificus* can adapt to a wide range of ecological relationships, including favorable partnerships with other bacteria or hosts (Thompson et al., 2004). It is feasible that pathogenic hybrid strains emerge particularly in the advantageous conditions of aquaculture settings (e.g., high nutrient loads and high host density), which may serve as reservoirs, where strains such as those in cluster II diverge from more typical environmental strains (Ben-Haim et al., 2003; Thompson et al., 2004). In this regard, Cohen et al. (2007) showed that cluster II strains (described as lineage I in their paper) are associated with a 33-kb GI (region XII), one of three regions identified by genome comparisons as unique to the species. This region contained a cassette of genes absent from most of cluster I strains (described as lineage II in their paper); it included an arylsulfatase gene cluster and a sulfate reduction system, which were suggested to play a role in pathogenesis and in improving survival in the human host. The authors suggested that this region might have provided cluster II strains with some selective advantage in the human host or in the aquatic environment.

A recent comparative genomic analysis that compared the complete genome sequences of two clinical strains of *V. vulnificus*, CMCP6 and YJ016, identified 14 regions that had the characteristics of GIs (Quirke et al., 2006). The authors identified nine GIs present in YJ016 but absent from CMCP6. Some GIs were involved in sugar transport and metabolism; some encoded possible virulence genes (pathogenicity islands), while others encoded multidrug resistance genes (putative resistance genes). When they examined the distribution of these GIs among 27 clinical and environmental strains, none of these GIs marked any clinical, environmental, or biotype-specific strains, in contrast to observations among *V. cholerae* (Karaolis et al., 1998) and *V. parahaemolyticus* (Hurley et al., 2006).

18.5 CONCLUSIONS

Populations of *V. cholerae*, *V. parahaemolyticus*, and *V. vulnificus* are highly diverse recombining populations from which novel human-pathogenic clones emerge. This is best illustrated by the emergence of pandemic *V. cholerae* O1 and O139, and most recently by the emergence of the O3:K6 clone of *V. parahaemolyticus* and *V. vulnificus* biotype 3. Available data from molecular studies of large and global populations of *V. cholerae* allowed a better understanding of its epidemiology. There is evidence for frequent intragenic and assortative recombination events in housekeeping genes. In spite of the frequent horizontal gene transfer, clonal lineages of *V. cholerae* emerged and persisted for decades, as seen by the emergence and persistence of epidemic and pandemic clones. Given the frequency and magnitude of horizontal gene transfer, it is almost certain that new pathogenic clones of *V. cholerae*, possibly with epidemic or pandemic potential, will arise in the future. The global burden of the disease caused by *V. parahaemolyticus* and *V. vulnificus* is far less severe than that caused by *V. cholerae*; yet, new human-pathogenic clones could arise in the future possibly altering the global epidemiology.

REFERENCES

Albert, M. J., Siddique, A. K., Islam, M. S. et al. (1993) Large outbreak of clinical cholera due to *Vibrio cholerae* non-O1 in Bangladesh. *Lancet* **341,** 704.

Amita, K., Chowdhury, S. R., Thungapathra, M. et al. (2003) Class I integrons and SXT elements in El Tor strains isolated before and after 1992 *Vibrio cholerae* O139 outbreak, Calcutta, India. *Emerg Infect Dis* **9,** 500–502.

Arias, C. R., Pujalte, M. J., Garay, E., and Aznar, R. (1998) Genetic relatedness among environmental, clinical, and diseased-eel *Vibrio vulnificus* isolates from different geographic regions by ribotyping and randomly amplified polymorphic DNA PCR. *Appl Environ Microbiol* **64,** 3403–3410.

Baba, K., Shirai, H., Terai, A. et al. (1991) Analysis of the tdh gene cloned from a tdh gene- and trh gene-positive strain of *Vibrio parahaemolyticus*. *Microbiol Immunol* **35,** 253–258.

Bag, P. K., Nandi, S., Bhadra, R. K. et al. (1999) Clonal diversity among recently emerged strains of *Vibrio parahaemolyticus* O3:K6 associated with pandemic spread. *J Clin Microbiol* **37,** 2354–2357.

Baross, J. A., Liston, J., and Morita, R. Y. (1978) Incidence of *Vibrio parahaemolyticus* bacteriophages and other *Vibrio* bacteriophages in marine samples. *Appl Environ Microbiol* **36,** 492–499.

Baselski, V. S., Medina, R. A., and Parker, C. D. (1979) In vivo and in vitro characterization of virulence-deficient mutants of *Vibrio cholerae*. *Infect Immun* **24,** 111–116.

Beltran, P., Delgado, G., Navarro, A. et al. (1999) Genetic diversity and population structure of *Vibrio cholerae*. *J Clin Microbiol* **37,** 581–590.

Ben-Haim, Y., Thompson, F. L., Thompson, C. C. et al. (2003) *Vibrio coralliilyticus* sp. nov., a temperature-dependent pathogen of the coral *Pocillopora damicornis*. *Int J Syst Evol Microbiol* **53,** 309–315.

Berche, P., Poyart, C., Abachin, E. et al. (1994) The novel epidemic strain O139 is closely related to the pandemic strain O1 of *Vibrio cholerae*. *J Infect Dis* **170,** 701–704.

Bik, E. M., Bunschoten, A. E., Gouw, R. D., and Mooi, F. R. (1995) Genesis of the novel epidemic *Vibrio cholerae* O139 strain: Evidence for horizontal transfer of genes involved in polysaccharide synthesis. *EMBO J* **14,** 209–216.

Bisharat, N., Agmon, V., Finkelstein, R. et al. (1999) Clinical, epidemiological, and microbiological features of *Vibrio vulnificus* biogroup 3 causing outbreaks of wound infection and bacteraemia in Israel. Israel Vibrio Study Group. *Lancet* **354,** 1421–1424.

Bisharat, N., Cohen, D. I., Harding, R. M. et al. (2005) Hybrid *Vibrio vulnificus*. *Emerg Infect Dis* **11,** 30–35.

Bisharat, N., Cohen, D. I., Maiden, M. C. et al. (2007) The evolution of genetic structure in the marine pathogen, *Vibrio vulnificus*. *Infect Genet Evol* **7,** 685–693.

Boucher, Y. and Stokes, H. (2006) The roles of lateral gene transfer and vertical descent in vibrio evolution. In *The Biology of Vibrios* (eds. F. Thompson, B. Austin, and J. Swings), pp. 84–94. ASM Press, Washington, DC.

Boyd, E. F., Cohen, A. L., Naughton, L. M. et al. (2008) Molecular analysis of the emergence of pandemic *Vibrio parahaemolyticus*. *BMC Microbiol* **8,** 110.

Broza, Y. Y., Danin-Poleg, Y., Lerner, L. et al. (2007) *Vibrio vulnificus* typing based on simple sequence repeats:

Insights into the biotype 3 group. *J Clin Microbiol* **45,** 2951–2959.

Byun, R., Elbourne, L. D., Lan, R., and Reeves, P. R. (1999) Evolutionary relationships of pathogenic clones of *Vibrio cholerae* by sequence analysis of four housekeeping genes. *Infect Immun*, **67,** 1116–24.

Calia, K. E., Waldor, M. K., and Calderwood, S. B. (1998) Use of representational difference analysis to identify genomic differences between pathogenic strains of *Vibrio cholerae*. *Infect Immun* **66,** 849–852.

Cameron, D. N., Khambaty, F. M., Wachsmuth, I. K. et al. (1994) Molecular characterization of *Vibrio cholerae* O1 strains by pulsed-field gel electrophoresis. *J Clin Microbiol* **32,** 1685–1690.

Centers for Disease Control and Prevention (1993) *Vibrio vulnificus* infections associated with raw oyster consumption-Florida 1981–1992. *MMWR Morb Mortal Wkly Rep* **42,** 405–407.

Chan, B., Miyamoto, H., Taniguchi, H., and Yoshida, S. (2002) Isolation and genetic characterization of a novel filamentous bacteriophage, a deleted form of phage f237, from a pandemic *Vibrio parahaemolyticus* O4:K68 strain. *Microbiol Immunol* **46,** 565–569.

Chen, F., Evins, G. M., Cook, W. L. et al. (1991) Genetic diversity among toxigenic and nontoxigenic *Vibrio cholerae* O1 isolated from the Western Hemisphere. *Epidemiol Infect* **107,** 225–233.

Chen, C. Y., Wu, K. M., Chang, Y. C. et al. (2003) Comparative genome analysis of *Vibrio vulnificus*, a marine pathogen. *Genome Res* **13,** 2577–2587.

Choudhury, S. R., Bhadra, R. K., and Das, J. (1994) Genome size and restriction fragment length polymorphism analysis of *Vibrio cholerae* strains belonging to different serovars and biotypes. *FEMS Microbiol Lett* **115,** 329–334.

Chowdhury, N. R., Chakraborty, S., Ramamurthy, T. et al. (2000) Molecular evidence of clonal *Vibrio parahaemolyticus* pandemic strains. *Emerg Infect Dis* **6,** 631–636.

Chowdhury, N. R., Stine, O. C., Morris, J. G., and Nair, G. B. (2004) Assessment of evolution of pandemic *Vibrio parahaemolyticus* by multilocus sequence typing. *J Clin Microbiol* **42,** 1280–1282.

Cohen, A. L., Oliver, J. D., DePaola, A. et al. (2007) Emergence of a virulent clade of *Vibrio vulnificus* and correlation with the presence of a 33-kilobase genomic island. *Appl Environ Microbiol* **73,** 5553–5565.

Colombo, M. M., Mastrandrea, S., Leite, F. et al. (1997) Tracking of clinical and environmental *Vibrio cholerae* O1 strains by combined analysis of the presence of toxin cassette, plasmid content and ERIC PCR. *FEMS Immunol Med Microbiol* **19,** 33–45.

Colwell, R. R. (1996) Global climate and infectious disease: The cholera paradigm. *Science* **274,** 2025–2031.

Colwell, R. R. and Spira, W.M. (1992) The ecology of *Vibrio cholerae*. In *Cholera* (eds. D. Brauna and W. B. Greenough III), pp. 107–127. Plenum, New York.

Dalsgaard, A., Forslund, A., Mortensen, H. F., and Shimada, T. (1998) Ribotypes of clinical *Vibrio cholerae* non-O1 non-O139 strains in relation to O-serotypes. *Epidemiol Infect* **121,** 535–545.

Dalsgaard, A., Forslund, A., Sandvang, D. et al. (2001) *Vibrio cholerae* O1 outbreak isolates in Mozambique and South Africa in 1998 are multiple-drug resistant, contain the SXT element and the aadA2 gene located on class 1 integrons. *J Antimicrob Chemother* **48,** 827–838.

Daniels, N. A., MacKinnon, L., Bishop, R. et al. (2000) *Vibrio parahaemolyticus* infections in the United States, 1973–1998. *J Infect Dis* **181,** 1661–1666.

DePaola, A., Kaysner, C. A., Bowers, J., and Cook, D. W. (2000) Environmental investigations of *Vibrio parahaemolyticus* in oysters after outbreaks in Washington, Texas, and New York (1997 and 1998). *Appl Environ Microbiol* **66,** 4649–4654.

Dobrindt, U., Hochhut, B., Hentschel, U., and Hacker, J. (2004) Genomic islands in pathogenic and environmental microorganisms. *Nat Rev Microbiol* **2,** 414–424.

Dziejman, M., Balon, E., Boyd, D. et al. (2002) Comparative genomic analysis of *Vibrio cholerae*: Genes that correlate with cholera endemic and pandemic disease. *Proc Natl Acad Sci U S A* **99,** 1556–1561.

Egan, E. S. and Waldor, M. K. (2003) Distinct replication requirements for the two *Vibrio cholerae* chromosomes. *Cell* **114,** 521–530.

Evins, G. M., Cameron, D. N., Wells, J. G. et al. (1995) The emerging diversity of the electrophoretic types of *Vibrio cholerae* in the Western Hemisphere. *J Infect Dis* **172,** 173–179.

Farfan, M., Minana, D., Fuste, M. C., and Loren, J. G. (2000) Genetic relationships between clinical and environmental *Vibrio cholerae* isolates based on multilocus enzyme electrophoresis. *Microbiology* **146**(Pt 10), 2613–2626.

Farfan, M., Minana-Galbis, D., Fuste, M. C., and Loren, J. G. (2002) Allelic diversity and population structure in *Vibrio cholerae* O139 Bengal based on nucleotide sequence analysis. *J Bacteriol* **184,** 1304–1313.

Farmer, J. J. and Hickman-Brenner FW. (1992) *Vibrio* and photobacterium. In *The Prokaryotes* (eds. A. Balows, H. Truper, M. Dworkin, W. Harder, and K. Schleifer), pp. 2952–3011. Springer-Verlag, Berlin.

Faruque, S. and Nair, B. (2006) Epidemiology. In *The Biology of Vibrios* (eds. F. Thompson, B. Austin, and J. Swings), pp. 385–398. ASM Press, Washington, DC.

Faruque, S. M., Albert, M. J., and Mekalanos, J. J. (1998) Epidemiology, genetics, and ecology of toxigenic *Vibrio cholerae*. *Microbiol Mol Biol Rev* **62,** 1301–1314.

Faruque, S. M., Asadulghani, M., Kamruzzaman, M. et al. (2002) RS1 element of *Vibrio cholerae* can propagate horizontally as a filamentous phage exploiting the morphogenesis genes of CTXphi. *Infect Immun* **70,** 163–170.

Faruque, S. M., Chowdhury, N., Kamruzzaman, M. et al. (2004) Genetic diversity and virulence potential of environmental *Vibrio cholerae* population in a cholera-endemic area. *Proc Natl Acad Sci U S A* **101,** 2123–2128.

Faruque, S. M. and Mekalanos, J. J. (2003) Pathogenicity islands and phages in *Vibrio cholerae* evolution. *Trends Microbiol* **11,** 505–510.

Garg, P., Aydanian, A., Smith, D. et al. (2003) Molecular epidemiology of O139 *Vibrio cholerae*: Mutation, lateral gene transfer, and founder flush. *Emerg Infect Dis* **9,** 810–814.

Gevers, D. and Van de Peer, Y. (2006) Gene duplicates in vibrio genomes. In *The Biology of Vibrios* (eds. F. Thompson, B. Austin, and J. Swings), pp. 76–83. ASM Press, Washington, DC.

Goldberg, S. and Murphy, J. R. (1983) Molecular epidemiological studies of United States Gulf Coast *Vibrio cholerae* strains: Integration site of mutator vibriophage VcA-3. *Infect Immun* **42,** 224–230.

Gonzalez-Escalona, N., Cachicas, V., Acevedo, C. et al. (2005) *Vibrio parahaemolyticus* diarrhea, Chile, 1998 and 2004. *Emerg Infect Dis* **11,** 129–131.

Gonzalez-Escalona, N., Martinez-Urtaza, J., Romero, J. et al. (2008) Determination of molecular phylogenetics of *Vibrio parahaemolyticus* strains by multilocus sequence typing. *J Bacteriol* **190,** 2831–2840.

Gutacker, M., Conza, N., Benagli, C. et al. (2003) Population genetics of *Vibrio vulnificus*: Identification of two divisions and a distinct eel-pathogenic clone. *Appl Environ Microbiol* **69,** 3203–3212.

Hall, R. M. (2002) Gene cassettes and integrons; moving single genes. In *Horizontal Gene Transfer* (eds. M. Syvanan and C. Kado), pp. 19–28. Harcourt Publishing, London.

Hayat, U., Reddy, G. P., Bush, C. A. et al. (1993) Capsular types of *Vibrio vulnificus*: An analysis of strains from clinical and environmental sources. *J Infect Dis* **168,** 758–762.

Hazen, T. H., Wu, D., Eisen, J. A., and Sobecky, P. A. (2007) Sequence characterization and comparative analysis of three plasmids isolated from environmental *Vibrio* spp. *Appl Environ Microbiol* **73,** 7703–7710.

Heidelberg, J. F., Eisen, J. A., Nelson, W. C. et al. (2000) DNA sequence of both chromosomes of the cholera pathogen *Vibrio cholerae*. *Nature* **406,** 477–483.

Hoi, L., Dalsgaard, I., DePaola, A. et al. (1998) Heterogeneity among isolates of *Vibrio vulnificus* recovered from eels (*Anguilla anguilla*) in Denmark. *Appl Environ Microbiol* **64,** 4676–4682.

Hueck, C. J. (1998) Type III protein secretion systems in bacterial pathogens of animals and plants. *Microbiol Mol Biol Rev* **62,** 379–433.

Hurley, C. C., Quirke, A., Reen, F. J., and Boyd, E. F. (2006) Four genomic islands that mark post-1995 pandemic *Vibrio parahaemolyticus* isolates. *BMC Genomics* **7,** 104.

Iida, T., Hattori, A., Tagomori, K. et al. (2001) Filamentous phage associated with recent pandemic strains of *Vibrio parahaemolyticus*. *Emerg Infect Dis* **7,** 477–478.

Iida, T., Park, K., Honda T. (2006). Vibrio parahaemolyticus. In *The Biology of Vibrios* (eds. F. Thompson, B. Austin, and J. Swings), pp. 340–348. ASM Press, Washington, DC.

Jackson, J. K., Murphree, R. L., and Tamplin, M. L. (1997) Evidence that mortality from *Vibrio vulnificus* infection results from single strains among heterogeneous populations in shellfish. *J Clin Microbiol* **35,** 2098–2101.

Jermyn, W. S. and Boyd, E. F. (2002) Characterization of a novel vibrio pathogenicity island (VPI-2) encoding neuraminidase (nanH) among toxigenic *Vibrio cholerae* isolates. *Microbiology* **148,** 3681–3693.

Jiang, S. C., Louis, V., Choopun, N. et al. (2000) Genetic diversity of *Vibrio cholerae* in Chesapeake Bay determined by amplified fragment length polymorphism fingerprinting. *Appl Environ Microbiol* **66,** 140–147.

Joseph, S. W., Colwell, R. R., and Kaper, J. B. (1982) *Vibrio parahaemolyticus* and related halophilic Vibrios. *Crit Rev Microbiol* **10,** 77–124.

Kaper, J. B., Bradford, H. B., Roberts, N. C., and Falkow, S. (1982) Molecular epidemiology of *Vibrio cholerae* in the U.S. Gulf Coast. *J Clin Microbiol* **16,** 129–134.

Kaper, J. B., Morris, J. G. Jr., and Levine, M. M. (1995) Cholera. *Clin Microbiol Rev* **8,** 48–86.

Karaolis, D. K., Johnson, J. A., Bailey, C. C. et al. (1998) A *Vibrio cholerae* pathogenicity island associated with epidemic and pandemic strains. *Proc Natl Acad Sci U S A* **95,** 3134–3139.

Karaolis, D. K., Lan, R., and Reeves, P. R. (1994) Molecular evolution of the seventh-pandemic clone of *Vibrio cholerae* and its relationship to other pandemic and epidemic *V. cholerae* isolates. *J Bacteriol* **176,** 6199–6206.

Karaolis, D. K., Lan, R., and Reeves, P. R. (1995) The sixth and seventh cholera pandemics are due to independent clones separately derived from environmental, nontoxigenic, non-O1 *Vibrio cholerae*. *J Bacteriol* **177,** 3191–3198.

Keymer, D. P., Miller, M. C., Schoolnik, G. K., and Boehm, A. B. (2007) Genomic and phenotypic diversity of coastal *Vibrio cholerae* strains is linked to environmental factors. *Appl Environ Microbiol* **73,** 3705–3714.

Koblavi, S., Grimont, F., and Grimont, P. A. (1990) Clonal diversity of *Vibrio cholerae* O1 evidenced by rRNA gene restriction patterns. *Res Microbiol* **141,** 645–657.

Kotetishvili, M., Stine, O. C., Chen, Y. et al. (2003) Multilocus sequence typing has better discriminatory ability for typing *Vibrio cholerae* than does pulsed-field gel electrophoresis and provides a measure of phylogenetic relatedness. *J Clin Microbiol* **41,** 2191–2196.

Kovach, M. E., Shaffer, M. D., and Peterson, K. M. (1996) A putative integrase gene defines the distal end of a large cluster of ToxR-regulated colonization genes in *Vibrio cholerae*. *Microbiology* **142**(Pt 8), 2165–2174.

Labbate, M., Boucher, Y., Joss, M. J. et al. (2007) Use of chromosomal integron arrays as a phylogenetic typing system for *Vibrio cholerae* pandemic strains. *Microbiology* **153,** 1488–1498.

Laohaprertthisan, V., Chowdhury, A., Kongmuang, U. et al. (2003) Prevalence and serodiversity of the pandemic clone among the clinical strains of *Vibrio parahaemolyticus* isolated in southern Thailand. *Epidemiol Infect* **130,** 395–406.

Levine, W. C. and Griffin, P. M. (1993) *Vibrio infections* on the Gulf Coast: Results of first year of regional surveillance. Gulf Coast Vibrio Working Group. *J Infect Dis* **167,** 479–483.

Lin, M., Payne, D. A., and Schwarz, J. R. (2003) Intraspecific diversity of *Vibrio vulnificus* in Galveston Bay water and oysters as determined by randomly amplified polymorphic DNA PCR. *Appl Environ Microbiol* **69,** 3170–3175.

Lin, M. and Schwarz, J. (2003) Seasonal shifts in population structure of *Vibrio vulnificus* in an estuarine environment as revealed by partial 16S ribosomal DNA sequencing. *FEMS Microbiol Ecol* **45,** 23–27.

Maiden, M. C., Bygraves, J. A., Feil, E. et al. (1998) Multilocus sequence typing: A portable approach to the identification of clones within populations of pathogenic microorganisms. *Proc Natl Acad Sci U S A* **95,** 3140–3145.

Makino, K., Oshima, K., Kurokawa, K. et al. (2003) Genome sequence of *Vibrio parahaemolyticus*: A pathogenic mechanism distinct from that of *V. cholerae*. *Lancet* **361,** 743–749.

Matsumoto, C., Okuda, J., Ishibashi, M. et al. (2000) Pandemic spread of an O3:K6 clone of *Vibrio parahaemolyticus* and emergence of related strains evidenced by arbitrarily primed PCR and toxRS sequence analyses. *J Clin Microbiol* **38,** 578–585.

McCarter, L. (1999) The multiple identities of *Vibrio parahaemolyticus*. *J Mol Microbiol Biotechnol* **1,** 51–57.

Mekalanos, J. J. (1983) Duplication and amplification of toxin genes in *Vibrio cholerae*. *Cell* **35,** 253–263.

Mekalanos, J. J., Collier, R. J., and Romig, W. R. (1978) Affinity filters, a new approach to the isolation of tox mutants of *Vibrio cholerae*. *Proc Natl Acad Sci U S A* **75,** 941–945.

Mohapatra, S. S., Ramachandran, D., Mantri, C. K. et al. (2009) Determination of relationships among non-toxigenic *Vibrio cholerae* O1 biotype El Tor strains from housekeeping gene sequences and ribotype patterns. *Res Microbiol* **160,** 57–62.

Nair, G. B., Ramamurthy, T., Bhattacharya, S. K. et al. (2007) Global dissemination of *Vibrio parahaemolyticus* serotype O3:K6 and its serovariants. *Clin Microbiol Rev* **20,** 39–48.

Nasu, H., Iida, T., Sugahara, T. et al. (2000) A filamentous phage associated with recent pandemic *Vibrio parahaemolyticus* O3:K6 strains. *J Clin Microbiol* **38,** 2156–2161.

Nilsson, W. B., Paranjype, R. N., DePaola, A., and Strom, M. S. (2003) Sequence polymorphism of the 16S rRNA gene of *Vibrio vulnificus* is a possible indicator of strain virulence. *J Clin Microbiol* **41,** 442–446.

Nishibuchi, M. and Kaper, J. B. (1990) Duplication and variation of the thermostable direct haemolysin (tdh) gene in *Vibrio parahaemolyticus*. *Mol Microbiol* **4,** 87–99.

Okada, K., Iida, T., Kita-Tsukamoto, K., and Honda, T. (2005) Vibrios commonly possess two chromosomes. *J Bacteriol* **187,** 752–757.

Okuda, J., Ishibashi, M., Hayakawa, E. et al. (1997) Emergence of a unique O3:K6 clone of *Vibrio parahaemolyticus* in Calcutta, India, and isolation of strains from the same clonal group from Southeast Asian travelers arriving in Japan. *J Clin Microbiol* **35,** 3150–3155.

Okura, M., Osawa, R., Iguchi, A. et al. (2003) Genotypic analyses of *Vibrio parahaemolyticus* and development of a pandemic group-specific multiplex PCR assay. *J Clin Microbiol* **41,** 4676–4682.

Oliver, J. D. (1989) *Vibrio vulnificus*. In *Food-borne Bacterial Pathogens* (ed. M. P. Doyle), pp. 569–600. Marcel Dekker, New York.

Oliver, J. D. (2006) *Vibrio vulnificus*. In *The Biology of Vibrios* (eds. F. L. Thompson, B. Austin, and J. G. Swings), pp. 349–366. ASM Press, Washington, DC.

Oliver, J. D., Warner, R. A., and Cleland, D. R. (1982) Distribution and ecology of *Vibrio vulnificus* and other lactose-fermenting marine vibrios in coastal waters of the Southeastern United States. *Appl Environ Microbiol* **44,** 1404–1414.

O'Shea, Y. A., Reen, F. J., Quirke, A. M., and Boyd, E. F. (2004) Evolutionary genetic analysis of the emergence of epidemic *Vibrio cholerae* isolates on the basis of comparative nucleotide sequence analysis and multilocus virulence gene profiles. *J Clin Microbiol* **42,** 4657–4671.

Pan, T. M., Wang, T. K., Lee, C. L. et al. (1997) Food-borne disease outbreaks due to bacteria in Taiwan, 1986 to 1995. *J Clin Microbiol* **35,** 1260–1262.

Popovic, T., Bopp, C., Olsvik, O., and Wachsmuth, K. (1993) Epidemiologic application of a standardized ribotype scheme for *Vibrio cholerae* O1. *J Clin Microbiol* **31,** 2474–2482.

Quilici, M. L., Robert-Pillot, A., Picart, J., and Fournier, J. M. (2005) Pandemic *Vibrio parahaemolyticus* O3:K6 spread, France. *Emerg Infect Dis* **11,** 1148–1149.

Quirke, A. M., Reen, F. J., Claesson, M. J., and Boyd, E. F. (2006) Genomic island identification in *Vibrio vulnificus* reveals significant genome plasticity in this human pathogen. *Bioinformatics* **22,** 905–910.

Ramamurthy, T., Garg, S., Sharma, R. et al. (1993) Emergence of novel strain of *Vibrio cholerae* with epidemic potential in southern and eastern India. *Lancet* **341,** 703–704.

Reen, F. J., Almagro-Moreno, S., Ussery, D., and Boyd, E. F. (2006) The genomic code: Inferring Vibrionaceae niche specialization. *Nat Rev Microbiol* **4,** 697–704.

Reen, F. J. and Boyd, E. F. (2005) Molecular typing of epidemic and nonepidemic *Vibrio cholerae* isolates and differentiation of *V. cholerae* and *V. mimicus* isolates by PCR-single-strand conformation polymorphism analysis. *J Appl Microbiol* **98,** 544–555.

RIES, A. A., VUGIA, D. J., BEINGOLEA, L. et al. (1992) Cholera in Piura, Peru: A modern urban epidemic. *J Infect Dis* **166,** 1429–1433.

ROSCHE, T. M., YANO, Y., and OLIVER, J. D. (2005) A rapid and simple PCR analysis indicates there are two subgroups of *Vibrio vulnificus* which correlate with clinical or environmental isolation. *Microbiol Immunol* **49,** 381–389.

ROSENBERG, E. and BEN-HAIM, Y. (2002) Microbial diseases of corals and global warming. *Environ Microbiol* **4,** 318–326.

ROWE-MAGNUS, D. A., ZOUINE, M., and MAZEL, D. (2006) The adaptive genetic arsenal of pathogenic *Vibrio* species: The role of integrons. In *The Biology of Vibrios* (eds. F. Thompson, B. Austin, and J. Swings), pp. 95–111. ASM Press, Washington, DC.

RUBIN, E., WALDOR, M., and MEKALANOS, J. (1998) Mobile genetic elements and the evolution of new epidemic strains of *Vibrio cholerae*. In *Emerging Infections* (ed. R. Krause), pp. 147–161. Academic Press, San Diego, CA.

RUBY, E. G., URBANOWSKI, M., CAMPBELL, J. et al. (2005) Complete genome sequence of *Vibrio fischeri*: A symbiotic bacterium with pathogenic congeners. *Proc Natl Acad Sci U S A* **102,** 3004–3009.

SACK, D. A., SACK, R. B., NAIR, G. B., and SIDDIQUE, A. K. (2004) Cholera. *Lancet* **363,** 223–233.

SCHOOLNIK, G. K. and YILDIZ, F. H. (2000) The complete genome sequence of *Vibrio cholerae*: A tale of two chromosomes and of two lifestyles. *Genome Biol* **1,** 1016.1–1016.3.

SHARMA, C., THUNGAPATHRA, M., GHOSH, A. et al. (1998) Molecular analysis of non-O1, non-O139 *Vibrio cholerae* associated with an unusual upsurge in the incidence of cholera-like disease in Calcutta, India. *J Clin Microbiol* **36,** 756–763.

STINE, O. C., SOZHAMANNAN, S., GOU, Q. et al. (2000) Phylogeny of *Vibrio cholerae* based on recA sequence. *Infect Immun* **68,** 7180–7185.

STROEHER, U. H., JEDANI, K. E., DREDGE, B. K. et al. (1995) Genetic rearrangements in the rfb regions of *Vibrio cholerae* O1 and O139. *Proc Natl Acad Sci U S A* **92,** 10374–10378.

SWERDLOW, D. L. and RIES, A. A. (1993) *Vibrio cholerae* non-O1–the eighth pandemic? *Lancet* **342,** 382–383.

TAMPLIN, M. L., JACKSON, J. K., BUCHRIESER, C. et al. (1996) Pulsed-field gel electrophoresis and ribotype profiles of clinical and environmental *Vibrio vulnificus* isolates. *Appl Environ Microbiol* **62,** 3572–3580.

TAUXE, R. L., SEMINARIO, L., TAPIA, R., and LIBEL, M. (1994) The Latin America epidemic. In *Vibrio Cholerae and Cholera: Molecular to Global Perspectives* (eds. I. K. Waschmuth, P. A. Black, and O. Olsvik), pp. 321–344. ASM Press, Washington, DC.

TAUXE, R. V. and BLAKE, P. A. (1992) Epidemic cholera in Latin America. *JAMA* **267,** 1388–1390.

TAYLOR, R. K., MILLER, V. L., FURLONG, D. B., and MEKALANOS, J. J. (1987) Use of phoA gene fusions to identify a pilus colonization factor coordinately regulated with cholera toxin. *Proc Natl Acad Sci U S A* **84,** 2833–2837.

THOMPSON, F. L., IIDA, T., and SWINGS, J. (2004) Biodiversity of vibrios. *Microbiol Mol Biol Rev* **68,** 403–431.

THOMPSON, F. L., THOMPSON, C. C., VICENTE, A. C. et al. (2003) Genomic diversity of clinical and environmental *Vibrio cholerae* strains isolated in Brazil between 1991 and 2001 as revealed by fluorescent amplified fragment length polymorphism analysis. *J Clin Microbiol* **41,** 1946–1950.

TISON, D. L., NISHIBUCHI, M., GREENWOOD, J. D., and SEIDLER, R. J. (1982) *Vibrio vulnificus* biogroup 2: New biogroup pathogenic for eels. *Appl Environ Microbiol* **44,** 640–646.

VICKERY, M. C., HAROLD, N., and BEJ, A. K. (2000) Cluster analysis of AP-PCR generated DNA fingerprints of *Vibrio vulnificus* isolates from patients fatally infected after consumption of raw oysters. *Lett Appl Microbiol* **30,** 258–262.

WACHSMUTH, I. K., EVINS, G. M., FIELDS, P. I. et al. (1993) The molecular epidemiology of cholera in Latin America. *J Infect Dis* **167,** 621–626.

WALDOR, M. K. and MEKALANOS, J. J. (1996) Lysogenic conversion by a filamentous phage encoding cholera toxin. *Science* **272,** 1910–1914.

WALDOR, M. K. and RAYCHAUDHURI, D. (2000) Treasure trove for cholera research. *Nature* **406,** 469–470.

WALDOR, M. K., TSCHAPE, H., and MEKALANOS, J. J. (1996) A new type of conjugative transposon encodes resistance to sulfamethoxazole, trimethoprim, and streptomycin in *Vibrio cholerae* O139. *J Bacteriol* **178,** 4157–4165.

WARNER, J. M. and OLIVER, J. D. (1999) Randomly amplified polymorphic DNA analysis of clinical and environmental isolates of *Vibrio vulnificus* and other vibrio species. *Appl Environ Microbiol* **65,** 1141–1144.

WOMMACK, K. E. and COLWELL, R. R. (2000) Virioplankton: Viruses in aquatic ecosystems. *Microbiol Mol Biol Rev* **64,** 69–114.

WONG, H. C., LIU, S. H., WANG, T. K. et al. (2000) Characteristics of *Vibrio parahaemolyticus* O3:K6 from Asia. *Appl Environ Microbiol* **66,** 3981–3986.

YAMAI, S., OKITSU, T., SHIMADA, T., and KATSUBE, Y. (1997) [Distribution of serogroups of *Vibrio cholerae* non-O1 non-O139 with specific reference to their ability to produce cholera toxin, and addition of novel serogroups]. *J Jpn Assoc Infect Dis* **71,** 1037–1045.

YAMAICHI, Y., IIDA, T., PARK, K. S. et al. (1999) Physical and genetic map of the genome of *Vibrio parahaemolyticus*: Presence of two chromosomes in *Vibrio* species. *Mol Microbiol* **31,** 1513–1521.

YEUNG, P. S. and BOOR, K. J. (2004) Epidemiology, pathogenesis, and prevention of foodborne *Vibrio parahaemolyticus* infections. *Foodborne Pathog Dis* **1,** 74–88.

Yoh, M., Miwatani, T., and Honda, T. (1992) Comparison of *Vibrio parahaemolyticus* hemolysin (Vp-TRH) produced by environmental and clinical isolates. *FEMS Microbiol Lett* **71,** 157–161.

Zaidenstein, R., Sadik, C., Lerner, L. et al. (2008) Clinical characteristics and molecular subtyping of *Vibrio vulnificus* illnesses, Israel. *Emerg Infect Dis* **14,** 1875–1882.

Zo, Y. G., Rivera, I. N., Russek-Cohen, E. et al. (2002) Genomic profiles of clinical and environmental isolates of *Vibrio cholerae* O1 in cholera-endemic areas of Bangladesh. *Proc Natl Acad Sci U S A* **99,** 12409–12414.

Index

Note: Page numbers in *italics* refer to Figures; those in **bold** to Tables.